WITHDRAWN

Biotechnology for Fuels and Chemicals

*Proceedings of the Eighteenth Symposium
on Biotechnology for Fuels and Chemicals
Held May 5–9, 1996, at Gatlinburg, Tennessee*

Sponsored by
U. S. Department of Energy's Biofuels Energy Systems Division
and the Biological and Chemical Technology Research Program (DOE)
Oak Ridge National Laboratory
National Renewable Energy Laboratory
Idaho National Laboratory
Lockheed Martin Energy Systems, Inc.
A. E. Staley Manufacturing Company
Archer Daniels Midland Company
Bio-Technical Resources, L. P.
Chronopol, Inc.
ConAgra Grain Processing Companies
Enzyme Bio-Systems, Ltd.
E. I. Du Pont de Nemours and Company
Grain Processing Corporation
Raphael Katzen Associates International, Inc.
Weyerhaeuser Company
American Chemical Society's Division of Biochemical Technology

Editors
Brian H. Davison
Oak Ridge National Laboratory
Charles E. Wyman
National Renewable Energy Laboratory
Mark Finkelstein
National Renewable Energy Laboratory

Humana Press • Totowa, New Jersey

Applied Biochemistry and Biotechnology

Volumes 63–65, Complete, Spring 1997

Copyright ©1997 Humana Press Inc.

All Rights Reserved.

No part of this publication may be reproduced or transmitted in any form or by any means, electronic or mechanical, including photocopy, recording, or any information storage and retrieval system, without permission in writing from the copyright owner.

Applied Biochemistry and Biotechnology is abstracted or indexed regularly in *Chemical Abstracts, Biological Abstracts, Current Contents, Science Citation Index, Excerpta Medica, Index Medicus,* and appropriate related compendia.

Introduction

BRIAN H. DAVISON
Oak Ridge National Laboratory

MARK FINKELSTEIN
National Renewable Energy Laboratory

CHARLES E. WYMAN
Oak Ridge National Laboratory

The Eighteenth Symposium on Biotechnology for Fuels and Chemicals continues to provide a forum for the presentation of research results and the exchange of ideas on advances in biotechnology for the production of fuels and chemicals. Although the emphasis is on utilization of renewable resources, the scope of the Symposium is broader than this and includes bioconversion of fossil fuels and syngas and the new area of conversions in nonaqueous environments; these areas were discussed in Session 5 and in a Special Topic Discussion Group at the Symposium. In addition, recent developments in bioremediation were well represented in Session 6 and in the poster session.

The Symposium involved both the development of new biological agents (such as enzymes or microbes) to carry out targeted conversions as well as bioprocess development. The first area covered improvements in enzymes as well as fundamental insights into substrate–enzyme interactions and photosynthesis. The latter area focused on converting one material into another using biological agents through combinations of chemical engineering, biological sciences, and fermentation technology. This area also refers to an overall processing involving at least one biologically catalyzed step in combination with other physical and/or chemical processing operations. Agricultural crops, such as corn and corn fiber as well as woody biomass and lignocellulosic wastes, are emphasized for process feedstocks and their pretreatment investigated.

This area is gaining increased interest as some processes are moving to commercialization. Along with continuing advances in ethanol production, both from corn and from lignocellulosics, technology for lactic acid production by fermentation processes is being improved and commercialized. Papers were also presented on other potential products including fumaric acid, succinic acid, methane, enzymes, glucuronic acid, and biodiesel. Industrial issues and needs for commercialization were addressed in a new Session 4 that proved exciting, provocative, and well-attended. The International Energy Agency cosponsored a special discussion on "Technoeconomic Modeling of Lignocellulosic Conversion to Ethanol" during the meeting.

The papers in this volume were drawn from the 37 oral and 103 poster presentations made to the approximately 180 attendees in Gatlinburg, TN. Overall, we believe the Eighteenth Symposium continued the tradition established by its founder, Charles D. Scott, of providing both technical and informal interactions between representatives of industry, academia, and the government research laboratories during the sessions, banquets, and tours, including a tour of Oak Ridge National Laboratory.

The Eighteenth Symposium was sponsored by the U. S. Department of Energy's Biofuels Energy Systems Division and the Biological and Chemical Technology Research Program (DOE), Oak Ridge National Laboratory, National Renewable Energy Laboratory, Idaho National Laboratory, Lockheed Martin Energy Systems, Inc., A. E. Staley Manufacturing Company, Archer Daniels Midland Company, Bio-Technical Resources, L. P., Chronopol, Inc., ConAgra Grain Processing Companies, Enzyme Bio-Systems, Ltd., E. I. DuPont de Nemours and Company, Grain Processing Corporation, Raphael Katzen Associates International, Inc., Weyerhauser Company, American Chemical Society's Division of Biochemical Technology.

Organization of the Symposium was as follows:

Organizing Committee
Brian H. Davison, *Cochairman, Oak Ridge National Laboratory*
Charles E. Wyman, *Chairman, National Renewable Energy Laboratory*
Bill Apel, *Idaho National Engineering Laboratory*
Rakesh Bajpai, *University of Missouri-Columbia*
David Boron, *U. S. Department of Energy*
Ting Carlson, *Cargill, Inc.*
James A. Doncheck, *Bio-Technical Resources, L. P.*
Mark Finkelstein, *National Renewable Energy Laboratory*
Donald L. Johnson, *Grain Processing Corporation*
Raphael Katzen, *Raphael Katzen Associates International, Inc.*
Lee R. Lynd, *Dartmouth College*
Valerie Sarisky-Reed, *U. S. Department of Energy*
Jonathan Woodward, *Oak Ridge National Laboratory*

Session Chairpersons and Cochairpersons
Session 1: Thermal, Chemical, and Biological Processing
Mark T. Holtzapple, *Texas A&M University*
Robert Torget, *National Renewable Energy Laboratory*

Session 2: Applied Biological Research
Valerie Sarisky-Reed, *U. S. Department of Energy*
Jonathan Woodward, *Oak Ridge National Laboratory*

Session 3: Bioprocessing Research
Robert R. Dorsch, *DuPont*
Christos Hatzis, *National Renewable Energy Laboratory*

Session 4: Industrial Needs for Commercialization
Dale Monceaux, *Raphael Katzen Associates International, Inc.*
James L. Gaddy, *Bioengineering Resources, Inc.*

Session 5: Emerging Topics in Industrial Biotechnology
Bruce Dale, *Michigan State University*
Eric Kaufman, *Oak Ridge National Laboratory*

Session 6: Environmental Biotechnology
Mary Jim Beck, *Tennessee Valley Authority*
Joni M. Barnes, *Idaho National Engineering Laboratory*

Acknowledgments

The able assistance of Anne Greenbaum as Symposium Secretary, assisted by Lois Hamm, Marsha Savage, Liz Willson, Joan Taylor, and Norma Caldwell. Liz Willson and Renae Humphrey assisted with the proceedings.

Oak Ridge National Laboratory is managed by Lockheed Martin Energy Systems, Inc., for the US Department of Energy under contract DE-AC05-96OR22464.

National Renewable Energy Laboratory is managed by the Midwest Research Institute, for the US Department of Energy under contract AC36-83CH190093.

"The submitted manuscript has been authored by a contractor of the US Government under contract DE-AC05-96OR22464. Accordingly, the US Government retains a nonexclusive, royalty-free license to publish or reproduce the published form of this contribution, or allow others to do so, for US Government purposes."

Other Proceedings in This Series

1. "Proceedings of the First Symposium on Biotechnology in Energy Production and Conservation" (1978), *Biotechnol. Bioeng. Symp.* **8**.
2. "Proceedings of the Second Symposium on Biotechnology in Energy Production and Conservation" (1980), *Biotechnol. Bioeng. Symp.* **10**.
3. "Proceedings of the Third Symposium on Biotechnology in Energy Production and Conservation" (1981), *Biotechnol. Bioeng. Symp.* **11**.
4. "Proceedings of the Fourth Symposium on Biotechnology in Energy Production and Conservation" (1982), *Biotechnol. Bioeng. Symp.* **12**.
5. "Proceedings of the Fifth Symposium on Biotechnology for Fuels and Chemicals" (1983), *Biotechnol. Bioeng. Symp.* **13**.

6. "Proceedings of the Sixth Symposium on Biotechnology for Fuels and Chemicals" (1984), *Biotechnol. Bioeng. Symp.* **14.**
7. "Proceedings of the Seventh Symposium on Biotechnology for Fuels and Chemicals" (1985), *Biotechnol. Bioeng. Symp.* **15.**
8. "Proceedings of the Eighth Symposium on Biotechnology for Fuels and Chemicals" (1986), *Biotechnol. Bioeng. Symp.* **17.**
9. "Proceedings of the Ninth Symposium on Biotechnology for Fuels and Chemicals" (1988), *Appl. Biochem. Biotechnol.* **17,18.**
10. "Proceedings of the Tenth Symposium on Biotechnology for Fuels and Chemicals" (1989), *Appl. Biochem. Biotechnol.* **20,21.**
11. "Proceedings of the Eleventh Symposium on Biotechnology for Fuels and Chemicals" (1990), *Appl. Biochem. Biotechnol.* **24,25.**
12. "Proceedings of the Twelfth Symposium on Biotechnology for Fuels and Chemicals" (1991), *Appl. Biochem. Biotechnol.* **28,29.**
13. "Proceedings of the Thirteenth Symposium on Biotechnology for Fuels and Chemicals" (1992), *Appl. Biochem. Biotechnol.* **34,35.**
14. "Proceedings of the Fourteenth Symposium on Biotechnology for Fuels and Chemicals" (1993), *Appl. Biochem. Biotechnol.* **39,40.**
15. "Proceedings of the Fifteenth Symposium on Biotechnology for Fuels and Chemicals" (1994), *Appl. Biochem. Biotechnol.* **45,46.**
16. "Proceedings of the Sixteenth Symposium on Biotechnology for Fuels and Chemicals" (1995), *Appl. Biochem. Biotechnol.* **51/52.**
17. "Proceedings of the Seventeenth Symposium on Biotechnology for Fuels and Chemicals" (1996), *Appl. Biochem. Biotechnol.* **57/58.**

This symposium has been held annually since 1978. We are pleased to have the proceedings of the Eighteenth Symposium currently published in this special issue to continue the tradition of providing a record of the contributions made.

The Nineteenth Symposium is planned for May 4–8, 1997 in Colorado Springs, Colorado and the Twentieth Symposium is planned for May 3–7, 1998 in Gatlinburg, Tennessee. We encourage comments or discussions relevant to the format or content of these meetings.

Contents

Introduction
Brian H. Davison, Charles E. Wyman, and Mark Finkelstein iii

Session 1—Thermal, Chemical, and Biological Processing

Introduction to Session 1
Mark T. Holtzapple and Robert Torget 1

Lime Pretreatment of Switchgrass
Vincent S. Chang, Barry Burr, and Mark T. Holtzapple* 3

Ammonia Recycled Percolation as a Complementary Pretreatment to the Dilute-Acid Process
Zhangwen Wu and Y. Y. Lee* 21

Preliminary Study of the Pyrolysis of Steam Classified Municipal Solid Waste
John M. Sebghati and Michael H. Eley* 35

Two-Phase Model of Hydrolysis Kinetics and Its Applications to Anaerobic Degradation of Particulate Organic Matter
Vasily A. Vavilin,* Sergei V. Rytov, and Ljudmila Ya. Lokshina 45

Preprocessed Barley, Rye, and Triticale as a Feedstock for an Integrated Fuel Ethanol-Feedlot Plant
Krystyna Sosulski,* Sunmin Wang, W. M. Ingledew, Frank W. Sosulski, and Juming Tang 59

Session 2—Biological Research

Introduction to Session 2
Valerie Sarisky-Reed and Jonathan Woodward 71

A Stable Lipase from *Candida lipolytica*: Cultivation Conditions and Crude Enzyme Characteristics
Fatima Ventura Pereira-Meirelles, Maria Helena Miguez Rocha-Leão, and Geraldo Lippel Sant' Anna, Jr.* 73

Glucoamylase Isoenzymes Tailoring Through Medium Composition
José G. Silva, Jr., Hilton J. Nascimento, Valíria F. Soares, and Elba P. S. Bon* 87

Regulation of Phosphotransferases in Glucose- and Xylose-Fermenting Yeasts
Vina W. Yang and Thomas W. Jeffries* 97

Diminished Respirative Growth and Enhanced Assimilative Sugar Uptake Result in Higher Specific Fermentation Rates by the Mutant *Pichia stipitis* FPL-061
Hassan K. Sreenath and Thomas W. Jeffries* 109

---

*For papers with multiple authorship, the asterisk identifies the author to whom correspondence and reprint requests should be addressed.

Production of Xylitol from D-Xylose by *Debaryomyces hansenii*
  Jose M. Dominguez,* Cheng S. Gong, and George T. Tsao 117

Production of 2,3-Butanediol from Pretreated Corn Cob by *Klebsiella oxytoca* in the Presence of Fungal Cellulase
  Ningjun Cao,* Youkun Xia, Cheng S. Gong, and George T. Tsao 129

Oxygen Sensitivity of Algal H_2-Production
  Maria L. Ghirardi,* Robert K. Togasaki, and Michael Seibert 141

Expression of *Ascaris suum* Malic Enzyme in a Mutant *Escherichia coli* Allows Production of Succinic Acid from Glucose
  Lucy Stols, Gopal Kulkarni, Ben G. Harris, and Mark I. Donnelly* 153

Reaction Engineering Aspects of α-1,4-D-Glucan Phosphorylase Catalysis: *Comparison of Plant and Bacterial Enzymes for the Continuous Synthesis of D-Glucose-1-Phosphate*
  Bernd Nidetzky,* Richard Griessler, Andreas Weinhäusel, Dietmar Haltrich, and Klaus D. Kulbe 159

Simultaneous Enzymatic Synthesis of Gluconic Acid and Sorbitol: *Production, Purification, and Application of Glucose-Fructose Oxidoreductase and Gluconolactonase*
  Bernd Nidetzky,* Monika Fürlinger, Dorothee Gollhofer, Iris Haug, Dietmar Haltrich, and Klaus D. Kulbe 173

Production of Hemicellulose- and Cellulose-Degrading Enzymes by Various Strains of *Sclerotium Rolfsii*
  Alois Sachslehner, Dietmar Haltrich,* Bernd Nidetzky, and Klaus D. Kulbe 189

Asparaginase II of *Saccharomyces cerevisiae*: GLN3/URE2 Regulation of a Periplasmic Enzyme
  Elba P. S. Bon,* Elvira Carvajal, Mike Stanbrough, Donald Rowen, and Boris Magasanik 203

Production of α-Terpineol from *Escherichia coli* Cells Expressing Thermostable Limonene Hydratase
  Natarajan Savithiry, Tae Kyou Cheong, and Patrick Oriel* 213

Fermentation of Biomass-Derived Glucuronic Acid by *pet* Expressing Recombinants of *E. coli* B
  Hugh G. Lawford* and Joyce D. Rousseau 221

Enhanced Cofermentation of Glucose and Xylose by Recombinant *Saccharomyces* Yeast Strains in Batch and Continuous Operating Modes
  Susan T. Toon, George P. Philippidis, Nancy W. Y. Ho, ZhengDao Chen, Adam Brainard, Robert E. Lumpkin, and Cynthia J. Riley* 243

Stabilization and Reutilization of *Bacillus megaterium* Glucose
    Dehydrogenase by Immobilization
  Madalena Baron, José D. Fontana, Manoel F. Guimarães,
    and Jonathan Woodward* 257

Optimization of Seed Production for a Simultaneous Saccharification
    Cofermentation Biomass-to-Ethanol Process Using Recombinant
    Zymomonas
  Hugh G. Lawford,* Joyce D. Rousseau, and James D. McMillan 269

Corn Steep Liquor as a Cost-Effective Nutrition Adjunct
    in High-Performance *Zymomonas* Ethanol Fermentations
  Hugh G. Lawford* and Joyce D. Rousseau 287

Astaxanthinogenesis in the Yeast *Phaffia rhodozyma: Optimization
    of Low-Cost Culture Media and Yeast Cell-Wall Lysis*
  José D. Fontana, Miriam B. Chocial, Madalena Baron,
    Manoel F. Guimaraes, Marcelo Maraschin, Cirano Ulhoa,
    Jose A. Florêncio, and Tania M. B. Bonfim* 305

Polysaccharide Hydrolase Folds Diversity of Structure and Convergence
    of Function
  Michael E. Himmel, P. Andrew Karplus, Joshua Sakon,
    William S. Adney, John O. Baker, and Steven R. Thomas* 315

Acetobacter Cellulosic Biofilms Search for New Modulators
    of Cellulogenesis and Native Membrane Treatments
  José D. Fontana, Cassandra G. Joerke, Madalena Baron,
    Marcelo Maraschin, Antonio G. Ferreira, Iris Torriani,
    A. M. Souza, Marisa B. Soares, Milene A. Fontana,
    and Manoel F. Guimaraes* 327

Experimental Data Analysis: *An Algorithm for Determining Rates
    and Smoothing Data*
  K. Thomas Klasson 339

SESSION 3—BIOPROCESSING RESEARCH

Introduction to Session 3
  Robert R. Dorsch and Christos Hatzis 349

Cellulase Production Based on Hemicellulose Hydrolysate
    from Steam-Pretreated Willow
  Zsolt Szengyel,* Guido Zacchi, and Kati Réczey 351

Effect of Impeller Geometry on Gas-Liquid Mass Transfer Coefficients
    in Filamentous Suspensions
  Sundeep N. Dronawat, C. Kurt Svihla, and Thomas R. Hanley* 363

Measurement of the Steady-State Shear Characteristics of Filamentous
    Suspensions Using Turbine, Vane, and Helical Impellers
    *C. Kurt Svihla, Sundeep N. Dronawat, Jennifer A. Donnelly,
    Thomas C. Rieth, and Thomas R. Hanley*** 375

Production of Fumaric Acid by Immobilized *Rhizopus* Using Rotary
    Biofilm Contactor
    Ningjun Cao, Jianxin Du, Cheeshan S. Chen, Cheng S. Gong,
    and George T. Tsao* 387

The Effect of Pectinase on the Bubble Fractionation of Invertase
    from α-Amylase
    Veara Loha, Robert D. Tanner,* and Ales Prokop 395

Lipase Production by *Penicillium restrictum* in a Bench-Scale Fermenter:
    Effect of Carbon and Nitrogen Nutrition, Agitation, and Aeration
    *Denise M. Freire, Elaine M. F. Teles, Elba P. S. Bon,
    and Geraldo Lippel Sant' Anna, Jr.** 409

Potassium Acetate by Fermentation with *Clostridium thermoaceticum*
    Minish M. Shah, Fola Akanbi, and Munir Cheryan* 423

Membrane-Mediated Extractive Fermentation for Lactic Acid
    Production from Cellulosic Biomass
    Rongfu Chen and Y. Y. Lee* 435

Enzyme-Supported Oil Extraction from *Jatropha curcas* Seeds
    Elisabeth Winkler, Nikolaus Foidl, Georg M. Gübitz,
    Ruth Staubmann, and Walter Steiner* 449

Biogas Production from *Jatropha curcas* Press-Cake
    Ruth Staubmann, Gabriele Foidl, Nikolaus Foidl, Georg M. Gübitz,
    Robert M. Lafferty, Victoria M. Valencia Arbizu,
    and Walter Steiner* 457

Evaluation of PTMSP Membranes in Achieving Enhanced Ethanol
    Removal from Fermentations by Pervaporation
    *Sherry L. Schmidt, Michele D. Myers, Stephen S. Kelley,
    James D. McMillan, and Nandan Padukone** 469

Performance of Coimmobilized Yeast and Amyloglucosidase
    in a Fluidized Bed Reactor for Fuel Ethanol Production
    *May Y. Sun, Paul R. Bienkowski, Brian H. Davison,
    Merry A. Spurrier, and Oren F. Webb** 483

A Mathematical Model of Ethanol Fermentation from Cheese Whey:
    I: Model Development and Parameter Estimation
    Chen-Jen Wang and Rakesh K. Bajpai* 495

A Mathematical Model of Ethanol Fermentation from Cheese Whey:
  II: Simulation and Comparison with Experimental Data
  Chen-Jen Wang and Rakesh K. Bajpai* 511

Modeling Fixed and Fluidized Reactors for Cassava Starch
  Saccharification with Immobilized Enzyme
  Gisella M. Zanin* and Flávio F. De Moraes 527

Fumaric Acid Production in Airlift Loop Reactor with Porous Sparger
  Jianxin Du,* Ningjun Cao, Cheng S. Gong, George T. Tsao,
  and Naiju Yuan 541

Maximizing the Xylitol Production from Sugar Cane Bagasse
  Hydrolysate by Controlling the Aeration Rate
  Silvio S. Silva,* João D. Ribeiro, Maria G. A. Felipe,
  and Michele Vitolo 557

Production of Succinic Acid by *Anaerobiospirillum succiniciproducens*
  Nhuan P. Nghiem,* Brian H. Davison, Bruce E. Suttle,
  and Gerald R. Richardson 565

Spiral Tubular Bioreactors for Hydrogen Production by Photosynthetic
  Microorganisms: Design and Operation
  Sergei A. Markov,* Paul F. Weaver, and Michael Seibert 577

Use of a New Membrane-Reactor Saccharification Assay to Evaluate
  the Performance of Cellulases Under Simulated SSF Conditions:
  Effect on Enzyme Quality of Growing Trichoderma reesei
  in the Presence of Targeted Lignocellulosic Substrate
  John O. Baker,* Todd B. Vinzant, Christine I. Ehrman,
  William S. Adney, and Michael E. Himmel 585

SESSION 4—INDUSTRIAL NEEDS FOR COMMERCIALIZATION

Introduction to Session 4
  Dale Monceaux and James L. Gaddy 597

Net Present Value Analysis to Select Public R&D Programs and Valuate
  Expected Private Sector Participation
  Norman D. Hinman* and Mark A. Yancey 599

A Review of Techno-Economic Modeling Methodology
  for a Wood-to-Ethanol Process
  David J. Gregg and John N. Saddler* 609

SESSION 5—EMERGING TOPICS IN INDUSTRIAL BIOTECHNOLOGY

Introduction to Session 5
  Bruce E. Dale and Eric N. Kaufman 625

Coupling of Waste Water Treatment with Storage Polymer Production
H. Chua,* P. H. F. Yu, and L. Y. Ho 627

Effect of Surfactants on Carbon Monoxide Fermentations
by *Butyribacterium methylotrophicum*
M. D. Bredwell, M. D. Telgenhoff, S. Barnard, and R. M. Worden* 637

Principles for Efficient Utilization of Light for Mass Production
of Photoautotrophic Microorganisms
Amos Richmond* and Hu Qiang 649

An Optical Resolution of Racemic Organophosphorous Esters
by Phosphotriesterase-Catalyzing Hydrolysis
*Shokichi Ohuchi, Hiroyuki Nakamura, Hiroto Sugiura,
Mitsuaki Narita, and Koji Sode** 659

SESSION 6—ENVIRONMENTAL BIOTECHNOLOGY

Introduction to Session 6
Mary J. Beck and Joni M. Barnes 667

Hydrodynamic Characteristics in Aerobic Biofilm Reactor Treating
High-Strength Trade Effluent
H. Chua* and P. H. F. Yu 669

A Biological Process for the Reclamation of Flue Gas Desulfurization Gypsum
Using Mixed Sulfate-Reducing Bacteria with Inexpensive Carbon Sources
Eric N. Kaufman,* Mark H. Little, and Punjai T. Selvaraj 677

A Preliminary Cost Analysis of the Biotreatment of Refinery
Spent-Sulfidic Caustic
Kerry L. Sublette 695

The Degradation of L-Tyrosine to Phenol and Benzoate in Pig Manure:
The Role of 4-Hydroxy-Benzoate
P. Antoine,* X. Taillieu, and P. Thonart 707

The Potential for Intrinsic Bioremediation of BTEX Hydrocarbons
in Soil/Ground Water Contaminated with Gas Condensate
Abhijeet P. Borole, Kerry L. Sublette, Kevin T. Raterman,
Minoo Javanmardian, and J. Berton Fisher* 719

Adsorption of Heavy Metal Ions by Immobilized Phytic Acid
George T. Tsao, Yizhou Zheng, Jean Lu, and Cheng S. Gong* 731

Observations of Metabolite Formation and Variable Yield
in Thiodiglycol Biodegradation Process: *Impact on Reactor Design*
Tsu-Shun Lee, William A. Weigand, and William E. Bentley* 743

Pilot-Scale Bioremediation of PAH-Contaminated Soils
S. P. Pradhan, J. R. Paterek, B. Y. Liu, J. R. Conrad,
and V. J. Srivastava* 759

Relating Ground Water and Sediment Chemistry to Microbial
     Characterization at a BTEX-Contaminated Site
  S. M. Pfiffner, A. V. Palumbo,* T. Gibson, D. B. Ringelberg,
     and J. F. McCarthy 775

Retaining and Recovering Enzyme Activity During Degradation of TCE
     by Methanotrophs
  A. V. Palumbo,* J. M. Strong-Gunderson, and S. Carroll 789

Spatial and Temporal Variations of Microbial Properties at Different
     Scales in Shallow Subsurface Sediments
  Chuanlun Zhang, Richard M. Lehman, Susan M. Pfiffner,
     Shirley P. Scarborough, Anthony V. Palumbo,*
     Tommy J. Phelps, John J. Beauchamp, and Frederick S. Colwell 797

Development of a Membrane-Based Vapor-Phase Bioreactor
  Nathalie Rouhana, Naresh Handagama, and Paul R. Bienkowski* 809

Intrinsic Bioremediation of Gas Condensate Hydrocarbons: *Results
     of Over Two Years of Ground Water and Soil Core Analysis and Monitoring*
  Kerry L. Sublette,* Ravindra V. Kolhatkar, Abhijeet Borole,
     Kevin T. Raterman, Gary L. Trent, Minoo Javanmardian,
     and J. Berton Fisher 823

Effects of a Nutrient-Surfactant Compound on Solubilization Rates of TCE
  M. T. Gillespie and J. M. Strong-Gunderson* 835

Porphyrin-Catalyzed Oxidation of Trichlorophenol
  Saleem Hasan and Kerry L. Sublette* 845

Bacterial Reduction of Chromium
  Eric A. Schmieman, James N. Peterson,* David R. Yonge,
     Donald L. Johnstone, Yared Bereded-Samuel, William A. Apel,
     and Charles E. Turick 855

Degradation of Polycyclic Aromatic Hydrocarbons (PAHs) by Indigenous
     Mixed and Pure Cultures Isolated from Coastal Sediments
  Mahasin G. Tadros* and Joseph B. Hughes 865

Reduction of $Cr(6^+)$ to $Cr(3^+)$ in a Packed-Bed Bioreactor
  Charles E. Turick,* Carl E. Camp, and William A. Apel 871

Managed Bioremediation of Soil Contaminated with Crude Oil:
     Soil Chemistry and Microbial Ecology Three Years Later
  Kathleen Duncan, Estelle Levetin, Harrington Wells,
     Eleanor Jennings, Susan Hettenbach, Scott Bailey,
     Kevin Lawlor, Kerry Sublette,* and J. Berton Fisher 879

Author Index 891

Subject Index 895

Copyright © 1997 by Humana Press Inc.
All rights of any nature whatsoever reserved.
0273-2289/97/63-65—0001$8.50

Session 1

Thermal, Chemical, and Biological Processing

MARK T. HOLTZAPPLE[1] AND ROBERT TORGET[2]

[1]*Texas A&M University, College Station, TX;*
and [2]*National Renewable Energy Laboratory, Golden, CO*

This session focuses on processing steps required to convert the carbohydrates of biomass into fuels and chemicals. For illustrative purposes, Fig. 1 shows a representative block flow diagram for a bioethanol process. (Of course, other approaches may be taken as well.) First, the biomass is pretreated to render it more digestible. Then enzymes are added that hydrolyze cellulose and hemicellulose into sugars. Finally, microorganisms (e.g., yeast) ferment these sugars to ethanol, which is recovered by distillation. The papers in this session describe various aspects of the process shown in Fig. 1, such as new developments in pretreatments, models of the saccharification and fermentation steps, enzyme recovery, combined enzyme/ethanol production, and a process that allows ethanol to be produced from biomass contaminated with radioactivity.

The Ammonia Fiber Explosion (AFEX) process has been optimized to pretreat corn fiber, a residue from corn wet-milling operations. The corn fiber was treated with liquid ammonia at 90°C; then the pressure was instantaneously released causing the corn fiber to explode. The resulting material was very digestible. Recombinant *Saccharomyces* strain (1400pLNH32) fermented both hexose and xyloses to ethanol with high efficiency.

Dilute-acid pretreatment allows fairly selective removal of hemicellulose from biomass, but the remaining solids are high in lignin. By removing the lignin with ammonia recycled percolation (ARP), the remaining solids are relatively pure cellulose. Thus, the three dominant components of biomass (cellulose, hemicellulose, lignin) are separated into relatively pure fractions. Because the lignin is removed, enzymatic saccharification of the relatively pure cellulose requires less enzyme and allows for better enzyme recovery. When conditions are properly optimized, lime pretreatment significantly increases the digestibility of herbaceous biomass. Many previous studies of lime pretreatment used suboptimal conditions; therefore, it has unfairly acquired a reputation for being a poor pretreatment agent compared to other alkalis (e.g., sodium hydroxide, ammonia). After the pretreatment, spent lime can be recovered by carbonating the wash water to form calcium carbonate, which is subsequently recovered in a lime kiln.

Kinetic models of the saccharification and fermentation portions of biomass processes are important design tools; in addition, models can bring fundamental insights into the process. A two-phase dynamic model accounts for the heterogeneous nature of biomass. This model describes enzyme adsorption with subsequent hydrolysis of the biomass and allows for a variety of particle geometries. It successfully describes the anaerobic digestion of cellulose and sludge.

Enzymes are a significant cost in the process shown in Fig. 1. A novel process allows the enzymes to be recovered and recycled that potentially will lower costs. This process precipitates the enzymes with subsequent recovery using continuous-column flotation.

Another approach to lowering the cost of enzymes is to consolidate enzyme production and fermentation into a single step. An important challenge of this approach is to obtain high enzyme productivity in the presence of high ethanol concentrations. New developments have shown that inhibition of *Clostridium thermosaccharolyticum*, which is generally attributed to ethanol, may actually be due to an insufficient nutrient supply or the accumulation of ions used to maintain a neutral pH.

The recent release of radioactivity from Chernobyl has contaminated vast regions of Russia. The biomass from these regions is radioactive, and hence, has little value. A process such as that shown in Fig. 1 allows the radioactive ash to be separated from the digestible portions of the biomass, thus having the double benefit of removing radioactivity from the land and producing useful fuels and chemicals from the biomass.

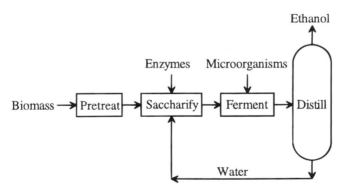

Fig. 1. A process for converting biomass to ethanol.

Lime Pretreatment of Switchgrass

VINCENT S. CHANG, BARRY BURR,
AND MARK T. HOLTZAPPLE*

Department of Chemical Engineering, Texas A&M University, College Station, TX 77843

ABSTRACT

Lime (calcium hydroxide) was used as a pretreatment agent to enhance the enzymatic digestibility of switchgrass. After studying many conditions, the recommended pretreatment conditions are: time = 2 h, temperature = 100°C and 120°C, lime loading = 0.1 g $Ca(OH)_2$/g dry biomass, water loading = 9 mL/g dry biomass. Studies on the effect of particle size indicate that there was little benefit of grinding below 20 mesh; even coarse particles (4–10 mesh) digested well. Using the recommended pretreatment conditions, the 3-d reducing sugar yield was five times that of untreated switchgrass, the 3-d total sugar (glucose + xylose) yield was seven times, the 3-d glucose yield was five times, and the 3-d xylose yield was 21 times. A material balance study showed that little glucan (approx 10%) was solubilized as a result of the lime pretreatment, whereas about 26% of xylan and 29% of lignin became solubilized.

Index Entries: Lignocellulose; pretreatment; lime; cellulase; sugar.

INTRODUCTION

Lignocellulose is a valuable alternative energy source because it is widely available and can be converted to various organic compounds such as sugars and alcohols *(1)*. The susceptibility of lignocellulosic biomass to enzymatic hydrolysis is constrained because of structural characteristics such as cellulose crystallinity, hemicellulose acetylation, inaccessible surface area, and lignin content *(2,3)*. To enhance enzymatic hydrolysis, pretreatment is essential. In general, pretreatment methods are of four types: physical, chemical, multiple (i.e., physical + chemical), and biological. Excellent reviews on pretreatments have been published by Lin et al. *(4)*, Fan et al. *(5)*, Chang et al. *(6)*, and Weil et al. *(7)*.

*Author to whom all correspondence and reprint requests should be addressed.

Of the chemicals used as pretreatment agents, alkalis have received much attention (8). According to fundamental studies by Kong et al. (3), alkalis remove acetate groups from hemicellulose, thereby reducing the steric hindrance of hydrolytic enzymes and greatly enhancing carbohydrate digestibility. Sodium hydroxide effectively enhances lignocellulose digestibility (9–11), but it has several disadvantages; it is expensive ($0.62/kg) (12), dangerous to handle, and difficult to recover. Ammonia pretreatment has also received much attention (13,14) because it is easy to recover; unfortunately, it is moderately expensive ($0.12/kg) (12) and requires careful handling. In contrast, lime (calcium hydroxide) has many advantages; it is very inexpensive ($0.04/kg) (12), is safe, and can be recovered by carbonating wash water (15). Table 1 summarizes the results of previous lime pretreatment studies (16–29). Although these results showed that lignocellulose digestibility improved by lime pretreatment, authors usually concluded that calcium hydroxide was less effective than other alkalis such as sodium hydroxide (16–18,22,23,25,29), ammonia (20,25,29), and potassium hydroxide (22,23). However, many of these comparisons employed the same pretreatment conditions, i.e., regardless of the alkali studied, equal amounts of water and alkali were used. Because calcium hydroxide is a weak alkali and poorly soluble in water, these studies put lime at a disadvantage. By modifying the pretreatment conditions to be compatible with lime, it is as effective as other alkalis in enhancing lignocellulose digestibility (28).

The bioconversion of crop residues (e.g., wheat straw, barley straw, corn cob, etc.) is widely studied; however, relatively few studies have been performed on forage grasses. Switchgrass (*Panicum virgatum*) is a good forage species that tolerates a wide variety of environmental conditions (30). It grows in all regions of the United States, except the Northwest and California (31). Because of its excellent growth, switchgrass is a potential renewable energy source that deserves further investigation. In this research, the effects of lime pretreatment conditions were studied, the effect of cellulase loading was determined, hydrolysis profiles were performed, and material balances were made to determine how much biomass is solubilized as a result of lime pretreatment.

MATERIALS AND METHODS

Lime Pretreatment

Switchgrass was pretreated with lime (calcium hydroxide) in the presence of water. Six 3.8-cm I.D. × 12.7-cm long, 304 stainless steel-capped pipe nipples were used as the pretreatment reactors. To ensure thorough mixing of contents inside, the pipe nipples were mounted on a rotating shaft inside an oven providing the desired pretreatment temperature. To perform the pretreatment, the oven was heated in advance to reach the desired temperature. A measured amount of switchgrass (7.5 g dry weight) and calcium

Table 1
Summary of Lime Pretreatment Conditions and Results

Biomass	Temp. (°C)	Time (h)	Lime Loading (g Ca(OH)$_2$/g dry biomass)	Water Loading (mL/g dry biomass)	Particle Size	Effect on Digestibility (%)	Reference
in vivo							
Corn stalks	amb.	24	0.12	8	2-4 cm	51.55 to 85.59[a]	(16)
Sorghum stalks	amb.	24	0.12	8	2-4 cm	49.93 to 83.64[a]	(16)
Sweet potato vines	amb.	24	0.12	8	2-4 cm	38.59 to 61.28[a]	(16)
Corn cobs	amb.	336	0.04	1.5	ground	improved digestibility	(17)
Barley straw	amb.	2160	0.056	0.9	chopped	54.8 to 56.9[b]	(18)
Wheat straw	amb.	24	0.09	11	3 cm	54.1 to 61.9[b]	(19)
Corn stover	21	336	0.04	0.54	0.95 cm	53.2 to 54.5[b]	(20)
in vitro							
Poplar bark	amb.	24, 3600	5.3, 10.6, 15.9, **21.1**	1.5	0.95 cm	38.6 to 52.0[c]	(21)
Corn Cobs	amb.	24	0.074, 0.0925	1	20 mesh	-----	(22)
Barley straw	amb.	2160	0.025, 0.05, **0.1**	1.2	5 cm	47.6 to 64.5[b]	(23)
Corn cobs	amb.	24 to 360	0.05	0.25, **0.67**, 1.5	-----	52 to 70[d]	(24)
Barley straw	amb.	24	0.05, **0.1**, 0.15	1, **20**	1.5 cm	37.9 to 52.5[e]	(25)
Pea straw	amb.	24	**0.05**, 0.1, 0.15	**1**, 20	1.5 cm	41.7 to 37.6[e]	(25)
Bagasse	amb.	24	0.05, **0.1**, 0.15	1, **20**	particulate	32.8 to 43.0[e]	(25)
Sunflower hulls	amb.	24	0.05, **0.1**, 0.15	1, **20**	particulate	16.6 to 20.3[e]	(25)
Bagasse	20	192	0.06, 0.09, 0.12, 0.15, 0.18, 0.24, **0.30**	1.74	0.225 cm	19.7 to 72.4[b]	(26)
Rice straw	amb.	24	0.05, 0.1, **0.2**	4, **8**, 12	-----	40 to 59[e]	(27)
Corn cobs	45	>6	0.05	1.5	-----	43 to 54[f]	(28)
Soya-bean straw	amb.	720	0.02, 0.03, **0.04**, 0.05	0.65	chopped	35.8 to 41.3[e]	(29)

amb. = ambient temperature
Bold conditions correspond to the best lime pretreatment results.
[a] Crude fibre digestion coefficient
[b] Organic matter digestibility (OMD)
[c] *in vitro* true digestibility (IVTD)
[d] *in vitro* dry matter digestibility (IVDMD)
[e] *in vitro* organic matter digestibility (IVOMD)
[f] *in vitro* digestibility (IVD)

hydroxide (according to the desired lime loading) were placed in each reactor and thoroughly mixed using a spatula. Distilled water (according to the desired water loading) was then added to the dry mixture. After tightly capping the reactors, they were placed in boiling water for 10 min to quickly reach a high temperature and then put into the oven. The motor was turned on to start the rotating shaft and the reactors were left for the desired pretreatment time. After the pretreatment time elapsed, the reactors were moved out of the oven and immersed in cold water to cool them to ambient temperature. Samples were then removed for enzymatic hydrolysis.

Enzymatic Hydrolysis

Lime-pretreated switchgrass was transferred from the reactors to 500-mL Erlenmeyer flasks with distilled water. Citrate buffer (7.5 mL, 1.0M, pH 4.8) and sodium azide (5 mL, 0.01 g/mL) were added to the slurry to keep a constant pH and prevent microbial growth, respectively. Glacial acetic acid was added to reduce the pH from about 11.5 to 4.8. Then, the total volume of the slurry was adjusted to 150 mL by adding distilled water. The flasks were placed in a 100-rpm shaking air bath. When the temperature reached 50°C, cellulase (5 FPU/g dry biomass) and cellobiase (28.4 CbU/g dry biomass) were added to the flasks. The activity of the cellulase (Cytolase CL enzyme, lot no. 17-92262-09, Environmental BioTechnologies, Santa Rosa, CA) was 91 FPU/mL, as determined using the filter paper assay [32]. The activity of cellobiase (Novozym 188, batch no. DCN0015, Novo Nordisk Bioindustrials, Franklinton, NC) was 250 CbU/g.

To investigate lime treatment conditions, enzyme hydrolysis samples (approx 4 mL) were withdrawn after 3 d and then boiled for 15 min in sealed tubes to denature the enzymes and thus prevent further hydrolysis; then reducing sugars were measured. When the hydrolysis profiles were performed, samples were withdrawn as a function of time (i.e., 0, 1, 3, 6, 10, 16, 24, 36, 48, and 72 h) and boiled for 15 min in sealed tubes; then glucose, xylose, and reducing sugars were measured at each time point. The same procedure was also applied to untreated switchgrass.

Sugar Measurement

Reducing sugars were measured using the dinitrosalicylic acid (DNS) assay [33]. A 200-mg/dL glucose standard solution (Yellow Springs Instruments, Yellow Springs, OH) was used for the calibration, thus the reducing sugars were measured as "equivalent glucose." The sugar content in the enzymes (approx 4.2 mg eq. glucose/g dry biomass) was subtracted from the original reducing sugar yields to determine the actual amounts of reducing sugar produced from the biomass. After subtracting the enzyme sugars, the yields were multiplied by a correction factor to account for calcium acetate inhibition and were called "corrected" reducing sugar yields. The correction factor depends upon lime loading (see Table 2).

Table 2
Effects of Calcium Acetate Inhibition and Correction Factors

Lime Addition (g Ca(OH)$_2$/g dry biomass)	3-d Reducing Sugar Yield (mg eq. glucose/g dry biomass)	Correction Factor for Calcium Acetate Inhibition
0	465	1.000
0.02	491	0.948
0.05	464	1.002
0.1	458	1.015
0.2	442	1.051
0.3	419	1.110

Glucose and xylose were measured using high performance liquid chromatography (HPLC). A Bio-Rad (Cambridge, MA) Aminex HPX-87P column was used for carbohydrate separation; a refractive index detector (LDC/Milton Roy, refractoMonitor III, Riviera Beach, FL) was used to detect sugars; a Spectra-Physics (San Jose, CA) integrator (SP4270) was used for integration. Throughout the paper, total sugar denotes the summation of glucose and xylose because no other carbohydrates were detected.

There are discrepancies between the reducing sugar measurements and the total sugar measurements because of inaccuracies associated with expressing xylose as equivalent glucose; nonetheless, the DNS assay was accurate enough to rapidly screen the pretreatment conditions in the early stages of this study.

Dry Weight and Composition

Throughout the paper, the sugar yields were calculated based on the dry weight of biomass that was determined by drying the biomass at 105°C for 8 h. However, when performing material balances, the biomass samples (raw, washed only, and pretreated and washed) were dried at 45°C to prevent carbohydrate destruction. A part of the 45°C-dried samples was then dried at 105°C to determine its dry weight, whereas the rest was used to determine its composition (i.e., lignin, glucan, xylan, crude protein, and ash) (32).

RESULTS AND DISCUSSION

Effects of Calcium Acetate Inhibition

After biomass is treated with lime, the pH is as high as 11.5, which is incompatible with cellulase. For convenience in a laboratory setting, the lime was neutralized with acetic acid. (Industrially, the lime would be removed by washing the lime-treated biomass with water that is subsequently carbonated to precipitate calcium carbonate. To regenerate the lime from calcium carbonate, a lime kiln would be employed.) However,

Table 3
Lime Pretreatment Conditions Explored for Switchgrass

	Time (h)	Temperature (°C)	Lime Loading (g Ca(OH)$_2$/g dry biomass)	Water Loading (mL/g dry biomass)	Particle Size (Mesh)
Study 1	1 to 24	60 to 130	0.1	10	-40
Study 2	3	100 and 120	0.01 to 0.30	10	-40
Study 3	3	100 and 120	0.1	5 to 15	-40
Study 4	3	100	0.1	9	5 to -80
Study 5	1 to 3	100	0.1	9	-40

the resulting calcium acetate may inhibit cellulase. To measure the effects of calcium acetate inhibition, the following experiment was conducted:

A large sample (approx 200 g) of switchgrass (–40 mesh) was treated with lime. The pretreatment was performed using recommended conditions by Nagwani (15) (temperature = 100°C, time = 1 h, lime loading = 0.1 g Ca(OH)$_2$/g dry biomass, and water loading = 10 mL/g dry biomass). The pretreated switchgrass was washed with fresh distilled water 10 times to remove the lime. Then, the pretreated-and-washed switchgrass was air-dried and divided into 12 flasks that contained citrate buffer and various lime additions (0, 0.02, 0.05, 0.1, 0.2, and 0.3 g Ca(OH)$_2$/g dry biomass, each in duplicate). The lime added to each flask was neutralized by adding various amounts of acetic acid such that the pH of each flask was 4.8. Then enzymatic hydrolysis was performed at 50°C for 3 d. The reducing sugar yields at each lime loading were measured as a function of time using the DNS assay. The reducing sugar yields of the time-zero samples determined the sugar content of the enzymes and were subtracted from the sugar yields at other time points.

Table 2 summarizes the 3-d reducing sugar yields. As anticipated, calcium acetate inhibited the enzyme which caused about 1.5% loss of sugar yields at recommended lime loadings. The correction factors reported in Table 2 were used in the subsequent experiments to correct the 3-d reducing sugar yields.

Effects of Pretreatment Conditions

The pretreatment conditions were systematically varied to explore the effects of process variables (i.e., time, temperature, lime loading, water loading, and biomass particle size) on digestibility. Although some researchers have explored the effects of lime loading (21–23,25–27,29) and water loading (24,25,27), little has been done on the effects of pretreatment time (21) or temperature. Recently, Nagwani (15) determined that time and temperature had the greatest impact on biomass digestibility. Lime load-

Lime Pretreatment 9

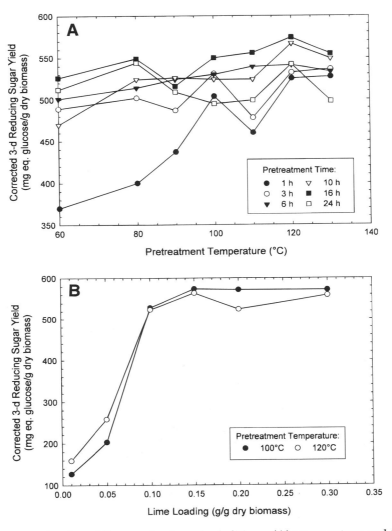

Fig. 1. Investigation of lime pretreatment conditions: **(A)** temperature and time, **(B)** lime loading, **(C)** water loading, **(D)** particle size.

ing generally had a critical value (approx 0.1 g Ca(OH)$_2$/g dry biomass) below which the digestibility greatly declined, and above which the digestibility only increased slightly. Water loading had little effect on the digestibility. Therefore, this study was conducted to hold the low-impact variables (e.g., lime loading, water loading, and particle size) constant while systematically varying the high-impact variables (e.g., time and temperature). Table 3 shows the range of conditions explored.

Figure 1A shows the 3-d reducing sugar yields as a function of pretreatment temperature at various reaction times. The best temperature lies between 100 and 120°C. The best pretreatment resulted after 16 h, but this

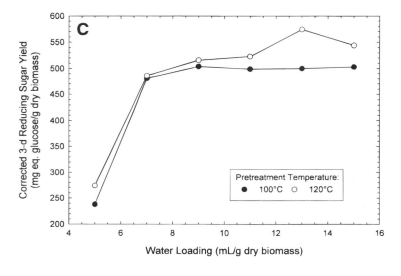

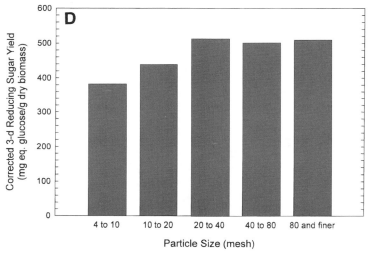

Fig. 1. (continued)

is excessively long from an economic viewpoint. One hour of reaction time was obviously insufficient to achieve good sugar yields, whereas pretreatment times longer than 3 h had little additional benefit. Therefore, 3 h was temporarily selected as a standard pretreatment time with which other investigations were conducted.

Figure 1B shows the effect of lime loading at two pretreatment temperatures (100 and 120°C). The most effective lime loading was 0.1 g $Ca(OH)_2$/g dry biomass. Although slightly greater sugar yields were obtained at a lime loading of 0.15 g $Ca(OH)_2$/g dry biomass, the 50% increase in lime consumption cannot be justified from an economic viewpoint. The results at 100 and 120°C were very similar.

Table 4
Further Investigation of Pretreatment Time

Pretreatment Time (h)	3-d Corrected Reducing Sugar Yield (mg eq. glucose/g dry biomass)
1	447
1.5	469
2[a]	483
3	486

[a]Recommended pretreatment time.

Figure 1C shows the effect of water loading at two pretreatment temperatures (100 and 120°C). Although water loadings as low as 7 mL/g dry biomass are effective, there is little economic incentive to reduce the water loadings to a bare minimum. Therefore, a water loading of 9 mL/g dry biomass, which has a slight extra benefit (approx 5% increase in sugar yields), can be used.

Figure 1D shows the effect of biomass particle size on the digestibility. Five different particle sizes were studied. Grinding to less than 20 mesh is sufficient for lime pretreatment. Even though it was not necessary to grind biomass below 40 mesh, subsequent experiments were performed using particle sizes less than 40 mesh because there was a large quantity of this material available. Also, it is of a more uniform particle size that reduces variability between experiments.

A higher resolution study of pretreatment times was performed to focus in the range of 1 to 3 h. The reducing sugar yields were measured for biomass samples that had been pretreated for 1, 1.5, 2, and 3 h. Table 4 shows that full pretreatment likely occurs after 2 h; therefore, it was selected as the standard pretreatment time.

Enzyme Loading Studies

A cellulase loading of 5 FPU/g dry biomass was used in the studies presented in Fig. 1 and Table 2. Here, cellulase loading is studied to determine if there are yield benefits from loadings higher than 5 FPU/g dry biomass or if cellulase loadings less than 5 FPU/g dry biomass are acceptable.

Approximately 90 g of switchgrass (–40 mesh) was pretreated using the recommended conditions (i.e., time = 2 h, temperature = 120°C, lime loading = 0.1 g $Ca(OH)_2$/g dry biomass, water loading = 9 mL/g dry biomass). The pretreated and untreated switchgrass (concentration = 50 g/L) were hydrolyzed at 50°C, pH 4.8 for 3 d in a 100-rpm air shaker, using an excess cellobiase loading (i.e., 28.4 CbU/g dry biomass) and various cellulase loadings (i.e., 0, 1, 3, 5, 10, 25, 50, 75, 100 FPU/g dry biomass).

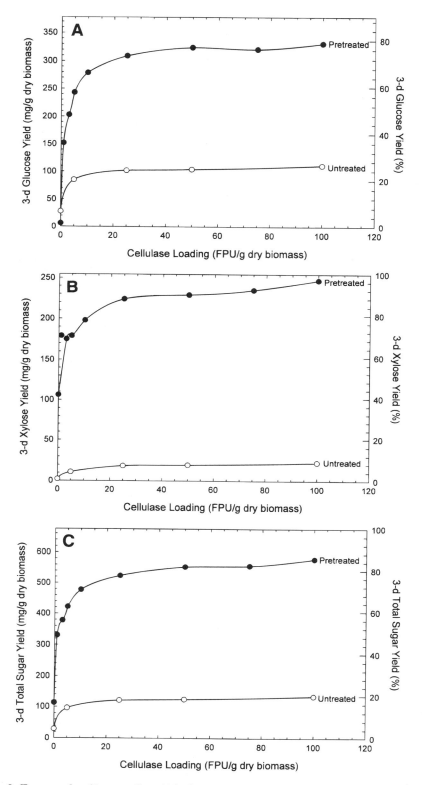

Fig. 2. Enzyme loading studies: **(A)** glucose yields, **(B)** xylose yields, **(C)** total sugar yields. (Pretreatment conditions: 120°C, 2 h, 0.1 g Ca(OH)$_2$/g dry biomass, 9 mL water/g dry biomass. Hydrolysis conditions: 50°C, pH 4.8, 28.4 CbU cellobiase/g dry biomass.)

The 3-d glucose, xylose, and "total sugar" (glucose + xylose) yields of untreated and pretreated switchgrass at various cellulase loadings are shown in Fig. 2. For cellulase loadings larger than 25 FPU/g dry biomass, the sugar yields remain essentially constant because the cellulose sites are likely saturated by the enzyme. The maximum total sugar yield was 85%. A cellulase loading of 5 FPU/g dry biomass is the "shoulder" of the curve. Although a cellulase loading of 10 FPU/g dry biomass increases total sugar yield by about 13%, doubling enzyme usage cannot be justified economically. (For example, assuming an enzyme cost of $9.4/million FPU *(34)* and an ethanol yield of 100 gal/t dry biomass, an enzyme loading of 5 FPU/g dry biomass corresponds to $0.47/gal ethanol, which is about 43% of its sales price; doubling the enzyme loading is impossible unless enzyme cost is reduced significantly.)

Hydrolysis Profiles

Instead of measuring just 3-d reducing sugar yields, a complete hydrolysis profile was measured so the sugar yields were determined as a function of time. Switchgrass (–40 mesh) was pretreated at 120°C for 2 h in the presence of 0.1 g $Ca(OH)_2$/g dry biomass and 9 mL water/g dry biomass. The pretreated switchgrass was then transferred from the reactors to Erlenmeyer flasks for enzymatic hydrolysis. This experiment was performed in triplicate.

Using cellulase loadings of 5 FPU/g dry biomass, Fig. 3 shows the yields of reducing sugars, glucose, xylose, and "total sugar" (glucose + xylose). (Note: Because of high cellobiase activity, cellobiose concentrations were negligible.) The sugar yields from pretreated switchgrass are significantly higher than from untreated switchgrass. The 3-d reducing sugar yield of pretreated switchgrass increased about five times (i.e., from 102 to 538 mg eq. glucose/g dry biomass), the 3-d total sugar yield increased about seven times (i.e., from 8.7% to 58.1%), the 3-d glucose yield increased about five times (i.e., from 12.3% to 58.0%), and the 3-d xylose yield increased about 21 times (i.e., from 2.8% to 58.1%). This dramatic increase in xylose yield indicates that lime has a selective effect on hemicellulose. The likely mechanism is that lime removes acetate groups from hemicellulose rendering it more accessible to hydrolytic enzymes *(3)*.

Material Balances

To remove solubles, switchgrass (either untreated or pretreated with the recommended conditions) was repeatedly washed with fresh distilled water until the decanted water became colorless. The total dry weight of the sample was measured before and after the pretreatment and wash. The compositions (i.e., glucan, xylan, lignin, crude protein, and ash) of raw, washed only, and pretreated and washed switchgrass were then determined (*see* Fig. 4).

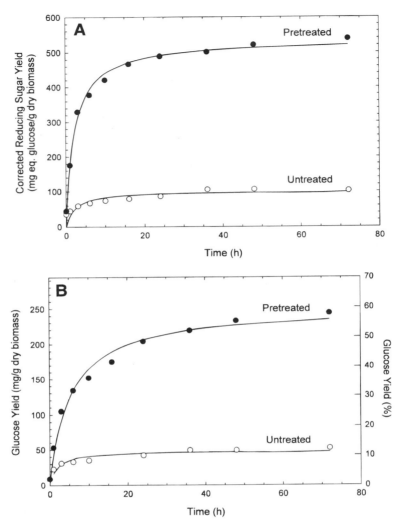

Fig. 3. 3-d hydrolysis profiles: (**A**) corrected reducing sugar, (**B**) glucose, (**C**) xylose, (**D**) total sugar. (Pretreatment conditions: 120°C, 2 h, 0.1 g Ca(OH)$_2$/g dry biomass, 9 mL water/g dry biomass. Hydrolysis conditions: 50°C, pH 4.8, 5 FPU cellulase/g dry biomass, 28.4 CbU cellobiase/g dry biomass.)

Table 5 summarizes the losses of each component before and after pretreatment. All the components became more water soluble because of the lime pretreatment, except ash. Fairly large quantities of lignin, xylan (hemicellulose), crude protein, and other components (e.g., extractives) were removed by the lime pretreatment, whereas little glucan (cellulose) was removed; hence, the pretreated and washed biomass was slightly enriched in cellulose. The results suggest that the removal of lignin and hemicellulose both contribute to the increase of the biomass digestibility.

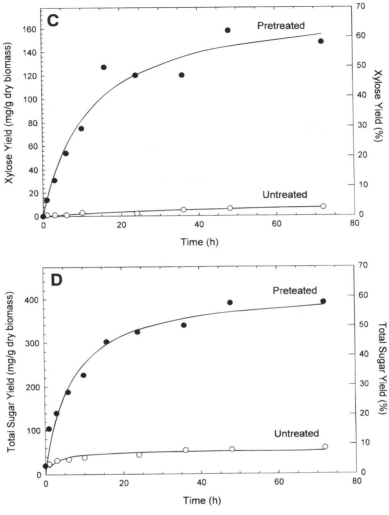

Fig. 3. (continued)

Lime Recovery and Recycle

Lime, although inexpensive, still requires recovery and recycle for it to be an economically viable pretreatment agent. The costs of various recovery alternatives have been evaluated (35). For example, the pretreated biomass can be washed with water to remove the lime. The wash water can be contacted with carbon dioxide to precipitate calcium carbonate, which subsequently may be separated from the liquid and converted to lime using a lime kiln. The lime kiln operates at a high temperature (900°C), so the exhaust gases are hot enough to generate steam for motive power or process heat; thus, there is not a significant energy penalty asso-

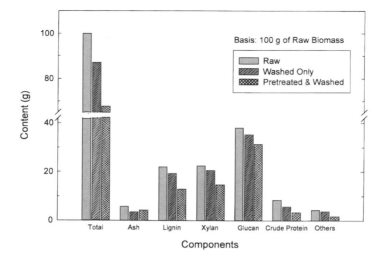

Fig. 4. Material balances for raw, washed only, and pretreated and washed switchgrass. (Pretreatment conditions: 120°C, 2 h, 0.1 g Ca(OH)$_2$/g dry biomass, 9 mL water/g dry biomass.)

Table 5
Summary of Water-Solubility of Switchgrass Components
Before and After Lime Pretreatment

Components	Raw Composition (g component/g total)	Weight Loss Percentage[a]		Amount Removed by Lime Pretreatment
		Washed Only	Pretreated & Washed	
Total	-----	13.0%	32.4%	19.4%
Ash	0.06	40.6%	27.6%	-13.0%
Lignin	0.22	11.4%	40.8%	29.4%
Xylan	0.22	8.5%	35.0%	26.5%
Glucan	0.38	7.7%	17.3%	9.7%
Crude Protein	0.08	31.8%	59.7%	27.9%
Others	0.04	19.2%	63.3%	44.1%

[a]Weight percentage based on the initial weight of each component.

ciated with operating the lime kiln. This process allows lime to be recycled so that the cost of lime consumption is effectively minimized.

CONCLUSIONS

The recommended conditions for lime pretreatment of switchgrass are: temperature = 100–120°C, time = 2 h, lime loading = 0.1 g Ca(OH)$_2$/g dry biomass, water loading = 9 mL/g dry biomass, and particle size = 20 mesh and finer. Under these conditions, the 3-d total sugar yield of pre-

Table 6
Summary of Effects of Various Alkaline Pretreatment Agents

Pretreatment Agent	Biomass	Untreated Digestibility (%)	Treated Digestibility[a] (%)	Ratio of Digestibility	Reference
NaOH	Corn stalks	51.55[b]	91.84[b]	1.78	(16)
	Sorghum stalks	49.93[b]	93.64[b]	1.88	(16)
	Sweet potato vine	38.59[b]	78.05[b]	2.02	(16)
	Barley straw	54.8[c]	67.4[c]	1.23	(18)
	Barley straw	47.6[c]	75.7[c]	1.59	(23)
	Barley straw	37.9[d]	59.7[d]	1.58	(25)
	Pea straw	41.7[d]	54.6[d]	1.31	(25)
	Bagasse	32.8[d]	51.7[d]	1.58	(25)
	Sunflower hulls	16.6[d]	23.1[d]	1.39	(25)
	Soya-bean straw	35.8[d]	44.2[d]	1.23	(29)
NH$_3$ or NH$_4$OH	Corn stover	53.2[c]	62.8[c]	1.18	(20)
	Barley straw	37.9[d]	59.7[d]	1.58	(25)
	Pea straw	41.7[d]	54.6[d]	1.31	(25)
	Bagasse	32.8[d]	51.7[d]	1.58	(25)
	Sunflower hulls	16.6[d]	23.1[d]	1.39	(25)
	Soya-bean straw	35.8[d]	43.3[d]	1.21	(29)
KOH	Barley straw	47.6[c]	69.7[c]	1.46	(23)
Ca(OH)$_2$	Switchgrass	19.8[e]	85.6[e]	4.32	This work

[a]Best result reported in the literature.
[b]Crude fibre digestion coefficient.
[c]Organic matter digestibility (OMD).
[d]*in vitro* organic matter digestibility (IVOMD).
[e]3-d total sugar yield at 100 FPU cellulase/g dry biomass.

treated switchgrass increases from 19.8 to 85.6% at an excess enzyme loading. Table 6 summarizes the effects of pretreatments using various alkalis as pretreatment agents. In most cases, the authors concluded that calcium hydroxide is not as effective as other alkalis such as sodium hydroxide, ammonia, and potassium hydroxide. However, when proper conditions are employed, the digestibility improvement resulting from lime pretreatment is significant.

For industrial applications, a pretreatment agent must be effective, economical, safe, environmentally friendly, easy-to-use, and easy-to-recover. Lime meets these objectives and warrants further research.

ACKNOWLEDGMENTS

The research is supported by NREL subcontract No. XAW-3-11181-03.

REFERENCES

1. Klyosov, A. A. (1986), *Appl. Biochem. Biotechnol.* **12**, 249–300.
2. Fan, L. T., Lee, Y.-H., and Gharpuray, M. M. (1982), *Adv. Biochem. Eng.* **23**, 157–187.
3. Kong, R., Engler, C. R. and Soltes, E. J. (1992), *Appl. Biochem. Biotechnol.* **34**, 23–35.
4. Lin, K. W., Ladisch, M. R., Schaefer, D. M., Noller, C. H., Lechtenberg, V., and Tsao, G. T. (1981), *AIChE Symposium Series* **207**, vol. 77, 102–106.
5. Fan, L. T., Gharpuray, M. M., and Lee, Y.-H., (1987), in *Cellulose Hydrolysis*, Springer-Verlag, Berlin, pp. 52–69.
6. Chang, M. M., Chou, T. Y. C., and Tsao, G. T. (1982), *Adv. Biochem. Eng.* **20**, 15–42.
7. Weil, J., Westgate, P., Kohlmann, K., and Ladisch, M. R. (1994), *Enzyme Microb. Technol.* **16**, 1002–1004.
8. Jackson, M. G. (1977), *Anim. Feed Sci. Technol.* **2**, 105–130.
9. Anderson, D. C. and Ralston, A. T. (1973), *J. Anim. Sci.* **37**, no. 1, 148–152.
10. Lesoing, G., Klopfenstein, T., Rush, I., and Ward, J. (1981), *J. Anim. Sci.* **51**, no. 2, 263–269.
11. Clunningham, R. L. and Carr, M. E. (1984), *Biotechnol. Bioeng. Symp.* **14**, 95–103.
12. Peters, M. S. and Timmerhaus, K. D. (1991), in *Plant Design and Economics for Chemical Engineers*, McGraw-Hill, New York, NY, p. 816.
13. Holtzapple, M. T., Jun, J.-H., Ashok, G., Patibandla, S. L., and Dale, B. E. (1991), *Appl. Biochem. Biotechnol.* **28**, 59–74.
14. Holtzapple, M. T., Lundeen, J. E., Sturgis, R., Lewis, J. E., and Dale, B. E. (1992), *Appl. Biochem. Biotechnol.* **34**, 5–21.
15. Nagwani, M. (1992), M.S. thesis, Texas A&M University, College Station, Texas.
16. Abou-Raya, A. K., Abou-Hussein, E. R. M., Ghoneim, A., Raafat, M. A., and Mohamed, A. A. (1964), *J. Anim. Prod. U. A. R.* **4**, no. 1, 55–65.
17. Waller, J. C. and Klopfenstein, T. (1975), *J. Anim. Sci.* **41**, no. 1, 242–425 (abstract).
18. Owen, E. and Nwadukwe, B. S. (1980), *Anim. Prod.* **30**, 489 (abstract).
19. Djajanegara, A., Molina, B. T., and Doyle, P. T. (1984), *Anim. Feed Sci. Technol.* **12**, 141–150.
20. Oliveros, B. A., Britton, R. A., and Klopfenstein, T. J. (1993), *Anim. Feed Sci. Technol.* **44**, 59–72.
21. Gharib, F. H., Meiske, J. C., Goodrich, R. D., and Serafy, A. M. (1975), *J. Anim. Sci.* **40**, no. 4, 734–742.
22. Rounds, W., Klopfenstein, T., Waller, J., and Messersmith, T. (1976), *J. Anim. Sci.* **43**, no. 2, 478–482.
23. Wilkinson, J. M. and Santillana, R. G. (1978), *Anim. Feed Sci. Technol.* **3**, 117–132.
24. Paterson, J. A., Stock, A. R., and Klopfenstein, T. J. (1980), in *Calcium Hydroxide Treatment of Crop Residues*, Nebraska Beef Cattle Rep., EC 80-218, 21.
25. Ibrahim, M. N. M. and Pearce, G. R. (1983), *Agricultural Wastes* **5**, 135–156.
26. Playne, M. J. (1984), *Biotechnol. Bioeng.* **26**, 426–433.
27. Winugroho, M., Ibrahim, M. N. M., and Pearce, G. R. (1984), *Agricultural Wastes* **9**, 87–89.
28. Owen, E., Klopfenstein, T., and Urio, N. A. (1984), in *Straw and Other Fibrous By-products as Feed*, Sundstøl, F. and Owen, E., eds., Elsevier Science Publishers B. V., Amsterdam, Netherlands, pp. 248–275.
29. Felix, A., Hill, R. A., and Diarra, B. (1990), *Anim. Prod.* **51**, 47–61.
30. Mohlenbrock, R. H. (1973), in *The Illustrated Flora of Illinois: Grasses Panicum to Danthonia*, Feffer & Simons, Inc., London and Armsterdam, pp. 67, 68.

31. Gould, F. W. (1968), in *Grass Systematics*, McGraw-Hill, New York, p. 214.
32. *Chemical Analysis & Testing Standard Procedure*, National Renewable Energy Laboratory.
33. Miller, G. L. (1959), *Anal. Chem.* **31**, 426–428.
34. Holtzapple, M. T., Ripley, E. P., and Nikolaou, M. (1994), *Biotechnol. Bioeng.* **44**, 1122–1131.
35. Burr, B., Chang, V. S., Holtzapple, M. T., and Davidson, R. (1996), in *Development of Alternative Pretreatment and Biomass Fractionation Processes: Lime Pretreatment*, Final Report, Part II, Subcontract XAW-3-11181-03, National Renewable Energy Laboratory.

Ammonia Recycled Percolation as a Complementary Pretreatment to the Dilute-Acid Process

ZHANGWEN WU AND Y. Y. LEE*

Chemical Engineering Department, Auburn University, Auburn, AL 36849

ABSTRACT

A two-stage dilute-acid percolation (DA) was investigated as a pretreatment method for switchgrass. With use of extremely low acid (0.078 wt% sulfuric acid) under moderate temperature (145–170°C), hemicellulose in switchgrass was completely solubilized showing no sugar decomposition. The treated switchgrass contained about 70% glucan and 30% lignin. The high lignin content in the treated feedstock raises a concern that it may cause a high enzyme consumption because of irreversible adsorption of cellulase enzymes to lignin. This problem may be amplified in the SSF operation since it is usually run in fed-batch mode and the residual lignin is accumulated. The DA pretreatment was, therefore, combined with the ammonia recycled percolation (ARP) process that has been proven to be effective in delignification. The combined pretreatment essentially fractionated the switchgrass into three major components. The treated feedstock contained about 90% glucan and 10% lignin. The digestibility of these samples was consistently higher that that of DA treated samples. Further study on the interaction of cellulase with xylan and that with lignin has shown that the enzymatic hydrolysis of cellulose is inhibited by lignin as well as xylan. The external xylan was found to be a noncompetitive inhibitor to cellulose hydrolysis. The cellulase used in this study was proven to have the xylanase activity.

Index Entries: Pretreatment; delignification; dilute-acid; cellulase adsorption; xylan hydrolysis.

---

*Author to whom all correspondence and reprint requests should be addressed.

INTRODUCTION

The primary purpose of pretreatment is to make the lignocellulosic substrate amenable to the action of cellulase. As such, the pretreatments are evaluated on the basis of the initial rate and the extent of cellulose hydrolysis. It is known that cellulase is adsorbed onto isolated lignin (1), or the lignaceous residues even after complete hydrolysis of the cellulose component (2–6). As a result, the cellulase enzyme becomes less efficient when it is applied to lignocellulosic substrates. The irreversible adsorption of cellulase to the lignin also makes it difficult to recover the enzyme. The cost of cellulase enzyme is one of the major cost items in the overall bioconversion process (7). Obviously, removal of lignin and/or hemicellulose is an important factor that needs to be considered in the pretreatment. Although there has been some dispute over the relative influence of lignin and hemicellulose in enzymatic hydrolysis of lignocellulosic materials (8,9), it is believed that lignin and hemicellulose are two main factors influencing the cellulose hydrolysis. They physically block the access of cellulase enzyme, adsorb it, and may even inhibit the reaction. Hence, removal of lignin and hemicellulose is a major occurrence in a number of known pretreatment methods. As one of such, a pretreatment method based on aqueous ammonia termed Ammonia Recycled Percolation (ARP) process was recently developed in the laboratory at Auburn University (10). In this process, aqueous ammonia is used as a pretreatment in a flow-through packed-bed reactor (percolation reactor). Recent results on the ARP show that it is highly effective in removing lignin (80–85% of lignin removal for herbaceous biomass). It also meets other pretreatment criteria: 90–95% digestibility and a high retention of glucan (11). Pretreatment by dilute sulfuric acid is also an established process. It is known for its effectiveness in removing hemicellulose (12,13). The recent modeling work and process study suggest that the hemicellulose can be efficiently removed through a two-stage dilute acid percolation process (14,15). These results further suggest that if the ARP and dilute-acid (DA) processes are combined, it can be improved to the point where the biomass is fractionated into the three main components. This investigation was undertaken to assess the effectiveness of the combined pretreatment of ARP and DA, and to elucidate the role of lignin and hemicellulose in the enzymatic hydrolysis of cellulose.

MATERIALS AND METHODS

Materials

Dry switchgrass milled and screened to 10–40 mesh was supplied by the National Renewable Energy Laboratory (NREL) and used as the lignocellulosic substrate. The composition of switchgrass as determined by

NREL Standard Procedures is as follows (% oven-dry biomass): glucan 35.23, xylan 17.26, galactan 1.41, arabinan 3.54, klason lignin 20.30, acid soluble lignin 3.30, ash 6.03, extractives 10.53, and other 1.9. The cellulase enzyme, Spezyme-CP, Lot No. 41-95034-004, was obtained from Environmental Biotechnologies, Menlo Park, CA. The specific activity of the enzyme as determined by the supplier is as follows: Filter paper activity = 64.5 FPU/mL, β-glucosidase activity = 57.6 p-NPGU/mL. Birch wood xylan (Sigma, St. Louis, MO) was used in hydrolysis experiment.

Experimental Setup and Operation

The details of the experimental apparatus for the pretreatment and the operation of ARP process have been previously reported *(11)*. A constant acid concentration (0.0784 wt%) and two different temperatures were applied in the two-stage dilute-acid process. A slight modification was made in that two additional feed reservoirs were added to the previous setup. One was for dilute acid and the other was for the deionized water. The latter was used to wash out the pretreatment agent remaining in the solid substrate.

Digestibility Test

Enzymatic hydrolysis of pretreated substrates was performed in 250 mL glass bottles at 50°C, pH 4.8, with a solid loading of 1% (w/v). It was agitated at 150 rpm on a Shaker Incubator. The enzyme loading of 60 IFPU/g glucan was applied. The enzymatic digestibility is defined as (total amount of glucose released) × 0.9/total glucan. A dehydration factor of 0.9 is used to convert the glucose to glucan.

Analytical Methods

The biomass samples were analyzed for sugar and lignin content following the procedure described in NREL-CAT Standard Procedures (No. 002–005 and LAP 010). Bio-Rad Aminex HPX-87H and HPX-87P HPLC columns were used for analysis of sugars and decomposition products. The sugar content in liquid sample was determined after the liquid sample was subjected to a secondary acid hydrolysis. The conditions in the secondary hydrolysis were: 4 wt% sulfuric acid, 121°C, and 1 h.

Enzyme Adsorption Test

The enzyme adsorption experiments were carried at 5°C to suppress the hydrolysis reaction. Sodium citrate (0.05M) was used as a buffer to keep the pH at 4.8. The protein in the solution was determined by the Bradford colorimetric method. Bovine serum albumin was used as the protein standard. The amount of adsorbed protein was calculated from the difference between the initial and final protein concentration in the supernatant.

RESULTS AND DISCUSSION

Effect of Hemicellulose on Cellulose Hydrolysis

Cellulose in lignocellulosic biomass is embedded in a sheath of hemicellulose and lignin. In order for the enzyme to attack the cellulose, it has to be first adsorbed on the surface of cellulose. In a previous study on the ARP it was found that the enzymatic digestibility of the lignocellulosic biomass is greatly enhanced by removal of the hemicellulose and lignin (10,11). Since the hemicellulose and lignin were removed simultaneously in the ARP process, the isolated effect of them could not be identified. The role of the lignin in the enzymatic hydrolysis of lignocellulosic biomass has long been recognized. However, the role of the hemicellulose in hydrolysis of lignocellulosic biomass is still uncertain. Unlike cellulose, hemicellulose is a heteropolymer. It is distributed in each layer of the cell wall and functions as supporting material in the cell walls (16). Grohmann et al. (17) reported that ester groups play an important role in the mechanism of plant cell wall resistance to enzyme hydrolysis. Kong et al. (18) examined the effects of cell-wall acetate, xylan backbone, and lignin on enzymatic hydrolysis of aspen wood. They found that acetyl groups and lignin were important barriers to enzyme hydrolysis, but the xylan backbone was not.

In this work, we studied the effect of the hemicellulose on the enzymatic digestibility of dilute-acid pretreated switchgrass. The pretreatment was carried out in a batch reactor with 0.0784 wt% sulfuric acid at 180°C applying five different levels of reaction time ranging from 5 to 45 min. The amount of hemicellulose solubilized increased with the reaction time whereas the amount of lignin solubilized stayed relatively constant. Removal of hemicellulose ranged from 30 to 88%. The plot of digestibility vs the percent hemicellulose removed (Fig. 1) indicates that the hemicellulose is an important factor influencing the hydrolysis of the lignocellulosic biomass. It is well known that the "cellulase" exhibits xylanase activity. However, it is unclear at this time whether the xylanase activity of cellulase comes from a separate enzyme(s) or from the cellulase itself. A recent study on cellulase protein indicates that a large number of the families of cellulases are polyfunctional, meaning that there are more than one specific catalytic domain in cellulase enzymes (19). At least two different enzymes have been verified to possess activities on both CMC and xylan (20,21). Johnston and Shoemaker (22) also reported that the endoglucanse has xylanase activity.

Shown in Fig. 2 is the effect of external xylan on cellulose hydrolysis at various levels of enzyme loading. The hydrolysis was conducted using the two substrates simultaneously with xylan to cellulose ratio of 0.4:1. It is clearly shown that with the enzyme loadings of 10 and 20 IFPU/g glucan, the cellulose hydrolysis is suppressed in the presence of the xylan. As the enzyme loading was raised to 40 and 60 IFPU/g glucan, the effect of xylan diminished.

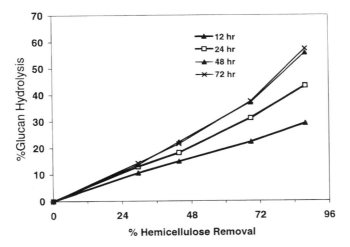

Fig. 1. Effect of hemicellulose content on enzymatic hydrolysis of switchgrass after dilute-acid batch pretreatment. Pretreatment condition: 0.05%(w/w) sulfuric acid, 180°C, reaction time 5–45 min.

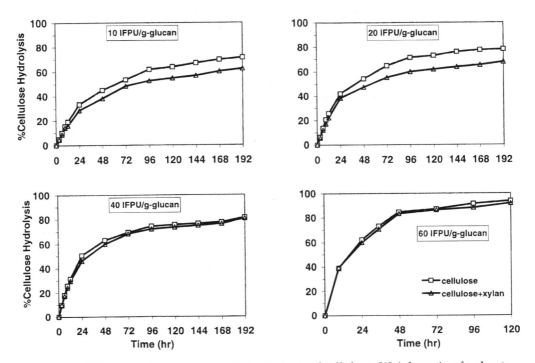

Fig. 2. Effect of xylan on enzymatic hydrolysis of cellulose. Weight ratio of xylan to cellulose = 0.4:1.

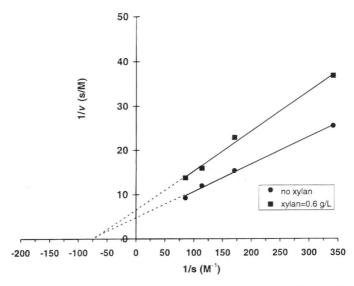

Fig. 3. Lineweaver–Burk plot of enzymatic hydrolysis of cellulose in presence of xylan. Hydrolysis conditions: 50°C, 40 IFPU/g glucan, pH 4.8, 1% (w/v) filter paper, time = 1.5 h.

The nature of the interaction between xylan and cellulase was further examined from a kinetic standpoint. An inhibition experiment was thus carried out using filter paper as a substrate. The enzymatic hydrolysis conditions were: 50°C, pH 4.8, 1% (w/v) filter paper, and 40 IFPU of cellulase/g glucan. The total amount of glucose and celloboise released from the filter paper at 90 min was used to calculate the initial rate of hydrolysis. The Lineweaver-Burk plot prepared from these results indicates that xylan inhibits cellulose hydrolysis in a noncompetitive inhibition mode (the vertical intercept and the slope increase in the presence of xylan) (Fig. 3). Under this kinetic pattern xylan and cellulose do not compete for the same active sites. The cellulase enzyme, most likely the endoglucanase, has active sites for xylan hydrolysis. It seems that adsorption of xylan onto the active sites somehow inhibits the cellulase activity.

Xylanase activity of cellulase was further tested using xylan as the only substrate (Xylan was from Sigma and its purity was not measured when this experiment was conducted). A peculiar reaction pattern was observed that the hydrolysis was very rapid for the first 12 h giving 40% conversion. The reaction, however, almost ceased after that period. It is unclear why there is a sudden stoppage of the reaction. One may speculate that xylose is a strong inhibitor to the enzyme or the substrate (commercial xylan) has a structure that limits the terminal digestibility. Addition of excessive amount of enzyme (total loading of 673 IFPU/g xylan) and a 140 h extension of reaction time increased the conversion

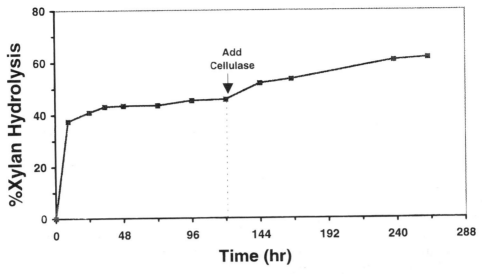

Fig. 4. Enzymatic hydrolysis of xylan by cellulase. Hydrolysis condition: 50°C, 20 IFPU/g-xylan, pH 4.8, 1% (w/v) xylan. The arrow indicates an additional 50 IFPU cellulase was added to the solution at 120 h. The total enzyme loading at this point is 673 IFPU/g xylan.

only by 15%. It appears that xylan plays a negative role in cellulose hydrolysis in two different ways. First, it adsorbs cellulase enzyme (formation of enzyme-substrate complex). This fraction of cellulase is, therefore, unavailable for cellulose hydrolysis. This effect applies for both internal xylan (Fig. 1) and external xylan (Fig. 2). Second, xylan or hemicellulose existing in a lignocellulosic substrate physically blocks the contact between cellulase and cellulose. It is our contention that hemicellulose has a profound effect on cellulose hydrolysis and its removal is a prerequisite for complete hydrolysis of cellulose in biomass.

Two-Stage Dilute-Acid Pretreatment

Dilute-acid treatment is known to be effective in removing hemicellulose from biomass. Recent studies along these lines including the ones from our laboratory suggest that two-stage treatment is highly effective in recovering hemicellulose sugars without decomposition (14,15). The two-stage dilute-acid pretreatment of this work employed extremely low sulfuric acid (0.0784 wt%) with a reaction time of 10 min per stage. The results are summarized in Table 1. The solubilization of hemicellulose, cellulose, and lignin generally increase with temperature. When a low side temperature above 135°C and a high side temperature above 165°C were applied, the hemicellulose was completely removed. The treated solid thus contained lignin and glucan only. In addition to the solubilization of the hemicellulose, 40–48% of lignin and 16–35% glucan were also solubilized. The

Table 1
Two-Stage Dilute-Acid Pretreatment on Switchgrass[a]

Two-Stage Reaction Temperature	Solid Residue			Primary Hydrolyzate			pH
	% of Original Amount Removed			Hemicellulose Sugars			
	Lignin	Glucan	Hemicellulose	%Monomer	%Oligomer	%Furfural	
165°C + 195°C	48.23	35.00	100.00	43.51	55.28	trace	2.60
155°C + 185°C	47.38	18.83	100.00	39.63	54.47	trace	2.40
145°C + 175°C	43.21	17.47	100.00	37.10	61.90	none	2.37
135°C + 165°C	41.52	16.08	100.00	34.63	63.26	none	2.38
125°C + 155°C	43.90	14.57	65.11	36.40	63.61	none	2.23

[a]Data in the table based on the oven-dry untreated biomass.
Pretreatment condition: 0.0784 w/v% sulfuric acid, reaction time of 10 min for each stage, flow rate = 4.0 mL/min.

hemicellulose hydrolyzate contained both oligomers and monomers, the former being more than 50%. The oligomer content decreased as the reaction temperature was increased. Below 185°C, no furfural or hydroxymethylfurfural (HMF) was detected in the hydrolyzate. Two-stage process was applied because the hemicellulose in herbaceous biomass is biphasic (14,15,23). In this experiment, we found that most of the hemicellulose (85% of original) and lignin (35 out of 45% total removal) were removed during the first stage.

Combined Pretreatment of ARP and DA

The ARP has been proven to be a highly effective pretreatment for switchgrass especially in deligninfication. In this work, we have shown that DA treatment is most effective in solubilizing hemicellulose. It is conceivable that combination of these two can fractionate biomass into the three major constituent polymers. The investigation of combined pretreatment of ARP and DA thus ensued. It was studied in both directions: DA followed by ARP (DA-ARP), and ARP followed by DA (ARP-DA). In the DA-ARP process, the two-stage DA conditions were: 0.0784 wt% sulfuric acid, low reaction temperature 140°C, high reaction temperature 170°C, reaction time of 10 min for each stage, flow rate of 4 mL/min. After the DA prehydrolysis, the reactor was washed on line with water, and then 10 wt% ammonia was pumped in and left for 4 h at room temperature for presoaking prior to the ARP operation. The ARP conditions were: 170°C, 10 wt% ammonia, 20 min. These reaction conditions were selected since they were the optimum conditions for the individual pretreatment. Similar reaction conditions and operation were applied to the ARP-DA process. The sugar and lignin balances in the combined pretreatment are shown in Tables 2 and 3. The hemicellulose is completely removed from the solid in both combined processes leaving the treated solid to contain glucan and lignin. In the DA-ARP process, all of hemicellulose is removed during the DA step whereas lignin is solubilized in both steps (37% in DA step and 45% in ARP step for a total of 82%). In the ARP-DA process, the two-thirds of the hemi-

Table 2
Sugar Balance in Switchgrass after the Combined Pretreatment of DA and ARP[a]

Pretreatment	Solid Residue						Acid Hydrolyzate or ARP Effluent			
	%wt	%glucan	%xylan	%galactan	%arabinan		%glucan	%xylan	%galactan	%arabinan
Untreated Switchgrass	100	35.23	17.76	1.41	3.54		0	0	0	0
DA-ARP	30.24	26.30	trace	0	0	DA	4.37	17.33	1.53	2.96
						ARP	0.24	0.24	0	0
ARP-DA	33.84	30.09	trace	0	0	ARP	2.44	11.05	1.19	2.78
						DA	2.43	6.59	0.17	0.91

[a]Data in the table based on the oven-dry untreated biomass. DA-ARP: dilute-acid pretreatment followed by the ammonia recycled percolation process. Pretreatment condition: DA: 140°C, 10 min + 170°C, 10 min, 0.0784 wt% sulfuric acid, 4 mL/min. ARP: 170°C, 10 wt% ammonia, 20 min, 4 mL/min. ARP-DA: ammonia recycle percolation process followed by the dilute-acid pretreatment. Pretreatment condition: ARP: 175°C, 10 wt% ammonia, 20 min, 4 mL/min. DA: 175°C, 0.0784 wt% sulfuric acid, 20 min 4 mL/min.

Table 3
Lignin Balance in Switchgrass after the Combined Pretreatment of DA and ARP[a]

Pretreatment		% Klason Lignin	% Acid Soluble Lignin	Total
Untreated Switchgrass		20.30	3.30	23.60
DA-ARP	Solid Residue	3.81	0.10	3.91
	DA Hydrolyzate	5.66*	3.08	8.74
	ARP Effluent	9.02*	1.54	10.56
ARP-DA	Solid Residue	3.78	0.07	3.85
	ARP Effluent	12.50*	4.67	17.17
	DA Hydrolyzate	0.00	0.36	0.36

[a]Data in the table based on the oven-dry untreated biomass. Pretreatment conditions same as Table 2.
*Klason lignin of DA hydrolyzate or ARP effluent is the lignin precipitated during the secondary acid hydrolysis.

cellulose is removed in ARP step and one-third hemicellulose is removed in the DA step. As for the lignin, a total of 83% of the initial lignin is removed, almost all of it is in the ARP step. In both combined pretreatment processes, the hemicellulose recovery was about 96%. Sugar decomposition was negligible in the ARP-DA process. However, about 4% of glucan is not accounted for in the DA-ARP process most likely by decomposition. One would thus prefer the pretreatment sequence of ARP followed by DA process to that of DA followed by ARP.

The enzymatic digestibility were determined for these samples. Results are shown in Fig. 5 and 6. A sample treated with DA only was included as a reference substrate. The composition of the DA treated sample was: 69.82% glucan and 30.28% lignin. At each level of enzyme loading, the digestibility of swichgrass after the combined pretreatment is seen to

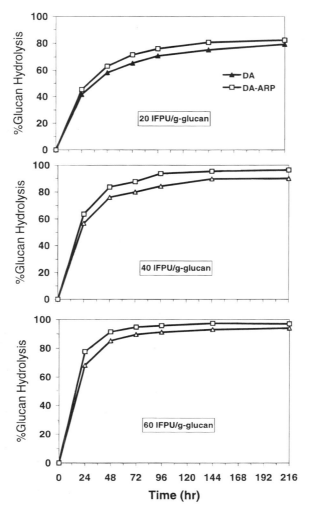

Fig. 5. Enzymatic hydrolysis of switchgrass after DA and DA-ARP pretreatment.

be higher than that of the sample treated with DA only. The digestibility results were quite similar in either sequence of the combined processes. We believe that the enhanced efficiency of cellulase enzymes observed here is directly linked to the low lignin content in the substrates that have been subjected to the combined pretreatment.

Enzyme Adsorption

Adsorption of enzyme onto lignocellulosic substrate is a required step in cellulose hydrolysis. Since we found some evidence that residual lignin affects the digestibility, it became in our interest to verify how the adsorption phenomena affects the enzymatic hydrolysis of cellulose. The following experiments address this issue.

Ammonia Recycled Percolation

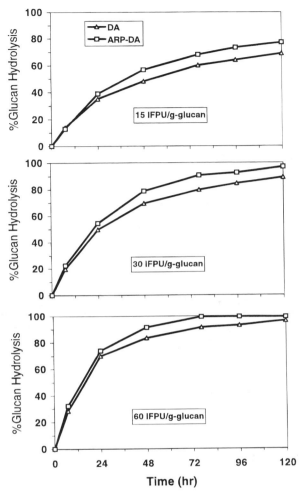

Fig. 6. Enzymatic hydrolysis of switchgrass after DA and ARP-DA pretreatment.

The enzyme adsorption capacity on cellulose was first studied covering the enzyme concentrations of 0.028–19.40 mg/mL. Figure 7 shows the enzyme (measured as protein) adsorption isotherm at 5°C. The enzyme adsorption on cellulose appears to follow the Langmuir multimolecular layer adsorption pattern. There are two stages of leveling off, one at the enzyme concentration of 0.056 to 0.084 mg/mL, the other at 5.89 to 12.10 mg/mL. It is followed by a steady increase not showing an upper limit of enzyme adsorption within this experiment range. Adsorption of cellulase enzyme on various substrates is presented in Table 4. The enzyme adsorption varies widely among the substrates. The amount of enzymes adsorbed is dependent on the nature and surface structure of substrates. Switchgrass upon pretreatment adsorbs more enzyme than the lignin residues and α-cellulose.

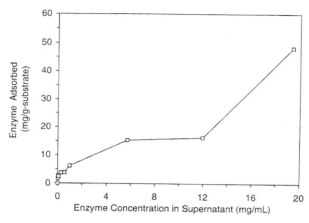

Fig. 7. Cellulase adsorption isotherm. Adsorption condition: 5°C, pH 4.8, 1% (w/v) cellulose, adsorption time = 1 h.

Table 4
Enzyme Adsorption on Various Substrates

ARP-DA treated switchgrass Substrates	Cellulase Adsorption (mg Protrein/g substrate)
DA treated switchgrass	14.51
ARP treated switchgrass	12.74
ARP-DA treated switchgrass	15.83
Cellulose*	3.52
Lignin from DA treatment	8.83
Lignin from ARP-DA Process	10.76
Indulin*	1.37

*All samples except those noted are kept wet before they are put into test. Adsorption condition: 5°C, 1 h 1% (w/v) substrate.

The effect of the isolated lignin on the cellulose hydrolysis is shown in Fig. 8. Commercial lignin (Indulin AT, Westvaco, North Charleston, SC), a dry lignin from the ARP process, and a wet lignin residue from the dilute-acid process were studied. The digestibility is seen to decrease when external lignin is added to the solution.

Adsorption of enzyme was further examined from an activity standpoint. The substrate used in the test was the switchgrass treated by dilute-acid in a batch mode. It contained 50% glucan, 7% xylan, and 38% lignin. After the substrate was suspended in the enzyme solution (1 IFPU/mL) for 2 h at 5°C, the solid substrate was separated from supernatant by vacuum filtration. Seventy five percent of total protein was found to be adsorbed to the solid substrate. The activity of cellulase in the supernatant and that of original enzyme solution were then measured by the enzymatic digestibility

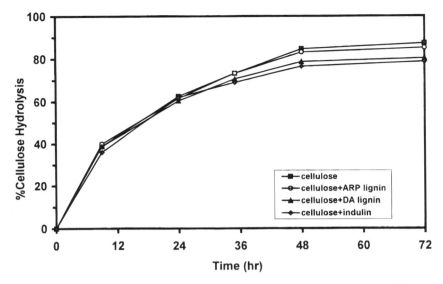

Fig. 8. Effect of external lignin on enzymatic hydrolysis of cellulose DA-lignin refers to the lignin residue of dilute-acid treated switchgrass after enzymatic hydrolysis. ARP-lignin refers to the lignin of switchgrass produced from the ARP process.

test. The data indicated that the supernatant had far less specific cellulase activity than that of original enzyme solution. Since 75% of the total protein is adsorbed onto the substrate, one expects to see 25% of enzyme activity retention in the supernatant, However, the supernatant showed only 15% of the initial activity. This result indicates that cellulase-substrate affinity promotes adsorption of cellulase enzymes much more so than the noncellulase proteins. This agrees with a recent finding by Lee et al. (3) that the activity of the recovered cellulase enzyme is much less than the value expected from the total recovered protein.

CONCLUSIONS

Hemicellulose in switchgrass can be completely solubilized and recovered through a two-stage dilute-acid percolation treatment conducted under an extremely low acid concentration (0.078 wt% sulfuric acid) and moderate reaction temperature (140–170°C). The combined pretreatment of ARP-DA on switchgrass has essentially fractionated the biomass into three major components. The treated feedstock contained glucan and a small amount of lignin. The combined sequence of ARP-DA showed better performance than DA-ARP in sugar decomposition. The cellulase enzyme used in this work (Spezyme-CP) was found to have activities on xylan as well as cellulose. However, it does not have sufficient xylanase activity to completely hydrolyze xylan. Both internal and external xylan have been proven to have a negative effect on cellulose hydrolysis. The xylan inhibits the cellulose hydrolysis in a noncompetitive mode. Active cellulase is adsorbed more strongly onto the cellulosic substrates than inactive cellulase or noncellulase protein.

ACKNOWLEDGMENT

This research was partially supported by the National Renewable Energy Laboratory through a subcontract NREL-XAW-3-11181-02. Additional equipment support was provided by College of Engineering, Auburn University.

REFERENCES

1. Chernaglazov, V. M., Ermolova, O. V., and Klyosv, A. A. (1988), *Enzyme Microb. Technol.* **10,** 503–507.
2. Tanka, M., Fukui, M., and Mastsuno, R. (1988), *Biotechnol. Bioeng.* **32,** 897–902.
3. Lee, D., Yu, A. H. C., and Saddler, J. N. (1995), *Biotechnol. Bioeng.* **45,** 328–336.
4. Castellanos, O. F., Sinitsyn, A. P., and Vlasenko, E. Y. (1995), *Biorsouce Technol.* **52,** 109–117.
5. Giard, D. J. and Converse, A. O. (1993), *Appl. Biochem. Biotechnol.* **39/40,** 521–533.
6. Ooshima, H., Burns, D. S., and Converse, A. O. (1990), *Biotechnol. Bioeng.* **36,** 446–452.
7. Ngugen, Q. A., and Saddler, J. N. (1991), *Bioresearch and Technol.* **35,** 275–282.
8. Stone, J., Scallan, A., Donefer, E., and Ahlgren, E. (1969), *Adv. Chem. Seri.* **95,** 219–241.
9. Ramos, L. P., Breuil, C., and Saddler, J. N. (1992), *Appl. Biochem. Biotechnol.* **34/35,** 37–48.
10. Yoon, H. H., Wu, Z. W., Kim, S. B., and Lee, Y. Y. (1995), *Appl. Biochem. Biotechnol.* **51/52,** 5–19.
11. Iyer, P., Wu, Z. W., and Lee, Y. Y. (1996), *Appl. Biochem. Biotechnol.* **57/58,** 121–132.
12. Lee, Y. Y., Lin, C. M., and Chambers, R. P. (1979), *Biotechnol. Bioeng. Symp.* **8,** 75–88.
13. Torget, R., Walter, P., Himmel, M., and Grohmann, K. (1991), *Appl. Biochem. Biotechnol.* **28/29,** 75–86.
14. Chen, R., Torget, R., and Lee, Y. Y. (1996), *Appl. Biochem. Biotechnol.* **57/58,** 133–146.
15. Torget, R., Hatzis, C., Hayward, T. K., Hsu, T., and Philippidis, G. P. (1996) *Appl. Biochem. Biotechnol.* **57/58,** (in press).
16. Sjöström, E. (1993), *Wood Chemistry: Fundamentals and Applications*, 2nd edition, Academic Press, New York, NY.
17. Grohmann, K., Mitchell, D. J., Himmel, M. E., Dale, B. E., and Schroeder, H. A. (1989), *Appl. Biochem. Biotechnol.* **20/21,** 45–61.
18. Kong, F., Engler, C. R., and Soltes, E. J. (1992), *Appl. Biochem. Biotechnol.* **34/35,** 23–35.
19. Wilson, D. B. (1992), *Critical Review in Biotechnology*, **12(1/2),** 45–63.
20. Matsushita, O., Russell, J. B., and Wilson, D. B. (1990), *J. Bacteriol.*, **17,** 3620–3630.
21. Yagüe, E., Béguin, P., and Aubert, J.-P. (1990), *Gene* **89,** 61–67.
22. Johnston, D. B. and Sheomaker, S. P. (1996), Assessment of Enzyme Kinetics Using Purified Endoglucanases and Insoluble Substrates, presented at 18th Symposium of Biotechnology for Chemicals and Fuel, Gatlinburg, TN.
23. Kim, B. J., Lee, Y. Y., and Torget, R. (1994), *Appl. Biochem. Biotechnol.* **45/46,** 113–129.

Preliminary Study of the Pyrolysis of Steam Classified Municipal Solid Waste

JOHN M. SEBGHATI AND MICHAEL H. ELEY*

Johnson Research Center, The University of Alabama in Huntsville, Huntsville, AL 35899

ABSTRACT

Steam classified municipal solid waste (MSW) has been studied for use as a combustion fuel, feedstock for composting, and cellulytic enzyme hydrolysis. A preliminary study has been conducted using a prototype plasma arc pyrolysis system (in cooperation with Plasma Energy Applied Technology Inc., Huntsville, AL) to convert the steam classified MSW into a pyrolysis gas and vitrified material. Using a feed rate of 50 lbs/h, 300 lbs of the material was pyrolysized. The major components of this pyrolysis gas were H_2, CO, and CO_2. A detailed presentation of the emission data along with details on the system used will be presented.

Index Entries: Municipal solid waste; gasification; pyrolysis; biomass; plasma arc.

INTRODUCTION

In 1990, the United States generated an estimated 195 million tons of municipal solid waste (MSW) *(1)*. The primary method used for disposal of MSW is landfilling. About 75% of existing landfills are expected to close in the next 10 to 15 y. New environmental laws have made landfilling much more expensive *(2)*. Incineration, another method of MSW disposal, is often more expensive than landfilling and is not very popular with the public because of the pollution concerns.

A method called "Steam Classification" has been developed at the University of Alabama in Huntsville (UAH) to treat MSW *(3,4)*. With the addition of moisture, the waste is exposed to saturated steam under pressure for approx 1 h with constant mixing. The biomass components are predominantly transformed into a reasonably uniform material that can be isolated from most of the glass, metals, textiles, and plastic through the use

---

*Author to whom all correspondence and reprint requests should be addressed.

of a trommel and air classifier. This material has been studied in conjunction with composting, enzyme hydrolysis, and incineration *(5,6)*. Steam classification has been found to produce a better product than conventional methods of refuse-derived fuel (RDF) preparation.

The purpose of this study is to examine the gasification of steam classified MSW. This was achieved using a plasma arc pyrolysis unit at Plasma Energy Applied Technology (PEAT), in Huntsville, Alabama, that was originally developed for disposal of hazardous waste. High temperatures under controlled process conditions to promote physical and chemical changes in material are used. The waste is converted to a pyrolytic gas and a vitrified solid. A plasma arc pyrolysis system was used because it was the only gasification system available at the time. Because of the preliminary nature of this study, a cost analysis has yet to be done.

METHODS

Steam Classification

Steam classification is a method whereby essentially unsorted MSW can be processed and subsequently separated into biomass and nonbiomass fractions. A combination of continuous mixing with added moisture as well as pressurized with saturated steam basically pulps or repulps cellulosic materials to a relatively uniform material by size and density. After processing, this biomass material can be separated from nonbiomass components for use as a chemical feedstock. The nonbiomass components can be separated into recyclable metals, glass, textiles, and plastics with only a minor fraction requiring disposal *(4)*.

The steam classified material used for this study was produced in Walton, KY. Approximately seven metric tons of ordinary municipal solid waste was placed inside a vessel along with some water. The vessel was thereupon closed and pressurized to about 400 kPa with saturated steam (150°C). The vessel was rotated, thus mixing the waste with internal helical flighting, for about 1 h. The vessel was depressurized and the contents were emptied onto a conveyor that transferred the material to a rotary trommel. The trommel separated out the material that was less than 3 cm from larger components. Recyclable materials were manually sorted before the remainder was landfilled. A composite sample of about 5000 kg of the under 3 cm material was transported to UAH. The material was then screened on a vibratory screener to recover the <1.3-cm fraction. This fraction was then air classified to remove most of the glass and remaining dense contaminants.

Plasma Arc Pyrolysis

Plasma, the fourth state of matter, is the universe's most common form of matter. It is basically an energetically charged gas. Fundamentally, the plasma arc torch consists of an electrically neutral gas forced through

Study of the Pyrolysis

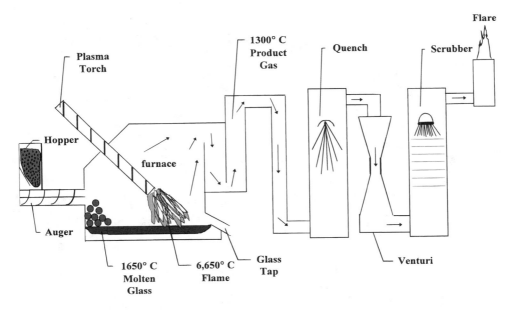

Fig. 1. Schematic representation of the Plasma Energy Applied Technology, plasma arc pyrolysis pilot plant.

the electric arc created between two electrodes in a metallic tube. It is possible to split complex materials into their basic compounds and individual elements because the torch reaches temperatures in excess of 6650°C heating the pyrolysis vessel to over 1650°C. Oxygen is restricted to the minimum necessary to accomplish gasification by utilizing steam as the oxidant, thus the gaseous atmosphere remains very reducing and has an abundance of hydrogen. The volume of off-gas is also reduced significantly compared to combustion, thus minimizing any cleanup of the gas (7).

The final biomass material prepared at UAH was transported to the PEAT facilities for the pyrolysis test. Batches of the material were loaded into a feed hopper and continuously fed via an auger into the pyrolysis unit. The feed rate had to be varied somewhat from 19 to 23 kg/h in order to prevent a buildup of unreacted material inside the unit. Early in the test a buildup of a white powder was observed that was determined to be kaolin from the paper products in the feed material. To eliminate the powder, soda ash (Na_2CO_3) was added with each batch of feed material.

For our test, the unit was heated to between 1260°C and 1315°C. Figure 1 shows a schematic of the PEAT pyrolysis process system.

The steam classified MSW was added to the hopper in measured batches. As the lid was closed the hopper is pressurized with N_2 driving air out and keeping a positive pressure. The feed screw fed the material into the furnace at a measured rate. As the material is consumed it is trans-

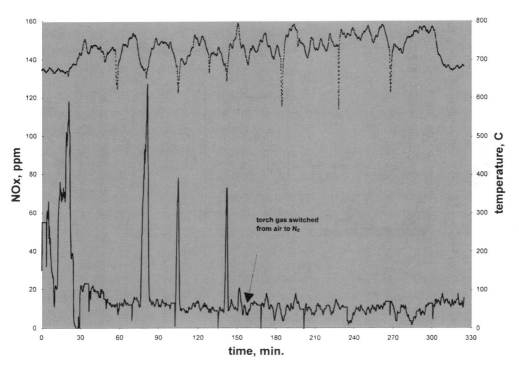

Fig. 2. Greenline Combustion Gas Analyzer data for NO_x and temperature measured after the flare. Note the change from air to N_2 as the torch gas at 160 min into the test. —— NO_x, ppm; - - - - flare temp., °C.

formed to either a gas or a vitrified slag. The slag is periodically tapped. The acid gases were measured at the rear of the pyrolysis unit as the gases exited the furnace. After exiting the furnace, the gases entered the quench that consists of a spray mist of city water at tap temperature that cooled the gases. The cooled gases pass into a venturi that includes an alkaline spray mist to extract most of the particulates and partially neutralize the acid gases. A scrubber was then used to scrub any remaining acid gases from the off-gas. A sodium hydroxide (NaOH) solution was used in both the venturi and scrubber. Particulates from the water quench, the venturi, and the scrubber are combined, recovered in a filter press, and reprocessed in the furnace. After exiting the scrubber, the product gas was saturated with water. The gas concentration was monitored at this point. The CO and CO_2 were measured using a Teledyne Model 731 nondispersal infrared analyzer. A Model 235 thermal conductivity analyzer monitored H_2. The CH_4 infrared sensor did not work in our test. The product gas entered a flare fueled by propane and air, and was combusted before being exhausted to the atmosphere. Before leaving the stack a number of measurements were taken. Using a Eurotron "Greenline" Combustion Gas Analyzer, the O_2, CO, CO_2, NO, NO_x, and SO_2 were measured. The SO_2 analyzer did not work for this test. Also measured were the metals, dioxins/furans, and particulates.

Study of the Pyrolysis

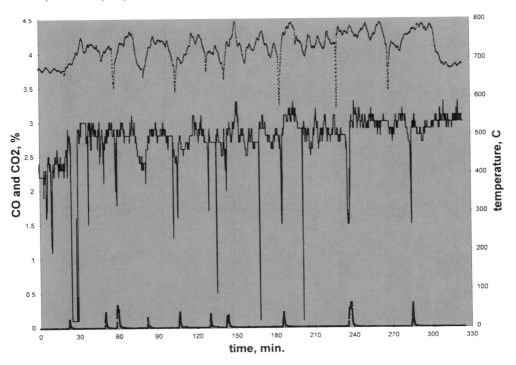

Fig. 3. Greenline Combustion Gas Analyzer data for CO and CO_2, after the flare and the flare temperature. Note that the CO_2 levels at this location would also include the combustion of propane. — —, CO, %; ——, CO_2, %; - - - -, flare temp., °C.

RESULTS AND DISCUSSION

Feedstock material was fed into the pyrolysis furnace over a period of 5 h 10 min, except for an occasional interruption to prevent accumulation of unreacted material in the furnace chamber. In the first 10 min a white powdery material was noted to accumulate in the furnace. It was determined to be kaolin, which is an additive of paper products. Soda ash (Na_2CO_3) was added as a flux agent to reduce or eliminate the kaolin floc.

The Greenline gas analyzer continuously monitored the gases at the flare exhaust. Since air was admitted at the base of the flare, the oxygen was in excess to allow the complete combustion of the propane and product gas before release to the atmosphere. During the first 2 ½ h it was noted that the NO and NO_x levels fluctuated excessively. The whole time the torch gas had been compressed air. It should be noted the quantity of air was insufficient to support combustion. It was then decided to switch the torch gas to compressed nitrogen, which considerably dampened the fluctuations in the NO/NO_x regions (*see* Fig. 2). The difference in the NO and NO_x readings were not significant. The flare temperature, plotted in Fig. 2, with only a few exceptions was maintained above 700°C. Except for a few spikes

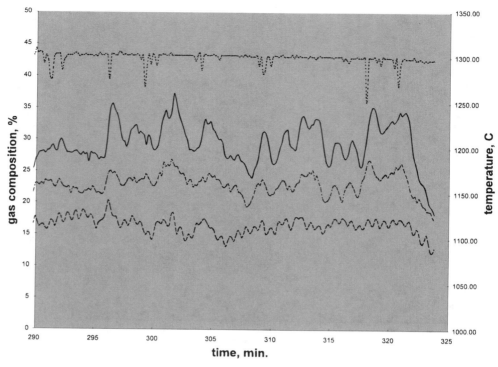

Fig. 4. Teledyne Model 731 Nondispersal Infrared Analyzer data for CO and CO_2 and Model 235 Thermal Conductivity Analyzer data for H_2 measured at the furnace exit. Furnace gas temperature is also included. Note that this data was collected only during the last 30 min of the test because of a gas sampling line blockage. ——, H_2; — - - —, CO; - - -, CO_2; - - - -, furn. gas temp., °C.

Table 1
Ultimate Analysis

Parameter	Dry Basis	As Received
% Moisture	0	50.54
% Carbon	39.79	19.68
% Hydrogen	5.47	2.71
% Nitrogen	0.83	0.41
% Sulfur	2.18	1.08
% Chlorine	0.77	0.38
% Ash	17.16	8.49
% Oxygen	33.8	16.72
fuel value (kJ/kg)	18,530	9,160

before the torch gas was changed from air to N_2, the NO_x level did not exceed the EPA limit of 180 ppm when corrected to 7% oxygen (8,9).

Although the CO concentration of the product gas was a major constituent of the product gas prior to combustion in the flare, the postflare CO

Table 2
Product Gas Composition and Temperatures

	H_2	CO	CO_2	Gas Temp., °C	Furn. Gas Temp., °C
average	29.49	22.97	16.39	34.56	1299.43
standard dev	3.23	1.69	1.26	0.40	6.48

Table 3
Acid Gas Analyses

Gas	measurements
	ppb
H_2S	104
Sulphate	9.042
Nitrate	0.319
Fluoride	0.28
Chloride*	0.205
Nitride	0.015
* EPA Limit - 35 ppm	

levels were reduced to almost zero. It should be noted that air is being admitted in excess at the base of the flare stack, thus promoting combustion. A few spikes of less than a half percent were noted shortly after each feed interruptions. The same figure also shows concentrations of CO_2 mostly in the range of 2.5–3%, resulting from the combustion of the product gas and propane in the flare (see Fig. 3).

Figure 4 shows the composition of the product gas prior to combustion in the flare only during the last half hour of the run. Previous to this time, inadequate readings indicated a low gas yield that was inconsistent with the quantity of feed material having a composition as noted in Table 1. Although a number of troubleshooting measures were taken during the entire run to identify the problem, it was not until the last half hour that the problem was identified and corrected. The gas sampling line to the Teledyne analyzer was partially blocked, and this blockage was removed by forcing air through the line. Table 2 shows the average composition and standard deviations after the correction. The significant quantities of H_2 and CO present indicated a combustible gas. The unaccounted portion of the gas was predominantly moisture since the gas was saturated during clean-up. The temperature of the product gas as it leaves the furnace is also indicated in the figure with the average also found in Table 2.

The exhaust gas from the furnace was sampled prior to the water quench for acid gas components. Table 3 shows the results of these analyses. Although in ppb the only acid gases of note were those containing sulfur, which probably results from the relatively high (approx 2% dry

Table 4
Metal Analysis

Metals	Analysis ppm	EPA Limit ppm
Copper	0.09	
Lead	0.73	5.0
Chromium	<0.02	5.0
Selenium	<0.2	1.0
Arsenic	<0.08	5.0
Mercury	0.14	0.2
Barium	<0.01	100.0
Silver	<0.01	5.0
Nickel	<0.05	
Potassium	<0.83	
Sodium	2.22	
Manganese	<0.01	
Cobalt	<0.07	
Zinc	0.19	
Iron	<0.07	
Cadmium	0.02	1.0
Magnesium	0.05	
Aluminum	0.08	
Calcium	0.45	
Beryllium	<0.01	

Table 5
Particulates*

Location	emissions ppb
Post Quench	25871
Post Quench	12504
Post Quench	37339
Post Flare	55.4
Post Flare	177.2
Post Flare	83.0
* EPA Limit - 34 ppb	

weight) of sulfur noted in the ultimate analysis (Table 1). The HCl level measured, 0.21 ppb, did not approach the EPA limit of 35 ppm (8). Table 4 shows the results from a metals analysis collected after combustion of the product gas. The only metal found in significant quantity was sodium,

Table 6
Dioxins and Furans*

Analytes	Sample 1 ng/m³	Sample 2 ng/m³
1234678-HpCDD	{0.02}	0.03
OCDD	0.16	{0.14}
1234678-HpCDF	0.03	{0.03}
OCDF	{0.06}	0.36
TOTAL HpCDD	{0.04}	0.05
TOTAL TCDF	{0.01}	
TOTAL HpCDF	0.04	0.01
{Est. Maximum Possible Concentration}		
* EPA Limit - 30 ng/m³		

which is indicative of added soda ash. It should be noted that all EPA-regulated metals are below acceptable emission levels.

Table 5 shows the particulate levels before and after the venturi and gas scrubbing treatment, which indicates a significant reduction in the quantities of particulates by such treatment. The postflare levels are approx 2–5 times the allowable EPA emission level of 34 ppb *(8)*, thus indicating additional gas cleaning would be necessary, such as a larger scrubber or a baghouse. The particulates were measured in accordance with EPA Method 5, Appendix A. The size and solubility were not measured. Semivolatiles, including dioxins and furans, were also collected after combustion. Table 6 shows that the dioxins and furans are well below the EPA limit of 30 ng/m³ *(8)*. Other semivolatiles detected at levels of parts-per-trillion were Phenol, Benzoic acid, Di-n-butylphthalate, and bis(2-Ethylhexyl)phthalate.

ACKNOWLEDGMENTS

This research was supported in part by funds from the Alabama Legacy for Environmental Research Trust Fund (ALERT) and the National Science Foundation—Experimental Program to Stimulate Competitive Research (NSF-EPSCOR). We wish to thank Plasma Energy Applied Technologies, Inc. (PEAT), Huntsville, Alabama, for their in-kind services and participation in this project. Special thanks to Dr. Marlin Springer and Tom Barkley for their technical assistance. We also wish to thank Ben Johnston, Joyita Bagchi, and Tom Carrington of the University of Alabama in Huntsville Environmental Laboratory for their participation in the laboratory analyses. We also appreciate Dr. Gerald R. Guinn of the UAH Johnson Research Center for his guidance and insight into the project planning and data analysis.

REFERENCES

1. Franklin, W. E. and Franklin, M. A. (1992), *EPA J.* **18(3),** 7–13.
2. *Federal Register*, 40CFR, part 258, October 9, 1991.
3. Eley, M. H. and Holloway, C. C. (1988), *Applied Biochem. Biotech.* **17,** 125–135.
4. Eley, M. H. (1994), *Applied Biochem. Biotech.* **45/46,** 69–79.
5. Eley, M. H., Guinn, G. R., and Bagchi, J. (1995) *Applied Biochem. Biotech.* **51/52,** 387–397.
6. Eley, M. H. and Guinn, G. R. (1994), *National Waste Processing Conference ASME*, pp. 283–291.
7. Plasma Energy Applied Technology, Inc. Huntsville, AL, patent pending.
8. *Air Pollution Control—BNA Policy and Practice Series*, The Bureau of National Affairs, Inc., Washington, DC latest updates.
9. Mahan, S. E. (1990), *Hazardous Waste Chemistry, Toxicology and Treatment*, Lewis Publishers, Chelsea, MI, p. 334.

Two-Phase Model of Hydrolysis Kinetics and Its Applications to Anaerobic Degradation of Particulate Organic Matter

VASILY A. VAVILIN,*,[1] SERGEI V. RYTOV,[1] AND LJUDMILA YA. LOKSHINA[2]

[1]*Water Problems Institute of the Russian Academy of Sciences, Novaja Basmannaja str. 10, P.O. Box 231, 107078 Moscow, Russia;*
[2]*Moscow Institute of Physics and Technology, MIPT-8, Dolgoprudny Region, Moscow 141700, Russia*

ABSTRACT

The various equations of hydrolysis kinetics included into the generalized simulation model ⟨METHANE⟩ were tested on the anaerobic digestion of cellulose, sludge, and cattle manure. The good agreement between the model simulation results and experimental data was obtained. The Contois equation, taking into account a hydrolytic biomass, and the first-order equation with respect to the particulate substrate only, were shown to be the approximations of two-phase hydrolysis kinetics.

Index Entries: Anaerobic digestion; hydrolytic bacteria; simulation model; sludge; cattle manure; cellulose.

INTRODUCTION

During anaerobic digestion the hydrolysis is normally rate-limiting if the substrate is in particulate form (1). The equations traditionally used for solids kinetics are the first-order equation in respect to the particulate substrate only, and the Contois equation that takes into account a hydrolytic biomass. According to Eastman and Ferguson (2), the first-order hydrolysis function is an empirical expression that reflects the cumulative effects of many processes. Large particles with a low surface-to-volume ratio would be hydrolyzed more slowly than small particles.

*Author to whom all correspondence and reprint requests should be addressed.

The Contois equation has been more or less successfully tested on the anaerobic degradation of suspended solids. However, it was developed originally for dissolved substrates. For hydrolysis process a heterogeneous reaction system, in which a particulate substrate contacts with microbial cells and related enzymes, must be taken into consideration (3).

A structured approach to the modeling of anaerobic reactors has provided an important tool for the integral study of microbial ecology as well as the investigation of advanced control strategies of high-rate anaerobic reactors. There are few simulation models of anaerobic digestion that take into account hydrolysis of particulate organic matter (4,5). The simulation model ⟨METHANE⟩ (6,7) is among them.

In this article, the various versions of hydrolysis kinetics included into the model ⟨METHANE⟩ were tested on anaerobic digestion of sludge, cellulose, and cattle manure.

SIMULATION MODEL OF ANAEROBIC DIGESTION

The following groups of model variables were included in the generalized simulation model of anaerobic digestion <METHANE>.

1. suspended solids concentration (X_k, k = 1,2,3);
2. active biomass concentration (B_i, i = 1,2, ..., 10);
3. dissolved substrate concentration (S_j, j = 1,2, ..., 15);
4. partial gas pressure (G_l, l = 1,2, ..., 7).

The processes of hydrolysis, acidogenesis, acetogenesis, and methanogenesis conducted by various groups of micro-organisms are considered (Fig. 1). It is assumed in the model that dissolved organic substrate S_1 (DOS) is a mixture of carbohydrates (C), proteins (P), lipids (L), and an unknown organic substance (Z). The composition of the substrate S_1 in an anaerobic reactor is formed under the effect of the following processes (Fig. 2):

1. The input of P, L, C and Z, contained in the influent of wastewater, in a dissolved form S_F;
2. Hydrolysis of suspended organic matter X_F, coming into the reactor with the influent wastewater;
3. Lysis and hydrolysis of cell biomass B_{di}, formed in the reactor after the bacterial death.

A dissolved organic substrate S_1 transforms into methane through the intermediate products S_j.

In Fig. 2 the rectangles denote substrate concentrations, lines outgoing from the rectangles denote the fractions constituting a given compound; value *f* shows the relative content of the C, P, L, and Z in the compound, and value *p* shows the relative content of biodegradable fraction of biopolymers. The particulate organic matter are divided into volatile VX, VB, and mineral

Hydrolysis Kinetics

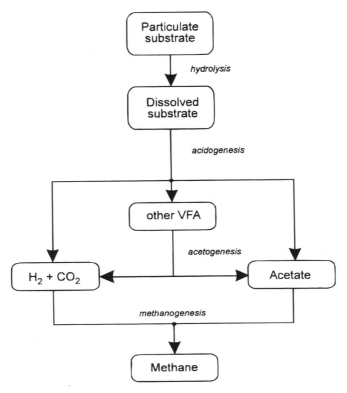

Fig. 1. Key stages of microbial transformation of complex organic matter.

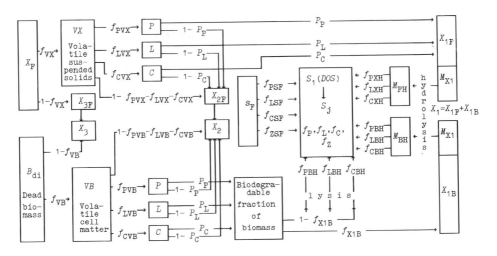

Fig. 2. Ways of formation of suspended and dissolved components in the anaerobic system.

X_3 fractions. Volatile organic matter VX and VB comprises P, L, C, and the nonbiodegradable substrates X_2. As result of hydrolysis of monomers of the degradable fraction of the influent suspended solids X_{1F} and the matter of the dead cells X_{1B}, the total mass of hydrolysis products will be equal to M_{FH} in the first case and M_{BH} in the second case (for simplicity below X_1 will be written as X and the hydrolytic biomass will be written as B).

Various versions of hydrolysis kinetics including the simplest first-order kinetics and the Contois equation are considered in the model <METHANE> for description of degradation of suspended solids:

$$v_H = k_H \cdot X \qquad \text{(First-order)} \quad (1)$$

$$v_H = v_{mH} \cdot B \cdot \frac{X/B}{\tilde{K}_X + X/B} \qquad \text{(Contois)} \quad (2)$$

where v_H is the rate of solids degradation, v_{mH} is the maximum specific hydrolysis rate, X and B are the concentrations of biodegradable suspended solids and hydrolytic biomass, respectively, k_H is the first-order kinetics constant, $\tilde{K}_X$ is the half-saturation constant.

RESULTS AND DISCUSSION

N-Order Solids Degradation Kinetics

During the hydrolysis the particulate substrates contact with hydrolytic microbial cells and the released enzymes (8). Two main phases must be taken into account for a description of the hydrolysis kinetics. The first phase is bacterial colonization during which the hydrolytic bacteria cover a surface of the solids. The bacteria located on or near the particle surface release the enzymes and produce monomers, which can be utilized by the hydrolytic bacteria themselves, as well as by other bacteria. The daughter cells that fall off into the liquid phase are trying to attach to some new place of a particle surface. According to Hobson and Wheatley (3), when an available surface is covered with bacteria, the surface area will be changed at a constant depth per unit of time (second phase).

Following a two-phase approach let us write the equations for the concentration of hydrolytic biomass B and the concentration of suspended solids X in the continuous-flow stirred-tank digester, respectively:

$$\begin{aligned} \frac{dB}{dt} &= Y \cdot v_H(B, X) - k_d \cdot B - \frac{1}{\theta} \cdot B \\ \frac{dX}{dt} &= -v_H(B, X) + \frac{X_F - X}{\theta} \end{aligned} \qquad (3)$$

where k_d is the rate coefficient of biomass decay, Y is the yield coefficient, X_F is the influent suspended solids concentration, θ is the solids retention time, and the $v_H(B, X)$ is the hydrolysis rate function that equals to

Hydrolysis Kinetics

$$\upsilon_H(B,X) = \begin{cases} \upsilon_{mH} \cdot B, & \text{if } B < \tilde{B} \quad \text{(A)} \\ \upsilon_{mH} \cdot \tilde{B}, & \text{if } B \geq \tilde{B} \quad \text{(B)} \end{cases} \tag{4}$$

where $\tilde{B}$ is the saturated biomass concentration covering the contact surface of the hydrolyzed particles, and υ_{mH} is the maximum specific hydrolysis rate.

Vavilin et al. *(9)* obtained the following generalized-type function that describes the rate of degradation of solids surface limited by a contact area between particles and bacterial mass

$$\upsilon_H = \upsilon_{mH} \cdot K \cdot X_F^{1-n} \cdot X^n \tag{5}$$

where K is the dimensionless constant and n is the degree coefficient. For the spherically symmetrical particles, it was obtained:

$$\tilde{B} = K \cdot X_F^{1/3} \cdot X^{2/3} \tag{6}$$

$$K = 6 \cdot \frac{\rho_B}{\rho_X} \cdot \frac{\delta}{d_{XF}} \tag{7}$$

$$\upsilon_H = \upsilon_{mH} \cdot K \cdot X_F^{1/3} \cdot X^{2/3} \tag{8}$$

where ρ_B, ρ_X are the bacterial and solids density, respectively, δ is the depth of bacterial layer; d_{XF} is the average input diameter of the particles hydrolyzed. It was assumed that the hydrolysis rate is limited by a contact area between particles and bacterial mass, and the size of the hydrolyzed particle is much greater than the depth of bacterial layer.

For the cylindrical symmetrical particles it was obtained:

$$\tilde{B} = K \cdot X_F^{1/2} \cdot X^{1/2} \tag{9}$$

$$K = 4 \cdot \frac{\rho_B}{\rho_X} \cdot \frac{\delta}{d_{XF}} \tag{10}$$

$$\upsilon_H = \upsilon_{mH} \cdot K \cdot X_F^{1/2} \cdot X^{1/2} \tag{11}$$

The case with $n = 0$ corresponds to a disk-like particle. The first-order kinetics following from equation 5 at $n = 1$ can be considered as an approximation of equation 8 or equation 11. Therefore, the hydrolysis constant $k_H = \upsilon_{mH} \cdot K$ is the function of the ratio between the characteristic sizes of bacterial layer and solids hydrolyzed. Shimuzu et al. *(10)* have reported that after ultrasonic lysis of the biopolymers of waste sludge, the hydrolysis constant of the first order kinetics k_H increases from 0.16 day^{-1} to 1.2 day^{-1}. It is evident that the characteristic size of suspended solids decreases, but the contact area between particulate organic material and bacteria increases substantially in this case.

Obviously, the two-phase model (3) that takes into account a colonization phase describes a biomass washing phenomenon for a continuos-flow system if

$$\theta < \theta_{cr} = 1/(Y \cdot \upsilon_{mH} - k_d) \qquad (12)$$

The simplest equation, 1 is unable to describe the biomass washing. It gives lower concentration of effluent suspended solids at low θ. The n-order kinetics fits well experimental data of sludge and cellulose mesophylic digestion over wide range of SRT (Fig. 3).

It is easy to see that at $n = 1$ a piece-wise (discrete) function (4) has the same limiting cases as the continuous Contois function (2). The ratio between the solids and biomass concentrations $\xi = X/B$ regulates a degradation rate, which is proportional to the biomass concentration B at high ξ and to the biodegradable solids concentration X at low ξ.

A generalized continuous function for the model (3) may be written in the form

$$\upsilon_H = \upsilon_{max} \cdot \sigma(B) \cdot A(X_F, X) \qquad (13)$$

where $\sigma(B)$ is the fraction of the surface occupied by bacteria or released enzymes as a function of the concentration of hydrolytic bacteria, $A(X_F, X)$ is the relative surface area subjected to degradation as a function of the influent and effluent solids concentrations.

Assuming the Langmuir function for an adsorption kinetics of hydrolytic biomass and the Monod function for a surface degradation kinetics, the function (13) can be written as

$$\upsilon_H = \upsilon_{max} \cdot \frac{\lambda B}{1 + \lambda B} \cdot \frac{X}{K_X + X} \qquad (14)$$

where λ is the equilibrium constant equal to the ratio between the adsorption and desorption rates constants.

Balance Between Hydrolysis and Methanogenesis

If the hydrolysis step is rate limiting, using the first-order equation (1) for a batch process, the degradable suspended solids concentration X and the methane volume released Q are given as

$$X = X_0 \cdot e^{-k_H t} \qquad (15)$$

$$Q = M_{Ym} \cdot X_0 \cdot (1 - e^{-k_H t}) \qquad (16)$$

where M_{Ym} is the methane to substrate conversion factor, which takes into account substrate converted to biogas; X_0 is the initial concentration of degradable organic matter. The values of M_{Ym} and k_H for different wastes can be found in the literature (14).

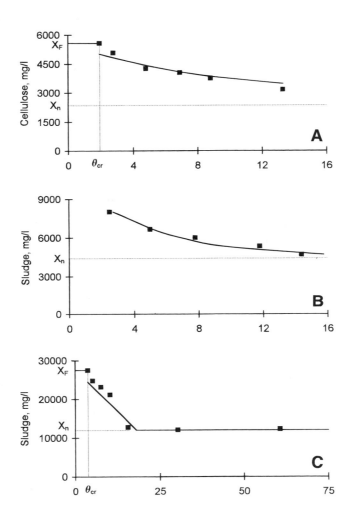

Fig. 3. Steady-state concentrations of cellulose (**A**) and sludge (**B, C**) with the n-order hydrolysis kinetics as a function of solids retention time:

A) $n = 0.5$, $v_{mH} \cdot K = 0.19$ day^{-1}, $X_n = 2360$ mg/L;
B) $n = 0.3$, $v_{mH} \cdot K = 0.15$ day^{-1}, $X_n = 4400$ mg/L;
C) $n = 0$, $v_{mH} \cdot K = 0.06$ day^{-1}, $X_n = 12000$ mg/L;

where X_n = nondegradable suspended solids concentration, θ_{cr} = critical (biomass washing) value of solids retention time.

Symbols: experimental values; lines: model prediction. Experimental data are taken from refs. *11–13*.

Table 1
The Characteristic Values of Maximum Growth Rate Constants of Biomass and Hydrolysis Rate Constants Used in ⟨Methane⟩ Model

Process	Feed	Temperature	Figure	Constants[1]	Source of experimental data
Hydrolysis	Sludge	35°C	4	k_H=0.25 day^{-1} (Eq.1)	5
	Cellulose	28°C	5	k_H=0.15 day^{-1} (Eq.1)	15
		15°C	5	υ_{mH}=0.45 g/g·day, $\widetilde{K}_X$=1.5 g/g (Eq.2)	15
		6°C	5	υ_{mH}=0.11 g/g·day, $\widetilde{K}_X$=1.8 g/g (Eq.2)	15
	Cattle manure	6°C	6	k_H=0.13 day^{-1} (Eq.1)	17
Acidogenesis	Sludge	35°C	4	5.0 day^{-1}	5
	Cellulose	28°C	5	10.0 day^{-1}	15
		15°C	5	4.0 day^{-1}	15
		6°C	5	2.0 day^{-1}	15
	Cattle manure	6°C	6	2.0 day^{-1}	17
Acetogenesis	Sludge	35°C	4	0.8 day^{-1}	5
	Cellulose	28°C	5	0.28 day^{-1}	15
		15°C	5	0.12 day^{-1}	15
		6°C	5	0.07 day^{-1}	15
	Cattle manure	6°C	6	0.11 day^{-1}	17
Methanogenesis	Sludge	35°C	4	0.5 day^{-1} (A) 2.0 day^{-1} (H)	5
	Cellulose	28°C	5	0.2 day^{-1} (A) 0.45 day^{-1} (H)	15
		15°C	5	0.096 day^{-1} (A) 0.18 day^{-1} (H)	15
		6°C	5	0.048 day^{-1} (A) 0.13 day^{-1} (H)	15
	Cattle manure	6°C	6	0.07 day^{-1} (A) 0.28 day^{-1} (H)	17

[1] A = acetate utilizing methanogens, H = hydrogens utilizing methanogens. For cattle manure the constants of the single pathway are presented.

The ⟨METHANE⟩ model was used to analyze the dynamics of sludge degradation after the sewage sludge feed of continuously fed labor digester was stopped (5). Some constants of the model for the case-studies simulated are summarized in Table 1. From Fig. 4 it can be concluded that

Hydrolysis Kinetics

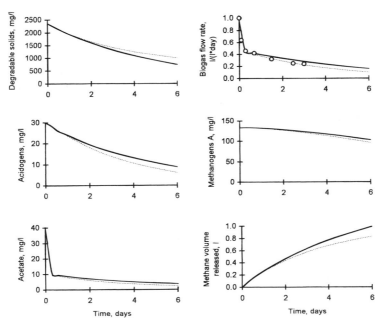

Fig. 4. Simulation of the sludge mesophylic digestion dynamics after stopping of sewage sludge feed of a continuously fed digester.
Symbols: experimental values; lines: model prediction. Experimental data are taken from ref. 5. ——— First order equation; - - - - Contois equation; ○ Experimental data.

a high initial biomass concentration does not limit the process that can be described by the simple equations (15), (16).

During the colonization phase the hydrolytic bacteria cover the surface of the solids. The rate of this process is proportional to the concentration of hydrolytic biomass. Figure 5 shows one-year graphs of cellulose degradation at the different temperature with micro-organism inoculum taken from the sediments of highly polluted treatment pond (15). In this case the dynamics of released methane volume can not be described by the simple equation (16) because the low initial microorganism concentrations. The first-order equation (1) of hydrolysis kinetics was applied in the ⟨METHANE⟩ model for 28°C only and the Contois equation (2) was applied for 15°C and 6°C.

Two Parallel Pathways of Degradation of Cattle Manure

It was mentioned above that the particle surface decreases at a constant depth per unit of time. However, availability of the particle surface for enzyme attack may change during the hydrolysis. Hobson (16) described the particle degradation of pig and cattle wastes by two growth rates in Monod equations for hydrolytic bacteria.

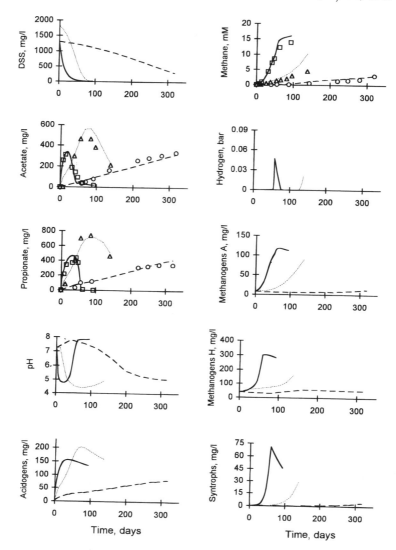

Fig. 5. Dynamics of anaerobic conversion of cellulose at different temperatures. Symbols: experimental values; lines: model prediction. Experimental data are taken from ref. 15. − − − 6°C ○; - - - - 15°C △; —— 28°C □.

As result of a long-time (2.5 yr) selection, Kotsyurbenko et al. *(17)* have obtained an acclimated microbial consortium fermenting cattle manure at 6°C with methane production. Using this consortium as a seed material, a cattle manure degradation was studied at 6°C. No methane production was obtained during manure degradation without addition of inoculum and the final products were volatile fatty acids.

Figure 6 shows the results of simulations of the system with inoculum. It can be concluded that there are two parallel pathways of cattle

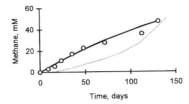
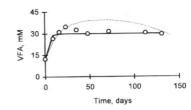

Fig. 6. Dynamics of anaerobic conversion of cattle manure at 6°C described by a single and two parallel pathways.

The first-order kinetics equation 16 was used for description of two parallel pathways of manure degradation:

$k_H = 0.005$ day^{-1}, $M_{Ym} = 2.2$ mM/mM, $X_0 = 45$ mg/L (CH$_4$)

$k_H = 0.03$ day^{-1}, $M_{Ym} = 2.5$ mM/mM, $X_0 = 7$ mg/L (VFA)

where M_{Ym} = conversion factor of methane or VFA to particulate substrate. Some constants of ⟨METHANE⟩ model used for description of a single pathway are presented in Table 1.

Symbols: experimental values; lines: model prediction. Experimental data are taken from ref. 17. ——— Two pathways, - - - - Single pathway, ○ Experimental data.

manure degradation with the final products of volatile fatty acids and a predominance of acetate (i) and methane (ii), produced without acetate consumption. The first pathway is conducted by micro-organisms presented in raw cattle manure and the second pathway is conducted by micro-organisms presented in inoculum. A much worse fitness of the model to experimental data was obtained assuming as usually the single pathway of solids degradation and methane production through acetotrophic and hydrogenotrophic methanogenesis. Some model constants are presented in Table 1.

CONCLUSIONS

N-order kinetics with the rate constant dependable on a surface-to-volume ratio should be applied for description of solids degradation.

A delay of methane production during batch test happens usually because of the low initial biomass concentrations.

At low temperature the cattle manure is transformed into methane without acetate consumption.

ACKNOWLEDGMENTS

This research was supported by the Russian Foundation for Basic Research (Project No. 96-01-00893).

NOMENCLATURE

B_i	concentration of biomass of i-th group of microorganisms
B	concentration of biomass of hydrolytic microorganisms
$\tilde{B}$	saturated hydrolytic biomass concentration covering the particle surface
B_{di}	concentration of dead biomass of i-th group
C	carbohydrate fraction of organic matter
d_X	current diameter of particles
d_{XF}	input diameter of particles
f	relative contents of carbohydrates, proteins, lipids and undefined organic matter
G_i	partial pressure of gas components
M_{Ym}	methane to substrate conversion factor
M_{BH}	monomers of the biodegradable fraction of dead biomass
M_{FH}	monomers of the biodegradable fraction of suspended solids
K	dimensionless constant
k_H	kinetic rate constant
k_d	biomass decay-rate coefficient
K_X	half-saturation coefficient for the generalized function (14)
$\tilde{K}_X$	half-saturation coefficient for the Contois function (2)
L	lipid fraction of organic matter
n	degree coefficient
P	protein fraction of organic matter
p	relative content of biodegradable fraction of biopolymers
Q	methane volume released
S_j	concentration of dissolved substrate components
S_1	DOS = concentration of dissolved carbohydrates, proteins and lipids
t	current time
VB	volatile cell matter
VFA	volatile fatty acids
VX	volatile suspended solids
X	$X_1 = DSS$ = concentration of biodegradable suspended solids
X_2	concentration of nonbiodegradable fraction of suspended solids and biomass
X_3	concentration of mineral fraction of suspended solids and biomass
X_F	concentration of influent biodegradable suspended solids
X_0	concentration of initial biodegradable suspended solids
Y	yield coefficient
δ	depth of bacteria layer on particles

θ	solids retention time
θ_{cr}	critical (biomass washing) value of solids retention time
λ	equilibrium constant equal to the ratio between the adsorption and desorption rate constants
ξ	parameter equal to the ratio between biomass and solids concentration
ρ_B	bacterial density
ρ_X	solids density
σ	the fraction of the surface occupied by bacteria
υ_H	rate of solids degradation
υ_{max}	maximal rate of solids degradation
υ_{mH}	maximal specific hydrolysis rate

REFERENCES

1. Pavlostathis, S. G. and Giraldo-Gomez, E. (1991), *Crit. Rev. Env. Contr.* **21**, 411–490.
2. Eastman, J. A. and Ferguson, J. F. (1981), *J. WPCF*, **53**, 352–366.
3. Hobson, P. N. and Wheatley, A. (1992), *Anaerobic Digestion: Modern Theory and Practice*, Elsevier, Oxford.
4. Angelidaki, I., Ellegaard L., and Ahring B. K. (1993), *Biotechnol. Bioengn.* **42**, 159–166.
5. Siergist, H., Renggli, D., and Gujer, W. (1992), in *Proc. Intern. Symp. on Anaerobic Digestion of Solid Waste*. Venice, 14–17 April, 1992. Stamperia di Venezia, pp. 51–64.
6. Vasiliev, V. B., Vavilin, V. A., Rytov, S. V., and Ponomarev, A. V. (1993), *Water Resources* **20**, 633–643.
7. Vavilin, V. A., Vasiliev, V. B., and Rytov, S. V. (1993), In *Modeling of Organic Matter Destruction by Microorganism Community*, Nauka Publishers, Moscow (in Russian).
8. Lee, Y. H. and Fan, L. T. (1982), *Biotech. Bioengn.* **24**, 2383–2406.
9. Vavilin, V. A., Rytov, S. V., and Lokshina, L. Ya. (1996), *Bioresource Technol.* **56**, 229–237.
10. Shimuzu, T., Kudo, K., and Nasu, Y. (1993), *Biotechnol. Bioeng.* **41**, 1082–1091.
11. Noike, T., Endo, G., Chang, J.-E., and Matsumoto, J. I. (1985), *Biotechnol. Bioeng.* **27**, 1482–1489.
12. O'Rourke, J. R. (1968), Ph.D. thesis, Standford University, Stanford, CA.
13. Chen, Y. R. and Hashimoto, A. G. (1980), *Biotechnol. Bioeng.* **22**, 2081–2095.
14. Owens, J. M. and Chynoweth, D. P. (1992), in *Proc. of Inter. Symposium on Anaerobic Digestion of Solid Waste*. Venice, 14–17 April, 1992, Stamperia di Venezia, pp. 29–42.
15. Kotsyurbenko, O. R., Nozhevnikova, A. N., and Zavarzin, G. A. (1992), *J. General Biol.* **53**, 159–175 (in Russian).
16. Hobson, P. N. (1983), *J. Chem. Tech. Biotechn.*, **33B**, 1–20.
17. Kotsyurbenko, O. R. Nozhevnikova, A. N., Kalyuzhny and Zavarzin, G. A. (1993), *J. General Biol.* **54**, 761–772 (in Russian).

Preprocessed Barley, Rye, and Triticale as a Feedstock for an Integrated Fuel Ethanol-Feedlot Plant

KRYSTYNA SOSULSKI,*,[1] SUNMIN WANG,[2] W. M. INGLEDEW,[2] FRANK W. SOSULSKI,[2] AND JUMING TANG[3]

[1]*Saskatchewan Research Council, 15 Innovation Blvd., Saskatoon, SK S7N 2X8, Canada;* [2]*University of Saskatchewan, Saskatoon, SK; and* [3]*Washington State University, Pullman, WA*

ABSTRACT

Rye, triticale, and barley were evaluated as starch feedstock to replace wheat for ethanol production. Preprocessing of grain by abrasion on a Satake mill reduced fiber and increased starch concentrations in feedstock for fermentations. Higher concentrations of starch in flours from preprocessed cereal grains would increase plant throughput by 8–23% since more starch is processed in the same weight of feedstock. Increased concentrations of starch for fermentation resulted in higher concentrations of ethanol in beer. Energy requirements to produce one L of ethanol from preprocessed grains were reduced, the natural gas by 3.5–11.4%, whereas power consumption was reduced by 5.2–15.6%.

Index Entries: Cereal grains; preprocessing by abrasion; ethanol yield; mass balance; energy.

INTRODUCTION

High-yielding Canadian Prairie Spring (CPS) wheat varieties are used as raw material for fuel ethanol production in Western Canada. Local availability and a high starch content make CPS wheat the feedstock of choice. However, in 1995–96, the decline in the world wheat reserves and predictions of low winter wheat yields in the United States pushed prices of wheat to US $4.70/bu (1). Thus, wheat became too expensive for fuel ethanol production.

*Author to whom all correspondence and reprint requests should be addressed.

Other cereal grains, triticale, rye, and barley, are traditionally lower priced than wheat; their 1995–96 prices ranged from US $110 to $120/tonne (1). Two of these grains, triticale and rye, are comparable in starch content to wheat, whereas barley is substantially lower because of the presence of hull (2).

Preprocessing of wheat was reported to remove bran fiber and to concentrate starch for fermentation (3). The present investigation describes preprocessing of barley, rye, and triticale by abrasion and the effect of fiber removal on concentration of starch in flours from the preprocessed grains. Ethanol concentrations and yields, mass balance and composition of products, and the influence of preprocessing on plant throughput and energy requirements are described.

MATERIALS AND METHODS

Cultivars selected for the study included cereal grains commonly grown in Western Canada and available for processing to ethanol. The grains were: Prima fall rye, CDC Dolly 2-row barley and AC Copia triticale.

A Satake abrasive rice mill, model TM05 (Satake Co., Japan), equipped with a medium coarse-grain stone, was used to evaluate the influence of grain preprocessing on chemical composition, ethanol yield, plant throughput, and energy requirements. Cereal grains, conditioned to 12.5% moisture content, were abraded on the Satake mill for 55 s at stone speed of 1450 rpm. The preprocessed grains (containing mainly endosperm) were separated from fines that were further separated on screens into bran and flour for rye and triticale, and into hulls—bran and flour for barley. The separated flours were combined with preprocessed endosperms.

At each stage of preprocessing samples of grain, preprocessed endosperms, bran and hull-bran fractions were collected, ground, and passed through a 60 mesh screen before chemical analysis and fermentations. All chemical analyses and fermentation experiments were duplicated.

Ground grains or preprocessed flours (Fig. 1) were dispersed in water, ratio of grain or flour to water being 1:3. The alpha-amylase enzyme, Maxaliq (International Bio-Synthetics, Charlotte, NC), was added to the mash in two portions, each at a concentration 0.4% (v/w enzyme to grain/flour). Starch hydrolysis with the first portion of enzyme was carried out at 95°C for 45 min. After lowering the temperature to 80°C, the second portion of enzyme was added and liquefaction was continued for an additional 30 min. Grain or flour mash was then transferred into a jacketed fermenter (Wheaton Scientific, Millville, NJ) containing 0.4% yeast extract (AYE 2200). Saccharification with Allcoholase II (Alltech Inc., Nicholasville, KY) was carried out at 30°C for 30 min at an enzyme concentration of 0.8% (v/w enzyme to grain/flour). Inoculum (10^7 cells/mL mash) of *Saccharomyces cerevisiae* yeast (Alltech Inc., Nicholasville, KY) was added to mash cooled to 27°C. Fermentations were carried out for 72 h.

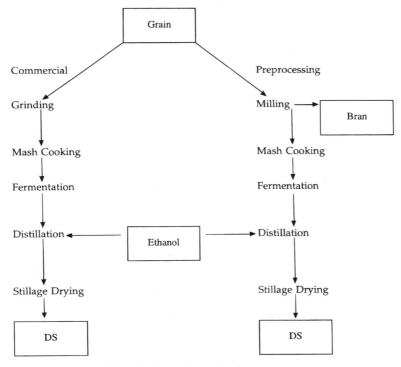

Fig. 1. Ethanol production process.

Ethanol was distilled from beer and the concentration of ethanol in the distillate was measured by an alcohol dehydrogenase assay, Kit # 332-A (Sigma Chemical Co., St. Louis, MO). Stillage was freeze-dried for further analysis.

Samples of grain, preprocessed flours, bran or hull-bran fractions, and dried stillage (DS) were analyzed by standard AACC (4) procedures for moisture (Method 44-15A), crude protein (Method 46-13) and crude fat (Method 30-25). Protein contents of samples were calculated using the nitrogen-to-protein conversion factor of 5.7 (5). Starch in grains, flours, and dried stillage was measured as glucose after hydrolysis with alpha-amylase and amyloglucosidase, following the procedure of Budke (6). Total dietary fiber (TDF) was quantitated by the enzymatic gravimetric method of Prosky et al. (7).

Economic evaluation of ethanol production from the preprocessed cereal grains and energy requirements for the process were carried out using a computer program developed for a small-scale ethanol plant. Data to create the computer program included equipment energy requirements, commercial and experimental process conditions, e.g., temperature, time of treatment, amounts of grain and water, and so on.

Table 1
The Influence of Grain Class and Grain Preprocessing
on Ethanol Concentration in the Beer

Grain class and product	Protein	Starch	Ethanol concentration		Fermentation efficiency
	--- % as is ---		% w/v	% v/v	%
Grain					
Fall rye	8.8 ± 0.1	56.1 ± 0.6	9.5 ± 0.6	12.1 ± 0.7	91.6 ± 0.7
Triticale	11.4 ± 0.2	55.6 ± 0.4	9.5 ± 1.0	12.0 ± 1.1	90.8 ± 1.1
2-row barley	10.2 ± 0.3	51.7 ± 0.5	8.8 ± 0.8	11.1 ± 0.9	89.2 ± 0.8
Flour					
Fall rye	8.2 ± 0.2	60.1 ± 0.2	10.2 ± 0.8	12.9 ± 0.8	93.6 ± 0.8
Triticale	11.4 ± 0.1	61.1 ± 0.4	10.4 ± 0.8	13.2 ± 0.9	94.6 ± 0.9
2-row barley	9.9 ± 0.2	61.6 ± 0.5	10.5 ± 0.8	13.3 ± 0.9	92.3 ± 0.9

*Ratio of grain/flour to water 1:3.

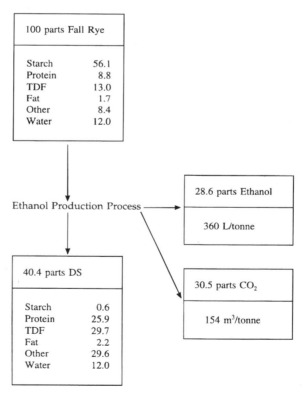

Fig. 2. Mass balance, ethanol yield, and dried stillage composition for Prima fall rye after commercial processing.

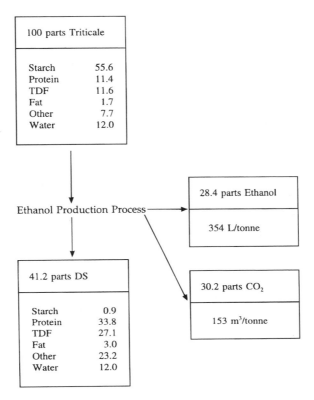

Fig. 3. Mass balance, ethanol yield, and dried stillage composition for AC Copia triticale after commercial processing.

RESULTS AND DISCUSSION

Fall rye contained the lowest protein and the highest starch contents (Table 1). Triticale was the highest in protein among the three grains, but was comparable to rye in starch content. Thus both grains should yield similar volumes of ethanol. Although intermediate in protein, barley was about 5% lower in starch than rye. As concentration of ethanol in the beer depended directly on the starch content in the feedstock, rye, and triticale yielded beers that were almost 1% higher in ethanol than barley.

Preprocessing of grain reduced protein and increased starch contents in the flours (Table 1). The increased concentrations of starch, from 51–56% in grains to 60–62% in the flours, increased concentrations of ethanol in the beer from an average 9.2% (w/v) to 10.3% (w/v), respectively. This represents an 11.9% increase in ethanol concentration in the beer that will significantly reduce distillation cost. The greatest increase in ethanol concentration, because of preprocessing, was observed for barley.

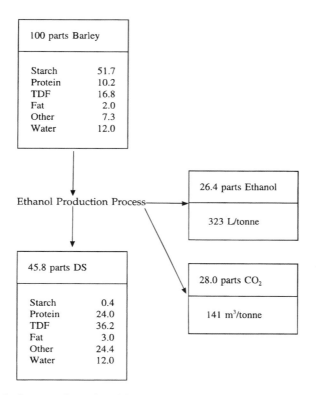

Fig. 4. Mass balance, ethanol yield, and dried stillage composition for CDC Dolly 2-row barley after commercial processing.

The fermentation efficiencies, calculated as the ratio between experimental and theoretical ethanol yields, ranged from 89.2–91.6% for grain to 92.3–94.6% for preprocessed grain samples (Table 1).

The volumes of ethanol produced from cereal grains corresponded to their starch contents. Fall rye and triticale, which were similar in starch contents, yielded 360 L/tonne and 354 L/tonne, respectively (Figs. 2 and 3). Barley, low in starch but high in TDF and protein as compared to rye and triticale, produced 323 L of ethanol per tonne of grain (Fig. 4). Ethanol represented only 26.4–28.6% of total product yields (Figs. 2–4). Carbon dioxide, which is not utilized in small ethanol plants, accounted for 28.0–30.5% of total products. Dry stillage (DS) represented 40.4–45.8% of total products from ethanol production. More stillage was produced from barley than from rye or triticale. And the stillage from barley contained the lowest level of protein and the highest level of fiber.

In addition to protein (24.0–33.8%), TDF (27.1–36.2%), and fat (2.2–3.0%), DS contained 23.2–29.6% of other constituents, including pentosans, beta-glucans, ash, phytate, and fermentation by-products such as

Integrated Fuel Ethanol-Feedlot Plant

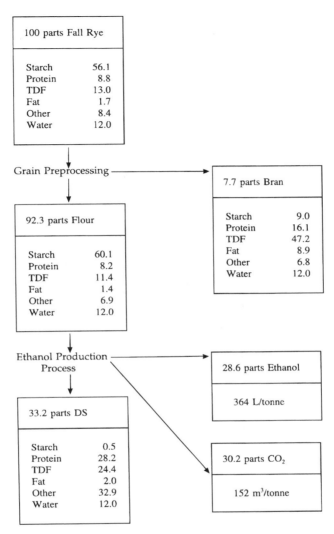

Fig. 5. Mass balance, ethanol yield, and product compositions for preprocessed Prima fall rye.

glycerol and organic acids (Figs. 2–4). The residual starch was below 1%. No free glucose was determined in stillage.

Preprocessing of grain removed from 7.7–21.7% of grain weight into bran fractions (Figs. 5–7). TDF was the major component of bran. However, bran contained 9.0 to 23.8% starch that represented a 2.0–9.7% loss of the total grain starch into bran. The most starch was lost during preprocessing of soft-kerneled triticale, followed by barley and rye. The starch lost to bran, although not available for conversion to ethanol, would

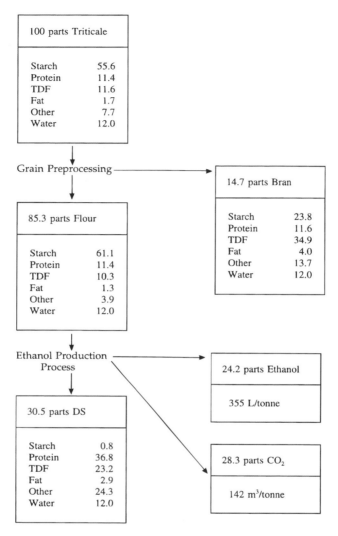

Fig. 6. Mass balance, ethanol yield, and product compositions for preprocessed AC Copia triticale.

increase bran quality as animal feed in an integrated fuel ethanol-feedlot operation.

The flours represented 78.3–92.7% of grain weights and contained 60.1–61.6% starch (Figs. 5–7). Preprocessing of rye and triticale increased starch concentration in the flour by 4–5%, which corresponded to low or intermediate removal of grain dry matter to bran fractions. Removal of hull and bran from barley, which represented 21.7% of grain, increased starch concentration in the flour by 10%. Increased starch concentrations in flours as compared to grain yielded greater concentrations of ethanol in the beer (Table 1).

Integrated Fuel Ethanol-Feedlot Plant

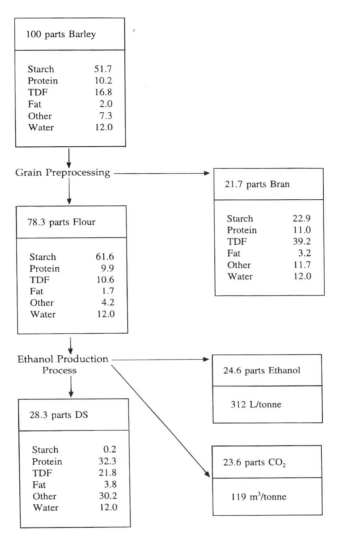

Fig. 7. Mass balance, ethanol yield, and product compositions for preprocessed Dolly 2-row barley.

Protein concentrations in the preprocessed flours were only slightly lower as compared to grains, but TDF contents were substantially reduced (Figs. 5–7). Reduction of TDF in flours, which became the feedstocks for ethanol fermentation, reduced TDF in DS with corresponding increases in protein contents, as compared to DS from grains (Figs. 2–4). No residual glucose and lower residual starch in DS from flours, as compared to DS from grains, indicate that preprocessed rye, triticale, and barley flours can be successfully fermented to ethanol.

Table 2
Effect of Preprocessing on Product Yields and Energy
Required During Fermentation, Based on a 10 Million L Ethanol
Plant Processing Prima Fall Rye

	Prima fall rye	
	Whole grain	Preprocessed
Material balance:		
Grain processed (tonnes)	27,778	30,095
Flour (tonnes)	--	27,778
Bran (tonnes)	--	2317
DS (tonnes)	11,112	9179
Ethanol (million L)	10.00	10.80
Ethanol production increase (%)	--	8.0
Energy balance:		
Gas:		
Cooking (MJ/L)	2.06	1.97
Distillation (MJ/L)	7.70	7.58
Drying (MJ/L)	19.60	18.83
Total gas (MJ/L)	29.36	28.35
Gas use decrease (%)	--	3.5
Electricity:		
Preprocessing (MJ/L)	--	0.09
Grinding (MJ/L)	0.16	0.15
Ethanol production	2.54	2.32
Total electricity (MJ/L)	2.70	2.56
Electricity use decrease (%)	--	5.2

The results from the experimental laboratory fermentations were used to evaluate the benefits of preprocessing of cereal grains for a theoretical 10 million L/y ethanol plant. On a constant throughput basis for the fermenters, preprocessing increased ethanol production capacity of this theoretical plant by 8–23%, depending on the preprocessed cereal grain (Tables 2–4). The best results were obtained with barley.

Preprocessing of grain to remove bran, or hull and bran in barley, and thus reducing nonfermentable constituents in feedstocks for the fermentation process, reduced material weights for distillation and drying to DS. Both units of operation showed decreased energy requirements per one L of ethanol (Tables 2–4). Combined use of natural gas was reduced by 3.5–11.4%. Reduction in power consumption ranged from 5.2 to 15.6%.

Table 3
Effect of Preprocessing on Product Yields and Energy
Required During Fermentation, Based on a 10 Million L Ethanol
Plant Processing AC Copia Triticale

	AC Copia triticale	
	Whole grain	Preprocessed
Mass balance:		
Grain processed (tonnes)	28,248	33,116
Flour (tonnes)	--	28,248
Bran (tonnes)	--	4868
DS (tonnes)	11,638	8616
Ethanol (million L)	10.00	11.16
Ethanol production increase (%)	--	11.6
Energy balance:		
Gas:		
Cooking (MJ/L)	2.03	1.93
Distillation (MJ/L)	7.67	7.54
Drying (MJ/L)	19.40	18.60
Total gas (MJ/L)	29.10	28.08
Gas use decrease (%)	--	3.5
Electricity:		
Preprocessing (MJ/L)	--	0.09
Grinding (MJ/L)	0.16	0.16
Ethanol production (MJ/L)	2.58	2.20
Total electricity (MJ/L)	2.74	2.44
Electicity use decrease (%)	--	10.9

CONCLUSIONS

1. Preprocessing of grain reduced fiber and increased starch contents in feedstock for ethanol production, yielding beers with increased concentrations of ethanol. Preprocessing was especially beneficial for barley where removal of 22% grain weight increased starch content by 10%.
2. Grain preprocessing would increase the capacity of an ethanol plant, since more starch is processed in the same weight of feedstock. The increase in plant capacity would be in the range 8–23%, depending on grain, with the highest value being for barley flour.
3. Grain preprocessing reduced energy requirements in ethanol production. Total gas use would be reduced by 3.5–11.4%, whereas power consumption would be reduced by 5.2–15.6%.

Table 4
Effect of Preprocessing on Product Yields and Energy
Required During Fermentation, Based on a 10 Million L Ethanol
Plant Processing CDC Dolly 2-row Barley

	CDC Dolly 2-row barley	
	Whole grain	Preprocessed
Mass balance:		
Grain processed (tonnes)	30,950	39,527
Flour (tonnes)	--	30,950
Bran (tonnes)	--	8577
DS (tonnes)	14,175	8759
Ethanol (million L)	10.00	12.33
Ethanol production increase (%)	--	23.3
Energy balance:		
Gas:		
Cooking (MJ/L)	2.20	1.91
Distillation (MJ/L)	7.92	7.49
Drying (MJ/L)	21.00	18.20
Total gas (MJ/L)	31.12	27.60
Gas use decrease (%)	--	11.4
Electricity:		
Preprocessing (MJ/L)	--	0.10
Grinding (MJ/L)	0.18	0.15
Ethanol production (MJ/L)	2.83	2.29
Total electricity (MJ/L)	3.01	2.54
Electricity use decrease (%)	--	15.6

REFERENCES

1. Anonymous. (1996), *Market Trends*, Saskatchewan Agriculture and Food. April 24.
2. Sosulski, K. and Tarasoff, L. (1996), Report G680-1C.96, Saskatchewan Research Council, Saskatoon, SK, Canada.
3. Sosulski, K. and Sosulski, F. (1994), *Applied Biochem. Biotechnol.* **45/46,** 169–180.
4. American Association of Cereal Chemists (1995), Approved Methods of the AACC, 9th ed., St. Paul, MN.
5. Sosulski, F. and Imafidon, G. I. (1990), *J. Agric. Food Chem.* **38,** 1352–1358.
6. Budke, C. C. (1984), *J. Agric. Food Chem.* **32,** 34–37.
7. Prosky, L., Asp, N. G., Furda, I., DeVries, J. W., Schweizer, T. F., and Harland, B. A. (1985), *J. Assoc. Off. Anal. Chem.* **68,** 677–685.

Session 2

Biological Research

VALERIE SARISKY-REED[1] AND JONATHAN WOODWARD[2]

*[1]US Department of Energy, Washington, D.C.;
and [2]Oak Ridge National Laboratory, Oak Ridge, TN*

Biotechnology is not a new phenomenon; its techniques have been used throughout history. Microorganisms have been used since early times to make bread and beer. These processes were the beginnings of the microbial fermentation technologies that continue to be practiced today. It was not until the 1970s, however, that scientists learned to reach into living organisms and precisely alter their genetic structures, leading to the development of metabolic engineering, which takes advantage of advances in DNA cloning and sequencing technologies, synthesis and amplification techniques, arming scientists with the ability to target and purposefully alter metabolic pathways within organisms. This will result in a better understanding and utilization of the cellular pathways for chemical transformation, energy transduction, and supermolecular assembly.

Bioprocessing is the ability to use living organisms or their components to produce marketable materials. Chemicals and fuels can be produced through bioprocesses, eliminating the need for energy-intensive chemical processes and reducing our national dependence on petroleum. In the case of fuel production, ethanol produced from lignocellulosic biomass, which includes agricultural wastes, such as rice straw and corn fiber, represents an opportunity to clean up waste products that have polluted our environment and turn them into a valuable commodity.

The work presented in this session is representative research in chemical and fuels production using biological organisms in place of chemical methods. These biological systems, whether they are *E. coli* engineered with the capability to grow on limonene, or yeasts that can ferment five and six carbon sugars, will be valuable to the future of the chemical and fuels industry. As scientists learn more about our environment and the need to reduce petroleum-based chemistry, more and more opportunities arise for novel bioprocesses, and with the development of each process, there is an overwhelming need to learn more about the systems under development.

A Stable Lipase from *Candida lipolytica*

Cultivation Conditions and Crude Enzyme Characteristics

FATIMA VENTURA PEREIRA-MEIRELLES, MARIA HELENA MIGUEZ ROCHA-LEÃO, AND GERALDO LIPPEL SANT' ANNA, JR.*

Instituto de Quimica and COPPE, Universidade Federal do Rio de Janeiro, P.O. Box 68502, CEP 21945-970, Rio de Janeiro, RJ, Brazil

ABSTRACT

Although lipases have been intensively studied, some aspects of enzyme production like substrate uptake, catabolite repression, and enzyme stability under long storage periods are seldom discussed in the literature. This work deals with the production of lipase by a new selected strain of *Candida lipolytica*. Concerning nutrition, it was observed that inorganic nitrogen sources were not as effective as peptone, and that oleic acid or triacylglycerides (TAG) were essential carbon sources. Repression by glucose and stimulation by oleic acid and long chain TAG (triolein and olive oil) were observed. Extracellular lipase activity was only observed at high levels at late stationary phase, whereas intracellular lipase levels were constant and almost undetectable during the cultivation period, suggesting that the produced enzyme was attached to the cell wall, mainly at the beginning of cultivation. The crude lipase produced by this yeast strain shows the following optima conditions: pH 8.0–10.0, temperature of 55°C. Moreover, this preparation maintains its full activity for at least 370 d at 5°C.

Index Entries: *Candida lipolytica*; *Candida*; lipase; lipase production; lipolytic activity.

INTRODUCTION

There is a growing interest in microbial lipases (acylglycerol hydrolases, E.C.3.1.1.3), because of their substrate specificity, which can be exploited for fine organic synthesis and related applications *(1–5)*.

Although lipase utilization, mainly for chiral compound synthesis, is a field of growing interest and investigation, lipase production and char-

---

*Author to whom all correspondence and reprint requests should be addressed.

acterization are not well studied from a biochemical and physiological point of view.

Yeasts are interesting lipase producers because they have short generation times. Consequently, the cultivation of yeasts may be performed in shorter periods of time, using simple control procedures. Additionally, many yeasts have a generally recognized as safe (GRAS) status. These characteristics make yeasts good candidates to be genetically manipulated aiming at the overexpression of heterologous proteins.

Production of lipases by yeasts was reviewed (6) with most published works on the *Candida* species. Few yeast lipases are commercially available, however, there is an increasing interest in these enzymes. Commercialization will require a better understanding of the production process as well as novel and well-characterized lipases. Many aspects of the lipase production process are not well understood and questions related to substrate consumption and induction or repression are not elucidated.

The aim of this work was to study the production of lipase by a new *C. lipolytica* strain, to determine the best conditions for enzyme production by this strain. This work attempts to address issues concerning intra and extracellular lipase levels and substrate (olive oil) consumption during cultivation and its relationship with lipolytic activity.

MATERIALS AND METHODS

Materials

Triolein, oleic acid, tributyrin, and azocasein were obtained from Sigma Chemical (St Louis, MO). Peptone and yeast extract were obtained from Difco (Detroit, MI). Acid alumina was obtained from BDH Chemicals (UK) and all the other analytical-grade materials are from Riede-de Haen (Germany).

Microorganism

A wild type strain of *Candida lipolytica* was selected from an estuary in the vicinity of Rio de Janeiro, Brazil (7).

Media Composition (all w/v):

SPC (sucrose 2%, bac. peptone 0.64%), TBS (the same as SPC plus tributyrin 1%), TBT (bac. peptone 0.64%, tributyrin 1%), OO (olive oil 1%, bac. peptone 0.64%), SA (olive oil 1%, ammonium sulfate 0.28%), SAT (the same as SA buffered with KH_2PO_4/K_2HPO_4, pH 6.0), U (olive oil 1%, urea 0.14%), UT (the same as U, buffered as SAT), G (glucose 2%, bac. peptone 0.64%), GOO (the same as G plus olive oil 1%), TO (triolein 1%, bac peptone 0.64%), AO (oleic acid 0.96%, bac. peptone 0.64%), OL (glycerol 0.1%, bac. peptone 0.64%). All the media were supplemented with yeast extract (0.1%) and the pH was adjusted to pH 6.0.

Inoculum and Cultivation Conditions

Incubations were carried out in 2000-mL Erlenmeyer flasks, containing 400 mL of culture medium, in a rotary shaker (160 rpm) at 29°C. A mass of cells leading to an initial concentration of 0.5 mg dry weight cell per mL (mg d.w.mL^{-1}) was used as inoculum.

Olive Oil Purification

To remove free fatty acids and other impurities from the commercial olive oil, a purification procedure was performed as follows: olive oil was mixed with a solvent mixture (1:1 v/v) containing petroleum ether and ethyl ether (10:1 v/v), and added to a column (2.4 cm diameter; 21 cm height) containing acid alumina. The solvent from the eluate was evaporated under mild stirring at room temperature. Purified olive oil was used in further experiments.

Analytical Methods

Cell growth was followed by optical density measurements at 570 nm and those values were converted to mg d.w.mL^{-1} using a standard curve relating those variables. Glucose concentration in the cultivation medium was determined at selected time intervals by using the classical 3,5 dinitrosalicilic acid (DNS) method *(8)* (olive oil does not interfere with the assay). Sucrose was determined by Anthrone method *(9)*. Olive oil content was determined according to Frings and Dunn *(10)* after three extractions with the same volume of chloroform, as established by previous experiments carried out at our laboratory.

Enzymatic Activity Assays

Lipolytic activity was determined by three different procedures, as follows:

1. Spectrophotometric method—0.1 mL of the cultivation supernatant was added to a solution of 0.504 mM *p*-nitrophenyl laurate (*p*-NPL) in 50 mM phosphate buffer, pH 7.0. This mixture was incubated at 37°C in a Shimadzu (Mod UV2201) spectrophotometer cuvet. The production of *p*-nitrophenol was automatically monitored at 410 nm during the linear period of product accumulation *(11)*. One unit (U) of lipase activity was defined as the amount of enzyme that produces 1 µmol of product per minute.

2. Diffusion method—hydrolysis halo diameter was measured in a solid medium containing olive oil or babassu oil according to Sztajer and Maliszewska *(12)*. Lipase activity was expressed as the ratio between halo diameter and supernatant volume (mm:mL^{-1}) after plate incubation at 37°C for 45 h when the maximum diameter was reached.

3. Titrimetric method—this protocol was developed in our laboratory using olive oil as substrate. Cultivation supernatant (3 mL) was added to reaction flasks containing 17 mL of olive oil (5%)-arabic gum (5%) emulsion in phosphate buffer pH 7.0. Reactions were conducted at 37°C for 10 min and stopped by the addition of acetone:ethanol (1:1 v/v). The free fatty acids produced were titrated with 0.05 N NaOH. One unit (U) of lipolytic activity was defined as the am.ount of enzyme that produces 1 µmol of product per minute under the assay conditions.

Invertase was measured in a specific experiment incubating 3 mg d.w. cell with 1 mL of sucrose (100 mg mL^{-1}) and 0.5 mL of 50 mM acetate buffer pH 5.0 for 2 min at 30°C. Reducing sugars were measured according to Nelson *(13)*.

Protease activity was determined according to Charney and Tomarelli *(14)*.

Cell-Free Extract Preparation

Cells (50 mg mL^{-1}), washed twice with distilled water, were resuspended in 0.5 mL of 0.1M MOPS buffer, pH 7.0, with 2.0 mM EDTA and 5 mM β-mercaptoethanol. Disruption was obtained using glass beads as previously described *(15)*.

Enzyme Stability Assays

Lipase stability experiments were carried out at different temperatures and pH values. In each case the supernatant of centrifuged samples (7000g for 30 min), namely, crude preparation, was incubated at the selected conditions and its activity was measured at different time intervals by the spectrophotometric or the titrimetric method as indicated in each case.

RESULTS AND DISCUSSION

Lipolytic Activity Measurement

To select the best method to quantify lipase production by *C. lipolytica*, experiments comparing the three methods were performed and their results are presented in Fig. 1. Similar lipase activity profiles were obtained, independent of the method used, supporting our choice of using the spectrophotometric method to quantify *C. lipolytica* lipase activity. Data from Fig. 1 lead to the following coefficients for the correlation between the spectrophotometric and the other two methods (diffusion and titrimetric, respectively): 0.946 and 0.970.

Stable Lipase

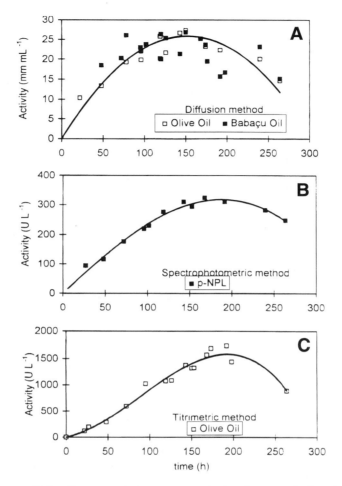

Fig. 1. Extracellular lipase activity measured simultaneously by three different methods during *C. lipolytica* cultivation in OO medium in a rotary shaker (160 rpm, 29°C). Lipolytic activity assay conditions are described in Materials and Methods.

Nutritional Conditions and Catabolite Repression

Some *Candida* species are usually cultivated in media containing a carbohydrate, a nitrogen source and an inducer molecule, generally, a triacylglyceride. In this work, several carbon and nitrogen sources were tested to produce lipase.

Results from Table 1 indicate that there is a correlation between maximum lipolytic activity and biomass concentration increase (ΔX) during cultivation ($r = 0.991$). However, no correlation was found between the specific growth rate (μ) and lipase activity. Glucose (G medium), but not sucrose (SPC medium), was effective for lipase production in the absence of TAG. This is explained by the absence of invertase activity in these cells. Invertase measurement in short-term

Table 1
Effect of Carbon Sources on Lipase Production

Main component	Code	ΔX $(mg.mL^{-1})^a$	μ (h^{-1})	Maximum activity $(U.L^{-1})$	Volumetric productivity $(U.L^{-1}.h^{-1})$
Carbohydrate	SPC	1.4	0.14	n.d.	—
	G	9.4	0.23	970	10.8
TAG	TBT	4.6	0.09	100	0.7
	OO	20.3	0.23	2700	17.7
	TO	20.0	0.42	3000	19.7
TAG + Carbohydrate	TBS	5.6	0.09	60	0.4
	GOO	19.1	—	3200^b	8.0
Others	AO	24.5	0.51	3800	25.0
	OL	4.5	0.32	80	0.5

^{a}Cell concentration increase expressed as dry weight per volume (ΔX is the difference between maximum and initial biomass concentrations.
^{b}Only reached 200 h after in comparison with other maximum values.
n.d. not detectable.
C. lipolytica cells were cultivated at 29°C in a rotary shaker at 160 rpm, on peptone and different carbon sources. Lipolytic activity was measured by the spectrophotometric method.

experiments with baker's yeast and *C. lipolytica* showed activity values of 11 and 0 µmol reducing sugars mL^{-1}, respectively. Residual sucrose after growth phase was 17 and 100% for baker's yeast and *C. lipolytica* cultivations, respectively. Another evidence that sucrose is not being consumed is the identical specific growth rate (0.09 h^{-1}) obtained in TBS and TBT media.

Although GOO medium lead to a high lipase level, maximum activity was attained late in comparison with medium OO, as illustrated by the results of volumetric productivity (Table 1) and long-term cultivation experiments (activity and specific activity are shown in Table 2). These results indicate that lipase from *C. lipolytica* undergoes glucose repression and that derepression does not depend on inducer presence. This fact is supported by the following:

1. Glucose was identically consumed in G and GOO media, indicating preferential glucose uptake (Fig. 2);
2. High extracellular lipase levels were only later detected in GOO media (Table 2);
3. Intracellular lipase levels were almost the same in G, GOO, and OO media without any internal accumulation (as illustrated in Fig. 3C for OO medium).

Table 2
Effect of Glucose on Extracellular Lipase Production

Time (h)	G Specific activity (U.g^{-1})^a	G Activity (U.mL^{-1})	GOO Specific activity (U.g^{-1})^a	GOO Activity (U.mL^{-1})	OO Specific activity (U.g^{-1})^a	OO Activity (U.mL^{-1})
26	n.d.	n.d.	n.d.	n.d.	n.d.	n.d.
74	9	90	2	40	7	60
113	22	260	1	30	51	760
139	6	80	2	50	42	510
186	3	30	10	220	100	1240
210	n.d.	n.d	2	30	18	230
214	n.d.	n.d	n.d	n.d.	5	40
336	—	—	17	340	—	—
410	—	—	53	860	—	—
432	—	—	22	360	—	—

^{a}Specific activity expressed as units per cell dry weight.
n.d. not detectable.
Time course activity results of three parallel cultivations carried out at 29°C in shaken flasks (160 rpm), on peptone and different carbon sources: glucose (G), glucose with olive oil (GOO) or olive oil (OO). Lipolytic activity was measured by the spectrophothometric method.

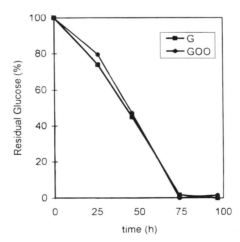

Fig. 2. Glucose consumption in presence (GOO media) or absence (G media) of olive oil. Experiments were carried out in parallel at 29°C and 160 rpm. Glucose concentration was measured by the DNS method.

It is important to note that previously reported results on glucose effect on lipase production are controversial. Muderhwa et al. *(16)* reported that lipase production by *Candida deformans* was repressed by

Table 3
Effect of Nitrogen Sources on Lipase Production

Code	ΔX (mg.mL^{-1})^a	μ (h^{-1})	Maximum activity (U.L^{-1})	Volumetric productivity (U.L^{-1}.h^{-1})	pH^{b,c}
OO	20.3	0.23	2700	17.7	5.8
U	7.1	—	60	1.3	3.0
UT	7.0	0.23	150	3.1	6.1
SA	10.1	0.10	20	0.1	4.0
SAT	5.0	0.17	130	0.6	5.2

^{a}Cell concentration increase expressed as dry weight per volume (ΔX is the difference between maximum and initial biomass concentrations.
^{b}Measured at the time of maximum lipolytic activity.
^{c}Initial pH=6.0.
C. lipolytica cells were cultivated at 29°C in a rotary shaker at 160 rpm, on olive oil and different nitrogen sources. Lipolytic activity was measured by the spectrophotometric method. OO medium contains peptone as nitrogen source.

glucose. On the other hand, no repression was observed with *Hansenula anomala (17)*.

As indicated in Table 1, pronounced lipase volumetric productivity was only observed when long chain TAG or oleic acid were used as carbon sources with peptone on absence of sucrose (OO, TO, and AO media). Glycerol and a short chain TAG like tributyrin were not effective (OL and TBT media, respectively).

The stimulation effect of unsaturated fatty acids on microbial lipase production has been reported *(18–20)*, however, its biochemical basis has not been elucidated. Our results show that oleic acid has an important role on the stimulation of lipase production by *C. lipolytica*, as this fatty acid led to the highest level of lipolytic activity, when compared with other carbon sources (Table 1).

Nitrogen effect on lipase production by *Candida* is a feature that is not much investigated. The results of cultivation experiments performed with different nitrogen sources are shown in Table 3. A pH decrease was always observed when urea (U medium) and ammonium sulfate (SA medium) were used in unbuffered media. When these media were buffered (UT and SAT media), lipase activity was increased by a factor of 2.5 and 6.5, respectively. It is worth nothing that although a nonbuffered media was used, pH was constant (5.8) during the cultivation period in OO medium.

The overall results shown in Tables 1 and 3 indicate that oleic acid and peptone were the best carbon and nitrogen sources, respectively, for lipase production by *C. lipolytica*. Considering that oleic acid and triolein are very expensive in comparison with the other tested nutrients, and that olive oil (a low-cost and available carbon source) leads to an appreciable lipase level, OO medium was used in further experiments.

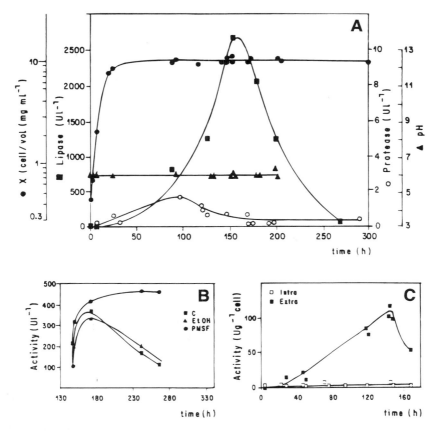

Fig. 3. Lipase production by *C. lipolytica* in OO medium. Cultivations were carried out in shaken-flasks at 29°C and 160 rpm. Lipolytic activity was measured by the spectrophotometric method. **(A)** Extracellular lipase and protease production, cell growth, and pH profile. **(B)** Extracellular lipase variation in OO medium (C-control experiment), OO medium with a serine protease inhibitor (PMSF-40 µg mL^{-1}) and OO medium with PMSF solvent (EtOH-ethanol). PMSF and/or ethanol were added after 142 hours of cultivation. **(C)** Intra- and extracellular lipolytic activity profiles (for comparison lipase levels were expressed as specific activity).

Enzyme Localization

A typical profile of cell concentration, pH, lipolytic, and proteolytic activities for the cultivation of the yeast in OO medium is shown in Fig. 3A. Maximum extracellular activity was only reached in the late stationary growth phase.

A sharp decrease on lipolytic activity occurs after 150 h of cultivation. In order to verify if this decay was caused by protease, a cultivation experiment was carried out by adding a serine protease inhibitor, phenylmethylsulphonyl fluoride (PMSF), to the culture medium. Figure 3B shows that the maximum level of lipase activity was maintained for an additional

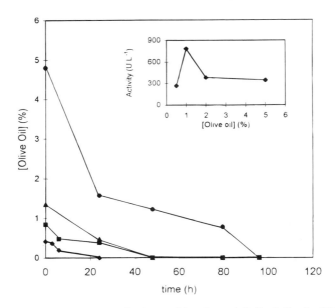

Fig. 4. Lipids content variation during cultivation of *C. lipolytica* in OO medium in a rotary shaker (160 rpm, 29°C). Olive oil initial concentration varied from 0.5 to 5% (w/v). Insert presents maximum extracellular lipolytic activity attained in each case, measured by the spectrophotometric method.

period of time (100 h) in comparison with the control experiment. Ethanol used as PMSF solvent had no effect on lipase production as also illustrated.

Intracellular lipase values for G and GOO media were similar to those presented in Fig. 3C (OO medium). A low intracellular lipase level was found during cultivation and no enzyme accumulation was observed within the cell. When lipolytic activity determinations (titrimetric method) were performed using no centrifuged cultivation broth samples (cells plus liquid medium) taken during the culture exponential growth phase (medium OO), a significant lipase activity value was obtained (2860 U L^{-1}). As neither intracellular nor extracellular lipase levels were detected at this cultivation time, our results suggest that lipase was cell-bounded. Emulsified hydrophobic substrates may attach to the cell wall retaining the enzyme in a supramolecular structure, containing lipolysaccharides, as observed by Kappeli et al *(21)*.

Substrate Consumption

Most of the published papers related to lipase production do not present profiles of oil exhaustion during cultivation, which might be a result of the difficulties in performing oil or lipids determinations in cultivation media. In this work, an attempt was made to quantify oil content, using chloroform to extract lipids from the supernatant. Oil was determined as described in the Materials and Methods section and typical results are shown in Fig. 4. This substrate practically vanished after 24 to

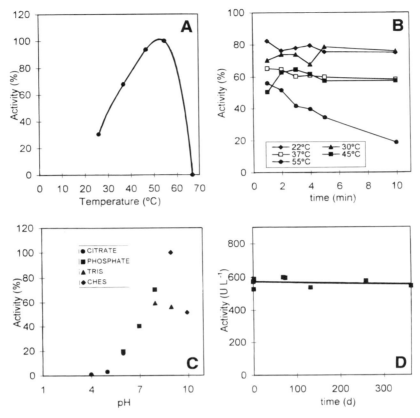

Fig. 5. Crude lipase preparation characteristics. Experiments were carried out with supernatants from *C. lipolytica* cultivations in OO medium. **(A)** Effect of temperature on extracellular lipolytic activity (spectrophotometric method). **(B)** Effect of temperature on extracellular lipase stability (spectrophotometric method). **(C)** Effect of pH on extracellular lipolytic activity (titrimetric method). **(D)** Long-term storage stability of the crude lipase preparation at 5°C (spectrophotometric method).

96 h of cultivation, depending on its initial concentration. The highest lipolytic activity was attained when 1% of olive oil was used.

Crude Enzyme Preparation Characteristics

After the selection of best conditions for lipase production by *C. lipolytica*, experiments were carried out to characterize the crude enzyme preparation.

The results concerning the effect of temperature on lipase activity are shown in Fig. 5A. The maximum activity was obtained at 55°C, but at this temperature the activity was quickly lost (Fig. 5B). However, the crude preparation was stable at 45°C and at lower temperatures (37, 30, and 22°C) as illustrated in Fig. 5B. Experiments carried out with different buffers (Fig. 5C) show that higher activities were observed in the pH range of 8.0 to 10.0. Figure 5D illustrates the enzyme stability profile. A singular high stability of the crude lipase preparation was found when it was stored at 5°C. The enzyme remained 100% stable for at least 370 d without any additives.

Comparison of stability results with published data is very difficult because of the scarcity of information, as well as the lack of definition of stability indicators (full stability value, percentage of loss as a function of time, etc.). To our knowledge this is the first time that long-term stability assays were presented with crude preparations showing such a promising result—that the enzyme could be stored for more than 1 yr without any activity loss.

In conclusion, our results indicate that the selected C. *lipolytica* strain produces lipase at appreciable levels when cultivated in media containing oleic acid or long chain TAG and peptone. Our data also suggest that lipase is produced since the beginning of the cultivation, but it remains cell bounded and is only significantly released to the medium at late stationary phase when the carbon source (olive oil) is practically exhausted. Besides the crude enzyme's interesting characteristics (optimum activity at high pH levels and high stability for long periods of time), which are suitable for future applications, this work also quantifies TAG content in the culture medium and relates substrate consumption with lipolytic activity, a feature that is not emphasized in the current literature.

ACKNOWLEDGMENTS

This project has been partially financed by CAPES, PADCT/CNPq (Proj. No 62.0160/91.8).

REFERENCES

1. Chen, P. Y., Wu, S. H., and Wang, K. T. (1993), *Biotechnol. Lett.* **15,** 181–184.
2. Pozo, M. and Gotor, V. (1993), *Tetrahedron* **49,** 10,725.
3. Mustranta, A., Forssell, P., and Poutanen, K. (1993), *Enz. .Microbiol. Technol.* **15,** 133–139.
4. Cernia, E., Delfini, M., Mgrii, A. D., and Palocci, C. (1994), *Cell. Mol. Biol.* **40,** 193–199.
5. Nakano, H., Kiki, Y., Ando, K., Kawashima, Y., Kitahata, S., Tominaga, Y., and Takenishi, S. (1994), *J. Ferm. Bioeng.* **78,** 70–73.
6. Hadeball, W. (1991), *Acta Biotechnol.* **11,** 159–167.
7. Hagler, A. N. and Mendonça-Hagler, L. C. (1981), *Appl. Environ. Microbiol.* **41,** 173–178.
8. Sumner, J. B. (1924), *J. Biol. Chem.* **62,** 287–290.
9. Brin, M. (1966), in *Methods in Enzymology*, vol 9, Colowick, S. P. and Kaplan N. O., eds., Academic Press, New York, NY, pp. 506–514.
10. Frings, C. S. and Dunn, R. T. (1970), *Am. J. Clin. Pathol.* **53,** 89–91.
11. Wills, E. D. (1965), *Adv. Lipid Res.* **3,** 197–240.
12. Sztajer, H. and Maliszewska, I. (1988), *Enz. Microbiol. Tecnol.* **10,** 492–497.
13. Nelson N. (1944), *J. Biol. Chem.* **153,** 357.
14. Charney, J. and Tomarelli, R. M. (1947), *J. Biol. Chem.* **171,** 501–505.
15. Panek, A. C., Araujo, P. S., Moura Neto, V., and Panek, A. D. (1987), *Curr. Genet.* **11,** 459–465.

16. Muderhwa, J. M., Ratomahenina, R., Pina, M., Graille, J., and Galzy, P. (1985), *J. Am. Oil. Chem. Soc.* **62,** 1031–1036.
17. Banerjee, M., Sengupta, I., and Majumdar, S. K. (1985), *J. Food Sci. Technol.* **22,** 137–139.
18. Gomi, K., Ota, Y., and Minoda, Y. (1984), *Agric. Biol. Chem.* **48,** 1061–1062.
19. Del Rio, J. L., Serra, P., Valero, F., Poch, M., and Solà, C. (1990), *Biotechnol. Lett.* **12,** 835–838.
20. Ohnishi, K., Yoshida, Y., and Sekiguchi, J. (1994), *J. Ferment. Bioeng.* **77,** 490–495.
21. Kappeli, O. and Flechter, M. (1977), *J. Bacteriol.* **131,** 917–923.

Glucoamylase Isoenzymes Tailoring Through Medium Composition

José G. Silva, Jr., Hilton J. Nascimento, Valíria F. Soares, and Elba P. S. Bon,*

Instituto de Química, Universidade Federal do Rio de Janeiro, CT, Bloco A, Ilha do Fundão, Rio de Janeiro, RJ, Brasil CEP:21949-900

ABSTRACT

Two major glucoamylase isoenzymes (GAI and GAII) have been identified in culture supernatants of *Aspergillus awamori*. It has been suggested that a stepwise degradation of a native enzyme during the fermentation by proteases and/or glucosidases results in the formation of isoenzymes that have different characteristics concerning substrate specificity and stability to pH and temperature. In this study, the glucoamylase isoenzymes produced by *Aspergillus awamori* using liquid media with C/N 10 (2.0% starch, 0.45% $(NH_4)_2 SO_4$) and C/N 26 (5.2% starch, 0.45% $(NH_4)_2 SO_4$) were analyzed. In both cases, GAI and GAII were characterized concerning its hydrolitic activities, mol wt, and isoeletric point. Using HPLC gel filtration and FPLC chromatofocusing, it was obtained for GAI a mol wt of 110,000 Da, pI 3.45 and for GAII a mol wt of 86,000 Da, pI 3.65. A different isoenzymes proportion was observed by the use of the two C/N ratios. In the lower carbohydrate content, fermentation of the GAI form predominated, whereas in the C/N 26 medium, GAII was prevalent. Gel eletrophoresis, amino acid analysis, and structural data confirmed that both preparations were glucoamylases with a high degree of homogeneity.

Index Entries: *Aspergillus awamori*; glucoamylase production; glucoamylase characterization; isoenzymes proportion; medium C/N ratio.

---

*Author to whom all correspondence and reprint requests should be addressed.

INTRODUCTION

Amylolytic enzymes from filamentous fungi have been extensively investigated because of their importance in the starch industry (1). Glucoamylase [α-(1–4) glucan glucohydrolase, [E.C.3.2.1.3] was identified in the early 1950s, and the existence of two isoenzymes produced by *Aspergillus niger* was firstly reported in 1959 (1,2). Since then the presence of two isoforms, glucoamylase I and II (GAI and GAII), have been consistently identified in fermentations carried out using several genera and species of fungi (3–12). Some authors, however, have also reported the existence of a higher number of glucoamylase isoenzymes that could be related to variable glycosylation patterns (13–20).

A great deal of work has been carried out aimed at the characterization of the two major isoenzymes, GAI and GAII concerning their physicochemical and catalytic properties (11,12,21,22). Both forms consist of a single glycosylated polypeptide chain with typical mol wt of 75,000 Da and 54,000 Da, respectively (7), although the molecular weights of these glycoproteins vary according to the analytical procedure used (21). The polypeptide chain of the two forms are very similar, however GAI and GAII have different C-terminal residues. GAII differs from GAI by lacking a COOH-terminal region of approx 100 amino acids residues (23). The isoforms can be distinguished in their function, the GAI (614–616 amino acids) having substantial activity toward raw starch in contrast to GAII (512 amino acids). This difference is atributted to the absence in the GAII molecule of a C-terminal peptide (glycopeptide Gp-1) that corresponds to a raw starch-binding domain (21,24–27). It is reported that the isoenzymes show the same activity towards soluble substrates (23).

The mechanism involved in the generation of the isoenzymes is debatable. Molecular cloning and characterization of the glucoamylase gene of *Aspergillus awamori* demonstrated that the glucoamylase gene exists as a single copy gene (28). Thus the mechanisms for the generation of the isoenzymes would be related to either differential mRNA splicing or posttranslational modifications. Two different glucoamylase cDNA were identified, a major form coding for GAI and a minor, smaller one, originating from an additional mRNA splicing within sequences encoding the C-terminal position of GAI (28,29). However, since the structure of the smaller form differed from its presumptive mRNA in the C-terminal sequence, it was suggested that GAII was generated by posttranslational proteolysis (30). This possibility was later addressed. The limited in vitro degradation of GAI using fungal acid proteases or subtilisin resulted in the production, by liberating the glycopeptide GpI, of a GAI' that was a raw starch nondigesting glucoamylase (31–33). These results were later confirmed by the determination of the amino acid sequence of both N- and C-terminal of Gp-I obtained by subtilisin cleavage of GAI and its insertion on the complete amino acid sequence of GAI (18).

The effect of the growth medium composition and the presence of extracellular proteases on the proportion of the isoenzymes was studied. A relationship was observed between proteolytic activity and the conversion of GAI into GAII during the course of the fermentation. Longer fermentations also favored the conversion process (15,34). Moreover, only one type of glucoamylase, GO-0, (raw starch digestive) was obtained from a protease-negative, glucosidase negative *Aspergillus awamori* mutant (33).

The main aim of the present work was to investigate the effect of the growth medium C/N ratio on the glucoamylase isoenzymes produced by *Aspergillus awamori* using ammonium sulfate as a nitrogen source and starch as a carbon source. Growth medium limited by carbon or nitrogen were designed by keeping the nitrogen source concentration constant and using a low and a high carbon source concentration (35). Under these conditions, shorter and longer fermentations were observed. The profile of glucoamylase isoenzymes from both conditions were compared. A higher proportion of GAI was observed under carbon limiting conditions while GAII prevailed on the nitrogen limited medium. These data suggest that the short fermentation condition diminished the processing of GAI by proteases in the culture medium, most probably because of a lower degree of cell lysis that dimished the enzyme liberation into the media. The repression of protease production in such an ammonium rich medium is also a possibility to be explored (36).

MATERIAL AND METHODS

Culture Maintenance and Propagation

All procedures were carried out according to previous work where the same strain, *Aspergillus awamori* 2.B.361 U2/1 was used (35,37).

Fermentations

Shaken flasks fermentations were carried out using growth medium containing 0.45% ammonium sulphate and 0.1% yeast extract plus starch 1.9% (medium A, C/N ratio 10) or 5.2% (medium B, C/N ratio 26). Medium A and B were carbon and nitrogen limited, respectively, as a balanced medium using ammonium sulfate as nitrogen source shows a C/N ratio of 15 (35). Shaken flasks containing 200 mL of the growth medium were incubated in a rotatory shaker at 30°C and 250 rpm. After glucose consumption, the mycelium was separated from the culture medium through filtration using a Buchner funnel and glass microfiber filter (Whatman GF/A). The culture supernatant was collected for further analysis.

Crude Glucoamylase Isolation

The culture supernatants (around 80 mL) from growth media A and B were concentrated by liophylization to half of its initial volume and chromatographed in a Biogel P-6 colunm (50 × 5 cm, eluant $0.1M$ acetic

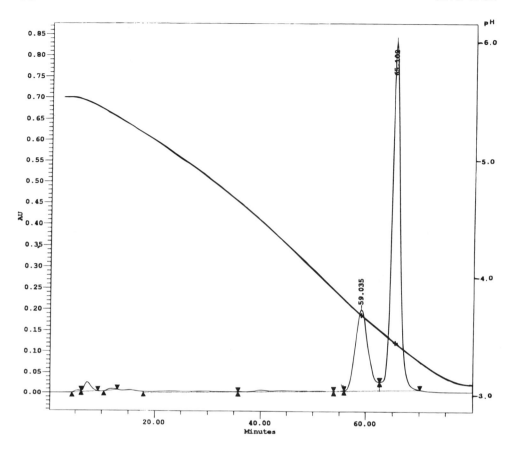

Fig. 1. FPLC-Chromatofocusing of 3.0 mg of crude glucoamylase preparation originating from medium A (C/N 10). Detection at 280 nm: Form I (pI = 3.45) and Form II (pI = 3.65).

acid and flow rate 35 mL/h) for the separation of the protein fraction from low-mol-wt compounds (MW < 6000 Da). The effluent was continuously monitored at 280 nm and the sole protein peak eluted in the column void volume, for medium A and B (crude glucoamylases), was collected and lyophilized.

Chromatofocusing of the Crude Glucoamylase Preparations

The two crude glucoamylase preparations were submitted to chromatofocusing in a mono P column (HR 5/20). The column was equilibrated with L-His 25 mM pH 5.5 and the pH gradient was established by the addition of polybuffer 74 (1/15, v/v in water), pH 3.1. The sample (3.0 mg dissolved in the start buffer, pH 5.5) fractionation was performed during 5 min in initial conditions followed by 85 min with the polybuffer solution, using a flow rate of 45 mL/h. The column efluent was used for pH

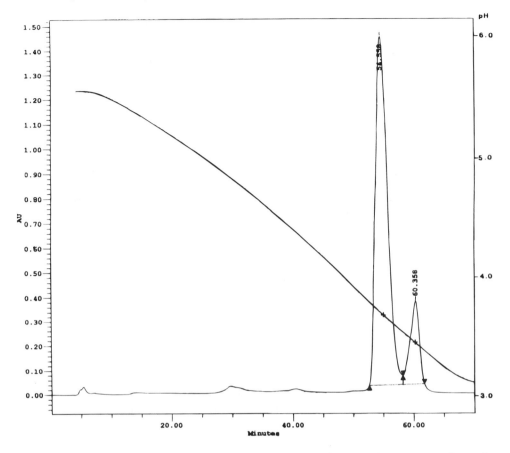

Fig. 2. FPLC-Chromatofocusing of 3.0 mg of crude glucoamylase preparation originating from medium B (C/N 26). Detection at 280 nm: Form I (pI = 3.45) and Form II (pI = 3.65).

and protein concentration determination (absorbance at 280 nm). Two main fractions, Form I, pI 3.45 (Fig. 1, retention time 65.102 min; Fig. 2, retention time 60.358 min) and Form II, pI 3.65 (Fig. 1, retention time 59.035 min; Fig. 2, retention time 54.558 min) were observed in the two crude glucoamylase preparations. Form I predominated in the preparation originating from medium A and Form II was the major component in the preparation from medium B. The fractions were concentrated in a speed vac and desalted in Biogel P-10. Both fractions from both growth media presented glucoamylase activity.

Native and Denaturing Polyacrylamide Gel Electrophoresis

The two crude glucoamylase preparations were submitted to native and denaturing discontinuous polyacrilamide gel electrophoresis (8–25% gradient gel) in a Phast System of Pharmacia using mol wt standards within the range of 20,000 to 200,000 Da for the calibration curve. In the denaturing gel electrophoresis the samples were previously incubated with SDS and dithiothreitol at 100°C for 5 min in 0.1M Tris-HCl buffer, pH 6.8. In all cases the gels were stained with CBR-250 and scanned in a LKB laser densitometer.

HPLC-Gel Permeation

The crude glucoamylases and Form I and Form II preparations were chromatographed in protein pak 300 SW (Waters). The column (30.0 × 7.8 mm) elution (30 mL/h) was performed using 0.2M phosphate buffer pH 6.0 and the effluent was monitored at 280 nm. Molecular weight standards within the range of 20,000 to 150,000 Da were used for the calibration curve.

Amino Acid Analysis

Form I chromatofocusing preparation from medium A and Form II from medium B were hydrolyzed in 5.8 N HCl vapor phase at 110°C for 22 h. The dried hydrolysate was derivatized with PITC, according to manufacturer, and analyzed in a PICO-TAG column (Waters). Each analysis was performed in duplicate.

Amino Acid Sequence

Glucoamylases Form I and II were submitted to N-terminal amino acid sequencing in a Shimadsu Sequenator. Approx 300 pmoles of the Form I and 250 pmoles of Form II were loaded into the sequenator.

Analytical

Protein concentration was measured using the SDS Folin-Lowry method. Enzyme and glucose concentrations were performed according to previous work *(35,37)*. The carbohydrate content of crude glucoamylases from medium A and B was determined using the phenol sulfuric method *(38)*.

RESULTS

Fermentations using medium A finished within 4 d whereas when medium B was used, glucose depletion ocurred within 10 d. A higher glucoamylase activity was observed in the culture supernatant of medium B, suggesting the presence of a more active molecule (Table 1). Medium B supernatant also presented a higher protein content: 261 mg of protein were recovered after Biogel P-6 chromatography in comparison to 40.6 mg from medium A. SDS-PAGE experiments showed one major peak (mol wt 110,000 Da) and three minor peaks, poorly resolved (mol wt 62,000; 65,000, and 70,000 Da) for medium A glucoamylase. An opposite pattern was observed for medium B glucoamylase, i.e., a minor component (mol wt 100,000 Da) and a protein family showing mol wt within 60,000–70,000 Da.

Table 1
Glucoamylase Activity of the Culture Supernants and Form I and II using
Maltose (a) and Starch (b) as Substrate[a]

Glucoamylase activity (U.mg^{-1})					
Culture Supernatants		Chromatofocusing Preparation			
Medium A (C/N 10)	Medium B (C/N 26)	Medium A (C/N 10)		Medium B (C/N 26)	
		Form I	Form II	Form I	Form II
4.0[(a)]	10.0[(a)]	2.0[(a)]	1.8[(a)]	4.2[(a)]	8.4[(a)]
		18.7[(b)]	13.1[(b)]	23.6[(b)]	64.0[(b)]

[a]Enzyme activity was expressed as µmol of glucose produced per minute per mg of protein (U.mg^{-1}) under inital rate conditions at 40°C.

Two major fractions with pH values 3.45 and 3.65 were observed after the chromatofocusing analysis of the crude glucoamylases. In preparation from medium A, the fraction with pI 3.45 (Form I) prevailed (80%) whereas the fraction with pI 3.65 (Form II) was predominant in preparations from medium B (85%), as shown in Figs. 1 and 2. The Form I and the Form II chromatographed by HPLC protein pak 300SW showed one peak with different retention times whose mol wt were 106,000 and 85,000 Da.

The amino acid composition of Form I and Form II (Table 2) relates closely to the reported composition of GAI and GAII (22,23). The partial N-terminal amino acid sequence of the two isoenzymes isolated by chromatofocusing gave "clean" sequence until the fifteenth amino acid residue, which was consistent with the sequence known for GAI and GAII (22,23).

According to the foregoing of the physicochemical characterization of the polypeptide chain of the glucoamylase preparations allowed, the identification of the two major isoenzymes, in different proportions, in the C and N nitrogen-limited medium. The isoforms activities varied according to the medium C/N ratio (Table 1). Isoenzymes from medium A (C/N ratio 10) presented equivalent activities toward maltose and soluble starch, whereas Form I and Form II from medium B (C/N ratio 26) were, respectively, two and four times more active in maltose in comparison to its medium A counterparts. Form II was, at least, three times more active in starch than all other isoforms.

The determined carbohydrate content of crude preparations from medium A and B were of 13.7 and 34.2%, respectively. Therefore, a higher degree of glycosylation occured in the isoforms from medium B that presented a higher carbohydrate surplus. Coincidently, crude glucoamylase B (medium C/N ratio 26) presented a carbohydrate content 2.5 times higher

Table 2
Amino Acid Composition (PICOTAG column, HCl 5.8N vapor phase hydrolysis) of Glucoamylases Preparations Form I and II, Isolated by FPLC-Chromatofocusing in mono P Column[a]

Amino acids	pmoles/µl (Form I)	pmoles/µl (Form II)	Reported compositions (GAI)[22]	Reported composition (GAII)[23]
D	95.34 (64)	72,15 (55)	66	56
E	59.57 (40)	40,89 (31)	41	31
S	107.43 (85)*	71,00 (63)*	87	76
G	68.09 (46)	47,70 (36)	45	40
H	7.40 (05)	5,54 (04)	04	04
R	27.48 (18/19)	19,51 (15)	18	15
T	91.41 (69)	61,63 (52)*	73	58
A	88.96 (60)	68,28 (51/52)	63	53
P	35.40 (24)	24,70 (19)	22	19
Y	40.90 (28)	25,08 (19)	27	21
V	53.98 (36)	44,05 (33)	41	33
M	4.85 (03)	4,31 (03)	02	02
C	10.46 (09)	6,41 (06)*	09	08
I	31.19 (21)	21,93 (16/17)	23	18
L	55.72 (37/38)	43,88 (33)	43	37
F	29.69 (20)	22,53 (17)	20	17
K	18.86 (13)	13,49 (10)	12	10
W	—	—	18	14
Number of residues	577-580	462-465	614	512

*Correction for acid destruction: 15% for serine (S), 10% for threonine (T), and 20% for cysteine (C).

Note: tryptophane (W) were not determined.

[a]In parenthesis are shown the number of amino acid residues/mol. The data are average values for two analyses of the same hydrolyzate.

in comparison to crude glucoamylase A (medium C/N ratio 10) and, therefore, protein glycosylation responded proportionally to the increase in the medium C/N ratio. As previously stated, the isoforms from medium B presented higher activity.

The similar patterns of activities toward maltose and starch presented by Form I and II from medium A, indicated that the lack of the raw starch binding domain did not have a major effect on the enzyme activity towards soluble substrates.

DISCUSSION

Structural and catalytic data allows the conclusion that the glucoamylases preparations, Form I and II, which were obtained by chromatofocusing, corresponds to the GAI and GAII isoenzymes molecules. The relative proportion of the isoenzymes were different prevailing GAI in medium A (C/N ratio 10) and GAII in medium B (C/N ratio 26). In this case, GAII showed to be two times more active in relation to GAI in maltose substrate and three times higher in starch (Table 1). The data concerning the eletrophoretic and chromatographic experiments are consistent with the hypothesis that extacellular proteases processes GAI, transforming it into GAII. This assumption is supported by the isoenzymes (Form I and II) amino acid compositions. Intracellular processing proteases are known to attack peptides linkages in dibasic sites. This is not the case for GAI transformation into GAII, as the C-terminal amino acid in GAII is alanine. Therefore, the GAI processing would involve an extracellular protease. The identification of the peptide Gp-1 in the extracellular environment would be a definitive argument for extracellular protease processing. In conclusion, media composition affected the isoenzymes proportion in the culture supernatant. The characterization of this effect can be used to obtain glucoamylases with particular characteristics. This possibility could be further explored as the presence of isoenzymes is a common pattern in extracellular fungi enzymes.

ACKNOWLEDGMENTS

This work was supported by the Brazilian Program for Science and Tchnology Development (PADCT) and the Brazilian Research Council (CNPq).

REFERENCES

1. Nikolov, Z. L. and Reilly, P. J. (1991), in *Biocatalysts for Industry*, Dordick, J. S., ed., Plenum Press, NY, London, pp. 37–62.
2. Pazur, J. H. and Ando, T. (1959), *J. Biol. Chem.* **234,** 1966–1970.
3. Watanabe, K. and Fukimbara, T. (1965), *J. Ferm. Tech.* **43,** 690–696.
4. Lineback, D. R., Russel, I. J., and Rasmussen, C. (1969), *Arch. Biochem. Biophys.* **134,** 539–553.
5. Lineback, D. R. and Baumann, W. E. (1970), *Carbohydr. Res.* **14,** 341–353.
6. Pazur, J. H., Knull, H. R., and Cepure, A. (1971), *Carbohydr. Res.* **20,** 83–96.
7. Lineback, D. R., Aira, L. A., and Horner, R. L. (1972), *Cereal Chem.* **49,** 283–298.
8. Yamasaki, Y., Suzuki, Y., and Ozawa, J. (1977), *Agric. Biol. Chem.* **41,** 755–762.
9. Yamasaki, Y., Tsuboi, A., and Suzuki, Y. (1977), *Agric. Biol. Chem.* **41,** 2139–2148.
10. Taylor, P. M., Napier, E. J., and Fleming, I. D. (1978), *Carbohydr. Res.* **61,** 301–308.
11. Alazard, D. and Baldensperger, J. F. (1982), *Carbohydr. Res.* **107,** 231–241
12. Ramasesh, N., Sreekantiah, K. R., and Murthi, V. S. (1982), *Stärke* **34,** 346–351.
13. Morita, Y., Shimizu, K., Ohga, M., and Korenaga, T. (1966), *Agric. Biol. Chem.* **30,** 114–121.
14. Ueda, S., Ohba, R., and Kano, S. (1974), *Die Stärke* **26,** 374–378.
15. Hayashida, S. (1975), *Agric. Biol. Chem.* **39,** 2093–2099.
16. Miah, M. N. N. and Ueda, S. (1977), *Die Stärke* **7,** 235–239.
17. Saha, B. C., Mitsue, T., and Ueda, S. (1979), *Stärke* **31,** 307–314.
18. Hayashida, S., Nakahara, K., Kuroda, K., Kamachi, T., Otha, K., Iwanaga, S., Miyata, T., and Sakaki, Y. (1988), *Agric. Biol. Chem.* **52,** 273–275.
19. Ono, K., Shintani, K., Shigeta, S., and Oka, S. (1988), *Agric. Biol. Chem.* **52,** 1689–1698.
20. Ono, K., Shintani, K., Shigeta, S., and Oka, S. (1988), *Agric. Biol. Chem.* **52,** 1699–1706.

21. Svensson, B., Pedersen, T. G., Svendsen, I., Sakai, T., and Ottesen, M. (1982), *Carlsberg Res. Commun.* **47,** 55–69.
22. Svensson, B., Larsen, K., and Svendsen, I. (1983), *Carlsberg Res. Commun.* **48,** 529–544.
23. Clarke, A. J. and Svensson, B. (1984), *Carlsberg Res. Commun.* **49,** 559–566.
24. Hayashida, S., Kunisaki, S., Nakao, M., and Flor, P. Q. (1982), *Agric. Biol. Chem.* **46,** 83–89.
25. Svensson, B., Jesperjen, H., Sierks, M. R., and MacGregor, E. A. (1989), *Biochem. J.* **264,** 309–311.
26. Evans, R., Ford, C., Sierks, M., Nikolov, Z., and Svensson, B. (1990), *Gene* **91,** 131–134.
27. Dalmia, B. K. and Nikolov, Z. L. (1991), *Enzyme Microb. Technol.* **13,** 982–989.
28. Nunberg, J. H., Meade, J. H., Cole, G., Lawyer, F. C., McCabe, Schweirckart, V., Tal, R., Wittman, V. P., Flatgaard, J. E., and Innis, M. A. (1984), *Mol. Cell. Biol.*, Nov., 2306–2315.
29. Boel, E., Hjort, I., Svensson, B., Norris, F., and Norris, K. E. (1984), *Embo J.* **3,** 1097–1102.
30. Svensson, B., Larsen, K., and Gunnarsson, A. (1986), *Eur. J. Biochem.* **154,** 497–499.
31. Yoshino, E. and Hayashida, S. (1978), *J. Ferment. Technol.* **56,** 289–295.
32. Hayashida, S. and Yoshino, E. (1978), *Agric. Biol. Chem.* **42,** 927–933.
33. Flor, P. Q. and Hayashida, S. (1983), *Appl. Environ. Microbiol.* **45,** 905–912.
34. Bartoszewicz, K. (1986), *Acta Biochim. Polonica* **23,** 17–29.
35. Bon, E. and Webb, C. (1993), *Appl. Biochem. Biotech.* **39/40,** 349–369.
36. Marzluf, G. A. and Fu, Y.-H. (1988), in *Nitrogen Source Control of Microbial Process*, Esquivel, S. S., ed., CRC Press, Boca Raton, FL., pp. 84–96.
37. Bon, E. and Webb, C. (1989), *Enzyme Microb. Technol.* **11,** 495–499.
38. Dubois, M., Gilles, K. A., Hamilton, J. K., Rebers, P. A., and Smith, F. (1956), *Anal. Chem.* **28,** 350–356.

Regulation of Phosphotransferases in Glucose- and Xylose-Fermenting Yeasts

VINA W. YANG AND THOMAS W. JEFFRIES*

Institute for Microbial and Biochemical Technology, Forest Products Laboratory,** One Gifford Pinchot Drive, Madison, WI 53705-2398

ABSTRACT

This research examined the titers of hexokinase (HK), phosphofructokinase (PFK), and xylulokinase (XUK) in *Saccharomyces cerevisiae* and two xylose fermenting yeasts, *Pachysolen tannophilus* and *Candida shehatae*, following shifts in carbon source and aeration. Xylose-grown *C. shehatae*, glucose-grown *P. tannophilus*, and glucose-grown *S. cerevisiae*, had the highest specific activities of XUK, HK, and PFK, respectively. XUK was induced by xylose to moderate levels in both *P. tannophilus* and *C. shehatae*, but was present only in trace levels in *S. cerevisiae*. HK activities in *P. tannophilus* were two to three fold higher when cells were grown on glucose than when grown on xylose, but HK levels were less inducible in *C. shehatae*. The PFK activities in *S. cerevisiae* were 1.5 to 2 times higher than in the two xylose-fermenting yeasts. Transfer from glucose to xylose rapidly inactivated HK in *P. tannophilus*, and transfer from xylose to glucose inactivated XUK in *C. shehatae*. The patterns of induction and inactivation indicate that the basic regulatory mechanisms differ in the two xylose fermenting yeasts.

Index Entries: *Saccharomyces cerevisiae*; *Pachysolen tannophilus*; *Candida shehatae*; 6-phosphofructokinase; hexokinase; D-xylulokinase; regulation; phosphotransferase.

---

*Author to whom all correspondence and reprint requests should be addressed.
**Maintained in cooperation with the University of Wisconsin, Madison, WI
This article was written and prepared by U.S. Government employees on official time, and it is therefore in the public domain and not subject to copyright.
The use of trade, firm or corporation names in this publication is for the information and convenience of the reader. Such use does not constitute an official endorsement or approval by the U.S. Department of Agriculture of any product or service to the exclusion of others which may be suitable.

INTRODUCTION

Aside from the documented induction of xylose reductase and xylitol dehydrogenase *(1)*, very little is known about regulation of metabolism in the fermentation of xylose or glucose by xylose-fermenting yeasts. In *Pichia stipitis, C. shehatae*, and most other yeasts, glucose represses the utilization of xylose and other sugars. This is a critical problem in the utilization of the mixture of glucose and other sugars found in hemicellulosic hydrolysates. The mechanism by which repression occurs in yeasts that rapidly metabolize xylose has not been elucidated. There is some indication that the basic regulatory mechanism in *P. tannophilus* might be similar to that of *S. cerevisiae* (2).

In *S. cerevisiae*, the *HXK2* gene product, hexokinase PII, EC 2.7.1.1, (HK PII) mediates uptake of hexoses and plays a role in carbon catabolite repression *(3,4)*. Hexokinase PII possesses both hexokinase and protein kinase activities, and the protein kinase activity of HK PII is also regulated by glucose *(5)*. In *S. cerevisiae*, mutants with reduced HK PII acvtivity have increased levels of glucose repressible enzymes such as invertase *(6)*. A regulatory role for HK is also seen in other yeasts. In *Kluyveromyces lactis*, the product of the HK gene, *RAG5*, controls the expression of *RAG1*, a low-affinity glucose-fructose transporter *(7)*. *P. tannophilus* mutants deficient in HK do not repress enzymes specific to xylose metabolism *(2)*, and they are able to ferment xylose in the presence of glucose *(8)*. Pardo et al. *(9)* isolated *P. stipitis* mutants defective in carbon catabolite repression by selecting for resistance to 2-deoxy-D-glucose, an antimetabolite known to select for HK-deficient cells. Therefore, HK appears to be important in regulating sugar uptake in many systems.

D-Xylulokinase (XUK), EC 2.7.1.17 is believed to be critical for the assimilation of pentose sugars *(10)*, but its regulation in xylose fermenting yeasts has not been reported previously. Phosphofructokinase (PFK), EC 2.7.1.11 is instrumental in the glycolytic flux of both pentoses and hexoses. In *S. cerevisiae*, allosteric inhibition of 6-phosphofructokinase (PFK) EC 2.7.1.11 kinetically regulates the glycolytic pathway *(11,12)*.

In *S. cerevisiae*, loss of respirative activity and onset of fermentation results from high glycolytic flux regardless of the carbon source *(13)*. In *Hanseniospora uvarum*, intracellular acetate accumulation appears to impede electron transport, thereby promoting the shift from respirative to fermentative metabolism *(14)*. In xylose-fermenting yeasts glycolysis does not appear to repress respirative activities *(15)*. Rather, fermentation is induced by restricting aeration *(16–20)*. It is possible that the glycolytic flux never becomes high enough to repress respiration in xylose-fermenting yeasts. A high glycolytic flux requires high phosphotransferase activities, so it is of interest to determine the relative activities of HK, XUK, and PFK in respirative and fermentative cells.

Biochemical studies with mutants of *P. tannophilus* and *C. shehatae* have shown that the level of D-xylulokinase (XUK, EC 2.7.1.17) is important in determining growth and xylose fermentation rates *(10,21)*. The possible roles of HK and PFK have not been previously studied in these organisms. The carbon source can be very important in determining the levels of glycolytic enzymes present in the cell *(22,23)*. Cells able to ferment both glucose and xylose should show adaptive responses when switched from one carbon source to the other, and cells grown on different carbon sources might demonstrate substantially different regulatory responses when metabolically perturbed.

P. tannophilus ferments glucose much faster than xylose *(24)*, whereas *C. shehatae* metabolizes glucose and xylose at similar rates. *C. shehatae* is a much better xylose fermenter than *P. tannophilus*. Such differences might be reflected in the relative activities of HK and XUK, so the phosphotransferase regulation in each of these species are of interest.

The objectives of this research were to compare the titers of HK, XUK, and PFK in yeasts capable of xylose and glucose fermentations, to determine whether these titers shifted significantly with growth, aeration, and anaerobiosis, and to see whether the carbon source employed for growth affected the regulatory patterns. A strain of *S. cerevisiae* was included in order to compare xylose-fermenting and non-xylose-fermenting yeasts.

MATERIALS AND METHODS

Micro-organisms

S. cerevisiae ATCC 26785 and *C. shehatae* ATCC 22984 were obtained from the American Type Culture Collection, Beltsville, MD. *P. tannophilus* NRRL Y-2460 was obtained from the Northern Regional Research Laboratory, Peoria, IL. All stock cultures were maintained on yeast malt agar (YMA, Difco).

Culture Methods

Cells were grown in 50 mL of 0.17% yeast nitrogen base without ammonium sulfate or amino acids (YB, Difco) using urea (2.27 g/L) as a nitrogen source and either glucose or xylose (90 g/L) as a carbon source. The pH was 4.6. Cells grown on fresh YPD medium plates (yeast extract, 6 g/L; peptone, 12 g/L; dextrose, 12 g/L), were washed and resuspended with sterile water and used as an inoculum. Initial cell densities were ≈ 2.0 OD at 525 nm. Cultures were incubated with shaking at 100 rpm, 30°C until midethanol production phase (midphase cells), which varied with the organism. For glucose-grown *S. cerevisiae, P. tannophilus,* or *C. shehatae* 12-, 18-, or 24-h-old cells were used, respectively. For xylose-grown *P. tannophilus,* or *C. shehatae,* 48- or 24-h-old cells were used, respectively. Continuous-culture studies were performed as previously described *(16)*.

Induction Studies

S. cerevisiae would not grow on xylose, so HK levels were assayed following incubation in the presence of xylose for 4.5 h. Cells from the three yeast species were cultivated on either glucose or xylose. Each batch of cells was then harvested, washed, and divided into 5-mL aliquots. One aliquot was immediately frozen (initial cells). Two aliquots were suspended in 50 mL YB with 5.0 g/L of $(NH_4)_2SO_4$ (YNB). These served as carbon starved controls. Two aliquots were suspended in 50 mL YNB medium along with 60 g/L glucose and two were suspended in 50 mL YNB plus 60 g/L xylose. Of each set of two aliquots, one was placed in a ventilated 125-mL Erlenmeyer flask (aerated), and the other was placed in a 100-mL serum vial, sealed with a rubber serum stopper and flushed with nitrogen (anaerobic). All flasks and serum vials were incubated with shaking at 200 rpm for 4.5 h at 30°C. Cells were then harvested, washed, and frozen in –80°C freezer until enzyme assays could be performed on the homogenates. Separate, longer-term induction studies were performed under fermentative conditions in a similar manner except that cycloheximide was added to determine how inhibition of protein synthesis would affect enzyme activity.

Preparation of Homogenates

Cells were thawed, washed, and suspended in a minimum volume of 0.1M 3-[*N*-morphilino]propanesulfonic acid (MOPS) buffer (pH 6.8). Cell slurries ($\approx$ 1.0 mL) were kept on ice and placed in a 13 mm (id) glass tube containing 1.0 g of 0.5 mm acid-washed glass beads and blended in a high speed vortex mixer for two 1-min bursts. Microscopic examination showed that approx 60% cell disruption was attained *(25)*. Cell homogenates were centrifuged at 3000g for 15 min, and supernatants were collected for enzyme assays.

Enzyme Assays

All assays were performed within 4 h of cell breakage. PFK was assayed by the method of Bruinenberg et al. *(26)*, but modified by the addition of 10 mM NH_4Cl, 1 mM 5'-AMP, and 0.02 mM fructose 2,6-bisphosphate. HK was assayed by the method of Bergmeyer *(27)* except that final concentrations of 11 mM glucose and 0.55 mM ATP were used. Activity was determined by observing the reduction of NADP of NADPH by glucose-6-phosphate dehydrogenase. XUK was assayed by a modification of the method of Simpson *(28)*. The reaction mixture contained 0.5M Tris/HCl buffer (with 0.01M EDTA), pH 7.8, 0.1 mL; freshly prepared 0.01M phospho-*enol*-pyruvate, 0.1 mL; 0.05M $MgCl_2$, 0.1 mL; 0.01M ATP 0.05 mL; 0.03M NADH, 0.03 mL; 0.01M D-xylulose, 0.1 mL; L-lactic acid dehydrogenase ($\approx$10 U), 0.03 mL; and diluted enzyme sample to a final volume of 1.0 mL. Addition of D-xylulose started the reaction. XUK activity was determined from the rate of NADH disappearance. Control assays for xylitol dehydrogenase (reaction mixture minus ATP) and NADH oxidation (reaction mixture minus ATP and D-xylulose) were performed, and the rates of these reactions were subtracted to obtain XUK activity.

Analytical Methods

Protein concentrations were determined by the method of Bradford (29) using bovine serum albumin as a standard. Specific activities are expressed in international units (µmoles of substrate consumed min^{-1}) per mg protein.

In this study, we cultivated all three organisms on glucose or xylose, induced them under six different conditions, prepared 21 different cell homogenates (including controls), and performed two to four enzyme assays of each homogenate. The experiment was repeated one to three times. Results were reported as average of the data obtained in these experiments, and we inferred effects of the variables only when the differences between the averages exceeded the sum of the observed ranges.

RESULTS AND DISCUSSION

Phosphofructokinase Activities

PFK activities were higher in *S. cerevisiae* than in the two xylose-fermenting yeasts (Table 1). The specific activities of PFK were similar in *P. tannophilus* and *C. shehatae*, when cells were grown on glucose. PFK levels were slightly higher in xylose-grown *C. shehatae* than on xylose-grown *P. tannophilus*. PFK levels in glucose-grown *P. tannophilus* were only one-third as high following anaerobic incubation in the presence of xylose as compared to glucose. Overall, the levels of PFK did not shift as much with *C. shehatae* as they did with *P. tannophilus*. These results indicate that the Embden-Meyerhoff Parnas pathway is probably functional in both glucose and xylose grown *P. tannophilus* and *C. shehatae*. This is consistent with the assumption that PFK plays a role in the metabolism of phosphorylated intermediates that are rearranged by the non-oxidative pentose phosphate pathway.

Hexokinase Activities

There was no significant difference between HK levels of *S. cerevisiae* and the two xylose fermenting yeasts when each was grown on glucose (Table 2). HK levels were higher in glucose-grown than in xylose-grown cells of *P. tannophilus* and. Less significant differences were seen with *C. shehatae*. Aerated incubation with xylose-strongly reduced HK titers in glucose grown *S. cerevisiae*, *P. tannophilus*, and *C. shehatae*. HK activity of xylose-grown cells of *C. shehatae* did not seem to be affected by this short-term incubation.

Table 1
Specific Activity of Phosphofructokinase[a] in *S. cerevisiae*, *P. tannophilus*, and *C. shehatae*

Incubation condition[b]	Saccharomyces cerevisiae	Pachysolen tannophilus		Candida shehatae	
	Glucose grown	Glucose grown	Xylose grown	Glucose grown	Xylose grown
Aerated					
YB + $(NH_4)_2SO_4$ + glucose	1.93 ±0.22	1.10 ±0.51	0.51 ±0.01	1.37 ±0.59	0.90 ±0.16
YB + $(NH_4)_2SO_4$ + xylose	1.99 ±0.12	1.07 ±0.18	0.59 ±0.10	1.17 ±0.22	1.05 ±0.01
Anaerobic					
YB + $(NH_4)_2SO_4$ + glucose	2.29 ±0.38	1.26 ±0.23	0.86 ±0.04	1.12 ±0.10	1.02 ±0.11
YB + $(NH_4)_2SO_4$ + xylose	1.86 ±0.34	0.42 ±0.15	0.89 ±0.09	1.24 ±0.16	1.21 ±0.12

[a]Specific activity IU/mg protein. Average ± range.
[b]YB = Yeast base.

Table 2
Specific Activity of Hexokinase[a] in *S. cerevisiae*, *P. tannophilus*, and *C. shehatae*

Incubation condition[b]	Saccharomyces cerevisiae	Pachysolen tannophilus		Candida shehatae	
	Glucose grown[c]	Glucose grown	Xylose grown	Glucose grown	Xylose grown
Aerated					
YB + $(NH_4)_2SO_4$ + glucose	2.23	2.98 ±0.90	1.05 ±0.03	1.99 ±0.35	1.49 ±0.43
YB + $(NH_4)_2SO_4$ + xylose	1.30	1.67 ±0.67	0.78 ±0.38	1.33 ±0.19	1.50 ±0.32
Anaerobic					
YB + $(NH_4)_2SO_4$ + glucose	1.90	2.65 ±1.00	0.96 ±0.28	1.91 ±0.23	1.63 ±0.22
YB + $(NH_4)_2SO_4$ + xylose	1.78	2.79 ±1.14	0.95 ±0.03	1.71 ±0.20	1.66 ±0.46

[a]Specific activity IU/mg protein. Average ± range.
[b]YB = Yeast base.
[c]Single determination.

Xylulokinase Activities

Xylose-induced XUK in *P. tannophilus* and *C. shehatae*. Specific activity was two to four times higher in xylose-grown cells than in glucose-grown cells (Table 3). Specific activity of XUK was only about one-third of HK (cf Table 2). Since no growth was obtained with *S. cerevisiae* in xylose, all induction experiments with this organism were performed with glucose-grown cells. XUK levels were barely detectable (Table 3). Xylose induced XUK activity in glucose-grown *P. tannophilus* and *C. shehatae* more under aerated conditions than under anaerobic conditions.

Time Course Studies

To examine induction and repression of HK and XUK in more detail, we grew cells of *C. shehatae* and *P. tannophilus* on either glucose or xylose to midphase then transferred the cells to the other carbon source.

Table 3
Specific Activity of D-xylulokinase[a] in *S. cerevisiae*, *P. tannophilus*, and *C. shehatae*

	Saccharomyces cerevisiae	Pachysolen tannophilus		Candida shehatae	
Incubation condition[b]	Glucose grown[c]	Glucose grown	Xylose grown	Glucose grown	Xylose grown
Aerated					
YB + (NH$_4$)$_2$SO$_4$ + glucose	0.023	0.09[c]	0.48 ±0.07	0.03[c]	0.59 ±0.08
YB + (NH$_4$)$_2$SO$_4$ + xylose	0.070	0.53±0.04	0.89 ±0.09	0.55±0.04	1.13 ±0.01
Anaerobic					
YB + (NH$_4$)$_2$SO$_4$ + glucose	0.029	0.25[c]	0.77 ±0.08	0 (nd)[d]	0.88 ±0.05
YB + (NH$_4$)$_2$SO$_4$ + xylose	0.026	0.27[c]	0.97 ±0.16	0.22[c]	1.03 ±0.05

[a]Specific activity IU/mg protein. Average ± range.
[b]YB = Yeast base.
[c]Single determination.
[d]Not detectable.

The specific activities of HK and XUK were periodically measured for up to 9 h.

Following a transfer of *C. shehatae* from glucose to xylose, XUK levels increased eight fold and HK levels decreased by half within 3 h (Fig. 1A). Following transfer from xylose to glucose, XUK was inactivated, but HK activity was essentially unchanged (Fig. 1B). When cycloheximide was added during induction by xylose, XUK activity did not increase, and HK activity dropped slowly with time (Fig. 1C). This indicated that in *C. shehatae*, xylose does not inactivate HK. When *C. shehatae* was transferred from xylose to glucose in the presence of cycloheximide, both XUK and HK activities dropped rapidly (Fig. 1D). The apparent loss of HK activity in the presence of cycloheximide (Fig. 1D), but not in its absence (Fig. 1B) indicates that new protein synthesis is necessary to maintain hexokinase activity in *C. shehatae* following transfer to glucose.

Different patterns of induction were observed with the HK and XUK titers of *P. tannophilus*. Following transfer from glucose to xylose, HK was rapidly (but incompletely) inactivated, and XUK activity was induced but to a lesser extent than in *C. shehatae* (cf Fig. 1A and 2A). In a transfer from xylose to glucose, XUK activity was not rapidly inactivated as in *C. shehatae* (cf Fig 1B and 2B). It did, however decrease gradually after several hours. This difference was also apparent when cells were transferred from xylose to glucose in the presence of cycloheximide. In marked contrast to *C. shehatae*, no inactivation of either HK or XUK was apparent in *P. tannophilus* (Fig. 2D).

Overall, the titers observed in the time-course induction experiments were consistent with those observed in earlier experiments with multiple variables (Tables 2 and 3).

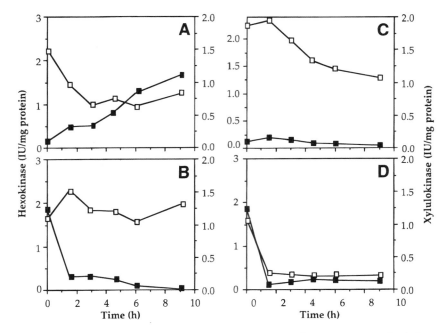

Fig. 1. Hexokinase (□) and D-xylulokinase (■) specific activities in *C. shehatae* following a shift from either **(A)** glucose to xylose or **(B)** xylose to glucose. Cells were grown to midethanol-production phase under aeration, then harvested, washed and transferred to fresh medium under similar conditions. HK and XUK specific activities in *C. shehatae* following a shift from either **(C)** glucose to xylose or **(D)** xylose to glucose with cycloheximide (final concentration, 6.8 µg/mL) added to the fresh medium. Time zero marks the transfer to fresh medium.

Continuous Culture Studies

To determine titers in steady-state cells, *C. shehatae* was cultivated under either fully aerobic or oxygen-limited (fermentative) conditions on xylose or glucose, and HK activities were measured. Four different dilution rates were examined, ranging between 0.05 and 0.22 h^{-1}, but no significant variation could be observed as a function of dilution rate, so the results are presented as averages. HK activity was slightly higher in glucose-grown than xylose-grown cells and somewhat higher in oxygen-limited than fully aerobic cells (Table 4). Studies with *S. cerevisiae* indicate that neither growth rate or cell cycle significantly affect HK or PFK specific activities *(30,31)*. Therefore, changes in enzyme titer likely represent regulatory adaptations to culture conditions.

Our data show that glucose induces HK activity and xylose induces XUK activity in both *P. tannophilus* and *C. shehatae* (Tables 2 and 3). There was no significant difference in PFK activity with these yeasts grown on glucose and incubated in glucose or xylose, and there was no difference when the yeasts were grown on xylose and incubated in glucose or xylose.

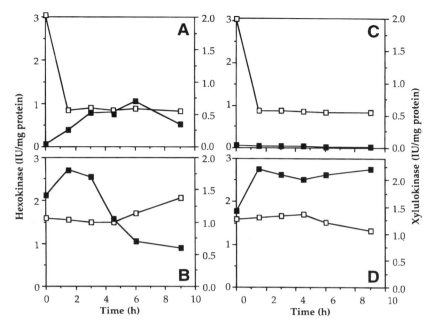

Fig. 2. Hexokinase (□) and D-xylulokinase (■) specific activities in *P. tannophilus* following a shift from either **(A)** glucose to xylose or **(B)** xylose to glucose. Cells were grown with aeration to midethanol production phase, then harvested, washed and transferred to fresh medium under similar conditions. HK and XUK specific activities in *P. tannophilus* following a shift from either **(C)** glucose to xylose or **(D)** xylose to glucose with cycloheximide added to the fresh medium. Time zero marks the transfer to fresh medium.

Table 4
Specific Activity of Hexokinase in *C. shehatae* Grown in Continuous Culture under Several Conditions

Growth condition	Carbon source	Hexokinase activity ave. ± SD (n)[a]
Fully-aerobic	xylose	0.72 ± 0.16 (4)
Fully-aerobic	glucose	0.92[b]
Oxygen-limited	xylose	1.07 ± 0.19 (11)
Oxygen-limited	glucose	1.84 ± 0.22 (4)

[a]Average ± standard deviation (number of replicates).
[b]Single determination.

C. shehatae HK activity observed in the presence and absence of cycloheximide, suggest the presence of two separate isozymes under different regulation in this organism. The total HK titer was maintained in growing cells by new protein synthesis, whereas in the presence of cycloheximide, the HK present during growth on xylose was inactivated, and new protein synthesis or the second HK could not occur to replace it. These results suggest that HK was synthesized in response to glucose, even though no net increase in total activity was observed.

C. tropicalis (32) and S. cerevisiae contain three glucose-phosphorylating enzymes, HK PI, HK PII, and glucokinase. Hexokinase PI is constitutive, whereas, HK PII and glucokinase are regulated by the carbon source. The addition of D-xylose to homogenates of glucose-grown cells causes an irreversible inactivation of all three enzymes in S. cerevisiae (33). In the study presented here, the HK of P. tannophilus was inactivated following transfer of glucose-grown cells to xylose, so these results are consistent with the previous findings with S. cerevisiae.

Hexokinase PII plays an important role in carbon catabolite repression in S. cerevisiae (2,34–36); in S. occidentalis, an unusual HK with two differing catalytic sites appears to have a similar function (37). Moreover, the xylose-induced decrease in HK PII activity confers resistance to carbon catabolite resistance in Saccharomyces carlsbergensis (38). In our study, XUK activity developed shortly following inactivation of HK, so the observations with P. tannophilus are consistent with the carbon catabolite repression mechanism mediated by HK PII described for S. cerevisiae (39). We do not know how important specific activities of these enzymes are to the overall metabolic activities of cells. The significance of PFK in determining the glycolytic flux has been demonstrated in erythrocytes (40) and more recently, in S. cerevisiae (41). In the latter case, PFK did not limit the fermentation rate, but it was not present in great excess either. About 80% of the PFK present was utilized by the glycolytic pathway. In the present studies of xylose-fermenting yeasts, the levels of PFK observed ranged between about 50 and 75% of those present in S. cerevisiae, so it is possible that this enzyme could limit glycolysis in the xylose fermenting yeasts. HK levels in P. tannophilus and C. shehatae were comparable to those observed in S. cerevisiae. Even though S. cerevisiae possesses several metabolic activities important to xylose metabolism (42), we suspect that the XUK levels present are insignificant, especially in light of the apparent importance of this enzyme in fermenting xylose (20).

In conclusion, the elevated levels of PFK observed with S. cerevisiae are consistent with its good fermentative activity on glucose. However, in comparison to xylose-fermenting organisms, S. cerevisiae does not possess significant XUK activity. Both HK and XUK are inductively regulated by glucose and xylose in P. tannophilus and C. shehatae, but significant levels of each enzyme are also present under fermentative conditions, regardless of the carbon source. The slightly higher level of XUK in xylose-grown C. shehatae and the higher level of HK in glucose-grown P. tannophilus might help account for the different rates at which these two organisms metabolize these sugars.

REFERENCES

1. Bolen, P. L. and Detroy, R. W. (1985), *Biotechnol. Bioengineer.* **27**, 302–307.
2. Wedlock, D. N. and Thornton, R. J. (1989), *J. Gen. Microbiol.* **135**, 2013–2018.
3. Entian, K.-D. and Fröhlich, K.-U. (1984), *J. Bact.* **158**, 29–35.
4. Trumbly, R. J. (1992), *Mol. Microbiol.* **6**, 15–22.
5. Herrero, P., Fernandez, R., and Moreno, F. (1989), *J. Gen. Microbiol.* **135**, 1209–1216.
6. Ma, H., Bloom, L. M., Walsh, C. T., and Botstein, D. (1989), *Mol. Cell. Biol.* **9**, 5643–5649.
7. Prior, C., Mamessier, P., Fukuhara, H., Chen, X. J., and Wesolowski, L. (1993), *Mol. Cell. Biol.* **13**, 3882–3889.
8. Wedlock, D. N., James, A. P., and Thornton, R. J. (1989), *J. Gen. Microbiol.* **135**, 2019–2026.
9. Pardo, E. H., Funayama, S., Pedrosa, F. O., and Rigo, L. U. (1992), *Can. J. Microbiol.* **38**, 417–422.
10. Lachke, A. H. and Jeffries, T. W. (1986), *Enzyme Microb. Technol.* **8**, 353–359.
11. Sols, A. (1967), in *Aspects of Yeast Metabolism*. Mills, A. K. and Krebs, H., eds., F. A. Davis Co. Philadelphia, pp. 47–66.
12. Turner, J. F. and Turner, D. H. (1975), *Ann. Rev. Plant Physiol.* **26**, 159–186.
13. Sierkstra, L. N., Nouwen, N. P., Verbakel, J. M. A., and Verrips, C. T. (1993), *Yeast* **9**, 787–795.
14. Venturin, C., Boze, H., Moulin, G., and Galzy, P. (1995), *Yeast* **11**, 327–336.
15. van Dijken, J. P. and Scheffers, W. A. (1986), *FEMS Microbiol. Rev.* **32**, 199–224.
16. Franzblau, S. G. and Sinclair, N. A. (1983), *Mycopathologia* **82**, 185–190.
17. Alexander, M. A., Chapman, T. W., and Jeffries, T. W. (1988), *Appl. Microbiol. Biotechnol.* **28**, 478–486.
18. Prior, B. A., Alexander, M. A., Yang, V., and Jeffries, T. W. (1988), *Biotechnol. Lett.* **10**, 37–42.
19. Alexander, M. A. and Jeffries, T. W. (1990), *Enzyme Microb. Technol.* **12**, 2–19.
20. Alexander, M. A., Yang, V., and Jeffries, T. W. (1988), *Appl. Microbiol. Biotechnol.* **29**, 282–288.
21. McCracken, L. D. and Gong, C.-S. (1983), *Adv. Biochem. Eng. Biotechnol.* **27**, 33–55.
22. Foy, J. J. and Bhattacharjee, J. K. (1978), *J. Bact.* **136**, 647–656.
23. Hommes, F. A. (1966), *Arch. Microbiol.* **58**, 296–301.
24. Jeffries, T. W. (1985), in *Energy Applications of Biomass*, Lownestein, M. Z. ed., Elsevier Applied Science Publishers, NY, pp. 231–252.
25. Ciriacy, M. (1975), *Mutation Research* **29**, 315–325.
26. Bruinenberg, P. M., van Dijken, J. P., and Scheffers, W. A. (1983), *J. Gen. Microbiol.* **129**, 965–971.
27. Bergmeyer, H. U., Gawehn, K., and Grassl, M. (1974), in *Methods of Enzymatic Analysis* 2nd ed., vol. 1., Bergmeyer, H. U. ed., Academic Press, Verlag Chemie, Weinheim/Bergstrasse, London, pp. 473, 474.
28. Simpson, F. J. (1966), *Meth. Enzymol.* **9**, 454–458.
29. Bradford, M. (1976), *Anal. Biochem.* **72**, 248–254.
30. Sierkstra, L. N., Verbakel, J. M. A., and Verrips, C. T. (1992), *J. Gen. Microbiol.* **138**, 2559–2566.
31. De Koning, W., Growneveld, K., Oehlen, L. J. W., Derden, J. A., and van Dam, K. (1991), *J. Gen. Microbiol.* **137**, 971–976.
32. Hirai, M., Ohtani, E., Tanaka, A., and Fukui, S. (1977), *Biochim. Biophys. Acta* **480**, 357–366.
33. Fernández, R., Herrero, P., and Moreno, F. (1985), *J. Gen. Microbiol.* **131**, 2705–2709.
34. Entian, K.-D. (1980), *Mol. Gen. Genet.* **178**, 633–637.
35. Entian, K.-D., Hilberg, F., Opitz, H., and Mecke, D. (1985), *Mol. Cell. Biol.* **5**, 3035–3040.
36. Entian, K.-D. (1986), *Microbiol. Sci.* **3**, 366–371.
37. McCann, A. K., Hilberg, F., Kenworthy, P., and Barnett, J. A. (1987), *J. Gen. Microbiol.* **133**, 381–389.

38. Fernández, R., Herrero, P., Gascón, S., and Moreno, F. (1984), *Arch. Microbiol.* **139,** 139–142.
39. Ma, H. and Botstein, D. (1986), *Mol. Cell. Biol.* **6,** 4046–4052.
40. Boscá, L. and Corredor, C. (1984), *Trends Biochem. Sci.* **9,** 372–373.
41. Liao, J. C., Lightfoot, E. N. Jr., Jolly, S. O., and Jacobson, G. K. (1988), *Biotechnol. Bioengineer.* **31,** 855–868.
42. Batt, C. A., Carvallo, S., Easson, D. D., Jr., Akedo, M., and Sinskey, A. J. (1986), *Biotechnol. Bioengineer.* **28,** 549–553.

Diminished Respirative Growth and Enhanced Assimilative Sugar Uptake Result in Higher Specific Fermentation Rates by the Mutant *Pichia stipitis* FPL-061

HASSAN K. SREENATH[1,2] AND THOMAS W. JEFFRIES*,[1,3]

[1]*Institute of Microbial Biochemical Technology, Forest Products Laboratory, Madison, WI 53705;* [2]*Visiting Scientist, Defence Food Research Laboratory, Mysore-570011, India;* [3]*Dept. Bacteriology, UW-Madison, Madison, WI 53706*

ABSTRACT

A mutant strain of *Pichia stipitis*, FPL-061, was obtained by selecting for growth on L-xylose in the presence of respiratory inhibitors. The specific fermentation rate of FPL-061, was higher than that of the parent, *Pichia stipitis* CBS 6054, because of its lower cell yield and growth rate and higher specific substrate uptake rate. With a mixture of glucose and xylose, the mutant strain FPL-061 produced 29.4 g ethanol/L with a yield of 0.42 g ethanol/g sugar consumed. By comparison, CBS 6054 produced 25.7 g ethanol/L with a yield of 0.35 g/g. The fermentation was most efficient at an aeration rate of 9.2 mmoles O_2 L^{-1} h^{-1}. At high aeration rates (22 mmoles O_2 L^{-1} h^{-1}), the mutant cell yield was less than that of the parent. At low aeration rates, (1.1 to 2.5 O_2 L^{-1} h^{-1}), cell yields were similar, the ethanol formation rates were low, and xylitol accumulation was observed in both the strains. Both strains respired the ethanol once sugar was exhausted. We infer from the results that the mutant, *P. stipitis* FPL-061, diverts a larger fraction of its metabolic energy from cell growth into ethanol production.

Index Entries: *Pichia stipitis*; respiration; ethanol production; specific productivity; mutation; salicyl hydroxamate; antimycin A.

---

*Author to whom all correspondence and reprint requests should be addressed.

INTRODUCTION

Pichia stipitis can convert xylose, glucose, mannose and galactose into ethanol *(1)*. However, glucose slows down its rate of xylose assimilation and ultimately results in lower productivity and yield *(2)*. This could be a result of catabolite repression or differential transport *(3–5)*. *P. stipitis* also has a lower ethanol yield at high aeration rates *(6–8)*. This appears to be caused by the respirative consumption of ethanol. Unfortunately, the sugar uptake and fermentation rates decrease as the aeration rate decreases *(9)*.

It is difficult to obtain improved fermentative mutants of *P. stipitis* because cell growth depends on aeration and fermentative activity occurs under oxygen-limiting conditions *(10,11)*. Strategies for obtaining improved yeast mutants with enhanced fermentation on mixed sugars suggest that derepression of enzymes for xylose metabolism enhance growth and fermentation rates *(12–15)*. Selection for mutants that can grow on carbon sources that do not induce pentose phosphate pathway enzymes, but which can be assimilated by that route can result in higher rates of xylose uptake. Moreover, by selecting for mutants that resist respiratory inhibitors, one can obtain lower respiratory activities *(16)*.

The objective of the present work was to characterize a mutant *P. stipitis* that had been selected for rapid growth on noninductive carbon sources in the presence of respiratory inhibitors. We have compared the *P. stipitis* parental strain with FPL-061 during fermentation of pentose and hexose sugar mixtures at various aeration rates.

MATERIALS AND METHODS

Yeast Strains and Inoculum Preparation

P. stipitis CBS 6054 was mutagenized with nitrosoguanidine (NG), and the cells were first screened by selecting for growth on L-xylose in the presence of salicylhydroxamic acid (SHAM, Sigma, St. Louis, MO) and antimycin A (AA, Sigma). FPL-061 was selected following screens of respiratory resistant mutants in microfuge tubes *(16)*. *P. stipitis* CBS 6054 and *P. stipitis* FPL-061 were cultivated on fresh plates of yeast extract peptone xylose (YEPX) agar at 31 to 32°C. Cells from a 48-h-old YEPX plate were washed with sterile water and used as the inoculum. The optical densities of cell suspensions were measured at 525 nm and adjusted to equivalent values by dilution with water.

Shake Flask Fermentation

Fermentation media contained 1.7 g/L filter sterilized yeast nitrogen base (YNB) without ammonium sulfate or amino acids (Difco). Urea, 2.27 g/L, and peptone, 6.56 g/L, were used as nitrogen sources. Fifty mL of medium containing xylose or sugar mixture in a 125-mL Erlenmeyer flask was shaken at 100 rpm, at 25 to 30°C. The fermentation was monitored in triplicate flasks for 3 to 10 d by removing 1.3 mL samples for sugar, ethanol, and cell analyses. The final pH was 4.0 to 4.5.

Fermentations used glucose, xylose, or arabinose or a 1:1 mixture of glucose and xylose. Final sugar concentrations ranged from 60 to 80 g/L. Individual sugars and sugar mixtures were autoclaved separately and added to media after they had cooled to room temperature. For aeration optimization, the dissolved oxygen transfer rate was estimated by the sulfite oxidation method *(17)*. For medium containing 70 g/L-xylose in various volumes of media 12.5 to 87.5 mL, oxygen transfer rates corresponding to 1.1 to 22.6 mmol of O_2 L^{-1} h^{-1} were used.

Analytical Methods

Cell densities were measured at 525 nm. An OD of 1.0 was equivalent to 0.21 mg dry weight of cells/mL. Sugars and fermentation products were determined by high performance liquid chromatography (HPLC) (Hewlett Packard series 1050) with an RI detector. Sugars and products were separated on an Aminex Carbohydrate HPX 87C, column (300 × 7.8 cm) maintained at 85°C *(18)*. The mobile phase was degassed distilled water at a flow rate of 0.5 mL/min at a pressure of 50 to 55 bar. Filtered clear samples (980 µL) were mixed with 20 µL of sucrose (500 g/L) as internal standard before injection.

RESULTS AND DISCUSSION

Selection of the FPL-061 Mutant

D-Xylose is the form found in nature; L-xylose is uncommon. Both, however, can be reduced to xylitol, which is neither D nor L. L-Xylose is not metabolized by *P. stipitis* CBS 6054. Presumably, this sugar does not induce assimilative enzymes, but this has not been confirmed. L-Xylose is, however, a (poor) substrate for aldose reductase *(18)*, and once reduced to xylitol, L-xylose can be metabolized. Growth on L-xylose, therefore, should select for strains constitutively derepressed for aldose reductase. It is not sufficient to identify rapid growers because such are usually poor fermenters. Hence, mutants were selected for growth in the presence of respiratory inhibitors. *P. stipitis* possesses both AA-sensitive and SHAM-sensitive respiratory pathways *(19)*. Blockage of either pathway alone is insufficient to inhibit growth, so both inhibitors were used. Mutants growing under these conditions were screened for fermentative activities in microfuge tubes *(20)*. Initial concentrations of cells from various strains were adjusted to a constant value so that strains with higher specific fermentation rates (rather than higher growth rates) could be identified. The properties of FPL-061 were then confirmed by trials in replicate shake flasks. Methods used to identify FPL-061 have been patented *(16)*.

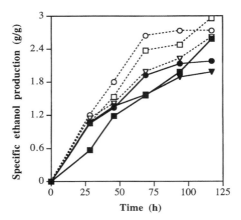

Fig. 1. Effect of xylose, glucose, and mixture of xylose and glucose on specific production in *Pichia stipitis* CBS 6054 (solid line) and *Pichia stipitis* FPL-061 (broken line). Symbols: (■), CBS 6054 with 74 g/L xylose; (▼), CBS 6054 with 70 g/L glucose; (●), CBS 6054 with 36 g/L xylose and 36 g/L glucose; (●), FPL 061 with 74 g/L xylose; (▼), FPL 061 with 70 g/L glucose; (○), FPL 061 with 36 g/L xylose and 36 g/L glucose.

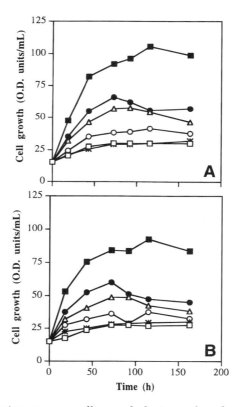

Fig. 2. Effect of aeration rates on cell growth during xylose fermentation in **(A)** *Pichia stipitis* CBS 6054 and **(B)** *Pichia stipitis* FPL-061. Symbols: aeration rate in mmol O_2 L^{-1} h^{-1}: (■), 22.5; (●), 11.8; (▽), 9.2; (○), 5.4; (□), 2.5; (X), 1.1.

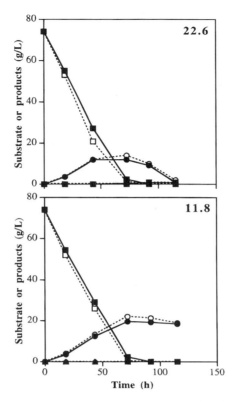

Fig. 3. Effect of aeration at 22.6 and 11.3 mmoles O_2 L^{-1} h^{-1} on ethanol production during xylose fermentation in *Pichia stipitis* CBS 6054 (solid line) and *Pichia stipitis* FPL-061 (broken line). Symbols: (■), CBS 6054 xylose; (♦), CBS 6054 ribitol; (▲), CBS 6054 xylitol; (●), CBS 6054 ethanol; (□), FPL 061 xylose; (○), FPL 061 ribitol; (△), FPL 061 xylitol; (○), FPL 061 ethanol.

Growth and Fermentation of Various Sugars and Sugar Mixtures

CBS 6054 grew faster than FPL-061 on a mixture of glucose and xylose, but on a specific cell basis, FPL-061 produced more ethanol in glucose, xylose, and sugar mixtures (Fig. 1). When grown on a 1:1 glucose:xylose mixture, FPL-061 assimilated xylose slightly faster than CBS 6054 (data not shown). FPL-061 produced a maximum ethanol concentration of 29.4 g/L with a yield of 0.42 g/g sugar consumed. By comparison, under the same conditions, CBS 6054 produced 25.7 g/L of ethanol with a yield of 0.35 g/g. Neither FPL-061 or CBS 6054, fermented L-arabinose, but both strains grew slowly on this sugar and converted it to arabitol.

Effect of Various Aeration Rates on Xylose Fermentation

The parental strain grew better than the mutant, particularly with high aeration. Growth rates of both strains were severely retarded with low aeration (Fig. 2). The mutant strain *P. stipitis* FPL-061 produced ethanol

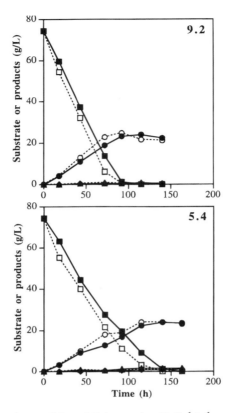

Fig. 4. Effect of aeration at 9.2 and 5.4 mmoles O_2 L^{-1} h^{-1} on ethanol production during xylose fermentation in *Pichia stipitis* CBS 6054 (solid line) and *Pichia stipitis* FPL-061 (broken line). Symbols same as Fig. 3.

from xylose faster than *P. stipitis* CBS 6054 when optimally aerated (Figs. 3 and 4). At the highest aeration rate, ethanol respiration was noted with both strains (Fig. 3). Reassimilation of ethanol at high aeration is well-recognized *(21)*. In contrast, at low aeration rates, we observed a low ethanol formation rate and xylitol accumulation by both strains (Fig. 5). FPL-061 and CBS 6054 both fermented xylose maximally at an aeration rate of 9.2 mmoles O_2 L^{-1} h^{-1}, but at that aeration rate, FPL-061 fermented faster (Fig. 4).

The mutant growth rates and cell yields were lower (Fig. 2), but its specific fermentation rates (Fig. 1) were higher than the parent, suggesting that FPL-061 has a deficient respirative capacity. This might have arisen when it gained resistance to AA or SHAM. For aeration rates of 5.4 to 9.2 mmol O_2 L^{-1} h^{-1}, substrate uptake by FPL-061 was more rapid than by CBS-6054, even though growth was less (cf. Fig. 1 and Fig. 3). However, for higher and lower aeration rates, no appreciable difference was observed. This suggests that substrate uptake is derepressed in the mutant. This conclusion was supported by the observation that FPL-061 ferments a mixture of glucose and xylose more rapidly than CBS 6054. Another possibility is

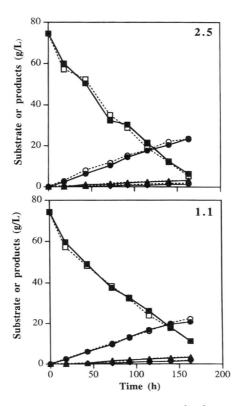

Fig. 5. Effect of aeration at 2.5 and 1.1 mmoles O_2 L^{-1} h^{-1} on ethanol production during xylose fermentation in *Pichia stipitis* CBS 6054 (solid line) and *Pichia stipitis* FPL-061 (broken line). Symbols same as Fig. 3.

that the mutant diverts a larger portion of metabolic energy into sugar uptake. In either case, of the sugar taken up, more is shunted into ethanol at intermediate aeration levels. At the lowest aeration rate, there was no discernible difference between the sugar uptake rates for the mutant and the parent (Fig. 5), and the cell growth rates were also similar (and very low). This indicates that the high affinity oxygen uptake systems in these two yeasts are similar.

Conclusions

The mutant strain *P. stipitis* FPL-061 produced ethanol at a higher yield as compared to its parental strain, *P. stipitis* CBS 6054. Elevated sugar uptake without concomitant growth suggests that the assimilative pathway is derepressed in the mutant. During xylose fermentation, an aeration rate of 9.2 mmol O_2 L^{-1} h^{-1} for ethanol production. Because the growth rate of the mutant was reduced at high aeration rates, it is probably defective in a low affinity oxygen uptake system that is responsible for energy production.

ACKNOWLEDGMENT

Hassan K. Sreenath is supported in part by the USDA, Forest Service International Forest Products Program. Additional support is provided by USAID Office of Research grant 10. 299.

REFERENCES

1. du Preez, J. C. (1994), *Enzyme Microb. Technol.* **16,** 944–956.
2. Panchal, C. J., Bast, L., Russel, I., and Stewart, G. G. (1988), *Can. J. Microbiol.* **34,** 1316–1320.
3. Jeffries, T. W. and Sreenath, H. K. (1988), *Biotechnol. Bioeng.* **31,** 502–506.
4. Slininger, P. J., Bolen, P. L., and Kurtzman, C. P. (1987), *Enzyme Microb. Technol.* **9,** 5–15.
5. Kilian, S. G. and van Uden, N. (1988), *Appl. Microbiol. Biotechnol.* **27,** 545–548.
6. Grootjen, D. R. J., Van, D. L. R., and Lubyen, K. C. A. (1990), *Enzyme Microb. Technol.* **12,** 20–23.
7. Gubel, D. V., Cordenons, A., Nudel, B. C., and Giulietti, A. M. (1991), *J. Indus. Microbiol.* **7,** 287–292.
8. Laplace, J. M., Delegenes, J. P., Moletta, R., and Navarro, J. M. (1991), *Appl. Microbiol. Biotechnol.* **36,** 158–162.
9. Boynton, B. L. and McMillan, J. D. (1994), *Appl. Biochem. Biotechnol.* **45/46,** 509–514.
10. Alexander, M. A., Yang, V. W., and Jeffries, T. W. (1988), *Appl. Microbiol. Biotechnol.* **29,** 282–288.
11. Passoth, V., Zimmerman, M., and Klinner, U. (1996), *Appl. Biochem. Biotechnol.* **57/58,** 201–212.
12. McCracken, L. D. and Gong, C. S. (1983), *Adv. Biochem. Bioeng.* **27,** 33–85.
13. Lachke, A. H. and Jeffries, T. W. (1986), *Enzyme Microb. Technol.* **8,** 353–359.
14. James, A. P., Zahab, D. M., Mahmourides, G., Maleszka, R., and Schneider, H. (1989), *Appl. Environ. Microbiol.* **55,** 2871–2876.
15. Laplace, J. M., Delgenes, J. P., Molleta, R., and Navarro, J. M. (1992), *Enzyme Microb. Technol.* **14,** 644–648.
16. Jeffries, T. W. and Livingston, P. L. (1992), U. S. Patent No. 5, 126, 266.
17. Sreenath, H. K., Chapman, T. W., and Jeffries, T. W. (1986), *Appl. Microbiol. Biotechnol.* **24,** 294–299.
18. Karassevitch, Y. N. (1976), *Biochimie* **58,** 239–242.
19. Jeppsson, H., Alexander, N. J., and Hahn-Hägerdal, B. (1995), *Appl. Environ. Microbiol.* **61,** 2596–2600.
20. Sreenath, H. K. and Jeffries, T. W. (1996), *Biotechnol. Tech.* **10,** 239–242.
21. Lohmeier-Vogel, E., Skoog, K., Vogel, H., and Hahn-Hägerdal, B. (1989), *Appl. Environ. Microbiol.* **55,** 1974–1980.

Production of Xylitol from D-Xylose by *Debaryomyces hansenii*

JOSE M. DOMINGUEZ,[1],* CHENG S. GONG,[2] AND GEORGE T. TSAO[2]

[1]*Department of Chemical Engineering, University of Vigo (Campus Orense), Las Lagunas, 32004 Orense, Spain;* [2]*Laboratory of Renewable Resources Engineering, Purdue University, West Lafayette, IN 47907*

ABSTRACT

Xylitol, a naturally occurring five-carbon sugar alcohol, can be produced from D-xylose through microbial hydrogenation. Xylitol has found increasing use in the food industries, especially in confectionary. It is the only so-called "second-generation polyol sweeteners" that is allowed to have the specific health claims in some world markets. In this study, the effect of cell density on the xylitol production by the yeast *Debaryomyces hansenii* NRRL Y-7426 from D-xylose under microaerobic conditions was examined. The rate of xylitol production increased with increasing yeast cell density to 3 g/L. Beyond this amount there was no increase in the xylitol production with increasing cell density. The optimal pH range for xylitol production was between 4.5 and 5.5. The optimal temperature was between 28 and 37°C, and the optimal shaking speed was 300 rpm. The rate of xylitol production increased linearly with increasing initial xylose concentration. A high concentration of xylose (279 g/L) was converted rapidly and efficiently to produce xylitol with a product concentration of 221 g/L was reached after 48 h of incubation under optimum conditions.

Index Entries: *Debaryomyces hansenii*; D-xylose; xylitol; biological hydrogenation; yeast.

INTRODUCTION

Sucrose is one of the most important ingredients of confectionary products, as it provides body, texture, and preservative properties, besides its sweetening effect. However, it is well known that consumption of

---

*Author to whom all correspondence and reprint requests should be addressed.

sucrose and fermentable carbohydrates facilitates the development of plaque, dental caries, and periodontal disease. To avoid these problems, noncariogenic polyol sweeteners are increasingly used. One of these, xylitol, may be considered as the best of all nutritive sweeteners because it has anticaries properties (1). Besides, it is tolerated by diabetics and it has a high negative heat of solution. For all these properties, xylitol is desirable for sugar-free confections (2). Unfortunately, it is one of the most expensive polyol sweeteners (3). Availability and cost of production are the obstacles impeding the increased use of xylitol.

Xylitol is a normal metabolic intermediate in animals. The human body produces 5–15 g of xylitol per day during normal metabolism (2). Xylitol occurs naturally in many fruits and vegetables (such as lettuce, cauliflower, strawberries) and constitutes part of the human diet. However, fruits and vegetables contain a small amount, usually less than 900 mg/100 g, rendering its extraction uneconomical. In industrial scale, it can be produced through chemical reduction of xylose derived from hemicellulosic hydrolyzate. This process includes extensive purification and separation steps to remove polymers of other sugars and other by-products present in the raw materials (4).

Xylitol can be formed too, as a metabolic intermediary product of D-xylose fermentation: D-xylose can be converted to xylitol by NADPH-dependent aldehyde reductase, or can be isomerized to D-xylulose by D-xylose isomerase, and then reduced to xylitol by NADH-dependent xylitol dehydrogenase (5). Many yeast strains have the ability to produce xylitol from xylose extracellularly as a normal metabolic activity (6). The prominent strains that produce xylitol include *Candida sp.* (7), *C. guillermondi* (8–10), *C. boidinii* (11), *C. tropicalis* (12), *C. parapsilosis* (13), and *D. hansenii* (14,15). However, D-xylose is an expensive substrate for xylitol production. Recent developments in obtaining xylose-rich hemicellulose hydrolyzates from lignocellulosic materials have identified economic source of xylose availability (16). As a result, xylitol can be produced from such materials as an option for effective utilization of lignocellulosic biomass (17).

In this report, we studied some characteristics of *D. hansenii* NRRL Y-7426 for xylitol production from xylose.

MATERIALS AND METHODS

Micro-organism

The yeast strain used in this study, *D. hansenii* NRRL Y-7426, was obtained from the Northern Regional Research Laboratory, (Peoria, IL). The yeast was grown for 3 d at 32°C in an incubator shaker at 200 rpm (New Brunswick), in a liquid media with 1% of glucose, 1% of xylose, 3 g/L of Bacto-yeast extract, 3 g/L of Bacto-malt extract, and 5 g/L of Bacto-peptone.

Fermentation

Shake flask fermentation experiments were carried out at 24°C, in 50-mL Erlenmeyer flasks (containing 10 mL of culture media) placed in a gyratory shaker at 180 rpm for 2 to 4 d. The sterile culture medium was prepared with pure xylose (12%) and supplemented with nutrients with the following composition per liter (pH 5.7): Bacto-yeast extract, 3 g; Bacto-malt extract, 3 g; and Bacto-peptone, 5 g.

Analytical Methods

Xylose and xylitol were analyzed using a Hitachi high-performance liquid chromatographic system consisting of an AS-4000 Intelligent Auto Sampler, a Hitachi L-3350 refractive index monitor, a Hitachi L-6000 pump, and a Hitachi D-2500 chromato-integrator. Separation was achieved using an organic acid column (Aminex HPX-87 H Ion Exclusion Column 300 × 7.8 mm, Bio-Rad, Hercules, CA) at 0.81 bar and 60°C with 0.01 N sulfuric acid as eluant at 0.8 mL/min over a 18-min period.

Determination of Cell Dry Weight

The cells were collected by centrifugation and dry weight was determined.

RESULTS AND DISCUSSION

Effect of the Cell Density

Figure 1 shows the effect of cell density on the xylitol production from D-xylose by *D. hansenii NRRL Y-7426*. When the initial yeast concentration increased gradually from 0.3 g/L to 3 g/L, the xylitol production increased from 60.1 g/L to 105.8 g/L after 72 h of fermentation. However, when the initial yeast concentration increased to 7.5 and 15 g/L, there was a little decrease in xylitol production. With 3 g/L of initial yeast concentration, xylose was rapidly converted to xylitol. In contrast, the rate of xylose utilization was slow and xylitol assimilation was lower at initial yeast concentration of 0.3 g/L. The highest xylitol productivity, 1.47 g/L/h, was obtained at 3 g/L of initial yeast concentration after 72 h of fermentation where 105.8 g/L of xylitol was produced. Similar results were reported by Cao et al. *(18)*, with *Candida* sp. B-22, the rate of xylitol production increased with increasing initial yeast concentration. After 72 h of fermentation, xylitol concentration decreased slowly as a result of lower residual xylose concentration and the assimilation of xylitol itself.

Effect of the Initial pH

The effect of the pH was studied using the same amount of yeast cells that led to the highest xylitol productions. Figure 2 shows the results after

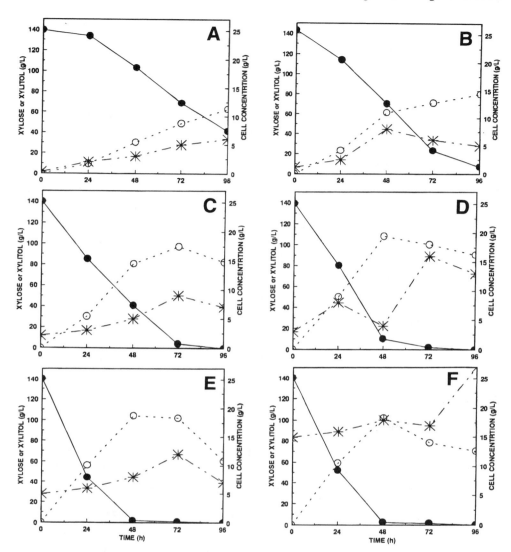

Fig. 1. Effect of the cell density on the production of xylitol from xylose by *D. hansenii*. Symbols: ●——●, xylose; ○- - -○, xylitol; ✱— ··✱, cell concentration. **(A)** 0.3 g/L, **(B)** 1.2 g/L, **(C)** 2.1 g/L, **(D)** 3 g/L, **(E)** 7.5 g/L, **(F)** 15 g/L.

48 h of fermentation under different initial pH. The optimal pH range for xylitol production was between 4.5 and 5.5. The xylitol concentration and the xylitol productivity were 86.29 g/L and 1.80 g/L/h at initial pH of 4.5. Whereas at initial pH of 5.5, xylitol concentration was 91.91 g/L and the productivity was 1.91 g/L/h. The xylitol concentrations at pH 3 and 6.5 were 69.91 g/L and 66.53 g/L, respectively. The xylitol concentration was lowest at both pH 8.0 (49.02 g/L) and pH 2.0 (53.32 g/L). The results indi-

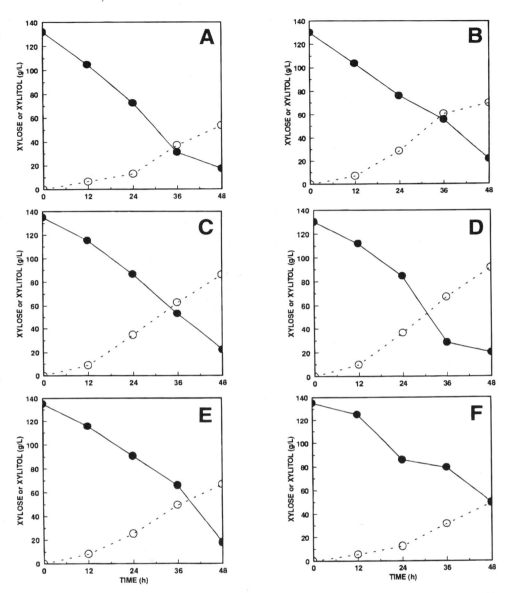

Fig. 2. Effect of the initial pH on the production of xylitol from xylose by *D. hansenii*. Symbols: ●——●, xylose; ○- - -○, xylitol. **(A)** pH = 2, **(B)** pH = 3, **(C)** pH = 4.5, **(D)** pH = 5.5, **(E)** pH = 6.5, **(F)** pH = 8.

cated that both the xylitol productivity and xylitol yields were influenced by initial pH. However, yeast cell growth is relatively resistant to pH change. This is indicated by the ability of *D. hansenii* to utilize xylose to produce xylitol at a relatively low pH environment.

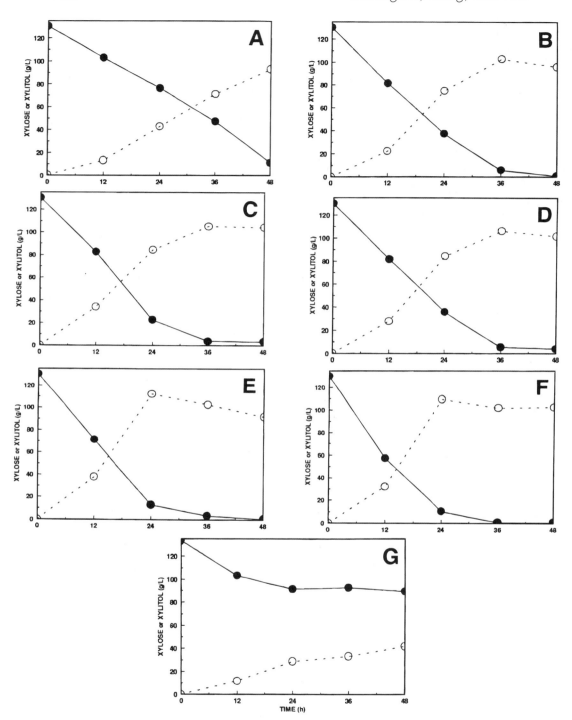

Fig. 3. Effect of the temperature on the production of xylitol from xylose by *D. hansenii*. Symbols: ●——●, xylose; O- - -O, xylitol. **(A)** 24°C, **(B)** 28°C, **(C)** 30°C, **(D)** 32°C, **(E)** 35°C, **(F)** 37°C, **(G)** 44°C.

Effect of the Temperature

The effect of temperature was studied at optimal pH 5.5 with 3 g/L initial yeast concentration for xylitol production from pure xylose. Figure 3 shows the fermentation carried out with the different temperatures chosen. The fermentation rate of xylose to xylitol was relatively constant over a temperature range of 28 to 37°C, and in this interval of temperatures, xylose was consumed rapidly and xylitol was produced efficiently (around 100 g/L). At 24°C, fermentation of xylose to xylitol was carried out slowly, although it reached a high xylitol concentration of 93.04 g/L. The decrease in fermentation rate was dramatic with a high temperature of 44°C as xylose utilization was slow, and xylitol production was lower (41.88 g/L). Cao et al. *(18)*, found that the initial fermentation rate of xylose to xylitol with *Candida* sp. B-22 was relatively constant over a temperature range of 35–40°C, however at incubation temperature of 45°C or higher, the fermentation rate was sharply reduced.

Effect of the Shaker Speed

The shaker speed is related to the concentration of dissolved oxygen that plays an important role in the fermentation of xylose into xylitol. With high aeration, the xylose fermentation shifted to growth phase, whereas under low speeds (anaerobic conditions) xylose could not be assimilated by yeast without NADH-linked xylose reductase. To improve the xylitol production, microaerobic conditions are required. Figure 4 shows the fermentation carried out at 32°C with three different shaker speeds during 48 h. When the shaker speed was 100 rpm, xylose was consumed slowly and the xylitol production was only 40.55 g/L. This is because of the low speed required to dissolve enough oxygen in the media. Working with 300 rpm, xylose was consumed rapidly to produce xylitol (99.67 g/L). Whereas at 200 rpm, xylose was consumed slowly, but reached a xylitol concentration of 113.54 g/L. Both 300 and 200 rpm can be considered microaerobic conditions. Parajó et al. *(15)* found aerobic condition in 300 rpm working with bigger flasks, this is because of the concentration of dissolved oxygen that was higher under these conditions. Figure 4 shows the yeast concentration under the different shaker speeds. At 100 rpm there was a little increase in yeast concentration. Whereas at 200 and 300 rpm, the increase in cell concentration was much higher.

Effect of the Initial Xylose Concentration

The effect of the initial xylose concentration was studied at shaker speed of 300 rpm. Figure 5 shows the xylose consumption and the xylitol production after 48 h of fermentation. Under these microaerobic conditions, all xylose was consumed after 48 h of incubation and xylitol production started without lag period. With low initial xylose concentrations,

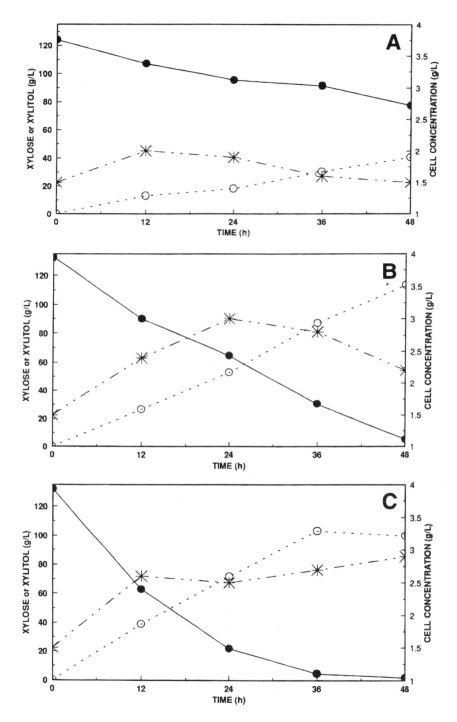

Fig. 4. Effect of the shaker speed on the production of xylitol from xylose by *D. hansenii*. Symbols: ●——●, xylose; ○- - -○, xylitol; ∗— ·· ∗, cell concentration. **(A)** Shaker Speed = 100 rpm; **(B)** Shaker Speed = 200 rpm; **(C)** Shaker Speed = 300 rpm.

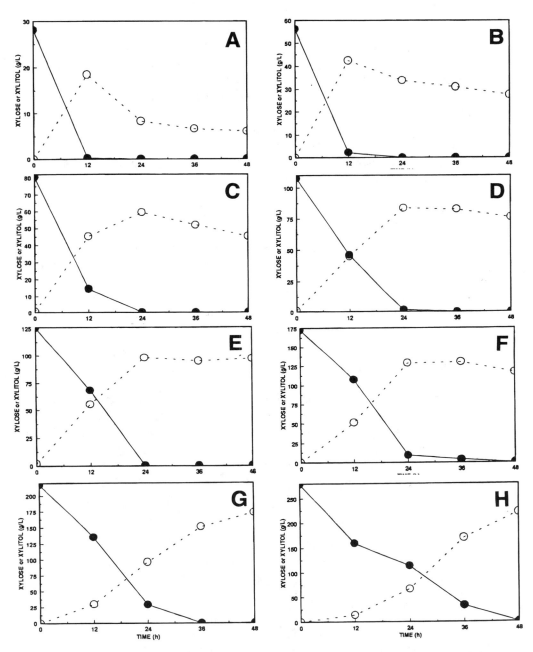

Fig. 5. Effect of the initial xylose concentration on the production of xylitol from xylose by *D. hansenii*. Symbols: ●——●, xylose; ○- - -○, xylitol. **(A)** X = 30 g/L; **(B)** X = 60 g/L; **(C)** X = 80 g/L; **(D)** X = 110 g/L; **(E)** X = 125 g/L; **(F)** X = 175 g/L; **(G)** X = 220 g/L; **(H)** X = 280 g/L.

xylose was converted rapidly and efficiently into xylitol. When the initial xylose concentration was very high at 217 and 279.24 g/L, xylose was consumed slowly at first, a result of the high osmotic pressure. After 24 h, when xylose decreased, xylose was consumed rapidly and reached a xylitol concentration of 173.08 g/L with a xylitol productivity of 3.6 g/L/h when initial xylose concentration was 217 g/L. At an initial xylose concentration of 279.24 g/L, final xylitol concentration was 221.12 g/L after 48 h of incubation with a productivity of 4.6 g/L/h. The xylitol weight yields were 79.76 and 79.19% for xylose concentration of 217 and 279.4 g/L, respectively. Similar effect of high initial xylose concentration on yeast productivity was observed by Meyrial et al. (10) with *C. guillermondii* at an initial xylose concentration of 300 g/L.

CONCLUSIONS

Xylose, the feedstock for xylitol, is abundant in nature in the form of hemicellulose as the major component of lignocellulose materials. It can be easily obtained by hydrolysis of hemicellulose. The use of this yeast or other high-xylitol producing yeast strains for xylitol production from biomass-derived xylose can provide a firm basis for expansion of supply of xylitol in food arena.

ACKNOWLEDGMENTS

This study was supported in part, through The Consortium for Plant Biotechnology Research by DOE cooperative agreement no. DE-FCO5-92OR22072. This support does not constitute an endorsement by DOE or by The Consortium for Plant Biotechnology Research of the view expressed in this article. J. M. D. was supported by Instituto Galego De Promoción Económica, Spain.

REFERENCES

1. Mäkinen, K. K. (1978), in *Biochemical Principles of the Use of Xylitol in Medicine and Nutrition with Special Consideration of Dental Aspects. Experientia:Suppl* **30,** 7–160. Birkhauser Verlag, Basel und Stuggart.
2. Pepper, T. and Olinger, P. M. (1988), *Food Technol.* **42,** 98–106.
3. Chemical Marketing Reporter (March 4, 1996).
4. Melaja, A. and Hämäläinen, L. (1977), US Patent 4,008,285.
5. Höfer, M., Betz, A., and Kotyk, A. (1971), *Biochem. Biophys. Acta.* **252,** 1–12.
6. Gong, C. S., Claypool, T. A., McCracken, L. D., Maun, C. M., Ueng, P. P., and Tsao, G. T. (1983), *Biotechnol. Bioeng.* **25,** 85–102.
7. Chen, L. F. and Gong, C. S. (1985), *J. Food Sci.* **50,** 226–228.
8. Barbosa, M. F. S., Medeiros, M. B., de Mancilha, I. M., Schneider, H., and Lee, H. (1988), *J. Ind. Microb.* **3,** 241–251.
9. Lee, H., Atkin, A., Barbosa, M. F. S., Dorscheid, D. R., and Schneider, H. (1988), *Enzyme Microb. Technol.* **10,** 81–84.

10. Meyrial, V., Delgenes, J. P., Moletta, R., and Navarro, J. M. (1991), *Biotech. Lett.* **13,** 281–286.
11. Vongsuvanlert, V. and Tani, Y. (1989), *J. Ferment. Bioeng.* **67,** 35–39.
12. Horitsu, H., Yahashi, Y., Takamizawa, K., Kawai, K., Suzuki. T., and Watanabe, N. (1992), *Biotech. Bioeng.* **40,** 1085–1091.
13. Furlan, S., Bouilloud, P., Strehaiano, P., and Riba, J. P. (1991), *Biotech. Lett.* **13,** 203–206.
14. Roseiro, J. C., Peito, M. A., Girio, F. M., and Amaral-Collaco, T. (1991), *Arch. Microbiol.* **156,** 484–490.
15. Parajó, J. C., Dominguez, H., and Dominguez, J. M. (1995), *Bioprocess Bioeng.* **13,** 125–131.
16. McMillan, J. D. (1994), in *Enzymic Conversion of Biomass for Fuels Production.* Himmel, M. E., Baker, J. O., and Overend, R. P., eds., ACS, Washington, DC, pp. 292–324.
17. Cao, N. J., Krishnan, M. S., Du, J. X., Gong, C. S., Ho, N. W. Y., Chen Z. D., and Tsao, G. T. (1996), *Biotechnol Lett.* **18,** 1013–1018.
18. Cao, N. J., Tang, R., Gong, C. S., and Chen, L. F. (1994), *Appl. Biochem. Biotechnol.* **45/46,** 515–519.

Production of 2,3-Butanediol from Pretreated Corn Cob by *Klebsiella oxytoca* in the Presence of Fungal Cellulase

NINGJUN CAO,* YOUKUN XIA, CHENG S. GONG, AND GEORGE T. TSAO

Laboratory of Renewable Resources Engineering, Purdue University, West Lafayette, IN 47907

ABSTRACT

A simple and effective method of treatment of lignocellulosic material was used for the preparation of corn cob for the production of 2,3-butanediol by *Klebsiella oxytoca* ATCC 8724 in a simultaneous saccharification and fermentation process. During the treatment, lignin, and alkaline extractives were solubilized and separated from cellulose and hemicellulose fractions by dilute ammonia (10%) steeping. Hemicellulose was then hydrolyzed by dilute hydrochloric acid (1%, w/v) hydrolysis at 100°C at atmospheric pressure and separated from cellulose fraction. The remaining solid, with 90% of cellulose, was then used as the substrate. A butanediol concentration of 25 g/L and an ethanol concentration of 7 g/L were produced by *K. oxytoca* from 80 g/L of corn cob cellulose with a cellulase dosage of 8.5 IFPU/g corn cob cellulose after 72 h of SSF. With only dilute acid hydrolysis, a butanediol production rate of 0.21 g/L/h was obtained that is much lower than the case in which corn cob was treated with ammonia steeping prior to acid hydrolysis. The butanediol production rate for the latter was 0.36 g/L/h.

Index Entries: Ammonia steeping; 2,3-butanediol; corn cob; *Klebsiella oxytoca*; simultaneous saccharification and fermentation (SSF).

INTRODUCTION

2,3-Butanediol is a colorless and odorless liquid having a high boiling point of 180–184°C and a low freezing point of –60°C. This petroleum-based product has diverse industrial use, particularly as a polymeric

*Author to whom all correspondence and reprint requests should be addressed.

substance in addition to its use for manufacturing butadiene or antifreeze. Currently, butanediol is enjoying 4–7% annual growth, buoyed by the increased demand for polybutylene terephthalate resins, gamma-butyrolactone, and spandex and its precursors *(1)*. Butanediol is also a potentially valuable fuel additive. The heating value of butanediol (27,198 J/g) is similar to other liquid fuels such as ethanol (29,055 J/g) and methanol (22,081 J/g). It can also be used as octane booster for gasoline or as high-grade aviation fuel after converting it to methyl ethyl ketone (MEK). Furthermore, butanediol can also be converted to diacetyl form for flavoring in food products *(2)*.

Industrially, butanediol is produced from petroleum-based feedstocks. It can also be produced from simple sugars such as glucose and xylose by bacterial fermentation. *Klebsiella oxytoca* (*K. pneumonia*), one of a few bacterial species that utilizes all the major sugars (hexoses and pentoses), produces butanediol in high yield and high concentration (up to 10%, w/v) under optimum conditions, e. g., temperature, dissolved oxygen, and so forth *(3,4)*. The production of butanediol from lignocellulosic materials has been considered as an alternative approach in the conversion of biomass substrates to liquid fuels and chemical feedstocks *(5,6)*.

Cellulosic biomass is a complex mixture of carbohydrate polymers from plant cell wall known as cellulose and hemicellulose, plus lignin and a smaller amount of minor compounds known as extractives. A key problem in utilization of cellulosic materials for fuel and chemical production is the poor yields of glucose from cellulose by acids or enzymes. Many pretreatment techniques have been used to improve cellulose hydrolysis. These techniques can be characterized as either chemical or physical in nature. The power required for physical treatments, including grinding and milling, is too costly. Chemical treatments with strong acids or bases are also expensive and compounded by the necessity of chemical recovery. Other treatment methods include the ammonia explosion *(7)*, steam explosion *(8,9)*, and ammonia-recycled percolation *(10)* processes. The ammonia explosion process allows the explosion of cellulosic materials at a relatively low temperature to avoid sugar decomposition. However, it is difficult to recover all the feed ammonia for reuse. The steam explosion process requires considerable thermal energy. And in the ammonia-recycled percolation method, the pretreatment is conducted at relatively high temperature (150°C) and pressure (325 psi), causing considerable degradation of hemicellulosic carbohydrates. However, the ammonia recovery rate can be as high as 99% *(10)*.

One method of producing butanediol from biomass is the simultaneous saccharification and fermentation (SSF). This process is similar in principle and in practice to that used in producing ethanol from cellulosic biomass. In this process, a cellulose hydrolyzing enzyme (cellulase) is combined with a butanediol producing organism to carry out simultaneous hydrolysis of cellulose and hemicellulose to glucose, xylose, and a mixture of minor sugars such as arabinose, and the conversion of sugars to butanediol. As a result, hydrolysis rates and yields of product are improved when compared to processes involving separate hydrolysis and fermentation steps *(11,12)*. As in the case of ethanol production by the SSF process, the cost of cellulase enzyme accounts for a large portion of the overall cost of conversion of biomass to butanediol. Therefore, a reduction in the cost of the enzyme usage would make this process more economically attractive. One way of reducing of enzyme usage is to remove lignin prior to SSF since it is known that lignin will adsorb cellulase and deactivate cellulase activity *(13–16)*. As a physical barrier, lignin also causes a higher cellulase dosage in order to achieve a reasonable cellulose hydrolysis rate. Another way is to recover and reuse cellulase. A prior removal of lignin will also allow a more complete cellulase recovery *(15,16)*. Likewise, the prior removal of hemicellulose will also enhance the reactivity of cellulose fraction *(17)*.

An effective utilization of xylose, arabinose, and other minor sugars in addition to glucose is important in the process economics. *K. oxytoca* ATCC 8724 has been studied the Laboratory of Renewable Resources Engineering *(3,4)*. This bacterial strain is capable to produce butanediol from both hexoses and pentoses with good yield. In this study, we combined *K. oxytoca* with a relatively easy and effective pretreatment technique to demonstrate the viability of producing butanediol from cellulosic biomass using ground corn cob as a model authentic lignocellulosic material.

MATERIALS AND METHODS

Materials

Ground corn cob (8% moisture) with an average particle diameter of 3.2 mm was purchased from Andersons Inc., Maumee, OH and was used for the experiments unless otherwise indicated. The corn cob (dry basis) has the following composition: 44.88% cellulose, 32.68% xylan, 7.41% lignin, and 2.51% acetate *(18)*. The liquid cellulase preparation with a specific activity of 170 IFPU/mL was provided by Iogen (Ottawa, Canada). Aqueous ammonia (30%) and hydrochloric acid (37%) were purchased from Mallinckrodt Chemical (Paris, KY). 2,3-Butanediol and other carbohydrates were purchased from Sigma Chemical (St. Louis, MO).

Organism and Medium

Klebsiella oxytoca ATCC 8724 was purchased from American Type Culture Collection, Rockville, MD and was maintained on YMA (Difco) slants. The medium (YMP) for cell growth contained the following: yeast extract (Difco), 3 g; malt extract (Difco), 3 g; peptone (Difco), 5 g; glucose, 10 g; and distilled water, 1 L. Sterilization was accomplished by autoclaving at 15 lb/in² for 15 min. Bacterial cells were grown aerobically in 250-mL Erlenmeyer flasks containing 100 mL growth medium at 25°C on a rotary shaker at 150 rpm for 24 h.

Methods

Ammonia Steeping to Remove Lignin

Corn cob (20 g) was mixed with 100 mL aqueous ammonia (10% ammonia) in a 250-mL Erlenmeyer flask and was incubated in a shaker for 24 h at ambient temperature. The mixture was then filtered to separate corn cob from ammonia solution. The corn cob was then washed twice by deionized water to remove residual ammonia from the surface of the particles. The delignined corn cob preparation was then obtained by vacuum evaporation to remove residual ammonia. The lignin fraction was precipitated out of solution upon ammonia removal and collected by filtration.

Dilute Acid Hydrolysis to Remove Hemicellulose

Delignined corn cob that comprises mainly cellulose and hemicellulose fractions was then treated with a 1% hydrochloric acid solution at 100–108°C for 1 h. The acidic hemicellulose hydrolyzate obtained can be neutralized as pentose-rich stream. The remaining solid (cellulose fraction) was then washed with deionized water to remove residual acid.

Enzymatic Hydrolysis of Cellulose Fraction

To the cellulose fraction (9.23 g) obtained from 20 g of corn cob, 50 mL water and 1.0 mL cellulase enzyme (equivalent to 8.5 IFPU/g corn cob cellulose) were added in a 250-mL flask. The saccharification was carried out at 50°C for 48 h.

Simultaneous Saccharification and Fermentation (SSF) of Cellulose Fraction

Pretreated corn cob (8 g, dry equivalent) was added to a 250-mL Erlenmeyer flask with 50 mL YMP medium. It was autoclaved for 15 min at 121°C. The total volume of water in the flask was 80 mL, after taking into account the water content of the pretreated corn cob. The initial pH was adjusted to 5.5 by phosphate buffer. Following this, 20 mL of actively growing bacterial cells that were previously grown in the YMP medium and cellulase (8.5 IFPU/g corn cob cellulose) were added to initiate SSF of the cellulose at 35°C in a shaker at 250 rpm. Samples (0.2 mL) were taken at 12 h intervals over 96 h of incubation.

Analysis

Glucose, xylose, 2,3-butanediol, ethanol, acetate, glycerol and xylitol were analyzed by HPLC (column: 300 × 7.8 mm HPX-87H, Bio-Rad, Richmond, CA). Lignin was determined as Klason lignin by the weight method (19). Cellulose and hemicellulose (as xylan) were determined

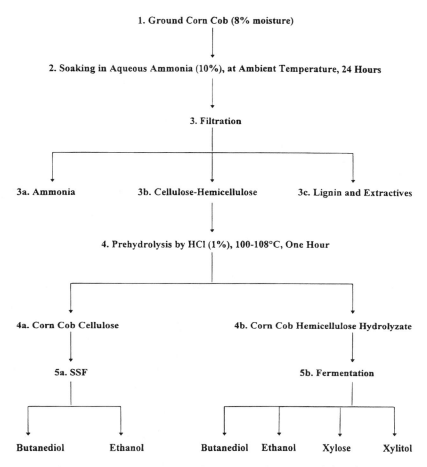

Fig. 1. A schematic representation of the procedures used for the preparation of substrate from ground corn cob.

according to the methods described (19). The residual ammonium ion in the solution was determined with a 05800-05 Solution Analyzer equipped with an ammonium electrode (Cole-Parmer Instrument, Niles, IL).

RESULTS AND DISCUSSION

The procedures for the substrate preparation for butanediol production by *K. oxytoca* are summarized in Fig. 1. In this pretreatment process, lignin and other alkaline extractives were solubilized and separated from corn cob after steeping in the dilute ammonia (10%) solution at ambient temperature. This was followed by hydrolysis using a dilute acid at 100–108°C to hydrolyze the hemicellulose. The changes in composition of corn cob after ammonia steeping are shown in Table 1. The original lignin content of corn cob was reduced by approx 90% from 0.074 to 0.0085 g/g after steeping. In addition, other alkaline extractives such as acetate and

Table 1
Composition of Corn Cob

Materials	Cellulose (%)	Xylan (%)	Lignin (%)	Acetate (%)	Other (%)
Original	44.88	32.68	7.41	2.51	12.53
After Ammonia Steeping	56.2	38.5	0.85	0.0	4.45
After Acid Hydrolysis	90.4	5.29	0.91	0.0	3.4

alkaline extractable materials were removed during steeping (Table 1). The content of hemicellulose, measured as xylan, remained more or less unchanged. Based on the composition of the treated corn cob dry weight, xylan content was increased from 32.68 to 38.5% after lignin removal. Likewise, cellulose content was increased from 44.88 to 56.2%. The retention of hemicellulose after ammonia steeping is in contrast to other treatment processes that resulted in the partial loss of the hemicellulose fraction. This is probably because of the mild condition (ambient temperature and atmospheric pressure) was employed during steeping.

The steeping process was followed by dilute acid hydrolysis. In this step, treated corn cob was incubated in the dilute acid solution at 100–108°C for 1 h to hydrolyze the hemicellulose. Hemicellulose solubilized by this treatment was about 87% since hemicellulose content dropped from 0.38 to 0.05 g/g dry corn cob (Table 1). A clean, light-amber colored hemicellulose hydrolyzate with carbohydrate concentration of 81.8 g/L was obtain with xylose comprising over 90% of the carbohydrates. This hemicellulose hydrolyzate had no acetate and alkali extractives present (Fig. 2). This is important for the utilization of hemicellulosic carbohydrates since acetate, even at a very low concentration of around 5 g/L, is known to inhibit ethanol *(20,21)* and butanediol fermentations *(22,23)*. The inhibition can be even more acute in the present of the fermentation product, ethanol, or butanediol (unpublished observation). This hemicellulose hydrolyzate has been used as a substrate for ethanol production by a xylose-fermenting yeast strain *(23)*. The same hydrolyzate has also been used as the substrate for the production of high-value sweetener, xylitol *(24)*, or as a raw material for xylose production.

The overall possible effect of sequential ammonia steeping and dilute acid treatment is to increase the exposure of cellulose to cellulase. The results from an increase in the surface area of cellulose available for the enzymatic hydrolysis because of the swelling of cellulose *(10,16)*, and also

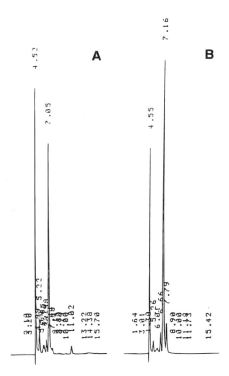

Fig. 2. LC profiles of dilute acid hydrolyzates before ammonia (A) and after ammonia steeping (B). Retention time (min): Acid (4.52); Xylose (7.05); Arabinose (7.79); Acetic Acid (11.02).

increase in the porosity, thereby enhanced the accessibility of cellulose by cellulase (25). Upon drying of the treated cellulose fraction, the beneficial effect of alkaline swelling of cellulose was reduced. This effect is shown in Fig. 3, dry cellulose no longer is as susceptible to cellulase as when it had not been dried. The material after lignin and hemicellulose removal has a cellulose content of 90.4% as based on dry weight (Table 2). This is an increase of cellulose content of over 100% from the original 0.45 g/g corn cob.

Since most of the lignin, acetate, alkali extractives, and hemicellulose have been removed by prior process steps, one would expect the remaining corn cob cellulose to be more reactive to cellulase hydrolysis. This is supported by the cellulase hydrolysis results. Figure 3 shows the effect of different stage of treatment on cellulase hydrolysis of corn cob when cellulase dosage of 8.5 IFPU per g corn cob cellulose was used. The results show that the combination of ammonia steeping and dilute acid hydrolysis gave the highest glucose yield of 91.8% based on dry cellulose within 48 h. The ammonia treated or dilute acid treated sample showed similar, but lower reactivity toward the cellulase. The sample without any treatment was least susceptible to cellulase hydrolysis.

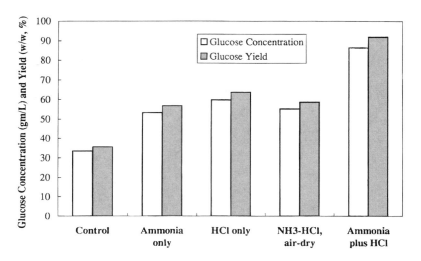

Fig. 3. Cellulase hydrolysis of corn cob after different treatment.

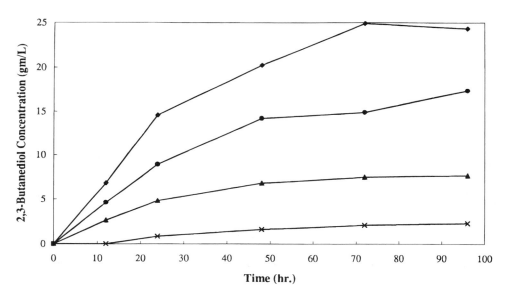

Fig. 4. Effect of different treatment on butanediol production from corn cob using SSF. ◆ Ammonia plus HCl, ● HCl only, ▲ Ammonia only, ✕ Control.

The SSF process is a viable option for conversion of biomass to ethanol (16) and butanediol (6). The SSF process improves the hydrolysis rates and product yields in comparison to processes involving separate hydrolysis and fermentation. To demonstrate the effectiveness of the current pretreatment process, SSF experiments were performed using bacterial culture *K. oxytoca* and corn cob samples after different steps of treatment. As shown in Fig. 4, the combination of ammonia steeping and

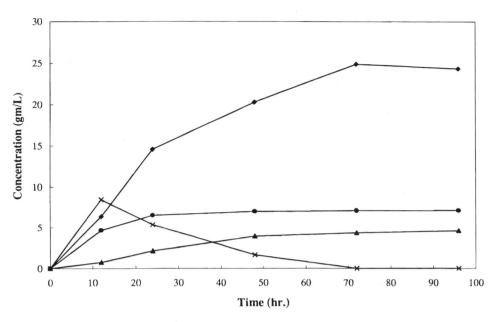

Fig. 5. Time course of SSF of cellulose fraction of corn cob. ♦ 2,3-Butanediol, ● Ethanol, ▲ Cellubiose, ✕ Glucose.

dilute acid hydrolysis gives the best results. The rate of butanediol production is at least threefold higher than those with ammonia, but without acid treated sample. Compared to the untreated sample, a tenfold increase in butanediol accumulation was obtained with ammonia-acid treated sample.

The time course of product formation during SSF with ammonia and acid treated sample is shown in Fig. 5. During 10 h of incubation, the glucose concentration increased to about 9 g/L and then declined as would be expected as a result of the presence of bacterial cells. The amount of cellobiose also increased slowly and reached the level of about 4 g/L at 48 h. Ethanol concentration increased to 7 g/L (accounting for about 17% substrate) by 24 h and stayed more or less unchanged throughout the remainder of the fermentation. The butanediol concentration increased almost linearly and reached a level of 25 g/L after 72 h. This is a yield of 62.5% based on total dry cellulose (80 g/L). The relatively low yield is due in part to coproduction of other products such as ethanol.

The stepwise removal of lignin and hemicellulose fractions from lignocellulose appears to be effective in the utilization of the cellulose fraction for butanediol production. The pretreatment can have the following beneficial effects: the solubilization of lignin and the separation of lignin from cellulose fraction, the chemical swelling of cellulose *(10,16)*, disruption of the crystalline structure of cellulose, and increase in the accessible surface area for cellulase *(25)*. Consequently, the cellulase dosage required for cellulose hydrolysis is 50% or less than those reported in the literature *(26)*.

Another advantage to be gained by the current procedure is the obtaining of a lignin- and acetate-free hemicellulose hydrolyzate. This fraction can be converted to products in the separate streams. This is significant in view of the fact that pentose utilization by micro-organisms is slower and its utilization by micro-organisms is often subjected to glucose inhibition even in the presence of low concentrations of glucose.

Unlike lignin recovered from acid hydrolysis or conventional pulping process, the lignin fraction removed by ammonia steeping remains chemically unchanged. This is because low temperatures and low pressures are used during ammonia steeping. The lignin residues after ammonia removal can be useful for synthesis of chemicals and polymers. Based on our estimation, at least 98% of ammonia is recoverable for reuse. The large scale ammonia recovery technology currently in practice in fertilizer and ammonia industries can be used for this purpose.

ACKNOWLEDGMENTS

This study was supported in part, through the Consortium for Plant Biotechnology Research, Inc. by DOE cooperative agreement no. DE-FCO5-92OR22072. This support does not constitute an endorsement by DOE or by The Consortium for Plant Biotechnology Research of the views expressed in this article.

REFERENCES

1. Chemical Marketing Reporter (July 24, 1995).
2. Garg, S. K. and Jain, A. (1995), *Biosource Technol.* **51,** 103–109.
3. Jansen, N. B. and Tsao, G. T. (1983), *Adv. Biochem. Eng. Biotechnol.* **27,** 85–100.
4. Jansen, N. B., Flickinger, M. C., and Tsao, G. T. (1984), *Biotechnol. Bioeng.* **26,** 362–369.
5. Laube, V. M., Groleau, D., and Martin, S. M. (1984), *Biotechnol. Lett.* **6,** 257–262.
6. Yu, E. K. C., Deschatelets, L., and Saddler, J. N. (1984), *Biotech. Bioeng. Symp.* **14,** 341–352.
7. Dale, B. E. and Moreira, M. J. (1982), *Biotech. Bioeng. Symp.* **12,** 31–43.
8. Grethlein, H. E. and Converse, A. O. (1991), *Bioresource Technol.* **36,** 77–82
9. McMillan, J. D. (1994), in *Enzymic Conversion of Biomass for Fuels Production.* Himmel, M. E., Baker, J. O., and Overend, R. P. eds., ACS, Washington, DC, pp. 292–324.
10. Yoon, H. H., Wu, Z. W., and Lee, Y. Y. (1995), *Appl. Biochem. Biotechnol.* **51/52,** 5–19.
11. Takagi, M., Abe, S., Suzuki, S., Emert, G. H., and Yata, N. (1977), in *Bioconversion Symposium Proceedings*, IIT, Delhi, pp. 551–571.
12. Wright, J. D., Wyman, C. E., and Grohmann, K. (1988), *Appl. Biochem. Biotechnol.* **18,** 75–90.
13. Sutcliffe, R. and Saddler, J. N. (1986), *Biotechnol. Bioeng. Symp.* **17,** 749–762.
14. Ooshima, H., Burns, D. S., and Converse, A. O. (1990), *Biotecnnol. Bioeng.* **36,** 446–452.
15. Elander, R. T. and Hsu, T. (1995), *Appl. Biochem. Biotechnol.* **51/52,** 463–478.
16. Philippidis, G. P. and Smith, T. K. (1995), *Appl. Biochem. Biotechnol.* **51/52,** 117–124.
17. Grohmann, K., Torget, R., and Himmel, M. (1985), *Biotech. Bioeng. Symp.* **15,** 59–80.
18. Foley, K. M. (1978), Chemical properties, physical properties and uses of the Andersons' corncob products. The Andersons, pp. 425.
19. Ghosh, S. (1989), *Biomass Handbook*, Gordon and Breach Science Publishers, New York, NY, pp. 395,

20. Lawford, H. G. and Rousseau, J. D. (1992), *Appl. Biochem. Biotechnol.* **34/35,** 185–204.
21. Lawford, H. G. and Rousseau, J. D. (1993), *Appl. Biochem. Biotechnol.* **39/40,** 301–322.
22. Frazer, F. R. and McCaskey, T. A. (1991), *Enzyme Micro. Technol.* **13,** 110–115.
23. Cao, N. J., Krishnan, M. S., Du, J. X., Gong, C. S., Ho, N. W. Y., Chen Z. D., and Tsao, G. T. (1996), *Biotechnol Lett.* **18,** 1013–1018.
24. Chen, L. F. and Gong, C. S. (1985), *J. Food Sci.* **50,** 226–228.
25. Burn, D. S., Ooshima, H., and Converse, A. O. (1989), *Appl. Biochem. Biotech.* **20/21,** 79–94.
26. Saddler, J. N., Mes-Hartree, M., Yu, E. K. C., and Browell, H. H. (1983), *Biotechnol. Bioeng. Symp.* **13,** 225–238.

Oxygen Sensitivity of Algal H_2-Production

MARIA L. GHIRARDI,*,[1] ROBERT K. TOGASAKI,[2] AND MICHAEL SEIBERT[1]

[1]*National Renewable Energy Laboratory, Golden, CO 80401; and*
[2]*Department of Biology, Indiana University, Bloomington, IN 47405*

ABSTRACT

Photoproduction of H_2 by green algae utilizes electrons originating from the photosynthetic oxidation of water and does not require metabolic intermediates. However, algal hydrogenases are extremely sensitive to O_2, which limits their usefulness in future commercial H_2-production systems. We designed an experimental technique for the selection of O_2-tolerant, H_2-producing variants of *Chlamydomonas reinhardtii* based on the ability of wild-type cells to survive a short (20 min) exposure to metronidazole in the presence of controlled concentrations of O_2. The number of survivors depends on the metronidazole concentration, light intensity, preinduction of the hydrogenase, and the presence or absence of O_2. Finally, we demonstrate that some of the selected survivors in fact exhibit H_2-production capacity that is less sensitive to O_2 than the original wild-type population.

Index Entries: Hydrogen; green algae; *Chlamydomonas*; oxygen; metronidazole.

Abbreviations: ATP, adenosine triphosphate; cw15, cell wall-less strain of *Chlamydomonas reinhardtii*; DCMU, 3-(3,4-dichlorophenyl)-1,1-dimethylurea; Fd, ferredoxin; I_{50}, inhibitor concentration that decreases the rate of an enzymatic reaction to 50% of the rate measured in the absence of the inhibitor; MNZ, metronidazole, [1-(2-hydroxyethyl)-2-methyl-5-nitroimidazole]; NADP, nicotinamide adenine dinucleotide phosphate; NADPH, reduced form of NADP; V_0, initial rate of an enzymatic reaction; WT, wild-type.

---

*Author to whom all correspondence and reprint requests should be addressed.

INTRODUCTION

There are three major classes of organisms that photoproduce H_2: photosynthetic bacteria, cyanobacteria, and green algae. Photosynthetic bacteria utilize reductants other than water to produce H_2, in a reaction catalyzed by the enzyme nitrogenase. Whereas this reaction is energy intensive, requiring at least 4 ATP/H_2 produced, it is also quite functional when waste reductant is available (1). Cyanobacteria also produce H_2 via a nitrogenase enzyme system. In this case the source of reductant is H_2O, but the reaction involves the formation of a metabolic intermediate. As a result, the quantum yield of H_2 production is rather low, in the order of one H_2/9–10 quanta (2). Green algae do not synthesize nitrogenase; rather, H_2 production in these organisms is catalyzed by the hydrogenase enzyme. Algal H_2 production does not require ATP input nor the generation of metabolic intermediates. Consequently, a higher theoretical quantum yield of one H_2/4 quanta is possible (2). Recent work with mutants of *Chlamydomonas reinhardtii* lacking Photosystem I suggests that an even higher quantum yield of one H_2/2 quanta may be attainable with green algae (3).

The desired characteristics of a photobiological H_2-production system are:

1. Use of water as the source of reductant,
2. Solar-driven,
3. High efficiency of solar energy conversion,
4. Durable and self-replicating,
5. H_2 production at high equilibrium pressure, and
6. Cost-competitiveness (4).

By these criteria, the use of green algae appears to be a promising alternative. On the other hand, the current practical limitations of green algae in a photobiological H_2-evolving system include:

1. The sensitivity of its hydrogenase to O_2,
2. The occurrence of a dark back reaction between O_2 and H_2 (i.e., the oxyhydrogen reaction),
3. Competition between the CO_2-fixation and the H_2-production pathways for electrons from H_2O,
4. The low equilibrium pressure of H_2 release, and
5. Saturation of H_2 production capacity at low light intensity.

A biological approach to address the hydrogenase O_2-sensitivity issue in green algae was developed about 20 yr ago by McBride et al. (5). The approach used a positive selection technique based on the reversible H_2 uptake reaction catalyzed by algal hydrogenases. Mutants were selected under this photoreductive pressure and increasing concentrations of O_2. These conditions require algal cells to utilize H_2 as a source of reducing equivalents to fix CO_2 in the presence of DCMU, an inhibitor of electron flow from Photosystem II and hence, from water. Consequently, cells with hydrogenase sensitive to low oxygen concentration starve to death, whereas cells with an O_2-tolerant hydrogenase grow. Oxygen-tolerant mutants of *C. reinhardtii* obtained by this technique could produce H_2 in the presence of up to 8% O_2. Unfortunately, the mutants exhibited high levels of the oxyhydrogen back reaction (5) and were not maintained in culture for further study.

It is clear that novel approaches will be required to obtain algal hydrogenases that are stable in the presence of O_2. We have addressed this problem by developing and examining a new technique based on the application of selective pressure under H_2-producing, rather than H_2-utilizing conditions. One of the potential advantages of this approach may be in elimination of the high rates of oxyhydrogen reaction observed in the old experiments. In this paper the authors describe the new selection procedure and report on the results of preliminary experiments done to test its validity in selecting for O_2-tolerant, H_2-producing organisms. Results suggest that application of this type of selective pressure, combined with mutagenesis, has the potential for yielding H_2-producing algal mutants with increased tolerance to O_2. These organisms may prove useful in future commercial photobioreactors for the continuous production of H_2 under aerobic conditions.

MATERIALS AND METHODS

Cell Growth

Wild-type (WT) *C. reinhardtii* (137mt$^+$) was obtained from Prof. S. Dutcher at the University of Colorado, Boulder. A cell wall-less strain, cw15 (CC-400 mt$^+$), was acquired from Dr. E. Harris at the Chlamydomonas Genetics Center, Duke University. Wild-type cells were grown photoautotrophically in Sager's minimal medium (6). The cw 15 strain required Sueoka's high salt medium, modified according to Vladimirova and Markelova (7). Both cultures were grown in a chamber at 25°C under 8 W/m^2 fluorescent illumination and continuous bubbling with a mixture of 1.7% CO_2 and air. Cells were harvested by centrifugation at 1000g for 10 min. Wild-type and cw15 cells were also grown on plates containing 1.5 and 0.8% agar, respectively.

H_2-Production Selection Technique

Harvested cells were resuspended in a small volume of resuspension buffer (50 mM potassium phosphate buffer, pH 7.2, containing 3 mM $MgCl_2$) (8) at a final concentration of about 200 µg Chl · mL^{-1} (9). Chlorophyll concentrations were determined by extracting the pigment using 95% ethanol and assaying spectrophotometrically (6). In order to induce the hydrogenase enzyme, the cell suspension was made anaerobic by bubbling argon for 30 min. Maintenance of anaerobic condition was insured by the addition of a glucose/catalase/glucose oxidase oxygen-scrubbing system *(10)*. This two-enzyme system reduces any available O_2 to water. *C. reinhardtii* cells cannot use glucose as a carbon source for growth *(6)*. The mixture was incubated at room temperature for 4 h in the dark *(11)* and transferred to 4°C for overnight storage.

For H_2-production selection, anaerobically-treated cells were added to a selective medium containing different concentrations of metronidazole and 1 mM sodium azide *(12)*. The azide inhibits endogenous catalase activity. All procedures were done under sterile conditions. The selection medium was also made anaerobic by argon bubbling before introduction of the cells. Oxygen was then added to the medium by syringe through a gas-tight septum to achieve final concentrations of O_2 in the gas phase (ranging 0–10%), as required. The final cell suspension was exposed to light of controlled intensity (Fiber-Lite High Intensity Illuminator, model 170-D, Dolan-Jenner Industries) for 20 min. The cells were pelleted out using an IEC clinical centrifuge, washed once with phosphate buffer, and then once with the resuspension buffer (5 mM potassium phosphate buffer containing 1 mM $CaCl_2$ and 1 mM Mg_2SO_4). Undiluted and sequential dilutions of each sample were plated on minimal medium and incubated in a growth chamber under low light levels. Survival rates were determined by counting the number of colonies detected on each plate following the treatment, and estimating the percentage of survivors with respect to the number of cells used in the MNZ treatment.

H_2-Evolution Measurements

Anaerobically-treated cells were added to a small volume of the assay buffer (50 mM MOPS, pH 6.8) *(9)* to a final concentration of about 15 µg Chl · mL^{-1}. The medium, in the assay chamber of a two-electrode apparatus (Clark-type, YSI 5331), was adjusted to different initial concentrations of O_2 before introduction of the cells. H_2 evolution was induced by illumination with saturating light from the same lamp described above except that a heat filter consisting of a 1% solution of $CuSO_4$ was used. The data were recorded on a strip-chart recorder and initial rates were calculated from the initial slopes of each curve. Initial O_2 concentrations were also determined from the recorded O_2 concentration measured at the time when the cells were introduced into the chamber. Gas concentrations were corrected for their decreased solubility in aqueous solution at Golden, CO, located 1580 m above sea level.

RESULTS AND DISCUSSION

The hydrogen-production selective pressure that we have employed in this work is based on the toxic effect of metronidazole (MNZ) on photosynthetic organisms. Metronidazole is a heterocyclic compound with a low redox potential (E_m = –325 mV at pH 6.9) *(13)* that is normally used to treat infections caused by protozoa and anaerobic bacteria *(14)*. In these organisms, MNZ is a strong oxidizer of ferredoxin, in a reaction catalyzed by the reversible hydrogenase *(14)*. Its toxic effect is a result of one of its reduced intermediate states. The site of action of MNZ in photosynthetic organisms is restricted to chloroplasts, where it oxidizes light-reduced ferredoxin *(13)*. Schmidt et al. *(13)* have proposed that the subsequent reoxidation of reduced MNZ by molecular O_2 yields a superoxide radical *(15)*, which then disproportionates into H_2O_2 *(16)*. Consequently, the toxicity of MNZ was attributed to the generation of H_2O_2. Long exposure (up to 24 h) to MNZ has been used to select for *Chlamydomonas* mutants defective in photosynthetic electron transport function *(13)*.

Ferredoxin (Fd) is a key electron carrier, located on the reducing side of photosystem I, and it provides the reducing equivalents to a variety of pathways. The most prevalent pathway, of course, is through NADP, which in turn is coupled to CO_2 fixation. Among the other Fd-dependent pathways is hydrogenase-catalyzed H_2 production. This pathway, of course, is inoperative during phototrophic growth because of inactivation of the enzyme by O_2 produced during the water-splitting process. It is reasoned that, since MNZ toxicity depends on the accumulation of reduced Fd, an organism with an active hydrogenase will be less sensitive to MNZ toxicity in the absence of CO_2, because an alternative pathway for electrons from Fd is available.

Based on the aforementioned information, a treatment to select for potential O_2-tolerant, H_2-producing organisms was designed. Cells were first incubated in the presence of MNZ at different concentrations of O_2 for a short period of time in the light and then grown on minimal agar medium to determine survival rates. The authors examined the effect of the following treatment parameters on the rate of survival: MNZ concentration, light intensity (which limits the rate of accumulation of reduced Fd), preinduction of the hydrogenase enzyme, and O_2 concentration in the selection medium. The authors tested both WT and a cell wall-less strain (cw15) of *C. reinhardtii*, taking into consideration that future work may require the use of genetic transformation, if they succeed in isolating a mutant containing an O_2-tolerant hydrogenase. The choice of the cw15 strain was based on the observation that the absence of the cell wall increases the efficiency of genetic transformation *(17)*.

Figure 1 shows the effect of increasing concentrations of MNZ and light intensity on the number of cw15 *Chlamydomonas* cells surviving the selective treatment. The light intensities examined were 17.5, 50 (Fig. 1A, first experiment), 200, and 400 (Fig. 1B, second experiment) W/m^2. The selective pressure was applied for 20 min to anaerobically-treated cells (containing an initially active hydrogenase), and O_2 was added at 2.8%. Oxygen inhibits most of the hydrogenase activity in both WT and cw15 cells at this concentration. The maximum concentration of MNZ used in the treatment (58 mM) was determined by the solubility of MNZ in aqueous

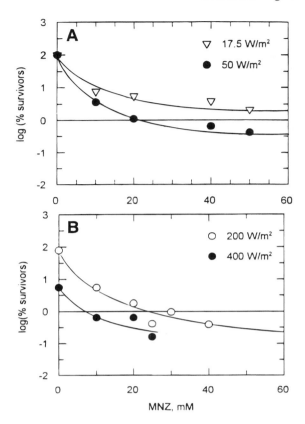

Fig. 1. Effect of MNZ concentration and light intensity on the number of survivors from two selective experiments. **(A)** Selection was done using cw15 cells in the presence of varying concentrations of MNZ and 2.8% O_2, at either 17.5 or 50 W/m². **(B)** Selection was done as above, under either 200 or 400 W/m². Symbols represent individual data points.

medium. It is clear from Fig. 1 that MNZ toxicity is a function of the light intensity, consistent with the theory that the extent of killing depends on the rate of electron transport. It is interesting that the curves in Fig. 1 do not fit an exponential function, which suggests that factors other than MNZ concentration are limiting the rate of killing (see the hydrogenase induction experiment as shown in Fig. 2). Figure 1 also shows the degree of variability of the results obtained from different experiments. The data obtained by exposure of cells to 50 W/m² in the first experiment (Fig. 1A) are very similar to the data obtained with 200 W/m² in the second experiment (Fig. 1B). It is possible that the phase of growth at which cells were harvested for each experiment is responsible for the variability. Treatment done in the presence of light intensities above 200 W/m² (Fig. 1B) caused an increase in temperature that may have affected the number of survivors detected even in control samples not exposed to MNZ.

Oxygen Sensitivity   147

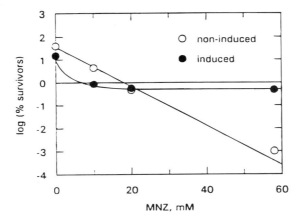

Fig. 2. Effect of anaerobic pre-induction of hydrogenase on the number of cw15 survivors. Cells submitted to the MNZ selective treatment were either preinduced by anaerobic incubation or not induced, as described in the Materials and Methods section. The treatment was done in the presence of 2.8% O_2 and 200 W/m².

The third parameter that we checked was the effect of hydrogenase induction on the number of survivors. The experiment was done with preinduced and uninduced cw15 *Chlamydomonas* cells subsequently treated with different concentrations of MNZ in the presence of 2.8% O_2 and illuminated for 20 min at 200 W/m². Figure 2 shows that MNZ killing in the absence of a preinduced hydrogenase is exponential. However, pre-induction of the hydrogenase causes a loss of exponential killing, consistent with the results in Fig. 1. This demonstrates that the effect of MNZ is limited in cells that have an active hydrogenase, confirming the hypothesis that hydrogenase-catalyzed H_2-production provides enough of an alternative sink for electrons from reduced Fd to mitigate the effects of MNZ.

Thus far the authors have found that their proposed selection technique depends on the concentration of MNZ, on the amount of accumulated reduced Fd (and thus on the electron transport rate), and on the presence of an active hydrogenase. They next examined the effect of O_2 concentration on the rate of survival. Oxygen is expected to inactivate the hydrogenase in a concentration-dependent manner, thus increasing the toxicity of MNZ. Figure 3 shows the result of adding 10% O_2 to the gas phase of the selection medium on the degree of survival of cw15 *Chlamydomonas* cells treated with MNZ and exposed to 200 W/m² light for 20 min. Whereas the increase in O_2 concentration in the gas phase appears to increase the sensitivity of MNZ-dependent killing, the magnitude of increase is not as pronounced as might be expected. One possibility is that internally generated photosynthetic O_2 production is affecting the MNZ treatment in part, and the impact of externally-set O_2 is superimposed on it.

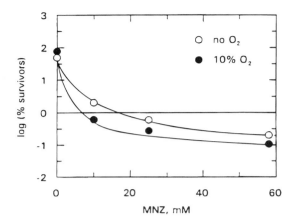

Fig. 3. Effect of O_2 concentration in the gas phase of the selective treatment vial on the number of survivors. MNZ incubation was done using cw15 cells, either under completely anaerobic conditions, or in the presence of 10% added O_2 in the gas phase of the selection medium. The light intensity was 200 W/m².

Finally, confirmation of the validity of the authors' proposed technique to select for O_2-tolerant, H_2-producing organisms comes from measurement of the O_2-sensitivity of H_2-production by some of the surviving isolates from two different selection experiments, compared to their parental strains. In order to establish baseline values, they initially determined the O_2 sensitivity of H_2 production by WT and cw15 cells. Initial rates of H_2 evolution were measured, as described in the Materials and Methods Section, as a function of the initial concentrations of O_2 in the assay medium. Given the variability of measuring rates of H_2 evolution from day to day (probably because of the condition of the cells at the time of harvest), the data from each experiment (from 5 to 10 points) were fitted to a single exponential decay equation from which the authors estimated V_0, the rate of H_2 evolution in the absence of O_2, and I_{50}, the initial O_2 concentration that causes a 50% decrease in the initial rate of H_2 evolution compared to V_0. The V_0s estimated from each experiment were then used to normalize the data obtained from all the experiments. The data in Fig. 4 show that as the initial concentration of O_2 increases, the rate of H_2 evolution by cw15 (Fig. 4A) or WT cells (Fig. 4B) decreases exponentially. The normalized V_0 values for the cw15 and WT strains were, respectively, 22 and 27 µmol $H_2 \cdot$ mg Chl$^{-1} \cdot$ h^{-1} and the estimated I_{50}s were 0.300% ± 0.028% and 0.394% ± 0.068% O_2.

Because of the variability in rates measured on different days, the authors always determined the O_2 sensitivity of H_2 production for survivors from different MNZ experiments and their respective parental strains on the same day. The results from two selection experiments are reported on Table 1. Experiment 1 was done with preinduced cw15 and WT cells treated with 58 mM MNZ in the presence of 2.8% O_2. There were

Oxygen Sensitivity

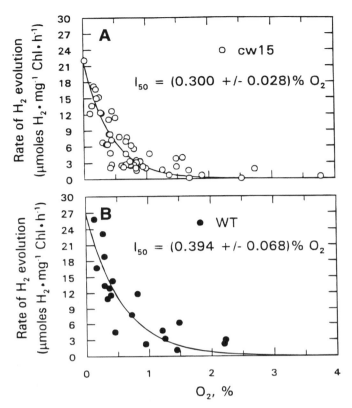

Fig. 4. Normalized rates of H_2 evolution measured with preinduced (**A**) cw 15 or (**B**) WT cells injected to an assay medium preset to the indicated initial concentrations of O_2. H_2 concentration in the assay medium was measured immediately after cell injection and the procedure included a 2-min interval prior to illumination to allow the cells to acclimate. No H_2 production was observed during this 2-min time. See the Materials and Methods section for a description of the experimental conditions and data analysis.

6 cw15 survivors, of which two were tested. Both showed increases in the I_{50} for O_2 compared to their parental strains. One particular survivor, identified as D5, had an estimated I_{50} 2.3 times higher than its parental strain. The other survivor, despite a higher I_{50}, had a decreased V_0 rate of H_2 evolution, perhaps indicating lower rates of electron transport. MNZ treatment of WT cells in Experiment 1 yielded five survivors. Two of them showed increases in I_{50} in the order of 50% compared to their parental strain. The rates of H_2 evolution by one of the survivors was very low, but the other had rates comparable to those of its parental strain. Experiment 2 was done only with WT cells following preinduction of the hydrogenase and MNZ treatment in the presence of 5% O_2. All of the four tested survivors out of a total of six had low rates of H_2 evolution. Two of them had I_{50}s in the order of their parental strain, whereas the other two exhibited small increases in their I_{50}s.

Table 1
H$_2$-Production Results from Two Selection Experiments done
with cw15 and WT *Chlamydomonas reinhardtii* cells[a]

Experiment	Organism	I$_{50}$ for O$_2$ % of control	V$_o$ % of control
Experiment 1 (2.8% O$_2$)	cw 15 (D5)	233	100
	cw15	153	20
	WT	152	100
	WT	163	12
Experiment 2 (5% O$_2$)	WT	100	23
	WT	138	44
	WT	168	21
	WT	100	5

[a]The cells were preinduced and subsequently treated with 58 mM MNZ for 20 min under 200 W/m^2 illumination in the presence of either 2.8% O$_2$ or 5% O$_2$. Survivors from the two experiments were plated, and selected clones were transferred to liquid medium and grown photoautotrophically for 3 d. Preinduced cells were used in measurements of H$_2$-evolution rates in the presence of different initial O$_2$ concentrations as in Fig. 4. Estimated V$_0$ and I$_{50}$ values for H$_2$ production obtained for each survivor were compared to values determined with control parental WT and cw15 cells on the same day.

The results using the selection technique described in this paper indicate that it is possible to select for *Chlamydomonas* isolates that exhibit H$_2$-production capability that is less sensitive to O$_2$ than the WT population. However, MNZ treatment can also select for mutants impaired in electron transport, as observed previously (5,13). Although H$_2$-production selective pressure is a valid option for reaching the long-term goal, the authors conclude that they need to improve the screening procedure to deselect for electron transport mutants. Their assay for O$_2$ sensitivity, based on measuring the I$_{50}$ of H$_2$ production in the presence of O$_2$ is time-consuming and subject to variability. Once this is done, H$_2$-production selection in combination with mutagenesis could lead to the identification of mutants that are tolerant to even higher concentrations of O$_2$. The ultimate goal is to develop a mutant that produces H$_2$ at ~1 atmosphere, with rates equivalent to the conversion of 10% of the incident light energy, and under atmospheric aerobic conditions. A mutant with these properties would demonstrate the commercial feasibility of this approach.

ACKNOWLEDGMENTS

This work was funded by the US Department of Energy Hydrogen Program. We thank Steve Toon for technical assistance.

REFERENCES

1. Weaver, P. F., Lien, S., and Seibert, M. (1979), in *Proceedings of the Fifth Joint US/USSR Conference of the Microbial Enzyme Reactions Project*, Jurmala, Latvia, USSR, pp. 461–479.
2. Benneman, J. R. (1994), in *Proceedings of the 10th World Hydrogen Energy Conference*, Cocoa Beach, FL.
3. Greenbaum, E., Lee, J. W., Tevault, C. V., Blankinship, S. L., and Mets, L. J. (1995), *Nature* **376**, 438–441.
4. Weaver, P. F., Lien, S., and Seibert, M. (1980), *Solar Energy* **24**, 3–45.
5. McBride, A. C., Lien, S., Togasaki, R. K., and San Pietro, A. (1977), in *Biological Solar Energy Conversion*, Mitsui, A., Miyachi, S., San Pietro, A., and Tamura, S. eds. Academic Press, New York, NY.
6. Harris, E. H. (1989), *The Chlamydomonas Sourcebook*, Academic Press, New York.
7. Vladimirova, M. G. and Markelova, A. G. (1980), *Sov. Plant Physiol.* **27**, 878–889.
8. Happe, T., Mosler, B., and Naber, J. D. (1994), *Eur. J. Biochem.* **222**, 769–774.
9. Roessler, P. and Lien, S. (1982), *Arch. Biochem. Biophys.* **213**, 37–44.
10. McTavish, H., Picorel, R., and Seibert, M. (1989), *Plant Physiol.* **89**, 452–456.
11. Roessler, P. G. and Lien, S. (1984), *Plant Physiol.* **76**, 1086–1089.
12. Asada, K. (1984), *Methods in Enzymology* vol. **105**, Academic Press, New York, NY, pp. 422–429.
13. Schmidt, G. W., Matlin, K. S., and Chua, N.-H. (1977), *Proc. Natl. Acad. Sci. USA* **74**, 610–614.
14. Church, D. L. and Laishley, E. J. (1995), *Anaerobe* **1**, 81–92.
15. Wardman, P. and Clarke, E. D. (1976), *Biochem. Biophys. Res. Comm.* **69**, 942–949.
16. Fridovich, I. (1974), in *Advances in Enzymology*, vol. **41**, Meister, A., ed., Wiley Interscience, New York, NY, pp. 35–97.
17. Kindle, K. L. (1990), *Proc Natl Acad Sci USA* **87**, 1228–1232.

Expression of *Ascaris suum* Malic Enzyme in a Mutant *Escherichia coli* Allows Production of Succinic Acid from Glucose

LUCY STOLS,[1] GOPAL KULKARNI,[2] BEN G. HARRIS,[2] AND MARK I. DONNELLY*,[1]

[1]*Environmental Research Division, Argonne National Laboratory, Argonne, IL 60439; and* [2]*Department of Biochemistry and Molecular Biology, University of North Texas, Fort Worth, TX 76107*

ABSTRACT

The malic enzyme gene of *Ascaris suum* was cloned into the vector pTRC99a in two forms encoding alternative amino-termini. The resulting plasmids, pMEA1 and pMEA2, were introduced into *Escherichia coli* NZN111, a strain that is unable to grow fermentatively because of inactivation of the genes encoding pyruvate dissimilation. Induction of pMEA1, which encodes the native animoterminus, gave better overexpression of malic enzyme, approx 12-fold compared to uninduced cells. Under the appropriate culture conditions, expression of malic enzyme allowed the fermentative dissimilation of glucose by NZN111. The major fermentation product formed in induced cultures was succinic acid.

Index Entries: Metabolic engineering; succinic acid; *Escherichia coli*; malic enzyme; *Ascaris suum*.

Succinic acid and other dicarboxylic acids can be produced as end products in microbial fermentations of renewable carbohydrate feedstocks and are potential intermediates in the synthesis of commodity chemicals *(1)*. Chemical conversion of succinic acid can yield established commodity chemicals, such as 1,4-butanediol and tetrahydrofuran *(1)*, or new products such as biodegradable solvents or polymers.

Escherichia coli normally produces a mixture of fermentation products of which succinic acid is a minor component *(2,3)*. Previously, we reported that overexpression of phosphoenolpyruvate carboxylase in *E. coli* results in increased succinic acid formation *(4)*. This manipulation involves

*Author to whom all correspondence and reprint requests should be addressed.

increasing the flux through the normal pathway to succinic acid. Other biochemically feasible routes to succinic acid do not occur naturally because of the regulation of crucial enzymes, either at the genetic or enzymatic level. Here we report that genetic manipulation of one such enzyme, malic enzyme, can at least in part overcome these regulatory limitations.

Physiologically malic enzyme catalyzes the conversion of malate and NAD^+ to pyruvate, NADH, and CO_2. However, their reverse reaction is favored thermodynamically (5). Malic enzyme is regulated genetically—its synthesis is induced in the presence of malate (6,7)—and at the the level of catalysis. Allosteric activation and inhibition have been reported for malic enzyme from various sources (8–10) and kinetic parameters favor the physiological reaction. The K_ms for malic enzyme from *Ascaris suum*, the only example for which K_ms were determined for the reaction in both directions, were 45 mM for pyruvate compared to 0.4 mM for malate (11). Under normal conditions, pyruvate would not accumulate to a sufficiently high concentration to allow malic enzyme to be effective in producing dicarboxylic acids.

Mutants of *E. coli* lacking pyruvate:formate lyase (*pfl*) and lactate dehydrogenase (*ldh*) are blocked in the metabolism of pyruvate and fail to grow fermentatively (12) (Fig. 1). Such strains accumulate pyruvate, excreting it into the medium to millimolar concentrations (13). The authors have previously shown that an enzyme with a poor K_m for pyruvate, a genetically engineered malate dehydrogenase with lactate dehydrogenase activity (13), can allow such a mutant to grow fermentatively, in that case by lactic acid fermentation. Here the authors evaluate the potential of malic enzyme expressed in the same strain to allow the production of succinic acid as the major fermentation product (Fig. 1).

The gene encoding the NAD^+-dependent malic enzyme of *A. suum* was cloned earlier under control of the *lac* promoter (14). Because this promoter is repressed by glucose, we recloned the gene into the vector pTRC99a (Pharmacia Biotech) that employs the *trc* promoter and permits induction by isopropyl-β-D-thiogalactopyranoside (IPTG) in the presence of glucose. The malic enzyme gene was amplified by the polymerase chain reaction (PCR) using plasmid pME-2 (14) as template and primers based on the published sequence of the *A. suum* malic enzyme gene (14). A truncated *N*-terminus was designed to match the *N*-terminus predicted for the *E. coli* enzyme based on its DNA sequence (15). The primers were:

C-terminus; ATTTAGGTACCTTAACCATCCATGCTGTCAT
N-terminus; TTTCCTCCATGGTTAAAAGTGTCGCTCATCAT
truncated N-terminus; TTAAATCCATGGACGAAAAAGAGATG

Primers were combined at 1 μM with approx 200 ng of pME-2 vector in a standard PCR using native Taq polymerase (Perkin-Elmer). Both sets of primers (generating products with alternate *N*-termini) gave products of 1.7 kb, the size expected for the malic enzyme gene (14). The fragment was verified to be the malic enzyme gene by digestion with restriction enzymes.

Malic Enzyme 155

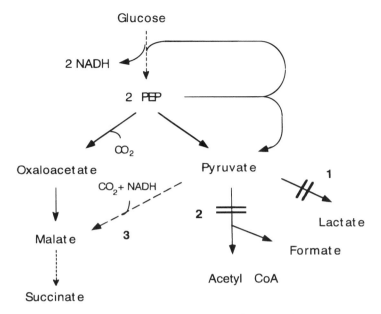

Fig. 1. Diversion of metabolites to succinic acid in E. coli NZN111. In the strain NZN111 fermentation is blocked by inactivation of the genes for lactate dehydrogenase (*ldh*, enzyme 1) and pyruvate:formate lyase (*pfl*, enzyme 2). In principle, introduction of malic enzyme (enzyme 3) can allow conversion of accumulated pyruvate to dicarboxylic acids.

The amplified gene was cloned into the NcoI and KpnI sites of pTRC99a by standard techniques (16). PCR products were digested with NcoI and KpnI (Promega), gel-isolated, purified with the Qiaex Gel Extraction Kit (Qiagen), and ligated into pTRC99a that had been cleaved with NcoI and KpnI and dephosphorylated with calf intestinal phosphatase (Promega).

The *E. coli* strain NZN111, a generous gift from Dr. David Clark, contains insertionally inactivated *pfl* and *ldh* genes and is incapable of fermentative growth. An aerobically grown culture was made competent and transformed with the above ligation mixtures by standard methods, and the resulting colonies were screened for the malic enzyme gene by restriction fragment analysis. Four colonies containing a plasmid encoding the mature malic enzyme (called pMEA1) and nine containing the truncated form (called pMEA2) were obtained. Representative colonies were evaluated for expression of malic enzyme induced by IPTG. Cells were lysed enzymatically and assayed using Mn^{2+} as the activating metal ion (11).

When induced at 37°C, the authors observed no evidence for malic enzyme overexpression. Denaturing polyacrylamide gel analysis of extracts confirmed that malic enzyme was not overexpressed. When grown at 30°C, on the other hand, moderate induction of malic enzyme activity occurred. Induction of a representative pMEA1- and pMEA2-containing strain with 1 m*M* IPTG resulted in maximal specific activities of 2.7 and 1.1

Table 1
Induction of *A. suum* Malic Enzyme in Transformants[a]

Strain	IPTG	Specific Activity (μmol/min/mg)	SD[b]	Induction (fold)
NZN111 (pMEA1)	–	0.44	0.13	
	+	5.26	1.49	11.9
NZN111 (pMEA2)	–	0.50	0.17	
	+	1.44	0.42	2.9

[a]Cultures were grown aerobically at 30°C in 25 mL LB medium containing 100 μg/mL ampicillin. At OD_{600} of 0.5, 1 mM IPTG was added. 2 mL aliquots were sampled and assayed for malic enzyme activity.
[b]Standard deviation.

μmol/min/mg, respectively, after 8 h compared to approx 0.3 μmol/min/mg for uninduced controls. Cells observed under the microscope were elongated and contained inclusion bodies. Denaturing gel electrophoresis of these samples revealed increased abundance of a protein of molecular weight 68,000, consistent with that expected for the *A. suum* malic enzyme (data not shown). Titration of the IPTG concentration indicated no difference in malic enzyme overexpression from pMEA1 between 0.25 and 4 mM IPTG. Representative transformants containing pMEA1 or pMEA2 were compared for overexpression induced by 1 mM IPTG in 6 h. Plasmid pMEA1 supported approx 12-fold overexpression of malic enzyme whereas pMEA2, which encodes a truncated form of the enzyme, consistently gave lower overexpression, approx threefold (Table 1).

The effect of overexpression of malic enzyme on fermentative growth was investigated using a transformant containing plasmid pMEA1. As a control we used NZN111 containing the parent vector, pTRC99a. Duplicate cultures of each strain were grown aerobically at 30°C in LB medium containing ampicillin and 10 g/L glucose. When the OD_{600} reached 0.5, one culture of each pair was induced with 1 mM IPTG. After 4 h (OD_{600} of 2–4), cells were centrifuged under sterile conditions, washed once with fresh medium and resuspended in sufficient medium to give an OD_{600} of 2.0 for each culture. One mL of these suspensions was immediately injected into sealed, stoppered serum tubes containing 10 mL of experimental medium—LB with 18 g/L glucose, 100 μg/mL ampicillin, 1 mM IPTG (for the induced cultures only), and 0.5 g of $MgCO_3$ to maintain pH—under an atmosphere of air:CO_2 (1:1) at 14 psi. The gas composition and pressure were established prior to inoculation by use of a gassing manifold *(17)*. Cultures were incubated on their side at 30°C and agitated at 100 rpm.

The metabolism of the cultures was analyzed by high-pressure liquid chromatography. Aliquots of 1 mL were removed with a syringe, centrifuged, and fractionated on a Bio-Rad Aminex HPX-87H column (7.8 × 300 mm) using a Shimadzu LC-10A chromatographic system with UV

Table 2
Effect of Expression of *A. suum* Malic Enzyme on Product Distribution[a]

Strain	IPTG	Amt (g/liter) of product					
		Glucose	Succinate	Pyruvate	Lactate	Acetate	Ethanol
NZN111 (pTRC99a)	−	2.58	2.63	4.02	0.34	0.72	2.18
	+	3.31	2.45	3.85	0.29	0.73	2.00
NZN111 (pMEA1)	−	5.50	2.06	2.50	0	0.61	1.11
	+	0	7.07	2.83	0	0.00	1.31

[a]Cultures were grown at 30°C in sealed serum tubes containing 10 mL of LB medium containing 18 g/L glucose, 100 µg/mL ampicillin, and, where indicated, 1 m*M* IPTG. The headspace was air:CO_2 (1:1) at 14 psi. Metabolites were analyzed by high pressure liquid chromatography after consumption of glucose ceased (24 h).

absorbance and refractive index detection. The column was eluted isocratically with 5 m*M* sulfuric acid, and data were analyzed with an EZChrom chromatographic data system (Scientific Software). Quantification was based on comparison to standards of known concentration.

After 24 h, metabolism of glucose had ceased. At this point, control cultures had partially metabolized glucose and generated small amounts of fermentation products (Table 2). The major fermentation product formed in these cultures was pyruvic acid, as expected for NZN111 because of the lack of functional pyruvate:formate lyase and lactate dehydrogenase. The authors assume that adequate pools of NADH were available for reduction of pyruvate; one NADH is produced per every three carbons metabolized via glycolysis and mutants blocked in pyruvate dissimilation fail to ferment glucose because of their inability to achieve electron balance *(12)*. Only the culture in which malic enzyme was induced—NZN111(pMEA1) containing IPTG—consumed all the glucose. Succinic acid was the major fermentation product in this culture, present at approximately three times the concentration observed in the control cultures. Acetic acid was present at lower levels. The amounts of other fermentation products were comparable to those in control cultures. The formation of succinic acid instead of malic or fumaric acid requires an additional reductive step and creates an electron imbalance relative to the NADH produced in glycolysis if succinic acid were the sole product. The production of acetate and pyruvate in this culture may be a consequence of the need to maintain electron balance in the fermentation *(2)*.

For all the cultures, part of the glucose consumption can be attributed to aerobic respiration using the air initially present in the culture tubes. If the authors provided a fully anaerobic atmosphere initially, none of the cultures grew. They attribute this failure to the demands put on the cells by simultaneous induction of malic enzyme, the transition to anaerobic growth, and the accumulation of pyruvate.

These results demonstrate both the potential and limitations of the use of malic enzyme to manipulate fermentative metabolism in *E. coli*. Whereas expression of malic enzyme does channel metabolites to succinic acid, it does so inefficiently. This inefficiency is a result in part because of the poor K_m of the *A. suum* malic enzyme for pyruvate, but may also be because of the poor expression obtained for the heterologous *A. suum* enzyme. Some of these issues can be addressed by the use of alternative malic enzymes or the development of improved enzymes through genetic engineering. The authors have initiated cloning of the NAD^+-dependent malic enzyme from *E. coli* in anticipation that it will give better overexpression and, perhaps, possess a better K_m for pyruvate.

ACKNOWLEDGMENTS

The submitted manuscript has been authored by a contractor of the U.S. Government under contract No. W-31-109-ENG-38. Accordingly, the U. S. Government retains a nonexclusive, royalty-free license to publish or reproduce the published form of this contribution, or allow others to do so, for U. S. Government purposes.

This work was supported by the Alternative Feedstock Program of the U.S. Department of Energy (DOE), Office of Industrial Technology, and by DOE's Assistant Secretary for Energy Efficiency and Renewable Energy, under contract W-31-109-Eng-38.

REFERENCES

1. Jain, M. K., Datta, R., and Zeikus, J. G. (1989), in *Bioprocess Engineering: The First Generation* Ghose, T. K., ed., Norwood, Chichester, UK, pp. 366–389.
2. Clark, D. P. (1989), *FEMS Microb. Rev.* **63**, 223–234.
3. Blackwood, A. C., Neish, A. C., and Ledingham, G. A. (1956), *J. Bacteriol.* **72**, 497–499.
4. Millard, C. S., Chao, Y.-P., Liao, J. C., and Donnelly, M. I. (1996), *Appl. Environ. Microbiol.* **62**, 1808–1810.
5. Thauer, R. K., Jungermann, K., and Decker, K. (1977), *Bacteriol. Rev.* **41**, 100–180.
6. Murai, T., Tokushige, M., Nagai, J., and Katsuki, H. (1971), *Biochem. Biophys. Res. Comm.* **43**, 875–881.
7. Murai, T., Tokushige, M., Nagai, J., and Katsuki, H. (1972), *J. Biochem.* **71**, 1015–1028.
8. Landsperger, W. J. and Harris, B. G. (1976), *J. Biol. Chem.* **251**, 3599–3602.
9. Takeo, K., Murai, T., Nagai, J., and Katsuki, H. (1967), *Biochem. Biophys. Res. Comm.* **29**, 717.
10. Sanwal, B. D. (1970), *J. Biol. Chem.* **245**, 1212–1216.
11. Mallick, S., Harris, B. G., and Cook, P. F. (1991), *J. Biol. Chem.* **266**, 2732–2738.
12. Mat-Jan, F., Kiswar, A. Y., and Clark, D. P. (1989), *J. Bacteriol.* **171**, 342–348.
13. Boernke, W. E., Millard, C. S., Stevens, P. W., Kakar, S. N., Stevens, F. J., and Donnelly, M. I. (1995), *Arch. Biochem. Biophys.* **322**, 43–52.
14. Kulkarni, G., Cook, P. F., and Harris, B. G. (1993), *Arch. Biochem. Biophys.* **300**, 231–237.
15. Mahajan, S. K., Chu, C. C., Willis, D. K., Templin, A., and Clark, A. J. (1990), *Genetics* **125**, 261–273.
16. Sambrook, J., Fritsch, E. F., and Maniatis, T. (1989), *Molecular Cloning: A Laboratory Manual, 2nd ed.*, 2nd ed., Cold Spring Harbor Press, Cold Spring Harbor, NY.
17. Balch, W. and Wolfe, R. S. (1976), *Appl. Environ. Microbiol.* **32**, 781–791.

Reaction Engineering Aspects of α-1,4-D-Glucan Phosphorylase Catalysis

Comparison of Plant and Bacterial Enzymes for the Continuous Synthesis of D-Glucose-1-Phosphate

BERND NIDETZKY,* RICHARD GRIESSLER, ANDREAS WEINHÄUSEL, DIETMAR HALTRICH, AND KLAUS D. KULBE

Division of Biochemical Engineering, Institute of Food Technology, Universität für Bodenkultur Wien, Muthgasse 18, A-1190 Vienna, Austria

ABSTRACT

Some important process properties of α-1,4-D-glucan phosphorylases isolated from the bacterium *Corynebacterium callunae* and potato tubers (*Solanum tuberosum*) were compared. Apart from minor differences in their stability and specificity (represented by the maximum degree of maltodextrin conversion) and a 10-fold higher affinity of the plant phosphorylase for maltodextrin (K_M of 1.3 g/L at 300 mM of orthophosphate), the performances of both enzymes in a continuous ultrafiltration membrane reactor were almost identical. Product synthesis was carried out over a time course of 300–400 h in the presence or absence of auxiliary pullulanase (increasing the accessibility of the glucan substrate for phosphorolytic attack up to 15–20%). The effect of varied dilution rate and reaction temperature on the resulting productivities was quantitated, and a maximum operational temperature of 40°C was identified.

Index Entries: α-1,4-D-Glucan phosphorylase; *Corynebacterium callunae*; *Solanum tuberosum*; α-D-glucose-1-phosphate; continuous production.

---

*Author to whom all correspondence and reprint requests should be addressed.

INTRODUCTION

α-1,4-D-Glucan phosphorylases are widespread in nature and catalyze the phosphorolytic degradation of α-1,4-linked oligo- or poly-D-glucose into α-D-glucose-1-phosphate (D-Glc-1-P). According to their natural substrates, the enzymes can be classified as glycogen, starch, or maltodextrin phosphorylases, and indeed phosphorylases isolated from various sources show striking differences in their specificity towards these substrates *(1,2)*. The phosphorolysis reaction is readily reversible in vitro as shown in equation 1 where N ($\approx$ 10–25) denotes the degree of polymerization of the available linear chains in the respective glucan *(3)*.

$$(\alpha\text{-}1,4\text{-}D\text{-glucan})_N + P_i \Leftrightarrow (\alpha\text{-}1,4\text{-}D\text{-glucan})_{N-1} + \alpha\text{-}D\text{-Glc-1-}P \qquad (1)$$

The equilibrium constant is pH-dependent, but even in alkaline regions between pH 8.0–9.0 polymer synthesis is favored. Glucan phosphorylases have attracted much attention with regard to their catalytic mechanism, regulation, and evolution *(4–6)*. Their use as biocatalysts in an applied field of research is, however, rarely documented, and with the exception of starch phosphorylase from *Solanum tuberosum* (potato), the process properties of these enzymes have not been studied in much detail *(7,8)*. We have recently become interested in possible applications of microbial phosphorylases lacking allosteric and covalent regulation of enzyme activity like the potato enzyme and have been studying the synthesis of D-Glc-1-P by phosphorylases from *Corynebacterium callunae* and *Escherichia coli* *(9–11)*.

D-Glc-1-P has limited applications as such. Likely fields for the direct use of this compound are mostly medically oriented *(12)*. However, as an activated or naturally C1-protected sugar it may serve as an important intermediate or starting material in the synthesis of simple and complex carbohydrates. The use of D-Glc-1-P for the production of glucuronic acid and α,α-trehalose, that both are fine chemical with large scale applications in food industries, has been recently demonstrated *(13,14)*. Oligosaccharide syntheses by chemoenzymatic approaches are often dependent of glycosyl transferases requiring UDP- or ADP-D-glucose as substrates. These compounds in turn are derived from D-Glc-1-P and the corresponding nucleotide triphosphates *(15)*.

Pertaining to process engineering aspect in continuous enzyme catalysis, the use of soluble biocatalysts is promising especially when the optimization of productivities and space time yields at concomitant high specificity and selectivity is considered. Immobilization of the biocatalysts in their native state by employing ultrafiltration membranes represents the tool of process technology to accomplish this objective *(16,17)*. We have recently demonstrated that microbial glucan phosphorylases are stable and well-suited for conversions of maltodextrins and orthophosphate in continuous ultrafiltration membrane reactors *(9–11)*. In this study the authors have compared the phosphorylase from *C. callunae* with the higher plant enzyme from *S. tuberosum*: The latter enzyme has already been used for the conversion of starchy material in soluble or solid-immobilized form *(7,8)*. Its application in membrane reactors, however, has so far not been studied.

MATERIALS AND METHODS

Chemicals

All chemicals were of reagent grade and obtained from Sigma (Deisenhofen, Germany) unless otherwise stated. The material for protein chromatography and electrophoresis was from Pharmacia (Uppsala, Sweden). Pullulanase (Promozyme) was from Novo (Bagsvaerd, Denmark) with a declared activity of 200 U per g liquid.

Enzyme Production and Isolation

Corynebacterium callunae DSM 20147 was used throughout this study and cultivated in shaken flasks or in a 10-L bioreactor system (MBR, Wetzikon, Switzerland) as previously described *(9)*. Glucose was used as a carbon source in the concentrations indicated. Maltodextrin or maltose were employed as the components inducing the synthesis of α-glucan phosphorylase. For monitoring the time course of enzyme production in bioprocess experiments, samples were taken at the times stated, and biomass formation and phosphorylase activity in cell extracts determined. The preparation of these extracts was as reported previously *(9)*.

The partial purification of *C. callunae* phosphorylase was accomplished by ammonium sulfate precipitation *(25% saturation)* followed by hydrophobic interaction chromatography on Phenylsepharose fast flow (decreasing linear gradient of 0 to 20% $(NH_4)_2SO_4$) to yield a stable enzyme preparation with a specific activity of 3.5 U/mg *(9,11)*.

The isolation of *S. tuberosum* phosphorylase was carried out as follows. Peeled and cut potato tubers were transferred into 10 mM P_i buffer pH 6.9 (supplemented with 0.1% sodium dithionite) and grounded for 1–2 min in a conventional kitchen mixer. The potato juice was centrifuged (7500 rpm, 15 min), and the slightly acidified (pH 5.8) supernatant treated with 5000 ppm biocryl processing aids (BPA 1050; Toso Haas, Stuttgart, Germany). The pelleted phosphorylase (6000 rpm; 15 min) was redissolved in 50 mM P_i pH 6.9 and brought to 25% saturation in $(NH_4)_2SO_4$. The supernatant recovered after ultracentrifugation (30,000 rpm, 40 min; 45.1 Ti rotor) was applied to a 25-mL XK 26 column of Phenylsepharose fast flow low sub. Bound protein was eluted at 2 mL/min using a decreasing step-gradient of 0, 7.5, 16, and 25% $(NH_4)_2SO_4$ in 50 mM P_i pH 6.9. Protein was detected at 280 nm, and phosphorylase was eluted at 7.5% $(NH_4)_2SO_4$. Concentration was carried out by crossflow (30 kDa Mini-Ultrasette;

Filtron, Northborough, MA) or dead-end ultrafiltration (Amicon stirred 50-mL cell equipped with 30 kDa membrane) as well as by using 30 kDa Centricon tubes (Millipore, Eschborn, Germany). Anion exchange chromatography of starch phosphorylase was carried out on DEAE membrane cartridges (Mem Sep 1000; Biorad, Hercules, CA). Phosphorylase was eluted by a linear gradient of 50–600 mM P_i pH 6.9. SDS electrophoresis was carried out on a Pharmacia Phast system using precast 8–25 gels and silver staining of protein bands.

Enzyme Characterization

Temperature optima, pH optima and the apparent kinetic constants (30°C) in the direction of α-glucan degradation (phosphorolysis, P-mode) were determined by reported methods using discontinuous assays and heat termination of reactions *(9,11)*. Incubations were carried out on a Thermomixer 5436 (Eppendorf, Munich, Germany) with gentle agitation at 5×100 min^{-1}. The mathematical analysis of the concentration dependence of the reaction rate was performed employing classic Michaelis-Menten models including terms for substrate inhibition when necessary and using non-linear regression for parameter estimation.

Assays

Phosphorylase activity was measured in P-mode at 30°C by a continuous coupled assay described recently *(9)* using 30 mg/mL maltodextrin DE 19.4 (Agrana, Vienna, Austria) and 50 mM orthophosphate as substrates. Accordingly, the concentration of D-Glc-1-*P* was quantitated in a discontinuous assay. One unit of phosphorylase activity refers to 1 µmol NADH formed per minute, and the concentration of D-Glc-1-*P* is determined from the amount of NADH produced in the discontinuous assays. Orthophosphate was quantitated spectrophotometrically using a commerical kit (Spectroquant; Merck, Darmstadt, Germany). Protein was measured according to Bradford *(18)*.

Discontinuous and Continuous Synthesis of D-Glc-1-P

Discontinuous syntheses of D-Glc-1-*P* were carried out at 30°C essentially as previously described *(9)*. The time-course of product formation was monitored until apparent equilibration was attained. Continuous experiments were carried out at temperatures from 25 to 45°C in a convective, well-mixed, stirred tank enzyme reactor with a total volume of 40 mL. A schematic representation of the reaction system employed in this study is shown in Fig. 1. The vessel (equipped with a heating mantle for temperature control using an external water bath) had a flat membrane configuration thus requiring dead-end filtration, and a 10 kDa ultrafiltration membrane (NMWL 10,000) was used to retain phosphorylase during continuous operation. In addition, the reactor was

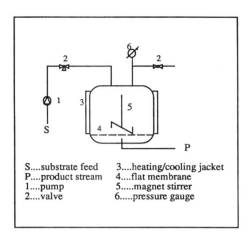

Fig. 1. Schematic representation of the enzyme reactor equipped with a flat 10 kD ultrafiltration membrane used in this study.

fitted with a conductivity electrode and a pressure sensor. Substrate was fed through a sterile filter at flow rates of 3 to 20 mL/h equivalent to average residence times of 2–13 h, and mixing was accomplished by magnetic stirring at 3–5 s^{-1} Equilibration with substrate in buffer was always allowed to proceed overnight, and the reaction was then started by the injection of a concentrated enzyme preparation (30–50 U/mL) through a septic seal. Samples were then periodically taken at the reactor outlet or, for determining the enzyme activity in the reactor, directly from the reactor through the seal. Samples were gelfiltered on NAP-5 desalting columns prior to measuring residual phosphorylase activity to remove D-Glc-1-*P* produced. Pretreatment of maltodextrin by pullulanase was accomplished by an enzyme dosage of 120 U/g dextrin *(9,10)*.

RESULTS AND DISCUSSION

Enzyme Production

Phosphorylase synthesis in *C. callunae* is inducible by maltodextrins or maltose and starts late in the exponential growth phase when the easily metabolizable carbon source such as glucose is already being depleted (Fig. 2). Table 1 summarizes results we have obtained in several bioprocess experiments aimed at the improvement and partial optimization of phosphorylase production by *C. callunae*. The data are results of bioreactor cultivations of the organism at 30°C, a constant pH of 7.4 and dissolved oxygen at 40% saturation *(9,11)*. Supplementation of the basal growth medium with glucose allows to increase the formation of wet biomass substantially with values in the range of 100 g/L of medium being achievable. The maximum phosphorylase activity produced by *C. callunae* is in the

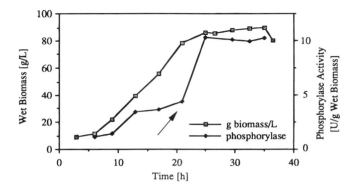

Fig. 2. Fermentation time course of *Corynebacterium callunae* on basal medium using 60 g/L glucose and 20 g/L maltose as the component inducing phosphorylase synthesis. The arrow indicates depletion of glucose. Other conditions: 30°C, 650 rpm, 40% oxygen saturation, pH 7.4.

Table 1
Production of α-1,4-D-Glucan Phosphorylase from *Corynebacterium callunae*

Glucose %	Inducing component %	Cultivation time h	Wet biomass g/L	Phosphorylase U/g biomass	Specific activity U/g protein
none	G_x, 1.6	10	15	10.5	0.25
none	G_x, 16.0	24	50	4.9	0.18
4	G_x, 16.0	36	98	6.7	0.16
12	G_x*, 3	45	110	1.8	n.d.
6	G_2, 2	36	92	10.2	0.24

Abbreviatons: G_x, maltodextrin; G_2, maltose; n.d., not determined; * induction at the end of the exponential growth phase.

range of 10 U/g biomass. In previous experiments it was found that the addition of a four fold excess of maltodextrin over glucose is necessary to obtain specific phosphorylase production of 6–8 U/g (wet) biomass (cf Table 1). The most likely explanation for this fact is that only certain limited components in a heterogeneous substrate such as a commercial maltodextrin are capable of inducing phosphorylase synthesis. Analysis of the fermentation medium by thin layer chromatography corroborates at least qualitatively, that the major fraction of the maltodextrin is not metabolized by the organism. Especially when high cell density cultivations of the organims are considered which require glucose concentrations of at least 40–60 g/L, the use of maltodextrin (in concentrations of 160–250 g/L) does not seem feasible and practical (because of the high viscosity of the fermentation medium). Induction of phosphorylase synthesis by maltose is thus more favorable, and using a cocentration of 20 g/L of maltose, the phosphorylase activities obtained per gram wet biomass are indeed in a range so far found to be maximal in *C. callunae* (Table 1). According to the data in Fig. 2 and Table 1, 1 L of fermentation medium yields approx 900 U of phosphorylase. When a comparison with potato tubers is made, these results show that 1–1.5 kg of the plant material need to be processed to obtain equal enzyme activities (0.60–0.75 U/g of tuber).

Enzyme Isolation

A minimal purification of *C. callunae* phosphorylase has recently been established that eliminates contaminating activities, especially that of phosphatase that hydrolyzes D-Glc-1-*P* to D-Glc and P_i. A strategy found to be efficient in processing starch phosphorylase is summarized in Table 2 that represents a considerable improvement of other literature protocols pertaining to the isolation of partially pure plant phosphorylase from potato tubers *(7)*. Hydrophobic followed by anion exchange chromatography is usually sufficient to remove all amylase, phosphoglucomutase, and phosphatase activities. As judged from SDS PAGE (not shown) the majority of contaminating proteins is removed after the ion exchange step resulting in an overall purification factor of approx 28–30 (Table 2). The yields in the purification sequences of both phosphorylases are comparable with approx 60–70%. If not indicated otherwise, a preparation of potato phosphorylase with a specific activity of 2.8 U/mg was used in the following conversion studies (Table 2).

Enzyme Characterization

Pertaining to their temperature optima the phosphorylases from *S. tuberosum* and *C. callunae* show very similar dependences of initial rate and stability on reaction temperature. The classic temperature optimum is found at 50–55°C (50 m*M* P_i, pH 7.5) certainly not coinciding with the operational optimum. The operational stability of both phosphorylases was studied in 300 and 600 m*M* P_i at 30 and 45°C. The corresponding half-lives are 13 and 20 d in case of starch phosphorylase from potato whereas values of 9 and 18 d were determined for the bacterial enzyme. Both phosphorylases are unstable in the absence of P_i, e.g., in TrisAc buffer the half-lives are <12 h at 4°C. Potato phosphorylase is destabilized by 10 m*M* chloride ions that in turn seem not to affect the stability of the bacterial enzyme at 4°C. The pH optimum of *S. tuberosum* phosphorylases is found at pH 7.8–8.0 and is shifted towards the alkaline region by nearly 1 U of pH when comparison is made with the *C. callunae* enzyme. The apparent kinetic constants for maltodextrin were determined at otherwise realistic (operational) conditions, i.e., 300 m*M* P_i and

Table 2
Partial Purification of Starch Phosphorylase from *Solanum tuberosum*

	Total activity (U)	Specific activity (U/mg)	Yield %	Purification factor (-fold)
Crude cell extract	860	0.24	100	1.0
BPA 1050 precipitation	743	0.77	86	3.3
HIC Phenyl-sepharose	610	2.76	71	11.7
DEAE ionexchange	525	6.50	61	27.6

Abbreviations: HIC, hydrophobic interaction chromatography; BPA, biocryl processing aids.

Table 3
Apparent Kinetic Constants for Maltodextrin Determined at Operationally Relevant Conditions: 300 mM P_i (saturation), 30°C, pH 7.5

Kinetic constant (range 0 - 100 g/L)	Starch phosphorylase (*S. tuberosum*)	Maltodextrin phosphorylase (*C. callunae*)
Maximum rate (V_{max}) (U/mg)	3.0	5.3
Apparent K_M (g maltodextrin/L)	1.29	12.0
Substrate inhibition K_{IS} (g/L)	131	none

pH 7.5 (Table 3) *(9)*. The K_M-values for maltodextrin differ by a factor of almost 10, and it seems thus likely that the potato phosphorylase will outperform the bacterial enzyme in terms of catalytic efficiency at low concentration of the glucan substrate. Substrate inhibition by maltodextrin was significant only in case of *S. tuberosum* phosphorylase. The affinities of both enzymes for untreated maltodextrin and maltodextrin that had been obtained after exhaustive treatment with pullulanase are identical as are the K_M values for orthophosphate (20–25 mM determined in P_i buffer using 80 g/L maltodextrin). Pretreatment was monitored by measurements of the formation of reducing sugars and by the iodine-starch reaction *(10)*.

Discontinuous Synthesis of D-Glc-1-P

The results of the kinetic characterization point to substrate inhibition of starch phosphorylase from *S. tuberosum* by maltodextrin. To determine whether this inhibition affects the synthesis of D-Glc-1-P, conversion experiments were carried out at varying concentrations of maltodextrin with orthophosphate kept constant at 600 mM. The amount of product formed was monitored after 90 and 240 min reaction time. It becomes obvious from the data in Figs. 3A and B that in case of potato phosphorylase the time-dependent product yields are decreased when the maltodextrin concentrations exceed 120 g/L. However, this effect does not reduce maximum attainable product concentrations in comparison to the results obtained with phosphorylase from *C. callunae* (Fig. 3A) and is of minor practical significance because the applied glucan concentrations will seldom be higher than 100 g/L.

In phosphorolysis direction of catalysis either P_i or the glucan substrate may be the limiting component determining the maximum concentration of the product D-Glc-1-P. In the presence of a molar excess of glucan (i.e., based on the concentration of anhydroglucose units), the attainable product yield is governed by the equilibrium constant $[P_{i,eq}/Glc\text{-}1\text{-}P_{eq}]$ *(9,10)*. In contrast, to study the maximum degradation of various α-glucans by the actions of plant and bacterial phosphorylase, an excess of orthophosphate had to be employed, usually 300 mM P_i at 10 g/L glucan, i.e., theoretically a complete degradation of the glucan should be possible even when the maximum conversion of the initial P_i in Glc-1-P cannot exceeed 18 to 20% (pH 7.5, 30°C). Among the substrates tested including various maltodextrins differing in their dextrose equivalent-values, soluble starch as well as pullulanase-pretreated material, all were more substantially (10 to 20%) degraded by starch phosphorylase from *S. tuberosum* as compared to the *Corynebacterium* enzyme. Typically the maximum degrees of glucan conversion are 30 to 45% for untreated- and 50 to 65% for pullulanase-treated substrates when starch phosphorylase is employed. The different maximal degrees of glucan conversion seen with bacterial and plant phosphorylase do not result from a more complete degradation of a linear α-glucan chain by the *S. tuberosum* enzyme. The limit dextrin of both phosphorylases, i.e., the smallest oligomeric substrate converted at significant rates in P-mode direction, is maltopentose (Grießler and Nidetzky, unpublished results).

Continuous Conversion in Membrane Reactors

The continuous synthesis of D-Glc-1-P in the reaction system represented in Fig. 1 was studied using partially purified *S. tuberosum* or *C. callunae* phosphorylase. In each experiment the attainable product concentration was limited by the amount of either orthophosphate or glucan in the substrate feed. Flow rates (dilution rates) were varied, and the

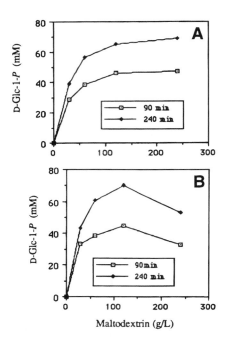

Fig. 3. Maltodextrin-dependent synthesis of D-Glc-1-P at 30°C and pH 7.5 using 600 mM P_i as a constant cosubstrate. **(A)** 0.8 U/mL *C. callunae* phosphorylase. **(B)** 0.8 U/mL *S. tuberosum* phosphorylase.

concomitant effect on the reactor productivity evaluated. The time courses of substrate conversion shown in Figs. 4 and 5 allow to compare the action of plant and bacterial phosphorylase with regard to the extent of maltodextrin degradation at several different operation conditions in the membrane reactor (glucan limit). The data indicate that the performances of both phosphorylases are very similar under the reaction conditions employed even though their kinetic constants, e.g., K_M-values, differ significantly. A more complete degradation of the glucan substrate by *S. tuberosum* phosphorylase noticed in the discontinuous experiments was not detected in the continuous conversions (cf Figs. 4 and 5), most probably because a maximum phosphorolytic degradation of the respective glucan can be achieved only at very low dilution rates. The pretreatment of maltodextrin by pullulanase has some effect on the attainable degrees of conversion (15 to 20% increase), and simultaneous or sequential enzyme action are equally effective (cf Figs. 4 and 5). When pretreatment and phosphorolysis are accomplished in a simultaneous manner, supplementation with fresh pullulanase approximately each 50 h is required (Fig. 5) pointing to some inactivation of the auxiliary biocatalyst. The dependence of the productivity on the applied dilution rate is roughly linear as shown in Fig. 6, and an increase of these values beyond 0.6 h^{-1} (corresponding to 25 mL/h) is expected to result in a further increase of productivity (at the cost

Glucan Phosphorylase

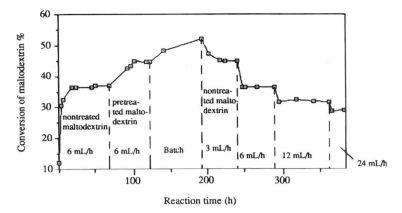

Fig. 4. Continuous conversion of 9.5 g/L maltodextrin at 30°C using 260 mM P_i (glucan limit) at varying operating conditions (pH 7.5). An initial *C. callunae* phosphorylase activity of 0.9 U/mL was employed.

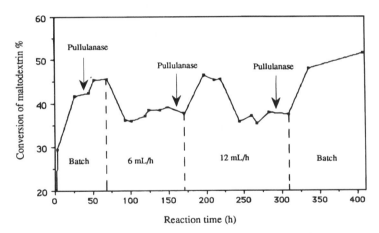

Fig. 5. Continuous conversion of 9.5 g/L maltodextrin at 30°C using 260 mM P_i (glucan limit) at varying operating conditions (pH 7.5). An initial *S. tuberosum* phosphorylase activity of 0.8 U/mL was employed.

of the extent of substrate conversion). Synthesis of D-Glc-1-P when employing P_i as the limiting substrate yields two to three-fold improved productivities with the maximum conversion of orthophosphate being in a range of 12 to 18% (Fig. 7). Again the results achieved, even on a quantitative basis, were nearly identical for plant and bacterial phosphorylase (not shown). The range of applicable flow rates in the reaction system used throughout this study was limited because of the retention of the polymeric substrate in dead-end ultrafiltration (cf Fig. 1). As possible alternatives other (external) membrane configurations are considered and immobilization of phosphorylase from *C. callunae* is currently being studied. Binding

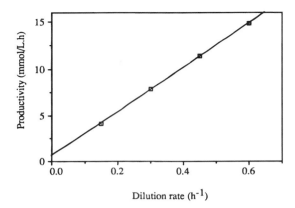

Fig. 6. Evaluation of reactor productivities at various dilution rates. Other conditions: 30°C, 260 mM P_i, 10 g/L maltodextrin, initial *C. callunae* phosphorylase activity of 0.9 U/mL.

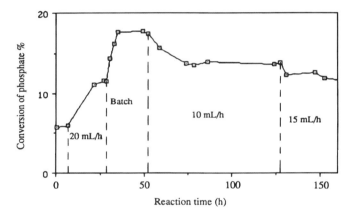

Fig. 7. Continuous conversion of 260 mM P_i at 30°C using 60 g/L maltodextrin (P_i limit) at varying operating conditions (pH 7.5). An initial *C. callunae* phosphorylase activity of 0.9 U/mL was employed.

of phosphorylase from *C. callunae* on weak or strong anion exchange resins has been accomplished, and production of D-Glc-1-P at substantially higher dilution rates than applied in the membrane reactor shown in Fig. 1 was indeed possible. However, preliminary results point to a rather low specific activity of the immobilized enzyme probably a result of intrinsic effects of matrix-binding of the phosphorylase (not shown).

Effect of Reaction Temperature

In the absence of rate limitations by external or internal mass transfer (which is assumed to be the case when soluble enzymes are employed), the reaction temperature is expected to be an important

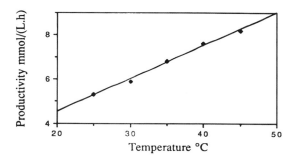

Fig. 8. Temperature effect on the continuous D-Glc-1-P synthesis by *C. callunae* phosphorylase. Other conditions: 600 mM P_i, 60 g/L maltodextrin (P_i limit), dilution rate 0.25 h^{-1}, initial *C. callunae* phosphorylase activity of 0.19 U/mL.

determinant of conversion rate and productivity. The continuous synthesis of D-Glc-1-P by *C. callunae* phosphorylase using orthophosphate-limited reaction conditions was followed in a temperature range of 25 to 45°C when applying a constant flow rate of 10 mL/h (0.25 h^{-1}). The enzymatic system was allowed to equilibrate at the corresponding temperature for at least 5 reactor cycles (i.e., 20 h), and during that period four to six samples were taken and product formation as well as enzyme activities quantitated. The product concentration in the filtrate remained fairly constant during each period of constant temperature as did the volumetric phosphorylase activity in the reactor. During the entire course of the experiment approx 20% of the initial activity were lost, and all productivities are corrected for the enzyme inactivation. The results shown in Fig. 8 clearly suggest that a well-defined linear relation holds for the dependence of productivity on the reaction temperature within the experimental range of 25 to 45°C. The operational (thermal) stability of both plant and bacterial phosphorylase allows their use at a reaction temperature up to 40 but not 45°C.

CONCLUSIONS

Starch phosphorylase from *S. tuberosum* and maltodextrin phosphorylase from *C. callunae* exhibit nearly identical process-relevant properties with regard to their application in a continuous synthesis of D-Glc-1-P. Despite some differences pertaining to thermal stability, kinetic properties (K_M-value for maltodextrin) and specificity (reflected by a different maximum extent of α-glucan substrate degradation), the performance of both enzymes during phosphorolytic conversion of maltodextrin and orthophosphate in an ultrafiltration membrane reactor are very well comparable. One additional discerning feature among both systems is the initially more than 10-fold higher phosphorylase activity in the bacterial biomass as compared to the plant material.

REFERENCES

1. Newgard, C. B., Hwang, P. K., and Fletterick, R. J. (1989), *Crit. Rev. Biochem. Mol. Biol.* **24,** 69–99.
2. Tanizawa, K., Mori, H., Tagaya, M., and Fukui, T. (1994), in *Molecular Aspects of Enzyme Catalysis*, Fukui, T. and Soda, K., eds., VCH, Weinheim, Germany, pp. 107–126.
3. Guilbot, A. and Mercier, C. (1985), in *The Polysaccharides*, Aspinall, G.O., ed., Academic Press, Orlando, FL, pp. 210–283.
4. Palm, D., Klein, H., Schinzel, R., Buchner, M., and Helmreich, E. J. M. (1990), *Biochemistry* **29,** 1099–1107.
5. Hudson, J. W., Golding, G. B., and Crerar, M. M. (1993), *J. Mol. Biol.* **234,** 700–721.
6. Barford, D. and Johnson, L. N. (1989), *Nature* **340,** 609–616.
7. Hollo, J., Laszlo, E., and Juhasz, J. (1967), *Plant α-Glucan Phosphorylase*, Akademiai Kiado, Budapest, Hungary.
8. Kayane, S., Kawai, T., Sakata, M., Imamura, T., Tanigaki, M., and Kurosaki, T. (1989), EP 0 305 981 A2.
9. Weinhäusel, A., Nidetzky, B., Rohrbach, M., Blauensteiner, B., and Kulbe, K. D. (1994), *Appl. Microbiol. Biotechnol.* **41,** 510–516.
10. Weinhäusel, A., Nidetzky, B., Kysela, C., and Kulbe, K. D. (1995), *Enzyme Microb. Technol.*, **17,** 130–135.
11. Nidetzky, B., Weinhäusel, A., Grießler, R., and Kulbe, K. D. (1995), *J. Carbohydr. Chem.* **14,** 1017–1028.
12. Vandamme, E., van Loo, J., Machtelinckx, L., and de Laporte, A. (1987), *Adv. Appl. Microbiol.* **32,** 163–201.
13. van Bekkum, H. (1991), in *Carbohydrates as Organic Raw Materials*. Lichtenthaler, F. W., ed., VCH, Weinheim, Germany, pp. 289–310.
14. Murao, S., Nagano, H., Ogura, S., and Nishino, T. (1985), *Agric. Biol. Chem.* **49,** 2113–2118.
15. Stabgier, P. and Thiem, J. (1991), in *Enzymes in Carbohydrate Synthesis* ACS Symp. Series 466, Bednarski, M. D. and Simon, E. S., eds., Washington, DC, pp. 63–78.
16. Prazeres, D. M. F. and Cabral, J. M. S. (1994), *Enzyme Microb. Technol.* **16,** 738–750.
17. Kula, M.-R. and Wandrey, C. (1987), *Meth. Enzymol.* **136,** 9–21.
18. Bradford, M. M. (1976), *Anal. Biochem.* **72,** 248–254.

Simultaneous Enzymatic Synthesis of Gluconic Acid and Sorbitol

Production, Purification, and Application of Glucose-Fructose Oxidoreductase and Gluconolactonase

Bernd Nidetzky,* Monika Fürlinger, Dorothee Gollhofer, Iris Haug, Dietmar Haltrich, and Klaus D. Kulbe

Division of Biochemical Engineering, Institute of Food Technology, Universität für Bodenkultur Wien, Muthgasse 18, A-1190 Vienna, Austria

ABSTRACT

With regard to the enzymatic synthesis of sorbitol and gluconic acid, a screening was carried out to identify promising producers of glucose-fructose oxidoreductase (GFOR) and gluconolactonase (GL). *Zymomonas mobilis* DSM 473 and *Rhodotorula rubra* DSM 70403 have been selected for the synthesis of GFOR and GL, respectively. Maximal enzyme production by these organisms has been achieved at D-glucose concentrations of 200 and 150 g/L, respectively. Both GFOR and GL were purified and characterized with respect to some of their catalytic properties. GL showed strict specificity for 1,5-(δ)-lactones and was activated by Mg^{2+} and Mn^{2+} ions. The potential use of soluble GFOR is limited by its inactivation during substrate conversion, and the effects of reaction temperature and pH on the "catalytic" stability of GFOR were evaluated. Exogenous adddition of auxiliary GL had no effect on oxidoreductase stability and did not improve productivities.

Index Entries: Glucose-fructose oxidoreductase; gluconolactonase; *Zymomonas mobilis*; *Rhodotorula rubra*; stability.

---

*Author to whom all correspondence and reprint requests should be addressed.

INTRODUCTION

Sorbitol and gluconic acid are bulk products with various applications in the food and pharmaceutical industry and have many other large scale uses in the chemical industry (1,2). An enzymic route for the synthesis of these two compounds, which operates under mild conditions and avoids by-product formation, could be a true alternative to the classic production processes (1,3). The ethanologenic bacterium *Zymomonas mobilis* was shown to synthesize sorbitol during growth on sucrose or on mixtures of fructose and glucose (4). The enzyme responsible for sorbitol formation by *Zymomonas* is glucose-fructose oxidoreductase (GFOR) that simultaneously converts mixtures of glucose and fructose into, respectively, glucono-δ-lactone and sorbitol. Tightly protein-bound NADP(H) serves as the cofactor in the oxidoreduction reaction (5). The simultaneous redox process shown in Scheme 1 is practically irreversible because gluconolactone is hydrolyzed to gluconic acid either in a spontaneous manner (6) or enzymatically by the action of gluconolactonase (GL) (5,7–10). Possible and promising applications of GFOR for the production of sorbitol and gluconic acid have been proposed soon after the first identification of the enzyme. Both isolated GFOR and permeabilized cells of *Z. mobilis* have been employed in several studies aimed at the optimization of the enzymic substrate conversion in discontinuous and continuous mode of operation (5,11–19). The principal advantages of GFOR over other enzyme/coenzyme systems for the production of gluconic acid and sorbitol are quite obvious: In contrast to other coupled enzyme systems, e.g., sorbitol-dehydrogenase/glucose-dehydrogenase or aldose-reductase/glucose-dehydrogenase (20), GFOR is a self-regenerating redox-enzyme-system whose activity does not depend on the efficiency of a (second) regenerating reaction system (21,22). Furthermore, the exogenous addition of the expensive and unstable NADP(H) is not necessary since the coenzyme in GFOR is bound to the protein in a nondissociable form. Consequently, the retention of the coenzyme in the reaction system during continuous operation employing specifically adapted techniques is not required, e.g., artificial enlargement of the coenzyme (23) or specific membrane technology (21,24–26).

The aim of this work was threefold. First the identification of not yet reported organisms, *Zymomonas* sp. or others, that are capable of expressing significant and possibly higher activities of GFOR and/or GL than those previously reported was taken into consideration. Although cloning and overexpression of the gene encoding GFOR and GL from *Z. mobilis* was successful (27) and a six to nine fold increased expression of GFOR in the native host organism was described (27,28), the identification of better producing wild type strains could be important. Second, the production, purification, and partial characterization of these enzymes was targeted. Finally, some aspects of the application of isolated GFOR with or without supplementation by GL for the synthesis of sorbitol and gluconic acid was studied.

Enzymatic Synthesis

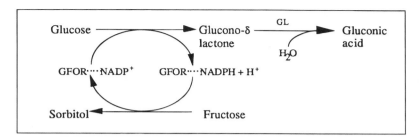

Scheme 1. Simultaneous conversion of glucose and fructose to sorbitol and gluconic acid by glucose-fructose oxidoreductase (GFOR) and gluconolactonase (GL)-an enzyme that catalyzes the hydrolysis of glucono-δ-lactone. The cofactor NADP(H) tightly bound to GFOR is indicated.

MATERIALS AND METHODS

Chemicals

Chemicals were reagent grade and purchased from Sigma (Deisenhofen, Germany) unless otherwise stated. The gel material for enzyme purification and electrophoresis was from Pharmacia (Uppsala, Sweden).

Screening

All micro-organisms were obtained from the German type culture collection DSM (Braunschweig, Germany) or the Centraalbureau voor Schimmelcultures (Baarn, The Netherlands) and were cultivated at 30°C according to the instructions of the suppliers using the following media components in g/L: *Zymomonas* (glucose [20.0], yeast extract [10.0], casein peptone [10.0]; pH 6.0, 60 rpm, 48 h, sealed serum flasks); *Pseudomonas* (casein peptone [5.0], meat extract [3.0], pH 7.0, 24 h, 100 rpm); *Gluconobacter, Acetobacter, Alteromonas* (glucose [25.0], yeast extract [5.0], casein peptone [3.0], pH 7.0, 48 h, 100 rpm); yeasts (peptone from soybean [3.0, Fluka], malt extract [30.0], pH 5.6, 24 h, 100 rpm). All aerobes were grown in baffled 250-mL Erlenmeyer flasks. Cells were harvested by centrifugation at 15,000g and 4°C for 20 min, washed twice, diluted 1:3 in 10 mM MES-buffer pH 6.4 and disrupted three times in a French pressure Mini cell (American Scientific Company, Silver Spring, MD) at 1200 psi. Debris and intact cells were removed by ultracentrifugation at 100,000g for 20 min (L8-70; Beckman, Fullerton, CA). The clear supernatants (cell extracts) were analyzed for enzyme activities and total protein as described in Enzyme Assays.

Production of GFOR and GL

Z. mobilis (DSM 473) was grown in shaken flask culture (serum bottles) for 48 h at 60 rpm on a reciprocal water-bath incubator (Infors HT; Infors, Bottmingen, Switzerland). The medium (pH 6.0) contained in g/L, yeast extract (5.0), KH_2PO_4 (0.5), $M_gSO_4 \times 7H_2O$ (0.02), $Fe(NH_4)_2(SO_4)_2 \times 6H_2O$ (0.02), biotin (0.001), calcium panthotenate (0.002), and carbohydrate component as indicated. *R. rubra* (DSM 70403) was cultivated at 30°C and 100 rpm for 24 h in complex medium or mineral medium (pH 5.6). The complex medium contained in g/L: glucose (20–200), malt extract (30), soja peptone (3). The mineral medium contained in g/L: glucose (20–200), KH_2PO_4 (0.88), K_2HPO_4 (0.13), NaCl (0.10), $(NH_4)_2SO_4$ (1.00), $M_gSO_4 \times 7H_2O$ (0.5), $CaCl_2 \times 2H_2O$ (0.1), biotin (0.001), 1 mL trace element solution ($Fe(NH_4)_2(SO_4)_2 \times 6H_2O$ [20 mg], $CuSO_4 \times 5\,H_2O$ [4 mg], KI [10 mg], $MnCl_2 \times 4H_2O$ [30 mg], $Na_2MoO_4 \times H_2O$ [25 mg], $ZnCl_2$ [20 mg]).

Alternatively, *Z. mobilis* was cultivated, without aeration, at 30°C and 150 rpm in a 20-L MBR fermentor with a working volume of 18 L (MBR, Wetzikon, Switzerland) fitted with pH and oxygen electrodes. 14 L of medium (with 200 g/L glucose) were inoculated with 1.4 L of preculture grown on the same medium for approx 50 h at 60 rpm. A constant pH value of 6.0 was achieved by the addition of 3*M* potassium hydroxide. *R. rubra* was grown in the same bioreactor system at 30°C, 300 rpm, a controlled pH of 5.0 and 0.5 vvm aeration rate using the specified complex medium containing 150 g/L glucose. Antifoam (polypropylenglycol) was added as required. Samples were periodically taken throughout the cultivations and for subsequent measurements centrifuged (3800–11,000*g*, 20–40 min, 4°C), washed and treated as described for the Screening Section.

Purification of GFOR and GL

Harvested cells of *R. rubra* were resuspended (1:1 w/v) in 10 m*M* MES buffer pH 6.4 containing 5 m*M* dithiothreitol (DTT) and disrupted at constant 4°C in a continuously operated Dyno Mill (Bachofen, Switzerland) using an average residence time of 7 min (glass beads of 0.5 mm diameter). Cells of *Z. mobilis* were two-fold diluted in 50 m*M* MES, pH 6.4 and subjected to a three-time disintegration in a 20K French pressure cell at 1200 psi. Following ultracentrifugation (30,000–100,000*g* for 40 min, 4°C) the supernatant was recovered and stored at –20°C until used for further purification.

All purification steps were caried out at room temperature except $(NH_4)_2SO_4$ precipitations at 4°C. Column chromatography was performed on a FPLC system (Pharmacia) using detection of eluting protein at 280 nm. Gel-filtration was accomplished using either Sephadex G-25 coarse (2.5 × 10 cm, 10 mL/min) or prefilled PD-10 and NAP 5 desalting columns. Concentration was carried out by cross-flow (10 kD Mini-Ultrasette; Filtron, Nothborough, MA) or dead-end ultrafiltration (Amicon-stirred 50-mL cell equipped with 10 kD membrane) as well as by using 10 kD Centricon tubes (Millipore, Eschborn, Germany). Hydrophobic interaction chromatography was carried out on Phenylsepharose fast flow (25 mL) equilibrated with 30% $(NH_4)_2SO_4$ pH 6.4 and eluting protein with a 5% step gradient (30–0%). In case of GL, 5 m*M* DTT was added to all buffers. Cation exchange chromatography of GFOR was carried out on a 1-mL column of Mono S HR 5/5 equilibrated with 10 mM MES, pH 6.4. Elution was accomplished at 0.5 mL/min using a linear gradient of 1M NaCl (0–150 mM in 60 min). Anion exchange chromatography of GL was performed on Mono Q HR 5/5 equilibrated with 20 mM piperazine, 1 mM DTT, 0.1% Tween, pH 6.0, and elution was at 0.5 mL/min with a linear gradient of 50–350 mM NaCl.

Enzyme Assays

GFOR activity was quantitated in a coupled assay together with *R. rubra* gluconolactonase by measuring the decolorization of a 0.29 mM *p*-nitro-phenol solution at 405 nm and 25°C. 400 mM Glucose and 800 mM fructose in 10 mM MES buffer pH 6.4 were used as substrates. It was proven (cf Results section) that 6–8 U/mL of lactonase are needed to avoid underestimation of GFOR activity because of rate-limitations in the coupled reaction (lactone hydrolysis). The activity of gluconolactonase was determined by the same method using 4 mM glucono-δ-lactone as a substrate. One unit of activity refers to 1 µmol gluconic acid formed per minute by the action of either GFOR/gluconolactonase or gluconolactonase alone. All rates are corrected for nonenzymic hydrolysis of gluconolactone in the absence of lactonase and, at 25°C, an equilibrium constant of 0.185 ± 0.05 of D-glucono-δ-lactone/D-gluconic acid is assumed *(6)*.

The pH and temperature optima of GFOR and GL were determined in discontinuous assays. In case of GFOR, but not GL, the reaction was terminated after 10–15 min by heating (10 min, ≈ 100°C) prior to employing HPLC analysis of product formation or substrate consumption (less than 15% conversion). Samples of GL-catalyzed conversions were analyzed immediately to reduce spontaneous hydrolysis of the lactone substrate. The substrate concentrations were 400 mM glucose/800 mM fructose (GFOR) and 100 mM glucono-δ-lactone (GL). The effect of various compounds on the stability and activity of both enzymes was tested in standard assays as indicated.

Simultaneous Substrate Conversions by GFOR/GL

Conversion of mixtures of fructose and glucose was carried out at the temperature indicated using varying activities of GFOR with or without supplementation by GL. GFOR was employed in varying degrees of purity whereas GL was used after ammonium sulfate precipitation (cf Results section). Incubation was in 100 mM potassium phosphate buffer at a constant pH value between 5.5–7.3 (maintained by the controlled addition of alkali). The reaction progress was monitored by reading either the volume or the weight of the consumed alkali. Samples (100–250 µL) were taken from the reaction mixture at times indicated. Approximately 100 µL were heat-inactivated (10 min; boiling water bath) and used for HPLC analysis (cf Analytical). The rest of the original sample was gel-filtered (NAP 5

Table 1
Screening for Microbial Producers of Glucose-Fructose Oxidoreductase (GFOR)
and Gluconolactonase (GL)

Organism	Strain	GFOR (U/mg)	GL (U/mg)	Wet biomass (g/L)
Z. mobilis	DSM 424	0.06 (0.08)*	n.d.	5.2
Z. mobilis	DSM 473	0.30 (1.51)*	0.81	7.8
Z. mobilis	DSM 762	n.d. (n.d.)	0.06	1.5
Z. mobilis	DSM 3580	0.32 (2.04)*	0.60	4.7
P. acidovorans	DSM 50251	n.d.	1.20	10.1
P. aeruginosa	DSM 1707	n.d.	1.62	5.6
P. fluorescens	DSM 50090	n.d.	2.42	4.8
P. chlororaphis	DSM 50083	n.d.	0.94	9.9
R. rubra	DSM 70403	n.d.	8.18	10.0
R. rubra	DSM 70404	n.d.	3.15	7.1
R. gracilis	CBS 6681	n.d.	2.70	8.5

*Values in parenthesis are measured in the presence of 8 U/mL gluconolactonase; n.d., not detectable.

colums) to remove substrates and products, and used for quantitation of residual GFOR activity.

Analytical

Glucose, fructose, sorbitol, and gluconic acid in reaction samples were base-line separated by HPLC using an Aminex HPX 87C (7.8 × 300 mm) ion exchange column (BioRad, Hercules, CA) at 85°C with 10 mM calcium nitrate as eluent. The flow rate was 0.7 mL/min, and detection was accomplished by refraction index. Gluconolactone was determined using a LiChrosorb NH_2 column (Merck, Darmstadt, Germany) with 77% acetonitrile/23% 10 mM potassium phosphate pH 5.5 as eluent (1.5 mL/min, 30°C). Protein was determined according to Bradford (29) using bovine serum albumin as standard.

RESULTS

Screening

Among 25 micro-organisms selected because their genera such as *Acetobacter* sp., *Gluconobacter* sp., and *Pseudomonas* sp. show some pheno- or genotypic similarity to *Zymomonas* (4), none were capable of producing detectable activities of GFOR. This result further supports the assumption that GFOR is an enzyme unique to *Z. mobilis* (4). The specific activities synthesized by different strains of *Z. mobilis* differed widely (Table 1), and the addition of exogenous GL activities (6–8 U/mL assay) was essential to

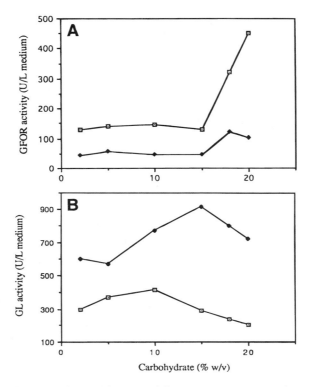

Fig. 1. **(A)** Production of GFOR by *Z. mobilis* DSM 473 using D-glucose (□) or D-fructose (◆) as a sole carbon source (30°C, 60 rpm, 48 h, microaerobically-sealed serum bottles). **(B)** Production of GL by *R. rubra* DSM 70403 using D-glucose as a carbon source in mineral (◆) medium and complex (□) medium (30 °C, 100 rpm, 24 h, baffled flasks).

avoid underestimation of actual GFOR activity especially when comparison of different strains had to be made. Since GL seemed to be an important enzyme at least from the analytical point of view, in addition to GFOR, the amount of GL produced by various organisms was assayed. Several bacteria (*Pseudomonas* sp.), but especially yeasts of *Rhodotorula* sp. were found to produce much higher activities than *Zymomonas*. In other yeast strains such as *Candida* sp. or *Pichia* sp. the volumetric and specific GL activities were low or even not detecable. From these preliminary screening studies *Z. mobilis* DSM 473 and *Rhodotorula rubra* DSM 70403 were selected for more detailed studies pertaining to GFOR and GL production, respectively. *Z. mobilis* DSM 473 was preferred over *Z. mobilis* DSM 3580 because of the better biomass formation of the former organism (cf. Table 1).

Enzyme Production

The effect of the type of the carbon source and its concentration on biomass formation and enzyme production by *Z. mobilis* and *R. rubra* was studied in shaken flask experiments, and results are shown in Figs. 1A and 1B.

The volumetric GFOR activities during growth on glucose were approx four fold higher than those obtained with fructose as a carbon source. The optimal glucose concentration with regard to biomass formation and enzyme production is different. The highest GFOR activities of approximately 450 U/L were obtained at 20% glucose (Fig. 1A) that is a sugar concentration where reduced growth of the organism was already observed (4.7 g/L). Maximal biomass yield of approx 8 g/L found between 100–150 g/L glucose. The use of fructose alone or equimolar mixtures of fructose and glucose reduced the biomass yields by approx 15–25%. Growth of R. rubra and concomitant GL production was monitored using glucose as a sole carbon source in concentrations up to 200 g/L. The optimal sugar concentrations for enzyme formation on mineral medium or on complex medium are distinct (100 and 150 g/L, respectively) and decrease when glucose is further increased beyond these concentrations (Fig. 1B).

In fermentor cultivations of Z. mobilis (200 g/L glucose) the formation of GFOR and of GL was found to be growth-associated and the production of these enzymes occured simultaneously (not shown). The stationary growth-phase was reached after 25–30 h cultivation time coinciding with depletion of glucose. After 30–35 h, the volumetric activity of GFOR started to decrease rapidly so that the time of cell harvest was important. In discontinuous experiments, a wet cell concentration of 17 g/L and enzyme activities of 800 U GFOR/L and 450 U GL/L were obtained. Fermentor cultivations of R. rubra on complex medium supplemented with 150 g/L glucose yielded approx 45 g/L wet biomass and 1200–1400 U GL/L medium after 20 h of growth. Enzyme production is growth-associated and, like in the case of GFOR, GL activities started to decrease by 10–20% when incubation was continued for further 5–10 h after the maximal activity had been attained.

Enzyme Purification

Since the addition of crude R. rubra GL was found to inhibit GFOR activity to some extent, a minimal purification of GL was required for its use in analysis and conversion studies. Table 2 presents a summary of the purification strategy and the results obtained. The fractions of GL achieved by the procedure outlined in Table 2 were added to a crude GFOR preparation, and their effect on the apparent oxidoreductase activity was determined. Ammonium sulfate precipitation of GL (45% saturation with a 90% saturated solution) followed by resuspension in 10 mM MES pH 6.4 was found to cause removal of all components inhibitory to GFOR. Consequently, for assaying GFOR and for supplementing the oxidoreductase during substrate conversions, this preparation of GL was useful after gel-filtration followed by four to five fold concentration (ultrafiltration). The Mono Q preparation was taken for the subsequent characterization of GL, although the isolation of the enzyme in homoge-

Table 2
Partial Purification of Guconolactonase from 70 mL Crude R. rubra Extract

	Total activity (U)	Total protein (mg)	Specific activity (U/mg)	Yield %	Purification factor (-fold)
Crude cell extract	1191	561	2.1	100	1.0
Ammonium sulfate (45%)	1406	483	2.9	118	1.4
Phenylsepharose fast flow	842	51	16.4	71	7.8
Mono Q	754	4.3	175.3	63	83.5

Table 3
Purification of Glucose-Fructose Oxidoreductase from 50 mL crude extract of Z. mobilis

	Total activity (U)	Total protein (mg)	Specific activity (U/mg)	Yield %	Purification factor (-fold)
Crude cell extract	2750	561	2.2	100	1.0
Ammonium sulfate (30%)	2842	483	3.3	≈100	1.4
Phenylsepharose fast flow	2236	51	20	81	7.8
Gel-filtration, ultrafiltration	1015		20-30	37	7.8 - 9.8
Mono S	563	2.3	245-310	20	114

neous form has not been fully accomplished by the ion exchange step (not shown).

GFOR was purified to apparent homogeneity by a similar procedure, and the results are shown in Table 3. The final cation exchange chromatographic step on Mono S yielded pure GFOR (SDS PAGE; not shown) although the specific activity of the resulting preparation could vary between 245–310 U/mg probably because of some inactivation of the enzyme during chromatography. The moderate yield of homogeneous enzyme is mainly a result of the concentration and desalting steps and the concomitant loss of enzyme activity.

Enzyme Characterization

GFOR and GL were characterized with regard to pH- and temperature-activity profiles, stability and specificity. The pH optimum of GFOR is between 6.5–7.0 whereas the pH dependence of GL catalysis could not be

Table 4
Activation and Inhibition of Glucose-fructose Oxidoreductase
and Gluconolactonase

GFOR activity			GL activity		
Compound	Concentration	Effect % *	Compound	Concentration	Effect % *
$NADP^+$	0.5; 2.5 mM	71; 40	$MnCl_2$	2.0 mM	156
NAD^+	0.5; 2.0 mM	93; 75	$MgCl_2$	2; 3; 5 mM	166; 181; 182
Tween 20	0.1 %	148	Tween 20	0.1 %	113
Triton X-100	0.1 %	146	Triton X-100	0.1 %	114
Sucrose	25 %	123	$MgAc_2$	2.0 mM	167
Trehalose	25 %	120	Trehalose	25 %	113
Glycerol	25 %	120	Glycerol	25 %	105
DMSO[a]	25 %	130	$CaAc_2$	0.5 mM	130
DTT[b]	5 mM	105	DTT	5 mM	≈100

*Based on standard assay at 25°C in the absence of additives; [a]dimethylsulfoxide; [b]dithiothreitol; Ac, acetate.

accurately quantitated because of increasing spontaneous hydrolysis of the lactone in alkaline regions, at least with the experimental assay lacking pH control that was employed throughout this study. The temperature optima of GFOR and GL were found to be 40°C and 32°C, respectively. The temperature effect on the GL-mediated hydrolysis probably also includes some minor pH effect because during the incubation period of 10 min a decrease in the initial pH of 6.5 to 5.5–6.0 was observed.

Tests of the substrate spectrum of GL revealed that the enzyme has a strict specificity for δ-lactones (D-glucono-[100%], D-cellobiono-[13%]; 4 mM) whereas all γ-lactones (D-glucono, D-glucurono-, D-galactono, D- and L-gulono-, L-arabono, D-ribono-; 4 mM) were not hydrolyzed. The only substrate accepted by GFOR in the reduction half-reaction is fructose that could not be replaced by other ketoses such as D-ribulose, D-xylulose, D-psicose, lactulose or L-sorbose (800 mM). Some activity was seen with 2-deoxy D-glucose (16% of the activity with the parent compound D-glucose; 400 mM). The apparent Michaelis constants of GFOR for fructose and glucose were determined as 400 mM and 20 mM, respectively.

The activity of GFOR and GL was probed in dependence of various additives (Table 4). Divalent cations especially Mn^{2+} and Mg^{2+} activated GL by more than 50%. The corresponding anion appears to be less important. GFOR was inhibited by free coenzyme $NAD(P)^+$ and activated by some nonionic detergents (Tween, Triton). For stability of GL at 25°C and 4°C (half lives 5 and >35 d, respectively), the addition of 0.1% Tween 20 and 5 mM DTT was required. NaCl (100 mM), K-gluconate (4 mM) or glycerol (25%) had slight additional stabilizing effects.

Table 5
First-order Deactivation Constants (k) of Glucose-Fructose
Oxidoreductase at 25°C

	k (h^{-1}) pH 5.5	k (h^{-1}) pH 6.4	k (h^{-1}) pH 7.3
Buffer	0.008	0.005	0.062
1 M Glucose	0.020	0.043	0.081
1 M Fructose	0.054	0.061	0.057

Stability of GFOR

The process-relevant stability of GFOR was studied with regard to two aspects considered especially important. The identification of a minimal required level of purity of the enzyme and the operational stability of this GFOR preparation during substrate conversion. The activity of various enzyme fractions differing in their specific activity according to Table 3 was monitored at 25°C and 45°C over a relevant time period in the absence of substrates or with one substrate added at a $1M$ concentration. The stability of GFOR in crude extracts of *Zymomonas* and that of the pure oxidoreductase were comparable, with a half-life of approx 55 h at 45°C and with no loss of enzyme activity at 25°C for a period of at least 2 wk. Since GFOR is an oligomeric protein that might dissociate upon dilution, its stability was assayed in dependence of the total enzyme or protein concentration (BSA added). In the concentration range of 50 µg/mL to 1 mg/mL, no effect on the stability of GFOR was observed using either crude or partially purified enzyme (20 U/mg). Substrates and products did not improve the stability of GFOR to any significant extent. In contrast, fructose and especially glucose decreased the stability of GFOR as did alkaline incubation conditions. Assuming first-order kinetics of enzyme denaturation, inactivation constants were calculated from plots of Ln (activity) vs incubation time and are summarized in Table 5. It is noteworthy that the destabilization by fructose showed no pH dependence whereas that by glucose did. Purification of GFOR to whatever extent did not change these stability characteristics significantly, and conversion studies were thus carried out with a crude *Zymomonas* enzyme preparation.

Enzymic Synthesis of Sorbitol and Gluconic Acid

The use of GFOR for the synthesis of sorbitol and gluconic acid appears to be promising because the enzyme is capable of fully converting extremely concentrated sugar solutions ($3M$ glucose, $3M$ fructose) whereas substrate inhibition is absent and product inhibition moderate *(12,18,19)*. However, in contrast to results achieved with "immobilized" GFOR, i.e., with the enzyme embedded into the matrix of permeabilized cells of

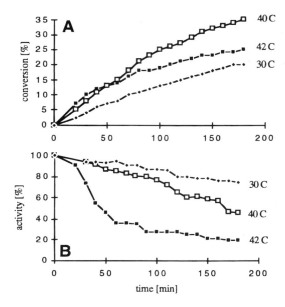

Fig. 2. Time-course of substrate conversion and residual GFOR activity in dependence of the reaction temperature. The conditions were: 1M fructose, 1M glucose, 100 mM P_i, pH 6.4 titrated with 1M KOH, 20 U/mL gluconolactonase from *R. rubra*. **(A)** Substrate conversion, **(B)** enzyme activity.

Zymomonas, the isolated enzyme is unstable during substrate conversion. The half-life of GFOR in buffer at 25–40°C is in the range of at least several days whereas substantial enzyme activity (30–40%) is lost within a few hours when the incubations are carried out in the presence of both substrates (Fig. 2A). Considering the time-courses for substrate conversion, an operational temperature optimum is difficult to define because inactivation occurs even at a reaction temperature of 30°C (Fig. 2B). However, GFOR loses activity at comparable rates at 30 and 35°C whereas substrate conversion is significantly higher at the latter temperature. All conversion/time profiles obtained at reaction temperatures above 35°C showed downward curvature indicating progressing loss of enzyme activity. The rapid inactivation of soluble GFOR is in marked contrast with the reported excellent process-stability of cell-bound GFOR at 39°C *(15)*. Destabilization by either fructose and glucose (cf Table 5) does not reflect this inactivation of the isolated enzyme. We have recently shown that the addition of 5–10 mM dithiothreitol as a protectant of sulfhydryl groups could stabilize GFOR to some extent *(18,19)*. The alkaline component employed for the titration of gluconic acid produced during reaction was also found to be important. Tris base could decrease but not completely abolish the rate of enzyme inactivation (Nidetzky and Fürlinger, unpublished results). Another important variable is the pH of the reaction mixture which is maintained by the addition of alkali during product formation (cf Table 5). At 30°C, a variation of

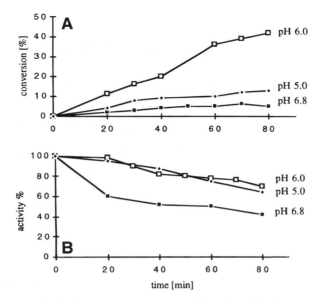

Fig. 3. Time-course of substrate conversion and residual GFOR activity at a reaction temperature of 40°C in dependence of the reaction pH. The conditions were: 1M fructose, 1M glucose, 100 mM P_i, pH 5.5–6.8 titrated with 1M KOH, 20 U/mL gluconolactonase from *R. rubra*. **(A)** Substrate conversion, **(B)** enzyme activity.

the controlled pH value in the range of 5.5–6.8 had only minor influence on the reaction rate reflecting the pH/activity profile of the enzyme and did not affect the stability of GFOR. At 40°C, however, the pH is extremely important as demonstrated by the data in Figs. 3A and B. When pH values above 7.0 were set, fast inactivation of GFOR was observed at all reaction temperatures applied. The ionic strength of the buffer system (0; 0.1; 1.0 M Pi) was found not to be a determinant of enzyme stability and activity.

Effect of Gluconolactonase

GL is useful as an auxiliary enzyme for assaying GFOR activity (cf Table 1) and, in vivo, it accelerates the hydrolysis of gluconolactone produced by GFOR. It is tempting to speculate that the presence of GL could have a beneficial effect on GFOR activity and, even more important, on its stability during substrate conversion in vitro. However, no such effect was observed when reactions were carried out in the presence or absence of GL and the resulting time-courses of product formation and inactivation of GFOR were compared. In a temperature range of 25–40°C, neither the stability of GFOR nor the space time yield of reaction was improved (when essentially complete substrate conversion was taken into consideration). For an efficient product formation in discontinuous reactions, the addition of exogenous GL is thus not required.

DISCUSSION

The results of the present work led to the identification of two previously uncharacterized strains, Z. mobilis DSM 473 and R. rubra DSM 70403, which are useful for the production of glucose-fructose oxidoreductase and gluconolactonase, respectively. According to the screening carried out here, the hypothesis of GFOR being an enzyme unique to Z. mobilis was corroborated (4). Not even traces of this oxidoreductase activity were found in any other species studied. Recently it became clear that GFOR has a very specialized physiological role in Zymomonas as it enables the organism to grow in concentrated sugar solutions while sorbitol (produced by GFOR) counteracts negative osmotic effects (30). Synthesis of GFOR by Z. mobilis DSM 473 showed a qualitatively similar dependence on the type and concentration of the carbon source as reported for Z. mobilis ATCC 29191 (5). The specific GFOR activities were comparable in both strains of Z. mobilis. The biomass formation by Z. mobilis DSM 473 was, however, two fold higher than those reported earlier (5). An alternative procedure for isolating GFOR in homogeneous form, based on hydrophobic and ion exchange chromatographies, was established, albeit the yields of pure enzyme need to be further improved. The results of a subsequent characterization of GFOR are in good accordance with the results of Zachariou & Scopes (5). When incubated in buffer with or without one of its substrates, GFOR exhibits stability characteristics typical for a mesophilic enzyme. The extreme instability of the oxidoreductase during its catalytic action remains still enigmatic, but was substantiated by analysing enzyme inactivation in dependence of the reaction temperature and pH. At least at present, the loss of GFOR activity during product synthesis certainly limits the applicability of the soluble enzyme.

R. rubra was identified as an outstanding natural producer of the auxiliary enzyme gluconolactonase. Partial purification of this enzyme was accomplished throughout this study that allowed characterization of the enzyme as well as its use for analytical purposes. The strict specificity of R. rubra GL for δ-lactones agrees with earlier reports on a gluconolactonase from Saccharomyces cerevisiae and Pseudomonas fluorescens (8,10), but distinguishes the new enzyme from other lactonases apparently hydrolyzing both δ- and γ-lactones (9). The activation by divalent cations is in accordance with data of metal analyses for a mammalian gluconolactonase indicating the presence of these ions bound to the native protein (7).

In the enzymatic synthesis of sorbitol and gluconic acid by GFOR, the presence of GL could be important. One tentative beneficial role of GL acting in combination with GFOR might be the removal of product inhibition because of the accelerated hydrolysis of the gluconolactone product in the presence of GL. However, no evidence was found in the course of the present study that would support any effect of GL during substrate conversions (either present in crude extracts of Zymomonas or added exogenously from a partially purified *R. rubra* preparation). We could detect no influence on the attainable space time yield and on GFOR stability.

In some cases the presence of GL may be even undesirable. During continuous conversions (where a certain substrate conversion has to be achieved) the average residence time in an enzyme reactor must be adjusted in accordance with the volumetric activity of the biocatalyst employed. One particular advantage of the homogeneous enzyme reactor is that high concentration of the biocatalyst can be used to guarantee product synthesis at efficient reaction rates. When taking into account that the formation of gluconic acid requires titration with alkali that in turn remains as a possible candidate responsible for some inactivation of GFOR, it might be desirable to carry out the enzymic conversions at a (dilution) rate that is significantly faster than that of the subsequent spontaneous hydrolysis of the formed glucono-δ-lactone. The first-order rate constants (h^{-1}) of the spontaneous decomposition of the 1,5-lactone at 25°C are 3.17 and 1.08 at pH 6.40 and 5.95, respectively *(6)*. These rate constants correspond to half-lives of 0.21 and 0.64 h. Considering volumetric activities of 5,000–10,000 U GFOR per-L, average residence times of 0.5 h or less seem to be feasible even when concentrated sugar solutions are to be converted. Under such conditions certainly the absence and not the presence of GL activity would be desirable and important.

ACKNOWLEDGMENTS

Part of this work was supported by the Jubiläumsfonds der Österreichischen Nationalbank, Grant 5089 to B. N.

REFERENCES

1. Milsom, P. E. and Meers, J. L. (1983), in *Comprehensive Biotechnology* vol. 3, Blanch, H., Drew, S., Wang, D. I. C., eds., Pergamon Press, Oxford, UK, pp. 681–700.
2. Schiweck, H., Vogel, R., Schwarz, E., and Kunz, M. (1994), in *Ullmann's Encyclopedia of Industrial Chemistry* 5th ed., Elvers, B., Hawkins, J., Russey, W., eds., vol. A25, VCH, Weinheim, Germany, pp. 413–437.
3. Kieboom, A. P. G. and van Bekkum, H. (1985), in *Starch Conversion Technology*, van Beynum, G. M. A. and Roels, J. A., eds., Marcel Dekker, New York, pp. 263–334.
4. Doelle, H. W., Kirk, L., Crittenden, R., Toh, H., and Doelle, M. B. (1993), *CRC Crit. Rev. Biotechnol.* **13**, 57–98.
5. Zachariou, M. and Scopes, R. K. (1986), *J. Bacteriol.* **167**, 863–869.
6. Mitchell, R. E. and Duke, F. R. (1970), *Ann. N. Y. Acad. Sci.* **172**, 129–138.
7. Carper, W. R., Mehra, A. S., Campbell, D. P., and Levisky, J. A. (1982), *Experientia* **38**, 1046–1047.
8. Constantinides, A., Myles, S. J., and Vieth, W. R. (1972), *Enzymologia* **43**, 121–128.
9. Yamada, K. (1959), *J. Biochem.* **46**, 361–372.
10. Jermyn, M. A. (1960), *Biochim. Biophys. Acta* **37**, 78–92.
11. Chun, U. H. and Rogers, P. L. (1988), *Appl. Microbiol. Biotechnol.* **29**, 19–24.
12. Scopes, R. K., Rogers, P. L., and Leigh, D. A (1988), US Patent 4755467.
13. Ichikawa, Y., Kitamoto, Y., Kato, N., and Mori, N. (1988), EP 0 322 723 A2.

14. Paterson, S. L., Fane, A. G., Fell, C. J. D., Chun, U. H., and Rogers, P. L. (1988), *Biocatalysis* **1,** 217–229.
15. Rehr, B., Wilhelm, C., and Sahm, H. (1991), *Appl. Microbiol. Biotechnol.* **35,** 144–148.
16. Roh, H.-S. and Kim, H.-S. (1992), *Enzyme Microb. Technol.* **13,** 920–924.
17. Kim, D.-M. and Kim, H.-S. (1992), *Biotechnol. Bioeng.* **39,** 336–342.
18. Gollhofer, D., Nidetzky, B., Fürlinger, M., and Kulbe, K. D. (1995), *Enzyme Microb. Technol.* **17,** 235–240.
19. Nidetzky, B., Gollhofer, D., Fürlinger, M., Haltrich, D., and Kulbe, K. D. (1995), in *Biochemical Engineering 3*, Schmid, R. D., ed., Kurz & Co., Stuttgart, Germany, pp.184–186.
20. Ikemi, M. and Ishimatsu, Y. (1990), *J. Biotechnol.* **14,** 211–220.
21. Howaldt, M. W., Kulbe, K. D. and Chmiel, H. (1990), *Ann. N. Y. Acad. Sci.* **589,** 253–260.
22. Chenault, H. K. and Whitesides, G. M. (1987) *Appl. Microbiol. Biotechnol.* **14,** 147–197.
23. Kula, M.-R. and Wandrey, C. (1987), *Meth. Enzymol.* **136,** 9–21.
24. Ikemi, M., Koizumi, N., and Ishimatsu, Y. (1990), *Biotechnol. Bioeng.* **36,** 149–154.
25. Ikemi, M., Ishimatsu, Y., and Kise, S. (1990), *Biotechnol. Bioeng.* **36,** 155–16526.
26. Nidetzky, B., Haltrich, D., and Kulbe, K. D. (1996), *Chemtech* **26,** 31–36.
27. Kanagasundaram, V. and Scopes, R. K. (1992), *J. Bacteriol.* **174,** 1439–1447.
28. Loos, H., Völler, M., Rehr, B., Stierhof, Y.-D., Sahm, H., and Sprenger, G. A. (1991), *FEMS Microbiol. Lett.* **84,** 211–216.
29. Bradford, M. M. (1976), *Anal Biochem* **72,** 248–254.
30. Loos, H., Krämer, M., Sahm, H., and Sprenger, G. A. (1994), *J. Bacteriol.* **176,** 7688–7693.

Production of Hemicellulose- and Cellulose-Degrading Enzymes by Various Strains of *Sclerotium Rolfsii*

ALOIS SACHSLEHNER, DIETMAR HALTRICH,* BERND NIDETZKY, AND KLAUS D. KULBE

Abteilung Biochemische Technologie, Institut für Lebensmitteltechnologie Universität für Bodenkultur Wien (BOKU), Muthgasse 18, A-1190 Wien, Austria

ABSTRACT

A number of wild-type isolates of *Sclerotium rolfsii* were screened for their capacity to produce lignocellulolytic enzymes when grown on a cellulose-based medium. *S. rolfsii* proved to be an efficient producer of hemicellulolytic enzymes under the conditions selected for this screening, although there was a great variability in enzyme activities formed by the different isolates. In addition to xylanase and mannanase, which were produced in remarkably high levels, a number of accessory enzymes, which are important for the complete degradation of substituted hemicelluloses and include α-arabinosidase, acetyl esterase, and α-galactosidase, are formed by *S. rolfsii*. Efficient production of xylanase and mannanase was achieved when cellulose-based media were used for growth. Under these conditions, enhanced levels of endoglucanase were formed as well. Formation of xylanase and mannanase could be more specifically induced when using xylan or mannan as growth substrates, although the enzyme activities thus obtained were significantly lower compared to cultivations on cellulose as main inducing substrate.

Index Entries: *Sclerotium rolfsii*; xylanase; mannanase; cellulase; screening.

*Author to whom all correspondence and reprint requests should be addressed.

INTRODUCTION

Plant cell walls are a major reservoir of fixed carbon in nature. In recent years there has been considerable interest in the utilization of plant material as a renewable source of fermentable sugars that could be subsequently converted into useful products such as liquid fuels, solvents, chemicals, food, or feed (1). Such bioconversion processes are particularly attractive for the elimination of residues and wastes produced by agriculture and forestry. Hemicellulose is one of the three main polymeric constituents of lignocellulose and comprises several heterogeneous groups of polysaccharides that are combined to this group on essentially practical and historical reasons such as solubility in alkali and application of chemical extraction procedures.

Xylans are the most abundant noncellulosic polysaccharides in hardwood and annual plants, where they account for 15–30% of the total dry weight. In softwoods, they are found in lesser quantities, accounting for approx 7–10% of the tissue dry weight. The basic structure of xylans is a main chain of (1→4)-linked β-D-xylopyranosyl residues. Typically, the linear chains of the xylans carry short side chains to a varying extent. Hardwood xylans from dicotyledonous angiosperms grown in temperate zones are all remarkably alike, showing only little variation. Typical substituents for hardwood xylans are (1→2)-linked 4-O-methyl-α-D-glucuronic acid residues and acetyl groups, which are attached at the O-2 or O-3 sites. Softwood xylans typically carry 4-O-methyl-α-D-glucuronic acid and L-arabinofuranose side groups. These side groups are linked at O-2 and O-3 of the xylose units in the main chain. Uronic acid residues seem to be rather localized on contiguous xylose residues than randomly dispersed over the molecule (2,3).

Mannans, or more appropriately galactoglucomannans, are the main hemicelluloses in softwoods, where their content varies between 15–20% of the total dry weight. In hardwood mannans are found in lesser quantities, constituting only up to 5%. The main chain of mannans is made up of β-(1→4)-linked D-mannopyranosyl and D-glucopyranosyl residues with widely varying proportions of these two types of sugar units. In softwood mannans α-D-galactosyl units are attached to the main chain units by (1→6)-bonds to a varying extent. The mannose and glucose units in the backbone are partially substituted at O-2 and O-3 by acetyl groups, whereas these side groups are not found in hardwood mannans (2–4).

Because of this complex structure of hemicelluloses, several different enzymes are necessary for their complete enzymatic degradation. The two main glycanases hydrolyzing the hemicellulose main chains are endo-(1→4)-β-D-xylanase and endo-(1→4)-β-D-mannanase. Small oligosaccharides formed by the action of these glycanases are further cleaved by β-D-xylosidase, β-D-mannosidase, and β-D-glucosidase. The side group substituents are liberated by α-L-arabinosidase, α-D-glucuronidase, α-D-galactosidase, and various esterases including acetyl esterase (5,6).

Sclerotium rolfsii (or *Athelia rolfsii*, which is used for the teleomorph) is an aggressive plant pathogen of many crop plants in the tropics and subtropics. The fungus colonizes organic matter in the soil from where it may parasitize certain plants. During its attack on plant material it forms large amounts of different enzymes that rapidly destroy host tissue and cell walls, thus enabling it to enter the host organism. *S. rolfsii* is known as an excellent producing strain of cellulolytic enzymes *(7,8)*. Furthermore, it forms high levels of a number of hemicellulolytic enzymes including xylanase and mannanase *(9)*.

It was the objective of our work to investigate the production of the spectrum of hemicellulose-degrading enzymes by different wild-type isolates of *S. rolfsii*. In addition, the effect of different carbon sources on the levels of xylanase and mannanase should be assessed. It was of special interest to investigate whether by selecting an appropriate inducing substrate high levels of hemicellulases with only low levels of concurrently produced cellulase could be attained, since hemicellulase preparations free of cellulase have gained significant interest during the last years because of their application in the pulp and paper industry *(10,11)*.

MATERIALS AND METHODS

Chemicals

α-Cellulose, *p*-nitrophenyl glycosides, α-naphthyl acetate, 2,6-dichlorophenol-indophenol, locust bean gum (a galactomannan from *Ceratonia siliqua* with a mannose-to-galactose ratio of 4:1), and guar gum (a galactomannan from *Cyamopsis tetragonobola* with a mannose-to-galactose ratio of 2:1) were from Sigma (St. Louis, MO); lactose, L-sorbose, D-xylose, and carboxymethylcellulose were from Fluka (Buchs, Switzerland). Konjac mannan, a glucomannan from *Amorphophallus konjac* with a mannose-to-glucose ratio of 1.8:1, was obtained from Arkopharma (Carros, France). Xylan from birchwood was from Roth (Karlsruhe, Germany) and peptone from meat was from Merck (Darmstadt, Germany). All other chemicals were analytical grade.

Organisms and Culture Conditions

Strains of *Sclerotium (Athelia) rolfsii* were obtained from Centraalbureau voor Schimmelcultures (CBS, Baarn, The Netherlands) or from the American Type Culture Collection (ATCC, Rockville, MD). Stock cultures were maintained on glucose-maltose Sabouraud agar and routinely subcultured every 4 wk. Inoculated plates were incubated at 30°C for 4–6 d and then stored at 4°C.

All strains were cultivated in unbaffled 300-mL Erlenmeyer flasks at 30°C for 13 d on a medium containing (in g/L): peptone from meat, 80; NH_4NO_3, 2.5; $MgSO_4 \cdot 7H_2O$, 1.5; KH_2PO_4, 1.2; KCl, 0.6, and trace element solution, 0.3 (mL/L) *(9)*. Unless otherwise indicated, 42.6 g/L α-cellulose were used as inducing substrate. All media were prepared with tap water. The pH-value was adjusted to 5.0 using phosphoric acid prior to sterilization. The trace element solution comprised (in g/L): $ZnSO_4 \cdot H_2O$, 1.0; $MnCl_2 \cdot 4H_2O$, 0.3; H_3BO_3, 3.0; $CoCl_2 \cdot 6H_2O$, 2.0; $CuSO_4 \cdot 5H_2O$, 0.1; $NiCl_2 \cdot 6H_2O$, 0.2, and H_2SO_4 conc., 4.0 (mL/L). Shake flasks were inoculated with a piece (1 cm^2) from an actively growing, 4–6-d-old colony of *S. rolfsii* on Sabouraud agar. The inoculated flasks were shaken continuously on an orbital shaker at 150 rpm (stroke 25 mm) and 30°C for 13 d. The culture was then centrifuged and the clear supernatant used for the estimation of enzyme activities.

Fermentation studies were carried out in a 20-L laboratory fermenter (MBR Bio Reactor, Wetzikon, Switzerland) with a working volume of 15 L and equipped with four disc turbine impellers, each with six flat blades.

Enzyme Activity Assays

All activity assays were carried out in 0.05*M* sodium citrate buffer, pH 4.5, unless otherwise stated. *Xylanase* (EC 3.2.1.8) activity was assayed using a 1% solution of xylan (4-*O*-methyl glucuronoxylan from birchwood; Roth Ltd., Karlsruhe, FRG) as a substrate *(12)*. The release of reducing sugars during 5 min at 50°C was measured as xylose equivalents using the dinitrosalicylic acid (DNS) method *(13)*. *Mannanase* (EC 3.2.1.78) and *endoglucanase* (carboxymethylcellulase, EC 3.2.1.4) activities were assayed similar to the determination of xylanase activity, using a 0.5% solution of locust bean gum galactomannan in 0.05*M* sodium citrate buffer, pH 4.0, or a 1% solution of carboxymethylcellulose (sodium salt, ultra low viscosity), respectively, as the substrates. Reducing sugars were assayed as mannose or glucose using the DNS method. Filter paper cellulase activity was measured according to IUPAC recommendations employing filter paper (Whatman No.1, Maidstone, UK) as a substrate *(14)*. One unit (IU) of enzyme activity is defined as the amount of enzyme liberating 1 μmol of xylose, mannose, or glucose equivalents per minute under the given conditions. 1 IU corresponds to 16.67 nkat.

α-*Arabinosidase* (α-L-arabinofuranosidase, EC 3.2.1.55), α-*galactosidase* (EC 3.2.1.22), β-*glucosidase* (EC 3.2.1.21), β-*mannosidase* (EC 3.2.1.25), and β-*xylosidase* (EC 3.2.1.37) were quantified in a similar manner, using the respective *p*-nitrophenyl-glycosides as substrates. Buffer (0.5 mL) was incubated with 0.25 mL of the appropriately diluted enzyme solution and 0.25 mL of substrate solution (8 m*M*) at 50°C for 10 min. The reaction was stopped by adding 2.0 mL of 1*M* Na_2CO_3 and the absorbance measured at 405 nm. Activities are expressed on the basis of the liberation of *p*-nitrophenol.

Acetyl esterase (EC 3.1.1.6) activity was determined using 1 m*M* α-naphthylacetat as the substrate *(15)*. One unit of enzyme activity is expressed as the amount of enzyme liberating 1μmol α-naphthol per minute.

The assay of *cellobiose dehydrogenase* (EC 1.1.99.18) activity was essentially as described by Sadana and Patil (16) using 1.8 mM cellobiose, 2.0 mM glucono-α-lactone, and 0.36 mM 2,6-dichlorophenol-indophenol in 100 mM-phosphate buffer, pH 6.3. Activities are expressed on the basis of the reduction of dichlorophenol-indophenol.

Protein Assays

Protein concentrations were determined according to the dye-binding method of Bradford (17) using bovine serum albumin (fraction V) as standard.

RESULTS

Screening of *S. rolfsii* Strains

Eight different strains of *S. rolfsii*, isolated worldwide at different geographic locations, were cultivated in shaken flasks using a cellulose-based medium. After 13 d of growth various hemicellulose- and cellulose-degrading enzyme activities produced by these different isolates were measured and compared. Results for xylanolytic, mannanolytic, and cellulolytic enzyme activities are shown in Figs. 1–3, respectively. The different strains produced all of the lignocellulose-degrading enzymes that were investigated in this study, albeit at greatly varying levels. Highest activities of xylanase were produced by *S. rolfsii* CBS 191.62 and *S. rolfsii* CBS 147.82. Furthermore, these two strains yielded the highest activities of β-xylosidase and acetyl esterase. On the other hand, xylanolytic enzyme activities formed by *S. rolfsii* CBS 149.82, which gave the lowest enzyme yields for most of the activities investigated, were only in the range of 15–25% of the maximal values produced by *S. rolfsii* CBS 191.62. Growth of *S. rolfsii* CBS 191.62 also resulted in the highest mannanase activity. Interestingly, the activity levels of the other mannan-degrading enzymes, i.e., β-mannosidase and α-galactosidase, produced by this strain were much lower in relation to the maximum values of 0.53 IU/mL for β-mannosidase formed by *S. rolfsii* CBS 151.31 and 26.4 IU/mL for α-galactosidase produced by *S. rolfsii* CBS 147.82. Highest activities of most cellulolytic enzymes were produced by *S. rolfsii* CBS 147.82. In addition to high activities of FP-cellulase (10.1 IU/mL) and endoglucanase (1800 IU/mL), a culture filtrate obtained after growth of this strain also contained high levels of β-glucosidase activity (21.2 IU/mL). Whereas the two former enzymes were produced in a similar range of activities (9.21 and 1640 IU/mL, respectively) by *S. rolfsii* CBS 191.62, β-glucosidase levels were considerably lower (2.8 IU/mL).

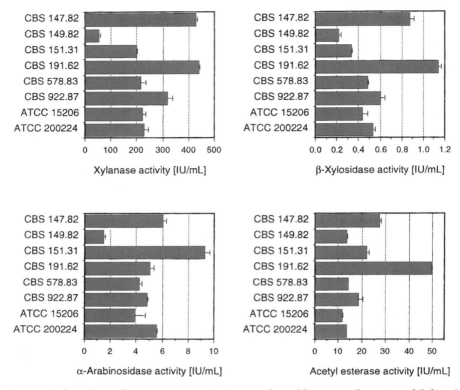

Fig. 1. Xylan-degrading enzyme activities produced by several strains of *Sclerotium rolfsii* when cultivated at 30°C on a cellulose-based medium for 13 d. Values are shown as the mean and standard deviation of two independent replicates.

Bioprocess Experiments

Strain *S. rolfsii* CBS 191.62 as the best producer of both xylanase and mannanase was selected for further studies. Production of xylanase, mannanase, and endoglucanase was followed in a 15-L laboratory fermentation using 42.6 g/L α-cellulose as the substrate. The inoculum was an 11-d-old shaken culture grown on the fermentation medium. The time course of this cultivation is shown in Fig. 4. Production of enzymes started after a lag of approx 30 h. This initial phase of growth was also accompanied by a rapid decrease in the pH from 4.8 to 3.3. Maximum xylanase and mannanase values of 192 and 568 IU/mL were reached after approx 165 h, corresponding to volumetric productivities of 1160 and 3440 IU/L · h, respectively. Thereafter, both enzyme activities remained constant for at least 3 d. Endoglucanase activity peaked after 213 h of cultivation reaching a value of 1290 IU/mL and then started to decrease slightly.

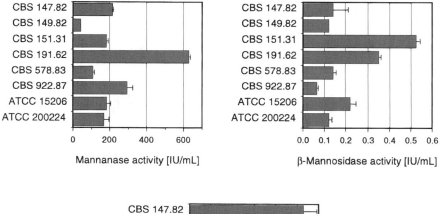

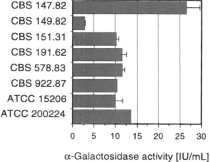

Fig. 2. Mannan-degrading enzyme activities produced by several strains of *Sclerotium rolfsii* when cultivated at 30°C on a cellulose-based medium for 13 d. Values are shown as the mean and standard deviation of two independent replicates.

Effect of Inducing Substrates

It was of special interest to investigate whether xylan or mannan when used as main carbon source could specifically provoke the synthesis of the enzyme activities necessary for the degradation of the respective polysaccharide by *S. rolfsii* CBS 191.62. Furthermore, small amounts of L-sorbose, D-xylose, or lactose, which were reported to enhance both xylanase and cellulase yields by *Trichoderma reesei* (18,19), were added to the cellulose-based medium. The level of extracellular protein as well as xylanase, mannanase, and endoglucanase activities obtained after 13 d of growth on these different media are shown in Table 1. Highest levels of all three enzyme activities were obtained when α-cellulose was the main inducing substrate. Replacing a minute part of this C source by L-sorbose (1.5 g/L) resulted in a small, yet significant increase in xylanase activity. Formation of endoglucanase was slightly lower after growth on the medium containing L-sorbose, whereas mannanase activity was decreased by approx 20%. Whereas the supplementation of 4.3 g/L of lactose had no

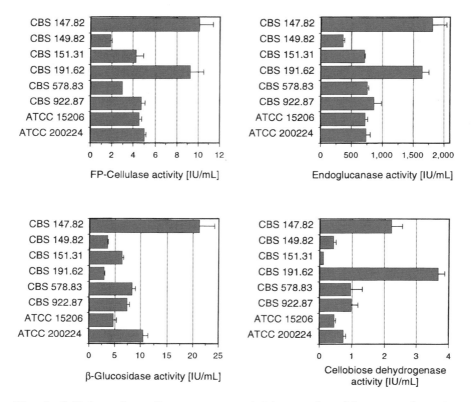

Fig. 3. Cellulose-degrading enzyme activities produced by several strains of *Sclerotium rolfsii* when cultivated at 30°C on a cellulose-based medium for 13 d. Values are shown as the mean and standard deviation of two independent replicates.

significant effect on the levels of enzymes formed by *S. rolfsii*, 4.3 g/L of D-xylose when added to the cellulose-based medium considerably reduced all three of the enzyme activities produced.

Production of mannanase could not be enhanced when several galacto- and glucomannans were used as sole inducing C sources or when these were added to cellulose. The mannans were each employed in a concentration of 21.3 g/L in these experiments. This resulted in a highly viscous culture medium that certainly caused severe mass transport limitations. However, this high viscosity was greatly reduced after only 2–3 d of cultivation because of the action of the endo-acting mannanases, which were excreted by the organism and cleaved the mannan main chains. In all cases, significantly higher mannanase activities were obtained when a mixture of mannan and α-cellulose (each at 21.3 g/L) was used as compared to a culture medium containing only mannan as main C source (Table 1). Similarly, the use of xylan did not yield higher levels of xylanase than a cultivation on an equal amount of α-cellulose.

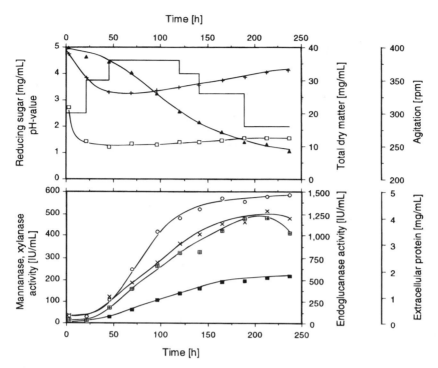

Fig. 4. Time course of xylanase, mannanase, and endoglucanase production by *Sclerotium rolfsii* CBS 191.62 in a 20-L stirred tank reactor (working volume 15 L). α-Cellulose (42.6 g/L) was used as the substrate (9). The temperature was controlled at 30°C and the pH, initially adjusted to 5.0, was allowed to float. Aeration was automatically varied from 0.1 to 1.0 vol of air per fluid vol per min to maintain a pO_2 of 40% of air saturation. Symbols : +, pH-value; ▲, total dry matter; □, reducing sugar; —, agitation; ⊞, extracellular protein; ○, mannanase; ■, xylanase; ×, endoglucanase.

Whereas xylan or mannan did not increase the production of xylanase or mannanase when employed as the main inducing substrates, these two polysaccharides could specifically induce the synthesis of the enzymes necessary for their hydrolysis. This is expressed by the ratios of xylanase, mannanase, and endoglucanase activities to each other that were calculated from the experimental data and are listed in Table 1. The ratio of xylanase to endoglucanase was fairly constant for the cellulose-based media (0.22–0.27). For the cultivation on xylan this value was significantly increased to 1.59, indicating that relatively more xylanase than endoglucanase is formed when *S. rolfsii* is grown on this substrate. In a similar manner, the ratio of mannanase to endoglucanase shows that mannans when employed as the only C source specifically provoke the synthesis of mannanase activity with only low levels of endoglucanase or xylanase formed. The ratio of mannanase to endoglucanase was increased to 4.06 and 5.49, respectively, when guar gum or locust bean gum galactomannan were

Table 1
Effect of Various Inducing Substrates used for Growth of *Sclerotium rolfsii* CBS 191.62 on the Formation of Extracellular Protein and Activities of Xylanase, Mannanase, and Endoglucanase[c]

Inducing substrate [g/L]	Extracellular protein [mg/mL]	Xylanase [IU/mL]	Mannanase [IU/mL]	Endo-glucanase [IU/mL]	Ratio xylanase to endoglucanase	Ratio mannanase to endoglucanase	Ratio mannanase to xylanase
Cellulose (42.6)	3.83 ± 0.11	257 ± 10	718 ± 9	1176 ± 59	0.22	0.61	2.79
Cellulose (38.3) + lactose (4.3)	4.04 ± 0.22	267 ± 7	724 ± 17	1165 ± 14	0.23	0.62	2.71
Cellulose (41.1) + sorbose (1.5)	3.83 ± 0.14	276 ± 18	580 ± 34	1104 ± 9	0.25	0.53	2.10
Cellulose (38.3) + xylose (4.3)	3.18 ± 0.10	181 ± 11	590 ± 11	765 ± 6	0.24	0.77	3.27
Cellulose (21.3) + guar gum (21.3)	3.24 ± 0.06	171 ± 6	528 ± 16	791 ± 38	0.22	0.67	3.09
α-Cellulose (21.3) + konjac GM[a] (21.3)	3.01 ± 0.30	176 ± 21	547 ± 28	645 ± 21	0.27	0.85	3.11
α-Cellulose (21.3) + LBG[b] (21.3)	2.98 ± 0.20	164 ± 18	535 ± 30	750 ± 124	0.22	0.71	3.26
Guar gum (21.3)	1.04 ± 0.07	12.2 ± 0.3	169 ± 17	41.6 ± 1.1	0.29	4.06	13.85
LBG[b] (21.3)	1.01 ± 0.06	7.8 ± 0.6	146 ± 6	26.6 ± 4.3	0.29	5.49	18.72
Xylan birchwood (42.6)	0.41 ± 0.04	4.6 ± 0.7	10.2 ± 2.5	2.9 ± 0.5	1.59	3.52	2.22

[a]Konjac glucomannan.
[b]Locust bean gum.
[c]Each value represents the mean ± standard deviation ($n = 2$). In addition, the ratio of these enzyme activities to each other are given.

used as the inducing substrates. This ratio was found to be in the range of 0.53–0.77 for the cellulose-based media, and even for a mixture of cellulose and mannan, both in equal concentrations, it was not significantly higher than this value.

DISCUSSION

Eight wild-type strains of *S. rolfsii*, which were screened for their capacity to produce various xylan-, mannan-, and cellulose-degrading enzymes, were all found to produce the enzyme activities investigated, although to a greatly varying extent. This variation, i.e., the ratio of the lowest to the highest values obtained, ranged from 4.3-fold for acetyl esterase to 38.5-fold for cellobiose dehydrogenase with the different strains studied.

Highest levels of cellulolytic enzyme activities were produced by *S. rolfsii* CBS 147.82. The ratio of β-glucosidase to FP-cellulase activity, which is important for efficient hydrolysis of cellulose, is significantly higher for an enzyme preparation obtained from this strain than that found for different *T. reesei* mutants *(20)*. It appears to be sufficient to hydrolyze cellulosic material at least at low substrate concentrations *(21)*, thus eliminating the need of supplementing additional β-glucosidase. The β-glucosidase to FP-cellulase ratio varied considerably with the different fungal isolates considered in this study. The lowest value of 0.3 was found for strain CBS 191.62, whereas this ratio was calculated to be 2.8 for strain CBS 578.83.

Strain *S. rolfsii* CBS 191.62 was superior in terms of both xylanase and mannanase activities produced. Especially the maximum value of 600–700 IU/mL of mannanase activity is remarkable, since this value is significantly higher than mannanase activities produced by most other microorganisms and appears to be among the highest mannanase values ever reported *(22–25)*. In addition to glycanases, this strain also formed high levels of accessory enzyme activities that are needed for a complete degradation of substituted xylans or mannans. Interestingly, production of glycanases, which hydrolyze the polysaccharide main chains, and of auxiliary enzymes did not correlate with the different strains of *S. rolfsii* investigated in the screening. Production of relatively high activities of these glycanases does not necessarily have to be accompanied by relatively high levels of auxiliary enzymes. In the case that these accessory enzymes are of primary interest, certain strains of *S. rolfsii* could be better suited for the production of these enzymes.

The type of inducing substrate had an important effect on the production of xylanase, mannanase, and endoglucanase by *S. rolfsii* CBS 191.62. Surprisingly, highest levels of all three enzyme activities were produced when the organism was grown on a medium containing cellulose as the main substrate, whereas xylan from birchwood as well as several gluco- and galactomannans as C source yielded only low levels of xylanase or mannanase. Similar observations pertaining to mannanase production have been reported for *Polyporus versicolor, Schizophyllum commune*, and *Trichoderma reesei*. Culture filtrates from the organisms grown on cellulose showed a considerable increase in mannanase activity over those obtained after growth on various mannans *(23,26,27)*. Whereas cellulose seems to be necessary for obtaining high levels of xylanase or mannanase activity, a more specific induction of the synthesis of these two enzymes by *S. rolfsii* could be achieved when xylan or mannan were used as main C source. When employing galactomannans as growth substrates, formation of mannanase activity was reduced as compared to the values obtained after growth on cellulose, yet the ratios of mannanase to endoglucanase or xylanase were increased by a factor of approx 8, indicating that relatively more mannanase was produced. A similar effect was observed for the ratio of xylanase to endoglucanase when comparing the values obtained after growth on xylan or cellulose.

A possible explanation for this could be a complex regulatory control of the synthesis of the multiple mannanases in *S. rolfsii*, where some mannanases are specifically induced in the presence of mannans, whereas the synthesis of other isoforms could be under a common control together with cellulases and possible xylanases. An alternative explanation could be the unspecificity of certain glycanases that not only show endoglucanase but mannanase and/or xylanase activity as well. This unspecificity of glycanases has been shown for certain enzymes from *Trichoderma harzianum* exhibiting mannanase and endoglucanase activities *(28)* as well as for *T. reesei* where unspecific glycanases hydrolyzed both xylan and cellulose *(29)*. However, at least one of the mannanases produced by *S. rolfsii* has recently been reported to be specific for various mannans *(30)*. The regulation of xylanase, mannanase, and endoglucanase synthesis in *S. rolfsii* is currently under investigation in the authors' laboratory.

ACKNOWLEDGMENT

The authors would like to thank Miss Marianne Prebio for linguistic corrections of the manuscript. This work was supported by grants from the Fonds zur Förderung der wissenschaftlichen Forschung P10753-MOB which the authors gratefully acknowledge.

REFERENCES

1. Kuhad, R. C. and Singh, A. (1993), *CRC Critical Rev. Biotechnol.* **13**, 151–172.
2. Stephen, A. M. (1983), in *The Polysaccharides*, vol. 3, Aspinall, G. O., ed., Academic Press, New York, London, pp. 97–193.
3. Eriksson, K.-E. L., Blanchette, R. A., and Ander, P. (1990), *Microbial and Enzymatic Degradation of Wood and Wood Components*, Springer, Berlin.
4. Ward, O. P. and Moo-Young, M. (1989), *CRC Critical Rev. Biotechnol.* **8**, 237–274.
5. Biely, P. (1985), *Trends Biotechnol.* **3**, 286–290.
6. Poutanen, K., Tenkanen, M., Korte, H., and Puls, J. (1991), in *Enzymes in Biomass Conversion*, ACS Symp. Ser. vol. 460, Leatham, G. F. and Himmel M. E., ed., American Chemical Society, Washington, pp. 426–436.
7. Lachke, A. H. and Deshpande, M. V. (1988), *FEMS Microbiol. Rev.* **54**, 177–194.
8. Kurosawa, K., Hosoguchi, M., Hariantono, J., Sasaki, H., and Takao, S. (1989), *Agric. Biol. Chem.* **53**, 931–937.
9. Haltrich, D., Laussamayer, B., and Steiner, W. (1994), *Appl. Microbiol. Biotechnol.* **42**, 522–530.
10. Biely, P. (1991), in *Enzymes in Biomass Conversion*, ACS Symp. Ser. vol. 460, Leatham, G. F. and Himmel M. E., ed., American Chemical Society, Washington, pp. 408–416.
11. Viikari, L., Kantelinen, A., Sundquist, J., and Linko, M. (1994), *FEMS Microbiol. Rev.* **13**, 335–350.
12. Bailey, M. J., Biely, P., and Poutanen, K. (1992), *J. Biotechnol.* **23**, 257–270.
13. Miller, G. L. (1959), *Anal. Chem.* **31**, 426–428.
14. Ghose, T. K. (1987), *Pure Appl. Chem.* **59**, 257–268.
15. Poutanen, K. and Puls, J. (1988), *Appl. Microbiol. Biotechnol.* **28**, 425–432.
16. Sadana, J. C. and Patil, R. V. (1985), *J. Gen. Microbiol.* **131**, 1917–1923.
17. Bradford, M. M. (1976), *Anal. Biochem.* **72**, 248–254.
18. Warzywoda, M., Larbre, E., and Pourquié, J. (1992), *Bioresource Technol.* **39**, 125–130.
19. Haapala, R., Parkkinen, E., Suominen, P., and Linko, S. (1995), *Appl. Microbiol. Biotechnol.* **43**, 815–821.
20. Esterbauer, H., Steiner, W., Labudova, I., Hermann, A., and Hayn, M. (1991), *Bioresource Technol.* **36**, 51–65.

21. Breuil, C., Chan, M., Gilbert, M., and Saddler, J. N. (1992), *Bioresource Technol.* **39,** 139–142.
22. Rättö, M. and Poutanen, K. (1988), *Biotechnol. Lett.* **10,** 661–664.
23. Johnson, K. G. (1990), *World J. Microbiol. Biotechnol.* **6,** 209–217.
24. Araujo, A. and Ward, O. W. (1990), *J. Ind. Microbiol.* **6,** 171–178.
25. Farrell, R. L., Biely, P., and McKay, D. L. (1996), in *Biotechnology in the Pulp and Paper Industry*, Srebotnik, E. and Messner K., ed., Facultas-Universitätsverlag, Vienna, pp. 485–489.
26. Arisan-Atac, I., Hodits, R., Kristufek, D., and Kubicek, C. P. (1993), *Appl. Microbiol. Biotechnol.* **39,** 58–62.
27. Haltrich, D. and Steiner, W. (1994), *Enzyme Microb. Technol.* **16,** 229–235.
28. Torrie, J. P., Senior, D. J., and Saddler, J. N. (1990), *Appl. Microbiol. Biotechnol.* **34,** 303–307.
29. Hrmová, M., Biely, P., and Vrsanská, M. (1986), *Arch. Microbiol.* **144,** 307–311.
30. Gübitz, G. M., Hayn, M., Urbanz, G., and Steiner, W. (1996), *J. Biotechnol.* **45,** 165–172.

Asparaginase II of *Saccharomyces cerevisiae*

GLN3/URE2 Regulation of a Periplasmic Enzyme

ELBA P. S. BON,*,[1] ELVIRA CARVAJAL,[2] MIKE STANBROUGH,[3] DONALD ROWEN,[3] AND BORIS MAGASANIK[3]

[1]*Instituto de Química, Universidade Federal do Rio de Janeiro;*
[2]*Departamento de Biologia Celular e Genética, Instituto de Biologia, Universidade Estadual do Rio de Janeiro, Rio de Janeiro, RJ, Brasil;*
and [3]*Department of Biology, Massachusetts Institute of Technology, Cambridge, MA*

ABSTRACT

The production of some extracellular enzymes is known to be negatively affected by readily metabolized nitrogen sources such as NH_4^+ although there is no consensus regarding the involved mechanisms. Asparaginase II is a periplasmic enzyme of *Saccharomyces cerevisiae* encoded by the *ASP3* gene. The enzyme activity is not found in cells grown in either ammonia, glutamine, or glutamate, but it is found in cells that have been subjected to nitrogen starvation or have been grown on a poor source of nitrogen such as proline. In this report it is shown that the formation of this enzyme is dependent upon the functional *GLN3* gene and that the response to nitrogen availability is under the control of the *URE2* gene product. In this respect the expression of ASP3 is similar to the system that regulates the *GLN1*, *GDH2*, *GAP1*, and *PUT4* genes that codes for glutamine synthetase, NAD-linked glutamate dehydrogenase, general amino-acid permease, and high affinity proline permease, respectively.

Index Entries: Asparaginase II; Nitrogen regulation; *GLN3*; *URE2*; Gene expression; *Saccharomyces cerevisiae*.

*Author to whom all correspondence and reprint requests should be addressed.

INTRODUCTION

The production of some extracellular enzymes such as proteases and ligninases is known to be negatively affected by eadily metabolized nitrogen sources such as NH_4^+. The majority of the data argues in favor of repressive effects rather than imbalance in the metabolic pool (1–8). At present there is substantial molecular data on yeast nitrogen regulation for some intracellular enzymes and permeases where the Ure2p/Gln3p transcription system plays a central role (9). The aim of the present study was to investigate if this same molecular system would regulate the formation of the extracellular yeast asparaginase II in a similar fashion. From the industrial and technological point of view, the understanding of the nitrogen regulatory mechanisms of gene expression for extracellular enzymes would be a tool for process optimization. The data for the nitrogen regulation of extracellular asparaginase could be further explored to provide adequate conditions for a range of extracellular enzymes production.

Asparaginase II (L-asparagine amidohydrolase, EC 3.5.1.1) located in the pericellular space of *Saccharomyces cerevisiae* is induced by nitrogen starvation, and not by asparagine (10). Growth on poor nitrogen substrates like proline results in intermediate to high enzyme activity, whereas complete nitrogen starvation for several hours in the presence of glucose results in the highest asparaginase II activity (10,11). *ASP3*, which encodes asparaginase II, has been cloned and was used to demonstrate that *ASP3* mRNA was absent from ammonia grown cells and that ASP3 mRNA accumulated for several hours during nitrogen starvation (12). The apparent correlation between the increase of asparaginase II activity and of *ASP3* mRNA during nitrogen starvation suggests that transcriptional control is the primary determinant of asparaginase II activity.

Two classes of mutants that affect the ability to produce asparaginase II have been described: mutants that produce high levels of asparaginase II during growth on repressing nitrogen sources such as asparagine or ammonia, and mutants that produce lower than normal amounts of asparaginase II activity during nitrogen starvation. The strongest of the derepressing class of mutants, *gdhCR* (also known as *and 4*), is known to be allelic to *ure2* (11,13,14). The URE2 product is a negative regulator of the transcriptional activator Gln3p such that *ure2* mutants produce high levels of mRNAs of nitrogen-regulated genes like *GLN1*, *GDH2*, *GAP1*, and *PUT4* that codes for glutamine synthetase NAD-linked glutamate dehydrogenase, general amino-acid permease and high affinity proline permease, respectively during growth on the repressing nitrogen source glutamine (15–19). The strongest of the repressing class of mutants, *asp6*, was shown to be independent of the structural gene *ASP3* and produced no asparaginase II activity during nitrogen starvation (20). The authors have now examined the formation of asparaginase II in mutant cells with insertions in *URE2* and/or in *GLN3* and that consequently lack the corresponding protein products. By introducing plasmids carrying the *GLN3* or *URE2* genes to these mutants, it is shown that the ability of cells to form asparaginase II requires a functional *GLN3* gene and that the response to the availability of nitrogen requires a functional *URE2* gene.

MATERIALS AND METHODS

Strains and Plasmids

The *Saccharomyces cerevisiae* strains and plasmids used in this work are listed in Table 1. Yeast transformation was performed by the method of Ito et al. *(21)*. DH5α *E. coli* cells transformation and plasmid isolation were carried out as described in Sambrook et al., 1989 *(22)*.

Minimal medium consisted of 20g/L of glucose, ammonium sulfate 9.9 g/L, and yeast nitrogen base without amino acids and ammonium sulfate (Difco) 6.7 g/L. Nutritional supplements (histidine, uracil, and adenine) were added at the concentrations specified by Sherman et al. *(23)* when required. Amino acid nitrogen sources (glutamine, glutamate, or proline) were added at 1.0 g/L in place of ammonia in some experiments.

Cells were grown at 30°C on a rotary shaker at 160 rev. min^{-1}. Growth was monitored by measurement of optical density at 600 nm in a Gilson Stasar II spectrophotometer.

Asparaginase II Induction and Assay

The effects of nitrogen limitation on the formation of the enzyme were determined in the following manner *(24,25)*. Two cultures were inoculated with washed cells from an overnight culture to an OD 600 nm of 0.1. The experimental culture was grown to an OD 600 nm of 0.7–0.8 and the cells were collected by centrifugation and washed with 20 mM potassium phosphate buffer pH 7.0 at 4°C. Half of these cells were stored in 1–2 mL of this buffer at 4°C to serve as a prestarvation control. Derepression of the other half of the experimental culture was accomplished by incubation of the cells in phosphate buffer containing 30 g/L glucose and no source of nitrogen for 3 h. The control culture was grown without interruption and was harvested at the same time as the nitrogen starved experimental culture. In all cases, the asparaginase II levels of cells from the control cultures and the prestarvation cultures were similar, therefore, data for the control cultures are not reported here. Cells from all three growth conditions were prepared for assay of asparaginase II activity by centrifugation, washing with 20 mM potassium phosphate buffer pH 7.0 at 4°C, and resuspending in 1–2 mL of the same buffer. At the time of assay, the cells were diluted in the same buffer. Asparaginase II activity was assayed in intact cells to ensure that intracellular asparaginase I enzyme did not contribute to the result *(10,26)*. It was not determined whether any of the strains employed produced asparaginase I activity. L-asparagine was added to the cell suspen-

Table 1
Strains and Plasmids

Strain	Genotype	Source
Σ1278b	MATα	Wiame(11)
X2180-1A	MATα, SUC2, mal, gal2, CUP1	YGSC
P40-2a	MATα, his4-619, leu2-3,112, ure2Δ11::LEU2	Coschigano (15)
BMY344	MATα, ade2-102, ura3-52, leu2-3,112, ure2Δ11::LEU2	Coschigano (15)
BMY224	MATα, ade2-102, ura3-52, leu2-3,112, ure2Δ11::LEU2, gln3Δ4::LEU2	Minehart (18)
MP38	MATα, ura3-52, leu2-3,112	Minehart (18)

Plasmid	Markers	Source
pPM7	AatII fragment of GLN3 cloned into AatII site of Yep24	Minehart (18)
p1-XS	XbaI-SalI fragment of URE2 cloned into NheI-SalI site of Yep24	Coschigano (15)
pEC22	BamHI-ClaI fragment of PCR amplified ASP3 cloned into BamHI-ClaI sites of pKS$^+$	This work
pEC23	BamHI-SalI fragment of pEC22 ASP3 cloned into BamHI-ClaI sites of pKP15 (18)	This work

sion (OD 600 nm of 0.6) to a final concentration of 500 mM and this reaction mixture was incubated at 23°C. At regular intervals up 30 min, 5 mL samples were withdrawn and filtered to stop the reaction. The ammonia concentration of the supernatant was measured. It was determined that these experimental conditions allowed the measurement of the enzyme activity under initial rate conditions.

The amount of ammonia in the supernatants was measured spectrophotometrically by coupling to glutamate dehydrogenase (Boehringer-Mannheim, Indianapolis, IN). The results are presented as Δ μg/mL NH$_3$.

Glutamine Synthetase Induction and Assay

The nitrogen regulation of the production of glutamine synthetase was tested in cells grown in glutamine or glutamate minimal medium to midlog phase. The glutamine grown cells were harvested, washed, and resuspended in various media or starved for nitrogen. After 3 h of incubation the cells were harvested and cell extracts were prepared for the enzyme assay as previously described *(15)*. Protein concentration of the extracts was measured by the Bradford assay using reagents from Bio-Rad Laboratories (Richmond, CA). Glutamine synthetase activity was measured as previously described *(15)*.

PCR Amplification, Cloning, and ASP3 Constructed Plasmids

A pair of custom 26-mers (5'-CTGGATCCCACCAACCTCCAAC-TATG and 5'-AGATCGATTGGCGTACTGTGGGGCAT) were used to amplify the ASP3 gene from genomic DNA of strain P40-3C. The 1003 bp PCR product from −23 position relative to ATG and 207 pb after the stop codon (GenBank/EMBL Data Bank, Kim et al., 1988) *(13)* was cloned in the BamHI-ClaI sites of the pBlueScript KS$^+$ (Stratagene). The resulting plasmid was denominated pEC22. The pEC23 was constructed by transferring the ASP3 fragment from pEC22 to the BamHI-SalI sites, being fused to the CYC1-UAS$_{GAL1-GAL10}$ region of plasmid pKP15 *(18)*.

DNA Manipulation

DNA digestions with restriction enzymes, ligation, and filling in reactions were carried out as suggested by the supplier of the enzymes (BioLabs).

RESULTS

Asparaginase II Activity in Wild Type Strains

The yeast strain Σ1278b, which was previously shown to be incapable of producing asparaginase II activity *(10)*, was used as negative control. No asparagine hydrolysis was detectable in suspensions of Σ1278b after growth on ammonia minimal medium or upon nitrogen starvation for 3 h. However, suspensions of strain X2180-1A did exhibit asparagine hydrolysis after nitrogen starvation. As expected from previous studies *(13)*, X2180-1A did not produce asparaginase II activity during growth on ammonia minimal medium. Strain X2180-1A also did not produce asparaginase II activity during growth on glutamine or glutamate minimal media, but again a shift to nitrogen-free medium from either of these media resulted in the appearance of significant enzyme activity. In contrast, growth of X2180-1A on proline minimal medium did result in the appearance of enzyme activity (Table 2).

Table 2
Asparaginase II Activities of Strain X2180-1A During Growth
on Various Nitrogen Sources and after Nitrogen Starvation

Nitrogen source	Enzyme activity (Δ µg NH_3/mL)	
	Pre-starvation	Post-starvation
NH3	0.1	2.4
glutamine	0.2	2.5
glutamate	0.25	4.0
proline	4.3	3.4

Table 3
Effects of Gln3p and of Ure2p on the Formation of Asparaginase II[a]

Strain	Genotype	Enzyme activity ((Δ µg NH_3/mL)	
		Pre-starvation	Post-starvation
BMY224	gln3, ure2	0.2	0.2
BMY224/p1-XS[b]	gln3, URE2	0.2	0.2
BMY224/pPM7	GLN3, ure2	2.7	4.3
BMY344	GLN3, ure2	1.7	2.2
BMY344/P1-XS	GLN3, URE2	0.2	1.4

[a]The cultures contained ammonium as source of nitrogen.
[b]This strain failed to produce asparaginase II when grown with proline as source of nitrogen.

Asparaginase II is Derepressed in *ure2* Strains

The product of the *URE2* gene has been shown to be required for the negative regulation of the transcriptional activator Gln3p during growth on repressing nitrogen sources such as glutamine (15). A strain in which the *URE2* gene was replaced by a *ure2::LEU2* null allele also produced high levels of asparaginase II activity during growth on ammonia (Table 3). When this strain was transformed with the *URE2* plasmid p1-XS, asparaginase II activity was not produced during growth on ammonia but was produced by nitrogen starvation.

Similar experiments were performed with strain BMY224 that carries the disrupted null alleles *ure2::LEU2* and *gln3::LEU2* (Table 3). Both BMY224 and BMY224 transformed with the *URE2* plasmid pl-XS were incapable of producing asparaginase II activity during nitrogen starvation. Very high asparaginase II activity was obtained during growth on ammonia and during nitrogen starvation in BMY224 transformed with the *GLN3* plasmid pPM7. These results are entirely consistent with those obtained in the study of the transcriptional regulation of the *GLN1* and *GDH2* genes *(15,16,18)*, indicating that the *ASP3* gene is transcriptionally regulated by the products of *URE2* and *GLN3*. In accord with previous studies, *gln3* is epistatic to *ure2* with respect to the production of asparaginase II activity and asparaginase II production is unregulated by nitrogen source in a *ure2 GLN3* strain *(15,27)*. It appears that the product of *URE2* is required to block transcriptional activation of *ASP3* by the product of *GLN3*.

Glutamine Synthetase is Derepressed by Nitrogen Starvation

The effects of nitrogen starvation on another gene that responds to activation by Gln3p were tested. Glutamine synthetase (GS), a product of *GLN1*, is found at low levels in a wild type strain during growth on glutamine and at approx 30-fold higher levels during growth on glutamate, whereas in a *gln3* strain the difference in GS levels between glutamate- and glutamine-grown cells is only threefold *(15)*. In the present work the level of GS of nitrogen-starved cells was compared to glutamine- or glutamate-grown cells. After 3 h of nitrogen starvation, the GS activity increased approx 20-fold compared to glutamine-grown cells (Table 4). Glutamate-grown cells produced about fourfold higher GS activity than nitrogen-starved cells. In order to directly compare the relative strength of cells grown on glutamate vs nitrogen starvation as signals for the production of GS, cells were shifted at midlog phase from minimal glutamine medium to either glutamine, glutamate, or nitrogen starvation medium for 3 h. GS activity increased about 20-fold in the nitrogen starved cells and about 39-fold in the cells shifted to glutamate medium, whereas no signifcant difference was observed in cells shifted from glutamine to glutamine (Table 4). GS levels are highest during growth on glutamate and intermediate after nitrogen starvation, whereas asparaginase II levels are highest after nitrogen starvation and are low during growth on glutamate. Thus, there appears to be a difference in the production of GS and asparaginase II in response to the same nitrogen signals and by the same transcriptional activator, Gln3p.

An Attempt to Overproduce Asparaginase II

Despite some strains, like BMY 344, show high levels of asparaginase II as compared to other strains (Table 3), we aimed to manipulate the *ASP3* gene in order to increase enzyme production. The pEC23 was used to

Table 4
Glutamine Synthetase Activity of Strain X2180-1A During Growth on Glutamine
or Glutamate and after Shift from Glutamine to other Media

Nitrogen source	Shifted to[a]	Specific activity (μmole/mg.min^{-1})
glutamate	--	2.43
glutamine	--	0.03
glutamine	glutamine	0.02
glutamine	glutamate	1.34
glutamine	no nitrogen	0.60

[a]Cells were collected by filtration, washed with water, resuspended in the indicated medium, and incubated under normal growth conditions for 3 h.

transform MP38, BMY224, and BMY344 strains. The selected transformants showed no asparaginase II activity related to the presence of the plasmid under the described conditions. This subject is currently under investigation to clear up these results.

DISCUSSION

Using a molecular approach it has been shown in this report that the mechanism of nitrogen control of the production of asparaginase II activity is mediated by the Ure2p/Gln3p transcriptional system. Interestingly, the pattern of nitrogen control on *ASP3* appears to be somewhat different than that on *GLN1*, which also responds to Ure2p/Gln3p. *GLN1* is most highly transcribed and GS activity is the highest when the nitrogen source is glutamate and GS activity is intermediate during nitrogen starvation. However, *ASP3* transcription during growth in glutamate (as judged by the production of asparaginase II activity) appears to be moderate *(11)* to low (Table 2). The disparity between the levels of GS and asparaginase II is even more striking if glutamate- vs proline-grown cells are compared. The amount of GS was sixfold higher in glutamate- than in proline-grown Σ1278b cells *(27)*, compared to 16-fold less asparaginase II in glutamate- than in proline-grown X2180-1A cells (Table 2). The possibility that the Ure2p/Gln3p transcription system produces different effects on different promoters under same nitrogen conditions is intriguing. Whether or not this effect is transcriptional must be clarified first because it is conceivable that asparaginase II activity or the secretion of asparaginase II is regulated posttranscriptionally by the nitrogen source, making it impossible to infer the transcriptional state of *ASP3* from asparaginase II activity.

It has been understood for some time that this asparaginase hydrolase is secreted into the pericellular space when the nitrogen source is poor or absent, and that growth on asparagine represses the production of asparaginase II activity *(10)*. This is the opposite of the expected pattern of regulation: high asparaginase II activity in the presence of high concentrations of asparagine. One possibility is that the intracellular asparaginase I, whose production does not appear to be regulated by nitrogen source, is present in sufficient quantities to allow good growth on high concentrations of asparagine, but that growth is limited by poor asparagine transport when the asparagine concentration is low. If the rate of ammonia or aspartate transport is higher than the rate of transport of equivalent, low concentrations of asparagine, then extracellular asparagine hydrolysis could compensate for poor transport. Another possibility is that nitrogen starvation allows *GLN3* to activate the production of a wide range of secreted enzymes. Interestingly, the primary nitrogen regulator of *Neurospora crassa*, the product of the *nit-2* gene and its analogous *areA* gene in *Aspergillus nidulans* encodes a metal finger domain homologous to that of the GLN3 protein *(28,29)*. In that respect similar patterns of nitrogen regulation in yeast and fungi could be envisaged. The use of yeast cells to study the subject is a tool to characterize the relevant genes and regulatory systems. These data have a potential use to engineer strains for increased extracellular enzymes production. Also, the evaluation of the effects of the nitrogen source on extracellular asparaginase II would allow the use of a scientific approach for fermentation medium design with obvious economic advantages, mainly taking into account the scale of extracellular enzymes production.

ACKNOWLEDGMENTS

This work was partially supported by the Brazilian Research Council (CNPq).

REFERENCES

1. Cohen, B. L. (1972), *J. Gen. Microbiol.* **71**, 293–299.
2. Cohen, B. L. and Drucker, H. (1977), *Archives Biochem. Biophys.* **182**, 601–613.
3. Keyser, P., Kirk, T. K., and Zeikus, J. G. (1978), *J. Bacteriol.* **135**, 790–797.
4. Kirk, T. K. and Fenn, P. (1981), *Arch. Microbiol.* **130**, 59–65.
5. Fenn, P., Choi, S., and Kirk T. K. (1981), *Arch. Microbiol.* **130**, 66–71.
6. Reid, I. D. (1983), *Appl. Environ. Microbiol.* **45**, 830–837.
7. Reid, I. D. (1983), *Appl. Environ. Microbiol.* **45**, 838–842.
8. Commanday, F. and Macy, J. M. (1985), *Arch. Microbiol.* **142**, 61–65.
9. Magasanik, B. (1992), in *The Molecular and Cellular Biology of the Yeast Saccharomyces: Gene Expression*, Vol.2, Cold Spring Harbor Laboratory Press, Cold Spring Harbor, NY, pp. 283–317.
10. Dunlop, P. C. and Ronn, R. J. (1975), *J. Bacteriol.* **122**, 1017–1024.
11. Wiame J. M., Grenson M., and Arst H. N. (1985), *Adv. Microb. Physiol.* **26**, 1–87.
12. Kim, K. W., Kamerud, J. Q., Livingston, D. M., and Roon, R. J. (1988), *J. Biol Chem.* **263**, 11,948–11,953.

13. Dunlop, P. C., Meyer, G. M., and Roon, R. J. (1980), *J. Bacteriol.* **143**, 422–426.
14. Kamerud, J. Q. and Roon, R. J. (1986), *J. Bacteriol.* **165**, 293–296.
15. Coschigano, P. W. and Magasanik, B. (1991), *Mol. Cell. Biol.* **11**, 822–832.
16. Miller, S. M. and Magasanik, B. (1991), *Mol. Cell. Biol.* **11**, 6229–6247.
17. Minehart, P. L. and Magasanik, B. (1991), *Mol. Cell. Biol.* **11**, 6216–6228.
18. Minehart, P. L. and Magasanik, B. (1992), *J. Bacteriol.* **174**, 1828–1836.
19. Daugherty, J. R., Rai, R., El Berry, J. M., and Cooper, T. G. (1993), *J. Bacteriol.* **175**, 64–73.
20. Kim, K. W. and Roon, R. J. (1984), *J. Bacteriol.* **157**, 958–961.
21. Ito, H., Fukada, Y., Murato, K., and Kimiru, A. (1983), *J. Bacteriol.* **153**, 163–168.
22. Sambroock, J., Fritsch, E. F., and Maniatis, T. (1989), *Molecular Cloning. A Laboratory Manual*, 2nd ed., Cold Spring Harbor Laboratory Press, Plainview, NY.
23. Sherman, F., Fink, G. R., and Lawrence, C. W. (1978), *Methods in Yeast Genetics.* Cold Spring Harbor Laboratory Press, Cold Spring Harbor, NY.
24. Pauling, K. D. and Jones, G. E. (1980), *Biochim. Biophys. Acta* **616**, 271–282.
25. Pauling, K. D. and Jones, G. E. (1980), *J. Gen. Microbiol.* **117**, 423–430.
26. Jones, G. E. (1977), *J. Bacteriol.* **130**, 128–130.
27. Courchesne, W. E. and Magasanik, B. (1988), *J. Bacteriol.* **170**, 708–713.
28. Fu, Y. and Marzluf, G. A. (1990), *Proc. Natl. Acad. Sci. USA* **87**, 5331–5335.
29. Kudla, B., Caddick, M. X., Langdon, T., Martinez-Rossi, N., Bennet, C. F., Silbley, S., Davies, R. W. and Arst, H. N. (1990), *EMBO J.* **9**, 1355–1364.

Production of α-Terpineol from *Escherichia coli* Cells Expressing Thermostable Limonene Hydratase

NATARAJAN SAVITHIRY, TAE KYOU CHEONG, AND PATRICK ORIEL*

Department of Microbiology, Michigan State University, East Lansing, MI 48824

ABSTRACT

The genes encoding a thermostable limonene hydratase have been located on a cloned fragment in *Escherichia coli* conferring growth on limonene and production of the monoterpenes perillyl alcohol and α-terpineol. Whole cell bioconversion studies at elevated temperature employing both an aqueous phase and neat limonene phase demonstrated significant production of α-terpineol with additional production of carvone.

Index Entries: Limonene; α-terpineol; thermostable; two-phase.

INTRODUCTION

Monoterpenes constitute a diverse group of C10-based plant secondary metabolites produced in part for defense against microbes and insects. Because of their unique organoleptic properties, certain monoterpenes utilized in fragrances and as food ingredients command some of the highest unit values among biotechnological products (1). Microbial conversion of low value monoterpenes to higher value derivatives has been recognized for some time as an attractive opportunity, but has been thwarted by the lack of knowledge of microbial monoterpene pathways leading to a multiplicity of monoterpene metabolites (for review, *see* ref. 2). Because of its low cost and extensive availability as a waste citrus product (3), the monoterpene R-(+)-limonene has been selected as a target for directed microbial bioconversions. To help avoid problems arising from

*Author to whom all correspondence and reprint requests should be addressed.

the microbial toxicity of this monoterpene, eubacterial thermophiles were targeted for investigation anticipating that their robust enzymes and growth ability in conditions favoring monoterpene volatilization might provide advantages in bioprocessing applications. In previous studies, a *Bacillus stearothermophilus* strain BR388 was isolated, which proved resistant to limonene toxicity, and which demonstrated production of perillyl alcohol and α-terpineol during growth on limonene *(4)*. The former compound has value as a flavorant *(2)*, whereas the latter is extensively utilized in perfume manufacture *(5)*. In order to study and control the pathway metabolites, the entire pathway was cloned into *Escherichia coli* as a 9.6-kb plasmid insert, conferring to the new host growth on limonene as a sole carbon source and production of perillyl alcohol and α-terpineol *(6)*. It was proposed that limonene degradation in both the thermophile and recombinant proceeded by oxidation of the C-1 methyl to perillic acid with further breakdown utilizing the β fatty acid pathway, whereas α-terpineol was formed as a hydratase-catalyzed reversible side product (Fig. 1). In this paper, information is presented on the cloned hydratase and the hydratase-catalyzed formation of α-terpineol and other products is examined in a two-phase bioreactor at elevated temperature.

MATERIALS AND METHODS

Growth of Microorganisms

E. coli recombinants carrying *B. stearothermophilus* inserts were grown in M9 salts medium containing either yeast extract or the selected monoterpene. M9 minimal medium *(7)* contains per liter: Na_2HPO_4, 6g; KH_2PO_4, 3g; NaCl, 0.5g; NH_4Cl, 1g; pH7.4. After autoclaving and cooling, 2 mL of $1M$ $MgSO_4$ and 0.1 mL of $1M$ $CaCl_2$ were added. Growth was carried out using 20 mL culture volumes in 40 mL serum bottles closed with Teflon-coated butyl stoppers and aluminum caps.

Limonene Hydratase Assay

Recombinant *E. coli* were grown in 50 mL LB medium containing 50 µg/mL ampicillin at 37°C overnight. Cells were centrifuged and resuspended in 5 mL of 50 mM sodium phosphate buffer, pH 7.0, and disrupted by 30-s bursts of sonication with cooling on ice for 3–5 min. The crude enzyme extract resulting from centrifugation at 35,000g for 30 min was assayed in a procedure adapted from Nagasawa et al. *(8)* using a 2 mL reaction mixture containing 1 mM 3-cyanopyridine in 50 mM phosphate buffer, pH 7.0. The reaction mixture was incubated at 55°C for 20 min and stopped by addition of 0.2 mL $1M$ HCl. Nicotinamide product was determined using HPLC analysis (Waters HP1050 with 3.9 mm × 15 cm NOVA C-18 column (Waters, Inc.) Peaks eluted using 60% acetonitrile in 5 mM

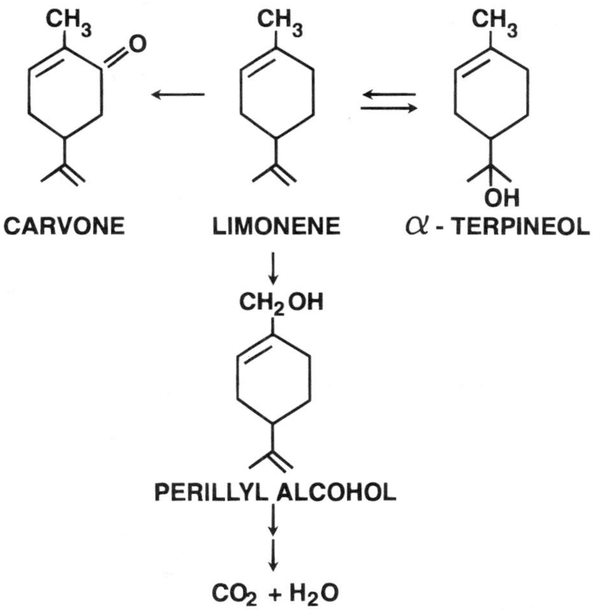

Fig. 1. Proposed pathway for limonene degradation and principal metabolic products of *E. coli* recombinants containing cloned DNA fragments from *B. stearothermophilus* BR388.

sodium phosphate buffer, pH 7.7, were determined at 230 nm. Peaks were identified and quantified using known standards. One unit of activity is defined as the amount of enzyme that catalyzes formation of 1 μM/min nicotinamide under the specified conditions.

GC/MS Product Analysis

Aqueous culture supernants were acidified to pH 2.0 with HCl, extracted three times with ether, evaporated, and analyzed using GC/MS equipment (Hewlett Packard model 5890) and procedures described previously *(4)*. Samples of the limonene phase from bioreactor studies were injected directly.

Subcloning of *E. coli* EC409A Insert

The limonene pathway encoded in the 9.6 kb insert of EC409A was subcloned utilizing HindIII digestion, separation of the 8.2, 3.8, and 0.6 kb fragments by agarose gel electrophoresis, fragment recovery using electroelution, ligation into the HindIII site of pBluescript (SK+), and transformation into *E. coli* DH5α using standard procedures *(7)*. Similar procedures were employed to obtain subcloned fragments utilizing other restriction sites.

Two-Phase Bioreactor Studies

To examine formation of α-terpineol in a whole-cell two phase bioreactor, 5 mL of neat limonene and 50 mL of EC423 cell suspensions (10^9 cells/mL) in M9 salts were shaken in a gyrotory water bath at 250 rpm in 125 mL screw cap bottles at varied temperatures. Samples of aqueous and limonene phases were taken for GC/MS analysis after various periods of incubation of the cell suspension.

RESULTS

Subcloning of the Limonene Hydratase Gene

Earlier work indicating that both α-terpineol and perillyl derivatives were metabolic products formed during growth of the recombinant *E. coli* EC409A on limonene suggested that both limonene hydratase and methyl oxidase activities were present on the cloned insert (6). In an effort to locate the genes encoding these activities, the transformant was subcloned into 8.2, 3.8, and 0.6 kb HindIII fragments. The transformant EC418 containing the 3.8 kb Hind III insert proved able to grow on limonene (Fig. 2), indicating that genes facilitating limonene catabolism in *E. coli* are retained on this fragment. GC/MS analysis of cell supernatants indicated that perillyl alcohol, α-terpineol, and lesser amounts of carveol were produced, suggesting that both methyl oxidase and hydratase activities were encoded (Table 1). The limonene hydratase gene was further subcloned as a 1.6-kb HindIII-Bg1II fragment as EC419 that expressed hydratase activity, but did not confer growth on limonene nor produce perillyl derivatives. Transformant EC421 carrying the adjacent 2.2-kb Bg1II-Hind III fragment, demonstrated production of perillyl alcohol and growth on limonene, indicating the presence of the gene(s) encoding limonene methyl oxidation. Surprisingly, limonene hydratase activity was also found with transformant EC423, indicating the presence of two distinct limonene hydratase genes in EC409A.

Preliminary Characterization of the Limonene Hydratase

The limonene hydratase in crude extracts demonstrated broad substrate specificity, in that in addition to hydration of limonene, the nitrile group of cyanopyridine could also be hydrated. Since the latter substrate is more soluble and less volatile than limonene, hydration of cyanopyridine to nicotinamide was utilized for standard assay using a published HPLC procedure (8). It was found that the hydratase enzyme expressed from EC419 was thermally unstable, demonstrating no enzymatic activity above 40°C whereas the hydratase expressed by EC423 exhibited an optimum temperature near 55°C which is close to the optimum growth temperature of the thermophile parent (data not shown). During preliminary

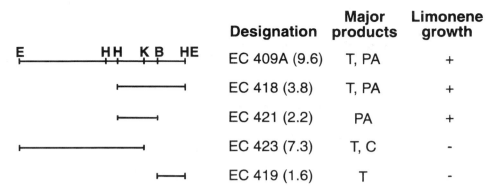

Fig. 2. Subcloning of the BR388 limonene degradation pathway. Product designations: T, α-terpineol; PA, perillyl alcohol. Enzyme designations: E, EcoRI; H, HindIII; K, KpnI; B, BglII. (+) Denotes ability to grow on limonene as sole carbon and energy source. Parentheses designate insert size.

Table 1
Monoterpene Products Produced by Recombinant
EC418 During Growth on Limonene

Growth stage	Metabolites and concentration (mg/l*)		
	α-terpineol	Carveol	Perillyl alcohol
Early log phase (6 h)	0.7	2.6	0.7
Late log phase (24 h)	1.1	1.7	2.5
Stationary phase (48 h)	29	0.9	72

attempts at purification, the EC423 enzyme could be pelleted by prolonged ultracentrifugation, suggesting that the enzyme may be membrane bound as previously reported for the limonene hydratase of *Pseudomonas gladioli* (9).

Whole Cell Bioreactor Studies

An attempt was made to utilize whole cells of *E. coli* EC423 in a two-phase bioreactor at elevated temperature as shown in Fig. 3. Neat limonene was utilized as the organic phase to maintain a saturated level of substrate and to facilitate product removal from the aqueous phase. Both actions serve to prevent dehydration of the α-terpineol product in this reversible reaction. As seen in Table 2, significant accumulation of α-terpineol was

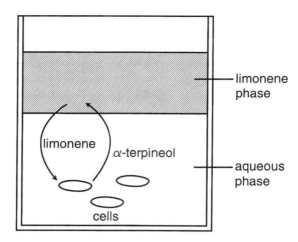

Fig. 3. Cartoon of two-phase whole-cell bioreactor utilizing neat limonene.

Table 2
Monoterpene Products Produced by Recombinant EC423 Utilizing Neat Limonene in a Two-Phase Bioreactor

Product	Temperature	Concentration (mg/l) Incubation Time (h)		
		24	48	72
α-terpineol	40°C	16	117	215
	50°C	19	168	235
	60°C	16	201	209
Carvone	40°C	3.9	20	23
	50°C	8	21	35
	60°C	17	19	28

Concentrations of product determined in the limonene phase are expressed per liter of aqueous cell suspension.

achieved with this simple bioreactor system. Surprisingly, the monoterpene carvone was also found in lesser, but significant amounts. Although minor amounts of carveol were produced during limonene utilization by EC418, production of carvone was not observed with either the parental thermophile or the recombinant EC409A containing the same DNA as part of a larger fragment. For both monoterpene products, optimal production occurred at 50°C, with carvone appearing at earlier times.

DISCUSSION

Cloning of the limonene degradative pathway provides an opportunity for separate examination and utilization of the conversion steps for production of valuable monoterpene metabolites. In this report, we have separated the limonene hydration and methyl oxidation steps, and have obtained additional evidence suggesting that the latter step participates in limonene utilization by the recombinant.

Although formation of carvone has been reported previously for other limonene-degrading bacteria (2), formation of this metabolite was not observed during studies of the parental thermophile, and was, therefore, unexpected. From subcloning studies, the gene encoding this ring oxidation activity appears to be distinct separate from that oxidizing the C-1 limonene methyl group. Since carvone is utilized as an important food flavoring (5), the enzyme and the encoding gene(s) merit further investigation.

To the authors' knowledge, this is the first attempt to utilize whole cells with thermostable enzymes in a two-phase bioreactor system at elevated temperature. Whereas significant further work is required for optimization and scaleup, the provision of excess monoterpene substrate and removal of reactive product facilitated by the separate phase exhibit significant promise for potential production of the specialty chemicals α-terpineol and carvone using elevated temperature. The introduction of thermostable enzymes catalyzing biotransformation at elevated temperature into a mesophilic bacterial host may also help to eliminate undesired side reactions catalyzed by host enzymes while retaining advantages of whole cell utilization.

ACKNOWLEDGMENTS

We gratefully acknowledge assistance in subcloning from M. R. Natarajan. This work was supported by the MSU Crop and Food Bioprocessing Center with funds provided by the State of Michigan Research Excellence Fund, and award 95-37500-1929 of the NRI Competitive Grants Program/USDA.

REFERENCES

1. Welsh, F. W., Murray, W. D., and Williams, R. E. (1989), *Crit. Rev. Biotechnol.* **9,** 105–169.
2. Krasnobajew, V. (1984), in *Biotechnology*, Vol. 6a. Lieslich, K. ed., Verlag Chemie, Weinheim.
3. Bowen, E. R. (1975), *Proc. Fla. State Hort. Soc.* **88,** 305–308.
4. Chang, H. C. and Oriel, P. (1994), *J. Food Sci.* **59,** 660–662.
5. Bauer, K., Garbe, D., and Surburg, H. (1990), *Common fragrance and flavor materials*, VCH, Weiheim.
6. Chang, H. C., Gage, D., and Oriel, P. (1995), *J. Food Sci.* **60,** 551, 552.

7. Maniatis, T., Fritsch E. F., and Sambrook, F. (1982), *Molecular Cloning*, Cold Spring Harbor Laboratory, New York.
8. Nagasawa, T., Mathew, C. D., Mauger, J., and Yamada, H. (1988), *Appl. Environ. Microbiol.* **54,** 1766–1769.
9. Cadwallader, K. R. and Braddock, R. J. (1992), in *Food Science and Human Nutrition* Charalambous, G., ed., Elsevier, New York, pp. 571–583.

Fermentation of Biomass-Derived Glucuronic Acid by *pet* Expressing Recombinants of *E. coli* B

HUGH G. LAWFORD* AND JOYCE D. ROUSSEAU

Bio-engineering Laboratory, Department of Biochemistry, University of Toronto, Toronto, Ontario, Canada M5S 1A8

ABSTRACT

The economics of large-scale production of fuel ethanol from biomass and wastes requires the efficient utilization of all the sugars derived from the hydrolysis of the heteropolymeric hemicellulose component of lignocellulosic feedstocks. Glucuronic and 4-O-methyl-glucuronic acids are major side chains in xylans of the grasses and hardwoods that have been targeted as potential feedstocks for the production of cellulosic ethanol. The amount of these acids is similar to that of arabinose, which is now being viewed as another potential substrate in the production of biomass-derived ethanol.

This study compared the end-product distribution associated with the fermentation of D-glucose (Glc) and D-glucuronic acid (GlcUA) (as sole carbon and energy sources) by *Escherichia coli* B (ATCC 11303) and two different ethanologenic recombinants—a strain in which *pet* expression was via a multicopy plasmid (pLOI297) and a chromosomally integrated construct, strain KO11. pH-stat batch fermentations were conducted using a modified LB medium with 2% (w/v) Glc or GlcUA with the set-point for pH control at either 6.3 or 7.0. The nontransformed host culture produced only lactic acid from glucose, but fermentation of GlcUA yielded a mixture of ethanol, acetic, and lactic acids, with acetic acid being the predominant end-product. The ethanol yield associated with GlcUA fermentation by both recombinants was similar, but acetic acid was a significant by-product. Increasing the pH from 6.3 to 7.0 increased the rate of glucuronate fermentation, but it also decreased the ethanol mass yield from 0.22 to 0.19 g/g primarily because of an increase in acetic acid production. In all fermentations there was good closure of

---

*Author to whom all correspondence and reprint requests should be addressed.

the carbon mass balance, the exception being the recombinant bearing plasmid pLOI297 that produced an unidentified product from GlcUA. The metabolism of GlcUA by this metabolically engineered construct remains unresolved. The results offered insights into metabolic fluxes and the regulation of pyruvate catabolism in the wild-type and engineered strains. End-product distribution for metabolism of glucuronic acid by the nontransformed, wild-type *E. coli* B and recombinant strain KO11 suggests that the enzyme pyruvate-formate lyase is not solely responsible for the production of acetylCoA from pyruvate and that derepressed pyruvate dehydrogenase may play a significant role in the metabolism of GlcUA.

Index Entries: Glucuronic acid; recombinant *E. coli* B; ethanol; pyruvate metabolism; acetic acid derepressed pyruvate dehydrogenase.

INTRODUCTION

Lignocellulosic biomass and wastes are being targeted as an economic alternative to agricultural food crops such as corn, cereal grains, and sugar cane, for the large-scale production of fermentation ethanol for use as an alternative liquid transportation fuel *(1–3)*. Fermentation feedstock costs dominate the economics of fuel ethanol production *(4,5)* and the efficient utilization of nonglucose sugars represents an opportunity to significantly reduce the cost of producing fuel ethanol from lignocellulosic biomass and wastes *(6,7)*.

Woody biomass consists primarily of three polymeric substances, cellulose, hemicellulose, and lignin *(8)*. Cellulose is a homopolymer of glucose and comprises about half of the dry mass; however, it is strongly resistant to depolymerization unless the lignocellulose is pretreated to remove the impediments to enzymic digestion that are caused by lignin and the hemicellulose fraction of biomass *(9–11)*. Unlike cellulose, hemicellulose is a heteropolymer with a structure and composition that is source dependent *(12,13)*. The term hemicellulose was introduced by Schulze in 1891, but it is non-descriptive, and in Europe the preferred term is "wood polyoses" *(12)*. Hemicellulose represents about one-third of the carbohydrate content of hardwood lignocellulosic biomass with the five-carbon sugar, D-xylose, being a major component. Thermochemical depolymerization of hemicellulose using dilute acid is efficient and cost effective *(10,14)*. The hemicellulose of temperate zone hardwoods (*Angiospermae*) such as aspen (poplar), beech, and oak is well conserved with a relatively invariant composition *(8,12,15)*. In chemical terms, hardwood hemicellulose consists of *O*-Acetyl-(4-*O*-methylglucurono)xylans accompanied by small proportions of galactomannan *(16)*. Hardwood (4-*O*-methylglucurono)xylan is completely devoid of arabinose and its presence in hydrolysates probably relates to the hydrolysis of other polysaccharide materials such as the pectic material of the primary cell wall *(12)*. The linear xylan backbone consists of

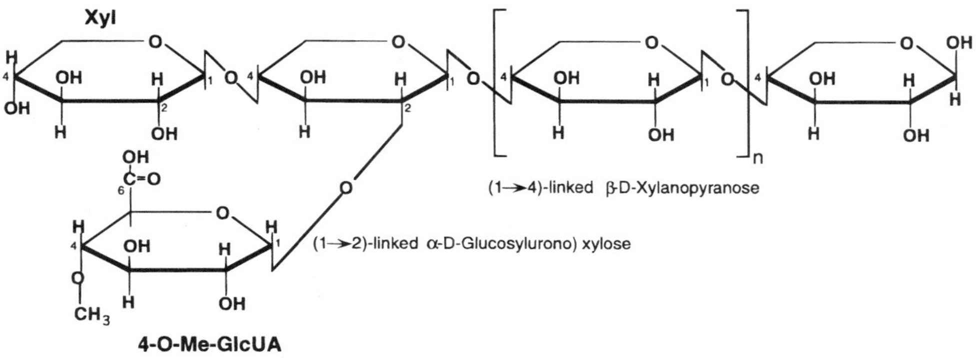

Fig. 1. Chemical structure of hardwood hemicellulose. The linear (1→4) β-D-xylopyranosyl backbone carries occasional substitutions at the C-2 position by 4-O-methyl-α-D-glucopyranosiduronic acid as well as randomly distributed acetyl groups (acetylation is not shown). The 4-O-methyl-α-D-glucopyranosiduronic acid linkage to xylose is highly resistant to acid hydrolysis and pretreatment of lignocellulosic hardwood biomass yields the disaccharide 2-O-(4-O-methyl-α-D-glucopyranosyluronic acid)-D-xylopyranose.

(1→4)-linked β-D-xylopyranose residues with the 4-O-methyl-glucuronic acid attached directly to the C-2 of (on average) approximately every tenth xylose residue (Fig. 1). X-ray analysis has revealed that hardwood xylan has a threefold screw axis with 120° for each xylose residue and a repeat length of 15å (cellulose has a twofold screw axis) *(12)*. The (4-O-methylglucurono)xylan is extensively acetylated in a random fashion, with acetyl groups (not shown in Fig. 1) amounting to about 3–5% of the wood substance *(8,17)*. In hardwood hemicellulose, the mole ratio of acetic acid to D-xylose is approx 7:10 (equivalent to a mass ratio of acetic acid to xylose of 0.28:1.0). It has been shown that ester groups play an important role in plant cell wall resistance to enzyme hydrolysis *(18)*.

The disaccharide that is formed at the branch in the hardwood xylan polymer is a (1→2)-linked (4-O-methyl-alpha-D-glucopyranosyluronic acid)-D-xylopyranose. This alpha-(1→2)-glycosidic bond between the 4-O-Me-GlcUA and xylose is the most acid stable bond found in woody biomass—being even more resistant to hydrolysis than the β-(1→4)-D-glucosidic bond in cellobiose *(12)*. Hence, this disaccharide is a by-product of hemicellulose acid hydrolysis. It is difficult to quantitate by the usual HPLC analysis (using a HPX-87H column) because it coelutes with several other di- and trimers. Other by-products of dilute-acid pretreatment include substances such as acetic acid, furfural, and lignin-derived phenolics *(19)* that are toxic to ethanologenic micro-organisms *(20–22)*. Recent advances in the area of the bioconversion of biomass hemicellulose to ethanol have been reviewed by McMillan *(23)*.

Considerable research has been directed to the search for organisms capable of high-performance fermentation of biomass prehydrolysates.

This search for xylose-fermenting ethanologenic micro-organisms has produced several alternatives including bacteria, yeasts, and fungi (for review *see* ref. *24*). In addition to natural isolates, several genetically engineered biocatalysts have been constructed for this purpose and prominent among these have been the patented ethanologenic *Escherichia coli* cultures that carry genes for ethanol production, namely pyruvate decarboxylase (*pdc*) and alcohol dehydrogenase II (*adhB*) cloned from *Zymomonas mobilis* CP4 *(25–27)*. Although *E. coli* is heterofermentative and produces primarily acid end-products *(28,29)*, it has been metabolically engineered to exhibit a very high degree of ethanol selectivity *(30)*. In the early stages of development, transformation of *E. coli* involved insertion of the ethanol production genes from *Zymomonas* (referred to as the *pet* operon) *(26)* on multicopy plasmids carrying marker genes responsible for resistance to tetracycline and ampicillin *(27,30)*. Although the pioneering work was done with *E. coli* K12 *(31–33)*, a subsequent physiological assessment of growth characteristics of several different potential host cultures of *E. coli* identified the wild-type Luria strain B (ATCC 11303) as a "hardy strain and a suitable host" *(34)* for *pet* transformation using the plasmid designated as pLOI297 *(30)*. For several years, we have been assessing the fermentation performance characteristics of this patented recombinant *E. coli* 11303:pLOI297 using both synthetic lab media *(35–42)* and biomass prehydrolysates prepared by different thermochemical processors from a variety of biomass/waste feedstocks, including both hardwood (aspen) *(43)* and softwood (pine) *(44)*, newsprint *(45)*, spent sulfite liquors *(46)*, and corn crop residues *(47)*.

Plasmid-bearing recombinants suffer from two limitations: firstly, they are inherently less stable than strains in which the foreign genes have been integrated into the host chromosome *(41,48)*, and secondly, high copy number plasmids are known to impose an energetic burden on the host *(49)*, which is often reflected in a reduced growth rate and yield *(40)*. With this in mind, Ingram and his associates engineered chromosomally integrated strains of *E. coli* B ATCC 11303 in which the *Zymomonas pdc* and *adhB* genes were inserted into the pyruvate-formate lyase gene (*pfl*) of the host *(48)*. However, it was discovered that single copy inserts of the *pdc* and *adhB* genes did not result in the same high level of activities of *Zymomonas* enzymes that had been achieved in multicopy plasmid-based recombinants *(48)*. In one series of constructs involving chromosomal integration, the transformation vector also contained the gene for chloramphenicol acetyl transferase (*cat*), which is responsible for conferring resistance to chloramphenicol (Cm). A spontaneous mutant, designated as strain KO11, was selected for resistance to high levels (600 µg/mL) of Cm and has been shown to express high levels of both the *Zymomonas* genes and *cat*. In addition, strain KO11 carries a mutation in its fumarate reductase gene that impairs its ability to produce succinate as a fermentation end-product *(48)*.

In addition to exhibiting a high level of conversion efficiency in laboratory media *(48)*, recombinant KO11 has been shown to ferment prehydrolysates prepared from pine *(50)*, and agricultural crop residues *(51,52)*. However, claims relating to its long-term stability *(48)* have recently been challenged in a study involving continuous culture *(41)*. Recently conducted comparative surveys of xylose-fermenting ethanologenic micro-organisms have concluded that recombinant *E. coli* strain KO11 is currently one of "the best candidates" for ethanol production from hemicellulosic hydrolysates *(23,53,54)*.

Filtered enzymic hydrolysates of peel and pulp wastes associated with the production of citrus fruit juices are a rich source of fermentable carbohydrates *(55)*. Almost one-third of the total mass of monosaccharides in an enzymic orange peel hydrolysate was shown to be galacturonic acid (GalUA), with the remainder being a mixture of glucose, fructose, galactose, and arabinose. In addition to the expected five-carbon and six-carbon neutral sugars, recombinant *E. coli* KO11 utilized the galacturonic acid. Using a nutrient-rich laboratory medium containing 2% (w/v) D-galacturonic acid, Grohmann et al. *(55)* showed that, at pH 7.0, *E. coli* KO11 produced equimolar amounts of acetic acid and ethanol, with carbon dioxide as the only other detectable fermentation product. Based on the proposed theoretical maximum ethanol yield of 0.237 g/g, the observed ethanol yield of 0.19 g/g *(55)* represents a conversion efficiency of 80%.

One objective of this work was to compare the ability of two high profile *E. coli* ethanologenic recombinants, specifically the plasmid recombinant 11303:pLOI297 and the chromosomal integrated strain KO11, to ferment D-glucuronic acid. Based on a knowledge of the chemical structure of hardwood hemicellulose, it is reasonable to assume that dilute-acid hydrolysis will produce 4-*O*-methyl-glucuronic acid and the C-2 xylose derivative disaccharide 2-*O*-(4-*O*-methyl-α-D-glucuronic acid)-D-xylose; however, in the absence of the commercial availability of either of these substances, our fermentation experiments were based on pure D-glucuronic acid as sole carbon (energy) source. Since end-product distribution has the potential to offer insights into how a substance is metabolized, a second objective of this work involved comparing the anaerobic catabolism of Glc and GlcUA by both the nontransformed wild-type culture and the two different metabolically engineered strains. Because the metabolic engineering was directed specifically toward alterations in pyruvate metabolism, it was hoped that the results of this study would shed light on metabolic fluxes and the regulation of pyruvate metabolism in *E. coli* B.

MATERIALS AND METHODS

Organisms

The wild-type, nontransformed host culture, *Escherichia coli* B (ATCC 11303) was obtained from The American Type Culture Collection (Rockland, MD). Recombinant *Escherichia coli* B (ATCC 11303 carrying the *pet* plasmid pLOI297) *(30)* and the chromosomally integrated strain KO11 *(48)* were received from L. O. Ingram (University of Florida, Gainesville, FL). Cultures grown from single colony isolates on selective antibiotic-containing agar medium were stored at –10°C in LB medium supplemented with glycerol (20 mL/dL) and sodium citrate (1.5 g/dL). Inocula were prepared using complex or defined media buffered with 100 mM phosphate (pH 7.0). Batch fermentations were inoculated by transferring approx 100 mL of an overnight flask culture directly to 1400 mL of medium in the stirred-tank bioreactor. The same sugar was used for preculture and fermentation. The initial cell density was monitored spectrophotometrically to give an OD_{550} in the range 0.1–0.2 corresponding to 30–50 mg dry cell mass (DCM)/L.

Culture Media

The nutrient-rich, complex culture medium Luria broth *(56)* was modified as described by Grohmann et al. *(55,57)* and contained 2.5 g Bacto Yeast Extract (Difco Laboratories, Detroit, MI) and 5 g Bacto Tryptone (Difco) per liter of distilled water. D-Glucuronic acid was obtained from Sigma Chemical (St. Louis, MO). The medium was sterilized by autoclaving. Stock sugar solutions were autoclaved separately and added at the concentration specified. When the *pet* transformed cultures were used, filter-sterilized antibiotics (final concentration of 40 mg/L ampicillin and 10 mg/L tetracycline for pLOI297 and 40 mg/L chloramphenicol for KO11) were added to the autoclaved fermentation media after cooling.

Fermentation Equipment

pH-stat batch fermentations were conducted in a volume of 1500 mL in MultiGen (model F2000) stirred-tank bioreactors fitted with agitation, pH, and temperature control (30°C) (New Brunswick Scientific, Edison, NJ). The pH was controlled either at 6.3 or 7.0 by the addition of 4N KOH.

Analytical Procedures

Growth was measured turbidometrically at 550 nm (1 cm lightpath) and dry cell mass (DCM) was measured by microfiltration as described previously *(40)*. Compositional analyses of culture media and cell-free spent broths were determined by HPLC using a HPX-87H column (Bio-Rad Labs, Richmond, CA) as described previously *(40)*. The concentration of metabolic end-products in spent fermentation broths was not corrected for the dilution caused by the addition of titrant during fermentation.

Determination of Fermentation Parameters

The molar growth yield coefficient with respect to carbon (energy) source was calculated by dividing the maximum cell density (g DCM/L) by the molar concentration of sugar added to the medium. The ethanol yield ($Y_{p/s}$) was calculated as the final mass concentration of ethanol divided by the initial sugar concentration. The average volumetric rate of sugar consumption ($_{av}Q_s$; g S/L.h) was determined by dividing the initial concentration of sugar (S) by the total time (post inoculation) required for the complete exhaustion of sugar from the medium. Carbon mass balances (expressed as percent carbon recovery) were calculated as described previously *(41)*. The carbon content of the *E. coli* dry cell mass was assumed constant at 47.6% carbon *(41)*.

RESULTS AND DISCUSSION

Figure 2 compares the growth and fermentation performance of recombinant *E. coli* 11303:pLOI297 in a nutrient-rich complex medium (mLB) with either 2% (w/v) Glc or GlcUA as sole carbon (energy) source. With GlcUA as carbon source, increasing the pH control set-point from 6.3 to 7.0 markedly improved both the sugar utilization rate and the ethanol productivity (Fig. 2), but both the growth yield and the rate of sugar consumption ($_{av}Q_s$) were decreased compared to Glc as substrate (Table 1). Recombinant 11303:pLOI297 exhibits a very high glucose-to-ethanol conversion efficiency (98%), with the ethanol yield being higher at pH 6.3 (0.50 g/g) than 7.0 (0.43 g/g) (Table 1). This observation confirms earlier observations *(30,34,35)*. There was good closure of the carbon mass balance and, at pH 7.0, the lower ethanol yield can be attributed to the formation of lactic acid (Table 2). Whereas Glc is converted primarily to ethanol, GlcUA is converted to acetic acid and ethanol (Table 2). On a weight basis, the ethanol yield (Yp/s/s) was 0.24 and 0.19 g/g at pH 6.3 and 7.0, respectively (Table 1). In the case of GlcUA at pH 6.3, the carbon mass balance did not exhibit closure (Table 1) and an unknown substance was detected in the spent broth from this fermentation. Attempts to positively identify this 'unknown' substance were unsuccessful, although under the conditions of operation of our HPLC system (*see* Materials and Methods), it eluted after acetic acid and before ethanol with a retention time of about 18 min. The only substance that exhibited a similar retention time was methylglyoxal *(29)*. Interestingly, Cooper and Anderson *(58)* have shown that *E. coli* B can synthesize methylglyoxal from the glycolytic intermediate dihydroxyacetone phosphate *(58)*.

The cell mass concentration in Glc and GlcUA fermentations is significantly different (Table 2) with the molar growth yield associated with Glc fermentation being about 1.7 times greater than with GlcUA as substrate (Table 1). The molar growth yield is a reflection of the net gain in energy (ATP) derived by substrate-level phosphorylation reactions associated with anaerobic sugar catabolism. Table 1 shows that the ATP gain (G_{ATP}, mol ATP/mol sugar) is greater with Glc compared to GlcUA as energy source. This bioenergetic parameter can be calculated based on the end-product distribution in combination with a knowledge of the metabolic pathway responsible for the metabolism of each sugar *(41)*. For Glc

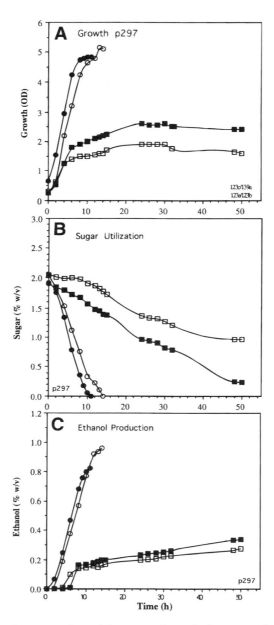

Fig. 2. Comparative growth and fermentation of glucose and glucuronic acid by recombinant *E. coli* B 11303:pLOI297. (**A**) growth, (**B**) sugar utilization, and (**C**) ethanol production. The medium was modified LB (mLB) with either approx 2% (w/v) glucose (Glc) or glucuronic acid (GlcUA) as sole carbon source (*see* Materials and Methods). Experimental data are summarized in Tables 1 and 2. ○ Glc pH 6.3; ● Glc pH 7.0; □ GlcUA pH 6.3; ■ GlcUA pH 7.0.

catabolism, the G_{ATP} per mole of ethanol, lactic acid, and succinate is 1.0 and 1.5 for acetic acid *(59)*. However, because GlcUA is metabolized differently *(60,61)* (*see* more detailed discussion following), the G_{ATP}/mol of ethanol, lactic acid, and succinate is only 0.5 and 1.0 for acetic acid.

Table 1
Summary of Growth and Fermentation Parameters

Culture	Sugar	pH	av Q_S (g S/L.h)	Growth Yield (g DCM/mol S)	G_{ATP} (mol ATP/mol S)	EtOH Yield (g EtOH/g S)	Carbon Recovery (%)
11303							
	Glc	6.3	0.83	11.5	2.00	-	105
		7.0	0.96	12.9	1.90	-	101
	GlcUA	6.3	0.37	8.7	1.59	0.12	114
		7.0	0.87	15.7	1.60	0.08	110
p297							
	Glc	6.3	1.36	16.9	2.01	0.50	113
		7.0	1.74	17.1	1.96	0.43	110
	GlcUA	6.3	0.22	10.9	0.83	0.24	77
		7.0	0.36	9.5	0.89	0.19	73
KO11							
	Glc	6.3	1.11	15.3	2.17	0.50	113
		7.0	1.41	12.3	2.38	0.44	113
	GlcUA	6.3	0.62	9.4	1.42	0.22	102
		7.0	1.45	17.6	1.62	0.19	114

$Y_{x/s}$ = molar yield (g DCM/mol S).

Furthermore, the calculated value of G_{ATP} for GlcUA fermentation by the plasmid-bearing recombinant is made lower by the production of the unidentified metabolic end-product for which an energy yield equivalence can not be assigned.

Acetic acid inhibits *E. coli* growth and fermentation *(62)*. The amount of acetic acid produced, and the sensitivity of *E. coli* to acetic acid inhibition, is known to be both pH and strain dependent *(63)*. In a previous study, we examined the sensitivity of *pet*-plasmid transformed *E. coli* B to acetic acid as a function of pH using different sugar substrates *(37)*. Because the undissociated (protonated) form of acetic acid is responsible for the inhibition, the inhibitory effect of acetic acid is decreased at pH 7.0 because of the lower concentration of the undissociated acid. However, the concentrations of acetic acid produced from GlcUA by the recombinant are well below the inhibitory threshold *(37)*.

Figure 3 compares the growth and fermentation performance of recombinant strain KO11 in mLB medium with either 2% (w/v) Glc or GlcUA as sole carbon (energy) source (note that for purposes of comparison the scales for the plot axes are the same in Figs. 2 and 3). With Glc as substrate, strain KO11 grows slower than the plasmid-bearing recombinant

Table 2
End-Product Distribution Associated with Glucose and Glucuronic Acid Fermentations by Wild-type *E. coli* B and *pet*-transformed Recombinants

	pH	Glc g/L	Glc mM	GlcUA g/L	GlcUA mM	Cell mass gDCM/L	EtOH mM	Succ. mM	Lactic mM	Acetic mM	(Total)
ATCC 11303											
	6.3	20.8	115.7			1.33	0.0	0.0	219.8 (1.90)	5.9 (0.05)	(1.95)
	7.0	26.9	149.3			1.92	0.4	0.0	266.0 (1.78)	9.6 (0.06)	(1.84)
	6.3			17.8	91.5	0.86	48.0 (0.52)	6.5 (0.07)	42.5 (0.46)	97.3 (1.06)	(2.11)
	7.0			19.1	98.6	1.55	32.1 (0.32)	6.8 (0.07)	32.5 (0.33)	122.4 (1.24)	(1.96)
11303:p297											
	6.3	19.0	105.3			1.78	208.4 (1.98)	3.5 (0.03)	0.0	0.0	(2.01)
	7.0	19.2	106.5			1.82	178.4 (1.68)	7.1 (0.07)	22.1 (0.21)	0.0	(1.96)
	6.3			10.8	55.9	0.61	57.0 (1.02)	2.9 (0.05)	0.0	16.5 (0.30)	(1.37)*
	7.0			18.0	92.8	0.88	74.2 (0.80)	6.8 (0.07)	0.0	41.6 (0.45)	(1.32)*
KO11											
	6.3	18.8	104.4			1.60	203.2 (1.95)	0.4	0.0	11.2 (0.11)	(2.06)
	7.0	22.6	125.7			1.54	216.4 (1.72)	0.0	12.1 (0.10)	35.0 (0.28)	(2.10)
	6.3			21.0	108.0	1.02	99.0 (0.92)	4.6 (0.04)	0.0	102.0 (0.94)	(1.90)
	7.0			17.5	90.0	1.58	73.6 (0.82)	0.0	4.3 (0.05)	106.0 (1.18)	(2.05)

Note: Formic acid was not detected in any of these expts. Medium = modified LB (mLB). Bracketed values represent molar yield of end-products (mole P/mole S).
*In Expt 123a and 123b an unknown peak on HPLC seems related to p297 metabolism of GlcUA.

(Fig. 3A). This slower growth is likely a result of the much higher level of inhibiting acetic acid (Table 2). The higher acetic acid concentration is reflected in the improved G_{ATP} (Table 1), but the energetic benefit is nullified at the lower pH by the energetic uncoupling effect of acetic acid (37). At pH 7.0, the molar growth yield with GlcUA surpasses that achieved with Glc (Table 1) and we have no explanation for this observation. With both recombinants, the ethanol yield from Glc was similar and was lower at pH 7 than at pH 6.3; although the rate of sugar consumption ($_{av}Q_s$) was higher for both recombinants at pH 7.0 (Table 1). With GlcUA as substrate, strain KO11 out performed the plasmid-bearing culture both with respect to growth rate (Fig. 3A) and the rate of sugar utilization (Fig. 3B). Unlike with strain 11303:pLOI297, there was good closure of the carbon balance for GlcUA metabolism by strain KO11 at both pH values (Table 1).

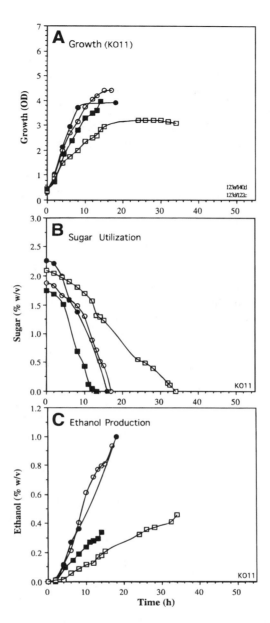

Fig. 3. Comparative growth and fermentation of glucose and glucuronic acid by recombinant *E. coli* B KO11. (**A**) growth, (**B**) sugar utilization, and (**C**) ethanol production. Experimental conditions are as described in legend to Fig. 2. Experimental data are summarized in Tables 1 and 2. ○ Glc pH 6.3; ● Glc pH 7.0; ☐ GlcUA pH 6.3; ■ GlcUA pH 7.0.

Grohmann et al. *(55,57)* have studied GalUA fermentation by recombinant KO11. Their experimental conditions were similar with respect to substrate concentration (2% w/v) and medium composition (mLB). With the pH controlled at 7.0, the ethanol mass yield from GalUA was 0.19 g/g *(55)*, which is identical to the ethanol yield from GlcUA (Table 1).

Furthermore, the molar yield for ethanol and acetic acid from GalUA was observed to be 0.80 and 0.78, respectively (55), which compares very favorably with the pattern for end-product distribution associated with GlcUA metabolism by KO11 under similar assay conditions, namely 0.82 and 1.18 for ethanol and acetic acid, respectively (Table 2). We observed that at pH 6.3, the distribution with respect to ethanol and acetic acid was equimolar, being 0.92 and 0.94, respectively (Table 2).

The pattern of end-product distribution led Grohmann et al. (55) to conclude that GalUA was metabolized according to the following relationship:

$$C_6H_{10}O_7 \rightarrow C_2H_{14}O_2 + C_2H_6O + 2CO_2$$
$$GlcUA \rightarrow \text{Acetic Acid} + EtOH + 2CO_2$$

From our observations with strain KO11 under similar conditions (pH 7.0), the same conclusion with respect to GlcUA metabolism would seem appropriate. The fact that this relationship was chemically balanced and that it coincided with their observations was sufficiently satisfying to Grohmann et al. (55) for them to suggest that it represented a "novel pattern of galacturonic acid fermentation" and these authors did not speculate regarding the metabolic mechanism responsible for the equimolar amounts of ethanol and acetic acid in recombinant E. coli KO11. However, the investigations by Grohmann et al. (55,57) were confined to recombinant KO11. In terms of end-product distribution, the results of our comparative study point to a difference in GlcUA metabolism between the two different metabolically engineered constructs that were examined.

In E. coli, the uptake and metabolism of both GalUA and GlcUA in terms of the conversion of these uronic acids to pyruvic acid is known to be similar (64). Figure 4 compares the catabolism of Glc and GlcUA. The following relationships represent the metabolic pathways depicted for the catabolism of Glc and GlcUA shown in Fig. 4.

For D-glucose (Glc)

$$Glc + 2\ NAD^+ + 2(ADP + P_i) \rightarrow 2\ \text{Pyruvic acid} + $$
$$2(NADH + H^+) + 2ATP$$
$$C_6H_{12}O_6 + 2\ NAD + 2(ADP + P_i) \rightarrow 2\ (C_3H_4O_3) + $$
$$2(NADH + H^+) + 2ATP$$

For D-glucuronic acid (GlcUA)

$$GlcUA + ADP + P_i \rightarrow 2\ \text{Pyruvic acid} + H_2O + ATP$$
$$C_6H_{10}O_7 + ADP + P_i \rightarrow 2(C_3H_4O_3) + H_2O + ATP$$

Succinic acid is sometimes observed as an end-product in E. coli fermentations (Table 2) and the pathway for its production is shown in Fig. 4. In the context of redox balancing, it is important to note that the production of succinic acid from phosphoenolpyruvate requires two pairs of

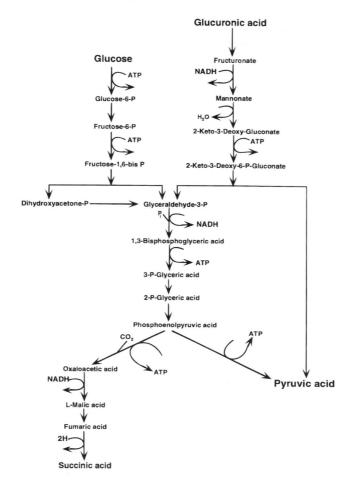

Fig. 4. Metabolic pathways for the partial catabolism of glucose and glucuronic acid by *Escherichia coli* B. The major metabolic pathways for the conversion of glucose and glucuronic acid to pyruvic acid are represented. Phosphoenolpyruvic acid can be converted to succinic acid, but this represents a minor pathway.

reducing equivalents (4H) (Fig. 4). However, the production of succinic acid represents a minor pathway, and furthermore, strain KO11 is supposedly incapable of producing succinate acid by virtue of the engineered interruption of the gene coding for fumarate reductase (48). It is clear from the relationships above that, apart from the participation of different enzymes, the major difference in catabolism of Glc and GlcUA relates to the production of two pairs of reducing equivalents (4H) from Glc, and the production of more energy (ATP) from Glc compared to GlcUA.

Figure 5 illustrates possible pathways for pyruvate metabolism by *E. coli* and includes pathways for both the wild-type, nontransformed, host culture ATCC 11303, and the *pet*-transformed, metabolically engineered, recombinants 11303:pLOI297 and KO11.

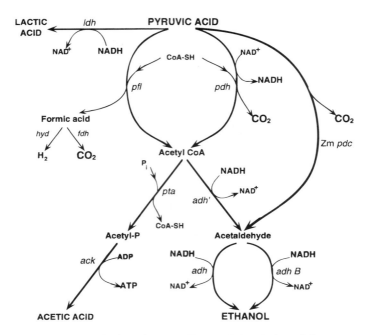

Fig. 5. Pyruvic acid metabolism by *E. coli* B ATCC 11303 and the *pet* recombinants 11303:pLOI297 and strain KO11. *adh* acetaldehyde dehydrogenase; *adh* alcohol dehydrogenase; *adh* B Zm alcohol dehydrogenase; *ack* acetate kinase; *fdh* formate dehydrogenase; *hyd* hydrogenase; *ldh* lactic acid dehydrogenase; pyruvate-formate lyase; *pdh* pyruvate dehydrogenase complex; *pta* phosphotransacetylase; ATP adenosine triphosphate; NADH nicotinamide adenine dinucleotide.

The commonly held view in the literature that deals with the subject of anaerobic glucose metabolism by *E. coli* is that pyruvic acid is reduced to acetyl coenzyme A (acetylCoA) + formic acid by pyruvate-formate lyase (*pfl*) (65) (for review *see* ref. 28) The level of *pfl* is known to be affected by both chemical and physical environmental conditions (66). Lactic acid can also be produced from pyruvic acid by lactic acid dehydrogenase (*ldh*). The metabolic engineering of *E. coli* for ethanol production (26,27) was based on this assumed metabolic fate of pyruvate, whereby the high level of expression of the *pet* operon enzymes from *Zymomonas*, namely pyruvate decarboxylase (*pdc*) and alcohol dehydrogenase II (*adh B*) resulted in a "redirection of pyruvate metabolism" (27) from a mixed acid fermentation to ethanol as the almost exclusive end-product (Fig. 5). If pyruvate were to be metabolized exclusively in this manner, the following end-product distribution, associated with GlcUA dissimilation by recombinant *E. coli*, might be predicted:

$$\begin{align} \text{GlcUA} &\rightarrow \text{Acetic acid} + \text{Formic acid} + \text{EtOH} + CO_2 \\ C_6H_{10}O_7 &\rightarrow C_2H_4O_2 + CH_2O_2 + C_2H_6O + CO_2 \\ &\rightarrow C_2H_4O_2 + C_2H_6O + 2CO_2 + H_2 \end{align}$$

However, this relationship produces an extra 2H and does not balance. From a theoretical perspective, if the recombinants were to produce ethanol through a utilization of the engineered ethanol production pathway from *Zymomonas* (or at least via the *Zymomonas pdc*), there would be a requirement for 2H (NADH). In searching for a possible source of this 2H (NADH), it was postulated that the native NAD-linked pyruvate dehydrogenase (*pdh*) is repressed by glucose, but in a nonrepressive growth environment, *pdh* would operate in competition with *pfl* for the conversion of pyruvate to AcCoA, thereby generating the necessary NADH for either the native or *Zymomonas*-derived alcohol dehydrogenase to convert acetaldehyde to ethanol (Fig. 5). The pyruvate dehydrogenase complex in *E. coli* is known to be subject to catabolite repression by glucose *(67)*, and Langley and Guest *(68)* have observed that with 50 m*M* glucose the levels of pdh were similar in *E. coli* strain H under both aerobic and anaerobic growth conditions. Whereas *pdc* is known to be absent from the nontransformed ATCC 11303 culture *(33)*, nevertheless there remains the potential for ethanol production via the combined action of the native NAD-linked enzymes, acetaldehyde dehydrogenase and alcohol dehydrogenase (Fig. 5). Therefore, it is conceivable that if *pdh* were solely responsible for the formation of AcCoA from pyruvate, equimolar amounts of ethanol and acetic acid could be produced from GlcUA in the recombinants totally independent of the existence (operation) of the engineered *pet* pathway. The balanced fermentation that would result from this proposed action is:

$$C_6H_{10}O_7 \rightarrow 2(C_3H_4O_3) + H_2O$$
Pyruvic acid
$$\rightarrow 2 \text{ AcetylCoA} + 2CO_2 + 2 (NADH + H^+)$$
$$\rightarrow C_2H_4O_2 + C_2H_6O + 2CO_2 + 2 \text{ CoA-SH} + 2NAD^+$$
Acetic acid   EtOH

In considering the fermentation of GlcUA, it was noted that the following relationship also balanced:

$$C_6H_{10}O_7 \rightarrow C_2H_4O_2 + C_3H_6O_3 + CO_2$$
GlcUA   Acetic acid   Lactic acid

However, this pattern of end-products is not consistent with the production of acetic acid by either *pfl* and/or *pdh* since both enzymes would generate an extra pair of reducing equivalents either as H_2 or as NADH.

The strategy for metabolically engineering *E. coli* for ethanol production was based on

1. The conversion of pyruvate to AcCoA exclusively by *pfl*, and
2. The absence of *pdc*, and
3. The ability of expressed *Zymomonas pdc* to effectively exclude the competition for pyruvate exerted by NAD-linked lactic acid dehydrogenase (*ldh*) *(26,27)*.

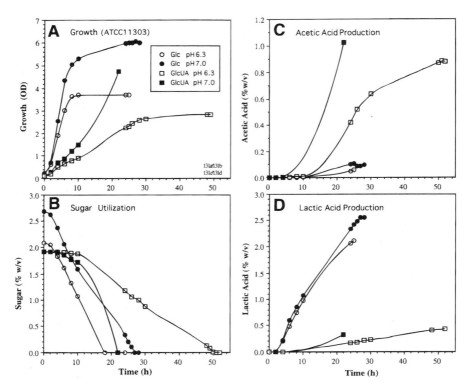

Fig. 6. Comparative growth and fermentation of glucose and glucuronic acid by the nontransformed host culture E. coli B ATCC 11303. (A) growth, (B) sugar utilization, (C) acetic acid production, and (D) lactic acid production. Experimental conditions are as described in legend to Fig. 2. Experimental data are summarized in Tables 1 and 2. ○ Glc pH 6.3; ● Glc pH 7.0.; □ GlcUA pH 6.3; ■ GlcUA pH 7.0.

However, if glucose acts to repress *pdh*, then under conditions where *pdh* is derepressed, the success of the competition for pyruvate between *pdh*, *pfl*, and *ldh*, will dictate the nature of end-product distribution. If GlcUA did not repress *pdh*, then it seems reasonable to expect that ethanol could be a product of GlcUA catabolism by the wild-type culture. To examine this possibility, the nature of end-product distribution in GlcUA fermentation by the nontransformed culture E. coli B ATCC 11303 was investigated. Figure 6 compares the growth and fermentation performance of wild-type E. coli B in mLB medium with either 2% (w/v) Glc or GlcUA as sole carbon (energy) source (note that for purposes of comparison the scales for the plot axes are the same as in Figs. 2 and 3).

With Glc as substrate, the higher cell density achieved at pH 7.0 compared to pH 6.3 (Fig. 6A) is a result of, in part, the higher sugar concentration (Fig. 6B and Table 2); however, the growth yield was only slightly higher at pH 7.0 (Table 1). The rate of Glc consumption ($_{av}Q_s$) was slower at both pH values compared to the recombinant cultures (Table 1), which may be related to the fact that lactic acid is the sole end-product of Glc catabolism by ATCC 11303 (Fig. 6D and Table 2). The high degree of selectivity for lactic acid associated with Glc fermentation by this culture confirms previous reports in the literature (31,35,39).

With GlcUA as substrate, the nontransformed culture produced acetic acid (Fig. 6C), lactic acid (Fig. 6D), and ethanol (not shown in Fig. 6). In all fermentations with ATCC 11303 (with both sugars) there was good closure of the carbon balance (Table 1). The concentrations of the different end-products associated with GlcUA fermentation are given in Table 2. The molar ratio of the end-products is not the same at pH 6.3 and pH 7.0 (Table 2). At pH 6.3 and 7.0, the yield of ethanol on a weight basis is 0.12 and 0.08 g/g, respectively (Table 1). This yield of ethanol is clearly inferior to that observed with recombinant strains, but the appearance of ethanol supports the proposed pathway for pyruvate metabolism with GlcUA as carbon source.

In the nontransformed culture, if the pyruvate derived from GlcUA were to be converted to ethanol, acetic acid and lactic acid by *pfl* and *ldh*, one could predict the following relationship:

$C_6H_{10}O_7$ --------> $2(C_3H_4O_3)$ + H_2O
GlcUA   Pyruvic acid
--------> $0.5 (C_3H_6O_3)$ + 1.5 AcetylCoA + 1.5 CH_2O_2
Lactic Acid
--------> $0.5 (C_3H_6O_3)$ + $C_2H_4O_2$ + $0.5 (C_2H_6O)$ + 1.5 CO_2
Lactic acid   Acetic acid   EtOH
+ 1.5 H_2

However, this relationship produces an extra 2H and does not balance. If the pyruvate derived from GlcUA were to be converted to ethanol, acetic acid and lactic acid by *ldh* and derepressed *pdh*, the fermentation balances as:

$C_6H_{10}O_7$ → $2(C_3H_4O_3)$ + H_2O
GlcUA   Pyruvic acid
→ $0.5 (C_3H_6O_3)$ + 1.5 AcetylCoA +
Lactic acid
1.5 CO_2 + 1.5 (NADH + H^+)
→ $0.5 (C_3H_6O_3)$ + $C_2H_4O_2$ + $0.5 (C_2H_6O)$ +
Lactic acid Acetic acid   EtOH
1.5 CO_2 + 1.5 NAD^+

This approximates closely the molar ratio of ethanol, acetic acid, and lactic acid that was observed for GlcUA fermentation by ATCC 11303 at pH 6.3 (Table 2). At pH 7.0, there is a shift in favor of acetic acid with the amounts both ethanol and lactic acid decreasing proportionately (Table 2).

To further support our hypothesis concerning a possible role for derepressed *pdh* in GlcUA metabolism, the literature was searched for reports concerning Glc fermentation by *E. coli*. A review of the literature revealed that the end-product distribution in anaerobic mixed acid fermentation by *E. coli* is strain-dependent. Too often generalizations concerning microbial metabolism are made based on the observations with only a single strain or with only a single substrate. Redox balance requires that the molar amount of acetic acid be equal to the sum of ethanol and succinic acid. Beläich and Beläich *(69)* observed that *E. coli* K-12 growing anaerobically in minimal medium at pH 7 produced no lactic acid and equimolar amounts of acetic acid + ethanol and succinic acid with a mass ethanol yield of 0.22 g/g. Varma and Palsson *(70)* observed a similar end-product distribution associated with anaerobic glucose dissimilation by *E. coli* strain W3110 (ATCC 27325) where the ethanol yield was 0.17 g/g. This pattern of end-product distribution is totally consistent with the catabolism of pyruvic acid by *pfl*. However, there are also reports in the literature where the amount of acetic acid produced during anaerobic glucose dissimilation by *E. coli* was not equal to the sum of the amounts of ethanol + succinic acid *(71,72)*. For example, Diaz-Ricci et al. *(72)* have studied the ethanologenicity of *pet* plasmid transformed *E. coli* strain HB101. Strain HB101 is a hybrid of *E. coli* K-12 and *E. coli* B *(73)* and the molar end-product distribution exhibited by the nontransformed culture is:

Glucose → 0.86 EtOH + 0.11 Succ + 0.75 Acetic acid
+0.01 Lactic acid + 1.25 Formic acid

The ethanol yield from glucose exhibited by the nontransformed culture was 0.22 g/g *(72)*. The labile nature of formic acid compromises interpretation based solely on the amount produced; however, pyruvate catabolism by *pfl* would yield an amount of formic acid equal to the sum of ethanol and acetic acid and clearly this is not the case in the work reported by Diaz-Ricci et al. *(72)*. Furthermore, the amount of ethanol + succinic acid is greater than the amount of acetic acid which suggests an additional source of reducing equivalents that would be required to form ethanol from acetylCoA by the combination of acetaldehyde dehydrogenase and alcohol dehydrogenase. It seems reasonable to assume that in this *E. coli* B hybrid strain *pdh* and *pfl* compete for pyruvic acid and that the extra NADH generated by *pdh* activity permits more ethanol to be synthesized. Collectively, these observations in the literature offer indirect support for our conclusion that *pdh* plays a role in GlcUA metabolism in recombinant *E. coli* B.

CONCLUSIONS

From this investigation it is concluded that GlcUA can be fermented by *E. coli* B ATCC 11303 and recombinants 11303:pLOI297 and KO11. Both recombinants produced ethanol from GlcUA in an amount and manner that was similar to the "novel fermentation" reported by Grohmann et al. (54,56) for GalUA with strain KO11. Acetic acid is a significant by-product of GlcUA fermentation by both recombinant cultures. The rate of fermentation is improved by increasing the pH-control set point from 6.3 to 7.0. The production of an unidentified metabolic end-product by the plasmid-bearing recombinant points to a different metabolic pathway operating in this metabolically engineered construct. Finally, our analysis of end-product distribution for metabolism of GlcUA by the non-transformed, wild-type *E. coli* B and recombinant strain KO11 suggests that pyruvate formate lyase is not solely responsible for the production of acetylCoA from pyruvate and that derepressed pyruvate dehydrogenase plays a role in the metabolism of GlcUA in these cultures.

ACKNOWLEDGMENTS

This research was internally funded by the University of Toronto. We are grateful to Lonnie Ingram for the gift of the recombinant *E. coli* cultures and to Karel Grohmann and Richard Helm for suggestions and advice.

REFERENCES

1. Lynd, L. R. (1990), *Appl. Biochem. Biotechnol.* **24/25,** 695–719.
2. Sheehan, J. J. (1994), in *Enzymatic Conversion of Biomass for Fuels Production*, Himmel, M. E., Baker, J. O., and Overend, R. P., eds., ACS Symposium Series 566, American Chemical Society, Washington, DC, pp. 1–53.
3. Lynd, L. R., Cushman, J. H., Nichols, R. J., and Wyman, C. E. (1991), *Science* **251,** 1318–1323.
4. Lynd, L. R., Elander, R. T., and Wyman, C. E. (1996), *Appl. Biochem. Biotechnol.* **57/58,** 741–762.
5. Hohman, N. and Rendleman, C. M. (1993), US Department of Agriculture Information Bulletin Number 663, pp. 1–17.
6. Hinman, N. D., Wright, J. D., Hoagland, W., and Wyman, C. E. (1989), *Appl. Biochem. Biotechnol.* **20/21,** 391–401.
7. Wyman, C. E. and Hinman, N. D. (1990), *Appl. Biochem. Biotechnol.* **24/25,** 735–753.
8. Timell, T. E. (1967), *Wood Sci. Technol.* **1,** 45–70.
9. Grohman, K., Himmel, M., Rivard, C., Tucker, M., Baker, T, Torget, R., and Graboski, M. (1984), *Biotechnol. Bioeng.* Symp. **14,** 139–157.
10. Grethlein, H. E. (1985), *Bio/Technolgy* **3,** 155–160.
11. Kong, F., Engler, C. R., and Soltes, E. (1992), *Appl. Biochem. Biotechnol.* **34/35,** 23–35.
12. Timell, T. E. (1964), *Adv. Carbohydrate Chem.* **19,** 247–302.
13. Timell, T. E. (1965), *Adv. Carbohydrate Chem.* **20,** 409–483.
14. Torget, R., Walter, P., Himmel, M., and Grohman, K. (1991), *Appl. Biochem. Biotechnol.* **28/29,** 75–86.
15. Aspinall, G. O. (1959), *Adv. Carbohydrate Chem.* **14,** 429–468.
16. Aspinall, G. O., Hirst, E. L., and Mahomed, R. S. (1954), *J. Chem. Soc.* **17,** 34–41.
17. Browning, B. L. (1967), in *Methods in Wood Chemistry*, Interscience Publishers, New York, NY.
18. Grohmann, K., Mitchell, D. J., Himmel, M. E., Dale, B. E., and Schroeder, H. A. (1989), *Appl. Biochem. Biotechnol.* **20/21,** 45–61.

19. Stanek, D. A. (1958), *TAPPI* **41,** 601–609.
20. Fein, J. E., Tallim, S. R., and Lawford, G. R. (1984), *Can. J. Microbiol.* **30,** 682–690.
21. Beck, M. J. (1986), *Biotechnol. Bioeng. Symp.* **17,** 617–627.
22. Frazer, F. R. and McCaskey, T. A. (1989), *Biomass,* **18,** 31–42.
23. McMillan, J. D. (1994), in *Enzymatic Conversion of Biomass for Fuels Production,* Himmel, M. E., Baker, J. O., and Overend, R. P., eds., ACS Symposium Series 566, American Chemical Society, Washington, DC, pp. 411–437.
24. McMillan, J. D. (1994), in *Enzymatic Conversion of Biomass for Fuels Production,* Himmel, M. E., Baker, J. O., and Overend, R. P., eds., ACS Symposium Series 566, American Chemical Society, Washington, DC, pp. 293–324.
25. Ingram, L. O., Conway, T., and Alterthum, F. (1991), United States Patent 5,000,000.
26. Ingram, L. O., Alterthum, F., Ohta, K., and Beall, D. S. (1990), in *Developments in Industrial Microbiology,* vol 31, Elsevier Science Publ., New York, NY, pp. 21–30.
27. Ingram, L. O. (1991), in *Energy from Biomass and Wastes XIV,* Klass, D. L., ed., Institute of Gas Technology, Chicago, IL, pp. 1105–1126.
28. Knappe, J. (1987), in *Escherichia coli and Salmonella typhimurium,* vol. 1, Neidhart, F. C., ed., Academic Press Inc., New York, pp. 151–155.
29. Gottschalk, G. (1986), in *Bacterial Metabolism,* 2nd ed., Springer-Verlag, New York, NY, pp. 208–282.
30. Alterthum, F. and Ingram, L. O. (1989), *Appl. Environ. Microbiol.* **55,** 1943–1948.
31. Ingram, L. O. and Conway, T. (1988), *Appl. Environ. Microbiol.* **54,** 397–404.
32. Brau, B. and Sahm, H. (1986), *Arch. Microbiol.* **144,** 296–301.
33. Neale, A. D., Scopes, R. K., Wettenhall, E. H., and Hoogenraad, N. J. (1987), *J. Bacteriol.* **169,** 1024–1028.
34. Beall, D. S., Ohta, K., and Ingram, L. O. (1991), *Biotechnol. Bioeng.* **38,** 296–303.
35. Lawford, H. G. and Rousseau, J. D. (1991), *Appl. Biochem. Biotechnol.* **28/29,** 221–236.
36. Lawford, H. G. and Rousseau, J. D. (1992), *Appl. Biochem. Biotechnol.* **34/35,** 185–204.
37. Lawford, H. G. and Rousseau, J. D. (1993), *Appl. Biochem. Biotechnol.* **39/40,** 301–322.
38. Lawford, H. G. and Rousseau, J. D. (1993), *Biotechnol. Letts.* **15,** 615–620.
39. Lawford, H. G. and Rousseau, J. D. (1991), in *Energy from Biomass & Wastes XV,* Klass, D. L., ed., Institute of Gas Technology, Chicago, IL, pp. 583–622.
40. Lawford, H. G. and Rousseau, J. D. (1993), *Appl. Biochem. Biotechnol.* **57/58,** 277–292.
41. Lawford, H. G. and Rousseau, J. D. (1993), *Appl. Biochem. Biotechnol.* **57/58,** 293–305.
42. Lawford, H. G. and Rousseau, J. D. (1993), *Appl. Biochem. Biotechnol.* **57/58,** 307–326.
43. Lawford, H. G. and Rousseau, J. D. (1991b), *Biotechnol. Letts.* **13,** 191–196.
44. Lawford, H. G. and Rousseau, J. D. (1993), in *Energy from Biomass and Wastes XVI,* Klass, D. L., ed., Institute of Gas Technology, Chicago, IL, pp. 559–597.
45. Lawford, H. G. and Rousseau, J. D. (1993), *Biotechnol. Letts.* **15,** 505–510.
46. Lawford, H. G. and Rousseau, J. D. (1993), *Appl. Biochem. Biotechnol.* **39/40,** 667–685.
47. Lawford, H. G. and Rousseau, J. D. (1992), *Biotechnol. Letts.* **14,** 421–426.
48. Ohta, K., Beall, D. S., Mejia, J. P., Shanmugam, K. T., and Ingram, L. O. (1991), *Appl. Environ. Microbiol.* **57,** 893–900.
49. DaSilva, N. A. and Bailey, J. E. (1986), *Biotechnol. Bioeng.* **28,** 741–746.
50. Barbosa, M. de F. S., Beck, M. J., Fein, J. E., Potts, D., and Ingram, L. O. (1992), *Appl. Environ. Microbiol.* **58,** 1182–1184.
51. Beall, D. S., Ingram, L. O., Ben-Bassat, A., Doran, J. B., Fowler, D. E., Hall, R. G., and Wood, D. E. (1992), *Biotechnol. Letts.* **14,** 857–862.
52. Asghari, A., Bothast, R. J., Doran, J. B., and Ingram, L. O. (1996), *J. Ind. Microbiol.* **16,** 42–47.
53. Hahn-Hägerdal, B., Jeppsson, H., Olsson, L., and Mohagheghi, A. (1994), *Appl. Microbiol. Biotechnol.* **41,** 62–72.
54. von Sivers, M., Zacchi, G., Olsson, L., and Hahn-Hägerdal, B. (1994), *Biotechnol. Prog.* **10,** 555–560.

55. Grohmann, K., Baldwin, E. A., Buslig, B. S., and Ingram, L. O. (1994), *Biotechnol. Letts.* **16,** 281–286.
56. Luria, S. E. and Delbruck, M. (1943), *Genetics* **28,** 491–511.
57. Grohmann, K., Cameron, R. G., and Buslig, B. S. (1995), *Appl. Biochem. Biotechnol.* **51/52,** 423–435.
58. Cooper, R. A. and Anderson, A. (1970), *FEBS Letts.* **11,** 273–276.
59. Tempest, D. W. and Neijssel, O. M. (1987), in *Escherichia coli and Salmonella typhimurium*, vol. 1, Neidhart, F.C., ed., Academic Press Inc., New York, Chap 52, pp. 797–806.
60. van Gijsegem, F., Hugouvieux-Cotte-Pattat, N., and Robert-Baudouy, J. (1985), *J. Bacteriol.* **161,** 702–708.
61. Portalier, R., Robert-Baudouy, J., and Nemoz, F. (1980), *J. Bacteriol.* **143,** 1095–1107.
62. Lui, G. W. and Strohl, W. R. (1990), *Appl. Environ. Microbiol.* **56,** 1004–1011.
63. Koh, B. T., Nakashimada, U., Pfeiffer, M., and Yap, M. G. S. (1992), *Biotechnol. Letts.* **14,** 1115–1118.
64. Lin (1987), in *Escherichia coli and Salmonella*, Neidhart, F. C., ed., Academic Press Inc., New York, Vol 1, Chap 18, pp. 244–284.
65. Clark, D. P., Cunningham, P. R., Reams, S. G., Mat-Jan, F., Mohammedkhani, R., and Williams, C. R. (1988), *Appl. Biochem. Biotechnol.* **17,** 163–173.
66. Rasmussen, L. J., Moller, P. L., and Atlung, T. (1991), *J. Bacteriol.* **173,** 6390–6397.
67. Nimmo, H. G. (1987), in *Escherichia coli and Salmonella*, Neidhart, F. C., ed., Academic Press Inc., New York, Vol 1, Chap 14, pp. 156–169.
68. Langley, D. and Guesty, J. R. (1978), *J. Gen. Microbiol.* **106,** 103–117.
69. Beläich, J. P. and Beläich, J. A. (1976), *J. Bacteriol.* **125,** 14–18.
70. Varma, A. and Palsson, B. O. (1994), *Appl. Environ. Microbiol.* **60,** 3724–3731.
71. Blackwood, A. C., Neish, A. C., and Ledingham, G. A. (1956), *J. Bacteriol.* **72,** 497–499.
72. Diaz-Ricci, J. C., Tsu, J. C., and Bailey, J. E. (1992), *Biotechnol. Bioeng.* **39,** 59–65.
73. Boyer, H. W. and Rouland-Dussioux, D. J. (1969), *J. Mol. Biol.* **41,** 459–472.

Enhanced Cofermentation of Glucose and Xylose by Recombinant *Saccharomyces* Yeast Strains in Batch and Continuous Operating Modes

SUSAN T. TOON,[1] GEORGE P. PHILIPPIDIS,[4]
NANCY W. Y. HO,[2] ZHENGDAO CHEN,[2] ADAM BRAINARD,[2]
ROBERT E. LUMPKIN,[3] AND CYNTHIA J. RILEY*,[1]

[1]*Biotechnology Center for Fuels and Chemicals, National Renewable Energy Laboratory (NREL), 1617 Cole Boulevard, Golden, CO 80401;*
[2]*Laboratory of Renewable Resources Engineering (LORRE), Purdue University, West Lafayette, IN 47907;* [3]*SWAN Biomass Company, Downers Grove, IL 60515; and* [4]*Thermo Fibergen, Inc., Bedford, MA 01730*

ABSTRACT

Agricultural residues, such as grain by-products, are rich in the hydrolyzable carbohydrate polymers hemicellulose and cellulose; hence, they represent a readily available source of the fermentable sugars xylose and glucose. The biomass-to-ethanol technology is now a step closer to commercialization because a stable recombinant yeast strain has been developed that can efficiently ferment glucose and xylose simultaneously (coferment) to ethanol. This strain, LNH-ST, is a derivative of *Saccharomyces* yeast strain 1400 that carries the xylose-catabolism encoding genes of *Pichia stipitis* in its chromosome. Continuous pure sugar cofermentation studies with this organism resulted in promising steady-state ethanol yields (70.4% of theoretical based on available sugars) at a residence time of 48 h. Further studies with corn biomass pretreated at the pilot scale confirmed the performance characteristics of the organism in a simultaneous saccharification and cofermentation (SSCF) process: LNH-ST converted 78.4% of the available glucose and 56.1% of the available xylose within 4 d, despite the presence of high levels of

---

*Author to whom all correspondence and reprint requests should be addressed.

metabolic inhibitors. These SSCF data were reproducible at the bench scale and verified in a 9000-L pilot scale bioreactor.

Index Entries: Simultaneous saccharification and cofermentation; glucose and xylose fermentation; cellulosic biomass; recombinant *Saccharomyces* sp.; fermentation scale-up.

INTRODUCTION

Agricultural residues represent a readily available source of cellulose and hemicellulose that can be converted to fuel ethanol *(1)*. The recent formation of the SWAN Biomass Company, a partnership between Amoco and Stone and Webster Engineering *(2)*, demonstrates the interest of the private sector in the commercialization of biomass-to-ethanol conversion technology developed by the National Renewable Energy Laboratory (NREL) and Amoco.

Cellulose conversion technology by the simultaneous saccharification and fermentation (SSF) process has made significant progress during the past ten years *(3)*. In contrast, the fermentation of hemicellulose-derived xylose to ethanol remains problematic because of the unavailability of efficient glucose- and xylose-cofermentation organisms *(4–6)*. As a result, the ethanol potential of biomass has not been fully realized and the cost of ethanol remains uncompetitive with gasoline. Fortunately, recent efforts in metabolic engineering have resulted in the development of organisms that can ferment glucose and xylose simultaneously, such as a recombinant *Zymomonas mobilis (7)* and recombinant derivatives of *Saccharomyces* yeasts *(8–12)*.

In 1993, Ho and coworkers first developed recombinant *Saccharomyces* yeasts that could effectively ferment xylose to ethanol. One such yeast is 1400(pLNH33), referred to as LNH33, which was developed by transforming *Saccharomyces* strain 1400, a fusion product of *Saccharomyces diastaticus* and *Saccharomyces uvarum (13)*, with the high-copy number yeast—*E. coli* shuttle vector pLNH33 containing the xylose reductase (XR), xylitol dehydrogenase (XD) (both from *Pichia stipitis*), and xylulokinase genes (XK) (from *S. cerevisiae*). These cloned genes were fused to promoters that are neither inhibited by the presence of glucose nor require the presence of xylose for induction *(8,10)*. As a result, LNH33 can ferment xylose to ethanol, as well as efficiently coferment glucose and xylose to ethanol. The xylose-fermenting ability of LNH33 can be maintained by growth on YEP (10 g/L yeast extract and 20 g/L peptone) medium supplemented with xylose (20 g/L) as the sole carbon source. However, like all plasmid-bearing transformants, LNH33 is not stable when successive generations are cultured in nonselective media. This instability could limit the use of LNH33 in a continuous fermentation process.

To overcome this problem, Ho and Chen *(12)* recently developed an improved recombinant *Saccharomyces* yeast, designated as LNH-ST, which contains multiple copies of the xylose metabolizing genes, XR-XD-XK, integrated into the chromosome of the host strain 1400. This strain coferments glucose and xylose with improved efficiencies, and is also stable. LNH-ST does not have to be cultured in any special medium to maintain the cloned genes; it can be cultured in nonselective media for an unlimited number of generations and still retain its full capability to ferment xylose to ethanol.

By using the recombinant yeast strains that utilize both glucose and xylose, the process becomes a SSCF. Higher ethanol yields and concentrations, in turn, minimize the risk of contamination in the SSCF bioreactors and further enhance the productivity and lower the cost of the technology. Hence, the current SWAN biomass conversion technology applied to grain by-products consists of three key steps:

1. The pretreatment process, which uses sulfuric acid and heat to hydrolyze hemicellulose to xylose and xylan oligomers and improves cellulose accessibility to enzymatic hydrolysis;
2. The SSCF process, in which cellulose is hydrolyzed to glucose and, at the same time, glucose and xylose are fermented to ethanol by the recombinant yeast; and
3. The downstream separation process, in which ethanol is recovered through distillation and the protein-rich solids are retrieved to be sold as a component of animal feed.

In the present work an evaluation of the cofermentation capabilities of the two recombinant yeast strains, LNH33 and LNH-ST is described, first using mixtures of pure sugars (glucose and xylose) and then pretreated corn biomass. To assist in the scale-up of the process, this study monitored the ethanol productivity of the organisms in both batch and continuous operating modes and compared the SSCF performances at the bench and pilot scales.

MATERIALS AND METHODS

Inoculum Generation

Frozen stock cultures of LNH33 and LNH-ST were prepared in YEPX [1% (w/v) yeast extract, 2% (w/v) peptone, 2% (w/v) xylose, pH 5.0], and 20% (w/w) glycerol solution. One-mL aliquots were dispensed into cryopreservation vials and stored at −70°C.

Seed cultures of LNH33 were prepared by inoculating 1 mL of LNH33 frozen stock into 50 mL of 1% (w/v) yeast extract, 2% (w/v) peptone, 2% (w/v) xylose, and 1% (w/v) corn steep liquor (CSL) at pH 5.0 in a 250-mL baffled Erlenmeyer flask. Cultures were incubated at 30°C with agitation at 150 rpm. After 24 h of growth, 10% (v/v) was transferred to 2% (w/v) CSL, 1% (w/v) yeast extract, and 2% (w/v) xylose at pH 5.0 for inoculum preparation.

LNH-ST inoculum was also prepared in two stages for the continuous fermentation studies, first by growth in YEPD [1% (w/v) yeast extract, 2% (w/v) peptone, and 2% (w/v) glucose, pH 5.0] and then in 1% (w/v) CSL, and 2% (w/v) glucose at pH 5.0. For comparative studies between bench-scale and the pilot-scale Process Development Unit (PDU), inoculum was prepared in five stages, to scale-up seed, in the presence of 1% (w/v) CSL and 2% (w/v) glucose at pH 5.0.

Cofermentation of Glucose and Xylose

Pure sugar cofermentation studies, with a mixture of glucose and xylose, were carried out in a two-stage continuous fermentation configuration at 30°C, 150 rpm, and pH 5.0. Two 1.7-L BioFlo III fermentors (New Brunswick Scientific) were employed at a working volume of 1 L per reactor. The first-stage fermentor was started in batch mode with 2% (w/v) CSL, 1% (w/v) yeast extract, 2.4% (w/v) glucose, and 3.4% (w/v) xylose at pH 5.0. Stock solutions of glucose and xylose were filter sterilized separately and added to the fermentor, along with 10% (v/v) inoculum. The first-stage fermentation was operated in batch mode for 24 h, then switched to continuous mode, and the effluent was directed to the second-stage fermentor. Feed for the first stage was prepared in 15-L batches and consisted of the same medium used in the batch phase. The dilution rate of the two fermentors was controlled by the feed rate to the first vessel and was monitored throughout the fermentation. The ethanol and by-product yields and the glucose and xylose conversions were calculated based on the steady-state concentrations. To minimize ethanol evaporation, the condensers on each fermentor were packed with l-mm glass beads for maximum surface area and equipped with 4°C water circulation. The pH was controlled at 5.0 with the addition of $3M$ sodium hydroxide. Neither air nor nitrogen were supplied to the fermentors.

Batch Simultaneous Saccharification and Cofermentation (SSCF)

Corn biomass was pretreated in a pilot-scale dilute-acid pretreatment reactor and the pH was adjusted to 5.0 with sodium hydroxide. SSCF batch experiments were performed at 20% (w/v) total solids concentration, 30°C, 150 rpm, and pH 5.0, controlled with the addition of $3M$ sodium hydroxide. CSL [1% (w/v)] was added as a nutrient source. The SSCF, as well as the continuous pure sugar fermentations, were carried out at 30°C because xylose fermentation is diminished at higher temperatures.

Cellulase enzyme (80 IFPU/mL), supplied by Iogen (Ottawa, Canada), was added to the SSCF at a level of 10 IFPU/g cellulose present in the raw corn biomass. Its activity was measured by the IUPAC methods *(14)*. Glucoamylase (285 IU/mL), from Enzyme Development Corporation (New York, NY), was added at 2 IU/g of starch present in the raw feedstock. Each enzyme preparation was filter-sterilized before addition to the fermentors.

Analysis

Samples were analyzed for ethanol and glucose with the Yellow Springs Instrument (YSI) Model 2700 Biochemistry Analyzer. Hexose and pentose sugars were analyzed by liquid chromatography using a Hewlett Packard 1090 HPLC unit equipped with an HP 1047 IR detector and an HPX87P column operating at 0.60 mL/min and 85°C. Ethanol and metabolic by-products, such as organic acids, glycerol, and xylitol, were quantified by HPLC using an HPX87H column operating at a flow rate of 0.60 mL/min and a temperature of 65°C. SSCF liquor residues were analyzed for oligomeric sugars, and a total compositional analysis was performed on the solids at the initial and final time points of each SSCF study to determine ethanol yields and to close the fermentation carbon balance (15).

RESULTS AND DISCUSSION

Two-Stage Continuous Cofermentation of Glucose and Xylose by LNH33

A two-stage continuous cofermentation was used to examine the ability of LNH33 to simultaneously ferment a mixture of xylose and glucose at levels representative of realistic pretreated corn biomass; and demonstrate its ability to grow in continuous culture under realistic process conditions prior to scale-up and commercialization efforts. The two-stage continuous cofermentation was operated in batch mode for 24 h. During the batch phase, all available glucose, as well as 38.6% of the xylose, was consumed (Fig. 1).

After the 24-h batch period, the fermentation was switched to continuous operation with a feed rate of 0.694 mL/min. This rate produced a 24-h residence time (0.042-h^{-1} dilution rate) in each stage and an overall process residence time of 48 h. After switching the system to continuous operation, the glucose concentration remained undetectable, but the xylose concentration increased during the 265-h duration of the continuous run to 31.2 g/L (Fig. 1), resulting in a xylose conversion of 11.4% in the first stage (Table 1). After the second stage was filled with effluent, the xylose concentration reached a minimum of 6.8 g/L. However, during the continuous phase, the xylose concentration in the second stage increased to 26 g/L (Fig. 1), resulting in a xylose conversion of 16.9%. Hence, the overall xylose conversion was 26.3% at a residence time of 48 h. Even after 11 residence times, the continuous system had not reached a steady state, presumably because of plasmid instability or a decrease in plasmid copy number.

The metabolic ethanol yields (based on the amount of consumed sugar) in the first and second stages were 84.3 and 85.2% of theoretical, respectively. The ethanol process yields (based on available fermentable sugars) were only 40.1 and 14.4%, respectively, because of the limited consumption of xylose, especially in the second stage (Table 1). For the over-

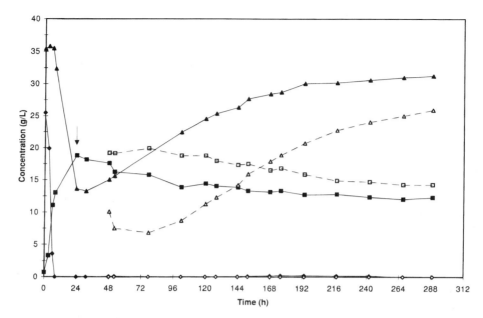

Fig. 1. Time course of the glucose (♦, ◇), xylose (▲, △), and ethanol (■, □) concentrations during the continuous, two-stage pure sugar cofermentation by LNH33. Closed symbols and continuous lines represent the first stage, whereas open symbols and dashed lines represent the second stage. The arrow indicates the switching point from batch to continuous operation.

Table 1
Pure Sugar Two-Stage Continuous Fermentation Performance of LNH33 and LNH-ST

Strain LNH33			
Stage	**1**	**2**	**Overall**
Residence Time (h)	24	24	48
Glucose Conversion	100.0%	-	100.0%
Xylose Conversion	11.4%	16.9%	26.3%
Ethanol Process Yield (% theoretical)	40.1%	14.4%	47.6%
Ethanol Metabolic Yield (% theoretical)	84.3%	85.2%	84.4%
Strain LNH-ST			
Stage	**1**	**2**	**Overall**
Residence Time (h)	23.3	23.3	46.6
Glucose Conversion:	99.5%	100.0%	100.0%
Xylose Conversion:	58.3%	67.4%	86.4%
Ethanol Process Yield (% theoretical)	58.7%	47.2%	70.4%
Ethanol Metabolic Yield (% theoretical)	78.0%	69.7%	76.5%

Table 2
Product Distribution during Pure Sugar Fermentation and Corn Biomass SSCF
(Expressed as Grams of Product per 100 g of Consumed Glucose and Xylose)

Product	Pure-Sugar Two-Stage		Corn Biomass SSCF by LNH-ST	
	LNH33	LNH-ST	Batch 1	Batch 2
Ethanol	43.14	39.09	41.14	45.00
Cell Mass	7.53	7.04	4.04	3.94
Carbon Dioxide	41.21	37.34	39.35	43.05
Glycerol	3.39	9.37	4.47	5.02
Xylitol	4.94	7.02	2.61	2.75
Succinic Acid	0.00	0.00	0.45	0.55
Total	100.21	99.86	92.05	100.31

all process, the ethanol metabolic yield was 84.4% and the ethanol process yield was 47.6%. The major by-products of the fermentation (expressed per gram of consumed glucose and xylose) were carbon dioxide (0.412 g, calculated on an equimolar basis with respect to produced ethanol), cell mass (0.075 g), xylitol (0.049 g), and glycerol (0.034 g). The overall carbon balance was 100.21% (Table 2).

Because of the poor xylose utilization by LNH33 in the two-stage continuous cofermentation study, strain LNH-ST was used in all subsequent studies.

Two-Stage Continuous Cofermentation of Glucose and Xylose by LNH-ST

The two-stage continuous cofermentation of glucose and xylose with LNH-ST was performed in the same manner as with LNH33, but the residence time was 23.3 h per stage (0.043-h^{-1} dilution rate), resulting in a slightly shorter overall process residence time of 46.6 h compared to 48 h for LNH33. The glucose disappeared within the first 24 h of the batch stage (Fig. 2). During the same period, residual xylose concentration decreased to 7.7 g/L from an original level of 32.1 g/L (76% consumption).

After 24 h of batch operation, the fermentation was switched to continuous mode with a feed rate of 0.716 mL/min. The glucose concentration remained at zero, and the xylose concentration increased from 7.7 g/L to a steady-state value of 13.4 g/L (Fig. 2), which represents a utilization of 58.3% of the feed xylose in the first stage (Table 1). In the second stage, the xylose concentration decreased further to a steady-state value of 4.4 g/L, representing an overall conversion of 86.4%. The overall xylose utilization of LNH-ST was significantly higher than that of LNH33 (86.4% vs 26.3%) under similar experimental conditions (Table 1).

The metabolic ethanol yields in the first and second stages were 78 and 69.7%, respectively, whereas the ethanol process yields were 58.7 and

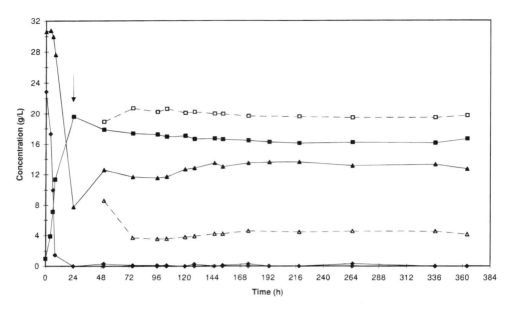

Fig. 2. Time course of the glucose (♦, ◊), xylose (▲, △), and ethanol (■, □) concentrations during the continuous two-stage pure sugar cofermentation by LNH-ST. Closed symbols and continuous lines represent the first stage, whereas open symbols and dashed lines represent the second stage. The arrow indicates the switching point from batch to continuous operation.

47.2%, respectively (Table 1). The lower metabolic yield of LNH-ST (76.5 vs 84.4% of LNH33) was a result of its higher production of glycerol and xylitol. This suggests that LNH33 was slightly more efficient, because it channels 10.5% more carbon into ethanol than LNH-ST. The major by-products (per g of consumed sugar) were carbon dioxide (0.373 g), glycerol (0.094 g), xylitol at (0.070 g), and cell mass (0.070 g) (Table 2). However, from a practical standpoint, LNH33 is a less efficient cofermenter than LNH-ST, because it fermented only 56.4% of the available sugar, whereas LNH-ST fermented 92.0%. The overall mass balance closure was 99.86% (Table 2).

SSCF of Corn Biomass by LNH-ST (Batch 1)

The pure sugar studies provided an indication of the ability of the organisms to use both xylose and glucose during the hydrolysis and fermentation of corn biomass. However, these studies were performed in the absence of metabolic inhibitors, such as acetic acid, lactic acid, hydroxymethyl furfural (HMF), furfural, and lignin-derived phenolics, which abound in biomass hydrolyzates. As the ultimate goal of this research is to examine the feasibility of large-scale conversion of pretreated corn biomass to ethanol, next studied was the ability of LNH-ST to produce ethanol during batch SSCF of pretreated corn biomass.

Cofermentation of Glucose and Xylose 251

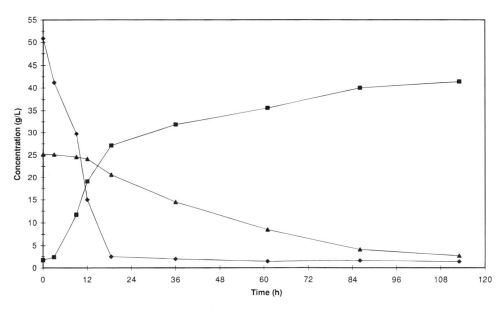

Fig. 3. Time course of the batch SSCF of pretreated corn biomass (batch 1) by LNH-ST. The symbols represent the concentrations of glucose (♦), xylose (▲), and ethanol (■).

Acetic acid (2.52 g/L) was the predominant inhibitor present in the pretreated biomass slurry. Other potential inhibitors included furfural (0.18 g/L) and HMF (0.22 g/L), both of which decreased in concentration over the time course of the fermentation. The addition of cellulase was needed for enzymatic hydrolysis of cellulose polymers and oligomers, whereas the addition of glucoamylase ensured the expedient conversion of residual starch to glucose. The pretreatment process liberated 62.3 g/L of soluble glucan (50.1 g/L glucose, 2.7 g/L cellobiose, and 9.5 g/L oligomers) and 39.3 g/L of soluble xylan (28 g/L xylose and 11.3 g/L oligomers) based on 20% total solids.

Within 24 h of inoculation, the available glucose was completely consumed (Fig. 3). Afterward the rate-limiting step to glucan utilization was the liberation of glucose from cellobiose and soluble oligomers in the liquid phase and from cellulose (28.7 g/L) in the solid phase by the catalytic action of the cellulase components endoglucanase, exoglucanase, and β-glucosidase (1). The overall glucan conversion, based on the soluble and insoluble fractions, was 74.1% after 113 h of fermentation (Table 3). The cellulose conversion was 75.4%. Of the 28 g/L monomeric xylose originally available, 26 g/L was consumed, corresponding to a conversion of 92.8% (62.8% conversion of total soluble [monomeric and oligomeric] and insoluble xylan). There was no detectable conversion of soluble oligomeric or insoluble xylan. The ethanol metabolic yield was quite high, 80.5% (Table 3), or 0.41 g/g of consumed sugars.

Table 3
Performance Parameters of Pretreated Corn Biomass Conversion
to Ethanol (SSCF) by LNH-ST

Parameter	Batch 1	Batch 2
Glucose Conversion[1]	74.1%	78.4%
Xylose Conversion[2]	62.8%	56.1%
Ethanol Process Yield (% theoretical)	56.9%	63.5%
Ethanol Metabolic Yield (% theoretical)	80.5%	88.1%

[1]Based on total glucan as equivalent glucose.
[2]Based on total xylan expressed as equivalent xylose.

The by-products were similar to those observed in the two-stage continuous cofermentation of pure sugar. As shown in Table 2, glycerol was the major by-product at 0.045 g/g of consumed sugars (glucose and xylose), whereas xylitol was synthesized at 0.026 g/g of consumed sugars. The overall mass balance was closed to 92.1%.

Reproducibility and Scaleability of Corn Biomass SSCF by LNH-ST (Batch 2)

The reliability and scaleability of LNH-ST is of particular importance to the commercialization of the biomass conversion technology, since it dictates the ethanol yield and production rate of ethanol, the desired product. It was, therefore, essential to examine whether the SSCF data could be reproduced and to assess how well the bench-scale performance of LNH-ST compared with that observed at the pilot scale.

Pretreated corn biomass already mixed at 20% solids loading with the appropriate amounts of cellulase and glucoamylase enzymes, CSL, and water (to resemble the batch study) was prepared in NREL's pilot plant and employed in both the bench scale 1.7-L fermentor and the 9000-L PDU fermentor to examine the process scaleability. Again, acetic acid (4.93 g/L) was the predominant inhibitor present in the pretreated biomass slurry, with only minor amounts of furfural (0.33 g/L) and HMF (0.27 g/L). The higher level of acetic acid liberated during the dilute-acid pretreatment step, resulted from slightly more severe conditions used in the PDU pretreatment reactor. The pretreatment generated 63.1 g/L of soluble glucan (51.4 g/L glucose, 5.6 g/L cellobiose, and 6.1 g/L oligomers) and 34.6 g/L of soluble xylan (26.7 g/L xylose and 7.6 g/L oligomers) in the liquid phase, and 26.6 g/L of cellulose and 1.75 g/L of xylan remained in the solid biomass at a 20% solids level.

After 24 h of SSC in the 1.7-L fermentor, most of the available glucose was consumed (Fig. 4). The overall glucan conversion was 78.4% after 167 h of fermentation, similar to the 74.1% observed in the first batch study after 113 h of operation (Table 2). The cellulose conversion was 78.3%. Xylan utilization (56.1%) was slightly lower than that observed in the

Cofermentation of Glucose and Xylose

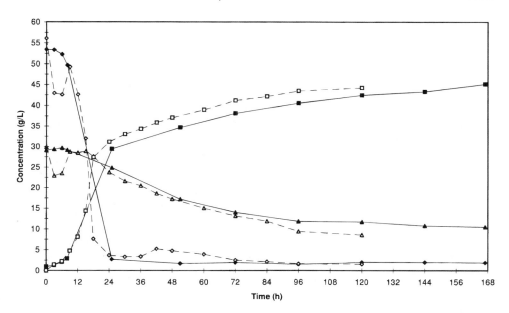

Fig. 4. Comparative study of the bench scale (closed symbols and continuous lines) and pilot scale (open symbols and dashed lines) performance of LNH-ST during the SSCF of pretreated corn biomass (batch 2). The symbols represent the concentrations of glucose (◆, ◇), xylose (▲, △), and ethanol (■, □).

previous study (62.8%). As the severity of pretreatment was the most notable difference between the two batch runs, the lower xylan utilization observed here may be because of the higher acetic acid concentration. The inhibitory effect of acetic acid on xylose utilization and ethanol production has been well documented (16,17). The ethanol process yield was 63.5%, whereas the ethanol metabolic yield was again very high at 88.1% (Table 2) or 0.45 g/g of consumed sugars. By-product formation resembled that in the first study: 5 g/L of glycerol and 2.7 g/L of xylitol. The overall mass balance closure was 100.3%.

The pretreated biomass was also subjected to SSCF in a 9000-L pilot plant fermentor under conditions similar to those at the bench scale. As seen in Fig. 4, the results obtained in the pilot plant run were similar to the bench-scale data. After 24 h of SSCF, the concentrations of glucose, xylose, and ethanol were 2.7 g/L, 24.9 g/L, and 29.4 g/L in the bench-scale fermentor and 3.7 g/L, 23.8 g/L, and 31.1 g/L in the pilot fermentor. At 96 h, the concentrations of these same components were 1.5 g/L, 11.9 g/L, and 40.5 g/L in the bench reactor and 1.7 g/L, 9.4 g/L, and 43.4 g/L in the pilot-scale reactor. Similarly, by-product formation was in close agreement, as indicated by xylitol, which reached 2.4–2.5 g/L in both vessels (data not shown). The good correlation shows that despite the 5300-fold increase in fermentor size, the performance of LNH-ST remained unchanged and very promising.

CONCLUSION

Continuous and batch studies have demonstrated that the chromosome-integrated genes of LNH-ST enable the organism to broaden its substrate range to metabolize xylose in addition to glucose. The observed high levels of glucose and xylose conversion and ethanol yield were consistently reproducible on pure sugars and corn biomass prepared under realistic process conditions. The batch corn biomass SSCF data show that more than 78% of the available glucose and cellulose and more than 56% of the available xylose were fermented by the organism within 4 d. Despite the presence of high levels of acetic acid, a metabolic inhibitor, the overall ethanol process and metabolic yields were 63.5% and 88.1%, respectively, at both the bench and pilot scales. The scaleability and reproducibility of the LNH-ST SSCF performance justify further optimization and scale-up, studies in a continuous pilot plant operation for eventual application of the recombinant cofermenting yeast at the commercial scale.

ACKNOWLEDGMENTS

This work was funded by the Biochemical Conversion Element of the Biofuels Program of the US Department of Energy and Amoco Corporation. We would like to thank Nancy Dowe and the PDU team of NREL for the pilot scale data, Bill Adney of NREL for measuring the cellulase activity, and NREL's Chemical Analysis Team for analyzing the composition of biomass and fermentation samples.

REFERENCES

1. Philippidis, G. P. (1994), in *Enzymatic Conversion of Biomass for Fuels Production*, Himmel, M. E., Baker, J. O., and Overend, R. P., eds, American Chemical Society, Washington, DC, pp. 188–217.
2. Dow Jones News Service. Amoco, Stone & Webster in Biomass Conversion Partnership. Oct. 18, 1995.
3. Philippidis, G. P., Spindler, D. D., and Wyman, C. E. (1992), *Appl. Biochem. Biotechnol.* **34/35**, 543–556.
4. Laplace, J. M., Delgenes, J. P., Moletta, R., and Navarro, S. M. (1991), *Biotech. Let.*, **13**, 445–450.
5. Jeffries, T. W. and Kurtzman, C. P. (1994), *Enzyme Microb. Technol.* **16**, 922–932.
6. Olsson, L. and Hahn-Hägerdal, B. (1996), *Enzyme Microb. Technol.* **18**, 312–231.
7. Zhang, M., Eddy, C., Deanda, K., Finkelstein, M., and Picataggio, S. (1995), *Science* **267**, 240–243.
8. Ho, N. W. Y., Chen, Z. D. and Brainard, A. (1993), in *Proceedings of Tenth International Conference on Alcohols*, Colorado Springs, CO, p. 738.
9. Kötter, P. and Ciriacy, M. (1993), *Appl. Microbiol. Biotechnol.* **38**, 776–783.
10. Ho, N. W. Y. and Tsao, G. T. (1995), PCT patent No. WO 95/13362.
11. Walfridsson, M., Hallborn, J., Penttilä, M., Keränen, S., and Hahn-Hägerdal, B., (1995), *Appl. Environ. Microbiol.* **61**, 4181–4190.
12. Ho, N. W. Y. and Chen, Z. D. (1996), Patent pending.

13. Stewart, G. G., Panchal, C. J., and Rusell, I. (1982), *Brew. Distill. Int.* **12,** 33.
14. Ghose, T. K. (1987), *Pure Appl. Chem.* **59,** 257–268.
15. Hatzis, C., Riley, C., and Philippidis, G. P. (1996), *Appl. Biochem. Biotechnol.* **57/58,** 443–459.
16. Ramos, M. T. and Madeira-Lopes, A. (1990), *Biotech. Let.* **12,** 229–234.
17. Van Zyl, C., Prior, B. A., and Du Perez, J. (1991), *Enzyme Microb. Technol.* **13,** 82–86.

Stabilization and Reutilization of *Bacillus megaterium* Glucose Dehydrogenase by Immobilization

MADALENA BARON,*,[1,2] JOSÉ D. FONTANA,[1] MANOEL F. GUIMARÃES,[1] AND JONATHAN WOODWARD[2]

[1]Biomass Chemo/Biotechnology Laboratory, Department of Biochemistry, Federal University of Paraná, P. O. Box 19046 Curitiba, PR (81531-970) Brazil; [2]Chemical Technology Division; and [3]Oak Ridge National Laboratory

ABSTRACT

Glucose dehydrogenase (GDH) from *Bacillus megaterium* was immobilized using aminopropyl controlled-pore silica (CPS, average pore sizes of 170 and 500 Å) as a support and glutaraldehyde as a bifunctional crosslinking agent. The CPS-immobilized enzyme could be reused 12 times and the best results were obtained using aminopropyl CPS-500 and bovine serum albumin as a feeder for stabilizing the protein layer on the support. DEAE-Sephadex (A-25 and A-50) was also used as a support for immobilizing GDH, with yields of around 42% for A-25 and 25–30% for A-50. The effect of pH on the immobilization procedure showed pH 6.5 to be better than pH 7.5 with respect to the recovery of enzyme activity. Both preparations of DEAE-Sephadex immobilized GDH could be reused several times and were thermostable at 40°C for 7 h. The kinetic parameters as Michaelis constant and maximum rate were determined for the immobilized enzyme and compared with those for the freeform.

Index Entries: *Bacillus megaterium*; glucose dehydrogenase; aminopropyl controlled-pore silica; DEAE-Sephadex; immobilization.

INTRODUCTION

Glucose dehydrogenase (GDH) from *Bacillus megaterium* catalyzes the oxidation of β-D-glucose to D-glucono-1,5-lactone using NAD^+ or $NADP^+$ as the coenzyme *(1)*. Coupling GDH and hydrogenase from *Pyrococcus*

---

*Author to whom all correspondence and reprint requests should be addressed.

furiosus, which uses NADPH as an electron donor, may provide a convenient method for the biological production of hydrogen from glucose that in turn may derive from renewable sources as cellulose, starch, and lactose (2). When adsorbed on metal hydrides, hydrogen may be used in refrigerators and air conditioners, as a coolant in place of freon (3). The other product obtained in the GDH-catalyzed reaction, gluconic acid from the hydrolysis of D-glucono-1,5-lactone, is also a high value chemical for a variety of industries including alkaline detergents for metal cleaning, pulp and paper, textile, glass, and pharmaceuticals (4). GDH is a tetrameric protein displaying molecular weight of 116,000 by gel permeation chromatography. In the presence of 0.1% sodium dodecylsulfate and $8M$ urea, the enzyme dissociates into four identical subunits, each one containing 262 amino acid residues (5), with molecular weight around 30,000 as determined by dodecylsulfate gel electrophoresis. Unfolding of the enzyme in $8M$ urea is strongly inhibited by high concentrations of NaCl (6). At 67 mM phosphate buffer, pH 6.5, in the presence of $3M$ NaCl, GDH shows optimal stability against thermal inactivation, and solutions of the enzyme may be stored for months at room temperature without loss of activity. The addition of polyethyleneimine (PEI), a water-soluble cationic polymer, to solutions of GDH at a molar ratio of PEI to GDH >10 also increases the thermal stability of GDH (7). In presence of PEI, the rate for the GDH-catalyzed oxidation of β-D-glucose increases in a low concentration range of NAD^+ and $NADP^+$ suggesting that negatively charged GDH interacts with cationic polymers by electrostatic attraction and the negatively charged coenzymes are adsorbed by the polymers, resulting in enrichment of the coenzymes in the vicinity of GDH.

Enzymatic stability may sometimes be enhanced by immobilization of the enzyme onto a solid support, where it can be recovered and continuously reused. The easiest method for immobilizing proteins is by adsorption involving noncovalent interactions as ionic, metal bridge, and hydrophobic binding (8). In this case, immobilization is normally achieved by simply incubating the support with a given amount of enzyme at a specific temperature, pH, and ionic strength. Alternatively, the proteins can be covalently attached to various support surfaces with a wide range of choices in selecting support materials and binding methods.

In this study, two methods for immobilizing GDH from *B. megaterium* were examined. Firstly, aminopropyl controlled-pore silica with average pore sizes of 170 and 500 Å was tested as an inorganic support, and glutaraldehyde was used as a bifunctional cross-linking agent in the covalent binding. Secondly, the anion exchanger DEAE-Sephadex was selected, that binds to the enzyme by electrostatic interactions. We compared the immobilization efficiencies for the covalent and non-covalent methods. Some enzymatic properties of the preparations and the possibility for reusing the immobilized biocatalyst are described and discussed.

MATERIALS AND METHODS

Materials

Glucose dehydrogenase (GDH, EC 1.1.1.47) from *Bacillus megaterium* (G-7653) with 250 U/mg protein (Biuret), β-nicotinamide adenine dinucleotide (β-NAD$^+$, 260-150, grade III), β-nicotinamide adenine dinucleotide phosphate (β-NADP$^+$, N-3886), bovine serum albumin (BSA, A-7638, essentially globulin free), glutaraldehyde (G-5882, grade I), aminopropyl controlled-pore silica (CPS, G-4643: average pore size = 500 Å, 200–400 mesh, amine content = 70 µmol/g silica and G-4518: average pore size = 170 Å, 200–400 mesh, amine content = 165 µmol/g silica) were purchased from Sigma Chemical Company, St. Louis, MO. DEAE-Sephadex A-50 (17-0180-01, useful MW range = 30,000–200,000) and A-25 (17-0170-01, useful MW range ≤30,000) were products from Pharmacia, Uppsala, Sweden. β-D-Glucose (34635) was obtained from Calbiochem, San Diego, CA and dye reagent concentrate (Coomassie Brilliant Blue G-250) for protein assay was from Bio-Rad Laboratories, Hercules, CA. All other reagents were of analytical grade.

Enzyme Assay

The enzyme activity was performed using β-D-glucose as substrate and NAD$^+$ or NADP$^+$ as coenzyme. Each assay contained 10 mM glucose and 1 mM NADP$^+$ or 2 mM NAD$^+$ prepared in 50 mM sodium phosphate buffer, pH 7.5. The reaction was initiated by the addition of enzyme (free or immobilized as indicated in each assay) and proceeded for 2 min or other indicated time, at room temperature (23°C). The production of NADH or NADPH was measured spectrophotometrically at 340 nm. One unit of enzyme activity (U) was defined as the amount of enzyme that oxidizes 1 µmol of β-D-glucose to D-glucono-δ lactone per minute at pH 7.5 at 23°C.

Immobilization of GDH Using Controlled-Pore Silica (CPS)

Two types of aminopropyl-CPS with average pore sizes of 170 and 500 Å (50–100 mg) were activated by addition of 2.5% (v/v) glutaraldehyde (0.25–0.50 mL) in 0.1M phosphate buffer, pH 7.5, with stirring for 1 h at 23°C, and the excess glutaraldehyde was washed off with deionized water. Enzyme solutions with 0.54 g% BSA (49.045 U/100 mg support) or without BSA (5.586 U/50 mg support) and the activated supports were shaken at 4°C for 41 h or at 22°C for 1 h, respectively. The immobilized enzyme was then washed with 50 mM phosphate buffer, pH 7.5, containing 1M NaCl until no enzymatic activity could be detected in the washings. The preparations were tested for enzyme activity as above, and after each assay, the immobilized enzyme was centrifuged and washed with phosphate-NaCl buffer in order to remove the residual products and reassayed to determine enzyme stability and reusability.

Immobilization of GDH Using DEAE-Sephadex

Effect of pH in the immobilization procedure: Two samples of DEAE-Sephadex A-50 (10 mg) were equilibrated with 50 mM phosphate buffer, pH 6.5 and pH 7.5, before enzyme immobilization. Enzyme solutions (0.419 U) prepared in phosphate buffer, pH 6.5 and pH 7.5, and the support in a total volume of 1 mL were stirred at 4°C for 15 h and the excess enzyme was washed off with the respective buffer solutions. The preparations were assayed as described in the Enyzme Assay Section, and after each assay recovered, washed and reassayed.

Effect of the type of DEAE-Sephadex in the immobilization procedure: Two types of DEAE-Sephadex A-25 (100 mg) and A-50 (100 mg) were equilibrated in 50 mM phosphate buffer, pH 6.5, and enzyme solutions (5.226 U), prepared in the same buffer, were added to the supports. The mixtures were shaken at 4°C for 1 h (DEAE-Sephadex A-50) or 5 h (DEAE-Sephadex A-25), and the excess enzyme was washed off with 50 mM phosphate buffer, pH 6.5. The preparations, after lyophilization, were assayed, as described in Enzyme Assay Section, recovered, and reassayed.

Determination of Kinetic Parameters for Free and DEAE-Sephadex Immobilized GDH

The kinetic parameters, Michaelis constant and maximum rate, for free GDH were determined by measuring the reaction rate at substrate concentrations between 1–30 mM whereas for immobilized GDH, concentrations of β-D-glucose between 2–160 mM were used.

Stability at 40° for Free and DEAE-Sephadex Immobilized GDH

Free GDH solutions (10.36 µg/mL) prepared in 50 mM phosphate buffer, pH 6.5 with NaCl (0.0857M), and without NaCl (after rapid desalting using Sephadex G-25 M), were heated at 40°C for 30 min, 1, 3, 5, and 7 h prior to the enzymatic activity measurement at 23°C. For the thermal stability of DEAE-Sephadex A-25 and A-50 immobilized GDH, samples of each preparation (2 mg) were heated at 40°C for the aforementioned periods of time and then assayed for enzymatic activity.

RESULTS AND DISCUSSION

Table 1 summarizes the results for covalent immobilization of GDH using aminopropyl CPS. In this method, glutaraldehyde was used to activate the amino groups on CPS and then to cross-link the enzyme to the support. In all cases, the covalent attachment of GDH resulted in most of the activity being lost (≥98.8%). Two types of aminopropyl-CPS with average pore sizes of 170 and 500 Å were used for immobilizing GDH, and the best results were obtained using CPS-500 with respect to the recovery of total activity after immobilization. Concerning the conditions for

Table 1
Covalent Immobilization of GDH Using CPS[a]

Immobilized GDH	Activity added (U)	Activity in washings (U)	Immobilized activity (U)		Immobilization yield (%) [(B/A) x 100]
			Theoretical (A)	Actual (B)	
CPS-170 (100 mg) With BSA (0.54g%)	49.05	2.32 (4.73)	46.72 (95.27)	0.03 (0.06)	0.06
CPS-500 (100 mg) With BSA (0.54g%)	49.04	8.22 (16.76)	40.82 (83.24)	0.31 (0.62)	0.75
CPS-170 (50 mg) Without BSA	5.59	0.07 (1.32)	5.51 (98.68)	0.04 (0.72)	0.73
CPS-500 (50 mg) Without BSA	5.59	0.11 (1.93)	5.48 (98.07)	0.06 (1.15)	1.18

[a]CPS = Controlled-pore silica; BSA = Bovine serum albumin; The values in parentheses refer to the activity retention (%). The conditions for immobilization were: 50 mM phosphate buffer, pH 7.5, containing 1M NaCl, shaking at 4°C for 41 h (with BSA) or shaking at 22°C for 1 h (without BSA).

immobilization, a shorter time (1 h instead of 41 h) and smaller enzyme:support ratio (0.11 U/mg instead 0.49 U/mg) were preferable for obtaining best recovery. BSA was used as a feeder (9) for stabilizing the enzymatic protein layer on the support and, after repeated assays with CPS-immobilized GDH, it was observed that the enzyme could be reused twelve times without total loss of activity (Fig. 1). Although, GDH immobilized on CPS could be reused several times, the immobilization yield was very low possibly because of exposure of the enzyme to harsh environments or toxic reagents such as glutaraldehyde. This crosslinking agent could act on amino groups located at or close to the active site [e.g., His-148 and Lys-201 are involved in catalysis or binding of ligands (5)] and could hinder the conformational adaptation of the enzyme to the substrate resulting in an inactivation of the enzyme molecules (9). The effect of pH on the immobilization of GDH using DEAE-Sephadex A-50 was examined in phosphate buffer solutions, pH 6.5 and pH 7.5. The reason for this is because pH 7.5 represents the optimum pH for determination of the GDH activity whereas pH 6.5 is more suitable for long-term storage of GDH (in presence of 3M NaCl) for months, without loss of activity (6). When the immobilized preparations made at pH 6.5 and pH 7.5 were subjected to repeated assays, it could be noted that pH 6.5 was better than pH 7.5 with respect to the recovery of enzyme activity (Fig. 2). Thus, pH 6.5 was chosen

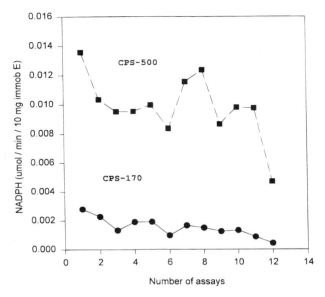

Fig. 1. Effect of repeated assays on the activity of CPS-170 and CPS-500 immobilized GDH/with BSA. The conditions for immobilization are in the footnote of Table 1. Incubation system: [glucose] = 10 µmol/mL, [NADP+] = 1 µmol/mL, [immobilized GDH] = 10 mg/mL, pH 7.5, room temperature, for 15 min.

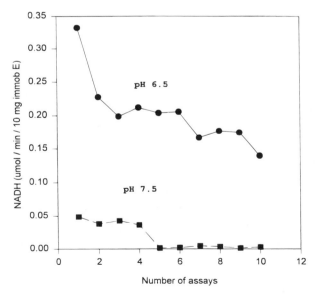

Fig. 2. Effect of repeated assays on the activity of DEAE-Sephadex A-50 immobilized GDH/pH 6.5 and pH 7.5. The conditions for immobilization were: 50 mM phosphate buffer, pH 6.5 or pH 7.5, shaking at 4°C for 15 h. Incubation system: [glucose] = 10 µmol/mL, [NAD+] = 2 µmol/mL, [GDH immobilized at pH 6.5] = 1.65 mg/mL, [GDH immobilized at pH 7.5] = 1.35 mg/mL, pH 7.5, room temperature for 2 min.

Table 2
Immobilization of GDH by Ionic Adsorption Using DEAE-Sephadex[a]

Immobilized GDH	Activity added (U)	Activity in washings (U)	Immobilized activity (U)		Immobilization yield (%) [(B/A) x 100]
			Theoretical (A)	Actual (B)	
DEAE-Sephadex A-25 (100 mg):					
Duplicate 1	5.23	0.00 (0)	5.23 (100)	2.23 (42.67)	42.67
Duplicate 2	5.23	0.00 (0)	5.23 (100)	2.22 (42.43)	42.43
DEAE-Sephadex A-50 (100 mg):					
Duplicate 1	5.23	1.49 (28.47)	3.74 (71.53)	0.96 (18.33)	25.63
Duplicate 2	5.23	1.50 (28.75)	3.72 (71.25)	1.14 (21.85)	30.67

[a]Actual activity was obtained after second reuse assay (see Figs. 3 and 4). The values in parentheses refer to the activity retention (%). The conditions for immobilization were: 50 mM phosphate buffer, pH 6.5, shaking at 4°C for 1 h (DEAE-Sephadex A-50) or 5 h (DEAE-Sephadex A-25).

for immobilizing GDH on DEAE-Sephadex. Two types of DEAE-Sephadex, namely A-25 and A-50, were tested as support in order to compare the immobilization efficiencies. The support A-25 presents useful MW range ≤30,000 and, in this case, GDH (MW = 116,000) is totally excluded from the gel while the support A-50 has useful MW range equal to 30,000–200,000 and there is some possibility for the inclusion of GDH inside the gel beads. Table 2 shows the results for immobilizing GDH on DEAE-Sephadex and the support A-25 was more efficient for obtaining superior activity recovery (42%) when compared to the support A-50 in which the recovery was around 25–30%. Both immobilized enzymes could be reused, as shown in Fig. 3 for support A-50 and Fig. 4 for support A-25, and both enzymes maintained their activities after several recycles. It is important to observe that the maximum activity for DEAE-Sephadex immobilized GDH, using $NADP^+$ as coenzyme (Figs. 3 and 4), was only obtained after the second reutilization assay. In comparison to the result shown in Fig. 2 using NAD^+ as cofactor, it is possible that the additional negative charge from phosphate group in the $NADP^+$ molecule plays an important role in its adsorption to the support. Indeed, NADPH was demonstrated to adsorb to DEAE-Sephadex (results not shown). It is possible, therefore, that the concentration of $NADP^+$ in the first use was effectively reduced in the assay by this nonspecific adsorption. In the second and subsequent assays, most if not all the $NADP^+$ would be available to the enzyme and not subject to further adsorption.

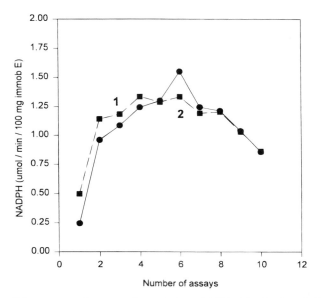

Fig. 3. Effect of the repeated use on the activity of DEAE-Sephadex A-50 immobilized GDH/pH 6.5. The conditions for immobilization were in the footnote of the Table 2. Incubation system: [glucose] = 10 μmol/mL, [NADP+] = 1 μmol/mL, [immobilized GDH] = 5 mg/mL, pH 7.5, room temperature for 2 min.

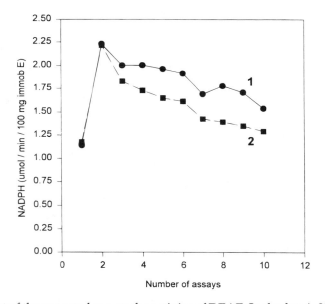

Fig. 4. Effect of the repeated use on the activity of DEAE-Sephadex A-25 immobilized GDH/pH 6.5. The conditions for immobilization were in the footnote of the Table 2. Incubation system = [glucose] = 10 μmol/mL, [NADP+] = 1 μmol/mL, [immobilized GDH] = 3.33 mg/mL, pH 7.5, room temperature for 2 min.

Table 3
Kinetic Parameters for Free and DEAE-Sephadex Immobilized GDH

Kinetic parameter	Free GDH	Immobilized GDH	
		DEAE-Sephadex A-25	DEAE-Sephadex A-50
Km(a)	3.83	3.84	12.11
Vmax	5.77(b)	2.60(c)	1.50(c)
r^2(d)	0.99	0.85	0.98

(a) Michaelis constant in mM.
(b) Maximum rate in μmol NADPH/min/0.2 mL (volume of free enzyme used in the immobilization procedure).
(c) Maximum rate in μmol NADPH/min/100 mg immobilized enzyme.
(d) Regression coefficient for Lineweaver–Burk plot.
The incubation conditions for assay with free enzyme were: [GDH]free = 0.36 μg/mL, [glucose] = 1–30 mM, [NADP$^+$] = 1 mM, 50 mM phosphate buffer, pH 7.5, room temperature for 2 min whereas for immobilized enzyme were: [GDH]immobilized A-25 = 2 mg/mL or [GDH]immobilized A-50 = 1 mg/mL, [glucose] = 2–160 mM, [NADP$^+$] = 1 mM, in the same conditions of pH, temperature, and time.

The kinetic parameters for the preparations of DEAE-Sephadex immobilized GDH are shown in the Table 3 and the apparent Km for DEAE-Sephadex A-50 immobilized GDH was calculated as being 12.11 mM, which is higher than the Km value for free GDH. According to the literature (7), Km values for free GDH at pH 7.0 and pH 7.8, using NADP$^+$ as cofactor, were respectively 2 mM and 4.55 mM and, in our procedure, the Km value at pH 7.5 was found as 3.83 mM. The apparent Km for DEAE-Sephadex A-25 immobilized GDH was 3.84 mM and it is similar to that for free GDH. The stability for GDH is an important study because it is a stable tetramer at pH 6.5 but is readily dissociated into four inactive protomers by shifting the pH to 8.5 (6). Rapid reactivation can be achieved by readjustment to the original pH value. Adding NaCl to the GDH solutions and the presence of the coenzyme stabilize the quaternary structure of the enzyme. For thermal stability of free GDH at 40°C, the effect of NaCl can be observed in the Fig. 5. When the free enzyme was preheated at 40°C, without presence of NaCl, the enzyme activity dropped very quickly after heating of 30 min. If the enzyme solution contained NaCl (0.086M), it maintained its activity after heating at 40°C for 7 h. In the Fig. 6, the thermostability at 40°C for DEAE-Sephadex immobilized GDH was analyzed and after preheating for 7 h, the immobilized enzyme was thermostable without loss of activity. In this case, the presence of NaCl was not required for stability of GDH. It can be

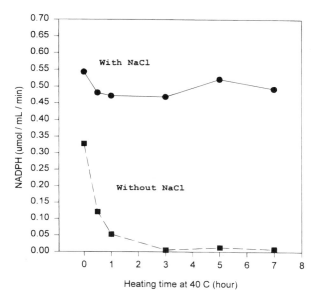

Fig. 5. Effect of sodium chloride for thermal stability of free GDH at 40°C. Incubation system: [glucose] = 10 µmol/mL, [NADP+] = 1 µmol/mL, [GDH solution, previously heated at 40°C, with NaCl (0.086M)] = 10.36 µg/mL, [GDH solution, previously heated at 40°C, without NaCl] = amount above after rapid desalting using Sephadex G-25 M, pH 7.5, room temperature for 2 min.

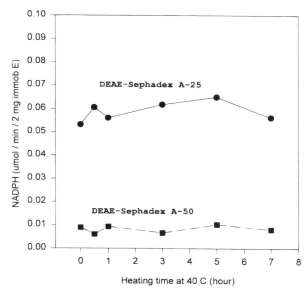

Fig. 6. Stability at 40°C for DEAE-Sephadex A-25 and A-50 immobilized GDH. Incubation system: [glucose] = 10 µmol/mL, [NADP+] = 1 µmol/mL, [immobilized GDH] = 2 mg/mL, pH 7.5, room temperature, for 2 min.

concluded that the immobilization of GDH using DEAE-Sephadex is a method to obtain stability of GDH that could be then reused several times with maintenance of its enzymatic activity. The stability of GDH is also necessary for coupling it with other enzymes as hydrogenase from *Pyrococcus furiosus* (2) since GDH is an useful enzyme for generating reduced coenzyme as NADPH, which in turn can be used as an electron donor for the hydrogenase activity in the production of hydrogen. Moreover, the coimmobilization of hydrogenase and GDH will also deserve attention in ongoing work in order to obtain an enzymatic system more stable than that formed with the free enzymes. The enzymatic pathway for the conversion of biomass-derived glucose to hydrogen is advantageous by requiring relatively mild conditions without the formation of waste gases as carbon dioxide and carbon monoxide when compared with other routes to hydrogen production as gasification, pyrolysis, and fermentation of biomass.

CONCLUSIONS

GDH from *Bacillus megaterium* was immobilized using aminopropyl controlled-pore silica as support and glutaraldehyde as bifunctional crosslinking agent. This method resulted in most of the activity being lost even so the immobilized enzyme activity was reusable. GDH immobilized on CPS-500 was reused twelve times without total loss of activity; BSA, as a feeder in the covalent interaction, stabilized the enzyme protein layer on the support. DEAE-Sephadex was also used as support for immobilizing GDH, through electrostatic binding, resulting in recovered activities around 42% (A-25) and 25–30% (A-50); pH 6.5 in the immobilization procedure was better than pH 7.5 with respect to the enzymatic activity obtained. The preparations of DEAE-Sephadex immobilized GDH could be reused several times with a good maintenance of activity and were thermostable at 40°C for 7 h.

ACKNOWLEDGMENTS

The authors would like to thank Barbara Evans, Michael Hu, and Vinod Shah for all suggestions and Maria Blanco-Rivera for practical assistance. This work was supported by CNPq (Brazilian Financial Agency) and the Advanced Utility Concepts Division, U.S. Department of Energy. Oak Ridge National Laboratory is managed by Lockheed Martin Energy Research Corporation for the U.S. Department of Energy under contract DE-AC05-96OR22464.

REFERENCES

1. Pauly, H. E. and Pfleiderer, G. (1975), *Hoppe-Seyler's Z. Physiol. Chem.* **356**, 1613–1623.
2. Woodward, J., Mattingly, S. M., Danson, M. J., Hough, D. W., Ward, N., and Adams, M. W. W. (1996), *Nature Biotechnol.* **14**, 872–874.

3. Jones, J. (1995), *DOE This Month* **18,** 14.
4. Sodium gluconate handbook (1994), PMP Fermentation Products, Inc., Chicago, IL.
5. Jany, K., Ulmer, W., Fröschle, M., and Pfleiderer, G. (1984), *FEBS* **165,** 6–10.
6. Pauly, H. E. and Pfleiderer, G. (1977), *Biochemistry* **16,** 4599–4604.
7. Teramoto, M., Nishibue, H., Okuhara, K., Ogawa, H., Kozomo, H., Matsuyama, H., and Kajiwara, K. (1992), *Appl. Microbiol. Biotechnol.* **38,** 203–208.
8. Bernath, F. R. and Venkatasubramanian, K. (1986), in *Manual of Industrial Microbiology and Biotechnology*, Demain, A. L. and Solomon, N. A., eds., American Society for Microbiology, Washington, ch. 19. pp. 230–247.
9. Broun, G. B. (1976), *Methods Enzymol.* **44,** 263–280.

Optimization of Seed Production for a Simultaneous Saccharification Cofermentation Biomass-to-Ethanol Process Using Recombinant *Zymomonas*

HUGH G. LAWFORD,*,[1] JOYCE D. ROUSSEAU,[1] AND JAMES D. MCMILLAN[2]

[1]*Bio-engineering Laboratory, Department of Biochemistry, University of Toronto, Toronto, Ontario, Canada M5S 1A8; and* [2]*Biotechnology Center for Fuels and Chemicals, National Renewable Energy Laboratory, 1617 Cole Boulevard, Golden, CO 80401-3393*

ABSTRACT

The five-carbon sugar D-xylose is a major component of hemicellulose and accounts for roughly one-third of the carbohydrate content of many lignocellulosic materials. The efficient fermentation of xylose-rich hemicellulose hydrolyzates (prehydrolyzates) represents an opportunity to improve significantly the economics of large-scale fuel ethanol production from lignocellulosic feedstocks. The National Renewable Energy Laboratory (NREL) is currently investigating a simultaneous saccharification and cofermentation (SSCF) process for ethanol production from biomass that uses a dilute-acid pretreatment and a metabolically engineered strain of *Zymomonas mobilis* that can coferment glucose and xylose. The objective of this study was to establish optimal conditions for cost-effective seed production that are compatible with the SSCF process design.

Two-level and three-level full factorial experimental designs were employed to characterize efficiently the growth performance of recombinant *Z. mobilis* CP4:pZB5 as a function of nutrient level, pH, and acetic acid concentration using a synthetic hardwood hemicellulose hydrolyzate containing 4% (w/v) xylose and 0.8% (w/v) glucose. Fermentations were run batchwise and were pH-controlled at low levels

*Author to whom all correspondence and reprint requests should be addressed.

of clarified corn steep liquor (cCSL, 1–2% v/v), which were used as the sole source of nutrients. For the purpose of assessing comparative fermentation performance, seed production was also carried out using a "benchmark" yeast extract-based laboratory medium. Analysis of variance (ANOVA) of experimental results was performed to determine the main effects and possible interactive effects of nutrient (cCSL) level, pH, and acetic acid concentration on the rate of xylose utilization and the extent of cell mass production. Results indicate that the concentration of acetic acid is the most significant limiting factor for the xylose utilization rate and the extent of cell mass production; nutrient level and pH exerted weaker, but statistically significant effects. At pH 6.0, in the absence of acetic acid, the final cell mass concentration was 1.4 g dry cell mass/L (g DCM/L), but decreased to 0.92 and 0.64 g DCM/L in the presence of 0.5 and 1.0% (w/v) acetic acid, respectively. At concentrations of acetic acid of 0.75 (w/v) or lower, fermentation was complete within 1.5 d. In contrast, in the presence of 1.0% (w/v) acetic acid, 25% of the xylose remained after 2 d. At a volumetric supplementation level of 1.5–2.0% (v/v), cCSL proved to be a cost-effective single-source nutritional adjunct that can support growth and fermentation performance at levels comparable to those achieved using the expensive yeast extract-based laboratory reference medium.

Index Entries: Recombinant *Zymomonas*, seed production via cofermentation of glucose and xylose, corn steep liquor, pH, acetic acid, synthetic hemicellulose hydrolyzate.

INTRODUCTION

Fermentation ethanol is currently produced from six-carbon hexose sugars derived either from starch or sucrose; however, the value of these carbohydrates as potential food (feed) resources seriously restricts fermentation ethanol from cost-effective competition in the alternative transportation fuels market *(1–3)*. Therefore, cost reduction is the driving force for R&D directed toward the use of alternative fermentation feedstocks. Lignocellulosic biomass (including short rotation energy crops, agricultural, forestry, and municipal wastes) is considered an excellent alternative fermentation feedstock, because it is inexpensive, plentiful, and renewable *(4,5)*.

Although *Saccharomyces* yeast currently enjoys a monopoly as the fermentation process biocatalyst in the fuel ethanol industry *(6)*, it is not the only ethanol-producing microorganism. Furthermore, the yeasts currently used in starch and sucrose-based fermentations cannot ferment pentose sugars *(7)*. One biological aspect of engineering a process that uses alternative fermentation feedstocks involves the use of an alternative process organism (fermentation biocatalyst) that has been either selected for, or tailored to, the specific requirements of the biomass-to-ethanol process. Research in this area has produced a variety of pentose-utilizing ethanologenic organisms, including yeasts, molds, and bacteria. Some of these are natural isolates; others represent genetically engineered recombinant variants *(8–12)*.

The National Renewable Energy Laboratory (NREL) is considering a variety of bioconversion processes for converting lignocellulosic biomass to ethanol on an industrial scale. In the general process design, feedstock comminution is followed by a dilute-acid pretreatment process. Economic sensitivity analysis of several process designs has demonstrated the substantial cost reduction that accompanies modifying the design from one of sequential hydrolysis and fermentation (SHF) to a simultaneous saccharification and fermentation (SSF) process *(1,11–13)*. Furthermore, there is potential for additional cost reduction (capital and operating costs) by combining into a single-unit operation the pentose fermentation and cellulose conversion (SSF) unit operations of the process *(14)*. However, such a design would require a biocatalyst that can coferment xylose and glucose. NREL has surveyed numerous potential industrial ethanol fermentation biocatalysts in a comprehensive study that compared known metabolic characteristics to a weighted list of fermentation performance criteria (including yield, ethanol tolerance, specific productivity, generally recognized as safe (GRAS) status, and sensitivity to inhibitory compounds typically present in biomass hydrolyzates) *(15)*. Using a nutrient-supplemented, dilute-acid hardwood prehydrolyzate as a screening medium, several strains of *Zymomonas* were selected as targets for improvement by metabolic engineering *(16)*.

The anaerobic growth characteristics of *Saccharomyces* yeast and the ethanologenic bacterium *Zymomonas mobilis* are similar, and the ethanol tolerance of both organisms is comparable *(17–19)*. Several comprehensive reviews in the literature pertain to the biology and physiology of *Zymomonas (20,21)* as well as biochemical and bioengineering aspects relating to its fermentation performance, and its potential as a process organism for producing fuel ethanol both in batch and continuous processes *(17,18,22)*.

By virtue of its demonstrated superior fermentation performance characteristics, *Zymomonas* offers an opportunity for process improvement with respect to both conversion efficiency (yield) and productivity *(17,19,22–24)*. It has the potential to revolutionize the fuel ethanol industry, and although not currently used commercially (for fermentations trials at industrial scale, *see* ref. *25*; for pilot-scale trials, *see* review by Doelle et al., ref. *22*), laboratory- and pilot-scale operations indicate it can generate near-theoretical maximum yields from several feedstocks, including sugar cane *(26)*, molasses *(27)*, saccharified starch from corn *(28)*, wheat *(25)*, cassava and sago *(29)*, and an enzymatic hydrolyzate of wood-derived cellulose *(30,31)*.

A serious limitation to the potential of *Zymomonas* as a universal biocatalyst in the fuel ethanol industry is its capacity to utilize only glucose, fructose, or sucrose. It lacks a complete pentose metabolism pathway necessary for xylose fermentation. Previous attempts to engineer metabolically *Zymomonas* for xylose fermentation were not entirely successful, as evidenced by the failure of the recombinants to grow on xylose as a sole carbon source *(32,33)*. In addition to a marker gene for tetracycline resistance, the plasmid constructed by NREL (designated as pZB5) carries genes, cloned from *Escherichia coli*, for four enzymes (xylose isomerase, xylulose kinase, transketolase, and transaldolase) that were required to create a functional xylose metabolism pathway *(34)*. *Z. mobilis* CP4, transformed with pZB5, grows on xylose as sole carbon source, and coferments xylose and glucose to ethanol in high yield *(34)*.

One of several biomass-to-ethanol processes currently under investigation by NREL is the simultaneous saccharification and cofermentation (SSCF) process that is based on the use of an NREL-proprietary genetically engineered strain of *Z. mobilis* CP4:pZB5 *(14)*. Encouraging cofermentation performance data (yield and productivity) have been obtained for this organism using laboratory media *(14,34)*. However, prior to the present investigation, the potential to achieve high ethanol yield on cost-effective media formulations and to perform efficiently in synthetic prehydrolyzate media that contain inhibitory substances, such as acetic acid, had not been demonstrated. Fermentation nutrient costs, for both seed production and SSF, can be a significant contributor to the overall cost of producing ethanol *(35)*.

The metabolic engineering of *Zymomonas* to ferment xylose to ethanol in high yield represents a major step forward in the economic production of fuel ethanol from biomass, but other parameters, such as the nutritional and physical environment to which the recombinant will be exposed, could significantly alter both yield and productivity. Culture media used to screen microorganisms for process-related characteristics seldom become incorporated into the industrial process simply because they include economically unattractive nutrients. Several reports in the literature concern the use of defined and minimal media formulations that are commensurate with high-performance fermentation by *Zymomonas (23,36–40)*; however, the influence of specific medium components is not fully understood. Furthermore, previous work in this area has been largely restricted to wild-type cultures, and little is known concerning the nutritional requirements of xylose-fermenting recombinant *Zymomonas*.

The objectives of this investigation were to gain insight into the relative importance of three key process variables—nutrient level, pH, and acetic acid concentration—anticipated to affect growth of recombinant *Zymomonas* and to identify any interactions among these variables that significantly affect cell mass production. This information is prerequisite to optimizing seed production (i.e., maximizing the cell mass yield on substrate and minimizing nutrient costs) in the context of the proposed SSCF process design. Statistical experimental design concepts were applied to enable us to characterize efficiently the dependence of seed production on these variables. Both two-level and three-level full-factorial designs with duplicate centerpoints were employed. A series of batch pH-controlled fermentations was carried out to complete each design. The fermentation medium consisted of a synthetic hardwood hemicellulose hydrolyzate (prehydrolyzate) with a xylose-to-glucose mass ratio of approximately 5:1. Analysis of variance (ANOVA) was used to determine the main and interactive variable effects significantly affecting cell mass concentration and the rate of xylose utilization.

MATERIALS AND METHODS

Organism

The recombinant *Z. mobilis* strain CP4 carrying the plasmid pZB5 (designated as Zm CP4:pZB5) (*34*) was received from M. Zhang (NREL, Golden, CO).

Long-Term Storage and Maintenance of Organism

Plasmid-bearing cultures, grown from single-colony isolates on selective agar medium (xylose + tetracycline), were stored at −70°C in RM medium supplemented with antifreeze (glycerol at 15 mL/dL). The phenotypic characteristics of the recombinant culture were related to growth in the presence of tetracycline and the production of ethanol from D-xylose. Cultures were generally revived in RM medium that contained 2% (w/v) glucose and 2% (w/v) xylose.

Fermentation Equipment

Batch fermentations were conducted in 1- or 2-L stirred-tank bioreactors (STR). pH-stat STR batch fermentations were conducted in a volume of 500 or 1500 mL in MultiGen (New Brunswick Scientific, Edison, NJ) bioreactors fitted with agitation, pH, and temperature control (30°C). The pH was monitored using a sterilizable combination pH electrode (Ingold) and was controlled by adding 4N KOH (NBS model pH-40 controller).

Methods of Preculture and Inoculation Procedures

A 1-mL aliquot of a glycerol-preserved culture was removed from cold storage (freezer) and transferred to about 100 mL of complex medium (RM), containing about 2% (w/v) xylose and 2% (w/v) glucose supplemented with tetracycline (Tc) (20 mg/L) in 125 mL screw-cap Erlenmeyer flasks, and grown overnight at 30°C. Alternatively, the inoculum was prepared by transferring an aliquot of a glycerol-preserved culture to a clarified corn steep liquor (cCSL) medium containing 2% (w/v) xylose, 2% (w/v) glucose, 0.2% (w/v) KH_2PO_4, and 20 mg/L tetracycline.

STR batch fermentations were inoculated by transferring approx 10% (v/v) of the overnight flask culture directly to the medium in the bioreactor. For STR fermentations, the initial cell density was monitored spectrophotometrically to give an OD_{600} in the range 0.1–0.2 corresponding to 30–50 mg dry wt cells/L.

Fermentation Media

For comparative purposes, the "benchmark" or reference medium for fermentation performance testing was the nutrient-rich, complex culture medium designated as RM (36). RM medium consists of 1% (w/v) Difco yeast extract (YE) (Difco Laboratories, Detroit, MI), and 0.2% (w/v) KH_2PO_4/distilled water. A CSL medium, which consisted of autoclaved tap water (TW) supplemented with centrifugally clarified CSL (range 1–2% v/v), was added at the time of inoculation.

A synthetic "biomass prehydrolyzate" (BPH) was formulated to model the sugar concentration in the NREL hardwood dilute-acid prehydrolyzate. The synthetic BPH medium was made with tap water, and contained 4% (w/v) xylose and 0.8% (w/v) glucose; it was supplemented with either 1% (w/v) YE and 0.2% (w/v) KH_2PO_4 or cCSL. All media contained 20 mg/L Tc.

Analytical Procedures

Growth was measured turbidometrically at 600 nm (1 cm lightpath) (Unicam spectrophotometer, model SP1800). In all cases, the blank cuvet contained distilled water. Dry cell mass (DCM) was determined by microfiltration of an aliquot of culture followed by washing and drying of the filter to constant weight under an infrared heat lamp. Fermentation media and cell-free spent media were compositionally analyzed by HPLC with a refractive index monitor and computer-interfaced controller/integrator (Bio-Rad Labs, Richmond, CA). Separations were performed at 65°C using an HPX-87H column (300 × 7.8 mm) (Bio-Rad Labs).

Statistical Analysis

ANOVA on results of the 2^3 full-factorial design was performed using an estimate for standard error based on a pooled standard deviation based on the two sets of duplicate centerpoints. ANOVA on the results of the 3^2 design was performed using Design-Expert™ version 4.0.5c software (Stat-Ease, Inc., Minneapolis, MN).

Determining Growth and Fermentation Parameters

The ethanol yield ($Y_{p/s}$) was calculated as the mass of ethanol produced (final concentration) per mass of sugar added to the medium, and was not corrected either for the dilution caused by adding alkali during the fermentation or for the contribution from fermentable components other than xylose and glucose. The average volumetric rate of xylose utilization ($_{av}Q_sxyl$) was determined by dividing the initial sugar concentration by the total time required to deplete sugar from the medium completely.

RESULTS AND DISCUSSION

Growth and Fermentation Characteristics of Recombinant ZmCP4:pZB5 in a Synthetic Hardwood Prehydrolyzate

The pioneering work of Zhang and her colleagues at NREL with recombinant Zm CP4:pZB5 involved the use of a nutrient-rich laboratory medium (RM), and experiments were conducted in the absence of pH control *(34)*. All batch fermentations in our investigation were conducted in bench-scale bioreactors fitted with controllers for agitation, pH, and temperature. For comparative fermentation performance testing, the RM medium *(36)* used in this study as the benchmark or reference medium was the same as that used previously *(34)*.

Figure 1 shows a typical growth and fermentation time-course using the recombinant culture pZB5 in the nutrient-rich RM medium. With the pH controlled at 6.0, the specific maximum growth rate (μ_{max}) in 4% (w/v) xylose was 0.23/h (Fig. 1), whereas in the absence of pH control, it was previously much slower (0.057/h) *(34)*. In another experiment (results not shown), we observed that use of a seed culture that had been grown in xylose as sole carbon (energy) source led to very slow growth (μ_{max} = 0.032/h) following subculturing into fresh medium of the same composition. Therefore, for seed production, a medium containing both glucose and xylose is preferred. With 4% (w/v) xylose as sole carbon and energy source, the ethanol yield was 0.48 g/g, which represents a conversion efficiency of 94%, compared to 86% without pH control *(34)*. Without pH control, the recombinant bacterium exhibited an average volumetric productivity of 0.23 g/L/h *(34)*, but with the pH controlled at 6.0, the productivity is increased almost threefold to 0.62 g/L/h (Fig. 1).

Addition of 0.8% (w/v) glucose to the 4% xylose-RM medium profoundly enhanced performance of recombinant Zm CP4:pZB5 (Fig. 1). These concentrations were selected to mimic the composition of a dilute-acid hardwood hemicellulose hydrolyzate *(14)*, and as such, this medium represented a synthetic BPH medium. Apart from phosphate, the sole source of nutrients was provided by 1% (w/v) Difco YE. In this medium at pH 6.0, both the specific growth rate and final cell mass concentration were increased about twofold to 0.43/h and 1.46 g DCM/L, respectively (Fig. 1 and Table 1). At 0.52 g/g, the ethanol yield is greater than the theoretical maximum of 0.51, which probably reflects the potential for some minor contribution from noncarbohydrate elements in the complex medium (Table 1). Compared to the medium with 4% xylose as the sole carbon source, the rate of xylose utilization in the synthetic BPH medium is

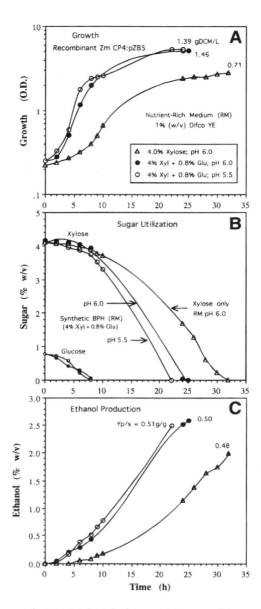

Fig. 1. Time-course of pH-stat batch fermentations with recombinant Z. *mobilis* CP4:pZB5 in yeast extract medium. **(A)** Growth, **(B)** glucose and xylose utilization, and **(C)** ethanol production. The medium was RM, and was supplemented with either 4% (w/v) xylose or a combination of 4% (w/v) xylose and 0.8% (w/v) glucose. The temperature was kept constant at 30°C. The pH-control set point was either 5.5 or 6.0. In panel A, the final DCM concentration (as determined directly by ultrafiltration) is indicated, and the ethanol yield ($Y_{p/s}$) is shown in panel C.

increased from 1.28–1.71 g/L/h (Table 1). The pH optimum for both growth rate and cell mass yield for wild-type Z. *mobilis* is close to 6.0 *(24,41)*. In the SSCF process design proposed by NREL *(14)*, the saccharifying enzymes operate at higher efficiency at pH <6.0 Figure 1 shows little effect

Table 1
Summary of Growth and Fermentation Parameters
for Recombinant Zm CP4:pZB5

Medium composition[o]	pH	Cell Mass (g DCM/L)	$Y_{p/s}$ (g/g)	av Q_s (xyl) (g Xyl/L.h)
RM (4% xylose)	6.0	0.71	0.48	1.28
RM (4% xyl + 0.8% glu)	6.0	1.46	0.52	1.71
RM (4% xyl + 0.8% glu)	5.5	1.39	0.51	1.86
TW + 1.0% cCSL	6.0	0.81	0.48	1.25
TW + 2.0% cCSL	6.0	1.40	0.50	1.56
TW + 2% cCSL + 0.5% Ac	6.0	0.92	0.46	1.24
TW + 2% cCSL + 1.5% Ac	6.0	0.43	0.49*	0.30*

[o]All media contain 20 µg/mL Tc and all fermentations were conducted at 30°C
*Based on sugar consumed at T = 54 h.
RM = rich medium; xyl = xylose (%w/v); glu = glucose (%w/v); TW = tap water; cCSL = clarified corn steep liquor (%v/v); Ac = acetic acid (%w/v).

on growth and fermentation performance of recombinant Zm CP4:pZB5 when the pH control set point was decreased from 6.0–5.5. At pH 5.5, the rate of xylose utilization increased from 1.71–1.86 g/L/h, representing a 4-h decrease in fermentation time from 26 to 22 h (Fig. 1B).

CSL has been shown to be a cost-effective means of nutrient supplementation in fermentations with ethanologenic recombinant *E. coli* (42,43). Lawford and Rousseau (44) have reported at this symposium that cCSL (i.e., with insolubles removed by centrifugation) cost-effectively satisfies the nutritional requirements of wild-type *Z. mobilis* CP4. Figure 2 shows the comparative growth and fermentation performance of recombinant Zm CP4:pZB5 in a synthetic BPH medium that consists of TW and 2% (v/v) cCSL. With the pH controlled at 6.0, the values for the final cell mass concentration and ethanol yield were very similar to those observed for supplementation with 1% YE (Table 1). The small decrease in the rate of xylose utilization probably reflects the proportional decrease in cell mass concentration (Table 1). A 50% reduction in the level of supplementation by cCSL produced a fermentation performance that was similar to that observed with the 4% xylose-RM medium (Fig. 2 and Table 1). At a cCSL supplementation rate of 1% (v/v), the final cell mass concentration was 0.81 g DCM/L, the ethanol yield was 0.48 g/g, and the rate of xylose utilization was 1.25 g/L/h (Fig. 2 and Table 1).

A well-known byproduct of dilute-acid pretreatment of lignocellulosic biomass is acetic acid (45,46) and, based on known structure and composition of the fermentation feedstock, the acetic acid concentration

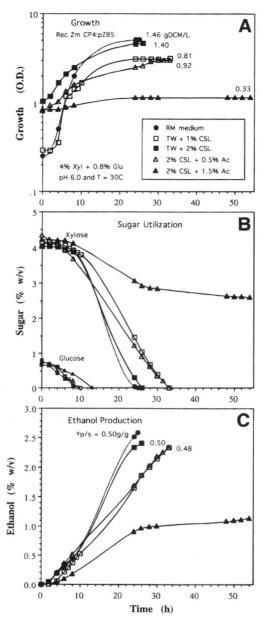

Fig. 2. Time-course of pH-stat batch fermentations with recombinant Z. *mobilis* CP4:pZB5 in CSL medium. **(A)** Growth, **(B)** glucose and xylose utilization, and **(C)** ethanol production. The medium consisted of TW supplemented with either 1 or 2% (v/v) cCSL. The triangle symbols represent fermentations with cCSL media supplemented with either 0.5 or 1.5% (w/v) acetic acid (Ac). The pH and temperature were controlled at 6.0 and 30°C, respectively. In all cases, the media contained 4% (w/v) xylose and 0.8% (w/v) glucose. The reference (control) medium was RM. In panel A, dry cell mass concentration is indicated, and the ethanol yield ($Y_{p/s}$) is shown in panel C.

can be predicted *(47)*. For hardwood biomass prehydrolyzate containing about 5% (w/v) sugars, the anticipated level of acetic acid is in the range 1.2–1.5% (w/v) *(47–49)*. The pH-dependent inhibitory effect of acetic acid on ethanologenic bacteria *(49,50)*, including *Zymomonas (51)*, is well documented. In terms of acetic acid toxicity, the undissociated protonated form of acetic acid is the causative agent, and the effect is potentiated at lower pH values *(50,51)*. The inhibitory effect of acetic acid is minimized by controlling the pH at 6.0 *(51)*. In a synthetic BPH medium with 2% (v/v) cCSL at pH 6.0, the presence of 0.5% (w/v) acetic acid results in growth and fermentation performance of Zm CP4:pZB5 that is very similar to the synthetic BPH medium in the absence of acetic acid and where the level of cCSL supplementation was only 1% (v/v) (Fig. 2 and Table 1). At a level of 1.5% (w/v) acetic acid, growth and fermentation kinetics are seriously compromised (Fig. 2). Although the ethanol yield (based on sugar consumed) remains high, about two-thirds of the xylose remains unfermented after 2 d (Fig. 2). From the perspective of seed production, the effect of acetic acid is most dramatic, and at the upper limit of 1.5% (w/v), the cell mass concentration decreases to 0.43 g DCM/L (Table 1).

Statistical Approach to Optimization

Optimization by the traditional method involves changing one independent variable while holding all others constant. This "one factor at a time" approach can be effective, but it is laborious and poorly suited for identifying interactions among multiple factors. A better approach is to use statistical experimental designs that enable the significance of multiple variables (or factors) to be assessed in an efficient manner *(52–55)*. Two-level full-factorial designs can be run, for example, to identify the most important process factors (variables) and factor–factor interactions (i.e., those that have the largest effect on the process). Then, once the most important variables are identified, higher-level designs—the simplest being a three-level factorial design—can be used to characterize better the response of the process within the experimental design domain.

Our earlier studies showed that batch seed growth was influenced by nutrient (cCSL) level, pH, and exogenous acetic acid concentration. Therefore, we designed a 2^3 full-factorial experiment (with duplicate centerpoints) to characterize efficiently the significance of these three factors on seed production performance and to determine whether there were any significant interactions among them. Their influence on cell production from glucose–xylose mixtures was examined using sugar loadings similar to those expected in hardwood hemicellulose hydrolyzates, (i.e., 4% w/v glucose + 0.8% w/v xylose). The overall design is depicted in Fig. 3A. which shows that the nutrient supplementation level varied from a low level of 1% (v/v) to a high level of 2% (v/v); pH ranged from a low level of 5.5 to a high level of 6.0; and acetic acid level varied from 0.5% (w/v) to 1.5% (w/v).

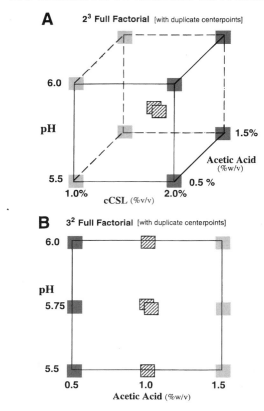

Fig. 3. Experimental design matrices. **(A)** 2^3 Full-factorial with duplicate center points and **(B)** 3^2 full-factorial with duplicate center points.

Results of the 2^3 full-factorial experiment are depicted in Fig. 4 and listed in Table 2, which shows that final cell mass concentration ranged from a low of 0.17 g DCM/L (at 1% [v/v]) cCSL, pH 5.5, and 1.5% [w/v] acetic acid to a high of 0.92 g DCM/L (at 2% [v/v] cCSL, pH 6.0, and 0.5% [w/v] acetic acid). Figure 4B shows that the average volumetric rate of xylose utilization behaved similarly, ranging from a low of 0.178 g/L/h (at 1% [v/v] cCSL, pH 5.5, and 1.5% [w/v] acetic acid) to a high of 1.24 g/L/h (at 2% [v/v] cCSL, pH 6.0, and 0.5% [w/v] acetic acid). Although more variable, the response in terms of cell mass yield as a function of factor levels also showed the same general trend (results not shown graphically, but values listed in Table 2). Figure 4 shows that cell mass production and xylose utilization exhibit similar behavior with respect to changes in pH and acetic acid concentration, but the influence of nutrient (cCSL) is different. Cell mass production shows a smaller dependence on the level of cCSL than the rate of xylose consumption, particularly at high pH. Figure 4B suggests that interactions may exist between the variables

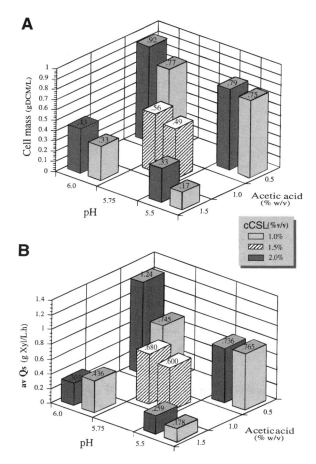

Fig. 4. Response profiles for 2^3 full-factorial study with Zm CP4:pZB5 using a synthetic biomass prehydrolyzate. **(A)** Cell mass concentration and **(B)** average volumetric rate of xylose utilization. The experimental design matrix is illustrated in Fig. 3A. The center bars in the 3D plot represent the duplicate center points. The base medium contained 4% xylose and 0.8% glucose. The three variables (levels in brackets) were pH (5.5 and 6.0), acetic acid (0.5 and 1.5% w/v), and cCSL (1 and 2% v/v).

with respect to xylose utilization. For example, at high pH and low acetic acid level, increasing cCSL level increases the utilization rate. Conversely, at high pH and high acetic acid level, increasing cCSL level results in a decrease.

ANOVA was performed to assess the statistical significance of these apparent effects. Results of the ANOVA on cell mass production show that at the 90% confidence level, all three factors significantly affects final cell mass concentration; the effects of acetic acid and nutrient are higher than the 95% confidence level (CL > 95%). Acetic acid concentration exerts the largest effect and negatively affects cell mass production, with nutrient level and pH exerting smaller, but positive effects on seed production. No

Table 2
Summary of Conditions and Responses for a 2^3 Factorial Study
with Recombinant pZB5 and a Synthetic BPH Medium[a]

Exp. #	ID #	Conditions			Responses		
		cCSL (% v/v)	pH	Acetic acid (% w/v)	Cell mass (g DCM/L)	Growth Yield (gDCM /g S used)	av Q_s (Xyl) (g Xyl/L.h)
1	37a	1.0	5.50	0.5	0.75	0.016	0.765
2	37c	2.0	5.50	0.5	0.79	0.017	0.736
3	37b	1.0	6.00	0.5	0.77	0.016	0.745
4	37d	2.0	6.00	0.5	0.92	0.018	1.238
5	44c	1.0	5.50	1.5	0.17	0.011	0.178
6	44e	2.0	5.50	1.5	0.33	0.018	0.259
7	44d	1.0	6.00	1.5	0.33	0.016	0.436
8	36b	2.0	6.00	1.5	0.43	0.023	0.303
9	39e	1.5	5.75	1.0	0.56	0.021	0.680
10	44f	1.5	5.75	1.0	0.49	0.018	0.600

[a]For description of experimental design matrix, see Fig. 3A. For graphical representation of the responses, see Fig. 4.

significant interactions are evident. Results of the ANOVA on xylose utilization rate (avQ_s) also indicate that all three factors significantly affect $_{av}Q_s$ (CL > 90%); the effects of acetic acid and pH (rather than nutrient) are higher than the 95% confidence level (CL > 95%). As with cell mass production, acetic acid concentration has the largest effect on avQ_s, and it is a negative effect, with the average rate of xylose consumption decreasing with increasing acetic acid concentration. pH again has a small, but statistically significant positive effect, with avQ_s increasing when seed growth is carried out at higher pH. Nutrient (cCSL) level also has an overall positive effect, although it is less pronounced than the effect of pH. ANOVA of avQ_s results also indicates that significant factor–factor interactions are present, with the nutrient × pH two-way interaction significant at above the 80% confidence level (CL > 80%) and the nutrient × acetic acid interaction significant at the 95% confidence level (CL > 95%). The three-way nutrient × pH × acetic acid interaction also appears to be highly significant (CL > 95%). Surprisingly, a significant interaction is not observed between pH and acetic acid for avQ_s.

The results of the 2^3 design conclusively demonstrate that low levels of inexpensive cCSL (1.5–2% v/v) cost-effectively supply nutrients for seed growth on synthetic hardwood hemicellulose hydrolyzates. The results of the 2^3 experiment also show that at acetic acid concentrations above 0.5% (w/v), seed production must be carried out at a pH above 5.5 to achieve robust cell growth. This finding contrasts with cell mass

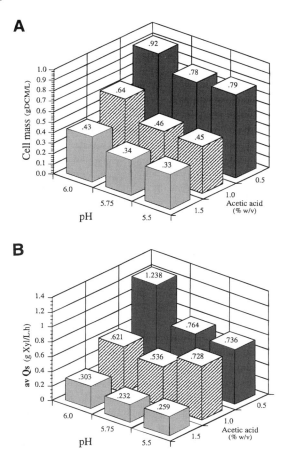

Fig. 5. Response profiles for 3^2 full-factorial study with Zm CP4:pZB5 using a synthetic biomass prehydrolyzate. **(A)** Cell mass concentration, and **(B)** average volumetric rate of xylose utilization. The experimental design matrix is illustrated in Fig. 3B. The center bar in the 3D plot represents the average of the duplicate center points. The two variables (levels in brackets) were pH (5.5, 5.75, and 6.0), acetic acid (0.5, 1.0, and 1.5% w/v). The medium contained 4% xylose and 0.8% glucose, and was supplemented with 2% (v/v) cCSL.

production in the absence of acetic acid, which exhibits no dependence on pH within the pH range of 5.5–6.0 (Fig. 1).

A follow-up 3^2 full-factorial experiment (with duplicate centerpoints) was designed to probe further for an anticipated pH × acetic acid interaction by characterizing, at a preliminary level, the response surfaces of cell mass production (as final cell mass concentration) and rate of xylose utilization (as avQ_s) in the acetic acid-pH design space. The overall experimental design is graphically depicted in Fig. 3B. Nutrient level was fixed at 2% (v/v) cCSL in this experiment to minimize the likelihood of nutrient limitation; sugar loading was the same as in the 2^3 experiment. Thus, an additional six fermen-

Table 3
Summary of Conditions and Responses for a 3^2 Factorial Study
with Recombinant pZB5 and a Synthetic BPH medium[a]

Exp. #	ID #	Conditions		Responses		
		Acetic acid (% w/v)	pH	Cell mass (g DCM/L)	Growth Yield (gDCM /g S used)	av Q_s (Xyl) (g Xyl/L.h)
1	37c	0.5	5.50	0.79	0.017	0.736
2	50a	1.0	5.50	0.45	0.017	0.728
3	44e	1.5	5.50	0.33	0.018	0.259
4	37d	0.5	6.00	0.92	0.018	1.238
5	50f	1.0	6.00	0.64	0.028	0.621
6	36b	1.5	6.00	0.43	0.023	0.303
7	50b	0.5	5.75	0.78	0.027	0.764
8	50c	1.0	5.75	0.45	0.021	0.550
9	50d	1.0	5.75	0.47	0.026	0.522
10	50e	1.5	5.75	0.34	0.021	0.232

[a]For description of experimental design matrix, see Fig. 3B. For graphical representation of the responses, see Fig. 5.

tations at the 2% (v/v) cCSL level beyond the four completed in the 2^3 design had to be run to complete this 3^2 full-factorial design.

Results of the 3^2 response surface study are shown in Fig. 5 and values are listed in Table 3. As Fig. 5A shows, there is not much curvature evident in the final cell mass concentration response. Dependence on acetic acid level is pronounced compared to dependence on pH. Values are highest at the lowest acid acid level regardless of pH, although at any particular acetic acid level, cell growth is highest at pH 6.0. The avQ_s response surface is more complicated and exhibits more curvature. The rate of xylose utilization (avQ_s) at the lowest acetic acid concentration and highest pH is significantly greater than at any other condition.

ANOVA calculations performed on the 3^2 results indicate that the final cell mass concentration response exhibits curvature with respect to both acetic acid level and pH (CL > 95%), but that no interaction between acetic acid and pH is present. In contrast, ANOVA for avQ_s indicates that the response of average xylose utilization rate has significant curvature with respect to pH (CL > 85%), but is linear with respect to acetic acid concentration (CL > 95%). However, the response of avQ_s exhibits a significant pH x acetic acid interaction (CL > 90%). These results indicate that the pH x acetic acid two-way interaction primarily influences the rate of xylose utilization. Thus, knowledge of this two-way interaction is necessary to maximize the rate of xylose utilization, but not to maximize the extent of cell mass production.

The results shown in Fig. 5 can be used to determine the approximate optimum pH to maximize seed production for a given (known) concentration of acetic acid. This is useful because acetic acid levels in hemicellulose hydrolyzates vary for different feedstocks. Alternatively, the results shown in Fig. 5 can be used to motivate the need to maintain acetic acid concentrations below 1.0% (w/v) to achieve robust cell production.

ACKNOWLEDGMENTS

This work was funded by the Biochemical Conversion Element of the Office of Fuels Development of the US Department of Energy. Research conducted at the University of Toronto was part of Phase II of a Subcontract AAP-4-11195-03 from NREL. The authors are grateful to D. Mandell for assistance in preparing some of the graphics presentations of the data. We also wish to thank M. Zhang for the recombinant Z. mobilis CP4:pZB5 and S. Picataggio for helpful discussions.

REFERENCES

1. Wright, J. D. (1988), *Chem. Eng. Prog.* **84,** 62–74.
2. Wright, J. D., Wyman, C. E., and Grohman, K. (1988), *Appl. Biochem. Biotechnol.* **18,** 75–90.
3. Wyman, C. E. and Hinman, N. D. (1990), *Appl. Biochem. Biotechnol.* **24/25,** 735–753.
4. Lynd, L. R. (1989), *Adv. Biochem. Eng. Biotechnol.* **38,** 1–52.
5. Lynd, L. R., Cushman, J. H., Nichols, R. J. and Wyman, C. E. (1991), *Science* **251,** 1318–1323.
6. Keim, C. R. (1983), *Enzyme Microbiol. Technol.* **5,** 103–114.
7. Jeffries, T. W. (1981), *Biotechnol. Bioeng., Symp.* **11,** 315–324.
8. Jeffries, T. W. (1983), *Adv. Biochem. Eng. Biotechnol.* **27,** 1–32.
9. Jeffries, T. W. (1990), in *Yeast: Biotechnology and Biocatalysis*, Verachtert, H. and De Mot, R., eds., Marcel Dekker, New York, pp. 349–394.
10. Hahn-Hägerdahl, B., Hallborn, J., Jeppson, H., Olsson, L., Skoog, K., and Walfridson, M. (1993), in *Bioconversion of Forest and Agricultural Plant Residues* Saddler, J. N., ed., C.A.B. International, Wallingford, UK, pp. 231–290.
11. Hinman, N. D., Wright, J. D., Hoagland, W., and Wyman, C. E. (1989), *Appl. Biochem. Biotechnol.* **20/21,** 391–401.
12. Hinman, N. D., Schell, D. J, Riley, C. J., Bergeron, P. W., and Walter, P. J. (1992), *Appl. Biochem. Biotechnol.* **34/35,** 639–649.
13. Lynd, L. R. (1990), *Appl. Biochem. Biotechnol.* **24/25,** 695–719.
14. Picataggio, S. K., Eddy, C., Deanda, K., Franden, M. A., Finkelstein, M., and Zhang, M. (1996), *Seventeenth Symposium on Biotechnology for Fuels & Chemicals* (Paper #9).
15. Picataggio, S. K., Zhang, M., and Finkelstein, M. (1994), in: *Enzymatic Conversion of Biomass for Fuels Production*, Himmel, M. E., Baker, J. O., and Overend, R. A., eds., American Chemical Society, Washington, DC, *ACS Symposium Series* 566, pp. 342–362.
16. Zhang, M., Franden, M. A., Newman, M., McMillan, J., Finkelstein, M., and Picataggio, S. K. (1995), *Appl. Biochem. Biotechnol.* **51/52,** 527–536.
17. Rogers, P. L., Lee, K. J., Skotnicki, M. L., and Tribe, D. E. (1982), *Adv. Biochem. Eng.* **23,** 37–84.
18. Lawford, G. R., Lavers, B. H., Good, D., Charley, R., Fein, J., and Lawford, H. G. (1982), in *International Symposium on Ethanol from Biomass*, Duckworth, H. E. and Thompson, E. A., eds., Royal Society of Canada, Winnipeg, Canada, pp. 482–507.

19. Lawford, H. G. (1988), *Proc. VIII Int'l Symp. on Alcohol Fuels*, New Energy Development Organization, Tokyo, pp. 21–27.
20. Swings, J. and DeLey, J. (1977), *Bacteriol. Rev.* **41**, 1–46.
21. Montenecourt, B. S. (1985), in *Biology of Industrial Microorganisms*, Demain, A. L. and Simon, N. A., eds., Benjamin/Cummings, Meno Park, CA, pp. 216–287.
22. Doelle, H. W., Kirk, L., Crittenden, R., Toh, H., and Doelle, M. (1993), *Crit. Rev. Biotechnol.* **13**, 57–98.
23. Lawford, H. G. (1988), *Appl. Biochem. Biotechnol.* **17**, 203–219.
24. Lawford, H. G. and Ruggiero, A. (1990), *Biotechnol. Appl. Biochem.* **12**, 206–211.
25. Bringer, S., Sahm, H., and Swyzen, W. (1984), *Biotechnol. Bioeng. Symp.* **No. 14**, 311–319.
26. Rodriguez, E. and Callieri, D. A. S. (1986), *Biotechnol. Lett.* **8**, 745–748.
27. Doelle, M. B., Greenfield, P. F., and Doelle, H. W. (1990), *Proc. Biochem.* **25**, 151–156.
28. Beavan, M., Zawadzki, B., Droiniuk, R., Fein, J. E., and Lawford, H. G. (1989), *Appl. Biochem. Biotechnol.* **20/21**, 319–326.
29. Lee, G. M., Kim, C. H., Lee, K. J., Zainal Abidin Mohd, Y., Han, M. H., and Rhee, S. K. (1986), *J. Ferment Technol.* **64**, 293–297.
30. Parekh, S. R., Parekh, R. S., and Wayman, M. (1989), *Proc. Biochem.* **24**, 85–91.
31. Park, S. C., Kademi, A., and Baratti, J. C. (1993), *Biotechnol. Lett.* **15(11)**, 1179–1184.
32. Liu, C.-Q., Goodman, A. E., and Dunn, N. W. (1988), *J. Biotechnol.* **7**, 61.
33. Feldman, S. D., Sahm, H., and Sprenger, G. A. (1992), *Appl. Microbiol. Biotechnol.* **38**, 354.
34. Zhang, M., Eddy, C., Deanda, K., Finkelstein, M., and Picataggio, S. K. (1995), *Science* **267**, 240–243.
35. Lynd, L. R., Elander, R. T., and Wyman, C. E. (1996), *Appl. Biochem. Biotechnol.* **57/58**, 641–661.
36. Goodman, A. E., Rogers, P. L., and Skotnicki, M. L. (1982), *Appl. Environ. Microbiol.* **44(2)**, 496–498.
37. Fein, J. E., Charley, R. C., Hopkins, K. A., Lavers, B., and Lawford, H. G. (1983), *Biotechnol Lett.* **5**, 1–6.
38. Nipkow, A., Beyeler, W., and Feichter, A. (1984), *Appl. Microbiol. Biotechnol.* **19**, 237–240.
39. Galani, I., Drainas, C., and Typas, M. A. (1985), *Biotechnol. Lett.* **7**, 673–678.
40. Baratti, J., Varma, R., and Bu'Lock, J. D. (1986), *Biotechnol. Lett.* **8**, 175–180.
41. Lawford, H. G., Holloway, P., and Ruggiero, A. (1988), *Biotechnol. Lett.* **10**, 809–814.
42. Lawford, H. G. and Rousseau, J. D. (1996), *Appl. Biochem. Biotechnol.* **57/58**, 307–326.
43. Asghari, A., Bothast, R. J., Doran, J. B., and Ingram, L. O. (1996), *J. Ind. Microbiol.* **16**, 42–47.
44. Lawford, H. G. and Rousseau, J. D. (1997), *Appl. Biochem. Biotechnol.* (18th Symp.), **63–65**, 287.
45. Grohman, K., Himmel, M., Rivard, C., Tucker, M., Baker, T., Torget, R., and Graboski, M. (1984), *Biotechnol. Bioeng. Symp.* **14**, 139–157.
46. Kong, F., Engler, C. R., and Soltes, E. (1992), *Appl. Biochem. Biotechnol.* **34/35**, 23–35.
47. Timell, T. E. (1964), *Adv. Carbohydrate Chem.* **19**, 247–302.
48. Lawford, H. G. and Rousseau, J. D. (1993), in *Energy from Biomass and Wastes XVI* (March 1992), Klass, D. L., ed., Institute of Gas Technology, Chicago, IL, pp. 559–597.
49. McMillan, J. D. (1994), in *Enzymatic Conversion of Biomass for Fuels Production*, Himmel, M. E., Baker, J. O., and Overend, R. A., eds., American Chemical Society, Washington, DC, *ACS Symposium Series* **566**, pp. 411–437.
50. Lawford, H. G. and Rousseau, J. D. (1993), *Appl. Biochem. Biotechnol.* **39/40**, 301–322.
51. Lawford, H. G. and Rousseau, J. D. (1994), *Appl. Biochem. Biotechnol.* **45/46**, 437–448.
52. Box, G. E. P., Hunter, W. G., and Hunter, J. S. (1978), in: *Statistics for Experimenters* Wiley, New York.
53. Davies, O. L. (1967), in *Design and Analysis of Industrial Experiments*, Hafneri, New York.
54. Maddox, I. S. and Richert, S. H. (1977), *J. Appl. Bacteriol.* **43**, 197–204.
55. Myers, R. H. (1971), in *Response Surface Methodology*, Allyn and Bacon, Boston.

Corn Steep Liquor as a Cost-Effective Nutrition Adjunct in High-Performance *Zymomonas* Ethanol Fermentations

HUGH G. LAWFORD* AND JOYCE D. ROUSSEAU

Bio-engineering Laboratory, Department of Biochemistry, University of Toronto, Toronto, Ontario, Canada M5S 1A8

ABSTRACT

The ethanologenic bacterium *Zymomonas mobilis* has been demonstrated to possess several fermentation performance characteristics that are superior to yeast. In a recent survey conducted by the National Renewable Energy Laboratory (NREL), *Zymomonas* was selected as the most promising host for improvement by genetic engineering directed to pentose metabolism for the production of ethanol from lignocellulosic biomass and wastes. Minimization of costs associated with nutritional supplements and seed production is essential for economic large-scale production of fuel ethanol. Corn steep liquor (CSL) is a byproduct of corn wet-milling and has been used as a fermentation nutrient supplement in several different fermentations. This study employed pH-controlled batch fermenters to compare the growth and fermentation performance of *Z. mobilis* in glucose media with whole and clarified corn steep liquor as sole nutrient source, and to determine minimal amounts of CSL required to sustain high-performance fermentation.

It was concluded that CSL can be used as a cost-effective single-source nutrition adjunct for *Zymomonas* fermentations. Supplementation with inorganic nitrogen significantly reduced the requirement for CSL. Depending on the type of process and mode of operation, there can be a significant contribution of nutrients from the seed culture, and this would also reduce the requirement for CSL. Removal of the insolubles (40% of the total solids) from CSL did not detract significantly from its nutritional effectiveness. On an equal-volume basis, clarified CSL was 1.33 times more "effective" (in terms of cell mass yield and fermentation time) than whole CSL. For fermentations at sugar loading of >5% (w/v), the

*Author to whom all correspondence and reprint requests should be addressed.

recommended level of supplementation with clarified CSL is 1.0% (v/v). Based on CSL at US $50/t, the cost associated with using clarified CSL at 1.0% (v/v) is 88¢/1000 L of medium and 5.3¢/gal of undenatured ethanol for fermentation of 10% (w/v) glucose. This cost compares favorably to estimates for using inorganic nutrients. The cost impact is reduced to 3.1¢/gal if there is a byproduct credit for selling the insolubles as animal feed at a price of about US $100/t. Therefore, the disposition of the CSL insolubles can significantly impact the calculations of cost associated with the use of CSL as a nutritional adjunct in large-scale fermentations.

Index Entries: *Zymomonas*; clarified corn steep liquor; whole CSL; nutrition; ethanol; economic impact; cell yield; high-performance fermentation; insolubles; defined medium.

INTRODUCTION

In North America, the increasing practice of a legislated requirement for oxygenate additives in gasoline offers a growth opportunity for the fuel ethanol industry. At the present time, this industry relies primarily on the fermentation of glucose derived from corn or cereal grain starch, and to a much lesser extent, fructose (sucrose) from cane and beet sugar *(1,2)*. For the most part, the technology used in the large-scale production of fermentation fuel ethanol is the same as that currently practiced in the beverage alcohol industry where yeast is used to convert sugar to ethanol *(1,2)*.

For the fuel ethanol industry to expand and remain economically viable, there is a requirement for alternative sources of fermentable carbohydrates *(3,4)*, and the "yeast monopoly" *(1,2)* is now being seriously challenged by other ethanologenic biocatalysts that have been either selected or engineered for their improved fermentation efficiency *(5,6)*. The bacterium *Zymomonas* (for reviews, see *7,8*) has proven fermentation performance characteristics that are superior to yeast *(9–14)*, and this biocatalyst has the potential to "revolutionize the industry" *(15)*. In a survey of several microorganisms that was recently conducted by the National Renewable Energy Laboratory (NREL), *Zymomonas* was selected as the most promising host for improvement by genetic engineering directed to pentose metabolism *(16,17)*.

The biological approach to process improvement is directed toward the performance characteristics of the biocatalyst *(18)*, and although major developments are usually credited to the selection or engineering of a superior strain, other parameters, such as the nutritional and physical environment to which the biocatalyst will be exposed, are also known to have a significant potential influence on yield and productivity both with respect to growth and fermentation *(19)*. Minimization of costs associated with nutritional supplements and seed production is essential for economic large-scale production of fuel ethanol *(20)*.

Most *Zymomonas* strains are known to be auxotrophic for the vitamin pantothenic acid *(7,21–24)*, and although there are several reports in the literature of defined and minimal media formulations commensurate with high-performance fermentation by *Zymomonas (11,18,25–28)*, relatively little is understood concerning the specific influence on growth and fermentation of the individual medium components. Furthermore, these defined media are not useful in an industrial context owing to the prohibitive cost of the vitamin supplements.

Corn steep liquor (CSL) is a byproduct of corn wet-milling, and contains a rich complement of important nutrients to support robust microbial growth and fermentation *(29)*. It was first used as a nutrient source in the 1940s in the development of large-scale penicillin fermentations and continues to be used extensively today in diverse industrial fermentation processes. The process (light) steep water (LSW) is concentrated about 10-fold by evaporation to 45–55% solids to produce heavy steep water (HSW) or CSL. CSL is sold into the animal feed market and has a protein value judged to be equivalent to gluten feed selling for US $100/t. Since on average CSL is about 50% dry substance *(29)*, the selling price is US $50/t. The protein content is estimated from the determination of the total Kjeldahl nitrogen *(29)*. As a fermentation medium supplement, CSL can be viewed either as a complete source of nutrients or as a source of vitamins and other trace elements (growth factors) *(19,30)*. Since *Zymomonas* can assimilate inorganic nitrogen *(18)*, supplementation with inorganic nitrogen potentially could significantly reduce the level of CSL required to support growth and fermentation.

Previous work conducted in our lab *(12,15,31,32)* as well as that of others *(33,34)* has showed that LSW and HSW from corn wet-milling were effective nutritional supplements for *Zymomonas* fermentations. CSL has also been investigated in terms of its equivalence to yeast extract as a complex nutritional supplement for xylose-fermenting yeasts *(35)* and used in the development of a yeast-based simultaneous saccharification and fermentation biomass-to-ethanol process *(36)*. CSL has also been used in ag-waste and wood biomass-processes that propose to use recombinant *Escherichia coli (37–40)*.

The purpose of this study was fourfold:

1. To compare the fermentation performance of a wild-type strain of *Zymomonas mobilis* CP4 in media with whole and clarified CSL as sole nutrient source;
2. To determine minimal amounts of CSL required to sustain high-performance fermentation;
3. To examine the potential for reducing the amount of CSL through supplementation with inorganic nitrogen; and
4. To estimate the economic impact of using CSL on the cost on producing fuel ethanol.

MATERIALS AND METHODS

Organisms

The wild-type strain *Z. mobilis* CP4 was received from M. Zhang (National Renewable Energy Laboratory, Golden, CO). Cultures, grown from single-colony isolates on glucose agar medium, were stored at –70°C in a nutrient-rich complex medium supplemented with antifreeze (glycerol at 15 mL/dL).

Fermentation Media

The chemical composition of the different media used in this study is given in Table 1. For comparative purposes, the reference medium was a nutritionally rich, complex medium containing 3% (w/v) yeast extract (Difco Laboratories, Detroit, MI) and was designated as "ZM1" (Table 1). In the absence of yeast extract, the defined salts medium was designated as "DS" medium, and it was supplemented with the vitamins D-pantothenic acid (hemi-calcium salt, 1.0 mg/L) and biotin (1.0 mg/L) (Sigma Chemical, St. Louis, MO) (Table 1). D-Glucose was added (as specified) as the sole carbon (energy) source. The media were sterilized by autoclaving with the stock glucose solution being autoclaved separately to minimize browning.

Two samples of CSL were obtained from NACAN Products (Collingwood, Ontario, Canada) on separate occasions and stored in a refrigerator at 4°C. The "CSL medium" (Table 1) consisted of autoclaved tap water (TW) supplemented with either whole (wCSL) or centrifugally clarified corn steep liquor (cCSL), which was added at the time of inoculation. Following centrifugation ($10,000g$ for 10 min), the packed sediment represented 25% of the total volume.

Fermentation Equipment

Batch fermentations were conducted in 1- or 2-L stirred-tank bioreactors (STR) fitted with pH control *(30)*. The temperature was kept constant at 30°C. A bench-top chemostat with a working volume of 350 mL (Bioflo C-30, New Brunswick Scientific, Edison, NJ) was used to generate glucose-limited continuous cultures as described previously *(11)*.

Methods of Preculture and Inoculation Procedures

A 1-mL aliquot of a glycerol-preserved culture was removed from cold storage and transferred aseptically to a 125-mL screw-cap Erlenmyer flask containing about 100 mL of either ZM1 medium or 2% (v/v) cCSL medium with 2% (w/v) glucose. The CSL medium was supplemented with KH_2PO_4 (20 g/L), and the initial pH adjusted to 6.0. Seed flask cultures were grown statically overnight in an incubator (30°). In

Table 1
Fermentation Media Formulations[a]

Ingredient (g)	Medium Designation		
	ZM1	DS	CSL
D-Glucose	var	var	var
Yeast Extract (Difco)	3.00	0	0
NH_4Cl	0.81	1.6	var
cCSL (mL)	0	0	var
KH_2PO_4	3.48	3.48	0*
$MgSO_4$	0.49	0.49	0
$FeSO_4 \cdot 7H_2O$	0.01	0.01	0
Citric acid	0.21	0.21	0
Ca Pantothenate	0	0.001	0
Biotin	0	0.001	0
Distilled water (L)	1	1	-
Tap water (L)	-	-	1

[a]var = amount variable (as specified for each fermentation); DS = defined salts medium; cCSL = clarified CSL.

*KH_2PO_4 (2 g/L) added for pH buffering in seed flask cultures.

experiments where a chemostat culture was used as seed, the effluent was collected on ice overnight.

STR batch fermentations were inoculated by transferring approx 10% (v/v) of the overnight flask culture directly to the medium in the bioreactor. The initial cell density was monitored spectrophotometrically to give an OD_{600} in the range 0.1–0.2, corresponding to 30–50 mg dry cell mass (DCM)/L. An alternative inoculation procedure was developed that minimized the potential for transfer of nutrients from the preculture medium during inoculation. This method involved a typical centrifugal harvesting/washing procedure. Following overnight growth at 30°C, the culture was centrifuged at 10,000g for 10 min (Sorvall RC2B centrifuge). The cell pellet was resuspended in 0.1% (w/v) peptone (Difco) water, agitated to achieve uniformity and used to inoculate the STR at the desired initial cell density (OD). This procedure was designated as the "concentrated cell inoculation" (CCI) method.

Analytical Procedures

Growth was measured turbidometrically at 600 nm (1-cm lightpath) (Unicam spectrophotometer, model SP1800). In all cases, the blank cuvet contained distilled water. DCM was determined by microfiltration of an aliquot of culture, followed by washing and drying of the filter to constant weight under an infrared heat lamp. Compositional analyses of fermentation media and cell-free spent media were accomplished by HPLC as described previously (30).

RESULTS AND DISCUSSION

In this study, we adopted the strategy commonly practiced in microbial physiology for investigating microbial nutritional requirements through culture media formulation. Microorganisms exhibit growth and metabolism (fermentation) optima with respect to both physical and chemical environmental factors. A prerequisite to growth is that the medium supply the elements of carbon, nitrogen, phosphorous, and sulfur that, in addition to hydrogen and oxygen, are the major components of all biomolecules. In addition to sodium, potassium, magnesium, calcium, and iron, certain minerals are required in relatively much smaller amounts ("trace elements"). There is a low-level requirement for metals, such as manganese, copper, cobalt, molybdenum, zinc, and so forth, that act as enzyme co-factors. In addition, there can be certain other essential elements that, apart from the known "vitamins," are referred to collectively as "growth factors." The elemental composition of the organism reflects the mass ratio requirements of these various nutritional elements in the culture medium. However, in this context, it is important to bear in mind that cellular composition can be influenced by the chemical nature of the growth environment. Accordingly, our approach to formulating a cost-effective medium with minimal levels of CSL involved a balancing of the nitrogen content of the medium with the nitrogen requirement of the anticipated cell mass concentration. A prerequisite to this approach to medium formulation is a knowledge of the nitrogen content of both the nutritional supplement (in this case CSL) and the *Zymomonas* cell mass.

In a separate study that was reported at this meeting, we were interested in optimizing seed production for a biomass-to-ethanol process that proposes to use a recombinant *Zymomonas* (41). For the purpose of seed production, the objective is to maximize cell density, and in a pH-controlled batch culture, this can be accomplished using a semisynthetic medium, such as ZM1 (Table 1), where both yeast extract (YE) and inorganic nitrogen (NH_4Cl) act as sources of assimilable nitrogen (Fig. 1). In the plot of cell mass vs glucose concentration (Fig. 1), the slope of the tangent provides an estimation of the growth yield with respect to carbon (energy) source, which under these conditions, is 0.036 g DCM/g glucose. With the pH controlled at 5.0, a maximum cell mass concentration of 2.25 g DCM/L is achieved at a glucose level of about 65 g/L, with very little further increase in cell density at higher sugar loading (Fig. 1). The growth yield is known to be affected by pH (42,43), and we observed that the maximum cell mass concentration can be increased about 10% to

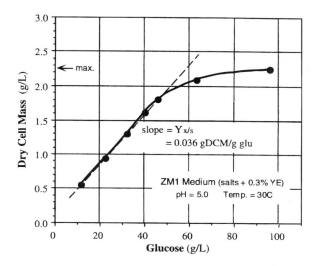

Fig. 1. Production of *Zymomonas* cell mass as a function of glucose concentration. Batch cultures of Z. *mobilis* CP4 were conducted in a semisynthetic medium (ZM1) containing glucose as sole carbon source (range approx 10–100 g/L). The pH and temperature were kept constant at 5.0 and 30°C, respectively. The y-axis arrow shows the observed maximum cell density of 2.25 g DCM/L.

about 2.6 g DCM/L by increasing the pH control set point from 5.0–6.0 (results not shown).

A Kjeldahl assay for total nitrogen (N) of the cell mass of Z. *mobilis* CP4 resulted in an average value of 13.6% dry basis (% db). This is equivalent to a growth yield with respect to N of 7.35 g DCM/g atom N, which agrees very closely with the value of 7.1 reported previously by Lawford and Stevnsborg *(11)* for a nitrogen-limited steady-state chemostat culture of Z. *mobilis* ATCC 29191 using a defined salts medium with ammonium chloride as the sole source of assimilable nitrogen. According to the concept of nitrogen balancing, the minimum level of N in the medium required to satisfy the requirement for the synthesis of 2.6 g DCM/L would be 0.354 g of assimilable N/L. This requirement could be supplied by 3.86 g/L of Bacto yeast extract (Difco) (total N = 9.18% db) or alternatively by a source of inorganic nitrogen, such as ammonium chloride, at a level of 1.35 g/L. It should be noted that these levels of N supplementation by yeast extract or NH_4Cl represent the minimum amounts required to satisfy the N requirement for a cell concentration of 2.6 g DCM/L, and assume an equivalence between total N and assimilable N.

This study is an extension of previous work that involved testing the fermentation performance response of a recombinant ethanologenic *E. coli* with a view toward designing a cost-effective, nutritionally lean medium for large-scale cellulosic ethanol production *(30)*. However, whereas previously we relied on CSL product specifications given to us by the supplier, in this study, we analyzed the two samples of CSL that were

Table 2
Composition of CSL[a]

Composition	cCSL	cCSL	wCSL
Batch No.	1	2	2
Density	1.12	1.13	1.20
Percent Solids (w/w)	46.1	41.4	44.2
Insolubles (% db)	ND	ND	40.0
Ash (% db)	15.8	ND	ND
Protein (% db) Total Kjeldahl N x 6.25	45.8	44.4	42.5
Carbohydrate (% db) (by difference)	16.5	ND	ND
Glucose (% db)	ND	0.19	ND
Lactic acid (% db)	ND	18.8	ND
Volumetric Total N (gN/mL)	0.0378	0.0333	0.0364

[a]CSL was from Nacan Products Ltd. (Collingwood, ON, Canada). wCSL = whole CSL; cCSL = centrifugally clarified CSL; % db = percent dry basis; ND = not determined.

provided to us at different times from a local corn wet-milling operation. The results of our compositional analysis of CSL are summarized in Table 2. Another distinguishing feature of this study was the use of cCSL. The high turbidity of the cell-free medium that is caused by the solids (insolubles) of the CSL supplement compromises the turbidometric measurement of growth, and the use of cCSL overcame this problem. The wCSL contained 44.2% solids (i.e., 44.2% by weight is dry substance) and 40% (db) insolubles that could be removed by centrifugation (Table 2)—the packed volume of the insoluble matter was 25%. In experiments designed to test for minimal levels of nutritional supplementation by a complex adjunct, such as CSL, it is important to restrict (minimize) the transfer of nutrients during inoculation. For this reason, in this work, we adopted a procedure of using centrifugally harvested cells as inoculum (*see* Materials and Methods).

Figure 2 shows typical time-courses for growth and glucose utilization by *Z. mobilis* CP4 in a medium consisting solely of TW and 4% (w/v) glucose that was amended with different amounts of cCSL over the range 4–20 mL/L. The pH was controlled at 5.0, and the temperature was 30°C. For comparative purposes, ZM1 medium (4% glucose) was used as the "control"

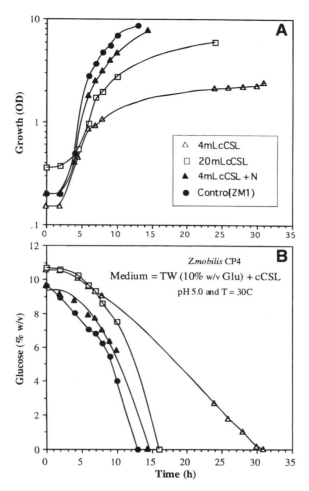

Fig. 2. Time-course of batch fermentations with Z. mobilis CP4. **(A)** Growth and **(B)** glucose utilization. The reference (control) medium was ZM1. The CSL medium contained either 4 or 10 mL of cCSL/L TW. In one case, ammonium chloride (1.6 g/L) was used as a N supplement. All media contained about 10% (w/v) glucose. The pH and temperature were kept constant at 5.0 and 30°C, respectively.

(Fig. 2). In all cases, the glucose-to-ethanol conversion efficiency was ≥98% theoretical maximum (results not shown). At a volumetric supplementation of 2%, the fermentation performance was similar to that achieved in the semisynthetic reference medium (Fig. 3). Addition of inorganic nitrogen (NH_4Cl) proved to be an effective means of decreasing the level cCSL required for growth and fermentation (Fig. 2). In a similar fashion, Fig. 3 shows the response by Z. mobilis CP4, in terms of cell mass and rate of glucose utilization, to the amount (mL) of cCSL added to a medium consisting solely of autoclaved TW and either 4 or 10% glucose. At the lower glucose concentration, the cell mass reaches a plateau of 1.6 g DCM/L (correspond-

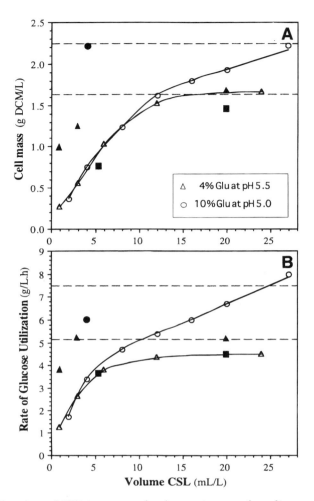

Fig. 3. Effectiveness of CSL in terms of volumetric rate of medium amendment. **(A)** Cell mass and **(B)** rate of glucose utilization. The media consisted TW with either 4% (w/v) or 10% (w/v) glucose with pH controlled at 5.5 and 5.0, respectively. The filled square symbols represent fermentations conducted with wCSL, and in all other cases, CSL was used. In some fermentations, the medium was supplemented with ammonium chloride: the filled triangles represent addition of 1.2 g/L to the 4% glucose-cCSL medium, and the filled circle represents the addition of 1.6 g/L to the 10% glucose-cCSL medium. The dashed lines represent the levels observed with the semisynthetic reference medium (ZM1) with either 4 or 10% glucose.

ing to the cell density achieved with the semisynthetic reference medium) at about 15 mL cCSL (1.5% v/v) (Fig. 3A). In the case of the rate of glucose utilization, the maximal rate is achieved at about half the amount of cCSL, but is less than the rate observed with the ZM1 medium (Fig. 3B). The solid square symbols in Fig. 3 represent the addition of 5.5 and 20 mL of nonclarified CSL (wCSL) to a TW (4% glucose) medium. Although the cell mass level using wCSL is lower than for comparable volumes of cCSL (Fig. 3A), the rate

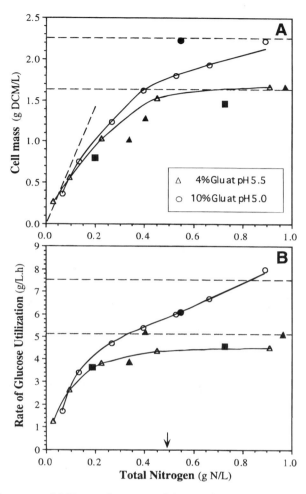

Fig. 4. Effectiveness of CSL as a function of the total nitrogen content of the medium. **(A)** Cell mass and **(B)** rate of glucose utilization. The conditions and symbols are the same as for Fig. 3. The x-axis arrow indicates the total N content of the reference medium (ZM1). The tangential dashed line in panel A shows the predicted response in terms of cell mass and is based on 100% assimilation of the N in the medium and *Zymomonas* dry mass assayed at 13.6% N.

response appears indifferent (Fig. 3B). From Fig. 3 it can be estimated that it would take about 1.3 times more CSL to achieve the same cell density as cCSL. With 10% glucose, the amount of cCSL required to achieve a cell density and fermentation rate comparable to the reference medium is increased to >25 mL/L with an apparent linear dependency for both responses in the range 12–25 mL cCSL/L (Fig. 3). In the Fig. 3, the filled triangles and circle represent experiments where the medium was fortified with NH_4Cl. For low-level cCSL supplementation, addition of inorganic nitrogen increases both the cell density and the rate of glucose utilization (Fig. 3).

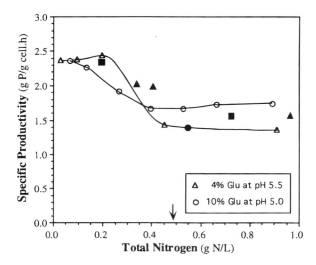

Fig. 5. Specific productivity as a function of the total N content of the medium. The conditions and symbols are the same as for Fig. 3. The *x*-axis arrow indicates the total N content of the reference medium (ZM1). The specific productivity was estimated as the ratio of the average rate of glucose utilization to the cell mass concentration, multiplied by the ethanol mass yield (which was relatively constant at 0.5 g/g).

The batch fermentations shown in Fig. 3 employed two different samples of CSL, and for the purpose of standardizing the data, the responses were plotted as a function of the N content of the medium (Fig. 4). The arrow on the abscissa indicates the N content of the reference medium ZM1 (Fig. 4B). In general, Fig. 4 shows that both the cell mass and rate of glucose utilization respond in accordance with the level of N in the medium up to a level of about 0.5 g N/L. The fact that maximal cell density is not achieved at this level of N with 10% glucose may relate more to a deficiency with respect to an essential growth element that is different from N—one possibility is pantothenic acid for which *Zymomonas* is known to have a specific growth requirement *(23)*. In all fermentations, the ethanol mass yield was ≥0.50 g ethanol/g glucose. Figure 5 shows the response of *Z. mobilis* CP4 in terms of specific productivity (g ethanol/g DCM/h) to the N content of the cCSL media. The increase in specific productivity at the growth-limiting levels of N is confirmation of previous observations with N-limited growth of other cultures of *Zymomonas (11,18)* where this condition leads to an energetic uncoupling of growth and glucose utilization.

Whereas for the purpose of defining minimal level of nutrients, it is imperative to minimize nutrient transfer for the seed culture medium at the time of inoculation, from a practical perspective, seed cultures are introduced to the batch fermenter at a volumetric "pitch" rate of about 5–10%. To determine the effect of nutrient transfer, we conducted a series of fermentations in which variously produced seed cultures were trans-

CSL as Nutrition Adjunct for Zymomonas

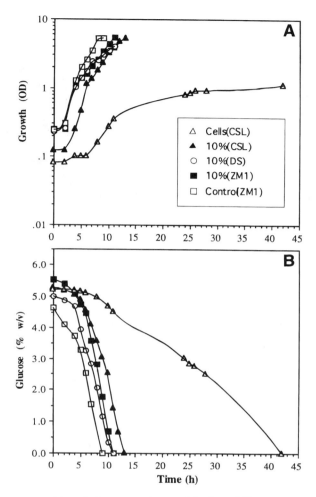

Fig. 6. Time-course of batch fermentations with Z. mobilis CP4. **(A)** Growth and **(B)** glucose utilization. The reference (control) medium was ZM1. The medium contained defined salts (DS), but was lacking the vitamins pantothenate and biotin (see Table 1). Three different media were used to generate the seed cultures for inoculation of the vitamin-deficient DS medium. For the purpose of seed production, the medium was (1) ZM1, (2) TW + 2% (v/v) cCSL, or (3) DS. A 10% (v/v) inoculum was used, except in one case where the seed culture grown in the CSL medium was harvested by centrifugation and the resuspended cells ("cells") were used as inoculum.

ferred into a minimal defined salts medium that did not contain the normal complement of vitamins. The following three media were used for the seed cultures:

1. TW with 20 mL/L cCSL;
2. A defined salts (DS) medium (Table 1); and
3. ZM1.

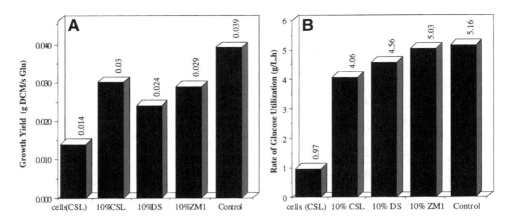

Fig. 7. Effect of nutrient transfer from seed culture on fermentation performance of *Zymomonas* in a defined salt medium lacking vitamins. **(A)** Growth yield and **(B)** rate of glucose utilization. The experimental design and conditions were as described in Fig. 6.

Figure 6 shows that surprisingly even the harvested cells could grow in the vitamin-deficient salts medium. The low level of growth may reflect a phenomenon known as "crossfeeding"—a form of microbial cannibalism. Where there was liquid culture used for inoculation, growth and fermentation were almost indistinguishable from the control using the relatively nutrient-rich ZM1 medium (Fig. 6). The responses in terms of cell mass concentration and glucose utilization rate are represented by the bar graphs in Fig. 7. These experiments clearly demonstrate that, at an inoculation ("pitch") rate of 10% (v/v), nutrient transfer is significant, and this practice could dramatically reduce the requirement for nutrient supplementation of the bulk fermentation medium.

Economic Impact of Using CSL as a Nutritional Adjunct in Large-Scale Fermentations

In terms of formulating a cost-effective fermentation medium that contains saturating yet minimal amounts of the essential elements for growth of the biocatalyst, cost reduction is a relatively facile exercise when the reference medium is comprised of research-grade lab chemicals. For example, the estimated cost of ZM1 medium, based solely on the use of industrial-grade yeast extract (US $3.30/lb), is US $21.80/1000 L. In a recent economic study based on a biomass-to-ethanol process using recombinant *E. coli*, it was suggested that inorganic chemicals could supply all the required elements for growth, and from the figures quoted in the report, the cost of the medium was determined to be 71¢/1000 L *(44)*. Although the efficacy of such a defined medium formulation was not tested, it still represents a cost of US $0.106/gal of denatured ethanol. According to current economic considerations for improving process efficiency, such a high cost for nutrients is viewed as unacceptable *(20)*.

Low-level use of CSL as a sole nutritional supplement in large-scale fermentation operations represents an opportunity for apparent significant cost reduction with recent independent estimates being US $0.042/gal ethanol in a biomass-to-ethanol process using hydrolyzates that contained 6–8% (w/v) fermentable sugar and a conversion efficiency of 90% *(30,39)*. However, both these estimates were erroneously based on CSL costing US $50/dry ton and consequently represent an underestimated cost impact associated with the use of CSL as a sole nutrient supplement. In reality, the cost impact should have been about double that estimated previously, namely, US $0.084/gal of ethanol.

Based on the nitrogen content of *E. coli*, Grethlein and Dill *(39)* have estimated that it would require 7 g dry wt (DW) CSL/L to achieve a cell density of 3 g DCM/L. At about 40–45% (db) protein; the N content of 7 g dry CSL would be about 0.45 g N ($7 \times 0.4 \times 0.16$), and at an assumed N content for the cell mass of 15% (w/w), the cell mass would represent 0.45 g N. For CSL costing US $50/t, supplementation at a rate of 7 g DW/L amounts to a cost of 77¢/1000 L of fermentation medium. The literature contains several reports of the fermentation media formulations involving CSL, but in the majority of these investigations, economics was not considered, and there was no attempt to define minimal levels compatible with high fermentation performance. For example, Amartey and Jeffries *(35)* used 28 g/L CSL to replace YE and other nutrients in xylose fermentations by *Pichia stipitis*. In fermentations of corn crop residues by recombinant *E. coli* KO11, the medium was supplemented with 2% (v/v) wCSL (equivalent to 24 g total mass of CSL/L) *(37)*, or alternatively, at a rate 1–5% (v/v) in conjunction with a crude yeast autolyzate as an additional nutritional supplement, the exact amounts of each component was not specified *(40)*. In all of these studies, it was concluded that CSL was an effective substitute for expensive YE and protein hydrolyzate additives.

For CSL selling at US $50/t, medium containing 1% (v/v) cCSL would cost 88¢/1000 L. The cost of using nonclarified (wCSL) at the same supplementation rate of 1% (v/v) would be 66¢/1000 L. However, we observed that equivalent performance is achieved with proportionately higher levels of CSL, and consequently, the cost of using wCSL at 1.3% (v/v) is the same. However, 40% of the solids are insoluble and could potentially be sold back into the animal feed market for a byproduct credit, thereby reducing the cost of using 1% cCSL to 52.8¢/1000 L. If wCSL is used, the protein content of the insolubles will contribute both to the mass and to the feed value of the fermentation residuals.

The economic impact that is associated with the use of CSL as a sole nutritional supplement in terms of cost per gallon of ethanol produced depends on the sugar loading (Fig. 8). At a sugar loading of about 10% (w/v), the cost is 5.2¢/gal undenatured ethanol (Fig. 8). This cost is based on:

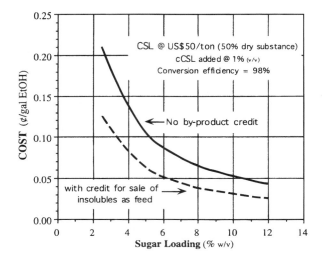

Fig. 8. Economic impact associated with use of CSL. Cost in terms of ¢/gal of undenatured ethanol was based on (1) using CSL as sole nutrient source, (2) CSL at US $50/t (50% solids), (3) supplementation rate using clarified CSL at 1% (v/v), (4) sugar-to-ethanol conversion efficiency of 98% by *Zymomonas*, (5) 100% product recovery, and (6) no by product credit for the sale of removed CSL solids (insolubles). The dashed line shows cost based on a by product credit for the sale of the CSL insolubles as feed at US $100/t.

1. CSL at US $50/t ("50% solids"—that is, 50% dry substance and therefore costing US $100/dry t);
2. The use of cCSL at a level of 1% (v/v);
3. Sugar-to-ethanol conversion efficiency by *Z. mobilis* of 98%; and
4. 100% product recovery.

An interesting aspect of this cost analysis is that although the cost of CSL is based on solids, 40% of these solids (i.e., the insolubles) are not used. Therefore, the cost impact is reduced by 40% (see dashed line in Fig. 8) if there is a byproduct credit for selling the insolubles as animal feed at a price of about US $100/t. Therefore, at a sugar loading of 10%, the cost of using 1% cCSL is reduced from 5.2 – 3.1¢/gal.

It can be concluded that low level CSL supplementation can supply the nutritional requirements compatible with growth and high-performance fermentation *of Zymomonas*, and that the disposition of the CSL insolubles can impact the calculations of cost associated with the use of CSL as a nutritional adjunct in large-scale fermentations.

ACKNOWLEDGMENTS

This work was funded by the University of Toronto. The authors are grateful to NACAN Products Limited for supplying samples of CSL.

REFERENCES

1. Keim, C. R. and Venkatasubramanian, K. (1989), *TIBTECH*, vol 7, Elsevier Science, UK, London, pp. 22–29.
2. Keim, C. R. (1983), *Enzyme Microbiol. Technol.* **5,** 103–114.
3. Lynd, L. R. (1990), *Appl. Biochem. Biotechnol.* **24/25,** 695–719.
4. Lynd, L. R., Cushman, J. H., Nichols, R. J., and Wyman, C. E. (1991), *Science* **251,** 1318–1323.
5. Skoog, K. and Hahn-Hägerdal, B. (1988), *Enzyme Microbiol. Technol.* **10,** 66–88.
6. McMillan, J. D. (1994), in *Enzymatic Conversion of Biomass for Fuels Production*, Himmel, M. E., Baker, J. O., and Overend, R. A., eds., American Chemical Society, Washington, DC., *ACS Symposium Series* 566, pp. 411–437.
7. Swings, J. and DeLey, J. (1977), *Bacteriol. Rev.* **41,** 1–46.
8. Montenecourt, B. S. (1985), in *Biology of Industrial Microorganisms*, Demain, A. L. and Simon, N. A., eds., Benjamin/Cummings, Meno Park, CA, pp. 216–287.
9. Baratti, J. C. and Bu'Lock, J. D. (1986), *Biotechnol. Adv.* **4,** 95–115.
10. Rogers, P. L., Lee, K. J., Skotnicki, M. L., and Tribe, D. E. (1982), *Adv. Biochem. Eng.* 37–84.
11. Lawford, H. G. and Stevnsborg, N. (1986), *Biotechnol. Bioeng.* **Symp. 17,** 209–219.
12. Lawford, H. G. and Ruggiero, A. (1990), in *Bioenergy* (Proceedings 7th Canadian Bioenergy R&D Seminar), Hogan, E., ed., National Research Council of Canada, Ottawa, Canada, pp. 401–410.
13. Busche, R., Scott, C. D. Davison, B. H., and Lynd, L. R. (1992), *Appl. Biochem. Biotechnol.* **34/35,** 395–417.
14. Doelle, H. W., Kirk, L., Crittenden, R., Toh, H., and Doelle, M. (1993), *Crit. Rev. Biotechnol.* **13,** 57–98.
15. Lawford, H. G. (1988), *Proceedings of VIII International Symposium on Alcohol Fuels*, New Energy Development Organization, Tokyo, Japan, (November 13–16), pp. 21–28.
16. Picataggio, S. K., Zhang, M., and Finkelstein, M. (1994), in *Enzymatic Conversion of Biomass for Fuels Production*, M. E. Himmel, J. O. Baker, and R. A. Overend, eds., American Chemical Society, Washington, DC, *ACS Symposium Series* **566,** pp. 342–362.
17. Zhang, M., Franden, M. A., Newman, M., McMillan, J., Finkelstein, M., and Picataggio, S. K. (1995), *Appl. Biochem. Biotechnol.* **51/52,** 527–536.
18. Lawford, H. G. (1988), *Appl. Biochem. Biotechnol.* **17,** 203–211.
19. Greasham, R. and Inamine, E. (1981), in *Manual of Industrial Microbiology and Biotechnology*, Demain, A. L. and Solomon, N. A, eds., American Society for Microbiology, Washington, DC, pp. 41–48.
20. Lynd, L. R., Elander, R. T., and Wyman, C. E. (1996), *Appl. Biochem. Biotechnol.* **57/58,** 741–761.
21. Belaïch, J. P. and Senez, J. C. (1965), *J. Bacteriol.* **89,** 1195–1200.
22. Belaïch, J. P., Belaïch, A., and Simonpietri, P. (1972), *J. Gen. Microbiol.* **70,** 179–185.
23. Lawford, H. G. and Stevnsborg, N. (1986), *Biotechnol. Lett.* **8,** 345–350.
24. Park, S. C., Kademi, A., and Baratti, J. C. (1993), *Biotechnol. Lett.* **15,** 1179–1184.
25. Goodman, A. E., Rogers, P. L., and Skotnicki, M. L. (1982), *Appl. Environ. Microbiol.* **44,** 496–498.
26. Fein, J. E., Charley, R. C., Hopkins, K. A., Lavers, B., and Lawford, H. G. (1983), *Biotechnol. Lett.* **5,** 1–6.
27. Nipkow, A., Beyeler, W., and Feichter, A. (1984), *Appl. Microbiol. Biotechnol.* **19,** 237–240.
28. Galani, I., Drainas, C., and Typas, M. A. (1985), *Biotechnol. Lett.* **7,** 673–678.
29. Anon (1975), "Properties and Uses of Feed Products from Corn Wet-Milling Operations." Corn Refiners Association Inc., Washington, DC.
30. Lawford, H. G. and Rousseau, J. D. (1996), *Appl. Biochem. Biotechnol.* **57/58,** 307–326.

31. Lawford, H. G. (1988), in *Canadian Power Alcohol Proceedings* (CANPAC'88), Biomass Energy Institute of Canada, Winnipeg, Manitoba, pp. 245–251.
32. Beavan, M., Zawadzki, B., Droiniuk, R., Fein J, E., and Lawford, H. G. (1989), *Appl. Biochem. Biotechnol.* **20/21,** 319–326.
33. Davison, B. H. and Scott, C. D. (1988), *Appl. Biochem. Biotechnol.* **18,** 19–34.
34. Webb, O. F., Davison, B. H., Scott, T. C., and Scott, C. D. (1994), *Appl. Biochem. Biotechnol.* **51/52,** 559–568.
35. Amartey, S. and Jeffries, T. W. (1994), *Biotechnol. Lett.* **16,** 211–214.
36. Kadam, K. L., Hayward, T. K., and Phillippidis, G. P. (1995), *ASME Solar Eng.* **1,** 339–347.
37. Beall, D. S., Ingram, L. O., Ben-Bassat, A., Doran, J. B., Fowler, D. E., Hall, R. G., and Wood, R. E. (1992), *Biotechnol. Lett.* **14,** 857–862.
38. Barbosa, M. de F. S., Beck, M. J., Fein, J. E., Potts, D., and Ingram, L. O. (1992), *Appl. Environ. Microbiol.* **58,** 1182–1184.
39. Grethlein, H. E. and Dill, T. (1993), SCA No. 58-1935-2-050, Agricultural Research Service, USDA, Philadelphia, PA.
40. Asghari, A., Bothast, R. J., Doran, J. B., and Ingram, L. O. (1996), *J. Ind. Microbiol.* **16,** 42–47.
41. Lawford, H. G., Rousseau, J. D., and McMillan, J. D. (1997), *Appl. Biochem. Biotechnol.* (18th Symp.), **63–65,** 269.
42. Lawford, H. G., Holloway, P., and Ruggiero, A. (1988), *Biotechnol. Lett.* **10,** 809–814.
43. Lawford, H. G. and Ruggiero, A. (1990), *Biotechnol. Appl. Biochem.* **12,** 206–211.
44. von Sivers, M., Zacchi, G., Olsson, L., and Hahn-Hägerdal, B. (1994), *Biotechnol. Prog.* **10,** 555–560.

Astaxanthinogenesis in the Yeast *Phaffia rhodozyma*

Optimization of Low-Cost Culture Media and Yeast Cell-Wall Lysis

JOSÉ D. FONTANA,*,[1] MIRIAM B. CHOCIAL,[2]
MADALENA BARON,[1] MANOEL F. GUIMARAES,[1]
MARCELO MARASCHIN,[1] CIRANO ULHOA,[3]
JOSE A. FLORÊNCIO,[1] AND TANIA M. B. BONFIM[2]

[1]*LQBB-Biomass Chemo Biotechnology Laboratory/Biochemistry/UFPR, P.O. Box 19046 (81531-990), Curitiba-PR-Brazil;* [2]*Pharmacy Course/UFPR; and* [3]*Celular Biology Department, UnB-Brasilia-DF-Brazil*

ABSTRACT

Astaxanthin is a diketo-dihydroxy-carotenoid produced by *Phaffia rhodozyma*, a basidiomicetous yeast. A low-cost fermentation medium consisting of raw sugarcane juice and urea was developed to exploit the active sucrolytic/urelolytic enzyme apparatus inherent to the yeast. As compared to the beneficial effect of 0.1 g% urea, a ready nitrogen source, mild phosphoric pre inversion of juice sucrose to glucose and fructose, promptly fermentable carbon sources, resulted in smaller benefits. Corn steep liquor (CSL) was found to be a valuable supplement for both yeast biomass yield (9.2 g dry cells/L) and astaxanthin production (1.3 mg/g cells). Distillery effluent (vinace), despite only a slightly positive effect on yeast growth, allowed for the highest pigment productivity (1.9 mg/g cells). Trace amounts of Ni^2 (1 mg/L, as a cofactor for urease) resulted in controversial effects, namely, biomass decrease and astaxanthin increase, with no effect on the release (and uptake) of ammonium ion from urea. Since the synthesized astaxanthin is associated with the yeast cell and the pigment requires facilitated release for aquaculture uses (farmed fish meat staining), an investigation of the yeast cell wall was undertaken using detergent-treated cells. The composition of the rigid yeast envelope

*Author to whom all correspondence and reprint requests should be addressed.

was found to be heterogeneous. Its partial acid or enzymatic depolymerization revealed glucose and xylose as common monomeric units of the cell-wall glycopolymers. Yeast cell-wall partial depolymerization with appropriate hydrolases may improve the pigment bioavailability for captive aquatic species and poultry.

Index Entries: *Phaffia*; astaxanthin; corn steep liquor; distillery effluent; cell wall.

INTRODUCTION

The basidiomicetous yeast *Phaffia rhodozyma* has gained an outstanding place in the industrial microbiology market (>100 million US$/year) solely by its peculiar biochemical ability to synthesize the highly oxygenated carotenoid astaxanthin *(1)*.

There are two main applications for this biotechnological product, aquaculture (farmed salmonid meat staining) *(2)* and orthomolecular medicine (as a quencher/scavenger of active oxygen species) *(3,4)*. Our first report on the subject dealt with the convenient and productive utilization of 1:10 diluted raw sugarcane juice and reduced supplementation with urea (0.1 g%) for *P. rhodozyma* growth and astaxanthin production *(5)*. The preceding report focused on a low-cost media improvement using soya meal and tannery shavings as alternative nitrogen sources to urea and enzymolyzed starches or ligno(hemi)cellulosics as disaccharide-enriched carbon sources. These substrates take advantage of the polyvalent fermentative capability of this microrganism *(6)*.

Several recent contributions on the basic and technological aspects of carotenoid production deserve mention. The astaxanthin deposition inside the yeast cell was clarified through cytofluorometry *(7)*. By comparison, a small protein (39 kD) was isolated as the wrapper element for the β-carotene globules observed in the halotolerant alga *Dunalliela bardawil (8)*. Genetic characterization and engineering of natural and cloned carotenoid producers are being examined. The electrophoretic karyotype patterns of three *P. rhodozyma* strains recalled its description as a single genus/species *(9)* and revealed 9–17 chromosomal bands. *Escherichia coli* recombinants were obtained after separated and sequential steps with the insertion of particular genes from *Erwinia* and then from the marine bacteria *Agrobacterium* or *Alcaligenes*. *E. coli*(r) incorporated the biochemical ability for the production of 303–494 µg of canthaxanthin (a diketo-carotenoid)/g cells *(10)*.

Two aspects of *P. rhodozyma* growth and astaxanthin production stand as valid prospects on carotenoid (bio)technology: culture media optimization owing to its alternative respiratory and fermentative metabolism *(1)* and yeast cell (envelope) lysis for the facilitated pigment release. New contributions for both aspects are presented in this article.

MATERIALS AND METHODS

Yeast Source, Maintenance, Growth in Shaken Cultures, and Routine Analytical Procedures

All small-scale procedures followed in this article use the guidelines already described in our previous publications *(5,6)*. For inversion, samples of sugarcane juice (1:10 dilution; total sugar content 2.0–2.5 g%) were adjusted to several pH values (4.5–1.5) with aqueous phosphoric acid, boiled for 30 min., and then titrated with aqueous ammonia to adjust the pH to 6.0 prior to yeast inoculation. The catalyst remained thus incorporated in the fermentation broth as ammonium phosphate. The control also received ammonia adjustment until pH 6.0. Corn steep liquor (CSL; 40 g% solids content) was provided by Refinacoes de Milho Brasil (Balsa Nova-PR, Brazil). Novo Nordisk hydrolases (Novozym 234 and SP-299 Mutanase) were a gift from E. Bordin. A *Trichoderma* sp. (ex-Brazilian "Cerrados") enzyme prepared by C. Ulhoa, UnB-Brasilia-DF, Brazil was also used. Gastric juice enzymes from the snail *Megalobulimus paranaguensis* were prepared at LQBB/UFPR.

Other Analytical Procedures

Ammonium ion, protein, and total carbohydrate were quantified by the Nessler, Coomassie (Bradford), and phenol-sulfuric reagents *(11)*, respectively. Optical and electronic micrography was carried out either in a Nikon/Labophot AFX-II (Tokyo, Japan) or a Philips SEM-505 apparatus (Amsterdam, The Netherlands). For the optical imaging, yeast cells or cell walls were prestained with 0.25% Coomassie blue, followed by destaining with methanol:acetic acid:glycerol:water (46:46:2:6) *(12)*. For scanning micrography, the biomaterial was fixed, buffered, and metalized (glutaraldehyde, Na^+ cacodilate, OsO_4, gold dust) according the classical technique. Acid hydrolysis of yeast cell walls or fractions was carried out with 2 or $4M$ trifluoroacetic acid (TFA) at 100°C from 1.5–5.0 h. The released sugars were analyzed by thin-layer chromatography (TLC) on silica-gel 60 plates (Merck, Darmstadt, Germany) using isopropanol:ethyl acetate:nitroethane:acetic acid:water (45:25:10:1:19) as mobile phase and staining with sulfuric orcinol or *p*-anisaldehyde *(11)*. Monosaccharide sylil derivatives were resolved by gas-liquid chromatography (GLC) on a HP-5 (Hewlett-Packard, Wilmington, DE) column programmed from 160–260°C (at 10°C/min) with an FID detector and using silylated *p*-nitro-Ø-D-galactopyranoside as an internal standard.

RESULTS AND DISCUSSION

P. rhodozyma growth in 1:10 diluted crude sugarcane juice alone (2.0–2.5 g% sugars) resulted in low biomass yield (<0.2 g cells dry wt/L) and intermediate astaxanthin content (0.5 mg/g cells) (Fig. 1; first bars of

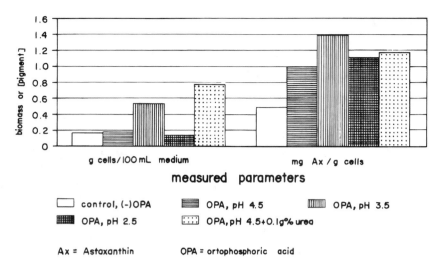

Fig. 1. Effect of OPA-mediated pre-inversion of raw diluted sugarcane on *P. rhodozyma* growth and astaxanthin production, OPA = orthophosphoric acid. All culture media, prior to yeast inoculation for growth at 25°C and 100 rpm for 96 h, were partially neutralized to pH 6.0 with aqueous ammonia. Yeast growth measured as dry wt biomass. Astaxanthin spectrophotometrically quantified at 478 nm.

both parameters). These control values were significantly increased as a result of a previous juice pretreatment, i.e., its mild inversion (sucrose → glucose + fructose) with drops of aqueous phosphoric acid (OPA), followed by a 30-min boiling. Of the pH values tested, best results for promoting growth were obtained at pH 3.5. This corresponded to biomass and astaxanthin yields of 0.54 g cells/L and 1.4 mg of pigment/g cells, respectively (Fig. 1; third bars). Deviation to either milder (pH 4.5) or stronger (pH 2.5) hydrolytic conditions resulted in diminished yields either to biomass or pigment, although a greater amount of reducing sugars was initially released at pH 2.5. Thus, a positive role for the catalyst itself (OPA → ammonium phosphate under neutralization with ammonia) cannot be ruled out. In addition, the supplementation of the mildest hydrolysate (pH 4.5) with 0.1 g% urea (last bars for both measured parameters) as compared to the urea-free control (second bars) implied an increase in both biomass and astaxanthin yields. This may be explained by the lesser amount of ammonia (a nitrogen source) employed for the neutralization of inversion done at pH 4.5. OPA inversion at pH 1.5 led to juice browning and yeast inhibition.

CSL (Fig. 2) was evaluated as an alternative to urea or ammonium phosphate, since the beneficial effect of complex nitrogen sources, like soya bean and tannery shavings proteolyzates (5), was already established. With respect to the control (first bars of the two blocks), and considering that all media were adjusted to pH with ammonia, CSL (0.2 mL% addition;

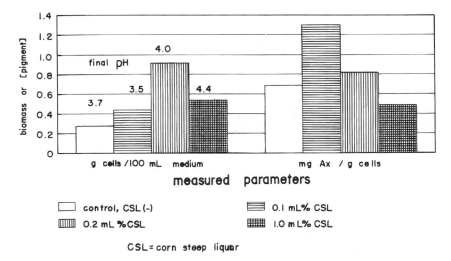

Fig. 2. Effect of CSL supplementation of sugarcane juice on *P. rhozoyma* growth and astaxanthin production (CSL = 40 g% solids; other details as in Fig. 1).

third bars) caused a maximum enhancement of 3.2-fold in biomass (almost 10 g cells/L) and 1.9-fold in astaxanthin (1.3 mg/g cells). Five to 10-fold increase CSL supplementation proved inhibitory for both yields. It may be also pointed out that moderate CSL addition (0.1 → 0.2 mL%) resulted in a progressively better consumption of the juice sugars by the yeast and better values for the biomass dry weights.

Distillery effluent (vinace), which is a major polluting industrial liquid, has found only limited application as a soil cofertilizer (K^+ supply for coffee and cane crops). It contains some mevalonic acid, a precursor for isoprenoid biosynthesis *(13)*. Addition of vinace (0.5%, v/v) to raw sugarcane juice resulted in improved astaxanthin productivity (1.9 mg/g cells; third bar, second block, Fig. 3), despite a smaller effect on biomass enhancement (1.6-fold; fourth bar, first block; Fig. 3) as compared to that obtained with CSL.

Nickel may be a bound metallic cofactor for urease. Since this enzyme plays an important role for nitrogen uptake by *Phaffia*, the effect of this cation was investigated in the 1–10 mg/L range. Experiments following the inclusion of dimethyl glyoxime (DMG), (stoichmetric amounts of 4–40 mg/L), an Ni^{2+} chelator *(11)* in the yeast culture media (irrespective the presence or absence of the cation), resulted in growth inhibition for the yeast. However, when Ni^{2+} was administered alone (1 mg/L), the biomass was reduced, but astaxanthin doubled (second bars, both blocks; Fig. 4). Addition of urea (0.1 g%) ensured an apparent better yeast performance for both parameters (third bars, both blocks; Fig. 4). This may be explainable by the effect of urea alone as indicated by the respective control (no Ni^{2+}

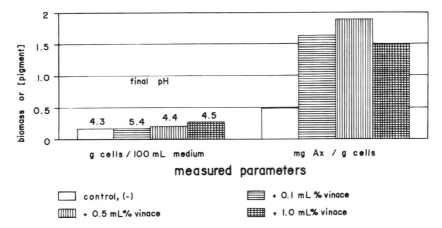

Fig. 3. Effect of distillery effluent (vinace) supplementation of sugarcane juice on *P. rhodozyma* growth and astaxanthin production. (Details as in Fig. 1.)

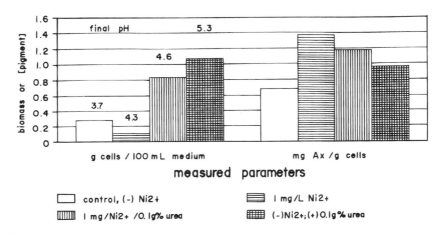

Fig. 4. Effect of Ni^{2+} supplementation of sugarcane juice containing or not containing urea on *P. rhodozyma* growth and astaxanthin production. (Details as in Fig. 1.)

addition; compare the first and last bars in each block; Fig. 4). Urea-derived ammonium ion was monitored (Nessler reagent) in the time-course of the fermentation from 2 until 96 h. There was only a small difference between the experiments with or without 1 mg/L Ni^{2+}. Urea hydrolysis and ammonium ion onset started in both cases at 6 h and reached a maximum at 48 h (30.9 mg% NH_4^+ for the experiment with 0.1% urea alone) and at 72 h (29.3 mg% NH_4^+ for the experiment with urea + Ni^{2+}). Both experiments finished (96 h) with an equivalent free NH^{4+} content in the culture media (17.6 and 17.9 mg%).

P. rhodozyma cells had previously proven very resistant to mechanical rupture (French Press or ultrasound procedures (5) for the purpose of cell lysis and invertase/urease release). One gram of yeast cells was rendered free of membrane and cytoplasmatic components (low-mol-wt metabolites, nucleic acids, proteins) through exhaustive washings with the detergent SDS (100 mL 2 g% sodium dodecylsulfate; threefold overnight agitated extractions). The extract was spectrophotometrically monitored at 260 (nucleic acids) and 280 nm (proteins). A reduction from 12–13 OD units (first cycle) to about 1/10 to 1/20 (third cycle) occurred. The aspect of fresh and SDS-washed (and also alkali-treated) cell walls can be seen in Fig. 5A and B. SDS cell walls were then fractionated by sequential 6-h extractions with half-saturated lithium isothiocyanate (a chaotropic agent) (I), and at room temperature (II) or boiling 5M KOH (III). All fractions produced polymeric material, which was then precipitated with 3 vol of ethanol. Yields were 125 mg (I), 155 mg (II), 115 mg (III), and final residue (405 mg), representing an 80% recovery. All of them also gave positive reactions for protein (Bradford reagent) and total carbohydrate (sulfuric phenol reagent). The chaotropic extractant yielded the richest protein fraction, and hot alkali produced the richest carbohydrate-containing preparation. The partial strong acid hydrolysis (TFA) of these materials indicated in every instance the dominant presence of xylose and glucose in the hydrolyzates, except for fraction III (hot alkali), which was richer in another hexose(s). GLC analyses (Fig. 6) confirmed the TLC initial data on sugar composition. The fraction arising from room temperature KOH extraction was preliminarily inspected by ^{13}C-nuclear magnetic resonance (NMR). Low-field signals at 105.1 and 104.5 ppm corresponded to to C-1 of β-D-glucopyranosyl and D-xylopyranosyl rings as nonreducing units (results not shown). Since in the time-course of TFA hydrolysis of whole SDS cell walls from 1.5 to 5.0 h, xylose + glucose were always detected together and in progressive amounts, a xyloglucan is indicated as a component of *P. rhodozyma* cell wall. SDS cell walls were also submitted to enzymatic attack with *Trichoderma* spp. (poly)hydrolases (*see* Materials and Methods; overnight incubation at 35°C). "Cerrados" *Trichoderma* sp. enzyme was more effective for sugar release on the final residue than on the original SDS cell walls. Glucose was found as the major released monosaccharide in all enzymatic incubations. Snail enzyme also released minor amounts of other sugars (mannose and less probably rhamnose). Neither xylose or *N*-acetyl-glucosamine was found by TLC in the enzymic digests. In *Phaffia* SDS cell walls chitin (resistant to TFA hydrolysis) remains to be characterized.

Since residual SDS cell walls remained pink-colored (despite the chaotropic and strong alkali pretreatments), a preliminary supercritical fluid extraction (SFE) was attempted on fresh yeast cells (about 1 mg astaxanthin/g cells) using a homemade steel reactor vessel and thermopressurized CO_2 just above the critical conditions for this gas (45°C and 80 atm). The extraction, under this condition, was not efficient. The procedure is being repeated under higher thermopressurization conditions

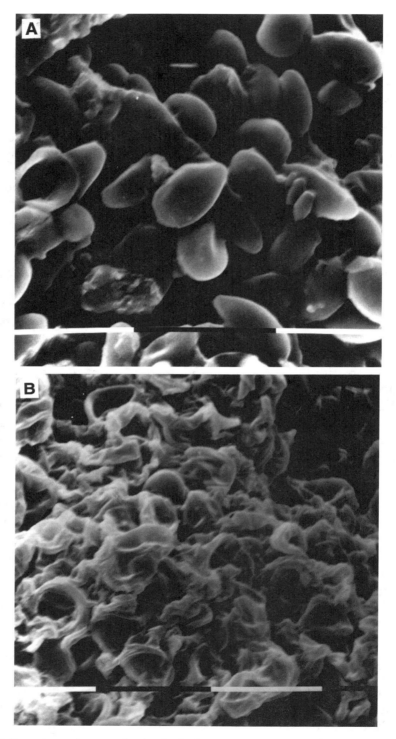

Fig. 5. Scanning micrography for *P. rhodozyma* fresh cells **(A)** and SDS/cold alkali-treated cell walls **(B)** (Nominal magnification: ×11,300).

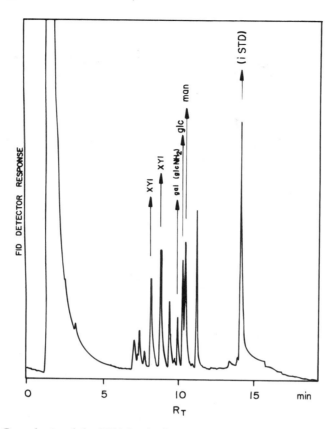

Fig. 6. GLC analysis of the TFA hydrolyzate of *P. rhodozyma* SDS cell walls (3.5 h of 2*M* TFA hydrolysis at 100°C; direct presilylated sugar derivatives; *see* Materials and Methods).

and including organosolvent moderators (e.g., methanol). The results will be published elsewhere.

CONCLUSIONS

The optimization of low-cost culture media for *P. rhodozyma* growth and astaxanthin production was attained through the addition of two new conutrients, CSL and distillery effluent. The superior sucrolytic/ureolytic activity possessed by this yeast, both genetically and physiologically well expressed, indicated that preinversion of sugarcane juice (sucrose) may not be essential for fermentation improvement. Purified yeast cell walls and their isolated subfractions revealed a heterogeneous nature concerning the polymeric carbohydrate architecture with an apparent dominance of xylosyl and glucosyl units. Cell-wall lysis and facilitated pigment release will require improvements concerning degradative enzyme induction/use or supercritical extraction procedures.

ACKNOWLEDGMENTS

The authors thank CNPq/PADCT-SBIO and CAPES (Brazilian funding agencies), J. A. Florencio (technical assistance), and M. Lima (drawings).

REFERENCES

1. Johnson, E. A. (1995), *Stud. Mycology* **38,** 81–90.
2. Starebakken, T. and No, N. K. (1992), *Aquaculture* **100,** 209–229.
3. Yokoyama, A., Izunida, H., and Miki, W. (1994), *Biosci. Biotechnol. Biochem.* **58(10),** 1842–1844.
4. Lambert, C. R. (1995), *Proc. SPIE—Int. Soc. Opt. Eng.* **2391** (Laser-Tissue Interaction VI), 218–224.
5. Fontana, J. D., Chocial, M. B., Baron, M., Guimaraes, M. F., Marines, N. T., Fontana, C. A., Maraschin, M., Ulhoa, C., Florêncio, J. A. and Bonfim, T. M. B. (1996), *Appl. Biochem. Biotechnol.* **57/58,** 413–422.
6. Fontana, J. D., Czeczuga, B., Bonfim, T. M. B., Chocial, M. B., Oliveira, B. H., Guimaraes, and M. F., Baron (1996), *Biores. Technol.* **58(2),** 121–125.
7. Johnson, E. A. (1995), *Proc. 1995 Intl. Chem. Congress of Pacific Basin Soc.* Honolulu, Hawaii, ref. 339 (Agrochemistry).
8. Katz, A., Jimenez, C., and Pick, U. (1995), *Plant Physiol.* **108(4),** 1657–1664.
9. Miller, M. W., Yokoyama, M., and Soneda, M. (1976), *Int. J. Syst. Bacteriol.* **26,** 73.
10. Misawa, N., Kajiwaya, S., Kondo, K., Yokoyama, A., Satomi, Y., Saito, T., Miki, W., and Ohtani, T. (1995), *Biophys. Biochem. Res. Commun.* **209(3),** 867–876.
11. Dawson, R. M. C., Elliot, D. C., and Jones, K. M. (1986), *Data for Biochemical Research*, Clarendon, Oxford.
12. Fontana, J. D., Souza, A. M., Fontana, C. K., Torriani, I., Moreschi, J. C., Gallotti, B. J., Souza, S. J., Narciso, G. P., Bichara, J. A., and Farah, L. F. X. (1990), *Appl. Biochem. Biotechnol.* **24/25,** 253–264.
13. Armstrong, G. A. and Hearst, J. E. (1996), *FASEB J.* **10,** 228–237.

Polysaccharide Hydrolase Folds Diversity of Structure and Convergence of Function

MICHAEL E. HIMMEL,*,[1] P. ANDREW KARPLUS,[2]
JOSHUA SAKON,[2] WILLAM S. ADNEY,[1] JOHN O. BAKER,[1]
AND STEVEN R. THOMAS[1]

[1]*Biotechnology Center for Renewable Fuels and Chemicals,
National Renewable Energy Laboratory, Golden, CO 80401;
and* [2]*Section of Biochemistry, Molecular and Cell Biology
Cornell University, Ithaca, NY 14853*

ABSTRACT

Polysaccharide glycosyl hydrolases are a group of enzymes that hydrolyze the glycosidic bond between carbohydrates or between a carbohydrate and a noncarbohydrate moiety. Here we illustrate that traditional schemes for grouping enzymes, such as by substrate specificity or by organism of origin, are not appropriate when thinking of structure–function relationships and protein engineering. Instead, sequence comparisons and structural studies reveal that enzymes with diverse specificities and from diverse organisms can be placed into groups among which mechanisms are largely conserved and insights are likely to be transferrable. In particular, we illustrate how enzymes have been grouped using protein sequence alignment algorithms and hydrophobic cluster analysis. Unfortunately for those who seek to improve cellulase function by design, cellulases are distributed throughout glycosyl hydrolase Families 1,5,6,7,9, and 45. These cellulase families include members from widely different fold types, i.e., the TIM-barrel, βαβ-barrel variant (a TIM-barrel-like structure that is imperfectly superimposable on the TIM-barrel template), β-sandwich, and α-helix circular array. This diversity in cellulase fold structure must be taken into account when considering the transfer and application of design strategies between various cellulases.

---

*Author to whom all correspondence and reprint requests should be addressed.

Index Entries: Cellulases; xylanases; amylases; glycosyl hydrolases; structural folds; X-ray structures; hydrophobic cluster families.

INTRODUCTION

In nature, the enzymatic degradation of cellulose is a fundamental mechanism of biomass conversion and carbon cycling in the biosphere. To produce alcohol fuels from lignocellulosic biomass, however, this process must be understood at the molecular level to develop highly efficient and cost-effective catalysts. One important step toward this goal is to determine key structure–function relationships of enzymes displaying activity on water-insoluble substrates, such as cellulose and other plant polysaccharides. In order to define and understand complex molecular mechanisms, detailed structural information, such as that determined from NMR or X-ray diffraction studies, is essential.

Understanding the evolutionary and mechanistic relationships of enzymes that catalyze similar reactions is, therefore, a highly desirable objective for those who design new strategies to improve enzyme function by site-directed mutagenesis (SDM). An examination of the SWISS-PROT data base by Orengo and coworkers *(1)* using sequence alignments (FASTA) revealed that the ~28,000 entries were reducible to no less than about 7700 unambiguously related groups. Groupings with such limited sensitivity tell us little about enzyme relatedness and are not a serviceable tool for understanding enzyme design. A more sensitive method is hydrophobic cluster analysis (HCA), which relies on the basic rules that underlie the folding of globular proteins and uses a two-dimensional plot to display the amino acid sequence of a protein depicted as an "unrolled longitudinal cut" of a cylinder *(2)*. The "helical net" produced by this graphical display allows the full sequence environment of each amino acid to be examined. HCA has been an extremely powerful method for classifying glycosyl hydrolases *(2)*. Gilkes and coworkers *(3)* originally proposed nine families, based on the glycosyl hydrolase sequences available at that time, and in the ensuing five years, these researchers have added substantially to the original classification list *(4,5)*. Development of glycosyl hydrolase classification is shown chronologically in Fig. 1. Today, Bairoch *(6)* has identified 56 families. This classification system provides a powerful tool for glycosyl hydrolase enzyme engineering studies, because many enzymes critical for industrial processes have not yet been crystallized or subjected to structure analysis. This article will review the correlations between polysaccharidase function and fold structure based on existing family assignments and reported macromolecular structures.

Enzyme Structure

Protein domains are grouped into four general structural categories (all-α, all-β, $\alpha + \beta$, and α/β) *(7,8)*. Proteins of the all-α class are usually comprised of multiple α-helices that may be oriented either along a common

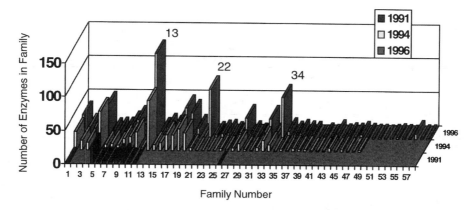

Fig. 1. The chronological development of O-glycosyl hydrolase classification by protein sequence alignment and hydrophobic cluster analysis (3,4,6).

bundle axis or randomly (9). Proteins of the all-β class contain β-strands that can be oriented either parallel or antiparallel, or in a mixture of the two (10). The α + β and α/β categories are distinguished by considering that α/β proteins have alternating β-strand and α-helical segments, whereas α + β proteins tend to contain regions definable as "mostly α" and "mostly β" (11). A common example of the α/β class is the TIM-barrel, named after the archetype of this fold, triose phosphate isomerase. In TIM-barrel proteins, the internal barrel is comprised of eight parallel β-strands, and the outer shell contains eight α-helices oriented with a cant relative to the axis of the barrel. Some protein domains do not fall into any of these categories and are grouped as irregular folds. Proteins representative of these domain (or fold) classes are myoglobin (all-α helix), immunoglobulin (all-β strand), cytochrome b5 (α + β), and triose phosphate isomerase (α/β) (7).

It is inferred that all proteins that have recognizable sequence similarity will have the same fold type. In many cases, the fold will be unique to that single family of proteins; such folds are known as structural singlets (1). In other cases, a domain structure (fold) may be shared by two or more proteins that appear unrelated by sequence and function. Such folds have been termed superfolds. Orengo and coworkers identified nine protein superfolds that were the α/β-doubly wound fold, the TIM-barrel, the split α/β-sandwich, the greek-key immunoglobulin fold, the up-down four-helix bundle, the globin fold, the β-jelly roll, the β-trefoil, and the ubiquitin αβ-roll (1).

Structure of Polysaccharide Hydrolases

A recent survey of the 3000 or more protein structures in the Brookhaven National Laboratory Protein Data Bank (PDB) revealed that only 6 cellulases, 8 xylanases, and 10 α-amylases, β-amylases, and glucoamylases have been deposited. A seventh cellulase structure was recently

Table 1
Distribution of Cellulases Among Glycosyl Hydrolase Families

Family	PDB file	Year filed	Res., Å	Fold class	Fold*	Ref.
Family 1	1cbg	1994	2.2	beta/alpha	TIM-barrel	16
Family 5	1cec	1995	2.2	beta/alpha	TIM-barrel	17
	1ece	1996	2.4	beta/alpha	TIM-barrel	12
Family 6	3cbh	1990	2	beta/alpha	TIM-barrel variant	18
	1tml	1993	1.8	beta/alpha	TIM-barrel variant	19
Family 7	1cel	1994	1.8	all beta	beta-sandwich, 12-14 strands	20
Family 9	1clc	1995	1.9	all alpha	6 alpha-helices/circular array	21
Family 45	1eng	1993	1.6	all beta	closed barrel, Barwin-like	22

*Assignment made by SCOP (15).

resolved for the endoglucanase EI from *Acidothermus cellulolyticus* (12). For the most part, the structural information available pertains to the catalytic domain only; however, two structures for cellulase cellulose binding domains (CBDs) have been filed with PDB (1cbh and 1exg). These structures are defined as small, all-β strand domains with 7 or 8 strands per molecule and were solved using 2D NMR techniques (13,14). This list will, of course, increase with time. Information regarding the structures of polysaccharide glycosyl hydrolases is shown in Tables 1–3. The structural classification given in these tables is based on recommendations from Structural Classification of Proteins (SCOP) (http://www.pdb.bnl.gov/scop) (15) and the recent glycosyl hydrolase family information taken from Bairoch, http://www.expasy.hcuge.ch/cgi-bin/lists?glycosid.txt (6).

Tertiary structure and key residues at active sites are generally better conserved than amino acid sequence, so it is no surprise that structural studies, combined with sequence comparisons directed at active site residues, have allowed many families to be grouped in clans that have a common fold and a common catalytic apparatus (39). Five such clans recently proposed for the glycosyl hydrolases are GH-A (including Families 1, 2, 5, 10, 17, 30, 35, 39, and 42); GH-B (Families 7 and 16); GH-C (Families 11 and 12); GH-D (Families 27 and 36); and GH-E (Families 33 and 34) (6). Among these, Clans GH-A and GH-B include cellulases.

DISCUSSION

The objective of this article is to compare the gross structural features recently made available for selected polysaccharide glycosyl hydrolases and to draw correlations with function. To accomplish this, the

Table 2
Distribution of Xylanases Among Glycosyl Hydrolase Families

Family	PDB file	Year file	Res.,Å	Fold class	Fold*	Ref.
Family 10	1xas	1994	2.6	beta/alpha	TIM-barrel	23
	1xyz	1995	1.4	beta/alpha	TIM-barrel	24
	1xys	1994	2.5	beta/alpha	TIM-barrel	25
	2exo	1994	1.8	beta/alpha	TIM-barrel	26
Family 11	1xyp	1994	1.5	all beta	beta-sandwich, 12-14 strand	27
	1xyn	1994	2	all beta	beta-sandwich, 12-14 strand	27
	1xnd	1994	1.8	all beta	beta-sandwich, 12-14 strand	28
	1bcx	1994	1.8	all beta	beta-sandwich, 12-14 strand	29

*Assignment made by SCOP (15).

Table 3
Distribution of Starch-Degrading Enzymes Among Glycosyl Hydrolase Families

Family	PDB file	Year filed	Res.,Å	Fold class	Fold*	Ref.
Family 13	2taa	1982	3	beta/alpha	TIM-barrel	30
	1cgt	1993	2	beta/alpha	TIM-barrel	31
	1cyg	1993	2.5	beta/alpha	TIM-barrel	32
	1ppi	1994	2.2	beta/alpha	TIM-barrel	33
	2aaa	1991	2.1	beta/alpha	TIM-barrel	34
	1amg	1993	2.2	beta/alpha	TIM-barrel	35
	1amy	1994	2.8	beta/alpha	TIM-barrel	36
Family 14	1byb	1994	1.9	beta/alpha	TIM-barrel	37
Family 15	3gly	1994	2.2	beta/alpha	6 alpha-helices/circular array	38

*Assignment made by SCOP (15).

families of glycosyl hydrolases classified by HCA and sequence alignment have been examined. Figure 2 shows the general substrate specificities of enzymes assigned to these families. Several families are highly conserved relative to substrate preference; these include the β-galactosidases (Family 2), β-glucosidases (Family 3), xylanases (Families 10 and 11), α-amylases (Family 13), β-amylases (Family 14), glucoamylases (Family 15), lichenases (Families 16 and 17), chitinases (Families 18 and 19), lysozymes (Families 21–24), neuraminidases (Families 33 and 34),

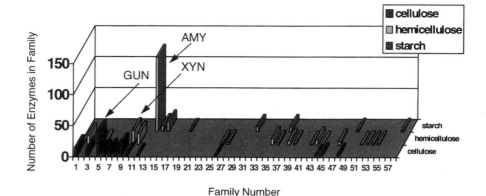

Fig. 2. The polysaccharide substrate specificity and distribution for families of O-glycosyl hydrolases.

and trehalases (Family 37). However, other families, such as Families 1, 4, 5, 39, 44, and others, include members that degrade different kinds of glycoside substrates.

Thus, within a single family, there may be diversity in the types of polysaccharides attacked, even though each family is expected to show a common tertiary structure and have a common mechanism of hydrolysis (4). Conversely, Tables 1–3 show that each general polysaccharide type may be attacked by hydrolases from various families, indeed with representation from widely different fold types. For instance, structurally known enzymes that degrade cellulose include two families with the TIM-barrel superfold (i.e., cyanogenic β-D-glucosidase [1cbg] from Family 1 and endoglucanase C from *Clostridium thermocellum* [1cec] and endoglucanase EI from *A. cellulolyticus* [1ece] from Family 5), one family with a modified TIM-barrel fold (i.e., cellobiohydrolase II from *Trichoderma reesei* [3cbh] and endoglucanase 2 from *Thermomonospora fusca* [1tml] from Family 6), one family with a β-strand sandwich fold (i.e., cellobiohydrolase I from *T. reesei* [1cel] from Family 7), one family with a six α-helix circular array fold (i.e., endoglucanase D from *C. thermocellum* [1c1c] from Family 9), and one family that has a Barwin-like, all β-strand closed barrel fold (i.e., endoglucanase V from *Humicula insolens* [1eng] from Family 45).

Table 2 shows a similar case for the eight xylanase structures reported, where one family with the TIM-barrel superfold (i.e., Family 10, xylanase A from *Streptomyces lividans* [1xas], xylanase Z from *C. thermocellum* [1xyz], xylanase A from *Pseudomonas fluorescens* var. *cellulosa* [1xys], and cellulase/xylanase from *Cellulomonas fimi* [2exo]) and a second family with a β-strand sandwich fold (i.e., Family 11, xylanase II from *T. reesei* [1xyp], xylanase I from *T. reesei* [1xyn], xylanase II from *Trichoderma harzianum* [1xnd], and xylanase II from *Bacillus circulans* [1bcx]) were

Polysaccharide Hydrolase Fold

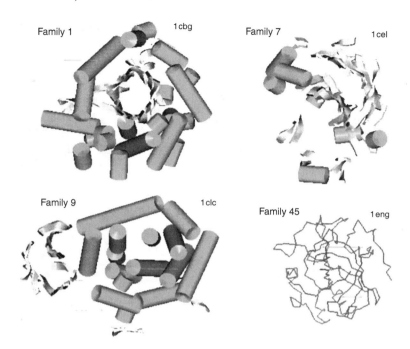

Fig. 3. Views of the secondary structures of cellulases and related enzymes from glycosyl hydrolase Families, 1, 7, 9, and 45. 1cbg is the cyanogenic-D-glucosidase from *Trifolia repens*, 1cel is the cellobiohydrolase I from *T. reesei*, 1c1c is the endoglucanase D from *C. thermocellum*, and 1eng is the endoglucanase V from *H. insolens*. Secondary assignments were made by Kabish-Sander algorithms. The structures were generated from PDB files (Brookhaven National Laboratory) using Biosym Version 94, Biosym Technologies, San Diego, CA.

found. Views of secondary structures for cellulase and xylanase enzymes generated from PDB files and assigned to glycosyl hydrolase families are shown in Figs. 3–5.

Table 3 shows less structural diversity for the starch-degrading enzymes studied so far, because all members of both Families 13 and 14, the α-amylases and β-amylases, respectively, contain TIM-barrel superfolds. The Family 15 glucoamylase, an enzyme from *Aspergillus awamori*, has a fold type much different from the TIM-barrel, i.e., a six α-helix circular array also found for endoglucanase D from *C. thermocellum*. This suggests that Family 15 enzymes may be mechanistically more closely related to the Family 9 cellulases than to the other amylases. Whereas the TIM-barrel superfold is a highly versatile and robust structure that dominates in the collection of glycosyl hydrolases as a whole (1,5), clearly many other folds are represented.

The high levels of structural and mechanistic similarity that often occur between enzymes showing preferences for different substrates also

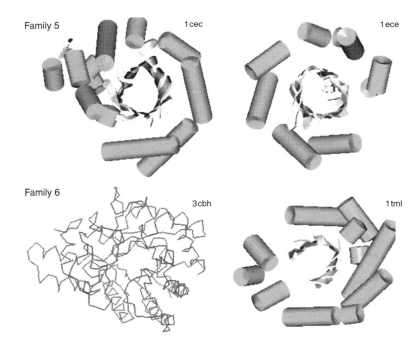

Fig. 4. Views of the secondary structures of cellulases from glycosyl hydrolase Families 5 and 6. 1cec is the endoglucanase C from *C. thermocellum*, 1ece is the endoglucanase EI from *A. cellulolyticus*, 3cbh is the cellobiohydrolase II from *T. reesei*, and 1tml is the endoglucanase 2 from *T. fusca*. The structures were generated from PDB files (Brookhaven National Laboratory) using Biosym Version 94, Biosym Technologies, San Diego, CA.

occur between enzymes from different environments and from different types of organisms. Within Family 5, for example, structural correlations were expected and have been recently confirmed experimentally between the endoglucanase C from *C. thermocellum* (1cec) and the endoglucanase EI from *A. cellulolyticus* (1ece). The α-carbon traces for these two enzymes are nearly superimposable *(12)*. This is intriguing, considering that endoglucanase C was produced as part of a mesophilic cellulosomal cellulase system and EI from *A. cellulolyticus* is a highly thermal tolerant endoglucanase (optimum temperature of 81°C) derived from a hot spring bacterium *(40)*. Furthermore, as shown by their grouping in the GH-A clan *(6)*, the Family 1 enzymes and the Family 10 xylanases are structurally and mechanistic cousins to the Family 5 cellulases. Structural and functional similarity superceding taxonomic boundaries is also common. For example, the Family 11 xylanases are all β-strand sandwich folds (Table 2), yet three are fungal (*Trichoderma*) and one is from a bacterium (*Bacillus*). Even cursory examination of Fig. 5 confirms the high degree of structural similarity for the three xylanases shown from Family 11.

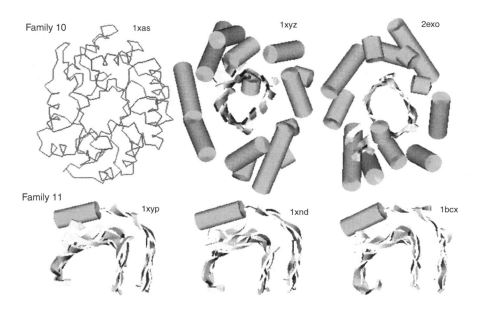

Fig. 5. Views of the secondary structures of xylanases from glycosyl hydrolase Families 10 and 11. 1xas is the xylanase A from *S. lividans*, 1xyz is the xylanase Z from *C. thermocellum*, 2exo is the cellulase/xylanase from *C. fimi*, 1xyp is the xylanase II from *T. reesei*, 1xnd is the xylanase II from *T. harzianum*, and 1bcx is the xylanase II from *B. circulans*. For clarity, structures for 1xys and 1xyn are not shown. The structures were generated from PDB files (Brookhaven National Laboratory) using Biosym Version 94, Biosym Technologies, San Diego, CA.

CONCLUSION

Ideally an experimental structure would be available for each glycosyl hydrolase, but we are limited by the size of the available structure data base at this time. Specifically, of the 811 glycosyl hydrolases listed by Bairoch (6), only 30 of those important for lignocellulosic biomass conversion have been subjected to X-ray diffraction and structure analysis. Still, nature has provided many independent structural schemes to bring a set of key active site amino acid residues into precise position to effect very similar or identical chemical reactions, in this case, transfer of the O-glycosyl bond to water. Thus, diversity of structure is a fact, and convergence of catalytic function is indicated for different types of glycosyl hydrolase folds. Glycosyl hydrolase Family 5 enzymes are especially important to researchers in the biomass conversion field, because one especially active and thermal tolerant endoglucanase, EI from *A. cellulolyticus*, belongs to Subclass 1 of this family. Although X-ray structures are currently available for only 2 of the 47 enzymes identified in Family 5 to date, endoglucanase C from *C. thermocellum* and endoglucanase EI from *A. cellulolyticus*, SDM work to improve Family 5 cellulases can proceed with a reasonable degree of confidence that lessons learned from modifying the structure of one family member will translate to other members *(41,42)*. In addition, because of the clear placement of Family 5 in the glycosyl hydrolase GH-A clan, insights based on studies of enzymes from glycosyl hydrolase Families 1, 2, 10, 17, 30, 35, 39, and 42 will also likely be relevant, even though those enzymes are not cellulases. In fact, such insights will probably be more relevant than those gleaned from the study of cellulases from the structurally unrelated Families 6, 7, 9, and 45.

ACKNOWLEDGMENT

This work was funded by the Biochemical Conversion Element of the Office of Fuels development of the US Department of Energy.

REFERENCES

1. Orengo, C. A., Jones, D. T., and Thornton, J. M. (1994), *Nature* **372**, 631–634.
2. Henrissat, B. (1991), *Biochem. J.* **280**, 309–316.
3. Gilkes, N. R., Henrissat, B., Kilburn, D. G., Miller, R. C., Jr., and Warren, R. A. J. (1991), *Microbiol. Rev.* **55**, 303–312.
4. Henrissat, B. and Bairoch, A. (1993), *Biochem. J.* **293**, 781–788.
5. Henrissat, B., Callebaut, I., Fabrega, S., Lehn, P., Mornon, J.-P., and Davies, G. (1995), *Proc. Natl. Acad. Sci.* **92**, 7090–7094.
6. Bairoch, A. (1996), SWISS-PROT Protein Sequence Data Bank (http://expasy.hcuge.ch/cgibin/lists?glycosid.text).
7. Levitt, M. and Chothia, C. (1976), *Nature* **261**, 552–557.
8. Efimov, A. V. (1994), *Structure* **2**, 999–1002.
9. Harris, N. L., Presnell, S. R., and Cohen, F. E. (1994), *J. Mol. Biol.* **236**, 1356–1368.
10. Chothia, C. and Janin, J. (1981), *Proc. Natl. Acad. Sci. USA* **78**, 4146–4150.
11. Orengo, C. A. and Thornton, J. M. (1993), *Structure* **1**, 105–120.
12. Sakon, J., Adney, W. S., Himmel, M. E., Thomas, S. R., and Karplus, P. A. (1996), *Biochemistry* **35**, 10648–10660.
13. Kraulis, P. J., Clore, G. M., Nilges, M., Jones, T. A., Pettersson, G., Knowles, J., and Gronenborn, A. M. (1989), *Biochemistry* **28**, 7241.
14. Xu, G.-Y., Ong, E., Gilkes, N. R., Kilburn, D. G., Muhandiram, D. R., Harris-Brandts, M., Carver, J. P., Kay, L. E., and Harvey, T. S. (1995), PDB entry 1exg.
15. Murzin, A. G., Brenner, S. E., Hubbard, T., and Chothia, C. (1995), *J. Mol. Biol.* **247**, 536–540.
16. Tolley, S. P., Barrett, T. E., Suresh, C. G., and Huges, M. A. (1993), *J. Mol. Biol.* **229**, 791.
17. Dominguez, R., Souchon, H., Spinelli, S., Dauter, Z., Wilson, K. S., Chauvaux, S., Beguin, P., and Alzari, P. M. (1995), *Nat. Struct. Biol.* **2**, 569.
18. Rouvinen, T., Rouvinen, J., Lehtovaara, P., Caldentey, X., Tomme, P., Claeyssens, M., Pettersson, G., and Teeri, T. (1989), *J. Mol. Biol.* **209**, 167.
19. Spezio, M., Wilson, D. B., and Karplus, P. A. (1993), *Biochemistry* **32**, 9906.
20. Divne, C., Stahlberg, J., Reinikainen, T., Ruohonen, L., Pettersson, G., Knowles, J. K. C., Teeri, T. T., and Jones, T. A. (1994), *Science* **265**, 524.
21. Alzari, P. M., Juy, M., and Souchon, H. (1993), *Biotechnol. Industrial Fermentation* **8**, 73.
22. Davies, G. J., Dodson, G. G., Hubbard, R. E., Tolley, S. P., Dauter, Z., Wilson, K. S., Hjort, C., Mikkelsen, J. M., Rasmussen, G., and Schulein, M. (1993), *Nature* **365**, 362.
23. Derewenda, U., Swenson, R., Green, R., Wei, Y., Morosoli, R., Shareck, F., Kluepfel, D., and Derewenda, Z. S. (1994), *J. Biol. Chem.* **269**, 20,811.

24. Dominguez, R., Souchon, H., Spinelli, S., Dauter, Z., Wilson, K. S., Chauvaux, S., Beguin, P., and Alzari, P. M. (1995), *Nat. Struct. Biol.* **2,** 569.
25. Harris, G. W., Jenkins, J. A., Connerton, I., Cummings, N., Lo Leggio, L., Scott, M., Hazlewood, G. P., Laurie, J. I., Gilbert, H. J., and Pickersgill, R. W. (1994), *Structure (Lond.)* **2,** 1107.
26. White, A., Withers, S. G., Gilkes, N. R., and Rose, D. R. (1994), *Biochemistry* **33,** 12,546.
27. Torronen, A. and Rouvinen, J. (1995), *Biochemistry* **34,** 847.
28. Campbell, R. L., Rose, D. R., Wakarchuk, W. W., To, R. J., Sung, W., and Yaguchi, M. (1994), PDB entry 1xnd.
29. Wakarchuk, W. W., Campbell, R. L., Sung, W. L., Davoodi, J., and Yaguchi, M. (1994), *Protein Sci.* **3,** 467.
30. Matsuura, Y., Kusunoki, M., Harada, W., and Kakudo, M. (1984), *J. Biochem. (Tokyo)* **95,** 697.
31. Klein, C. and Schulz, G. E. (1991), *J. Mol. Biol.* **217,** 737.
32. Kubota, M., Matsuura, Y., Sakai, S., and Katsube, Y. (1995), PDB entry 1cyg.
33. Qian, M., Haser, R., Buisson, G., Duee, E., and Payan, F. (1993), *J. Mol. Biol.* **231,** 785.
34. Boel, E., Brady, L., Brzozowski, A. M., Derewenda, Z., Dodson, G. G., Jensen, V. J., Petersen, S. B., Swift, H., Thim, L., and Woldike, H. F. (1990), *Biochemistry* **29,** 6244.
35. Morishita, Y., Matsuura, Y., Kubota, M., Sato, M., Sakai, S., and Katsube, Y. (1995), PDB entry 1amg.
36. Kadziola, A., Abe, J.-I., Svensson, B., and Haser, R. (1994), *J. Mol. Biol.* **239,** 104.
37. Mikami, B., Degano, M., Hehre, E. J., and Sacchettini, J. C. (1994), *Biochemistry* **33,** 7779.
38. Aleshin, A. E., Hoffman, C., Firsov, L. M., and Honzatko, R. B. (1994), *J. Mol. Biol.* **238,** 575.
39. Davies, G. and Henrissat, B. (1995), *Structure* **3,** 853–859.
40. Himmel, M. E., Adney, W. S., Grohmann, K., and Tucker, M. P. (1994), US Patent No. 5,275,944.
41. Wang, Q., Tull, D., Meinke, A., Gilkes, N. R., Warren, R. A. J., Aebersold, R., and Withers, S. G. (1993), *J. Biol. Chem.* **268,** 14,096–14,102.
42. Bortoli-German, I., Haiech, J., Chippaux, M., and Barras, F. (1995), *J. Mol. Biol.* **246,** 82–94.

Acetobacter Cellulosic Biofilms Search for New Modulators of Cellulogenesis and Native Membrane Treatments

JOSÉ D. FONTANA,*,1 CASSANDRA G. JOERKE,1
MADALENA BARON,1 MARCELO MARASCHIN,1
ANTONIO G. FERREIRA,2 IRIS TORRIANI,3 A. M. SOUZA,1
MARISA B. SOARES,1 MILENE A. FONTANA,4
AND MANOEL F. GUIMARAES1

[1]LQBB-Biomass Chemo/Biotechnology Laboratory, Department of Biochemistry, UFPR (Federal University of Paraná), PO Box 19046 (81531-990) CURITIBA-PR-Brazil; [2]NMR/Chemistry/UFSCar, Sao Carlos-SP; [3]Crystallography/UNICAMP-SP; and [4]N.S. Medianeira College, Curitiba-PR, Brazil

ABSTRACT

Since natural substances like pseudoxanthins exert a positive effect on the cellulogenic ability of *Acetobacter xylinum* when producing cellulosic pellicles suitable for skin burn therapy, new defined and complex modulators were sought. Ca^{2+} and Mg^{2+} (4 mM) were strongly stimulatory. Na^+ had no effect and K^+ was inhibitory. Ammonium dihydrogen phosphate (0.12 g/L) ensured the same nitrogen supply as the same concentration of yeast extract as measured by cellomembrane dry wt./yield albeit higher yeast extract supplies produced thicker membranes. Corn steep liquor (CSL) was also progressively beneficial from 0.125 to 0.5 mL/L, and this yield could be further improved by the combination of CSL with a tea infusion (source of caffeine). Uridine (precursor for UDP-Glc, sugar donor in cellulose biosynthesis), guanine, guanosine, and its butyrylated derivatives (precursors for the positive modulator of cellulose synthetase, di-cGMP) resulted in only moderate stimulation. Sodium phytate and betaine were also slightly stimulatory. The fibrilar product from

*Author to whom all correspondence and reprint requests should be addressed.

a new *Acetobacter* isolate (Ax-M) was characterized as cellulose by comparison with the solid-state ^{13}C-NMR of algal cellulose. Its X-ray diffractogram was a confirmatory analysis. After incorporation of tamarind xyloglucan to previously air-dried cellulosic pellicles, diffractometry displayed only slight differences. Mercerized (5M NaOH) fresh cellulosic biofilms underwent drastic size reduction (3.5-fold), turning compact nut still flexible if maintained wet.

Index Entries: *Acetobacter*; cellulosic biofilms; cellulogenesis modulation; bacterial cellulose.

INTRODUCTION

Bacterial cellulose has well-established applications as a (bio)technological tool. Examples are: artificial skin in the therapy of burns and other dermic injuries *(1)*, high-fidelity acoustic speakers *(2)*, high-quality paper *(3)*, ultrafiltration membranes, cover membrane for glucose biosensors, culture substrate for mammalian cells, binder/thickener for paint, ink, and adhesives, and finally, diet and dessert foods *(4)*. These applications are owing primarily to such properties as large surface area, high water-holding capacity, moldability, and strong shear resistance.

The scientific and technological research carried out on bacterial cellulogenesis has yielded some remarkable findings. A novel cyclic dinucleotide c(GMP)$_2$ was identified and characterized as the cellulose synthetase positive modulator *(5)*.

The positive metabolic role of pseudoxanthines from tea infusions (caffeine and theophylline), as potent stimulators of *Acetobacter xylinum* cellulosic biofilms by targeting the c(GMP)$_2$-destroying phosphodiesterases, has also been elucidated *(6)*.

Mixed fibers and composites from bacterial cellulose were first developed at the University of Texas at Austin *(7)*. A doubled yield of *A. pasteurianus* cellulose was achieved by including an oxygen-permeable membrane in the fermentation vessel *(4)*. New wild-type and mutant hyperproducing strains and a subsp. (*sucrofermentans*) of *A. xylinum* were recently described *(8,9)*. Culture media optimization for the production of bacterial cellulose was surprisingly achieved by adding endoglucanase, a cellulose depolymerizing enzyme *(10)*.

We routinely use static cultures of *Acetobacter* and collect the floating and progressively thickened cellulosic membrane at the medium surface line. The Rensseler Polytechnic Institute (Troy, NY) uses a rotating disk bioreactor *(11)*. Scaling up of bacterial cellulose production for several commercial applications was accomplished by a collaborative project between Weyerhaeuser Co. (Tacoma, WA) and Cetus Corporation, thus leading to a multipatent-protected product, Cellulon™ (Folster, H. G., lecture 28, 17th Symposium on Biotechnology for Fuels and Chemicals, Vail, CO, May 1995).

Following our previous contributions on the subject of bacterial cellulose biosynthesis *(1,6)*, a comparative study is now being undertaken of the responses of a new isolate (Ax-M) and a collection reference strain (Ax-G) of *A. xylinum* to several metabolic effectors of cellulogenesis.

MATERIALS AND METHODS

Strains—Origin and Maintenance

Ax-M, under PCR-based genetic characterization, is an isolate of *Acetobacter* sp. (presumably *xylinum*, collected by M A. Fontana) from a fermented coffee. Ax-G is a reference strain of *Acetobacter aceti* subsp. *xylinum* (DSM 2325; GBF, Braunschweig, Germany) received as a gift by M. Baron. Stock cultures were maintained as lyophilized mixture of cells and skim milk.

Routine Liquid Medium for Nonagitated Cultures

The nonstimulated medium (BM⁻) contained, per liter: glucose (40 g), yeast extract (0.1–0.2 g), ammonium dihydrogen phosphate (0.1 g), and 5 mL ethanol. Except in the experiments using *Camellia sinensis* (tea) infusions, caffeine, and theophylline (2 mg/L) were included as stimulators (BM⁺). All other additions are detailed in the text (Results and Discussion). Corn steep liquor (CSL) was provided by Refinaçoes de Milho Brasil (Balsa Nova, PR, Brazil) as a concentrate with ca. 40 g% solids. Its particulate matter was removed before use by centrifugation at 10 krpm (12,300 g). Liquid cultures were in a final volume of 100 mL in glass vessels of 250 mL with a slight reduction of the diameter from the top to bottom in order to avoid accidental submersion of the growing and floating cellulosic membrane. Fermentation vessels were covered with a double layer of micropierced napkin sheets in order to allow free air/gas exchange. Nystatin and benomyl (5 mg/L each) were added to all culture media in order to avoid yeast and/or mold contaminations. A suspension of bacterial cells (0.15–0.20 OD units at 650 nm) was used at 1/20 vol inoculum. Growth was at 28°C for 5 d without agitation. Cellulose pellicles, prior to gravimetric quantitation, were exhaustively washed with distilled water to remove all water-soluble contaminants and then lyophillized.

Physical Analyses

Diffractometry and electron microscopy were carried out as previously described *(1)*. For the diffractogram and solid-state ^{13}C-nuclear magnetic resonance (NMR) the cellulosic membranes were pretreated by extensive washing with 1% sodium dodecyl sulfate (SDS). Tamarind seed xyloglucan (Ts-XG) was obtained through five boiling cycles (microwave oven) of an aqueous suspension of seed endosperm, followed by centrifugation at 10

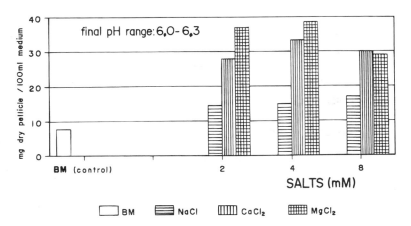

Fig. 1. Modulation of cellulosic membrane yields by strategic cations.

krpm (12,300 g) and lyophilization of the cleared and viscous supernatant. Guanosine (pyridine suspension) was derivatized with a stoichiometric amount of butyric anhydride for 12 h at room temperature and then dried in a Savant Speed vacuum machine with the application of heat.

RESULTS AND DISCUSSION

Culture media optimization was initially investigated via the dose–response for common, although strategic, cations owing to their important modulation of key enzymes of sugar activation/deactivation (e.g., phosphodiesterases). The best cellulosic pellicle mass increase was obtained with 2–4 mM calcium and magnesium chlorides (Fig. 1). Mixed at half-concentration, they act somewhat synergistically (results not shown), and were used in all subsequent experiments.

Ammonium dihydrogen phosphate (0.012 g/L) yielded the same cellulose equivalent to that obtained from a more expensive nitrogen source, namely, yeast extract (0.12 g/L) (Fig. 2). A better buffering condition against acetic acid action was also attained with this simpler nitrogen source. The particular effect of other sources like purines and pyrimidnes (Fig. 1) will be mentioned later.

CSL alone or in combination with tea (*C. sinensis*) infusions (Fig. 3) resulted in cellulosic membrane mass increases from 2- to 10-fold. Combinations of these complex nutrient sources resulted in additive effects (see the thicker and more uniform membrane in the fourth vessel from Fig. 4) and darker membranes, thus requiring more extensive washings.

An aqueous extract of previously organosolvent-extracted tea still stimulated cellulose production. Since pseudopurines were previously removed with the organosolvents, this gain is not explainable by caffeine/theophylline *(6)*. Therefore, beneficial tea components other than the pseudoxantines do exist and remain to be characterized.

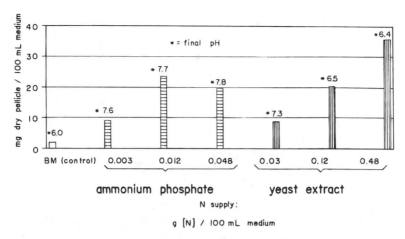

Fig. 2. The comparative effect of defined and undefined nitrogen sources for cellulose production.

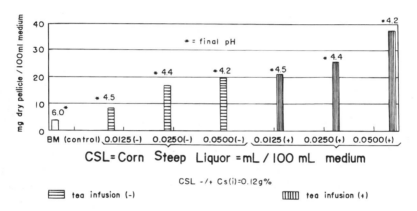

Fig. 3. Cellulosic membranes yield increase by CSL and/or tea infusion (CSi) additions.

CSL addition to the basic culture medium caused a severalfold increase in cellomembrane yield (Fig. 3). Interestingly, an anionic component from CSL (here recovered as its insoluble Ca^{2+} salt) produced a positive cellulogenic effect obtained from sodium phytate.

Since nucleotides play key roles in many biological processes (12), some precursors, such as nitrogen bases or natural and lipoderivatized nucleosides, were examined as selective and limited (up to 5 mg/L) additions. Precursors of metabolically important compounds were chosen: guanine or guanosine because of the crucial positive role of $c(GMP)_2$ on cellulose synthetase and uridine as part of UDP-glucose, the activated glucosyl donor for the same enzyme. A stimulated basic medium was used as control (Table 1). Data indicated a positive, although limited, effect on cel-

Fig. 4. Morphological aspect of cellomembranes from nonstimulated and stimulated culture media. (BM = basic medium; +Cs = addition of tea [*C. sinensis*] infusion; +CSL = addition of corn steep liquor; +Cs + CSL = both additions.)

lulose yield. Work is now in progress on long-chain derivatives (e.g., oleoyl) because of the possible toxic effects of free butyric acid.

The defined additions for other assays were phytic acid (sodium salt, pH adjusted to 6.8) and betaine, but the enhancement of cellulose yields was obtained only with higher inputs (0.5 and 0.05 g/L; Tables 1, 2, and 3). Interestingly, a fraction obtained from CSL as the precipitate from the addition of saturated $CaCl_2$ reproduced the effect of phytin. The particular effect of this cyclitol polyphosphate is difficult to explain in light of its chelating properties for Ca^{2+} and Mg^{2+}, but its potential benefit as precursor for phosphatidylinositol, an important component of biological membranes, cannot be discounted. Betaine may play the role of a membrane-reinforcer "osmolyte" *(13)*.

Since *Acetobacter* spp. may also biosynthesize other polysaccharides (e.g., the water-soluble acetan *14*), washing and/or pretreatment of the cellulosic membranes should resolve this issue. An alkaline bath treatment is the usual practice to obtain whitened membranes. Strong NaOH washing (5*M*, 6 h at room temperature with gentle agitation in a rotatory shaker) indeed produced such a desirable result. However, a dramatic effect of 3.5 times size reduction by this treatment (mass almost unaffected) was experienced by biofilms from both strains (Fig. 5). The new compact cellulose membrane architecture preserved some of its original flexibility, if maintained as hydrated material. Possible applications for these mercerized cellofilms are beyond the scope of this article, but ongoing physical analyses may shed some light on this aspect.

Table 1
Effect of Defined Supplements on *Acetobacter* Cellulosic Biofilm Production

mg dry pellicle/100 mL medium

		Acetobacter strain	
addition	concentration	"M"	"G"
BM #	(-)	119	74
BM + Na phytate	375 mg/L	148	123
Guanine	5 mg/L	129	93
Guanosine (GR)	5 mg/L	133	99
butyrylated GR *	0.25 mg/L	122	86
uridine	5 mg/L	125	89

#BM = basic medium = 40 g/L glucose, 1.25 g/L yeast extract, 0.25 g/L ammonium dihydrogen phosphate, 1 mg/L (each) caffeine + teophilline, 0.3 mM (each) Na^+, Ca^{2+}, Mg^{2+} chlorides, 5 mL/L ethanol.
*Derivative soluble in anhydrous methanol.

The expected nature of the polymer biosynthesized by both strains was then confirmed as cellulose using detergent-washed membranes in order to reduce any noise (entrapped cells and their multicomponents) in solid-state ^{13}C-NMR spectroscopy. The five-peak spectrum for Ax-M was similar to that previously reported for cellulose from the alga *Valonia* (15) and from crosspolarization/magic angle spinning (CP/MAS)-NMR (16). This spectrum was minimally altered (Fig. 6) when an air-dried membrane was previously equilibrated overnight with a 1 g% solution of tamarind seed xyloglucan (Ts-XG) followed by extensive washing (unbound polysaccharide removal). Except for a small new peak (asterisk label) at δ = 97.1 ppm from nonreducing -D-xylopyranosyl branching units of Ts-XG), no differences were noticeable since the Ts-XGT main -D-glucopyranosyl backbone is also cellulose.

SDS-washed cellulosic membranes, with (AxM-cel/Ts-XG) or without (Ax-M-cel) xyloglucan addition were also submitted to diffractometric analysis. The X-ray diffraction diagram of the control (Ax-M-cel) membrane is shown in Fig. 7. The concentric and complete-ring array

Table 2
Effect of Nondefined Supplements Related to Tea Infusion on
Acetobacter Cellulosic Biofilm Production

mg dry pellicle / 100 mL medium

		Acetobacter strain	
addition	concentration	"M"	"G"
BM #	(-)	30	14
BM + caffeine theophylline	2 mg/L (each)	60	30
BM + tea infusion	.2 g/L	135	33
BM + tea aq. fraction *	2.0 g/L	60	45
BM + betaine	50 mg/L	53	24

#BM = basic medium = 40 g/L glucose, 0.125 g/L yeast extract, 0.1 g/L ammonium dihydrogen phosphate, 0.3 mM (each) Na^+, Ca^{2+}, and Mg^{2+} chlorides, 5 mL/L ethanol.
*Whole tea was exhaustively extracted with organic solvents of increasing polarity (iso-octane to methanol) and the final residue then extracted with boiling water to render the assayed fraction.

indicated a high degree of crystallinity, but no preferential orientation for the cellulosic microfibril both kind of samples (control; Ax-M-cel and Ax-M-cel/Ts-XG) were compared by scanning and reflection geometry (θ-2 θ plot), a lower amorphous content (Fig. 8; asterisk label) was detected after Ts-XG binding, but the same peaks with no change in positions indicated the preservation of the same planar spacings in the crystalline regions. The diffractometry tracing on the two main peaks (2θ-degrees = 14.5 and 22.76, corresponding to the 101 and 002 nomenclature for cellulose crystallites) fitted closely to those reported for 1% NaOH-boiled bacterial cellulose, namely, 2θ-degrees = 14.52 and 22.67 *(17)*, except for their similar intensities in our diffractograms. The third diffractometric data (peak 101; 2θ-degrees = 16.8), although coincident in terms of position, was less distinguishable, but the different sample pretreatments may be recalled (SDS washing, our case; alkaline boiling for 8 h, ref. *17*).

Table 3
Effect of CSL and Derivatives on *Acetobacter* Cellulosic Biofilm Production

mg of dry pellicle/100 mL medium

Acetobacter strain

addition	concentration	"M"	"G"
BM	(-)	11	14
BM + caffeine (#) theophylline	2 mg/L (each)	17	18
BM + CSL	10 mL/L	88	103
BM# + CSL	10 mL/L	104	112
BM# + Na phytate	0.5 g/L	37	32
BM# + CSL Ca2+ pp *	0.5 g/L	47	35

#All experiments with CSL or fractions contained the basal (2 mg/L) addition of pseudopurines.
*Fraction from CSL precipitated by the addition of a saturated solution of calcium chloride.

CONCLUSIONS

Growth of wild-type *Acetobacter* strains for the production of floating cellulosic membranes could be improved through the incorporation of simple and complex conutrients in the media for nonagitated cultures (e.g., cellulose mass doubling owing to 2 mM calcium ion). CSL proved a cheap and very productive addition. Slight or drastic changes in the cellulosic pellicles resulted from the incorporation of either structurally related hemicelluloses (e.g., xyloglucan) or from strong alkali mercerization (e.g., 5M KOH), respectively. These modified products may generate new (bio)technological applications other than the reported one as a temporary skin substitute in the case of human burns and other dermic injuries.

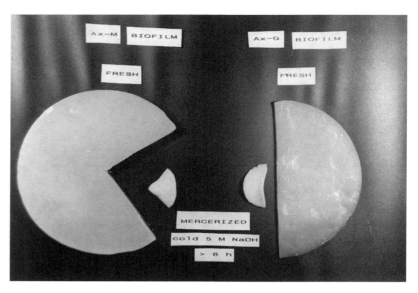

Fig. 5. Mercerization of fresh cellulosic biofilms (a quarter [Ax-M strain] or a half [Ax-G strain] of fresh cellulosic membranes was submerged in 5 M NaOH at room temperature for >6 h, and then thoroughly washed with distilled water before picturing).

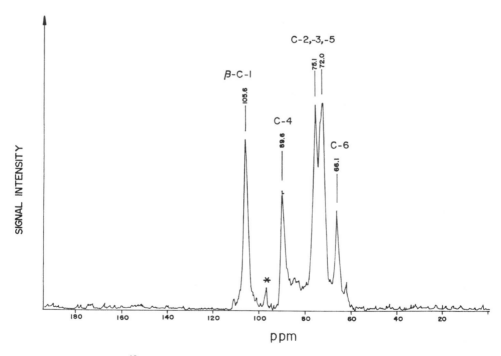

Fig. 6. Solid-state ^{13}C-NMR spectrum of an air-dried and powdered SDS-washed membrane (*) Before the spectral run, this particular air-dried membrane obtained from the Ax-M strain was equilibrated with xyloglucan and then thoroughly washed before comminution and drying.

Acetobacter Cellulosic Biofilms

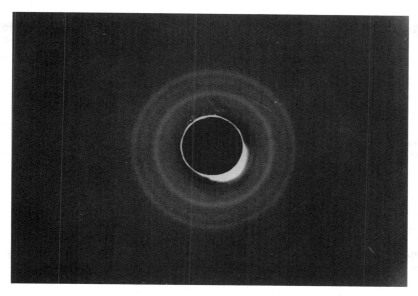

Fig. 7. X-ray transmission Laue diagram for an SDS-washed and air-dried membrane from the Ax-M strain.

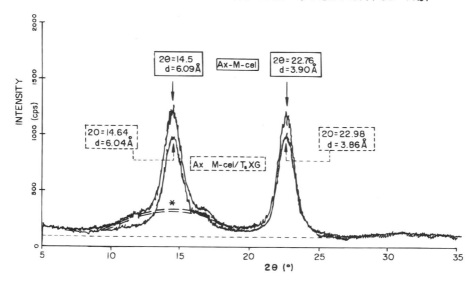

Fig. 8. Scanning and reflection geometry ($\theta 2\theta$ plot) from X-ray diffractometry of SDS-washed and air-dried membranes from Ax-M strain (upper tracing = membrane without addition of xyloglucan; lower tracing = membrane with addition of xyloglucan).

ACKNOWLEDGMENTS

Thanks are addressed to Brazilian funding agencies (CNPq/PADCT-SBIO and CAPES) and to A. C. Ramalho (UNICAMP; technical assistance with X-rays diffactometry), G. Yates (manuscript reading), and M. Lima (drawings).

REFERENCES

1. Fontana, J. D., Souza, A. M., Fontana, C. K., Torriani, I., Moreschi, J. C., Gallotti, B. J., Souza, S. J., Narciso, G. P., Bichara, J. A., and Farah, L. F. X. (1990), *Appl. Biochem. Biotechnol.* **24/25,** 253–264.
2. Yamanaka, S., Watanabe, K., Kitamura, N., Iguchi, M., Mitsuhashi, S., Nishi, Y., and Uryu, M. (1989), J. Mat. Sci. 24, 3141–3145.
3. Johnson, D. C. and Winslow, A. R. (1990), *Pulp & Paper* **May,** 105–107.
4. Yoshino, T., Asakura, T., and Toda, K. (1996), *J. Ferm. Bioeng.* **81(1),** 32–36.
5. Volman, G., Ohana, P. and Benziman, M. (1995), *Carbohydr. Europe* **12,** 17.
6. Fontana, J. D., Franco, V. C., Souza, S. J., Lyra, I. N., and Souza, A. M. (1991), *Appl. Biochem. Biotechnol.* **28/29,** 341–351.
7. Dean, K. L. (1993), *Ind. Bioprocessing* **15(4),** 2.
8. Toyosaki, H. (1995), *J. Gen. Appl. Microbiol.* **41(4),** 307–314.
9. Ishikawa, A., Matsuoka, M., Tsuchida, T., and Yoshinaga, F. (1995), *Biosci. Biotechnol. Biochem.* **59(12),** 2259–2262.
10. Tonouchi, N., Thara, N., Tsuchida, T., Yoshinaga, F., Beppu, T., and Hourinochi, S. (1995), *Biosci. Biotechnol. Biochem.* **59(5),** 805–811.
11. Honen, J. (1996), *Ind. Bioprocessing* **18(3),** 4–5.
12. Carver, J. D. and Walker, W. A. (1995), *Nutr. Biochem.* **6,** 58–72.
13. Coghlan, A. (1996), in *New Scientist* vol. 149, n. 2018, Anderson, A., ed., IPC Magazines, London, UK, p. 20.
14. Couso, R., Ielpi, L., and Dankert, M. (1987), *J. Gen. Microbiol.* **133,** 2123–2135.
15. Marchessault, R. H. and Sundararajan, P. R. (1983), in *The Polysaccharides* vol. 2, Aspinall, G. O., ed., Academic P, New York, p. 36.
16. Yamamoto, H. and Horii, F. (1994), *Cellulose* **1,** 57–66.
17. Uhlin, K. I., Atalla, R. H., and Thompson, N. S. (1995), *Cellulose* **2,** 129–144.

Experimental Data Analysis

An Algorithm for Determining Rates and Smoothing Data

K. THOMAS KLASSON

Chemical Technology Division, Oak Ridge National Laboratory,*
Oak Ridge, TN 37831-6044

ABSTRACT

Reaction rate determination from experimental data is generally an essential part of evaluating enzyme or microorganism growth kinetics and the effects on them. Commonly used methods include forward, centered, or backward finite difference equations using two or more data points. Another commonly applied method for determining rates is least-square regression techniques, and when the sought function is unknown, polynomials are often applied to represent the data. The cubic spline functions presented in this article represent a versatile method of evaluating rates. The advantage in using this method is that experimental error may be largely accounted for by the incorporation of a smoothing step of the experimental data without force-fitting of the data. It also works well when data are unevenly spaced (often the case for experiments running over long periods of time). The functions are easily manipulated, and the algorithm can be written concisely for computer programming. The development of spline functions to determine derivatives as well as integrals is presented.

Index Entries: Cubic splines; computer programming; derivatives; smoothing data.

INTRODUCTION

Reaction rate determination from experimental data is generally an essential part of evaluating enzyme or microorganism growth kinetics and effects on them. Commonly used methods include simple forward,

---

*Managed by Lockheed Martin Energy Research Corp. for the U.S. Department of Energy under contract DE-AC05-96OR22464. The submitted manuscript has been authored by a contractor of the U.S. Government. Accordingly, the U.S. Government retains a nonexclusive, royal-free license to publish or reproduce the published form of this contribution, or allow others to do so, for U.S. Government purposes.

centered, or backward finite difference equations using two or more data points. Another method was presented by LeDuy and Zajic (1), and involves the construction of a circle with the three neighboring points on the perimeter, finding the center coordinates, and the normal to the equation between the center coordinates and the second (middle) point. This method works well with well-behaved data evenly spaced. The advantage in using the method described in this article is that experimental error may be largely accounted for by the incorporation of a smoothing step of the experimental data. It also works well when data are unevenly spaced (often the case for experiments running over long periods of time). The difference between the numerical method proposed and the manual-graphical method of finding derivatives (dX/dt) may be represented as the following:

Manual or graphical method
1. Plot experimental data.
   (X as a function of t)
2. Draw a smooth curve through or close to the points.
3. Take many readings of curve and find derivatives at each of the original values of t using finite differences.

Numerical method
1. Construct cubic spline with a "smoothness factor."
2. Use spline functions to find derivatives at original t values.

The construction of cubic splines is simple with today's personal computers and available programming languages.

Cubic Spline Theory

A cubic spline joins two neighboring points with a third-order polynomial (2). Thus, if there are N experimental data points, the overall curve through the points may be described by $N - 1$ equations, each in the form of $f_i(t) = a_i t^3 + b_i t^2 + c_i t + d_i$ (see Fig. 1). The third-order equation contains four constants, and therefore, we need to determine a total of $4(N - 1)$ constants. In order to do this and to assure a continuous smooth curve without breaks and abrupt changes in the derivative at the points, a few criteria must be followed.

1. The equation values must be equal at the interior points ($2N - 4$ conditions). In other words, the spline must pass through the data points.
2. The first and last equations must pass through the end points (two conditions).
3. The first and second derivatives at the interior points must be equal ($2N - 4$ conditions). In other words, the spline must be smooth as it passes through the data points.
4. The second derivatives at the end points are zero (two conditions).

Experimental Data Analysis

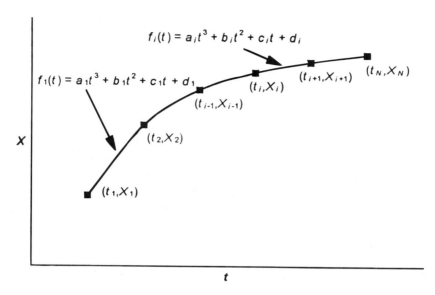

Fig. 1. Example of cubic spline construction.

Even though the above system is completely defined, it is more efficient to restructure the equations to contain fewer unknown constants (2). If the experimental data correspond to the points (t_1, X_1), (t_2, X_2), ... (t_1, X_1), ... (t_N, X_N), then the i:th polynomial between the points i and $i + 1$ may be described in the terms of surrounding points by:

$$f_i(t) = \frac{f''(t_i)}{6(t_{i+1} - t_i)}(t_{i+1} - t)^3 + \frac{f''(t_{i+1})}{6(t_{i+1} - t_i)}(t - t_i)^3$$
$$+ \left[\frac{X_i}{(t_{i+1} - t_i)} - \frac{f''(t_i)(t_{i+1} - t_i)}{6}\right](t_{i+1} - t) \qquad (1)$$
$$+ \left[\frac{X_{i+1}}{(t_{i+1} - t_i)} - \frac{f''(t_{i+1})(t_{i+1} - t_i)}{6}\right](t - t_i)$$

This equation contains only two unknowns, the second derivatives, which may be evaluated using:

$$(t_{i+1} - t_i)\, f''(t_i) + 2(t_{i+2} - t_i)\, f''(t_{i+1}) + (t_{i+2} - t_{i+1})\, f''(t_{i+2}) = \frac{6}{(t_{i+2} - t_{i+1})}(X_{i+2} - X_{i+1}) - \frac{6}{(t_{i+1} - t_i)}(X_{i+1} - X_i) \qquad (2)$$

Equation (2) was derived by taking the first derivative of Eq. (1) for both the i:th and $(i + 1)$:th polynomial and setting them equal at the shared point (t_{i+1}, X_{i+1}). When Eq. (2) is written for all interior points, we will have

$N-2$ simultaneous equations with $N-2$ unknown if criterion four above is incorporated. The equation system may be written in the following format:

$$a_2 f_1'' + b_2 f_2'' + c_2 f_3'' = r_2$$
$$a_3 f_2'' + b_3 f_3'' + c_3 f_4'' = r_3$$
$$\downarrow$$
$$a_{N-1} f_{N-2}'' + b_{N-1} f_{N-1}'' + c_{N-1} f_N'' = r_{N-1}$$

Since $f_1'' = [f''(t_1)] = 0$ and $f_N'' = [f''(t_N)] = 0$ (according to criterion four), the equation system will form a matrix with a tridiagonal band. This type of matrix may easily be solved using simplified Gaussian elimination with a few lines of computer programming. Once the second derivatives have been evaluated at the interior points, they may be used to define completely each of the polynomials, according to Eq. (1).

DETERMINATION OF DERIVATIVES AND INTEGRALS

Equation (1) may be differentiated to determine the derivative (e.g., when estimating rates) at any point on the splines yielding:

$$f_i'(t) = \frac{f''(t_{i+1})}{2(t_{i+1} - t_i)}(t - t_i)^2 - \frac{f''(t_i)}{2(t_{i+1} - t_i)}(t_{i+1} - t)^2 + \frac{(X_{i+1} - X_i)}{(t_{i+1} - t_i)} - \frac{(f''(t_{i+1}) - f''(t_i))}{6}(t_{i+1} - t_i) \quad (3)$$

Equation (3) may be used for the points 1 through $N-1$, and the derivative for the last point may be evaluated using the $(N-1)$:th equation. Note that Eq. (3) is further simplified, since t is substituted by t_i when evaluating derivatives at the data points.

Equation (1) may also be integrated to determine the integral under the data points. This may be desired when estimating cell and product yield uptake using an integrated equation vs a differentiated equation *(3)*. The integral between data points i and $i+1$ connected by the cubic spline function $f_i'(t)$ is given by:

$$\int_{t_i}^{t_{i+1}} f_i(t) dt = \tfrac{1}{2}(t_{i+1} - t_i)(X_i + X_{i+1}) - \tfrac{1}{24}(t_{i+1} - t_i)^3 (f''(t_i) + f''(t_{i+1})) \quad (4)$$

SMOOTHING OF DATA

The procedure for smoothing of data is based on the above cubic spline procedure with a few modifications and is presented by the following algorithm:

Experimental Data Analysis

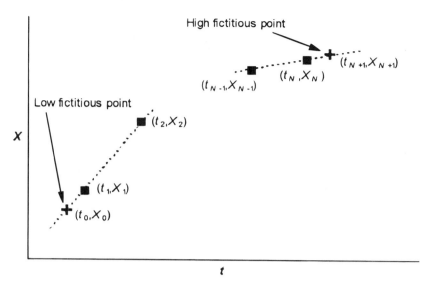

Fig. 2. Determination of fictitious points slightly outside data intervals.

1. Determine positions of two fictitious points outside the range of t by constructing a straight line between the first and second point, and finding a point on this line with a value of t slightly lower than t_1. Do the same for the fictitious point in the upper region of t (*see* Fig. 2).
2. Connect all points including fictitious points with straight lines, and find the center point for each interval of t equal to $(t_{i+1} + t_i)/2$. This will give $N + 1$ new values of t and X.
3. Create cubic splines through the new set of data (the midpoints).
4. Using the original values of t and the spline functions, calculate values of X for all the points. The values will correspond to smooth data and have slightly different values than the original "raw data."

The above procedure may be invoked iteratively by using the smooth data obtained in step 4 as starting values in step 1. One completed sequence may be seen as a smoothness factor (SF) of 1. Similarly, the original cubic spine has a SF of 0 (zero). Visually, one may see the smoothing of the original spline as a method of straightening the curve (*see* Fig. 3). An infinite SF will theoretically result in a straight line. The derivatives of the smooth data may of course be evaluated as described in the previous section.

AN EXAMPLE

In 1959, Luedeking and Piret published a bacterial product-formation equation (4), which has since been used repeatedly to describe production in various bioprocesses. In their original study, they used a graphical

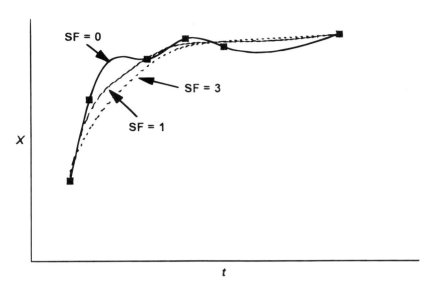

Fig. 3. Examples of cubic splines with different "smoothness" factor.

method, and reduced their 55 raw data points (time and cell concentration) to 25 data points by plotting their raw data, manually constructing a curve through the data, and finally taking readings off this curve. This graphical data were then the starting point for determination of growth rate (dN/dt) by graphical differentiation.

In Fig. 4, I have plotted Luedeking and Piret's determined growth rate (which they determined from 25 points) together with the growth rate calculated based on cubic splines with SFs of zero and one. As is noted when SF = 0 (cubic spline through the 55 raw data points), the calculated derivatives (growth rates) vary substantially from the graphically determined derivatives, but when the raw data are smoothed with an SF = 1, the derivatives compare well with Luedeking and Piret's. The reason that the cubic spline with SF = 0 produces a somewhat scattered result stems from the built-in strength of the spline to adjust itself to go through the points; this may result in an oscillating behavior (as noted in Fig. 3 with SF = 0) that is reflected in derivatives at the points. These oscillations are especially noted when there is experimental error in the data. Smoothing of data reduces oscillations and may be thought of as a method to adjust for experimental errors. Care must be taken not to overdo smoothing and jeopardize accurate representation of the data; the use of correlation coefficients, residuals, and visual inspection (as in Fig. 3) can help alleviate this concern.

Another commonly applied method for smoothing data is least-square regression techniques, and when the sought function is unknown, polynomials are often applied to represent the data. Derivatives (or rates) may then be analytically derived from the regression equation. In Fig. 5, the data from Fig 4. have been represented again, but this time, the deriv-

Experimental Data Analysis

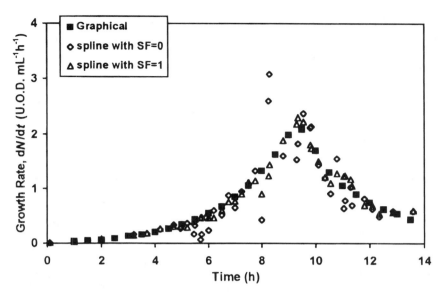

Fig. 4. Comparison between graphical and cubic spline methods when determining derivatives.

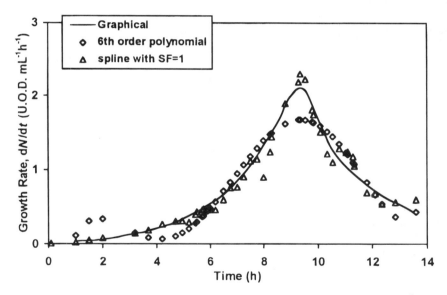

Fig. 5. Comparison of derivatives derived from cubic splines and a sixth-order polynomial.

atives from a spline with SF = 1 are compared with derivatives calculated from a sixth-order polynomial fitted (using method of least-squares) to the raw data. The polynomial fits the data quite well with a correlation coefficient of 0.998, but the calculated derivatives at the points are quite distant from the values obtained by Luedeking and Piret (represented with curve in Fig. 5). This indicates that it is important not to force-fit a set of data to any one function as a means to evaluate derivatives.

ERROR IN SMOOTHING

Since the smooth curve generated through the use of cubic splines with SF > 0 does not go through the data points, it is appropriate that we define a correlation coefficient as (5):

$$r^2 = \frac{\sum_{1}^{N}(X_i - \bar{X})^2 - \sum_{1}^{N}(X_i - \hat{X}_i)^2}{\sum_{1}^{N}(X_i - \bar{X})^2} = \frac{2\sum_{1}^{N} X_i \hat{X}_i - \sum_{1}^{N} \hat{X}_i^2 - \frac{1}{N}\left(\sum_{1}^{N} X_i\right)^2}{\sum_{1}^{N} X_i^2 - \frac{1}{N}\left(\sum_{1}^{N} X_i\right)^2} \quad (5)$$

where X_i is the i:th data value, $\hat{X}_i$ is the predicted value, and $\bar{X}$ is the average of X_i. The residual $(X_i - \hat{X}_i)$ is another useful parameter for estimating data correlation. In the example presented in Fig. 4, the correlation coefficient was equal to 1.00 for SF = 1; thus, the smoothed values were very close to the raw data.

COMPUTER PROGRAMMING

The programming was done in Microsoft Q-Basic for simplicity, but may easily be rewritten to any other programming language. The computer code is listed below and is available via electronic mail by contacting the author.

```
'Initiation
DECLARE SUB TRIMATRIX (t!(), X!(), i0!, N!, r!())
DIM t(50), X(50), tmid(50), Xmid(50), traw(50), Xraw(50), r(50), dXdt(50)
'
'Simple data input
INPUT "Number of points"; N
PRINT "Input in order of increasing t,"
FOR i = 1 TO N
    PRINT "(t,X) for point"; i;
    INPUT traw(i), Xraw(i)
    t(i) = traw(i): X(i) = Xraw(i)
NEXT i
'
'Smoothing of data
CLS
INPUT "Input Smoothness Factor (SF)"; SF
'
FOR SFcounter = 1 TO SF
    'Step 1. Find fictitious points
    t(0) = t(1) – .05 * (t(2) – t(1))
    X(0) = X(1) + (X(2) – X(1)) * (t(0) – t(1)) / (t(2) – t(1))
```

```
       t(N + 1) = t(N) + .05 * (t(N) – t(N – 1))
       X(N + 1) = X(N) + (X(N – 1) – X(N)) * (t(N + 1) – t(N)) / (t(N – 1) – t(N))
'
       'Step 2. Find mid-points on connecting lines
       FOR i = 0 TO N
           tmid(i) = (t(i + 1) + t(i))/2
           Xmid(i) = (X(i + 1) + X(i))/2
       NEXT i
'
       'Step 3. Create cubic splines with mid-points
       CALL TRIMATRIX(tmid(), Xmid(), 0, N, r())
'
       'Step 4. Find smooth X-values on splines at original t-values
       FOR i = 0 TO N – 1
           t = traw(i + 1)
           Xtemp = (r(i) * (tmid(i + 1) – t) ^ 3 + r(i + 1) * (t – tmid(i)) ^ 3) / 6 / (tmid(i + 1) –
           tmid(i))
           Xtemp = Xtemp + (Xmid(i) / (tmid(i + 1) – tmid(i)) – r(i) * (tmid(i + 1) – tmid(i))
           / 6) * (tmid(i + 1) – t)
           X(i + 1) = Xtemp + (Xmid(i + 1) / (tmid(i + 1) – tmid(i)) – r(i + 1) * (tmid(i + 1)
             – tmid(i)) / 6) * (t – tmid(i))
       NEXT i
   NEXT SFcounter
'
   CALL TRIMATRIX(t(), X(), 1, N, r())
'
   'Calculate derivatives based on spline equations
   FOR i = 1 TO N – 1
       dXdt(i) = –r(i) * (t(i + 1) – t(i)) / 2 + (X(i + 1) – X(i)) / (t(i + 1) – t(i)) – (r(i + 1) –
       r(i)) / 6 * (t(i + 1) – t(i))
   NEXT i
   'Derivative at last point (N) must be calculated separately
   dXdt(N) = (X(N) – X(N – 1)) / (t(N) – t(N – 1)) + r(N – 1) * (t(N) – t(N – 1)) / 6
'
   'Print smooth values and derivatives at smooth values
   PRINT "These are your raw and smooth values plus derivatives with SF="; SF
   PRINT "t-raw", "X-raw", "t-smooth", "X-smooth", "dX/dt"
   FOR i = 1 TO N
       PRINT traw(i), Xraw(i), t(i), X(i), dXdt(i)
       'Calculate the summations needed for r2 calculation
       sum1 = sum1 + X(i) * Xraw(i): sum2 = sum2 + X(i)^ 2: sum3 = sum3 + Xraw(i):
       sum4 = sum4 + Xraw(i)^ 2
   NEXT i
   PRINT "For SF="; SF; ", the correlation coefficient, r2 ="; (2 * sum1 – sum2 –
       sum3 ^ 2 / N) / (sum4 – sum3 ^ 2 / N)
   END
'
   SUB TRIMATRIX (t(), X(), i0, N, r())
       DIM a(50), b(50), c(50) 'Local variables
       'Create constants in Equation <2> based on points, numbered i0 to N
       FOR i = i0 TO N – 2
           a(i + 1) = t(i + 1) – t(i)
           b(i + 1) = 2 * (t(i + 2) – t(i))
```

```
        c(i + 1) = t(i + 2) - t(i + 1)
        r(i + 1) = 6 * (X(i + 2) - X(i + 1)) / (t(i + 2)
           - t(i + 1)) - 6 * (X(i + 1) - X(i)) / (t(i +
           1) - t(i))
    NEXT i
    r(i0) = 0: r(N) = 0 'Criterion 4
    '
    'Solve triagonal augmented matrix
    FOR i = i0 + 1 TO N - 2
        b(i + 1) = b(i + 1) - a(i + 1) / b(i) * c(i)
        r(i + 1) = r(i + 1) - a(i + 1) / b(i) * r(i)
    NEXT i
    FOR i = 1 TO N - 1 - i0
        r(N - i) = (r(N - i) - c(N - i) * r(N - i + 1))
           / b(N - i)
    NEXT i
    '
    'r() contains the solution = values for secondary derivatives
    END SUB
```

CONCLUSION

Cubic spline functions represent a powerful method of evaluating rates and/or smoothing data. The functions are easily manipulated, and the algorithm can be written concisely. The data smoothing can potentially adjust for experimental error and compares well to manual graphical methods. The method presented constitutes an improvement over force-fitting.

REFERENCES

1. LeDuy, A. and Zajic, J. E. (1973), *Biotechnol. Bioeng.* **15**, 805–810.
2. Cheney, W. and Kincaid, D. (1980), in *Numerical Mathematics and Computing*, Brooks/Cole, Monterey, CA, pp. 339–343.
3. Klasson, K. T., Clausen, E. C., and Gaddy, J. L. (1989), *Appl. Biochem. Biotechnol.* **20**, 491–509.
4. Luedeking, R. and Piret, E. L. (1959), *J. Biochem. Microbiol. Technol. Eng.* **1**, 393–412.
5. Walpole, R. E. and Myers, R. H. (1989), in *Probability and Statistics for Engineers and Scientists*, 4th ed., Macmillan, New York, p. 422.

Session 3

Bioprocessing Research

ROBERT R. DORSCH[1] AND CHRISTOS HATZIS[2]

[1]*DuPont, Wilmington, DE; and*
[2]*National Renewable Energy Laboratory, Golden, CO*

Bioprocessing can be described as the process whereby a material is converted into another using biological agents. This area combines chemical engineering, biology, and fermentations. This can also refer to an overall process in which one key step is carried out biologically (the other steps may be grinding, separations, purification, and so on). This session and the associated posters surveyed current efforts in the interest of taking a potentially valuable bioconversion and testing it as a process. The products included ethanol, fumaric and succinic acids, and the enzyme cellulase. The reactor schemes described included stirred batch reactors, slurry reactors, immobilized-cell fluidized-bed columnar reactors, and improved airlift columnar reactors.

Bioprocessing research also involves providing a better understanding of the operation of the reactor system. Even the basic stirred tank can provide challenges of oxygen transfer, especially into filamentous systems. Separations are a critical part of the bioprocess, and were represented by *in situ* gas stripping of a volatile product such as ethanol or the investigation of protein fouling of the membranes used to separate biomass and products.

Cellulase Production Based on Hemicellulose Hydrolysate from Steam-Pretreated Willow

ZSOLT SZENGYEL,*,[1] GUIDO ZACCHI,[1] AND KATI RÉCZEY[2]

[1] Department of Chemical Engineering 1, University of Lund, P.O. Box 124 S-22100, Lund, Sweden; and [2] Department of Agricultural Chemical Technology, Technical University of Budapest, Szent Gellért tér 4, Budapest, H-1521, Hungary

ABSTRACT

The production cost of cellulolytic enzymes is a major contributor to the high cost of ethanol production from lignocellulosics using enzymatic hydrolysis. The aim of the present study was to investigate the cellulolytic enzyme production of *Trichoderma reesei* Rut C 30, which is known as a good cellulase secreting micro-organism, using willow as the carbon source. The willow, which is a fast-growing energy crop in Sweden, was impregnated with 1–4% SO_2 and steam-pretreated for 5 min at 206°C. The pretreated willow was washed and the wash water, which contains several soluble sugars from the hemicellulose, was supplemented with fibrous pretreated willow and used for enzyme production. In addition to sugars, the liquid contains degradation products such as acetic acid, furfural, and 5-hydroxy-methylfurfural, which are inhibitory for microorganisms. The results showed that 50% of the cellulose can be replaced with sugars from the wash water. The highest enzyme activity, 1.79 FPU/mL and yield, 133 FPU/g carbohydrate, was obtained at pH 6.0 using 20 g/L carbon source concentration. At lower pHs, a total lack of growth and enzyme production was observed, which probably could be explained by furfural inhibition.

Index Entries: Cellulase enzyme production; *Trichoderma reesei* Rut C 30; lignocellulosics; furfural inhibition.

INTRODUCTION

Enzymatic conversion of willow, considered a promising raw material for large scale production of ethanol in Sweden, has been investigated during the last decade *(1,2)*. A typical process configuration consists of pre-

*Author to whom all correspondence and reprint requests should be addressed.

treatment, enzyme production, enzymatic hydrolysis, fermentation, and ethanol refining as the major process steps. The raw material is first steam exploded with high pressure saturated steam to make the cellulose more accessible to enzymatic attack. In the hydrolysis step the cellulose is converted to glucose with cellulolytic enzymes, which are produced from a part of the pretreated material using *T. reesei*. The hydrolysate is then separated from the solid residue and fermented with baker's yeast to ethanol. The solid residue, mostly lignin, can be used as a solid fuel for production of process steam. Finally the ethanol is refined using distillation.

The raw material constitutes a large fraction of the total production cost in the bioconversion of lignocellulosic materials to ethanol (3). A high degree of utilization of the raw material is thus required. When hardwood, such as willow, is used, the hemicellulose part, which constitutes 23% of the dry matter, is underutilized. During the steam-pretreatment of willow, the hemicelluloses are degraded mostly to monosaccharides of which 55% is xylose (4). In addition to soluble sugars, by-products are also formed, which are inhibitory to fermentation (5). When ordinary baker's yeast is used for alcoholic fermentation, the xylose fraction of the raw material leaves the process unchanged, which reduces the total yield of ethanol based on the amount of saccharides available in the raw material. There are also other yeasts, (such as *Pichia stipitis, Candida shehatae*), which are capable of converting the pentoses to ethanol, but they do not ferment well in undetoxified hydrolysates (6). Another possible way to utilize the pentoses is to use them for production of cellulolytic enzymes. This results in an increased ethanol yield as a part of the cellulose, which would be used for enzyme production that can be replaced with the xylose-rich liquid. The xylose utilization of *Trichoderma reesei* Rut C 30 was investigated by Mohagheghi et al. (7) using the mixture of Solka Floc and xylose in the medium. It was clearly shown that *T. reesei* can utilize the xylose, but for sufficient enzyme production the medium must be supplemented with either fibrous cellulose, or other inducers such as sorbose (8).

In a previous study, it was shown that steam-pretreated willow is a suitable carbon source for enzyme production using *T. reesei* Rut C 30 (9). The aim of the present work was to examine the enzyme production of the same strain on various mixtures of fibrous and liquid fractions from steam-pretreated willow as carbon sources. The effects of the composition, concentration of carbon source, stirring speed, and pH were investigated.

MATERIALS AND METHODS

Pretreatment

In the present study, *Salix caprea*, a fast-growing willow species, was used for production of cellulolytic enzymes. Chopped and screened willow was first presteamed for 40 min with saturated steam at 1 bar. After

Cellulase Production

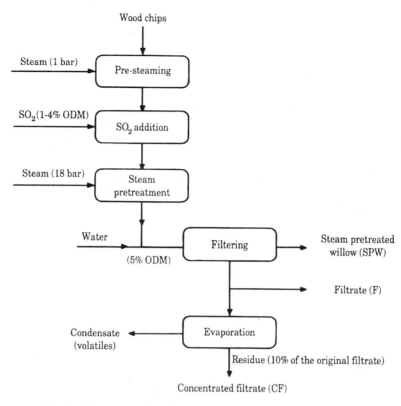

Fig. 1. Schematic flowsheet of carbon source preparation.

presteaming, the hot material was transferred into plastic bags and impregnated with 1–4% SO_2, based on oven-dried material (ODM), and stored overnight at 4°C. The steam treatment was performed at 206°C for 5 min *(1,2)*. The pretreated material was diluted with tap water to approx 5% ODM and filtered using a PF 0.1H2 (Larox OY, Finland) filter press unit *(10)*. The filtrate, i.e., the hydrolyzed hemicellulose part of the raw material, was divided into two fractions. One fraction was concentrated by removing 90% of the liquid by vacuum evaporation at 80°C and pH 2.8–3.0 using a Büchi RE 121 rotavapor. The concentrated filtrate (CF) had tenfold higher concentration of nonvolatiles, whereas most of the volatile compounds were removed. The fibrous material (SPW), the CF, and the unconcentrated filtrate (F) were all used as carbon sources for enzyme production. Figure 1 shows the schematic procedure for the preparation of the three fractions and the nomenclature that will be used. The composition of the pretreated willow was determined using Hägglund's method *(11)*, with the modification that the sugar content of the acid hydrolysate was also analyzed, and from that the cellulose content was calculated. The

SPW contained 48.2% cellulose based on ODM. The liquids were analyzed on HPLC for different sugars and inhibitory compounds.

Inoculum Preparation

The fungus *Trichoderma reseei* Rut C 30 was stored on agar slants containing (in g/L): 20 malt extract, 5 glucose, 1 proteose peptone, and 20 bacto agar, prior to usage. After 30 d at 30°C, the conidia were suspended in 5 mL of sterile water and 1.6 mL of the suspension was pipetted into a 1 L baffled E-flask containing 200 mL sterilized Mandels medium *(12)*, in which the concentrations of the nutrients in g/L were 0.3 urea, 1.4 $(NH_4)_2SO_4$, 2.0 KH_2PO_4, 0.3 $CaCl_2$, 0.3 $MgSO_4$, 0.25 yeast extract, and 0.75 proteose peptone together with 7.5 Solka Floc. Trace elements were also added (mg/L): 5 $FeSO_4 \cdot 7H_2O$, 20 $CoCl_2$, 1.6 $MnSO_4$, and 1.4 $ZnSO_4$, respectively. The pH before sterilization was adjusted to 5.4. After 4 d at 30°C and 300 rpm the inoculum was ready.

Enzyme Production in Shake Flasks

The mycelia from the inoculum was used to initiate growth in 1 L baffled flasks containing 200 mL of a modified Mandels medium where the yeast extract and the proteose peptone were replaced with 0.38 g/L dried yeast *(9)*. The inoculum constituted 10% of the total volume. The enzyme production was performed at two different carbon source concentrations corresponding to 10 and 20 g/L of total available carbohydrate. For the higher loading the amount of nutrients and inoculum were doubled. The enzyme production was performed in a rotary shake-incubator at 30°C and 300 rpm. The pH was adjusted if it was below 4.4 with addition of sterile 10 wt% NaOH solution. Samples were taken once a day and centrifuged at 2500g for 10 min using a Winfug (AB Winkelcentrifug, Sweden) centrifuge. The supernatant was then analyzed for enzyme activity and the concentration of sugars and some volatile inhibitors were also determined.

Analysis

The enzyme activity of the samples was determined both as filter paper activity (FPU) using Mandels procedure *(13)* and β-glucosidase activity using Berghem's method *(14)*.

The centrifuged samples from the enzyme production were filtered through 0.2 μm membrane filters (MFS-13, Micro Filtration System) and analyzed on an HPLC unit (Shimadzu, Japan) equipped with a refractive index detector. Cellobiose, glucose, xylose, acetic acid, 5-hydroxy-methyl-furfural, and furfural were separated on an Aminex HPX-87H column at 65°C using 5 mM H_2SO_4 as eluent, at a flow rate of 0.5 mL/min.

RESULTS AND DISCUSSION

Enzyme Production using Concentrated Hemicellulose Hydrolysate

Two main series of experiments were performed in shake flasks to investigate the cellulase production of *T. reesei* using the concentrated filtrate, CF, supplemented with fibrous pretreated willow, SPW. In the first series, the effect of the initial pH was studied. The initial pHs were set to 4.8, 5.3, and 5.8 before sterilization. The concentration of the sugars in the concentrated filtrate was set to 5 g/L by dilution with tap water, and the medium was supplemented with pretreated willow at an amount corresponding to a cellulose concentration of 5 g/L. The concentrated and rediluted filtrate to 5 g/L total sugar contained in g/L: 0.07 cellobiose, 1.48 glucose, 3.35 xylose, 0.63 acetic acid, and 0.13 5-hydroxi-methylfurfural. The concentration of furfural was below the detection limit.

In the second set of experiments, the effect of the concentration of carbon source and nonvolatile inhibitors were examined at two carbon source levels, 10 and 20 g/L of total carbohydrates. In both series the cellulose constituted 50% of the total carbon source. Control fermentations were run with SPW concentrations corresponding to 10 and 20 g/L of cellulose. For each conditions three fermentations were run in parallel and the mean values were calculated.

The enzyme activity and the total sugar concentration vs time are shown in Fig. 2 for the first series of fermentations. The mean value of the standard deviation was 0.04 FPU/mL, for the enzyme activity measurement and 0.30 g/L, for total sugar analysis in all experiments. The pH has a significant effect on enzyme production and in all cases the cellulase production was repressed while sugars were present in the medium. This can be explained by the presence of easily metabolisable sugars, which increase the intracellular level of glucose-6-phosphate or its analogues and repress the cellulase production *(15)*. Thus the cellulolytic enzyme production rate is a function of the sugar consumption (*see* Fig. 2A, B). At the lowest pH, 4.8, a very long lag phase can be seen, which is reduced by increasing pH. Also, the final enzyme yields differed although the initial carbon source concentration was the same in all experiments. The highest filter paper activity after 7 d cultivation, 0.89 FPU/mL, was obtained when the starting pH was 5.8, and the lowest 0.64 FPU/mL, for pH 4.8. It seemed that the pH had a double effect on cellulolytic enzyme production when the medium contained soluble sugars. Both the growth rate of the microorganism and the amount of secreted cellulases increased with increasing pH. The β-glucosidase production showed the same trend, with a final activity of 0.46 IU/mL obtained at pH 5.8 and 0.16 IU/mL at the lowest pH level.

The second series of fermentations were performed with the initial pH 5.8 based on the results of the previous experiments. Table 1 shows the final

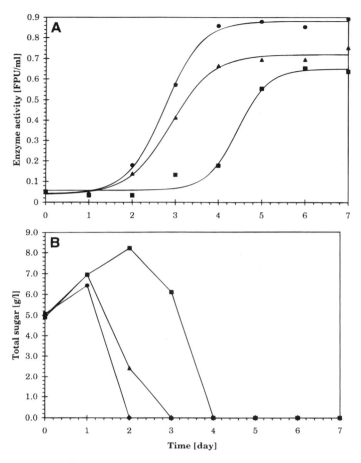

Fig. 2. Enzyme activity vs time for various staring pHs. **(A)** Cellulase activity; **(B)** total sugar concentration. ■ = pH 4.8, ▲ = pH 5.3, ● = pH 5.8.

Table 1
Enzyme Activities and Yields after 8 d Cultivation
for Concentrated Filtrate (CF) with Initial pH 5.8

No.	FPU [IU/ml]	Stdev$_{FPU}$ [IU/ml]	Yield [FPU/g carbohydrate]	ß-gluc. [IU/ml]	Stdev$_{ß-gluc.}$ [IU/ml]
A	1.04	0.03	104	0.31	0.01
B	1.33	0.08	133	0.47	0.06
C	1.58	0.06	79	0.44	0.03
D	1.79	0.06	86	0.43	0.02

A-10 g/L carbon source 100% SPW
B-10 g/L carbon source 50% SPW and 50% CF
C-20 g/L carbon source 100% SPW
D-20 g/L carbon source 50% SPW and 50% CF

enzyme activities and yields calculated on the total added carbon source (cellulose plus sugars). The FPU values were in the range 1.04 FPU/mL to 1.79 FPU/mL. The fermentations using a mixture of SPW and CF resulted in higher FPU activities compared with the fermentations where only SPW was used. One possible explanation is that the liquid (CF) contained some oligomeric saccharides, which were not considered, as only three carbohydrates were analyzed on HPLC. These media would then contain more carbon source than the media with SPW. The yields are lower for 20 g/L carbon source, than at 10 g/L concentration. The higest yield, 133 FPU/g added carbon source, was obtained with the medium containing the mixture of 5 g/L sugars from CF and 5 g/L cellulose from SPW. The lower yield at higher carbon source concentrations can be the effect of an increased mass transfer resistance when the concentration of the solids were doubled because of poor stirring in the shake flasks. The β-glucosidase activity varied less than the filter paper activity. It was in the range 0.43–0.47 IU/mL except for SPW with a cellulose content corresponding to 10 g/L, where a somewhat lower activity, 0.31 IU/mL, was obtained.

Enzyme Production Using Original Pretreated Filtrate

A central-composite face (CCF) experimental design was used to investigate the effect of three factors on enzyme production using the original F. The three factors were the pH, stirring speed, and the amount of soluble sugars in percent of the total carbon source (10 g/L), whereby the amount of inhibitors were also varied. Table 2 shows the levels of the three factors. The initial composition of the media are summarized in Table 3. The amount of added filtrate in the medium was calculated on the total amount of carbohydrates measured after enzymatic hydrolysis for 48 h with mixture of cellulolytic enzymes. This analysis showed that the total amount of the carbohydrates in the filtrate was 1.3 times higher than that of the amount of the sugars measured by HPLC. For each experimental setup three fermentations were run in parallel.

Unfortunately only six conditions, out of the 15, were successful, i.e., yielded a measurable amount of enzymes, thus the evaluation of the data using the CCF design was not possible, but general conclusions could be drawn by comparison of the individual fermentations. The final enzyme activities are shown in Table 4.

At the highest concentration of soluble sugars (8 g/L) no growth or enzyme production was observed even at the highest stirring speed and pH levels. The furfural concentration in the medium, which initially was around 1 g/L, decreased slightly during the cultivation, but was never reduced to 0.0 g/L. The concentration of sugars was increased during the fermentation, probably because of the action of the cellulases loaded with the inoculum. The level of acetic acid was unchanged during the whole cultivation.

Table 2
CCF Experimental Design Setup

Exp. No.	pH	Stirring speed [rpm]	Soluble carbon [% of total carbon source]
1	5.0	200	20
2	6.0	200	20
3	5.0	400	20
4	6.0	400	20
5	5.0	200	80
6	6.0	200	80
7	5.0	400	80
8	6.0	400	80
9	5.0	300	50
10	6.0	300	50
11	5.5	200	50
12	5.5	400	50
13	5.5	300	20
14	5.5	300	80
15	5.5	300	50

Table 3
Composition of Media Used in CCF Experimental Design*

Soluble carbon source [%]	CELL [g/l]	GLU [g/l]	XYL [g/l]	TS [g/l]	HAc [g/l]	HMF [g/l]	FUR [g/l]
20	0.05	0.32	1.18	1.55	0.72	0.06	0.27
50	0.12	0.91	3.11	4.14	1.78	0.16	0.73
80	0.15	1.35	4.65	6.15	2.63	0.23	1.08

*oligosaccharides are not shown
CELL = cellobiose, GLU = glucose, XYL = xylose, TS = total HPLC sugar, HAc = acetic acid, HMF = 5-hydroximethyl furfural, FUR = furfural

When the medium contained 2 g/L sugars, all conditions gave evaluable results. The enzyme activities were in the range 0.57–0.77 FPU/mL. As was stated before, there was no sense in evaluating the obtained data using the CCF design, but experiments 1–4 were calculated using a 2^2-type factorial design with the pH and stirring speed as variables, and the maximum

Cellulase Production

Table 4
Final FPU Activities for the Successful CCF Experiments

Exp. No.	FPU [IU/ml]	Stdev$_{FPU}$ [IU/ml]
1	0.57	0.05
2	0.77	0.04
3	0.63	0.02
4	0.76	0.03
10	0.59	0.04
13	0.64	0.04

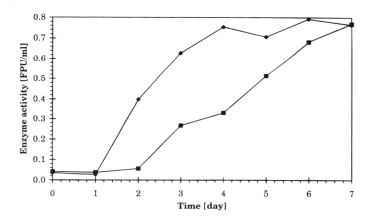

Fig. 3. CCF experimental design. Comparison of conditions No2. and No4. (*see* Table 2). ■ = FPU (No2., 200 rpm), ♦ = FPU (No4., 400 rpm).

FPU activity values obtained after 7-d cultivation as the dependent variable. The result of the regression analysis of the data comprising to fitting a polynomial equation and significance test of obtained parameters showed that the stirring has no significant effect on the final yield. The final equation obtained was $Y = 0.68 + 0.08x_1$, where Y is the response (FPU activity) and $x_1 = [\text{pH-5.5}]/0.5$. Although the stirring has no effect on the final yield as shown in Fig. 3, it has a significant effect on the enzyme production rate.

The effect of increased pH can be seen in Fig. 4, where the enzyme production, sugar concentration, and furfural concentration vs time are shown for conditions 1 and 2 (*see* Table 2). The final cellulolytic activity, after 7 d cultivation, increased from 0.57 FPU/mL at pH 5.0 to 0.77 FPU/mL at pH 6.0. In both cases the utilization of sugars was not started until the furfural was consumed, which can be considered to be a detoxification period. The soluble sugars were utilized to support the growth of the microorganism.

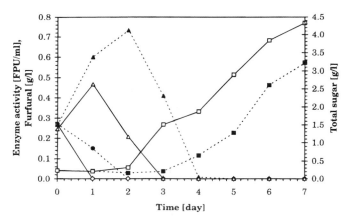

Fig. 4. CCF experimental design. Comparison of conditions No1. and No2. (*see* Table 2). ■ = FPU (No1.), ▲ = Total sugar (No1.), ● = Furfural (No1.), □ = FPU (No2.), △ = Total sugar (No2.), ○ = Furfural (No1.).

During this period no cellulase production was observed. This was followed by the cellulose utilization phase, with the effective cellulase production. At pH 6.0 the furfural was consumed already within 1 d, whereas it took about 2 d at pH 5.0. This gave a higher enzyme production rate although it seems that the production was not completed after 7 d in either of the two conditions. At pH 6.0 there is no difference between the final enzyme activities at various stirring speeds, but the fermentation performed at 400 rpm was finished after 4 d cultivation, whereas at 200 rpm, the FPU activity increased during the whole fermentation (Fig. 3).

There was only one fermentation with 50% F, which resulted in a measurable enzyme production (Exp. 10). This was performed at pH 6.0 and 300 rpm. The furfural concentration, which initially was around 0.70 g/L, decreased to 0.0 g/L during the first 2 d (*see* Fig. 5). The level of the acetic acid was constant for 2 d, but was then consumed completely within 4 d. The total sugar concentration increased during the first 2 d and then decreased to 0.0 g/L. After 2 d the cellulase production started and the final activity after 7 d cultivation was 0.59 FPU/mL. This clearly shows that the acetic acid is utilized together with the other soluble carbohydrates in the medium and that it has no effect on growth at least up to a concentration of 1.8 g/L and if the pH is high enough. The somewhat lower FPU activity, compared with the results at 20% soluble carbohydrates is probably the result of the higher concentration of furfural in the medium.

CONCLUSIONS

The experiments on concentrated filtrate show that 50% of the cellulose can be replaced with soluble sugars derived from the hemicellulose fraction of the willow. A yield of 133 FPU/g added carbohydrates was obtained using the mixture of 5 g/L SPW and 5 g/L sugars from CF,

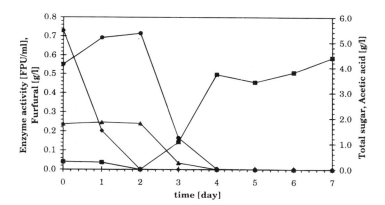

Fig. 5. CCF experimental design. Enzyme activity, acetic acid, total sugar and furfural concentration versus time for condition No10. (*see* Table 2). ■ = FPU, ◆ = Furfural, ▲ = Acetic acid, ● = Total sugar.

whereas the cultivation on 100% SPW gave 104 FPU/g. The yields at 20 g/L carbohydrates concentration were lower, 79 FPU/g for SPW and 86 FPU/g for the SPW and CF mixture. The yields obtained on media containing 50% soluble sugars of the total carbohydrates were higher than the yields on pure SPW. The yields at higher solids concentrations were lower than at lower solids concentrations, thus, it may be concluded that the improved mass transfer in the case of lower solids concentrations yielded an increased cellulase production.

It is evident that the evaporation of water solutions is expensive because of the high energy demand. From an economical point of view it is preferable to use the original filtrate. The effect of the pH, stirring speed, and amount of soluble sugars on enzyme production was investigated with a CCF experimental design. Only six conditions gave evaluable enzyme activities, thus the evaluation of the data with CCF model was not possible. There was no enzyme production when 8 g/L sugars were added to the medium, which corresponds to a furfural concentration of 1 g/L. Only one condition with 5 g/L sugars provided cellulases with an activity of 0.59 FPU/mL after 7 d cultivation. The experiments at a concentration of 2 g/L sugars indicate that furfural has an effect on growth and that is implied by a low pH. At pH 6.0 the furfural was rapidly consumed, yielding the highest cellulolytic activity, 0.77 FPU/mL. The result of the statistical analysis of data obtained at 2 g/L sugar concentration shows that only the pH has effect on cellulase yield at constant solids concentrations and the effect of stirring can be neglected.

ACKNOWLEDGMENT

The Swedish National Board for Industrial and Technical Development (NUTEK), and the National Research Fund of Hungary (OTKA T-017201) are gratefully acknowledged for their financial support.

REFERENCES

1. Galbe, M. (1994), PhD thesis, Lund Institute of Technology, Lund, Sweden.
2. Eklund, R. (1994), PhD thesis, Lund Institute of Technology, Lund, Sweden.
3. von Sivers, M. and Zacchi, G. (1995), *Bioresource Technol.* **51**, 43–52.
4. Eklund, R., Galbe, M., and Zacchi, G. (1995), *Bioresource Technol.* **51**, 225–229.
5. Palmqvist, E., Hahn-Hägerdal, B., Galbe, M., and Zacchi, G. (1996), *Enzyme Microb. Technol.* **19**, 470–476..
6. Olsson, L. and Hahn-Hägerdal, B. (1993), *Process Biochem.* **28**, 249–257.
7. Mohagheghi, A., Grohmann, K., and Wyman, C. E. (1988), *Appl. Biochem. Biotechnol.* **17**, 263–277.
8. Schaffner, D. W. and Toledo, R. T. (1991), *Biotehnol. Bioeng.* **37**, 12–14.
9. Réczey, K., Szengyel, Zs., Eklund, R., and Zacchi, G. (1996), *Bioresource Technol.* **57**, 25–30.
10. Palmqvist, E., Hahn-Hägerdal, B., Galbe, M., Larsson, M., Stenberg, K., Szengyel, Zs., Tengborg, C., and Zacchi, G. (1996), *Bioresource Technol*, **58**, 171–179.
11. Hägglund, E. (1951), in *Chemistry of Wood*, Academic Press, New York, NY.
12. Mandels, M. and Weber, J. (1969), *Adv. Chem. Ser.* **95**, 391–414.
13. Mandels, M., Andreotti, R., and Roche, C. (1976), *Biotechnol. Bioeng. Symp.* **6**, 21–33.
14. Berghem, L. E. R. and Petterson, L. G. (1974), *Eur. J. Biochem.* **46**, 295–305.
15. Sestak, S. and Farkas, V. (1993), *Can. J. Microbiol.* **39**, 342–347.

Effect of Impeller Geometry on Gas-Liquid Mass Transfer Coefficients in Filamentous Suspensions

SUNDEEP N. DRONAWAT, C. KURT SVIHLA, AND THOMAS R. HANLEY*

Department of Chemical Engineering, Speed Scientific School, University of Louisville, Louisville, KY 40292

ABSTRACT

Volumetric gas-liquid mass transfer coefficients were measured in suspensions of cellulose fibers with concentrations ranging from 0 to 20 g/L. The mass transfer coefficients were measured using the dynamic method. Results are presented for three different combinations of impellers at a variety of gassing rates and agitation speeds. Rheological properties of the cellulose fibers were also measured using the impeller viscometer method. Tests were conducted in a 20 L stirred-tank fermentor and in 65 L tank with a height to diameter ratio of 3:1. Power consumption was measured in both vessels. At low agitation rates, two Rushton turbines gave 20% better performance than the Rushton and hydrofoil combination and 40% better performance than the Rushton and propeller combination for oxygen transfer. At higher agitation rates, the Rushton and hydrofoil combination gave 14 and 25% better performance for oxygen transfer than two Rushton turbines and the Rushton and hydrofoil combination, respectively.

Index Entries: Filamentous suspensions; Gas-liquid mass transfer; non-Newtonian rheology.

INTRODUCTION

Submerged, aerobic industrial fermentations are usually carried out in sparged, agitated gas-liquid reactors. Mycelial fermentations, an important class of biochemical processes, present unique mixing prob-

---

*Author to whom all correspondence and reprint requests should be addressed.

lems due to non-Newtonian behavior, changes in the rate-controlling step, and relatively high oxygen demand (1). High dissolved gas demand and the shear insensitivity of the mycelial cultures make agitated tanks a popular reactor choice. A lack of understanding of gas-liquid mass transfer has hindered optimization for these processes (2). Few studies have been conducted to study the effect of impeller geometry on gas-liquid mass transfer in filamentous systems (3). Previous studies conducted by these authors focused on the effects of mixing on gluconic acid production by *Aspergillus niger* in submerged culture (4,5). In the present work, cellulose fibers were used to simulate the filamentous nature of *Aspergillus niger*. The aim of this study was to obtain experimental data for gas-liquid mass transfer coefficients in filamentous systems for three types of commonly used impellers.

Some previous studies have suggested that cellulose fiber suspensions exhibit rheological behavior similar to filamentous mycelial broths, although higher mass loadings are generally required (3,5). This work presents the results of mass transfer coefficient measurements in a mechanically agitated fermentor of 20 L capacity for a range of fiber concentrations, agitation speeds, and gassing rates using three types of impellers (a six-bladed Rushton turbine, a hydrofoil, and a marine propeller). Experiments were also conducted in a 65 L tank equipped with three Rushton turbine impellers to study the effect of scale-up. The rheological behavior of the fiber suspensions was characterized using the impeller method (5).

EXPERIMENTAL METHODS

The 20 L fermentor used in the tests was equipped with two agitators of diameter 0.069 m, four baffles, and a heating and cooling element. A liquid volume of 15 L was used in the fermentor. The 65 L Plexiglas vessel had a height to diameter ratio of 3:1 and a working volume of 63.7 liters. It was equipped with four baffles and three Rushton turbine impellers. Schematic diagrams of the two systems are shown in Fig. 1. Table 1 lists the dimensions of the impellers and tanks used.

The cellulose fibers used in the studies (Solka Floc, grade KS-1016 Fiber Sales and Development, Urbana, OH) had an average fiber length of 216 µm and were suspended in tap water. Fiber concentrations of 5, 10, 15, and 20 g/L were used at aeration rates of 5, 10, and 20 standard L per minute. Dissolved oxygen concentrations were measured with a polarographic electrode. Gas flow rates were measured using a mass flow meter and a rotameter. The dissolved oxygen probe was connected to a PC for data acquisition via a computer interface board.

A separate experimental apparatus was constructed to measure the dynamics of the electrode. The schematic of the experiment is given in Fig. 2. In this experiment the polarographic electrode was placed in the tube and an arrangement for instantaneous change from nitrogenated

Impeller Geometry

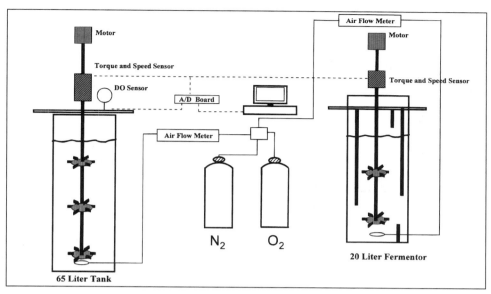

Fig. 1. Experimental setup for $k_L a$ and power measurements in the 20 L fermentor and 65 L tank.

Table 1
Dimensions of the Tanks and Impellers Used

20 Liter Fermentor		65 Liter Tank	
Diameter of Tank	22.5 cm	Diameter of Tank	30 cm
Height of Tank	50 cm	Height of Tank	106.7 cm
No of Baffles	4	No of Baffles	4
No. of Impellers	2	No. of Impellers	3
Baffle Width	2.0 cm	Baffle Width	2.85 cm
Impeller Diameter	6.9 cm	Impeller Diameter	9.9 cm

liquid to oxygenated liquid was constructed. The probe response was assumed to be linear and to be described by a second-order differential equation. A 386-microprocessor, 25 MHz computer was used for all trials and data analysis.

Volumetric gas-liquid mass transfer coefficients were measured using the polarographic electrode for which the dynamics were determined using the apparatus shown in Fig. 2. The dynamic method was used with an instantaneous interchange in inlet gas (6). The tank was initially sparged with nitrogen gas for sufficient amount of time to attain saturation. Oxygen was then sparged to the tank and the dissolved oxygen

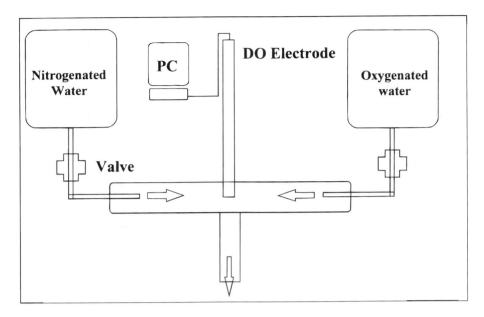

Fig. 2. Experimental setup for electrode dynamics.

concentration response measured at intervals of 0.10 s through the A/D interface. The measured dissolved oxygen concentration response curve was used to determine the value of $k_L a$ for the tests using the mathematical model discussed later in the paper. The temperature for all trials was maintained constant at 30°C.

The rheological behavior of the fiber suspensions were characterized using the impeller viscometer method for fiber concentrations varying from 10 to 30 g/L. The method for measurement of rheological properties of filamentous suspensions is described elsewhere (4,5,7).

MATHEMATICAL MODEL

In the model derivation, the liquid phase is assumed to be well-mixed and the gas interchange from nitrogen to oxygen is assumed to take place instantaneously. The initial period in which both oxygen and nitrogen are present in the fermentor is not considered. Henry's law is assumed to apply at the gas-liquid interface (3). The material balance for the liquid phase is:

$$\frac{dC_L}{dt} = k_L a (\alpha C_g - C_L) \tag{1}$$

where α is the Henry's law coefficient.

The electrode dynamics are assumed to be described by a second-order differential equation with time constants τ_1 and τ_2. When expressed in the dimensionless form:

$$\phi = \frac{C_E(T) - C_E}{C_E(T) - C_E(0)} \tag{2}$$

the probe response in the electrode dynamic tests depends only on the dynamic characteristics, τ_1 and τ_2.

A solution for the measured dissolved oxygen response to a step change in inlet gas concentration can be obtained using Laplace transforms:

$$\frac{C_E s}{C_E(T) - C_E(0)} = \frac{e^{-\tau_d s}}{\left(\frac{s}{k_L a} + 1\right)(\tau_1 s + 1)(\tau_2 s + 1)} \tag{3}$$

where a time lag τ_d has also been introduced. When this expression is inverted into the time domain, the resulting expression is a function only of the dynamic characteristics of the system.

The two time constants, τ_1 and τ_2, were determined by a nonlinear least squares regression of the electrode response curve. Since the probe dynamics did vary over time, the dynamics were determined just before or after a given set of trials. The dead time was obtained for each test by observing the plot of time vs the electrode response and the first observable electrode response was noted. The time axis was shifted by the value of the dead time determined for each trial. Since the electrode time constants τ_1 and τ_2 were determined in separate tests and τ_d found from direct inspection of the response curve; $k_L a$ is the only variable that is adjusted to fit the dynamic test data to the model.

RESULTS

The electrode time constants, τ_1 and τ_2, were typically in the range of 5.1 to 5.9 s, and 7.9 to 8.6 s, respectively. The largest $k_L a$ value for any absorption test was approx 0.036 s^{-1}, hence the electrode dynamics never dominated the overall response. The fit of the experimental data to the model was generally good as is shown in Fig. 3. The effect of cellulose fiber concentration on $k_L a$ for the Rushton and Rushton turbine combination for 10 L per minute air flow rate in the 20 L fermentor is presented in Fig. 4. It can be inferred from Fig. 4 that an increase in cellulose fiber concentration decreases the $k_L a$. This effect is more pronounced at higher agitation rates and higher concentrations of cellulose fiber. The effect of impeller geometry on the gas-liquid mass transfer coefficient is shown in Figs. 5 and 6 for cellulose fiber concentrations of 10 g/L and 20 g/L, respectively for 10 L per minute air-flow rate. The increase in cellulose fiber concentration decreases the gas-liquid mass transfer coefficient for all combinations of impellers. Similar behavior was observed at an air flow rate of 20 L per minute. It is evident that the gas-liquid mass transfer coefficient increases with an increase in aeration rate. Figures 7 and 8 show the effect of impeller geometry on power consumption at 10 L per minute and 20 L per minute

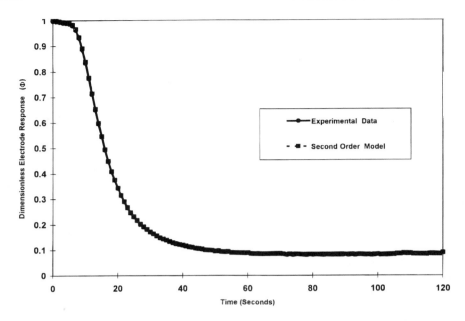

Fig. 3. Experimental electrode response fit to the model for k_La measurements.

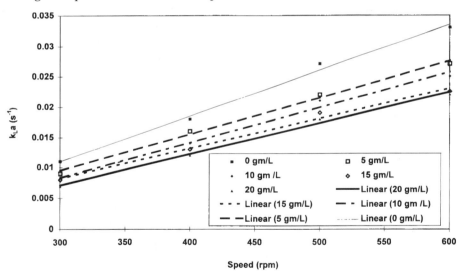

Fig. 4. Effect of fiber conc. on k_La at 10 lpm airflow rate for Rushton and Rushton (20 L fermentor).

flow rate, respectively with no cellulose fibers. Figures 9 and 10 show the effect of solids concentration on the gas-liquid mass transfer coefficient at air flow rates of 10 L per minute and 20 L per minute, respectively in the 65 L tank. Figure 11 shows the rheological properties of the cellulose fiber suspension measured using the impeller method. Table 2 lists the power law parameters for the cellulose fiber suspensions.

Table 2
Power Law Parameters for Cellulosic Fiber Suspensions

Fiber Concentration	n	kpl
10	0.458	1.24
20	0.401	1.82
30	0.283	3.09

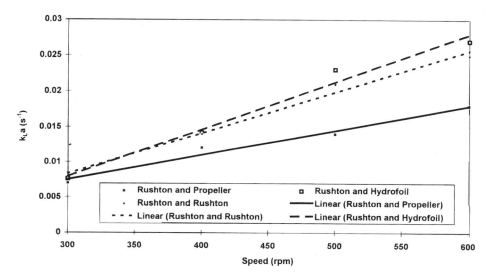

Fig. 5. Effect of impeller geometry on $k_L a$ at 10 lpm airflow rate and fiber conc. of 10 g/L (20 L fermentor).

DISCUSSION

The technique used to measure the gas-liquid mass transfer coefficient was consistent. The gas-liquid mass transfer coefficients decreased as the fiber concentrations increased, as expected.

At low agitation rates and with no fibers present, all three combinations of impellers performed similarly in the 20 L fermentor. The $k_L a$ values increased with increasing agitation for all three combinations of impellers. At higher fiber concentrations, the combination of a Rushton turbine and hydrofoil was more effective and had a higher $k_L a$ value than other combinations. At lower agitation rates the two Rushton turbines gave better performance for the conditions studied. The effect of solids concentration on $k_L a$ was also studied in the 65 L tank at 10 L per minute and 20 L per minute. From the results, it is evident that the $k_L a$ decreases with an increase in solids concentration.

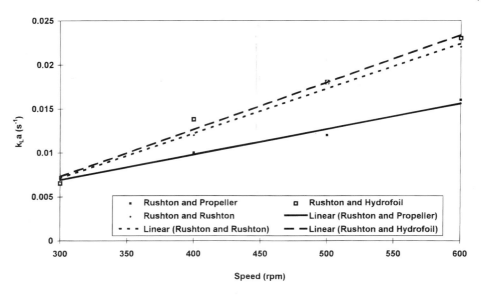

Fig. 6. Effect of impeller geometry on $k_L a$ at 10 lpm airflow rate and fiber conc. of 20 g/L (20L fermentor).

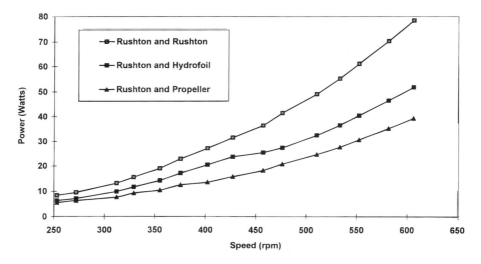

Fig. 7. Effect of impeller geometry on power at 10 lpm airflow rate (20 L fermentor).

The reduction in oxygen transfer that occurs in fermentations involving filamentous mycelial broths is postulated to be the result of viscous, shear-thinning broth rheology. The fibers used in these tests are straight and unbranched and hence are imperfect models of mycelial hypae. Although the measured viscosity of the suspensions increased with increasing fiber concentration, the effect was not nearly as large as that reported for filamentous mycelial broths at similar mass loadings. The presence of solids in the liquid phase can cause a decrease in the mass

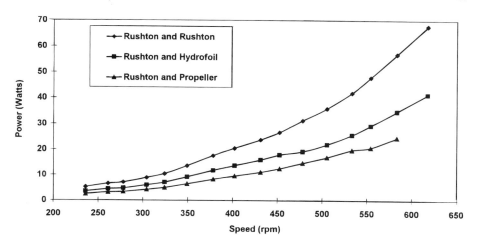

Fig. 8. Effect of impeller geometry on power at 20 lpm airflow rate (20 L fermentor).

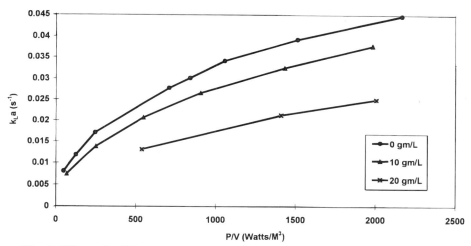

Fig. 9. Effect of solids concentration on $k_L a$ at 10 lpm airflow rate (65 L tank).

transfer coefficient even if the viscosity is similar to that of water, perhaps by acting as sites for bubble coalescence, thus decreasing the interfacial area (3), or by damping the level of turbulence in the liquid.

From the results of the power measurements, it is evident that the combination of two Rushton turbines consumes the most power in the 20 L fermentor. This effect is more evident at higher agitation rates. Similar effects were observed by many workers (7,8) who reported the power draw to double when impeller speed was doubled. The power draw increases since the energy is spent on liquid circulation and not for gas dispersion. Once the gas has been dispersed and is well-mixed in the fermentor, increasing the agitation will result in only a slight increase in the mixed fraction of the gas. At lower agitation rates, the gas tends to accumulate

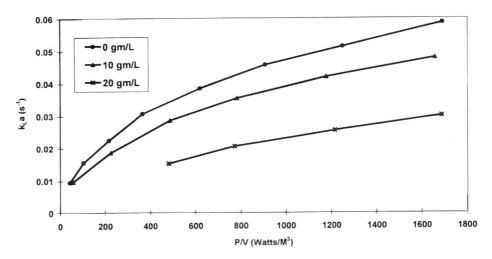

Fig. 10. Effect of cellulose fibers on $k_L a$ at 20 lpm airflow rate (65 L tank).

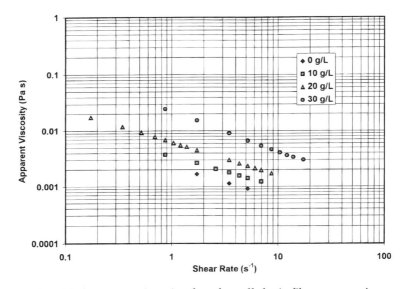

Fig. 11. Apparent viscosity data for cellulosic fiber suspensions.

near impellers with less circulation, which results in a decrease in the mixed fraction of gas. By increasing the gas flow rate from 10 to 20 L per minute, there was a slight decrease in power consumption, but the trends were similar to the results at 10 L per minute.

CONCLUSIONS

The conclusions made are limited to the two tank sizes, the three impellers, the fiber type and concentrations, and the aeration and agitation rates used in this investigation. In the 20 L fermentor at low agitation rates the two Rushton turbines gave 20% better performance than the Rushton and hydrofoil combination and 40% better performance than the Rushton and propeller combination. At higher agitation rates, the Rushton and hydrofoil combination gave 14% better performance than two Rushton turbines, and 25% better performance than the Rushton and hydrofoil combination. The power consumption with the Rushton and propeller combination was 57% less than the power consumption with two Rushton turbines. The power consumption with the Rushton and hydrofoil combination was 35% less than that of the two Rushton turbine combination.

NOTATION

$k_L a$ gas liquid mass transfer coefficient (s^{-1})
C_g gas side concentration
C_L liquid side concentration
$C_E(0)$ electrode response at time = 0
$C_E(T)$ electrode response at time = t
ϕ dimensionless electrode response
α Henrys law coefficient
τ_1 time constant
τ_2 time constant
τ_d dead time

REFERENCES

1. Moo-Young, M. and Blanch, H. W. (1981), *Adv. Biochem. Eng.* **19**, 1–69.
2. Chisti, M. Y. and Moo-Young, M. (1988), *Biotechnol. Bioeng.* **31**, 487.
3. Svihla, C. K. and Hanley, T. R. (1992), *AIChE Symposium Series*, **88**, 114–118.
4. Dronawat, S. N., Svihla, C. K., and Hanley, T. R. (1995), *Appl. Biochem. Biotechnol.* **51/52**, 347–354.
5. Svihla C. K., Dronawat, S. N., and Hanley, T. R. (1995), *Appl. Biochem. Biotechnol.* **51/52**, 355–366.
6. Linek, V., Vacek, V., and Benes, P. (1988), *Chem Eng. J.* **34**, 11.
7. Dronawat, S. N, (1996), Mixing in Fermentors: Analysis of Oxygen Transfer and Rheology, *Ph. D. Dissertation*, University of Louisville.
8. Laine, J. and Kuoppamaki, R. (1979), *Ind. Chem. Process Des. Dev.*, **18**, 501–550.

Measurement of the Steady-State Shear Characteristics of Filamentous Suspensions Using Turbine, Vane, and Helical Impellers

C. Kurt Svihla, Sundeep N. Dronawat,
Jennifer A. Donnelly, Thomas C. Rieth,
and Thomas R. Hanley*

*Department of Chemical Engineering, Speed Scientific School,
University of Louisville, Louisville, KY 40292*

ABSTRACT

The impeller viscometer technique is frequently used to characterize the rheology of filamentous suspensions in order to avoid difficulties encountered with conventional instruments. This work presents the results of experiments conducted with vane, turbine, and helical impellers. The validity of the assumptions made in the determination of the torque and shear-rate constants were assessed for each impeller type. For the turbine and vane impellers, an increase in the apparent torque constant c was observed with increasing Reynolds number even when measurements were confined to the viscous regime. The shear-rate constants determined for the vane and turbine impellers varied for different calibration fluids, which contradicts the assumptions usually invoked in the analysis of data for this technique. When the helical impeller was calibrated, consistent values for the torque and shear-rate constants were obtained. The three impeller types were also used to characterize the rheology of cellulose fiber suspensions and the results compared for consistency and reproducibility. The results have application in design of rheometers for use in process control and product quality assessment in the fermentation and pulp and paper industries.

Index Entries: Filamentous suspensions; helical impellers; non-Newtonian rheology.

---

*Author to whom all correspondence and reprint requests should be addressed.

INTRODUCTION

Knowledge of the rheological behavior of filamentous suspensions is necessary for the implementation of process and quality control for several processes in the biotechnology and pulp and paper industries. Filamentous suspensions can exhibit complex, non-Newtonian behavior, and characterization of their rheology has been a challenging problem. The use of the impeller viscometer for this purpose has become common practice (1–3), but the validity of a key assumption invoked in the analysis of data obtained from this technique has recently been questioned. Experiments were conducted to develop an impeller design that would yield consistent and reproducible results for non-Newtonian fluids and suspensions of widely varying properties. Turbine, vane, and helical impeller geometries were considered in this investigation.

METHODS

The vane and helical impellers used in the tests were fashioned from nylon using selective laser sintering technology. The six-bladed disk turbine used in the tests had a diameter of 0.0291 m with a blade width of 0.0079 m, a blade height of 0.0064 m, and a blade thickness of 0.0015 meters. The four-bladed vane impeller had a diameter of 0.0291 m and a height of 0.051 m. The helical impeller used in the tests had a single flight with an internal auger. It had a diameter of 0.04 m, a height of 0.055 m, and a pitch of 0.2. All tests were carried out using 800 mL samples in an unbaffled 1000 mL beaker (T = 0.104 m). Table 1 summarizes the pertinent dimensions of the various impeller and vessel combinations used in the tests.

Four different viscometers (Brookfield, Stoughton, MA) were used in the tests. The two dial-reading instruments (models LVT and LVF) had maximum spring torques of 673.7 dyno cm and maximum rotational speeds of 60 rpm. The two digital viscometers (models RV-DVIII and HB-DVIII) had maximum speeds of 250 rpm and full-scale spring torques of 7187 dyno cm and 57,496 dyno cm, respectively. Each viscometer had an experimental uncertainty of plus or minus 1% of its maximum reading. No data was recorded if the torque reading for a given instrument was less than 5% of its full-scale value. All experiments were conducted at a temperature of $25 \pm 0.1°C$.

The basis for the impeller viscometer technique has been well described elsewhere (2). The method relies on the use of Newtonian and non-Newtonian calibration fluids to determine constants relating the measured torque and speed to viscosity and shear rate. Newtonian silicone oils and solutions of glycerin and corn syrup were used to determine the impeller constant torque, c, while non-Newtonian carboxymethylcellulose (CMC), xanthan gum, and guar gum solutions were used to determine the shear rate constant, k. The properties of the Newtonian and non-

Table 1
Relative Dimensions of the Impeller Viscometer Apparatus

Impeller	Impeller Diameter D (m)	Impeller Height L/D	Liquid Height H_L/D	Off-Bottom Clearance C/D
Turbine	0.0291	0.22	3.30	1.96
Vane	0.0291	1.75	3.30	0.57
Helical	0.0400	1.37	2.40	0.16

Table 2
Properties of Newtonian Fluids Used in the Determination of the Torque Constant

Fluid	Viscosity (Pa-s)	Density (kg/m^3)
Glycerin (diluted)	0.175	1235
Corn Syrup (diluted)	0.205	1321
Silicone I	0.100	967
Silicone II	0.201	965

Table 3
Results of Shear Rate Constant Determination for the Turbine Impeller

Calibration Solution	Range of Power Law Index	Reynolds Number Range	Range of Values for k	k_{avg}
0.75% Xanthan	0.04 - 0.08	0.001 - 8.40	10.7 - 12.6	11.4
1.0% Xanthan	0.04 - 0.10	0.040 - 8.37	11.2 - 12.8	11.8
0.75% Guar	0.42 - 0.70	0.100 - 9.75	8.75 - 12.8	10.4
1.0% Guar	0.30 - 0.60	0.005 - 1.13	10.0 - 10.7	10.2
1.0% CMC	0.82 - 0.86	3.270 - 9.77	8.67 - 12.9	11.0
1.5% CMC	0.74 - 0.84	0.109 - 1.24	8.40 - 11.4	9.40

Table 4
Results of Shear Rate Constant Determination for the Vane Impeller

Calibration Solution	Range of Power Law Index	Reynolds Number Range	Range of Values for k	k_{avg}
0.75% Xanthan	0.04 - 0.08	0.001 - 9.00	9.41 - 11.1	10.2
1.0% Xanthan	0.03 - 0.10	0.004 - 7.93	9.79 - 11.7	10.4
0.75% Guar	0.58 - 0.70	0.100 - 1.38	9.19 - 10.4	9.56
1.0% Guar	0.34 - 0.70	0.046 - 1.82	9.66 - 12.6	10.6
1.0% CMC	0.85 - 0.86	0.400 - 7.59	8.87 - 9.48	9.20
1.5% CMC	0.70 - 0.84	0.440 - 1.80	12.4 - 19.2	14.2

Table 5
Results of Shear Rate Constant Determination for the Helical Impeller

Calibration Solution	Range of Power Law Index	Reynolds Number Range	Range of Values for k	k_{avg}
0.75% Guar	0.43 - 0.78	0.100 - 9.86	9.78 - 13.2	10.7
1.0% Guar	0.30 - 0.74	0.008 - 9.69	9.77 - 12.8	10.9

Newtonian calibration fluids were measured using a cone-and-plate attachment on the RV-DVIII instrument (Brookfield spindle CP-41). The properties of the Newtonian fluids used to determine the impeller torque constants are summarized in Table 2. Ranges for the power-law indices of the non-Newtonian calibration fluids can be found in Tables 3 through 5.

The fiber suspensions used in this study were composed of various concentrations of powdered cellulose (Solka-Floc, grade KS-1016, average fiber length = 290 µm) suspended in 0.05% and 0.10% xanthan gum solutions to retard settling.

RESULTS

Newtonian Calibration Fluids

In the viscous flow regime, the power number for an impeller is inversely proportional to the Reynolds number with the constant of proportionality defined as c. Thus, the measured torque is directly proportional to the rotational speed and the viscosity of the fluid:

Characteristics of Filamentous Suspensions

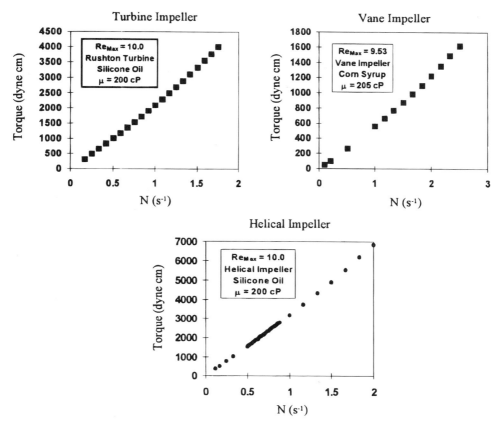

Fig. 1. Newtonian torque vs speed relationship for various impellers.

$$\Gamma = \frac{c}{2\pi}\mu ND^3 \qquad (1)$$

For a given rotational speed, the measured torque can thus be used along with the fluid viscosity and the impeller diameter to calculate the apparent impeller torque constant, c. Figure 1 illustrates typical Newtonian torque vs speed relationships for the three impellers used in this study. There is a pronounced curvature in the plots for the vane and turbine impellers, but the data for the helical ribbon seem to be linear in accordance with Equation 1. Figure 2 shows the same data plotted as the apparent value of c/c_0 where c_0 is the average value of the apparent impeller torque constant for Reynolds numbers less than 3. Based on the data of Donnelly (4), the following relationships for the apparent torque constants for the vane and turbine were determined:

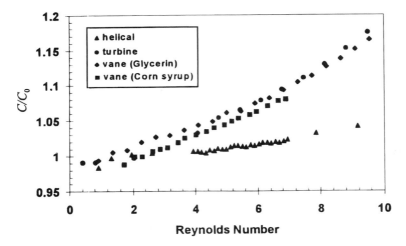

Fig. 2. Effect of Reynolds number on the apparent torque constant for various impellers.

            Turbine   Vane

$c = 67$ (Re < 2) $c = 244$ (Re < 2)
$c = 67 + 2\,\mathrm{Re}$ (2 < Re < 4) $c = 235 + 4.5\,\mathrm{Re}$ (2 < Re < 10)
$c = 64.4 + 1.65\,\mathrm{Re}$ (4 < Re < 10)

The data for the helical impeller can be adequately represented by a constant value of $c = 150$ in accordance with Equation 1.

Non-Newtonian Calibration Fluids

The impeller method is extended to non-Newtonian fluids by replacing the Newtonian viscosity μ in Equation 1 with the apparent viscosity η_a:

$$\Gamma = \frac{c}{2\pi}\eta_a N D^3 \qquad (2)$$

It is also assumed that the c value for the Newtonian case applies to nonNewtonian fluids as well. The apparent viscosity of a non-Newtonian fluid can, therefore, be determined from the measured torque and impeller speed once the impeller torque constant c has been determined. The question of what shear rate to associate with the apparent viscosity determined from application of Equation 2 remains, however. The procedure usually adopted is to make use of the Metzner-Otto assumption that the average shear rate near the impeller is related to the rotational speed by a fluid independent constant, k:

$$\gamma = kN \qquad (3)$$

If this approach is valid, the shear-rate constant can be determined from experimental measurements of torque vs impeller speed for non-Newtonian fluids of known properties. The apparent viscosity is determined from Equation 2, the value of shear rate that corresponds to this viscosity is obtained from the known rheogram for the non-Newtonian fluid, and the value of k then determined from Equation 3. If the Metzner-Otto assumption holds true, the same value of k will be found for all fluids, implies that the method can be applied with confidence to any non-Newtonian fluid or suspension of unknown properties.

Tables 3 through 5 summarize the results obtained from the shear rate constant calibration trials for the turbine, vane, and helical impellers. The average value of the shear-rate constant determined for the turbine was 11.0, for the vane 10.2, and for the helical impeller 10.8. The results from the CMC trials were not used in the calculation of the average shear rate constant for the turbine or the vane for reasons given in the discussion.

Filamentous Suspensions

The collection of rheological data for filamentous suspensions becomes complicated if the fibers have a tendency to settle. The fibers used in these tests tended to settle fairly rapidly in water, so small concentrations (0.05 and 0.10%) of xanthan gum were added to retard settling. The solutions were also mixed periodically throughout the trials.

Figure 3 demonstrates the ability of the Rushton turbine, vane impeller and helical impeller to measure the viscosity of the suspensions. It is evident from the figure that both the vane and Rushton turbine impellers return similar viscosities at a given shear rate. The viscosities measured by the helical impeller are slightly higher than the values given by the other impellers, but the general behavior is similar. The difference might be a result of a difference in the settling characteristics of the suspensions in the different systems since particle settling was generally not a problem with the helical impeller. Table 6 summarizes the results of the suspension tests.

DISCUSSION

Newtonian Calibration Fluids

The apparent Reynolds number dependence of the torque constants for the vane and turbine impeller is cause for some concern since this behavior has generally not been reported in the literature for Re < 10. This behavior was not observed when tests were conducted with a turbine impeller in a larger scale system (D = 0.0994 m, µ = 67.7 g/cm s) with a different torque measuring system. The possibility that the observed dependence may be an artifact because of some characteristic of the torque sensing devices built into the viscometers cannot be ruled out. The fact that

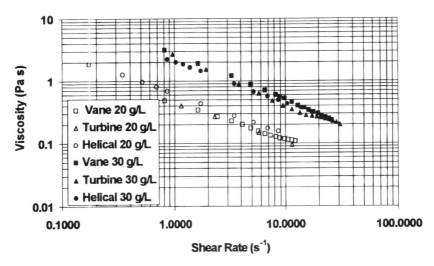

Fig. 3. Properties of cellulose fiber suspensions as measured by the different impellers.

the helical impeller gave c values that were very nearly constant for Reynolds numbers less than 10, may be a result of its larger diameter and the fact that the transition to turbulence for helical impellers occurs at higher Reynolds numbers than for turbines or vanes (5).

Non-Newtonian Calibration Fluids

The average shear rate constant did not appear to be independent of the fluid used to determine it for any of the impellers studied. Indeed, as Tables 3–5 indicate, there was considerable point to point variation in the k value for a given fluid. Nevertheless, use of the average k value determined for each impeller did not lead to more than about 10–15% error when the cone-and-plate data were compared with the values given by the respective impeller viscometers. The CMC results were not used to compute the average k values for the vane and turbine impellers because the power-law indices of the two solutions were relatively close to 1. In this range, the viscosity is relatively insensitive to shear rate, which is not a particularly desirable property of a non-Newtonian calibration fluid, since it will tend to magnify any experimental errors in the torque measurement. Nevertheless, the average k values for the CMC solutions are not much different from those determined for the more highly shear-thinning solutions.

Fiber Suspensions

The rheological data for 0.1 and 0.05% xanthan gum solutions are similar, which suggests that the addition of small amounts of xanthan to

Table 6
Rheological Parameters of Fiber Suspensions Determined Using Various Impellers

Fiber Concentration (kg/m³)	Xanthan Gum Concentration (kg/m³)	Impeller Type	K_{pl} (Pa s^n)	n	r²
20	1.0	Vane	0.398	0.507	0.966
30	1.0	Vane	3.04	0.228	0.999
40	1.0	Vane	13.6	0.255	0.9995
50	1.0	Vane	28.4	0.303	0.9998
20	1.0	Turbine	0.463	0.364	0.999
30	1.0	Turbine	2.45	0.263	0.988
40	1.0	Turbine	9.69	0.285	0.998
50	1.0	Turbine	27.5	0.258	0.999
20	0.5	Helical	0.624	0.361	0.999
25	0.5	Helical	0.814	0.330	0.999
30	0.5	Helical	2.062	0.337	0.999
35	0.5	Helical	5.967	0.246	0.996
10	1.0	Helical	0.347	0.458	0.9994
20	1.0	Helical	0.623	0.404	0.9997
30	1.0	Helical	2.200	0.283	0.9993

prevent particle settling did not significantly affect the measured torques and viscosities. In any case, the effect will be largest at low shear rates and will become progressively less important as the fiber concentration increases. The helical impeller was superior to the other two impellers since it allowed rheological measurements to be performed on suspensions with fiber concentrations as low as 10 g/L in 0.05% xanthan gum solution. In previous work conducted with *Aspergillus niger* broths (1), the rheology of pelletal suspensions could not be accurately characterized with a turbine impeller since the suspensions were insufficiently viscous and the resulting impeller torques were too small. The helical impeller should be better suited for measurements of this type and for less viscous suspensions in general.

The power-law parameters determined for the cellulose fiber suspensions can be compared with the results reported in the literature for the same fiber grade (3):

20 g/L: K (Pa s^n) = 1.46   n = 0.32
30 g/L: K (Pa s^n) = 6.13   n = 0.24

Comparison of the power-law parameters indicates that the suspensions studied here were less viscous and less-shear thinning than the suspensions used in the literature study. The differences might reflect a difference in the properties of the fibers, although the same fiber grade was used. The behavior of the suspensions was consistent with the literature results in that the increase in the consistency index as the power-law index increased was relatively moderate.

CONCLUSIONS

At a given speed, the vane and helical impellers draw more torque than the turbine, and were hence more useful for measuring the rheological properties of low viscosity fluids and suspensions. Of the three impellers studied, the helical impeller yielded results which appeared to best satisfy the assumptions inherent in the impeller viscometer approach.

NOMENCLATURE

c	impeller torque constant, dimensionless
C	off-bottom clearance of impeller, m
D	impeller diameter, m
H_L	liquid height, n
k	shear rate constant, dimensionless
K_{pl}	consistency index constant, Pa s^n
L	impeller height, m
n	power-law index
N	impeller speed, s^{-1}
Re	impeller Reynolds number, $\rho ND^2/\mu$
T	vessel diameter, m
Γ	impeller torque, Nm
γ	shear rate, s^{-1}
μ	Newtonian fluid viscosity, Pa-s
η_a	apparent viscosity of a non-Newtonian fluid, Pa-s
ρ	fluid density, kg/m^3

REFERENCES

1. Svihla, C. K., Dronawat, S. N., and Hanley, T. R. (1995), *Appl. Biochem. Biotechnol.* **51/52**, 355–366.
2. Allen, D. G. and Robinson, C. W. (1990), *Chem. Engng. Sci.* **45(1)**, 37–48.
3. Chisti, M. Y. and Moo-Young, M. (1988), *Biotechnol. Bioeng.* **31**, 487–494.
4. Donnelly, J. (1995), M. Eng. Thesis, University of Louisville, Louisville, Kentucky.
5. Kemblowski, Z., Sek, J., and Budzynski, P. (1988), *Rheological Acta* **27**, 82–91.

Production of Fumaric Acid by Immobilized Rhizopus Using Rotary Biofilm Contactor

NINGJUN CAO,*,[1] JIANXIN DU, CHEESHAN CHEN,[2] CHENG S. GONG,[1] AND GEORGE T. TSAO[1]

[1]*Laboratory of Renewable Resources Engineering, Purdue University, W. Lafayette, IN 47907; and* [2]*Department of Food Engineering, Da-Yeh Institute of Technology, Chang-Hwa, Taiwan 51505, R.O.C.*

ABSTRACT

Rotary biofilm contactor (RBC) is a reactor consisting of plastic discs that act as supports for micro-organisms. The discs are mounted on a horizontal shaft and placed in a medium-containing vessel. During nitrogen-rich growth phase, mycelia of *Rhizopus oryzae* ATCC 20344 grew on and around the discs and formed the "biofilm" of self-immobilized cells on the surface of the plastic discs. During the fermentation phase, the discs are slowly rotated, and the biofilms are exposed to the medium and the air space, alternately. With RBC, in the presence of $CaCO_3$, *Rhizopus* biofilm consumes glucose and produces fumaric acid with a volumetric productivity of 3.78 g/L/h within 24 h. The volumetric productivity is about threefolds higher with RBC than with a stirred-tank fermenter with $CaCO_3$. Furthermore, the duration of fermentation is one-third of the stirred-tank system. The immobilized biofilm is active for over a 2-wk period with repetitive use without loss of activity.

Index Entries: Biofilm; calcium carbonate; calcium fumarate; immobilized *Rhizopus*; fumaric acid; *Rhizopus oryzae*; rotary biofilm contactor (RBC).

INTRODUCTION

Fumaric acid, a naturally occurring four-carbon dicarboxylic acid, is a normal intermediate in cell metabolism. Because of its double bond and two carboxylic groups, fumaric acid has many potential industrial uses;

---

*Author to whom all correspondence and reprint requests should be addressed.

such as in the manufacturing of synthetic resins and biodegradable polymers *(1,2)*. It can also serve as an intermediate for chemical syntheses *(3)*. In addition to its use as feedstock for chemical synthesis, fumaric acid is commonly used as a food ingredient and beverage acidulant. Many studies show that fumaric acid reduces processing costs and improves the quality of many food products. There are only two major processes for fumaric acid production: chemical synthesis via the catalytic isomerization of petroleum-derived maleic anhydride or maleic acid *(3)*, and the biological production through the direct fermentation of glucose *(4,5)*. Currently, fumaric acid is produced exclusively using the synthetic route. Recently, however, the microbial production of fumaric acid has received increased attention because of its increased use in food industry.

Many species of micro-organisms produce small amounts of fumaric and other organic acids extracellularly as metabolic by-products during oxidative metabolism. In some cases, certain mycelial fungi are capable of producing significant quantities of fumaric acid from glucose. The most noticeable is the group of fungi belonging to the genus *Rhizopus (6)*. A few selected strains of *Rhizopus oryzae* (*R. arrhizus*) produce, almost exclusively, fumaric acid in high yield from glucose under specific cultural conditions *(7,8)*. The high yield of fumaric acid from glucose is believed to be a result of the ability of the fungi to fix CO_2 during the acid production stage *(9,10)*.

Currently, biological production of fumaric acid is too expensive to compete with the synthetic route *(11)*. Also, biological production is hindered by some technical drawbacks. For example, when *Rhizopus* produces fumaric acid, the pH of the fermentation broth decreases to a threshold point where the organism stops producing the acid and the metabolism shifts to produce ethanol and other products *(12)*. Final concentrations of free acid in the fermentation broth are typically very low. The most common approach to overcome the low concentration of acid in the fermentation broth is to continuously neutralize the acid produced by adding a base such as calcium carbonate. Consequently, the product that accumulates in the fermentation broth is calcium fumarate instead of fumaric acid. The recovery of calcium fumarate and the generation of free acid from calcium fumarate is difficult. Furthermore, when the fermentation is carried out with calcium carbonate as the neutralizing agent, calcium fumarate will precipitate because of its low solubility. It also causes an increase in liquid viscosity resulting in the stoppage of the agitator and thus, the entire fermentation operation *(8)*. If, however, sodium bicarbonate or other neutralizing agents were used in place of calcium carbonate, less fumarate is formed. This is because the accumulated sodium fumarate exhibited an inhibitory effect on fumaric acid production *(8,11)*. Another problem encountered is the need for a continuous supply of antifoam agent because of the tendency of excessive foaming during stirred tank fermentation.

In this research, self-immobilized *Rhizopus oryzae* mycelia were used to carry out repetitive fumaric acid fermentation in a rotary biofilm contactor (RBC) with the supplement of $CaCO_3$. The characteristics of biofilm during fermentation are also described.

MATERIALS AND METHODS

Micro-organism and Inoculum

Rhizopus oryzae ATCC 20344 was purchased from American Type Culture Collection (Rockville, MD). The culture was maintained on potato dextrose agar (Difco) slants and propagated by growing in Erlenmeyer flasks with potato dextrose agar to obtain sporangiospores. For fermentation studies, spores were obtained by washing spore-bearing cultures with sterile water, filtered through sterile filter paper and collected as spore suspension.

Medium

Growth medium contained: 10 g glucose, 2.0 g urea, 0.6 g KH_2PO_4, 0.25 g $MgSO_4 \cdot 7H_2O$, and 0.088 g $ZnSO_4 \cdot 7H_2O$ in one L distilled water (6). The fermentation medium consisted of 100 g glucose, 1.0 g yeast extract, 0.6 g KH_2PO_4, 0.25 g $MgSO_4 \cdot 7H_2O$, 0.088 g $ZnSO_4 \cdot 7H_2O$, in 1 L distilled water. $CaCO_3$ was added whenever necessary to maintain the pH at 5.0 (6).

Rotating Biofilm Contactor (RBC)

The reactor is a 2 L fermentor with 0.9 L working liquid volume (100 mm in diameter and 260 mm in length). The reactor contains six plastic discs (125 cm^2/disc) that act as supports for the micro-organism. The discs, with a total surface area of 750 cm^2, are mounted on a horizontal shaft and placed in the fermentor containing growth medium. Mycelial fungi showed a strong tendency to grow on and around the solid plastic surface of the discs and to form the "biofilm" and immobilized on the surface of the plastic discs. During the operation, air was flowed into and out of the upper portion of air space at an air-flow rate of 1.0 L/L/min. The discs were rotated at 22 rpm and the biofilms were exposed alternately to the fermentation medium and the air space. Figure 1 shows the schematic diagram of the RBC.

Growth of *Rhizopus* Biofilm

The spore suspension (100 mL with spore concentration of 1×10^6/mL) was inoculated into RBC containing 800 mL of the growth medium. The pH during growth stage was maintained at 5 with a small addition of $CaCO_3$. After 60 h of incubation at 35°C with a rotating speed of 14 rpm, biofilm was grown on and around the surface of the plastic

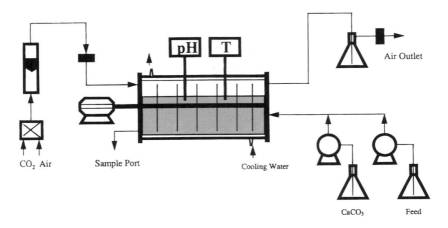

Fig. 1. Schematic diagram of rotary biofilm contactor.

discs. Then the broth in the reactor was changed from growth to nitrogen-deprived fermentation medium.

Analytical Methods

Fumaric acid, L-malic acid, succinic acid, and carbohydrates were determined and quantified by high-performance liquid chromatography (Hitachi Instrument, Tokyo, Japan, L-6200A) using a BioRad Aminex HPX-87H ion exclusion column (300 × 7.8 mm) with a refractive index Detector (Hitachi Instrument, L-3350 RI). The column was eluted with dilute sulfuric acid (0.005M) at a column temperature of 80°C and a flow rate of 0.8 mL/min over a 13-min period.

RESULTS AND DISCUSSION

Fumaric acid fermentation by *Rhizopus oryzae* can be separated into cell growth and product formation stages. During the aerobic growth in the presence of growth medium, *R. oryzae* grows with extended hyphae and forms large size pellets or mycelial aggregates. Once the cells are fully grown, the medium is changed to fermentation medium to encourage cells to produce fumaric acid from glucose. When fermentation was carried out in the conventional stirred tank fermentor, the growing mycelia adsorbed onto the heat exchanger and propellers and formed mycelial clumps. The fermentation time was long, and unwanted byproducts, particularly ethanol, were formed. The conversion yield was low because of localized anaerobic condition and the limitation of mass transfer. In addition, when $CaCO_3$ is added, it will intermingle with mycelial aggregates and further complicate the problem. One method that can alleviate this problem is to grow the cells in the specially formulated growth medium to prevent large mycelial pellet formation before subjecting the cells to the nongrowth fermentation medium (6). Another method is to use reactors, such as RBC.

The biofilm formation in the RBC was conducted after inoculating *Rhizopus* sporangiospores into liquid growth medium. During incubation, with slow rotation of the plastic discs, the germinated spores with hyphae became attached to the surface of the plastic discs upon contact. They grew on and around the discs and eventually covered the entire surface of the discs (Fig. 2A, B) and formed a mycelial film of not more than 2 mm in thickness (Fig. 2C, D). Typically, it takes from 48 to 72 h for growing mycelia to cover the entire surface of the discs. A small amount of $CaCO_3$ was added during the growth stage to maintain the pH of the medium at about 5.0. After the completion of the formation of biofilm, the fermentation medium was introduced to replace the growth medium in order to carry out fumaric acid production. During the fermentation stage, biofilm was exposed to sterile air in the head space of RBC that enhances oxygen exposure and opportunity for CO_2 fixation by biofilm.

Typical results of RBC fermentation are shown in Fig. 3. During the operation, the discs were rotated at 22 rpm, slightly faster than during growth stage. The slower rotating speed used during the biofilm formation stage was to allow the mycelia to attach itself to the disc without too much friction. In the absence of $CaCO_3$, the fumaric acid concentration reached to a level of 6 g/L, which is the solubility of fumaric acid at the incubation temperature of 35°C. At this concentration, the pH of the broth was at 2.9. In the presence of $CaCO_3$, the pH of the broth was at 5.0 and the soluble calcium fumarate concentration reached 30 g/L. The excess calcium salt was crystallized as calcium fumarate and settled to the bottom of RBC. After about 24 h of operation, the glucose present was consumed entirely to produce approx 75.7 g/L of calcium fumarate. During fermentation, ethanol was produced and reached a maximum of 12g/L and stayed more or less unchanged throughout the entire course of operation. A small amount of glycerol, approx 5 g/L, was produced by this strain of *R. oryzae*. Different strains of *Rhizopus* produced different amounts of glycerol. *R. oryzae* ATCC 12702 produced twice as much glycerol as *R. oryzae* ATCC 20344 under identical fermentation conditions (13).

An average weight yield of about 75% (74 g/L fumaric acid) was obtained from an initial average glucose concentration of 98.7 g/L. This is much higher than those in the stirred tank fermentation with $CaCO_3$. The weight yield of stirred tank fermentation was 65%. The time required for the completion of fermentation is 24 h, which is one-third the time required for the stirred tank fermentation. The most significant advantage of RBC fermentation is the increase of volumetric productivity and the weight yield of fumaric acid. The volumetric productivity (3.78 g/L/h) achieved using RBC is almost five times higher than that achieved using a stirred tank system (0.94 g/L/h) with neutralizing agents. Likewise, the weight yield is much higher in RBC than in a traditional fermentor (65%) with

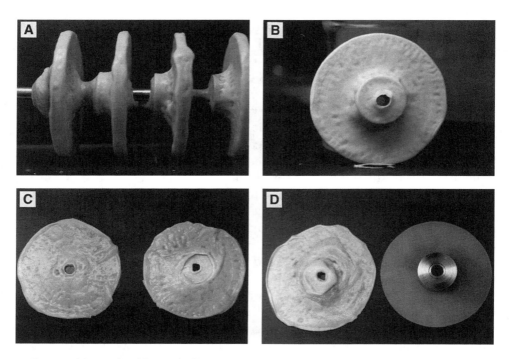

Fig. 2. *Rhizopus* biofilm. **a.** Fully grown biofilm on the discs that were attached to the horizontal shaft. **b.** Close-up of biofilm on a single disc. **c.** Peeled off biofilms. **d.** The naked disc with biofilm.

neutralizing agents *(8)*. Other significant advantages of RBC fermentation are the avoidance of foaming and the reusability of fungal mycelia. In stirred tank fermentation, the organisms can be used only once *(8)*.

Self-immobilized *Rhizopus* biofilm is stable during extended RBC operations. There is no observable deterioration of biofilm after 10 cycles of operation. Figure 4 shows the glucose consumption and fumaric acid production by biofilm after the repetitive fed-batch cycles. The glucose utilization rates were similar for the first three cycles of operation. During the fourth operation cycle, the glucose utilization activity was reduced by about 20%. The activity was restored to its full capacity after incubating the biofilm in growth medium. The biofilm activity after growth reactivation was shown in the fifth through seventh cycle. The same procedure was repeated with similar results.

The uses of RBC or the similar fermentor configuration for the biological applications have been reported. The similar reactor was described by Sudo and Aiba *(14)* in the treatment of polluted water with naturally occurring microflora, mainly protozoa. Likewise, Sublette et al. *(15)* described the treatment of munition waste water with RBC to reduce the concentrations of trinitrotoluene and cyclotrimethylene trinitramine using a mycelial fungus, *Phanerochaete chrysosporium*, as the source of biofilm.

Production of Fumaric Acid

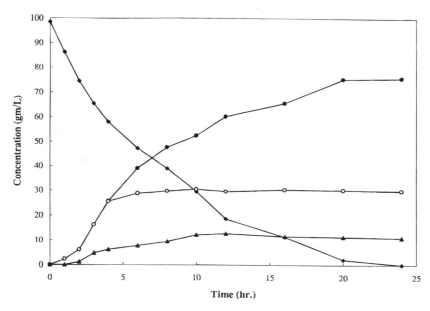

Fig. 3. Kinetics of fumaric acid fermentation by *Rhizopus oryzae* ATCC 20344 in rotary biofilm contactor. ◆ Glucose; ● Fumaric acid, total; ○ Fumaric acid, in solution; ▲ Ethanol.

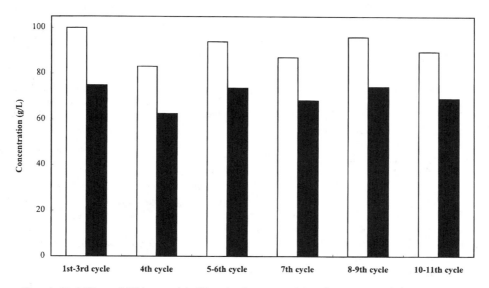

Fig. 4. Stability of *Rhizopus* biofilm during repetitive fumaric acid fermentation in RBC. ☐ Glucose utilized, ■ Fumaric acid produced.

Unfortunately, no detailed description of the reactor configuration was provided. Another example is the utilization of mixed-culture biofilm reactor for lactic acid production (16). In this case, a biofilm forming bacterium, *Streptomyces viridosporus*, was used for the biofilm formation on lignocellulosic materials. Another bacterium, *Lactobacillus casei*, was used for the lactic acid production. The system is more or less similar to the cell immobilization system. So far, we are not aware of any reports using RBC for the biological production of fumaric acid.

CONCLUSION

This fermentation system has several advantages including:

1. No direct physical contact of mycelia with $CaCO_3$, thus avoiding the mingling of $CaCO_3$ with mycelia that caused localized mass transfer interference;
2. The mycelia that formed the biofilm can be reused and revitalized *in situ*, whenever necessary, for reuse;
3. By periodically removing the calcium fumarate formed and by continuous feeding of fresh glucose, the production process can be kept almost continuously; and
4. Higher volumetric productivity and product yield.

ACKNOWLEDGMENTS

This research was supported in part by Division of Biological and Critical Systems, National Science Foundation grant BES-9412582.

REFERENCES

1. Robinson, W. D. and Mount R. A. (1981), in *Kirk-Othmer Encyclopedia of Chemical Technology, Vol. 14*, Grayson, M. and Eckroth, D., eds., Wiley, New York, NY, p. 770.
2. Tao, W. Y., Collier, J. R., Collier, B. J., and Negulescu, I. I. (1993), *Textile Res. J.* **63**, 162–170.
3. Ng, T. K., Hesser, R. J., Stieglitz, B., Griffiths, B. S., and Ling, L. B. (1986), *Biotechnol. Bioeng. Symp.* **17**, 355–363.
4. Foster, J. W. and Waksman, S. A. (1939), *J. Am. Chem. Soc.* **61**, 127–135.
5. Margulies, M. and Vishniac, W. (1981), *J. Bacteriol.* **81**, 1–9.
6. Yang, C. W., Lu, J. and Tsao, G. T. (1995), *Appl. Biochem. Biotechnol.* **51/52**, 57–71.
7. Rhodes, R. A., Moyer, A. J., Smith, M. L. and Kelley, S. E. (1959), *Appl. Microbiol.* **7**, 74–80.
8. Rhodes, R. A., Lagoda, A. A., Misenheimer, T. J., Smith, M. L., Anderson, R. F., and Jackson, R. W. (1962), *Appl. Microbiol.* **10**, 9–15.
9. Kenealy, W., Zaady, E., Du Preez, J. C., Stieglitz, B., and Goldberg, I. (1986), *Appl. Environ. Microbiol.* **52**, 128–133.
10. Osmani, S. A. and Scrutton, M. C. (1985), *Eur. J. Biochem.* **147**, 119–128.
11. Gangl, I. C., Weigand, W. A., and Keller, F. A. (1990), *Appl. Biochem. Biotechnol.* **24/25**, 663–677.
12. Romano, A. H., Bright, M. M., and Scott, W. E. (1967), *J. Bacteriol.* **93**, 600–604.
13. Lu, Z. J., Yang. C. W., and Tsao, G. T. (1995), *Appl. Biochem. Biotechnol.* **51/52**, 83–95.
14. Sudo, R. and Aiba, S. (1984), *Adv. Biochem. Eng. Biotechnol.* **29**, 117–141.
15. Sublette, K. L., Ganapathy, E. V., and Schwartz, S. (1992), *Appl. Biochem. Biotechnol.* **34/35**, 709–723.
16. Demirci, A., Pometto, A. L. III., and Johnson, K. E. (1993), *Appl. Environ. Microbiol.* **59**, 203–207.

The Effect of Pectinase on the Bubble Fractionation of Invertase from α-Amylase

VEARA LOHA, ROBERT D. TANNER*, AND ALES PROKOP

Department of Chemical Engineering, Vanderbilt University, Nashville, TN 37235

ABSTRACT

Fermentation broth normally contains many extracellular enzymes of industrial interest. To separate such enzymes on-line could be useful in reducing the cost of recovery as well as in keeping their yield at a maximum level by minimizing enzyme degradation from broth proteases (either the desired enzymes or the proteases could be removed selectively or both removed together and then separated). Several large-scale separation methods are candidates for such on-line recovery such as ultrafiltration, precipitation, and two-phase partitioning. Another promising technique for on-line recovery is adsorptive bubble fractionation, the subject of this study. Bubble fractionation, like ultrafiltration, does not require contaminating additives and can complement ultrafiltration by preconcentrating the enzymes using the gases normally present in a fermentation process. A mixture of enzymes in an aqueous bubble solution can, in principle, be separated by adjusting the pH of that solution to the isoelectric point (pI) of each enzyme as long as the enzymes have different pIs. The model system investigated here is comprised of three enzyme separations and the problem is posed as the effect of pectinase (a charged enzyme) on the bubble fractionation of invertase (a relatively hydrophilic enzyme) from α-amylase (a relatively hydrophobic enzyme).

The primary environmental variable studied, therefore, is the pH in the batch bubble fractionation column. Air was used as the carrier gas. This prototype mixture exemplifies an aerobic fungal fermentation process for producing enzymes. The enzyme concentration here is measured as total protein concentration by the Coomassie Blue (Bradford) solution method *(1)*, both as a function of time and column position for each batch run. Since, from a previous study *(2)*, it was found that inver-

---

*Author to whom all correspondence and reprint requests should be addressed.

tase and α-amylase in a two-enzyme system can be partially separated in favor of one vs the other at two different pHs (pH 5.0 and 9.0) with significant separation ratios, emphasis is placed on the effect of pectinase at these pHs. In this study, the addition of pectinase reduced the total separation ratio of the α-amylase-invertase mixture at both pHs.

Index Entries: Pectinase; invertase; α-amylase; enzymes; bubble fractionation; protein separation.

INTRODUCTION

Robert Lemlich (3) coined the expression "adsorptive bubble fractionation" to describe the separation process by which surface active solutes are separated from dilute solution by bubbling an inert gas through a column containing the solute–solution mixture. Generally, surface active solutes, such as enzymes and other proteins, strongly adsorb at air–liquid interfaces. A typical protein structure is comprised of hydrophobic and hydrophilic parts that contribute to high surface forces. In bubble fractionation, gas bubbles are generated by forcing the gas through a sparger, which is typically a nozzle, a porous ceramic plate, or a sintered metal distributor. As bubbles rise through the column, proteins move from the dilute bulk solution to those bubbles, attaching to an adsorptive layer at the gas bubble–liquid interface. In that bubbling process, the bubbles burst at the liquid surface at the top of the column, releasing the proteins. The surface protein concentration thus increases as the bubbles rise in the column, and this increase continues until the surface concentration reaches a maximum. A steady-state protein gradient along the axis of the column is reached after a given time, typically 15–20 min. At steady-state, the amount of released proteins transferred downward by bulk diffusion from the protein rich surface is exactly countered by the proteins rising with the bubbles. Thus, if the protein is actively adsorbed by the bubbles at a particular processing condition, the highest concentration of that recovered protein is typically reached at the liquid surface of the column. When proteins are removed by surface skimming, a continuously changing protein gradient profile will describe the shift in concentration below the original steady-state profile.

The notion of bubble fractionation occurring within a fermentor came about as a result of the observation by Park and coworkers (4), in which protein stratification was observed in the broth of a Baker's yeast fermentation process even though the broth was mixed vigorously. Effler et al. (5) replicated this phenomenon in a graduated cylinder, and Potter et al. (6) showed that this process could be modeled using the notion of bubble fractionation to affect protein stratification. The fermentation experiments (5) were performed at 32°C and pH 5.0 with and without sparging gases. These experiments, carried out within a graduated cylinder, were sampled at three positions along the axis and at 4, 7, and 10 h. Air and carbon dioxide were alternately used as the sparging gas. It was shown that the separation occurred for both the sparged and nonsparged gas systems. The highest axial protein concentration gradients along the column occurred when carbon dioxide, at levels above those naturally present in the fermentation processes, was used as the sparging gas.

To elucidate the protein separation phenomenon, DeSouza and coworkers (2) studied the effect of both pH and gas composition on the bubble fractionation of two enzymes in a nonfermentation system using the same graduated cylinder. Invertase (a highly glycosylated and relatively hydrophilic enzyme) and α-amylase (a relatively hydrophobic enzyme) were used individually and in combination. Air and carbon dioxide were used as the sparging gases. The experiments were conducted batchwise at ambient temperature. As expected, the highest separation ratios (defined as the ratio of the concentration of proteins at the top to the bulk protein concentration at the bottom of the column) reached locally maximum levels when the solution pH was positioned near the respective protein isoelectric point (pI). In the DeSouza et al. study (2), the apparent pIs were at pH 5.0 for invertase and pH 8.0 for α-amylase. Sparged carbon dioxide led to larger separation ratios than sparged air at these two pHs.

Mixtures of enzymes are normally found in living systems such as fermentation broths, plant, and animal cells. Fruits that contain the combination of α-amylase, invertase and pectinase, the mixture of enzymes used in this study, include apples, pears, and papayas. Such enzymes in fruits are usually found in dilute concentration and within a mixture of hundreds of enzymes. To even partially separate such enzymes from such fruits or a fermentation broth would be useful in reducing the separation cost of one or more of these enzymes. Generally, separation costs reach 80 to 90% of the total manufacturing cost of many enzymes. In a fermentation process, on-line separation of either proteases or other extracellular enzymes can contribute towards maximum yields of desired enzymes by lowering the contact time of enzyme degradation from broth proteases. One of the imposed constraints of this study is to not add any contaminating additives, such as surfactants or salts, because these additives usually must be removed in a subsequent time-consuming purification step. Enzymes such as α-amylase and invertase may be sufficiently surface-active to be bubble fractionated without the use of such additives. Bubble fractionation processes also have the advantage of requiring little additional energy above that normally needed by the fermentation process itself, when the fractionation is performed on-line. Bubble fractionation itself does not typically lead to highly purified protein products, but its simplicity in achieving a crude separation indicates that it may be a promising technique for the initial isolation and purification step of enzymes.

In this preliminary study of a three enzyme mixture, air alone was used as the carrier gas. Pectinase (a positively charged enzyme, since its pectin substrate is negatively charged), invertase, α-amylase, and various combinations of these three different type of proteins were selected for these experiments. The objective was to determine the effect of pectinase on the bubble fractionation of invertase from α-amylase by varying the pH and the concentration of pectinase in the bubble fractionation column.

MATERIALS AND METHODS

A schematic diagram of the bubble fractionation column previously developed (2) is depicted in Fig. 1. The bubble column is a Nalgene polycarbonate 1 L graduated cylinder of 6.2 cm diameter and 35 cm height. The column was modified to have three sample ports perpendicular to the column as shown: top, middle, and bottom positions. The bottom of the column contained a fritted glass sparger that was tightly fitted and glued to the inner diameter of the column. Compressed air, introduced below the column sparger, was prehumidified by bubbling it through water in order to minimize both column water loss and protein contamination from the influent air.

The bacterial (*B. subtilis*) crude α-amylase (Lot No. 113F-0516) and Baker's yeast crude invertase (Lot No. 101F-0147) used in the experiments were purchased in powder form from Sigma Chemical Company (St. Louis, MO). Solid powder pectinase produced by L.D. Carlson Company (Kent, OH) packaged as pectic enzyme, was purchased from a local winery store.

The Coomassie Blue (Bradford) method (1) was used to determine the total protein content in the solution samples. The Coomassie Blue binding reagent was prepared by dissolving 100 mg Coomassie Brilliant Blue G-250 (purchased from Bio-Rad Laboratories, Richmond, CA) in 50 mL of 95% ethanol (produced by Midwest Grain Products, Atchison, KS) and 100 mL of 85% phosphoric acid (purchased from Fisher Scientific, Fairlawn, NJ), then adding deionized water bringing the solution up to 1 L. The solution was filtered to remove any undissolved dye. Excess reagent was stored in the refrigerator at 4°C and used within 2 wk.

In all of these experiments, 2 mL of sample and 3 mL of Coomassie Blue binding agent were used for the total protein content analysis. Following complete mixing of the sample and the reagent, the absorbance of that mixture was read at 5 min (at a wavelength of 595 nm in a Bausch and Lomb Spectronic 20 spectophotometrer).

Calibration curves for α-amylase, invertase, and pectinase were developed by taking absorbance readings over a range of enzyme concentrations by using the Bradford procedure (1). For all three enzymes, the linear relationship, $A_i = K_i C_i$, i = 1, 2, 3, described the resulting calibration curves in the selected concentration ranges. At pH 6.0 these were: $K_1 = K_{amylase} = 1.95 \times 10^{-3}$ L/mg; $K_2 = K_{invertase} = 5.36 \times 10^{-4}$ L/mg; and $K_3 = K_{pectinase} = 6.99 \times 10^{-5}$ L/mg. Different calibration curves are

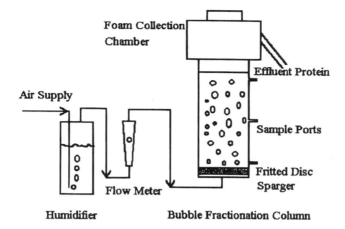

Fig. 1. Schematic diagram of the semibatch bubble fractionation process.

needed in general for each protein since each contains different amounts of arginine, the amino acid that is complexed with Coomassie Blue.

The impure invertase powder used dissolved only partially in water at room temperature. Invertase solution, therefore, was prepared by first dissolving invertase powder in deionized water, followed by filtering through Whatman No. 4 filter paper to remove the undissolved solids. In general, the concentration of this filtrate was then determined from its absorbance reading using $A_i = K_i C_i$. The K_i value in the $A_i = K_i C_i$ calibration curve was determined from a single concentration/absorbance (C_i, A_i) point where the absolute concentration was determined by taking the solution to dryness and then weighing it.

Both pectinase and α-amylase were weighed and mixed with deionized water before being used in the bubble fractionation experiments. Fresh mixtures used for the experiments were prepared in 1250 mL of deionized water without filtration and the pH was adjusted by using 0.1 N NaOH or HCl. No buffer was added. The pH of these three enzyme solutions was adjusted between 4.0 and 10.0 in the experiments. The air flow rate was held constant at 47 cm^3/min for all of the runs (ca. 0.05 vvm, where vvm = vol of air per vol of liquid solution, per minute). Bubble fractionation experiments were performed for one enzyme, two enzyme and then, the complete three enzyme system. The protein concentration conditions used in these experiments are listed in Table 1. The protein concentrations were selected so that the different enzyme components would each contribute about the same toward the solution absorbance reading. These concentrations differ from the previous experiment (2) where a 50–50% (100 mg/L each) mixture was used for the α-amylase-invertase system and may account for why the results for this mixture were somewhat different.

Samples for all of the enzyme cases listed in Table 1 were taken every 15 min up to 1 h at all three sampling ports. Since local maximum separa-

Table 1
Bubble Fractionation Conditions for the Experiments
using α-Amylase, Invertase, and Pectinase.

System	Protein solution	Protein concentration (mg/L)	pH range
Single enzyme	α-amylase	200	4-10
	invertase	600	4-10
	pectinase	2,000	4-10
Two enzymes	α-amylase and invertase	100 (α-amylase) and 300 (invertase)	4-10
	α-amylase and pectinase	100 (α-amylase) and 1,000 (pectinase)	4-10
	invertase and pectinase	300 (invertase) and 1,000 (pectinase)	4-10
Three enzymes	α-amylase, invertase and pectinase	100 (α-amylase), 300 (invertase) and 1,000 (pectinase)	4-10
	α-amylase, invertase and pectinase	100 (α-amylase), 300 (invertase) and 0-1,250 (pectinase)	5 and 9

tion ratios were obtained for α-amylase and invertase at pH 5.0 and 9.0, studies of the effect of pectinase on these other two enzymes were focused at these two pHs. The concentrations of pectinase were varied up to 2000 mg/L. The concentrations of α-amylase and invertase were kept constant at 100 mg/L and 300 mg/L, respectively, in the two and three enzyme experiments and at 200 and 600 mg/L in the single enzyme cases.

Solution samples (5 mL) for all cases were taken by syringe. The samples that were assigned time 0 were actually taken 2 min after pouring the solution into the bubble fractionation column. Sampling was taken every 15 min up to 60 min, from rubber septum sampling ports, located at the top, middle, and bottom of the column. For each pH profile case for a given enzyme system, 15 samples were analyzed for total protein content by the Coomassie Blue method (1).

RESULTS AND DISCUSSION

Air bubbles can cause proteins to rise in a dilute solution to the upper surface by adsorbing on the bubble-solution interface. Proteins concentrating on the bubbles lead to a protein concentration gradient from the top to the bottom of column as the bubbles rise up through the column. After the bubbles burst at the liquid-air surface of the bubble fractionation column, the released proteins build up to their highest concentration at that position.

Proteins tend to be hydrophobic and at their isoelectric point (where there is no charge in a water solution) they move toward the bubble—solution interface, thereby leaving the water solution by attaching to the gas bubbles. Once they latch onto a gas bubble, they are carried up to the top of the bulk solution, and thus lower the residual protein concentration of the remaining bulk solution. Since the protein-carrying bubbles collapse at the top surface, the protein concentration at that position reaches its highest level. The proteins, therefore, transfer downward and reflux by bulk diffusion. At steady state, the flux of proteins upward is equal to the flux of proteins downward. For the case of a mixed proteins, each of the enzyme molecules competes with each other to attach to adsorption sites on the bubbles. The more successful the protein is in competing for these interfacial sites, the better its separation within the bubble column. The separation ratio, defined as the ratio of a particular protein concentration at the top of the column to the bottom position, is used here to quantify the separation of a particular protein. In this study, the individual protein concentrations were not determined, in order to simplify the procedure. Here, the lumped protein optical absorbance ratio is used as a rough indicator of the partitioning. A mixed separation ratio, expressed as the ratio of optical absorbance at the top of column to the optical absorbance at the bottom of column, is used to infer the extent of the overall separation for the mixture of enzymes. Since certain proteins tend to favor particular pHs (at their isoelectric points) to achieve maximum separation as determined by their individual separation ratios, determining the optical absorbance ratio as a function of pH can provide clues for determining the best conditions to achieve maximum separations of components within a mixture.

Based on the experience gained in this study, steady state generally is achieved by 15–20 min. To ensure the samples were taken at steady-state, measurements were all taken at 60 min in the batch bubble fractionation process. Figure 2 shows the individual enzyme separation ratio vs pH trajectory for α-amylase, invertase, and pectinase solutions bubbled with air. α-amylase reaches a local maximum separation ratio at pH 8.0–9.0 and a bounded maximum at pH 4.0. Previous work (2) reported that a local max-

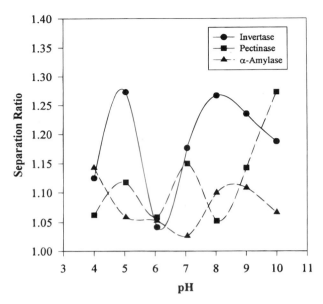

Fig. 2. Bubble fractionation of each of the three enzymes.

imum separation ratio of α-amylase, using either air or carbon dioxide as a carrier gas, was reached at pH 8.0. Local maximum separation ratios for invertase were obtained in these studies at pH 5.0 (close to the pH value used for invertase activity measurements by Vitolo and Borzani [7]) and pH 8.0. In previous experiments (2), with both air and carbon dioxide as the carrier gas, only one local maximum was reported at pH 5.0. The addition of a second peak may perhaps indicate that, under the conditions of this study, subunits of invertase (with pIs of 8) were released to the solution. The difference in air flow rate between previous (140 cm^3/min) experiments and these (47 cm^3/min) might cause the different degree of denaturation of proteins. Typically, the major cause of damage to proteins is from bursting bubbles (8). The longer sampling time for these experiments (60 min) may have led to more denaturation of enzymes than in previous experiments (20 min), but this longer time is countered by shorter residence times of oxygen contacting the recovered enzyme. The structure of denatured protein generally differs from the original protein. It is possible then that the denatured protein has a new isoelectric point from the undenatured protein that leads to a previously unobserved second peak. For invertase in these experiments, the second local maximum separation ratio appeared at pH 8.0. It is also noted that in these experiments the separation ratio trajectory decreases for α-amylase in the low pH range.

Pectinase seems to have the local maximum separation ratio at the pH bound of 10.0. Normally, pectinase is a mixture of three enzymes (9): pectinesterase, polygalacturonase, and pectate lyase. Multiple peaks of the pectinase separation ratio vs pH trajectory may indicate that three pectinase enzymes have separate and distinct isoelectric points. It is observed that the local maximum protein separations all occur in the pH range of 4.0 to 10.0: between pH 5.0 and 8.0 for invertase, between pH 5.0 and 7.0 for pectinase and between pH 8.0 and 9.0 for α-amylase.

The three combinations of two enzyme mixtures are comprised of the binary components of α-amylase-invertase, invertase-pectinase, and α-amylase-pectinase as listed in Table 1. The optical absorbances of the isolated α-amylase and invertase components in these binary combinations (prior to being combined) were chosen to be about 0.15, which corresponds to a concentration of 100 mg/L for α-amylase and 300 mg/L for invertase. Since pectinase is such a poor absorber of light in the Bradford procedure, the concentration of pectinase used in the binary enzyme combinations was raised to 1000 mg/L, which corresponds to an optical absorbance of 0.07. The optical absorbance of the α-amylase-invertase mixture is about 0.3, which is about twice that of the original optical absorbance of α-amylase and invertase, as expected, indicating that Beer's law holds in this concentration range. The optical absorbance of single enzyme systems (α-amylase and invertase) was selected to be the same as the α-amylase-invertase mixture at about 0.3 for α-amylase (at a doubled concentration of 200 mg/L) and for invertase (at a doubled concentration of 600 mg/L). Unfortunately, the optical absorbance of pectinase solution could only be raised to 0.14, even for a concentration of 2000 mg/L, the upper concentration limit explored in this study. Optical absorbances of 0.3, therefore, were not reached with pectinase mixtures.

Both the α-amylase-pectinase mixture and the invertase-pectinase mixture have optical absorbances around 0.22, inferring that Beer's law holds for these cases, as well as the α-amylase-invertase case. Like the α-amylase-invertase mixture, these experiments are designed to explore the behavior of two enzyme systems at an optical absorbance value in the range of 0.2 and 0.3. Figure 3 illustrates the mixed separation ratios of α-amylase-invertase, α-amylase-pectinase, and invertase-pectinase mixtures as a function of pH. The mixed separation ratio is the ratio of optical absorbances, A_{Top}/A_{Bottom}, for these systems. The absorbance, A, in dilute solutions can usually be represented by Beer's Law as $K_1C_1 + K_2C_2$. A can be used in place of C in this preliminary study in which only optical absorbance is measured. Since only one wavelength, λ, is appropriate for the Coomassie Blue method, the concentrations cannot be estimated from the linear relationship $A_{mixture} = \sum_{i}^{n} K_iC_i$, where $A_{mixture}$ and K_i are functions of the given λ. An additional procedure such as UV absorption needs to be used to provide extra information in order to estimate both C_1 and C_2 in this mixed enzyme system. It is noted that the mixed separation ratio, A_{Top}/A_{Bottom}, reduces to the regular separation ratio, C_{Top}/C_{Bottom}, for a single enzyme system, since A becomes KC for dilute systems and the K cancels out.

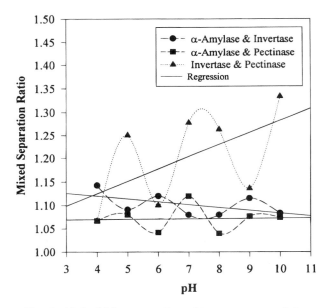

Fig. 3. Air bubble separation of two enzyme mixtures.

The mixed separation ratio in a multi-enzyme system is the ratio of the absorbance at the top position to the absorbance at the bottom position of the column. For dilute solutions this ratio may infer the weighted ratio of the total protein concentration at the top position to that at the bottom position since $A_{mixture} = \sum_{i=1}^{n} K_i C_i$, where $n = 3$. Here, the weighting factors are the absorbance coefficients, K_i. A value for total protein can be estimated if the values of $K_1 C_1 = K_2 C_2 = K_3 C_3 = K_i C_i = A_i$ are about equal. The mixed separation ratio using the absorbance ratio can be implied from the concentration ratio as follows:

$$A_{top} = \sum_{i=1}^{3} A_{i,Top}$$

where $A_{i,top} = [K_i C_i]_{Top}$ so that:

$$A_{top} = 3 A_{i,Top}$$

Similarly,

$$A_{bottom} = 3 A_{i,Bottom}$$

The mixed separation ratio (S.R.) becomes:

$$S.R. = \frac{A_{Top}}{A_{Bottom}} = \frac{A_{i,Top}}{A_{i,Bottom}} = \frac{[K_i C_i]_{Top}}{[K_i C_i]_{Bottom}}$$

If the additional approximation of $K_{i,Top} = K_{i,Bottom}$ is made for this "mean" value of K_i, then:

$$S.R. = = \frac{C_{i,Top}}{C_{i,Bottom}}$$

To qualitatively infer the separation tendency of these mixtures undergoing bubble fractionation, a linear regression of the mixed separation ratio, A_{Top}/A_{Bottom}, as a function of pH, was developed (in Fig. 3). It is observed that the mean value of the mixed separation ratio of the α-amylase-invertase mixture decreased slightly and the α-amylase-pectinase held constant as the pH of the mixture increased. It is also observed that the mean value of the mixed separation ratio of invertase-pectinase may increase when the pH of solution increases, albeit with significant oscillations. This seems to indicate that the standard deviation for the mixed separation ratio of invertase–pectinase is particularly large (more than double that observed in the other two binary systems). Comparing Fig. 2 and 3 it appears that pectinase does not seem to impair the separation of invertase (or at least the total protein) under the conditions studied, even though the pectinase concentration was more than three times that of the invertase concentration: enzyme activities were not measured in this study and should be checked in subsequent investigations to observe how they vary in response to the changing conditions. In contrast, the mixture of α-amylase-pectinase seems to track the pectinase pH profile more closely than the α-amylase profile, which is not surprising since the pectinase concentration is 10 times that of the α-amylase concentration. These experiments seem to indicate that pectinase can either cause the separation to remain unchanged (invertase) or enhance the separation (α-amylase). The effect of pectinase in the presence of another protein in bubble fractionation, therefore, depends on both the bubble adsorptive force of the protein being separated and the competition for the given bubble adsorption sites by the pectinase. The competition for adsorption sites depends on the protein structure of each enzyme, and their polarity, hydrophobicity, and concentration.

The mixture used in the three enzyme bubble fractionation experiment is comprised of 100 mg/L α-amylase, 300 mg/L invertase, and 1000 mg/L pectinase. As previously noted, these concentrations were selected to enable the optical absorbance of α-amylase and invertase components to each contribute about equally to the optical absorbance, whereas pectinase contributes to the optical absorbance only about one half of the other two enzymes. The resulting mixed separation ratio for the three enzyme system is shown as a function of pH in Fig. 4. The local maxima of this mixed separation ratio of the ternary mixture appear at pH 5.0, 7.0, and between 9.0 and 9.5. Unlike the single and binary cases shown in Figs. 2 and 3, the highest mixed separation ratio was reached at 1.10 (vs ca. 1.25 in the previous

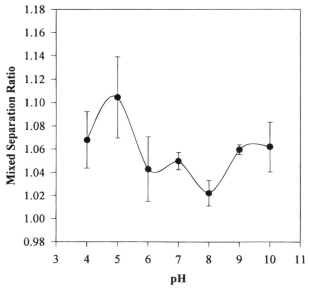

Fig. 4. Air bubble fractionation of the three enzyme mixture.

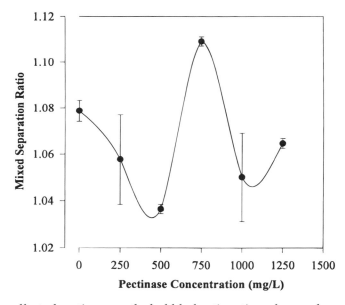

Fig. 5. The effect of pectinase on the bubble fractionation of α-amylase and invertase at pH 5.

cases). The largest two of the three local maxima located at pH 5.0 and 9.0 were thus selected (to effect the greatest resolution) for subsequent study of the effect of pectinase on the air bubble fractionation of α-amylase from invertase. These pHs also turn out to be close to the isoelectric values (pIs) of invertase and α-amylase, respectively (2).

Effect of Pectinase 407

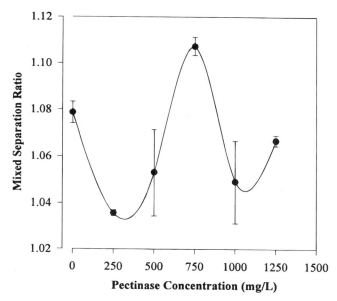

Fig. 6. The effect of pectinase on the bubble fractionation of α-amylase and invertase at pH 9.

Figs. 5 and 6 qualitatively depict the effect of various levels of pectinase on the separation of α-amylase from invertase at pH 5.0 and 9.0, respectively. Both graphs show similar curve patterns. The oscillation of these curves may be from the ionic charge on pectinase driving the separation ratio of the system to neutrality and polarity. For low pectinase concentration, $0 \leq C_{pectinase} \leq 500$ mg/L, the mixed separation ratio decreases as the pectinase concentration increases. Perhaps the ionic charges on pectinase neutralize the opposite charges on α-amylase and invertase to drive the system to neutrality. When the charges are balanced, the proteins generally tend to move freely and disperse evenly in the bubble column. This charge leads to a minimum in mixed separation ratio (at about 1.03) of this three enzyme mixture for a pectinase concentration of 500 mg/L for pH 5.0 and 300 mg/L for pH 9.0. When more pectinase is added beyond the neutral balance point, the ionic charges build up in the solution. An additional adsorptive effect of pectinase itself develops at the air-liquid interface leading to an enhanced pectinase layer that is brought up to the top by the air bubbles. The two pH systems reach a maximum of protein separation when the mixed separation ratio is about 1.11, at a pectinase concentration of 750 mg/L for both the pH 5.0 and 9.0 cases. Adding more pectinase beyond this 750 mg/L level, leads to excess ionic charges in the solution, and resulting repulsive forces. Therefore, it is difficult for proteins to form a layer on the bubbles and the resulting fractionation is degraded.

Since all of the mixed separation ratios determined in these experiments are greater than 1.0, this seems to indicate that some protein separation occurs over the entire pH range ($C_{i,Top} > C_{i,Bottom}$ for all i). Unfortunately, for the mixed enzyme mixture cases, the individual protein concentrations were not measured since the expected separation (which matched the individual protein cases at their isoelectric points) previously found by DeSouza et al. *(2)* for a 50–50% mixture of α-amylase and invertase, was not replicated here for any binary combination of enzymes. The total protein technique (Coomassie Blue method) used here only determines the concentration of individual enzymes in single protein systems. Polyacrylamine gel electrophoresis, HPLC, or a direct enzyme activity method must be employed in future work to sharpen these results. It seems clear, however, that a change in pectinase level has a strong effect on the apparent separation of the total protein mixture even when the pectinase concentration is less than 500 mg/L at pH 5.0 and less than 250 mg/L at pH 9.0.

CONCLUSIONS

In these preliminary experiments, pectinase does not seem to enhance the separation of an invertase-pectinase mixture (relative to the invertase system alone) and even seems to inhibit the bubble separation of α-amylase in an α-amylase-pectinase mixture. When pectinase was mixed with invertase and α-amylase to form a three enzyme system, the local maxima mixed separation ratios were reached at pH 5.0 and 9.0. The addition of pectinase to the mixture of α-amylase and invertase seemed to reduce the separation of these two enzymes by bubble fractionation at both pH 5.0 and 9.0.

REFERENCES

1. Bradford, M. M. (1976), *Anal. Biochem.* **72,** 248–254.
2. DeSouza, H. G. A., Tanner, R. D., and Effler, W. T. (1991), *Appl. Biochem. Biotechnol.* **28/29,** 655–666.
3. Lemlich, R. (1972), in *Adsorptive Bubble Separation Techniques*, Academic Press, New York, NY, pp. 133–143.
4. Park, D. H., Baker D. S., Brown K. G., Tanner R. D., and Malaney G. W. (1985), *J. Biotechnol.* **2,** 337–346.
5. Effler, W. T., Tanner, R. D., and Malaney, G. W. (1989), in *Bioproducts and Bioprocesses*, Fiechter, A., Okada, H., and Tanner, R. D., eds., Springer-Verlag, New York, NY, pp. 235–256.
6. Potter, F. J., DeSouza, H. G. A., Tanner, R. D., and Wilson, D. J. (1990), *Separation Sci. Technol.* **25,** p. 673–687.
7. Vitolo, M. and Borzani, W. (1983), *Analyt. Biochem.* **130,** 469, 470.
8. Wright, K. I. T. and Cui, Z. F. (1994), in *Separations for Biotechnology-3*, The Royal Society of Chemistry, London, pp. 577–585.
9. Whitaker, J. M. (1994), in *Principles of Enzymology for the Food Sciences*, Marcel Dekker, New York, NY, pp. 425–435.

Lipase Production by *Penicillium restrictum* in a Bench-Scale Fermenter
Effect of Carbon and Nitrogen Nutrition, Agitation, and Aeration

DENISE M. FREIRE, ELAINE M. F. TELES, ELBA P. S. BON, AND GERALDO LIPPEL SANT' ANNA, JR.*

Instituto de Química, Faculdade de Farmacia and COPPE, Universidade Federal do Rio de Janeiro, Rio de Janeiro, RJ, Brazil

ABSTRACT

A preliminary screening work selected *Penicillium restrictum* as a promising micro-organism for lipase production. The physiological response of the fungus towards cell growth and enzyme production upon variable carbon and nitrogen nutrition, specific air flow rate (Q_a) and agitation (N) was evaluated in a 5-L bench-scale fermenter. In optimized conditions for lipase production meat peptone at 2% (w/v) and olive oil at 1% (w/v) were used in a growth medium with a C/N ratio of 9.9. Higher C/N ratios favored cell growth in detriment of enzyme production. Low extracellular lipase activities were observed using glucose as carbon source suggesting glucose regulation. Final lipase accumulation of 13,000 U/L was obtained, using optimized specific air flow rate (Q_a) of 0.5 vvm and an impeller speed (N) of 200 rpm. Agitation showed to be an important parameter to ensure nutrient availability in a growth medium having olive oil as carbon source.

Index Entries: Lipase; lipase production; *Penicillium restrictum*; C and N nutrition; agitation and aeration.

INTRODUCTION

There is a growing interest on microbial lipases (acylglycerol hydrolases, E.C. 3.1.1.3) that are largely used in the detergent and food industries. Lipases with peculiar catalytic properties obtained from microbial

*Author to whom all correspondence and reprint requests should be addressed.

screening procedures may be used for the modification of fats and oils or synthesis of novel compounds that will probably have an important impact on a range of chemical industries (1).

Fungi from the genera *Mucor, Aspergillus, Rhizopus,* and *Penicillium* have been used in lipase production studies. Although the effects of media composition have been described (2–5), there is a general lack of discussion as to the causes of the micro-organism's physiological response. The use of triglycerides as carbon source is adequate for both cell growth and enzyme production, although the induction of lipase biosynthesis by free fatty acids or triglycerides is controversial (6–8). Consistent cell growth is likewise observed when glucose is used as a carbon source although lipase production by some *Penicillium* sp, may be repressed depending on the micro-organism studied (5). Concerning the nitrogen source, meat peptone is beneficial for lipase production by some *Penicillium* and other fungi species (4,7–11).

The reported agitation and aeration conditions for lipase production by fungi in bench-scale fermenters varies widely: 750 rpm and 2 vvm, 200 rpm and 1.0 vvm, and 80 rpm and 1.0 vvm were used in *Geotrichum candidum (12), Humicola lanuginosa (13),* and *Rhizopus* sp *(14)* fermentations, respectively.

Aiming at a better understanding of fungi lipase production, a stepwise study was carried out, focusing on the relationship between cell growth and enzyme production in response to variable carbon and nitrogen nutrition, specific air flow rate, and agitation in a 5-L instrumented fermenter.

MATERIALS AND METHODS

Micro-organism Maintenance and Propagation

Penicillium restrictum isolated from wastes of a Brazilian babassu coconut oil and identified as lipase producer was used in the experiments. For spores production the micro-organism was cultivated in agar slants (soluble starch 2%; olive oil 1%; yeast extract 0.1%; $MgSO_4 \cdot 7H_2O$ 0.025%; KH_2PO_4 0.05%; $CaCO_3$ 0.5%, and agar 1%) and incubated at 30°C for 1 wk.

Fermentations

Enzyme production was studied in shaken-flask fermentations: 120 mL of the growth media (olive oil 1%, meat peptone 2%, NaCl 0.5%, and yeast extract 0.1%, initial pH 5.5) was inoculated with 2×10^6 spores and incubated at 30°C and 160 rpm. Fermentations were also carried out in a 5-L fermenter (BIOFLO II-New Brunswick) containing 4000 mL of an optimized growth medium at 30°C, initial pH 5.5 (not controlled). The fermenter was inoculated with 2% (v/v) of its volume with 24 h grown cells. Mixing and aeration were investigated by varying N from 50 to 300 rpm and Qa from 0.2 to 1.0 vvm.

Effect of the Carbon and Nitrogen Sources

Fermentations were carried out as in Materials and Methods except for the olive oil, which was replaced by

1. glucose 2.2%
2. glucose 1.98% plus olive oil 0.1%
3. lactose 2.2% and
4. lactose 1.98% plus olive oil 0.1%.

In all media the same C/N ratio (total carbon to total nitrogen concentration) of 9.9 was used. In a second set of experiments the effect of olive oil in the basal media was investigated by varying its concentration within the range 0.1 to 2.0%. As a consequence of keeping the peptone concentration constant and increasing the olive oil, the media C/N ratio varied from 6.2 to 13.9. To study the effect of the nitrogen source, in preliminary experiments, the meat peptone in the growth media was replaced by 2% soya and 2% casein peptones. In a second set of experiments, the concentration of soya and casein peptones were raised to 7.0 and 4.3% respectively, to equalize the media nitrogen content to the meat peptone medium *(15)*. Inorganic nitrogen was also studied, replacing meat peptone by ammonium sulfate 0.46%, ammonium nitrate 0.28%, and ammonium chloride 0.38%. The salts concentration varied to keep a C/N ratio of 9.9 in all cases.

Analytical:

Biomass Concentration

At selected time intervals, samples were withdrawn and filtered through Whatman N°44 paper. The biomass was thoroughly washed and dried at 75°C for dry weight determination.

Lipids Assay

The culture supernatant was used for gravimetric determination of lipids according to Akhtar et al. *(2)*, replacing n-hexane by a mixture of petroleum-ether and ethyl-ether (1:1).

Lipase Assay

The reaction mixture consisted of 19 mL of olive oil/arabic gum emulsion (5% olive oil and 5% arabic gum) in 100 mM potassium phosphate buffer, pH 7.0. This mixture was homogenized in a blender for 3 min and the enzymatic reaction started by adding 1 mL of the culture supernatant. The assay was carried out at 37°C and 200 rpm for 30 min. After this time interval, the reaction was stopped by adding 20 mL of acetone-ethanol 1:1 (v/v) and the amount of fatty acids produced was then titrated with 0.05M NaOH until end-point 9.5 using an automatic titration apparatus (Mettler DL21). One unit of lipase activity was defined as the amount of enzyme

Protease Assay

Proteolytic activity was determined in the culture supernatant according to Charney and Tomarelli (16).

Relationship Between Cell Lysis and Protease

The relationship between cell lysis and protease concentration in the fermentation broth was evaluated in bench-scale fermenter experiments by plotting cellular decay rate $\Delta X/\Delta t$ (variation of biomass concentration since the beginning of the growth decay up to around 150 h of fermentation) against $\Delta P/\Delta t$ (variation of protease concentration in the same time interval). This determination was performed for all conditions of air supply (0.20 to 1.0 vvm) and agitation (50 to 300 rpm).

Yield Coefficient and Productivity

The overall yield coefficient (Yp/x) was defined as the maximum units of lipase produced per maximum biomass concentration. Productivity was the ratio between maximum lipase activity per fermentation time.

RESULTS AND DISCUSSION

Effect of Nitrogen Source

Penicillium restrictum was unable to grow in inorganic nitrogen, although it was reported that a strain of *P. roqueforti* was able to grow and produce lipase using ammonium sulfate as the only nitrogen source (7). This inability to grow on inorganic nitrogen has also been observed for *P. roqueforti (17,18)*, *P. verrucosum* var. *cyclopium* (8), and *P. citrinum* (5), and could be related to transport limitation or restrictions in the metabolic pathways to incorporate inorganic nitrogen.

Table 1 shows the highest values for cell and enzyme concentrations, and normalized enzyme production (units of lipase produced per mg of cell dry weight:Yp/x), in the culture media with meat, soya, and casein peptone at variable concentrations. Lipase concentration and the Yp/x parameter were more than two times higher in meat peptone, indicating that its amino acids and/or cofactors content matches the micro-organism physiological requirements for lipase biosynthesis. These results are in accordance to the literature concerning lipase production by *Penicillium* (5,7–9,11,17) and other fungi genera (4,10).

Effect of the Carbon Source

Table 1 also presents the peak data for biomass and lipase concentrations and normalized lipase production for different conditions of carbon nutrition. *Penicillium restrictum* was able to grow equally on glucose and

Table 1
Effect of the Nitrogen and Carbon Sources on Cell Growth, Overall
Yield Coefficient (Yp/x) and Lipase Production by *Penicillium restrictum*
in Shaken-Flask Fermentations Carried out at 30°C and 160 rpm

Nitrogen source[a]	Biomass concentration (mg/mL)	Activity (U/mL)	Yp/x (U/mg)
meat peptone (2.0%)	14.2	13.0	0.92
soya peptone (2.0%)	11.4	3.4	0.30
(7.0%)	16.5	6.6	0.40
casein (2.0%)	14.9	5.4	0.36
(4.3%)	17.5	6.2	0.35
Carbon source[b]	Biomass concentration (mg/mL)	Activity (U/mL)	Yp/x (U/mg)
glucose (2.2%)	14.1	1.9	0.14
olive oil (1.0%)	14.2	13.0	0.92
glucose (1.98%) + olive oil (0.1%)	13.9	1.8	0.13
lactose (2.2%)	4.3	1.7	0.40
lactose (1.98%) + olive oil (0.1%)	5.7	6.0	1.00

[a]olive oil 1% w/v was used as carbon source.
[b]meat peptone was used as nitrogen source (media with a constant ratio C/N of 9.9).

olive oil as similar biomass concentrations around 14.0 mg/mL were observed in both cases. Enzyme production however, was significantly higher (13.0 U/mL) when olive oil was used in comparison to glucose (1.9 U/mL) even when the glucose media was supplemented with 0.1% olive oil. Accordingly, lipase production by *P. restrictum* seems to be glucose regulated. Similar results were observed for *Penicillium citrinum* when glucose or sucrose were used even in the presence of triglycerides (5). These findings indicate that the avoidance of a repressive carbon source is of paramount relevance for lipase production.

The use of lactose resulted in poor cell growth (4.3 mg cell/mL) probably because of the inability of the micro-organism to metabolize lactose. Cell growth was, therefore, limited to the peptone availability. Concerning lipase production, although only basal levels were observed (1.7 U/mL), the normalized production (0.4 U/mg cell) reflected the absence of glucose repression. When lactose was supplemented with olive oil at 0.1%, an improvement in cell growth (5.7 mg/mL), enzyme production (6.0 U/mL), and normalized production (1.0 U/mg) were observed. This beneficial effect can be explained by the higher carbon availability to the cell metabolism or oil induction of enzyme biosynthesis *(3,6,19)*. In further experiments, similar enzyme levels were observed using peptone 2% w/v and olive oil 0.1% w/v (Fig. 1). In conclusion, olive oil showed to be more appropriate for both cell growth and lipase production, glucose repressed the enzyme synthesis, and lactose was poorly metabolized. Although the possible causes to these physiological responses were discussed, it is important to point out that the measurement of enzyme concentration in culture supernatant may not be directly related to its biosynthesis, thus, considerations related to regulation of gene expression are open to further discussions.

Effect of Olive Oil Concentration on Lipase Production and Cell Growth

Figure 1 compares the highest data for biomass concentrations and lipase accumulation for the range of oil concentrations studied. There is a clear positive effect of the concentration increase on cell growth indicating that the basal medium (1% olive oil, 2% peptone) was carbon limited. In response to the increase in oil availability from 0.1 to 2.0% (media C/N ratio from 6.2 to 13.9), biomass varied steadily from 5.5 to 22.4 mg/mL. The overall positive effect of oil concentration on cell growth was not likewise observed on lipase production as peak concentration of 13.0 U/mL were obtained with 0.5 and 1.0% olive oil decreasing afterwards. In conclusion, a threshold olive oil availability, related to the medium C/N ratio, would trigger a metabolic shift in *P. restrictum* towards cell growth in detriment of enzyme production. Lower enzyme activity at higher oil concentrations (C/N > 9.9) are explained by some authors as enzyme inhibition by free fatty acids. This effect, however, has not been properly characterized *(2,20)*.

Lipase and Protease Production in a Bench-Scale Fermenter

A typical fermentation time course for lipase production by *P. restrictum* is shown in Fig. 2. Using optimized inoculum and medium conditions, the experiments were carried out at 200 rpm and 0.5 vvm. The

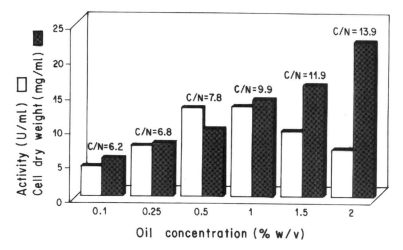

Fig. 1. Effect of olive oil concentration (w/v) on biomass and lipase accumulation in *Penicillium restrictum* shaken-flask fermentations (30°C and 160 rpm), having meat peptone 2% w/v as nitrogen source. Media carbon to nitrogen ratios (C/N) are presented for each case.

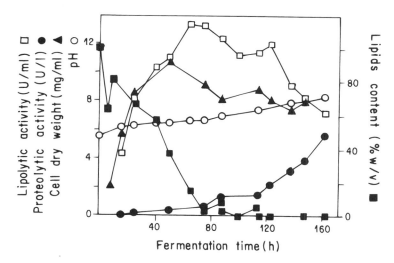

Fig. 2. Typical time course for cell growth, lipase production, protease accumulation, and lipids consumption by *Penicillium restrictum* in a 5-L bench-scale fermenter. Experiment carried out at 200 rpm, 0.5 vvm, 30°C, and initial pH 5.5.

maximum lipolytic activity (13.4 U/mL) was observed after 70 h of cultivation, which was coincident with the depletion of the carbon source (lipids). Lipase production showed to be growth associated, although a small increase on enzyme activity was observed after the end of the cell-growth phase. This increase could be related to the intracellular lipase release upon cell lysis that was observed after 40 h of fermentation. It was

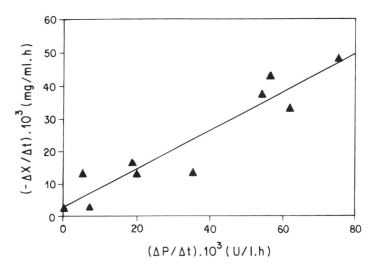

Fig. 3. Relationship between cellular decay rate ($-\Delta X/\Delta t$) and protease production rate ($\Delta P/\Delta t$) in *Penicillium restrictum* fermentations using a 5-L bench-scale fermenter. Working conditions according to text.

observed a direct relationship ($r = 0.92$) between cell lysis rate ($\Delta x/\Delta t$) and the rate of protease activity increase ($\Delta P/\Delta t$) at the later fermentation stage, indicating that protease activity increase was a result of the release of intracellular protease (Fig. 3). The decrease on lipase activity after 70 h of fermentation could be explained by pH inactivation, proteolysis, or both. Previous results showed that *P. restrictum* lipase is not stable at pH values above 8.0 and that the decrease on lipase activity was reduced when the serine protease inhibitor (PMSF) was added at the later fermentation stages (results not shown).

Effect of the Specific Air Flow Rate on Lipase, Protease, and Biomass Accumulation

The results presented in Table 2 and the profiles of lipase activity in experiments carried out using different specific air flow rates at 200 rpm presented in Fig. 4A, indicate that Qa variation had no significant effect on lipase production. However, Qa increase had a positive effect on cell growth and protease accumulation (Table 2 and Fig. 4B). A higher oxygen availability improved biomass accumulation and consequently the amount of protease released upon cell lysis. Isobe et al. (21), working with *P cyclopium*, utilized a high Qa value (1.0 vvm) to increase cell growth for the inoculum preparation and then a reduced Qa (0.25 vvm) during the fermentation. Such procedure is in accordance to our results since a higher oxygen availability promoted a metabolic shift toward cell growth in detriment of lipase production.

Table 2
Effect of the Impeller Speed and the Specific Air Flow Rate
on Lipolytic Activity, Biomass Concentration, Overall
Yield Coefficient (Yp/x), and Productivity[a]

N (rpm)	Qa (vvm)	Maximum lipolytic activity (U/mL)	Maximum biomass concentration (mg/mL)	Yp/x (U/mg)	Productivity (U/L.h)
50	0.20	3.5	5.6	0.63	30
50	0.50	4.9	5.9	0.83	42
100	0.20	8.9	6.0	1.48	75
100	0.50	9.3	8.5	1.09	78
100	1.00	9.1	11.5	0.79	82
200	0.20	12.3	7.2	1.71	141
200	0.50	13.4	9.3	1.44	209
200	0.75	13.3	11.7	1.14	147
200	1.00	11.4	13.5	0.84	208
300	0.20	11.3	13.0	0.87	246
300	0.50	12.2	13.9	0.87	306
300	1.00	13.0	15.1	0.86	263

[a]Maximum data for enzyme and biomass concentration in fermentations carried out in a 5-L bench-scale fermenter at 30°C and initial pH 5.5.

Effect of Agitation Speed on Lipase, Protease, and Biomass Accumulation

Figure 5A presents the effect of N on lipase production using 0.5 vvm. A positive effect of agitation on lipase production was observed up to 200 rpm. At this impeller speed, maximum enzyme activity was improved 1.4 and 2.7 times in comparison to the lipase concentration with 100 and 50 rpm, respectively. An acceleration was also observed on lipase accumulation and/or on its release to the medium as an anticipation of the activity peak was observed when N was increased from 50 to 300 rpm. A similar

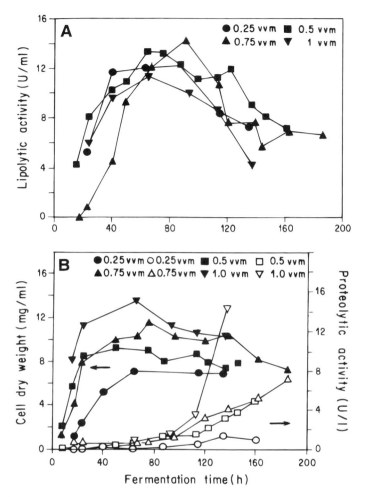

Fig. 4. Effect of the specific air flow rate on lipase production (**A**), cell growth and protease accumulation (**B**) by *Penicillium restrictum*. Experiments carried out in a 5-L instrumented fermenter (200 rpm, 30°C, and initial pH 5.5).

trend was observed for *Geotrichum candidum* cultivation in shaken flasks (22). This effect seems to be related to the improvement of the mass transfer conditions in the fermenter. As lipids are water nonmiscible carbon sources, their availability for microbial consumption is favored by the turbulence attained in the medium. At higher impeller speeds, the mixing improvement favors substrate availability being beneficial to a higher and faster cell growth and lipase production. It is well-established that N is a relevant operational parameter, mainly in highly viscous nonNewtonian fluids, as is the case of filamentous fungi fermentation broths. Figure 5B shows that protease activity increased upon increments on the impeller speed, which can be related to both higher biomass accumulation and cell lysis because of mycelium fragmentation.

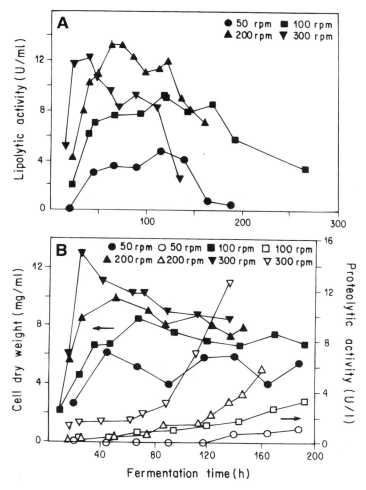

Fig. 5. Effect of the impeller speed on lipase production (**A**), cell growth and protease accumulation (**B**) by *Penicillium restrictum*. Experiments carried out in a 5-L instrumented fermenter (0.5 vvm, 30°C, and initial pH 5.5).

Effect of Air Supply and Agitation on the Yield Coefficient and Productivity

According to data presented in Table 2, productivity was not significantly affected by Qa. Figure 4A that compares enzyme concentration profiles for 200 rpm at variable Qa values shows that similar peak enzyme concentrations were observed in all cases at around 75 h of fermentation. Agitation showed a marked effect on productivity as by the use of 50, 100, 200, and 300 rpm, the average values obtained were of 36, 78, 176, and 272 U/L.h., respectively. This was a result of the anticipation of the activity peak from 120 h (50 rpm) to 30 h (300 rpm) and to the increase of the maximum activity value by a factor of 2.5 (Fig. 5A). The yield coefficient showed a decreasing trend with increase in the air flow rate, as observed in the experiments conducted at 100 and 200 rpm. For these two conditions the maximum lipolytic activity was not affected by Qa, although, maximum biomass concentration raised markedly. These results indicate that fungal growth was stimulated by a higher oxygen supply, and that lipase production was favored by oxygen limitation. At impeller speed of 300 rpm both cell growth and lipase production were not affected by air flow rate, indicating that a saturation condition was attained in terms of oxygen and nutrients availability.

A positive effect on the yield coefficient was observed with the increase on the impeller speed up to 200 rpm. In the range of 50 to 200 rpm (0.5 vvm), the maximum lipolytic activity increased by a factor of about 2.7, whereas the maximum biomass concentration increased 1.6 times. Consequently, the yield coefficient increased about 1.7 times. As previously commented, this result seems to be related to a better availability of nutrients caused by improved mixing conditions. Beyond 300 rpm no improvement on lipase production was observed and consequently the yield coefficient decreased.

In conclusion, the results hereby presented indicated that the best results concerning lipase production were achieved in fermentations conducted at 30°C when meat peptone (2% w/v) and olive oil (1% w/v) were used as nitrogen and carbon sources (medium with a C/N ratio of 9.9), initial pH 5.5 natural pH fermentations. Using these conditions, the optimized agitation and aeration in the bench-scale fermenter were 200 rpm and 0.5 vvm.

ACKNOWLEDGMENTS

This work was partially financed by CAPES and PADCT/CNPq (Proj. No. 62.0160/91.8)

REFERENCES

1. Björkling, F., Godtfredsen, S. E., Kirk, O. (1991), *TIBTECH* **9**, 360–363.
2. Akhtar, M. W., Mirza, A. Q., Nawazish, M. N., and Chughtai, M. I. D. (1983), *Can. J. Microbiol.* **29**, 664–669.
3. Pokorony, D., Friedrich, J., and Cimerman, S. (1994), *Biotechnol. Lett.* **16**, 363–366.
4. Salleh, A. B., Musani, R., Basri, M., Ampon, K., Yunus, W. M. Z., and Razak, C. N. A. (1993), *Can. J. Microbiol.* **39**, 978–981.
5. Sztajer, H. and Maliszewska, J. (1989), *Biotechnol. Lett.* **11**, 895–898.
6. Maliszewska, I. and Mastalerz, P. (1992), *Enzyme Microb. Technol.* **14**, 190–193.
7. Petrovic, S. E., Skrinjar, M., Becaveric, A., Vujicic, J. F., and Banka, L. (1990), *Biotechnol. Lett.* **12**, 299–304.
8. Glenza, A. and Jaballah, L. B. (1985), *Archs Inst. Pasteur Tunis* **62**, 69–89.
9. Chander, H., Sannabhati, S. S., Elias, J., and Ranganathan, B. (1977), *J. Food Sci.* **42**, 1677–1678.
10. Kennedy, J. F. (1987), in *Biotechnology: A comprehensive treatise in 8 vol.*, Rehm, H. J. and Reed, G., eds., pp. 113–121.

11. Druet, D., Abbadi, N. E., and Comeau, L. C. (1992), *Appl. Microbiol. Biotechnol.* **37,** 745–749.
12. Jacobsen, T., Olsen, J., and Allermann, K. (1989), *Enzyme Microb. Technol.* **11,** 90–95.
13. Ibrahim, O. C., Nishio, N., and Nagai, S. (1987), *Agric. Biol. Chem.* **51,** 2145–2151.
14. Upadhyay, C. M., Nehete, P. N., and Khotari, R. M. (1989), *Biotechnol. Lett.* **11,** 793–796.
15. Chopra, A. K., Chander, H., Batish, V. K., and Ranganathan, B. (1981), *J. Food Protection* **44,** 661–664.
16. Charney, J. and Tomarelli, R. M. (1947), *J. Biol. Chem.* **171,** 501–505.
17. Eitenmiller, R., Vakil, J. R., and Shahani, K. M. (1970), *J. Food Sci.* **35,** 130–133.
18. Menassa, A. and Lambert, G. (1982), *Le Lait* **62,** 32–43.
19. Okeke, C. N. and Okolo, B. N. (1990), *Biotechnol. Lett.* **12,** 747–750.
20. Akhtar, M. W., Mirza, A. Q., and Chughtai, M. I. D. (1980), *Appl. Environ. Microbiol.* **40,** 257–263.
21. Isobe, K., Akiba, T., and Yamaguchi, S. (1988), *Agric. Biol. Chem.* **52,** 41–47.
22. Baillargeon, M. W., Bistline, Jr., R. G., and Sonnet, P. E. (1989), *Appl. Microbiol. Biotechnol.* **30,** 92–96

Potassium Acetate by Fermentation with *Clostridium thermoaceticum*

Minish M. Shah, Fola Akanbi, and Munir Cheryan*

University of Illinois, Agricultural Bioprocess Laboratory, 1302 W. Pennsylvania Ave., Urbana, IL 61801

ABSTRACT

Potassium acetate is currently made by reacting petroleum-based acetic acid with potassium hydroxide. An alternate process, anaerobic fermentation of dextrose with *Clostridium thermoaceticum*, could be used and could possibly be cheaper. Growth characteristics and productivity of the fermentation were optimized to maximize acetate concentration in the broth. The effects of pH, type, and concentrations of nutrients and reducing agents were also evaluated. Corn steep liquor and stillage from an ethanol plant were effective and much cheaper substitutes for yeast extract. Preconcentrating the cells by ultrafiltration improved productivity, resulting in an acetic acid concentration of 53.6 g/L in 50 h at pH 6.5 using corn steep liquor. Sodium sulfide could be substituted for cysteine as the reducing agent with yields greater than 0.9 g acetic acid/g glucose.

Index Entries: *Clostridium thermoaceticum*; potassium acetate; acetic acid; fermentation.

INTRODUCTION

Potassium acetate has been approved by the US Federal Aviation Administration for deicing of airport runways and aircrafts, replacing urea and glycol (1). It is environmentally safe and easily biodegradable (2). It also can be used as a heat transfer fluid in antifreeze formulations and in heat pumps. Currently, potassium acetate is made by reacting petroleum-based acetic acid with potassium hydroxide. An alternate method is to manufacture the acetate from dextrose by fermentation.

Anaerobic fermentation for acetate production by *Clostridium thermoaceticum* has been extensively investigated (3–15). The most attractive fea-

*Author to whom all correspondence and reprint requests should be addressed.

ture of this fermentation is high acetate yield (theoretically, 1 g/g glucose), which is 50% higher than the two-step fermentation process used for vinegar production. Among the many improvements in the anaerobic fermentation are development of strains with tolerance to acetate concentrations as high as 6–10% *(3–6)*, reduction in nutrient costs *(7,8)* and higher productivity using membrane cell-recycle bioreactors *(9–11)* and immobilized cell bioreactors *(12)*. Much of the prior work focused on sodium acetate or calcium magnesium acetate *(5)*. However, potassium may be more toxic than sodium for strain DSM521 of this micro-organism *(13)*. This paper reports on the fermentation aspects of the process: effects of pH, nitrogen source, reducing agents, and cell concentration on production of potassium acetate with a mutant strain of *C.thermoaceticum*. Downstream processing is being investigated and will be published elsewhere.

MATERIALS AND METHODS

The mutant strain of *C. thermoaceticum* (ATCC 49707 and DSM 6867) was adapted to grow in media containing high concentrations of potassium by transferring it alternatively to the nutrient medium (described below) containing 5% potassium acetate and the medium without any acetate. The medium for culture adaptation and maintenance contained (in g/L) the following components: Glucose = 20; Buffering components ($KHCO_3$ = 9, KH_2PO_4 = 1.4, K_2HPO_4 = 1.1); yeast extract = 5; salts ($MgSO_4.7H_2O$ = 0.25, $(NH_4)_2SO_4$ = 1, $CoSO_4.7.5H_2O$ = 0.03, $Fe(NH_4)_2(SO_4)_2.6H_2O$ = 0.04, $Na_2WO_4.2H_2O$ = 0.0033, $Na_2MoO_4.2H_2O$ = 0.0024, $NiCl_2.6H_2O$ = 0.00024, $ZnSO_4.7H_2O$ = 0.00029, Na_2SeO_3 = 0.000017) and reducing agent (cysteine. $HCl.H_2O$) = 0.25. The medium for fermentation experiments was the same composition, except it contained only 1.8 g/L of $KHCO_3$. The concentrations of glucose, yeast extract (or other nitrogen source), and salts were varied in different experiments. The concentrations given above are referred to as X concentration. When concentrations of yeast extract and salts were two or three times the above concentration, they were referred to as 2X and 3X concentrations, respectively.

The nutrient medium for fermentation was prepared as described previously *(14)*. For batch experiments, the initial glucose concentration was 46–50 g/L. All fermentation experiments were carried out in a 2-L reactor with 1 L working volume. Temperature was controlled at 60°C and the pH was maintained at the desired value by addition of oxygen-free 8N KOH. The fermenter was overlaid with sterile CO_2. Fermentation was initiated by transferring 24-h old inoculum (10% v/v). Fed-batch experiments were conducted as described earlier *(10,17)*: Concentrated solutions of selected nutrients were added at various times during the fermentation.

Glucose and acetic acid were analyzed by HPLC using the Aminex 87H column (Bio-Rad, Hercules, CA). "Acetate" data in this paper refers to acetic acid with a molecular weight of 60. Thus 1 g of acetic acid = 1.635 g potassium acetate. Cell concentration was monitored by optical density at 600 nm wavelength.

Effect of pH

These experiments were conducted at pH 5.5, 6.0, 6.5, and 7.0. The concentration of yeast extract and salts were at the 2X level.

Effect of Reducing Agents

Various reducing agents were used at concentrations of 0 to 0.25 g/L. Yeast extract (Difco) was used as the complex nitrogen source and pH was kept at 6.5 with KOH. The concentrations of yeast extract and all salts were at the 2X level.

Low-Cost Nutrients

Corn steep liquor was obtained from A. E. Staley Manufacturing (Decatur, IL), as a 50% (w/w) solution. It contained 20% (w/w dry basis) lactic acid. Stillage was obtained from Pekin Energy (Pekin, IL). It is the aqueous suspension remaining after ethanol is steam-stripped away from an ethanol fermentation broth. It contained 7% total solids (mostly dead yeast cells) and 1.47% lactic acid (i.e., 21% lactic acid on a dry basis). Stillage was prefiltered using an ultrafiltration membrane (UFP500, A/G Technology, Needham, MA). The clear permeate from the membrane was used as a nutrient in the fermentation. Solids content was determined by air drying at 80°C for 24 h. Concentrations of nutrients reported are all on dry bases.

Fermentation with High Cell Density

The inoculum was grown in 3 L of medium. It was then concentrated by ultrafiltration to 800 mL. This inoculum with high cell concentration (OD of 15–19) was combined with 200 mL nutrient medium to initiate fermentation.

RESULTS AND DISCUSSION

Effect of pH

In all experiments, the typical fermentation pattern reported earlier (7,8,11) was observed: an increase in optical density until the end of the exponential phase, followed by a decrease (Fig. 1). Acetic acid production followed the cell growth curve, but decreased significantly after growth stopped. In some cases, fructose was produced as a by-product, especially when nutrients were insufficient. Table 1 summarizes results of two replicate experiments. With 2X yeast extract and 2X salts, pH 6.5 resulted in the

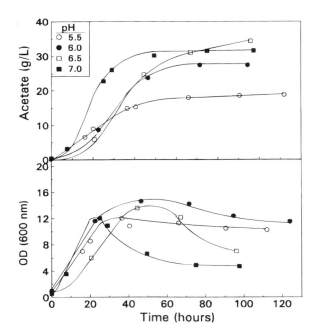

Fig. 1. Effect of pH on growth and acetate production by *C. thermoaceticum* in batch fermentation.

Table 1
Effect of pH on Potassium Acetate Fermentation*

pH	Glucose initial (g/L)	Glucose utilized (%)	Fermn. time (hours)	Acetate concn. (g/L)	Yield (g acetate/ g glucose)	Acetate productivity (g/L.h)	Max. OD
5.5	46.6	42.8	71	17.8	0.91	0.25	10.8
6.0	46.4	80.6	86	31.8	0.89	0.36	14.3
6.5	46.3	90.8	96	33.5	0.84	0.43	12.8
7.0	46.4	90.0	75	31.5	0.81	0.42	12.0

*Average of 2 experiments.

highest acetate concentration (33.5 g/L) and productivity (0.43 g/L.h). Acetate concentration was slightly lower at pH 7.0, but significantly lower at pH 5.5. This is probably because of the inhibitory effect of undissociated acetic acid on the micro-organism *(13)*; acetic acid is more dissociated at higher pH.

However, lower pH resulted in lower acetate concentration (Table 1), but more stable cell concentration (Fig. 1). The latter could be important for fermentation with cell recycle, e.g., a recent study showed that at pH 6.0, *C. thermoaceticum* cells remained active for extended periods and produced

Table 2
Effect of Reducing Agents (Nutrients were Yeast Extract and Salts at 2X Levels)

| Reducing agent | | Glucose | | Acetate | Acetate | Fermn. | Acetate | Max. |
Type	Conc. (g/L)	Conc. (g/L)	Utilized (%)	Conc. (g/L)	yield (g/g)	time (hours)	productivity (g/L.h)	OD
Cysteine	0.25	46.3	90.3	34.4	0.86	118	0.35	13.6
	0.20	48.9	96.7	37.9	0.85	119	0.32	13.7
	0.15	46.9	96.3	37.6	0.89	115	0.33	11.9
	0.10	46.2	97.7	40.7	0.96	123	0.33	8.4
	0.05	50.3	96.7	42.4	0.94	114	0.37	7.4
Sodium thioglycolate	0.25	47.2	87.4	36.8	0.91	119	0.30	6.1
	0.20	47.3	85.4	35.5	0.89	115	0.30	5.8
	0.15	48.6	92.9	35.3	0.82	125	0.28	6.7
	0.10	47.5	84.7	34.5	0.88	128	0.26	6.0
	0.05	48.6	91.0	40.5	0.93	140	0.27	4.8
Sodium sulfide	0.25	45.8	95.2	39.0	0.95	92	0.43	11.6
	0.10	47.2	97.5	38.5	0.89	146	0.26	6.1
None	0.00	48.4	80.2	29.5	0.80	92	0.32	7.3

up to 38 g/L acetate in continuous cell recycle fermentation (11). Since high acetate concentration is important in the economics of industrial production, pH 6.5 was selected as the optimum for potassium acetate production. The concentration of acetate obtained with KOH was slightly lower than that obtained with NaOH as a neutralizing agent (11).

Effect of Reducing Agents

In most of our prior research, cysteine·HCl·H_2O (0.25 g/L) was used as a reducing agent and as a source of sulfur that is required by *C. thermoaceticum* (15). Since cysteine is expensive, studies were conducted with alternate low-cost sulfur-containing reducing agents (Table 2). Interestingly, decreasing cysteine concentration from 0.25 to 0.05 g/L actually improved both the final acetate concentration in the broth (Fig. 2) and the yield (Table 2). At the same time, however, maximum OD in the fermentation broth decreased from 13.6 to 7.4. Thus the increase in acetate yield may be a result of decrease in cell mass yield that allowed more carbon to be channeled into acetate production. Koesnandar et al. (15) showed

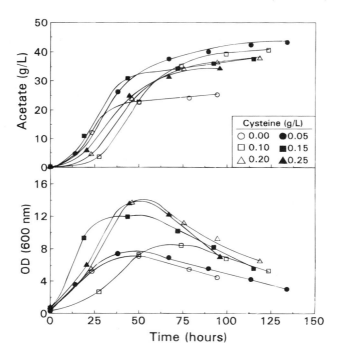

Fig. 2. Effect of cysteine HCl·H$_2$O concentration on growth and acetate production by *C. thermoaceticum* in batch fermentation.

that cysteine increased growth and acetate production when a minimal medium was used in the fermentation. In this study, the fermentation broth contained complex nitrogen sources such as yeast extract that may have made the presence of high cysteine levels unnecessary.

Similar effects were observed when sodium thioglycolate was used as the reducing agent (Table 2). The lowest sodium thioglycolate concentration resulted in the highest acetate yield and the lowest optical density. With sodium thioglycolate, cell growth was highest at 0.15 g/L and lowest at 0.05 g/L. However, with sodium sulfide (Na$_2$S·9H$_2$O), decreasing its level from 0.25 to 0.1 g/L resulted in lower cell growth and productivity, although acetate level was 38–39 g/L. Our results differ from those of Koesnandar et al. *(15)* who used a minimal medium consisting of mineral salts and vitamins, probably because our medium contained a complex nitrogen source. In addition, the mutant strain used in the present study appeared to be much better since it routinely achieved much higher acetate concentrations (35–40 g/L vs 15 g/L in their study).

A certain concentration of sulfur-containing reducing agent is necessary. With no reducing agent, acetate concentration was only 29.5 g/L, the yield was lower and the maximum OD was only 7.3. Since sodium sulfide is the cheapest of the three reducing agents studied, it could be used in commercial fermentation at a level of 0.1 g/L.

Table 3
Effect of Stillage Concentration

Stillage (g solids per liter)	Glucose initial (g/L)	Glucose utilized (%)	Fermn. time (hours)	Acetate Conc. (g/L)	Yield (g/g)	Productivity (g/Lh)	Max. OD
5	46.5	74.4	106	23.9	0.70	0.22	5.0
10	46.9	63.2	115	23.5	0.82	0.20	5.9
15	46.7	63.0	118	23.4	0.81	0.19	4.9
20	46.2	67.6	141	28.1	0.92	0.19	6.6
30	46.6	81.3	139	33.6	0.92	0.24	8.2

Low-Cost Nutrients

Because of the high cost of yeast extract, alternate low-cost nutrients were evaluated for the production of potassium acetate. Since most of the glucose in the United States is produced from corn, the focus was on byproducts from the corn wet milling industry.

Stillage

As shown in Table 3, stillage concentrations of 5–15 g/L (dry solids basis) resulted in. acetate concentrations of only 23–24 g/L. Higher levels, e.g., 30 g/L, resulted in higher acetate concentration of 33.6 g/L (Fig. 3). Traces of fructose were observed in all experiments. However, yields of acetate were good, especially at high stillage levels, probably because of the conversion of lactic acid in the stillage to acetic acid. Lactic acid constituted 21% w/w (dry basis) of the stillage solids. Thus, stillage is not only a good low-cost nutrient, but also provides an additional carbon source for producing acetic acid.

Corn Steep Liquor (CSL)

Corn steep liquor had previously been shown to be a good nutrient source for *C. thermoaceticum* (7,8), especially in combination with ammonium sulfate. Figure 4 shows fermentation patterns with CSL. With CSL at 5 g/L, increasing ammonium sulfate concentration from 3 to 5 g/L resulted in an increase in acetate concentration from 26.9 to 33 g/L (Table 4). In general, acetate productivities were higher with CSL (0.27–0.32 g/L·h) than with stillage, but not as good as with yeast extract.

The fermentation could be improved by using the fed-batch mode and an excess of nutrients. With CSL at 13.5 g/L and ammonium sulfate at 2.7 g/L, an acetate concentration of 38.8 g/L (Fig. 4B) and acetate yield of 1.02 g/g were obtained. The high yield is probably a result of utilization of lactate in CSL by the organism to produce additional acetate.

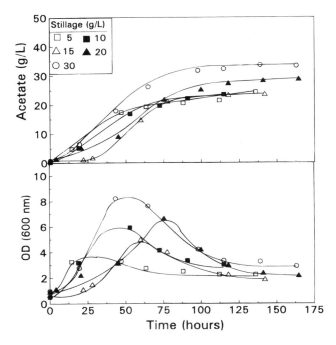

Fig. 3. Effect of filtered stillage concentration on growth and acetate production by *C. thermoaceticum* in batch fermentation.

Table 4
Batch and Fed-Batch Fermentation with Corn Steep Liquor (CSL)*

Experiment	CSL (g/L)	AS[a] (g/L)	Salts[b]	Glucose Conc. (g/L)	Glucose Utilized (%)	Fermn. time (hours)	Acetate Conc. (g/L)	Acetate yield (g/g)	Acetate productivity (g/L.h)	Max OD
Batch	5.0	3.0	2X	46.5	86.5	100	26.9	0.70	0.27	8.9
	5.0	4.0	2X	46.6	93.6	96	30.2	0.81	0.32	9.2
	5.0	5.0	2X	47.5	87.7	118	33.0	0.83	0.28	7.9
Fed-Batch	13.5	2.7	2.7X	46.3	95.1	129	38.8	1.02	0.30	10.5

*Average of 2 experiments.
[a]AS = Ammonium sulfate.
[b]Does not include ammonium sulfate.

Effect of Cell Density

Productivity can be improved by increasing cell density in the fermenter (10,11). This is shown in Table 5 with preconcentrated cells in media containing either yeast extract (YE) or CSL as the nutrient source. Productivities were higher (0.65–0.71 g/L.h) compared to fermentations

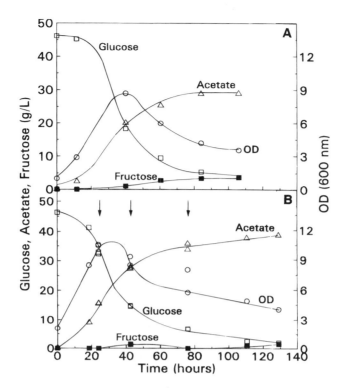

Fig. 4. **(A)** Batch fermentation with CSL = 5 g/L, salts = 2X, ammonium sulfate = 3 g/L. **(B)** Fed-batch fermentation with CSL = 13.5 g/L, salts = 2.7X, ammonium sulfate = 2.7 g/L. Arrows show the times of nutrient addition.

Table 5
Effect of Cell Density on Fermentation Parameters

Expt	Nutrient Type[a]	Conc. (g/L)	AS[a] (g/L)	Salts[b] (g/L)	Glucose Conc. (g/L)	Glucose Utilized (%)	Acetate Conc. (g/L)	Acetate yield (g/g)	Acetate productivity (g/L·h)	Max. OD
High cell density	Yeast extract	7.5	4.0	2.5X	71.6	86.1	50.3	0.74	0.67	22.1
		10.0	6.0	3X	67.8	87.2	48.9	0.82	0.67	19.1
	CSL	5.0	5.0	2X	61.5	84.2	46.4	0.82	0.50	27.7
		10.0	4.0	2X	83.7	81.3	53.6	0.81	0.72	19.4
Low cell density	CSL	5.0	5.0	2X	47.5	87.7	33.0	0.83	0.28	7.9
		10.0	4.0	2X	47.0	77.7	29.2	0.83	0.22	9.5

[a]AS = Ammonium sulfate; CSL = Corn steep liquor.
[b]Does not include ammonium sulfate.

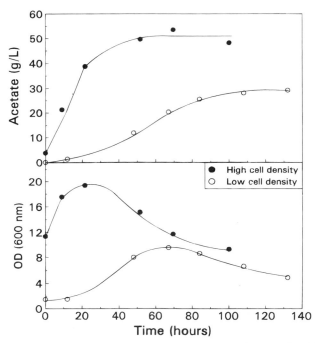

Fig. 5. Effect of cell density on growth and acetate production by *C. thermoaceticum* in batch fermentation. Medium contained CSL = 10 g/L, salts = 2X, ammonium sulfate = 4 g/L.

with normal cell levels (0.22–0.28 g/L.h). The maximum OD values and final acetate concentrations were also much higher. Figure 5 compares these fermentation patterns with the same nutrients but different cell densities. Another benefit of high cell density culture was elimination of the lag period at the start of the fermentation.

In summary, this study has identified the optimum levels of some of the important fermentation parameters for production of potassium acetate from dextrose by *C. thermoaceticum*. The materials cost can be substantially reduced by using sodium sulfide instead of cysteine as the reducing agent, and CSL or stillage as low-cost complex nutrient sources. Fermentation costs can be reduced by preconcentrating the cells, resulting in higher productivity, and higher acetate concentration.

ACKNOWLEDGMENTS

This research was supported by Illinois Corn Marketing Board, Minnesota Corn Promotion and Research Council, US Department of Agriculture through the NRICGP program, and the Illinois Agricultural Experiment Station.

REFERENCES

1. Johnson, K. (1993), Cryotech Deicing Technologies Corporation, Fort Madison, Iowa. *Personal communication.*
2. Sills, R. D. and Blakeslee, P. A. (1992), in *Chemical Deicers and the Environment*, D'Itri, F. M., ed., Lewis Publishers, Inc., MI, pp. 323–340.
3. Ljungdahl, L. G., Carreira, L. H., Garrison, R. J., Rabek, N. E., Gunter, L. F., and Weigel, J. (1986), *Final Report FHWA/RD-86/117*, Federal Highway Administration, Washington, DC.
4. Parekh, S. R. and Cheryan, M. (1990), *Appl. Microbiol. Biotechnol.* **36,** 384–387.
5. Parekh, S. R. and Cheryan, M. (1994). *Biotechnol. Lett.* **16,** 139–142.
6. Reed, W. M., Keller, F. A., Kite, F. E., Bogdan, M. E., and Ganoung, J. S. (1987), *Enz. Microbial. Technol.* **9,** 117–120.
7. Shah, M. M. and Cheryan, M. (1995), *J. Ind. Microbiol.* **15,** 424–428.
8. Witjitra K., Shah, M. M., and Cheryan, M. (1996), *Enz. Microbial Technol.* **19,** 322–327.
9. Reed, W. M. and Bogdan, M. E. (1985), *Biotechnol. Bioeng. Symp.* **15,** 641–647.
10. Parekh, S. R. and Cheryan. M. (1994), *Enz. Microbial Technol.* **16,** 104–109.
11. Shah, M. M. and Cheryan, M. (1995), *Appl. Biochem. Biotechnol.* **51/52,** 413–422.
12. Wang, G. and Wang, D. I. C. (1983), *Appl. Biochem. Biotechnol.* **8,** 491–503.
13. Wang, G. and Wang, D. I. C. (1984), *Appl. Environ. Microbiol.* **47,** 294–298.
14. Parekh, S. R. and Cheryan, M. (1990), *Process Biochem.* **25,** 117–121.
15. Koesnandar, Nishio, N., and Nagai, S. (1990), *Appl. Microbiol. Biotechnol.* **32,** 711–714.

Membrane-Mediated Extractive Fermentation for Lactic Acid Production from Cellulosic Biomass

RONGFU CHEN AND Y. Y. LEE*

Department of Chemical Engineering, Auburn University, Auburn, AL 36849

ABSTRACT

Lactic acid production from cellulosic biomass by cellulase and *Lactobacillus delbrueckii* was studied in a fermenter-extractor employing a microporous hollow fiber membrane (MHF). This bioreactor system was operated under a fed-batch mode with continuous removal of lactic acid by an *in situ* extraction. A tertiary amine (Alamine 336) was used as an extractant for lactic acid. The extraction capacity of Alamine 336 is greatly enhanced by addition of alcohol. Long-chain alcohols serve well for this purpose since they are less toxic to micro-organism. Addition of kerosene, a diluent, was necessary to reduce the solvent viscosity. A solvent mixture of 20% Alamine 336, 40% oleyl alcohol, and 40% kerosene was found to be most effective in the extraction of lactic acid. Progressive change of pH from an initial value of 5.0 down to 4.3 has significantly improved the overall performance of the simultaneous saccharification and extractive fermentation over that of constant pH operation. The change of pH was applied to promote cell growth in the early phase, and extraction in the latter phase.

Index Entries: Hollow fiber membrane; lactic acid; SSF; extraction; *in situ* separation.

INTRODUCTION

Lactic acid production from corn by fermentation routes is an established industrial practice. The current demand of lactic acid as a monomer (1) and the market of it as polymers (polylactic acid) (2,3) is bringing its status from a specialty chemical to a commodity chemical. In view of this growing demand, the lignocellulosic materials are currently being regarded as impor-

---

*Author to whom all correspondence and reprint requests should be addressed.

tant feedstocks for lactic acid production (4,5). The process schemes for the production of lactic acid from cellulosic biomass can share with those proposed for production of ethanol from biomass. One such scheme is simultaneous saccharification and fermentation (SSF). SSF is a bioprocess capable of directly converting lignocellulosic materials to end product. It has been extensively investigated in connection with ethanol production from cellulosic biomass (6,7). Recently it has also been brought up as a means of producing lactic acid (8). There is one technical aspect in lactic acid fermentation that makes it particularly suitable for SSF operation. Many of the lactic acid producing bacteria are thermotolerant. The operating temperature of the SSF can thus be brought to the level close to the optimum of the cellulase enzymes, making the overall process more efficient, especially in the use of enzymes. On the other hand, as is the case with most organic acid fermentations, the lactic acid fermentation is strongly inhibited by the end product (9). Therefore, it is necessary to separate lactic acid from broth during the fermentation.

In this work, we explored a method of removing lactic acid from the fermenter. This idea stems from recent research work concerning *in situ* removal of nonvolatile fermentation products. Various types of integrated fermentation-separation systems have indeed been developed and successfully applied for processes in which product inhibition is significant (10,11). In the recovery of carboxylic acids, extraction has been found to be most suitable (12). Kertes and King (13) have reviewed the extraction chemistry of carboxylic acids. The experimental data on extraction of citric acid using a tertiary amine has also been reported (14). For extraction of carboxylic acid, tertiary amides have been reported to be more suitable than the primary and secondary amines (15). All of these fermentation processes, however, dealt only with liquid substrates. With lactic acid production from cellulosic material, the traditional extraction methods become extremely difficult because of the emulsion formed when the particulate solid substrates come in contact with organic solvent. Use of hollow fiber membrane is, therefore, proposed here to resolve this problem. Insertion of a membrane between the two phases (broth and extractant) would certainly prevent solvent particle interaction. Proper choice of membrane can also prevent loss of the solvent as a result of partial dissolution. A clean and simple phase separation is therefore obtainable without having to deal with emulsion and coalescence. This investigation was undertaken to assess the technical feasibility of employing membrane-mediated extraction in the SSF to produce lactic acid from cellulosic biomass.

MATERIALS AND METHODS

Substrates

Two different substrates were employed in this study: α-cellulose (Sigma, St. Louis, MO) and switchgrass (Alamo species). The later was supplied by National Renewable Energy Laboratory (NREL) in the form of fine particles (20–60 mesh). The switchgrass was pretreated in a 2-gal batch reactor with 0.07% sulfuric acid at 175°C for 30 min, prior to fermentation.

Enzyme

The cellulase enzyme, *Cytolase CL* (Lot No. 17-92262-09) was obtained from Environmental Biotechnologies, Santa Rosa, CA. It has filter paper cellulase activity of 95.9 FPU/mL, β-glucosidase activity of 80.6 p-NPGU/mL, and endo-glucanase activity of 613 CMCU/mL.

Micro-organism and Medium

Lactobacillus delbrueckii (NRRL-B445) was used for fermentation. Elliker broth (Difco) was used as the culture medium. The culture was grown at 37°C for 36 h, and stored at 4°C in agar slants made of Elliker broth and 5% tomato juice agar (Difco). The fermentation medium contained (per liter): solid substrate (10–30 g), yeast extract (30 g), NaOH (1.25 g), K_2HPO_4 (0.2 g), KH_2PO_4 (0.2 g), $MgSO_4 \cdot 7H_2O$ (0.6 g), $MnSO_4 \cdot H_2O$ (0.03 g), and $FeSO_4 \cdot 7H_2O$ (0.03 g).

Solvents

Alamine 336 was purchased from Henkel Co. (Tucson, AZ). Oleyl alcohol (technical grade) and n-dodecanol (technical grade) were from Aldrich. Other solvents were from Fisher Scientific.

Microporous Hollow Fiber Membrane (MHF) and its Mediated Extraction

Liqui-Cel 5PCM-106, Hoechst Celanese, Charlotte, NC, was used as the MHF. The manufacturer's specifications are: Celgard X-10 microporous polypropylene hollow fiber, 2100 fibers, fiber internal diameter 240 µm, fiber wall thickness 30 µm, effective pore size 0.05 µm, porosity 30%, and effective surface area 0.23 m^2. The outer shell of the MHF was removed in order to bring the fibers in direct contact with the fermentation broth.

Experimental Setup

The experimental setup of the simultaneous saccharification and extractive fermentation (SSEF) is shown in Fig. 1. The SSEF system involved two processes, the SSF in the broth and the extraction across the MHF. The bioreactor (1.3-L working volume) is divided into two compartments separated by a partial glass wall, leaving openings at the top and bottom. The MHF unit is installed in one compartment, the extraction chamber. In the other fermentation chamber, an agitator is installed. The fermentation broth is driven downward by agitation in the fermentation chamber, and flows into the extraction chamber through the open area under the glass wall. It then flows upward in the extraction chamber, and returns to the fermentation chamber across the

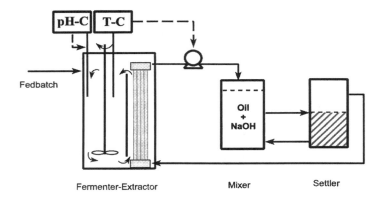

Fig. 1. Schematic of the SSEF.

top of the glass wall. The polypropylene membrane is hydrophobic. The organic solvent passes through the tube side of the fibers, and the broth is on the shell side. Since the membrane is hydrophobic the aqueous broth does not penetrate or wet the hollow fiber membrane pores. The organic solvent penetrates the pores and directly contacts the aqueous broth phase (Fig. 2). By applying a lower pressure on the solvent phase, a steady interface is formed at the pore entrances on the aqueous broth side. The temperature was controlled by a Proportional Temperature Controller (Cole-Parmer Versa-Therm). The pH was controlled by New Brunswick Model PH-22 pH controller. The pH of the fermentation broth was adjusted initially by adding NH_4OH (5 N) for promotion of cell growth. During the latter phase of the fermentation, the pH was controlled by on–off operation of lactic acid extraction. The pH was controlled in such a way that the production of lactic acid (downward shift of pH) initiates pumping of the extractant into the tube side of MHF. The extracted lactic acid was back-extracted by $5N$ NaOH. The mixed two phases were separated in a settler. The stripped solvent was then returned to the extractor.

Analytical Methods

The samples were analyzed for sugars, lactic acid, and acetic acid by HPLC (Water Associate), equipped with an RI detector. Bio-Rad's HPX-87H column was used at 65°C, with 0.005 M H_2SO_4 as mobile phase. The flow rate was set at 0.6 mL/min.

RESULTS AND DISCUSSION

Solvent Selection

There are a number of constraints in the selection of solvent for the MHF-mediated extraction. A solvent should have high extractive capacity and selectivity for lactic acid, low viscosity, and low toxicity to the micro-

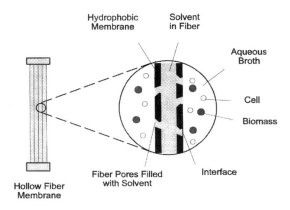

Fig. 2. Microporous hollow fiber mediated extractive fermentation.

organism. A series of extraction experiments was conducted to screen the solvent addressing this issue. High molecular weight aliphatic amines, such as Alamine 336 (trioctyl amine), have been reported to be effective in extraction of dilute carboxylic acids. Alamine 336 is a water-insoluble, saturated, straight-chain tertiary amine, with the alkyl groups of C_8–C_{10} (16). Its extraction capacity can be greatly enhanced when it is mixed with a polar solvent such as water insoluble alcohols. Since Alamine 336 and water insoluble alcohols are extremely viscous, and inoperable in a hollow fiber membrane system, kerosene was introduced as a diluent. For 20 (v/v)% Alamine 336 dissolved in kerosene alone without alcohol, the distribution coefficient (m) is only 0.1. When it is supplemented with 20% oleyl alcohol, the distribution coefficient (m) rose to 0.7 (Fig. 3). This improvement is a result of the fact that alcohols can interact with the acid-amine complex through hydrogen bonding and make the complex more stable in the solvent phase (17). Further experiments were carried out for final selection of alcohol. Four alcohols, 1-octanol, decyl alcohol, dodecanol, and oleyl alcohol, were examined for this purpose. As shown in Fig. 4, the order of the distribution coefficients from highest to lowest is 1-octanol→decyl alcohol→dodecanol→oleyl alcohol. It is shown that the distribution coefficient decreases with the chain length of alcohols. For a given alcohol, the higher the level of concentration, the more the hydrogen bonds are available, therefore, the higher the distribution coefficient observed.

For the solvents to be used in extractive fermentation, they should also be biocompatible. The same organic solvent system was tested for toxicity to the micro-organism. The solvents were washed with culture medium to remove water soluble impurities and saturated with mineral components of the medium. An autoclaved culture medium was added with 10% inoculum. For each sample, 10 mL culture medium containing seed bacteria was mixed with 1 mL prewashed solvent in a 20-mL vial. They were incubated at 37°C for 12 h in a shaker bath. A control test was

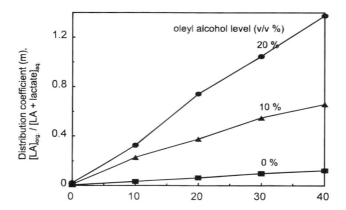

Fig. 3. Effect of alcohol modifier on distribution coefficient (solvent contains Alamine 336, alcohol, and kerosene. extraction conditions: solvent:lactic acid solution = 1:1, temperature = 26°C, initial pH = 2.3, and initial lactic acid conc. = 15.06).

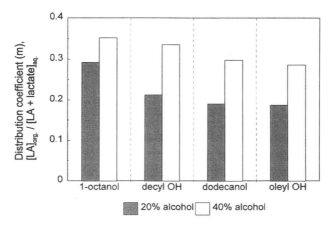

Fig. 4. Effect of alcohol type and concentration on distribution coefficient, (20% Alamine 336 + alcohol + balanced kerosene, extraction conditions: solvent:lactic acid solution = 1:1, temperature = 26°C, initial pH = 4.01, and initial lactic acid conc. = 15.86 g/L).

made with culture medium without solvent. Lactic acid was taken as the indicator in these tests. The results are summarized in Table 1. It is clearly shown that kerosene and oleyl alcohol are nontoxic to the micro-organism. From observation of the various mixtures, the solvent combination of 20% Alamine 336 + 40% oleyl alcohol + 40% kerosene was chosen for this work.

Effect of Temperature

The SSEF for lactic acid from cellulosic substrates involves three steps, saccharification, fermentation, and extraction. Unfortunately, the optimal temperature and pH differ significantly for the three processes.

Lactic Acid Production

Table 1
Biocompatibility of Solvent

Solvents	% of lactic acid produced
Control (medium only)	100
Pure solvents:	
Diluent: n-decane	71
n-dodecane	96
kerosene	100
Modifier: 1-octanol	5
decyl alcohol	16
oleyl alcohol	100
Solvent mixtures:	
20% Alamine 336 + 60% kerosene +:	
20% decyl alcohol	46
20% oleyl alcohol	84
20% Alamine 336 + 40% kerosene +:	
40% decyl alcohol	21
40% dodecanol	35
40% oleyl alcohol	92

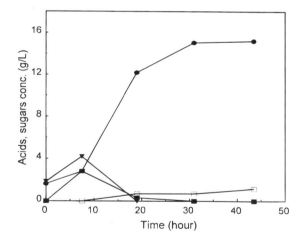

Fig. 5. Time course of a batch SSF (conditions: switchgrass 28.3 g/L, cellulase 25 IFPU/g substrate, pH 5.0, and temperature 46°C, symbol: ▼ glucose, ■ cellobiose, ● lactic acid, and □ acetic acid.)

Figure 5 shows a time course of an SSF run using pretreated switchgrass. Accumulation of glucose and cellobiose were seen in the initial phase. The fermentation then proceeded under glucose-limited conditions, indicating that the hydrolysis is the rate-limiting step in the SSF. The subsequent work in the SSF was, therefore, focused on the improvement of the hydrolysis process.

The optimum temperature for *Lactobacillus delbreuckii* is reported to be in the range of 42–60°C *(18)*. The optimum temperature for enzymatic hydrolysis is about 50°C. However, when a temperature above 47°C was applied in the SSF, we observed that the micro-organism activity was dramatically reduced and showed no lactic acid formation. The temperature of the SSF was chosen as the highest temperature the micro-organism can withstand. In this work, it was set at 46°C, one degree below the upper limit allowing a safety factor. The effect of temperature on extraction is shown in Fig. 6. The distribution coefficient (m) for lactic acid linearly decreases with temperature, low temperature favoring the extraction. The saccharification, fermentation, and extraction occur essentially in sequence. We have, therefore, chose the highest possible temperature as the optimum with the understanding that the enzymatic hydrolysis is the rate-limiting step in the SSEF.

Effect of pH

The effect of pH on enzymatic hydrolysis was studied. It has been reported that the optimum pH for lactic acid production by *L. delbreuckii* is between 5 and 6 *(18)*. The optimum pH of the enzyme has been found to be 4.5–5.0 in our study. If the two processes are combined (SSF), the optimum pH is 5.0. The effect of pH on extraction is shown in Fig. 7. The pH values were measured after extraction equilibrium was attained. The distribution coefficients (m) for the total lactic acid (lactate plus free lactic acid) were seen to linearly decrease as pH increases. It has been reported that Alamine 336 can extract the undissociated acid, but cannot extract organic acid under basic conditions *(19)*. To improve the extraction, it is desirable to raise the concentration of free lactic acid in the broth. This can be done by raising the total amount of lactic acid in the broth at a given pH. The concentration of the lactic acid, however, is limited by the micro-organism's tolerance. The only other option to increase free lactic acid in the broth is to reduce the pH. We have also observed that the lower limit of pH for lactic acid production is 4.1. From considerations of these findings we have chosen pH of 4.3 for the SSEF experiments.

However, employing a uniform pH of 4.3 for the entire SSF process resulted in low cell growth. The decreased cell activity was evidenced by low lactic acid production shown in Fig. 8. This indicates the process was limited by the fermentation. To overcome this problem, a progressive change of pH was attempted. The pH was first controlled at 5.0, an optimal pH for cell growth in the fermentation. It was maintained at this pH by addition of $5N$ NH_4OH solution. After sufficient bacterial growth, the alkali pump was stopped allowing pH to decrease to 4.3. The pH was then maintained at 4.3 for the remainder of the process. The results of the effect of pH on the SSF are shown in Fig. 8

Lactic Acid Production 443

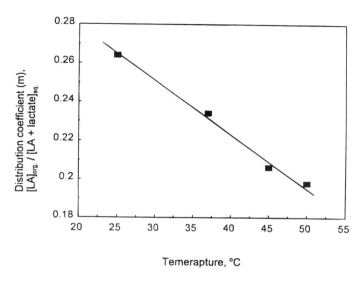

Fig. 6. Effect of temperature on distribution coefficient (extraction conditions: solvent: lactic acid solution = 1:1, initial aqueous pH = 4.09, and initial lactic acied conc. = 15.61 g/L).

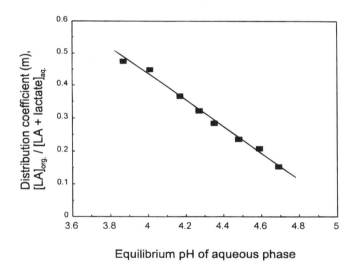

Fig. 7. Effect of pH on distribution coefficient (extraction conditions: solvent:lactic acid solution = 1:1, temperature 45°C, initial lactic acid conc. = 13.94 g/L).

for three different cases: constant pH at 4.3, constant pH at 5.0, and gradual change of pH from 5.0 to 4.3. With gradual change of pH (from 5.0 to 4.3), the lactic acid production was enhanced almost to the level of SSF at pH 5.0. There was no sugar accumulation in this run, another evidence of high cell activity.

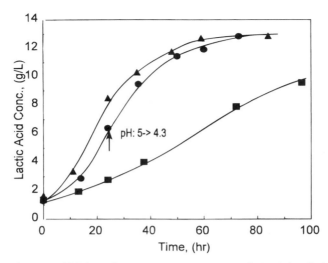

Fig. 8. Effect of pH on SSF (conditions: temperature 46°C, initial cellulose loading = 12.5 g/L, and enzyme load = 60 IFPU/g cellulose; symbols: ■ pH = 4.3, ▲pH = 5, and ● pH = 5 → 4.3).

Simultaneous Saccharification and Extractive Fermentation

The SSEF of this work is designed such that the fermentation products are continuously extracted through a membrane mediated extractor. The bioreactor system was sterilized with 0.1% NaOH solution for 2 h and washed with sterilized DI water. The solvent was prewashed with DI water to remove water soluble impurities and mixed with fermentation broth to saturate with minerals. For the extraction, 1300 mL solvent (20% Alamine 336, 40% oleyl alcohol, and 40% kerosene) was used, and 250 mL 5N NaOH for back extraction.

Two substrates, α-cellulose (76% glucan) and pretreated switchgrass (with 56.1% glucan and 26.2% lignin), were used as lignocellulosic biomass for lactic acid production. Fig. 9 shows the results of a SSEF run made in fed-batch mode using α-cellulose. The bioreactor was initially charged with 21.5 g/L substrate, 7% inoculum, 45 IFPU cellulase/g-substrate. Each additional feed in fed-batch operation contained 13 g α-cellulose with the same enzyme loading, and 6 g yeast extract. The SSEF was first run under the condition favoring the growth of the micro-organism. The pH was controlled at 5.0 with the use of $5N$ NH_4OH as a neutralizing agent for a period of 9 h. The pH was then reset to 4.4. When the pH reached 4.4, the extraction was initiated by the pH controller. This procedure of pH change is identical to that of SSF. The pH was then maintained by on–off control of an extraction recirculation pump. The glucose level in the broth dropped to almost zero when extraction began. This reaffirms that the hydrolysis is the rate-controlling step in the SSEF. Fed-batch feeding was applied twice at reaction times of 21.5 and 41 h. Glucose accumu-

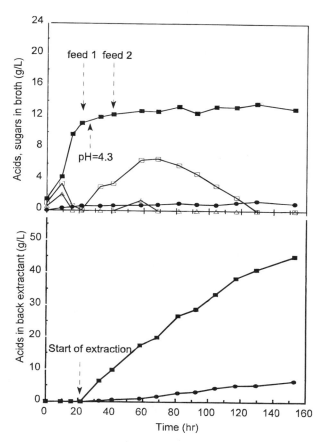

Fig. 9. SSEF of α-cellulose (conditions: fermentation volume = 1300 mL, temperature = 44°C, initial pH 5.0, pH 4.3 after start of extraction; inital loading: α-cellulose = 21.5 g/L, and enzyme = 45 IFPU/g sustrate; fed batch loading: α-cellulose = 13 g(dry), and enzyme = 45 IFPU/g substrate; symbols: ■ lactic acid, ● acetic acid, □ glucose, and △ cellobiose.)

lation was seen after fed-batch feeding (Fig. 9) during which the process was limited by the fermentation step. The near constant level of lactic acid in the broth indicates that the extraction is not the controlling step at any time during the SSEF. After 130 h, glucose was depleted indicating that the process was again limited by hydrolysis. The lactic acid concentration in the back extractant is seen to increase steadily showing stable and active SSEF process for the entire duration. It has been reported that about 1.1 g/L lactic acid formed from 30 g/L yeast extract by fermentation, in a control test at 45°C and pH 5.0 without using cellulose substrate (20). This lactic acid contribution from yeast extract was considered in our yield calculations. The observed overall yield of lactic acid (total of broth, extractant, and back extractant) in this run was calculated at 152 h, on the base of total substrate input, to be 67% of the theoretical maximum. At this point of the process, the bioreaction is incomplete leaving a significant

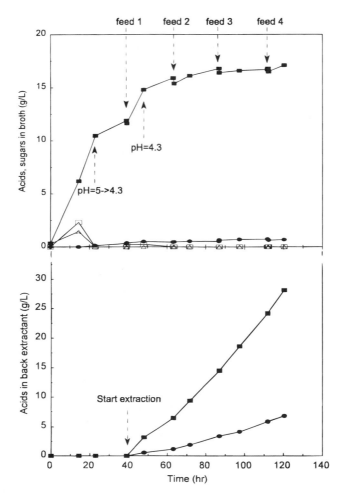

Fig. 10. SSEF of pretreated switchgrass (conditions: fermentation volume = 1300 mL, temperature = 43°C, initial pH 5.0, pH 4.3 after start of extraction; inital loading: switchgrass = 30 g/L, and enzyme = 25 IFPU/g sustrate; fed batch loading: switchgrass = 18 g(dry), and enzyme = 25 IFPU/g substrate; symbols: ■ lactic acid, ● acetic acid, □ glucose, and △ cellobiose.)

amount of unreacted substrate in the fermenter. Allowing more reaction time would therefore increase the overall yield.

The results of SSEF with pretreated switchgrass are shown in Fig. 10. The enzymatic digestibility of switchgrass after pretreatment was about 60%. The experiment was run with initial loading of 30 g/L substrate, 45 IFPU cellulase/g-glucan. Each additional fed-batch feed contained 18 g (dry base) substrate with 75% moisture content, 6 g yeast extract, and the same enzyme loading. The pH adjustment was same as that of α-cellulose run. Fed-batch feeding was applied four times at 24 h intervals. Hydrolysis is the rate-controlling step. The time-course data of Fig. 10 shows that lactic acid in the broth stayed at a relatively constant level once the extraction was applied. The lactic acid level in the back extractant again showed steady increase. Acetic acid was also extracted from the broth for both substrates. The lactic acid production was lower in the switchgrass run than in the α-cellulose run mainly because of lower digestibility of switchgrass.

In general, the SSEF approach works well with certain substrates in the form of fine particles, such as switchgrass sawdust, ground corn cobs/stover mixture, which can flow easily with fermentation broth. For those substrates with big size or pulp sludge, further modifications are needed. Yeast extract loading is not studied in this work. Further work on this matter needs be addressed.

CONCLUSIONS

In this work we have demonstrated that the extractive fermentation for lactic acid from lignocellulosic substrates can sustain a stable operation in fed-batch mode. The solvent extraction mediated by the microporous membrane has proven to be an effective *in situ* product separation scheme in the SSEF. The extraction capacity of Alamine 336 is greatly enhanced by addition of alcohol. Long-chain alcohols are preferred over short chain alcohols because of their low toxicity. Addition of a diluent (kerosene) is essential in reducing solvent viscosity. The solvent combination of 20% Alamine 336, 40% oleyl alcohol, and 40% kerosene was found to be most effective in the SSEF. For SSF the optimum pH is 5.0. For SSEF, a progressive change of pH, from 5.0 to 4.3 provides a better performance than a uniform pH condition because of a mismatch of pH optima for SSF and extraction. The SSEF run with α-cellulose gave higher overall product yield and productivity than with those switchgrass primarily because of the difference in digestibility.

ACKNOWLEDGMENT

Authors gratefully acknowledge the financial support from the National Science Foundation (NSF/EPSCOR-OSR-955-0480) and the Pulp and Paper Research and Education Center, Auburn University.

REFERENCES

1. Lepree, J. (1995), *Chemical Marketing Reporting*, March 13, p. 14.
2. Medisorb Technologies International L. P. (1995), "An Emerging Technology Takes Flight," Cincinnati, OH.
3. Lipinsky, E. S. and Sinclair, R. G. (1986), *Chem. Eng. Prog.* **82,** 26.
4. Wyman, C. E. (1990), "Ethanol from Biomass: Annual Review Meeting at Lincoln, NE," NREL, Golden, CO.
5. Montgomery Advertise (1994), "Gas from grass could be future," July 5, 1994.
6. Ooshima, H., Ishitani, Y., and Harano, Y. (1985), *Biotechnol. Bioeng.* **27,** 389.
7. Spangler, D. J. and Emert, G. H. (1986), *Biotechnol. Bioeng.* **28,** 115.
8. Abe, S. and Takagi, M. (1991), *Biotechnol. Bioeng.* **37,** 93.

9. Takagi, M. (1984), *Biotechnol. Bioeng.* **26,** 1507.
10. Daugulis, A. J. (1988), *Biotechnol. Prog.* **4,** 113.
11. Lewis P. V. and Yang, S. (1992), *Biotechnol. Prog.* **8,** 104.
12. Wang, C. J., Bajpai, R. K., and Iannotti, E. L. (1991), *Appl. Biochem. Biotechnol.* **28/29,** 1991.
13. Kertes, A. S. and King, C. J. (1986), *Biotechnol. Bioeng.* **28,** 269.
14. Baniel, A. M. (1982), Chem. Ab. **97,** 10,9557.
15. Pearson, R. G. and Vogelson, D. C. (1958), *J. Am. Chem. Soc.* **80,** 1038.
16. Henkel Corporation's Technical Bulletin, Blue Line—Alamine 336 (1994), Minerals Inductry Division, Tucson, AZ.
17. Tung, L. A. and King, C. J. (1994), *Ind. Eng. Chem. Res.* **33,** 3217.
18. Vickroy, T. B. (1985), in *Comprehensive Biotechnology*, vol. 3., Blanch, H. W., Drew, S., Wang, D. I. C., eds., Pergamon Press, New York, NY, pp. 761.
19. Yang, S., White, S. A., and Hsu, S. (1991), *Ind. Eng. Chem. Res.* **30,** 1335.
20. Thomas, S. (1995), Thesis, Auburn University, AL.

Enzyme-Supported Oil Extraction from *Jatropha curcas* Seeds

ELISABETH WINKLER,*,[1] NIKOLAUS FOIDL,[2] GEORG M. GÜBITZ,[1] RUTH STAUBMANN,[1] AND WALTER STEINER[1]

[1] Institute of Biotechnology, Graz University of Technology, Petersgasse 12, A-8010 Graz, Austria; and [2] Proyecto Biomasa, Managua University of Technology (UNI), AP 432, Managua, Nicaragua

ABSTRACT

Jatropha curcas is a tropical plant widely distributed in arid areas. The seeds contain about 55% of oil, which is mainly used for the production of soap as a fuel and after transesterification as biodiesel. Various methods for recovering of oil from the seeds, including extraction with organic solvents and water, have been investigated. Compared to hexane extraction (98%) the oil extraction using water only yielded 38% of the total oil content of the seeds. Using several cell wall degrading enzymes during aqueous extraction a maximum yield of 86% was obtained. The influence of cellulolytic, hemicellulolytic enzymes, as well as proteases was studied. The experiments were carried out at different pH-values and temperatures to find out the optimum for oil recovering using enzymes. Surprisingly, the best results (86%) were obtained using an alkaline protease. Combinations of proteases with hemicellulases and/or cellulases did not further increase the extraction yield. The enzyme-supported aqueous extraction offers a nontoxic alternative to common extraction methods using organic solvents with reasonable yields.

Index Entries: Enzyme-supported oil extraction; *Jatropha curcas*; oil seeds; aqueous oil extraction.

INTRODUCTION

Jatropha curcas Linn is a plant widely distributed in the arid regions of the hemispheres, mainly Central and South America. The high share of oil, but also the resistance to dryness and sterile soil make *Jatropha curcas* inter-

*Author to whom all correspondence and reprint requests should be addressed.

Table 1
Composition of *Jatropha curcas* Seeds

compounds	seeds with shells [%]	seeds without shells [%]
dry substance	94.23	92.00
rash	3.17	3.96
organic dry substance	91.06	88.16
protein	17.08	22.24
raw fat	34.38	54.38
raw fibre	22.96	2.21
starch	0.04	0.15
sugar	2.67	3.30
hemicellulose	3.22	0.18
raw cellulose	13.98	2.91
raw lignin	14.25	0.17

esting for afforestation and desirable as a source of alternative energy. Using the seeds that are rich in oil (55%) and proteins (22%) (Table 1) as a feed staff is not taken into account, because of contained phorbol ester and lectine (curcine). Therefore, the seeds are mainly used for the production of soap as a fuel and after transesterification as biodiesel.

Jatropha curcas fruits are as big as walnuts, divided in three parts containing three seeds formed like kidneys in a hard brown shell with a length of about 2 cm. Plant cell walls are unlignified and composed of cellulose fibers to which strands of hemicellulose are attached. The fibers are often embedded in a matrix of pectic substances linked to structural protein (1). The different possibilities of oil extraction are based upon the fact that the oil inside the cells is partly bounded to proteins and complex carbohydrates, like starch, pectin, cellulose, and xylan. During extraction with an organic solvent, the cell wall of cells not mechanically opened is distributed by osmotic pressure. The use of aqueous solutions has not played a part until now, because of the very bad solution quality of oil in water. This changed with the use of cell-degrading enzymes, mainly fungal preparations rich in xylanases, cellulases, and proteases, which have been interesting for the food processing industry to aid the isolation of proteinaceous rapeseed and other materials from plants (2,3). The degree of mechanical crushing is insofar important as a high crushing rate implies an emulsion of oil drops, cell remnants, proteins, and lipids (4). If the use of emulsion-breaking substances as detergents is not wanted, an optimal mechanical crushing degree must be determined. It should be remarked that a centrifugation can separate the emulsion partly. Since the composition of plant cell walls varies, individual enzyme combinations comprising xylanases, cellulases, pectinases, and proteinases are required for enzyme-supported oil extraction (5,6).

The aim of this work was to compare commercial available enzymes and to find the optimal enzyme combination, concentration, and conditions for enzyme-supported oil extraction of *Jatropha curcas* seeds.

MATERIAL AND METHODS

Substrate

Table 1 shows the composition of seeds with and without shells (7). For our experiments we used *Jatropha curcas* seeds, obtained from UNI Managua/Nicaragua with an average oil content of 59.8%. The pretreatment of the seeds included the removal of the external fruit flesh and the drying to an average water content of 8 to 10%. Before treatment with enzymes, the seeds were shelled and grinded, to a size of about 0.2 mm in a coffee mill.

Enzymes

Two protease preparations (Alcalase and Neutrase) and a hemicellulase/cellulase preparation (Viscozyme) were kindly provided from Novo Nordisk (Bagsvaerd, Denmark). The protease preparations BLAP and Corolase were obtained from COGNIS (Düsseldorf, Germany) and Röhm (Darmstadt, Germany), the hemicellulase LYX and the cellulase Cytolase CL from Gist-Brocades (Seclin, France), respectively.

Enzyme Activity Assay

Enzyme activities of the commercial preparations was measured at optimal pH and temperature as specified in the manufacturer's instructions. Protease activity was determined using the Anson/Kunitz method (8), xylanase activity using the DNS method (9), and cellulase activity (FPU) according to IUPAC recommendations (10).

Other Analytical Methods

The calculation of extraction yield is based on the initial oil content determined with soxhlet extraction and direct measurements of the weight of the oil after the extraction process. Free fatty acids were determined using the titration method with ethanol/diethyl ether and phenolphthalein as indicator to qualify the oil (11). The water content of the oil was quantified using the Karl Fischer method.

Table 2
Optimal pH and Temperature for Enzymatic Activity with Casein (Proteases),
Filter Paper (FPU), and Xylane (XU) as Standard Substrates and
with *J. curcas* Seeds for Oil Extraction*

Enzyme	pH standard	T [°C] standard	Anson [U/g]	FPU filterpaper [U/g]	XU xylan [U/g]	pH J. curcas	T [°C] J. curcas	oil yield [%]
alcalase	8	50	1027	-------	-------	7	60	78
neutrase	6	45	852	-------	-------	6	45	61
corolase	9	45	99	-------	-------	9	45	59
BLAP	11	65	623	-------	-------	11,5	60	60
cytolase cl	5	45	-------	175	37	4,5	45	40
LYX	5	40	-------	23	165	5	45	42
viscozyme	4	45	-------	25	225	4,5	45	68

*Incubation system in case of *Jatropha curcas* as substrate: 0.5M citrat or phosphate buffer (depends on optimal pH), 5 Anson U/g seed or 3 XU/g seed for 2 h.

Enzyme Treatment

A suspension of *Jatropha curcas* seeds (460 g/L, dry weight) was incubated with different enzymes for 2 h at different pH (0.05M citrate buffer [pH ≤ 7] 0.05M phosphate buffer [pH > 7]) and temperatures as indicated in Table 2. Thereafter, the suspension was centrifuged at 10,000 rpm for 10 min yielding a four-phase system (Fig. 1). The oil phase was separated after dilution with n-hexane (100% hexane v/v), which was removed by distillation afterwards. The oil content was quantified gravimetrically after drying at 105°C for 1 h.

RESULTS AND DISCUSSION

First, the correlation between substrate concentration and oil extraction yield was studied. A suspension of the seeds was incubated with 3 XU/g seed for Viscocyme activity or 5 Anson U/g seed for protease activity (Neutrase, Alcalase, Corolase, and BLAP) and the extraction yield was determined. For all enzyme preparations, the best results were obtained at a substrate concentration around 46% w/w, dryweight, which is shown in Fig. 2 for Alcalase and Viscozyme. These findings are in contrast to observations from Barrios et al. (12) for coconuts of 25% w/w and from

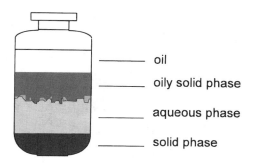

Fig. 1. Four-phase system after centrifugation of the enzyme-treated seed-water solution.

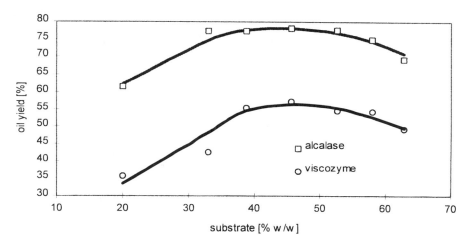

Fig. 2. Effect of substrate concentration (w/w) on *J. curcas* seed extraction yield with Alcalase (5 Anson U/g seed, pH: 7) and Viscozyme (3 XU/g seed, pH: 4.5); incubation time: 2 h, temperature: 60°C (Alcalase), 45°C (Neutrase).

Buenrostro *(13)* for avocados of 20% w/w. All further experiments were carried out at a substrate concentration of 460 g/L.

In a second step, the optimal pH and temperature for the enzyme treatment of *J. curcas* seeds were determined. The cellulase-hemicellulase gave the best results at pH around 5.0, whereas proteases preferred higher pH between 6.0 and 11.5 (Table 2). All preparations showed different optimum temperature ranging from 45 to 60°C. Optimal pH and temperature measured on *J. curcas* seeds did not vary significantly from those reported in the manufacturer's instructions for standard substrates. The oil extraction yields obtained with the individual enzyme preparations under these conditions are also listed in Table 2.

Enzyme treatment was performed with different enzyme concentrations varying from 0.4 to 4 XU/g (Viscozyme) and from 0.3 to 30 Anson U/g for proteases (Neutrase, Alcalase, Corolase, and BLAP). The maximum

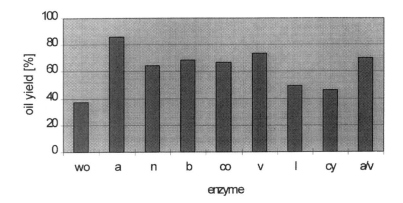

Fig. 3. Oil yields after enzymatic aqueous extraction with different enzymes at optimal conditions: 15 Anson units for proteases, 2 XU for hemicellulases/cellulases. (wo = without enzyme, a = Alcalase, n = Neutrase, b = BLAP, co = Corolase, v = Viscozyme, l = LYX, cy = cytolase and a/v = Alcalase/Viscozyme).

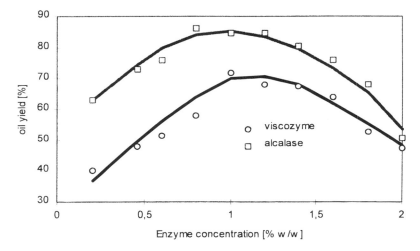

Fig. 4. Effect of enzyme concentration (w/w) on *J. curcas* seed extraction yield with Alcalase (1% w/w = 15 Anson U/g seed, pH: 7) and Viscozyme (1% w/w = 2 XU/g seed, pH, 4.5); incubation time: 2 h, temperature: 60°C (Alkalase), 45°C (Viscozyme).

extraction yield was reached with different activities of the individual enzymes as shown in Fig. 3. The highest extraction yield of 85.6% was obtained with the protease Alcalase. This is in a good agreement with data reported by Barrios (12) for avocados (80%), Buenrostro et al. (13) for coconuts (78%), Olsen (2) for rapeseed (95%), coconuts (95%), and flax seed (72%). In this case proteases seem to be more effective than cellulases or hemicellulases preparations that gave a maximum extraction yield of 73% in case of Viscozyme. Figure 4 shows the extraction yields obtained with different

Table 3
Content of Free Fatty Acids and Water in
the Oil after Enzymatic Treatment

Enzyme	free fatty acids [%]	water [%]
Alcalase	7.3	0.1621
BLAP	9.07	0.1690
Neutrase	11.18	0.1688
Viscozyme	7.5	0.1705
blank	2.5	0.1993
pressed oil	5.6	0.1677

enzyme concentrations using Viscozyme and Alcalase (Fig. 4). Interestingly, high enzyme concentrations resulted in a decrease of the extraction yield because of worse separation properties of the resulting suspension.

In order to study the effect of combinations of proteases with hemicellulases/cellulase, a common oil seed suspension (460 g/L) was incubated with Alcalase (15 U/g seed, dry weight, activity Alcalase: 1027 U/mL) and Viscozyme (2 XU/g seed, activity Viscozyme: 25 FPU/mL, 225 XU/mL). The optimal pH- and the temperature for combined enzyme treatment was determined, but an increase of the yield could not be obtained compared to separate treatment with Alcalase and Viscozyme. Under optimal conditions, an oil yield of 70% could be measured that is about the same as with Viscozyme alone. This is in contrast to results reported in the literature *(6,11,12,14)* where combinations of proteases, hemicellulases, and cellulases gave the best result.

Incubation of *J. curcas* seeds after milling for 1 h at 105°C before enzyme treatment decreased the yield about 10 to 15% in contrast to the results from Jensen et al. *(15)* for rapeseeds.

To characterize the oil quality with regard to followed esterification, the content of water and free fatty acids in the oil was determined (Table 3). The results were compared to data measured for oil extracted by the Soxhlet method with n-hexane. The content of free fatty acids was in all cases above the values of blank, and the contents of water were very similar. An aftertreatment of the oil seems to be necessary since fatty acids and water cause disturbing coreactions during esterification *(16)*.

CONCLUSIONS

When applying the enzyme-supported oil extraction of *Jatropha curcas* seeds described here, an oil yield of 86% is achieved, that is significant higher than aqueous extraction (38%). In general, the enzymatic process requires a considerably lower capital investment and consumes less energy, that makes this process interesting for developing countries, specially in case of *Jatropha curcas*, which is used for afforestation of arid areas.

ACKNOWLEDGMENTS

We would like to thank Sucher and Holzer, Graz, Austria, for financial support.

REFERENCES

1. Keegstra, K., Talmadge, K. W., Bauer, W. D., and Albertsheim, P. (1973), *Plant Physiol.* **51**, 188–196.
2. Olsen, H. S. (1988), *Lecture given at the workshop on Agricultural Refineries—a Bridge from Farm to Industry.*
3. Coll, L., Saura, D., Ruiz, J. M., Canovas, J. A., and Laencina, J. (1995), *Food Chem.* **53**, 27–434.
4. Flemming, M. C. (1991), *INFORM* **2**, 984–987.
5. Dominguez, H., Nunez, M. J., and Lema, J. M. (1995), *J. American Oil Chemists Soc.* **72**, 1409–1411.
6. Tanodebrah, K. and Ohta, Y. (1994), *Food Chem.* **49**, 271–286.
7. Böhme, H. (1988), *Möglichkeiten der Verwendung von Preβrückständen der Purgiernuβ in der Tierernährung auf den Cap Verden, Institut für Tierernährung der Bundesforschungsanstalt Braunschweig.*
8. Bergmeyer, (1984), *Methods Enzymatic Anal.* **5**, 258–277.
9. Bailey, M. J., Biely, P., and Poutanen, K. (1992), *J. Biotechnol.* **23**, 257–270.
10. IUPAC, (1987), *Pure Appl. Chem.* **59**, 257–268.
11. DIN, Deutsches Institut fuer Normung, e.V. (1994).
12. Barrios, V. A., Olmos, D. A., Noyola, R. A., and Lopez-Munguia, C. A. (1990), *Oléagineux* **45**, 35–42.
13. Buenrostro, M. and Lopez-Munguia, C. (1986), *Biotechnol. Lett.* **8**, 505, 506.
14. Lanzani, A., Petrini, M. C., Cozzoli, P., Gallavresi, C., Carola, G., and Jacini, G. (1974), *La rivista Italiana delle sostanze grasse* **11**, 226–229.
15. Jensen, S. K., Olsen, H. S., and Sorensen, H. (1990), *Commercial Processing New Dev.* 331–343.
16. Bockisch, M. (1993) *Handbuch der Lebensmitteltechnologie, Nahrungsfette und-öle* 412–426.

Biogas Production from *Jatropha curcas* Press-Cake

RUTH STAUBMANN,*,[1] GABRIELE FOIDL,[2] NIKOLAUS FOIDL,[2] GEORG M. GÜBITZ,[1] ROBERT M. LAFFERTY,[1] VICTORIA M. VALENCIA ARBIZU,[2] AND WALTER STEINER[1]

[1]*Institute of Biotechnology, Graz Technical University, A-8010 Graz, Austria; and* [2]*Proyecto Biomasa, Universidad Nacional de Ingenierìa, Managua, Nicaragua*

ABSTRACT

Seeds of the tropical plant *Jatropha curcas* (purge nut, physic nut) are used for the production of oil. Several methods for oil extraction have been developed. In all processes, about 50% of the weight of the seeds remain as a press cake containing mainly protein and carbohydrates. Investigations have shown that this residue contains toxic compounds and cannot be used as animal feed without further processing. Preliminary experiments have shown that the residue is a good substrate for biogas production. Biogas formation was studied using a semicontinous upflow anaerobic sludge blanket (UASB) reactor; a contact-process and an anaerobic filter each reactor having a total volume of 110 L. A maximum production rate of 3.5 m^3 m^{-3} d^{-1} was obtained in the anaerobic filter with a loading rate of 13 kg COD m^{-3} d^{-1}. However, the UASB reactor and the contact-process were not suitable for using this substrate. When using an anaerobic filter with *Jatropha curcas* seed cake as a substrate, 76% of the COD was degraded and 1 kg degraded COD yielded 355 L of biogas containing 70% methane.

Index Entries: Biogas; anaerobic filter; press cake; *Jatropha curcas*; methane production; renewable energy.

INTRODUCTION

Jatropha curcas (purge nut, physic nut) from the family of Euphorbiaceae is a common shrub of 3–6 m height. It originates in West India and is common in most arid areas of South America, Africa, and

*Author to whom all correspondence and reprint requests should be addressed.

Asia. Because of its extraordinary drought resistance, *Jatropha curcas* has attained economical importance in areas with extreme climates and soil conditions.

Different parts of the plant are used in nature medicine. The oil obtained from the seeds is used in the soap industry and also as source of energy (1). After transesterification of the oil it can be used as substitute for diesel oil, which is of interest for developing countries. This process is currently being carried out in Nicaragua as a development aid project.

The press cake of most oil seeds can be used as animal feed. Experiments have shown that the direct use of *Jatropha curcas* press cake for animal feed purposes is not possible because of the presence of toxic compounds such as curcin, a toxalbumin, and other equally negative substances such as phorbolic esters (2).

Another possibility of utilizing organic wastes is to convert these to biogas by means of an anaerobic fermentation. In this microbiological process, organic matter is converted to biogas with a high content of methane as a utilizable energy source. Biogas technology is very suitable for the partial treatment of all organic wastes (3). For example, in many food production processes, by-products cannot be used and recycled since they often contain organic and inorganic components. If directly discharged, the environment would be seriously polluted. In tropical countries, the anaerobic methane fermentation can be carried out at an ambient temperature, without any additional costs for heating. This relatively simple technology has made the biogas process interesting for developing countries.

The aim of this study was to find out the suitability of the press cake for anaerobic fermentation. The advantages of converting the press cake to biogas are a profitable removal of the press cake and the possibility to use the biogas directly within the process.

Environmental protection, improvements of hygiene and energy production are the relevant aspects of the biogas process. Biogas plants offering benefits for agriculture and the environment are used in Europe (4–6) as well as in Africa and Asia (7,8). The usefulness of decentralized family size biogas plants has been investigated in many different countries. Whereas studies of biogas plants in India showed very good results (9,10), other experiments in South Africa have pointed out some difficulties, such as the effects of shortages of water and manure, and the sceptisism of the users (11).

MATERIALS AND METHODS

Substrate

An aqueous suspension of press cake residue from *Jatropha curcas* without any additonal chemicals was used as substrate. The composition of press residues of *Jatropha curcas* seeds is given in Table 1. The actual

Table 1
Composition (%) of Seed Press Cake from *Jatropha curcas*

		a.		b.
dry weight		90.86		91.40
ash		6.03		6.55
org. dry weight		84.83		84.92
proteine		24.54		53.11
fat		6.40		6.32
fibre		32.26		5.60
starch		0.63		0.68
sugars		0.71		9.36
hemicellulose		5.55		1.94
cellulose		20.3		6.43
lignin		19.46		0.53

a. with Shells, b. without Shells

substrate suspension (30 g dry weight/L) was obtained after crushing the seed cake residue and seperation of the shells by sedimentation. Removal of the shells was necessary since in preliminary experiments the shells caused clogging in the pipes and were hardly fermentable. A composition of the substrate suspension is given in Table 2.

A COD:N:P ratio of 17.7:1:1 according to R. Braun *(12)* showed that a well-balanced nutrient composition was present and inhibition of methane fermentation because of ammonia formation was not expected.

Inoculum

For seeding of both the batch reactor and the semicontinuous reactors, an active slurry from a biogas plant running on pig manure was used. In case of the batch reactors, an inoculum concentration of 20% (v/v) and in case of the semicontinously operated reactors an inoculum concentration of 30% (v/v) was used.

Batch Fermentation

Preliminary batch experiments were carried out in 3-L glass fermenters to investigate the fermentability of the substrate. Seed cake residues were used at different concentrations. Fermentation experiments with the shells of the seeds that consist of about 90% lignin should indicate whether they are degradable or not. In the course of these experiments, volcanic stones that were used as a support in the anaerobic filter were tested for their toxicity with overall gas production as indication. All experiments were carried out at 37°C. Gas production and the pH value of the substrate were measured daily; solids, total nitrogen, and ammonium were measured at the beginning and at the end of each series of experiments.

Table 2
Composition of the actual suspension of the seed press cake
from *Jatropha curcas*

total nitrogen [g L^{-1}]	1.65
dry weight [g L^{-1}]	28.9
chemical oxygen demand (COD) [g L^{-1}]	29.2
organic dry material [g L^{-1}]	22.5
total phosphate [g L^{-1}]	1.73

Semicontinuous Fermentation

The formation of methane during the fermentation of seed cake was investigated in three different types of semicontinuously operated reactors utilizing the experimental data from the initial batch fermentation. All reactors were constructed from stainless steel tubing having a length of 2.87 m and a diameter of 0.25 m. The total volume of each bioreactor was 110 L. A schematic presentation of the reactor is given in Fig. 1. One reactor was operated as an anaerobic filter with a working volume of 73.8 L caused by partially filling with volcanic stones as a support. The second bioreactor was used as an UASB (upflow anaerobic sludge blanket)-reactor and the third as a contact-process type of bioreactor. The principal aim in all three reactor types was to increase the active biomass concentration in the fermenter either through growth of bacteria on the volcanic stones or by the formation of granular sludge or recirculation of biomass. In case of all three reactors, the substrate was added every 12 h.

Anaerobic Filter

The anaerobic filter was inoculated with 30% of its volume using digested sludge obtained from a biogas plant running on pig manure (13). The homogenized substrate was added at a concentration of 16 kg COD m^{-3}. During the first 5 d period, the high content of shells not being significantly degraded caused trouble in the tubes. Subsequently, shells in the substrate were first removed by sedimentation.

During the start-up period, substrate was added during 17 d at a concentration of 0.20 and for 6 d at a concentration of 0.30 kg COD m^{-3}d^{-1}. From twentyfifth day on, depending on process stability, the loading rate was increased daily by 10% so that after 61 d a loading rate of 10.57 kg COD m^{-3}d^{-1} was attained. In the following period, the loading rate was kept constant for 40 d since the pH value decreased and gas containing sludge came out of the gas outlet. This was probably caused by an increase of the solid concentration as a consequence of the short retention time. The loading was once again then slowly increased until a final value

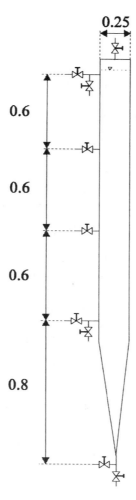

Fig. 1. Schematic presentation of the reactor (all dimensions in meters).

of 28 kg COD $m^{-3}d^{-1}$ was reached after 172 d. Because of the formation of gas containing sludge that could not be prevented and the decreasing pH the fermentation process broke down.

UASB-Reactor and the Contact Process

The start-up procedure for the UASB-reactor was the same as described for the anaerobic filter. After 37 d a loading rate of 1.42 kg COD $m^{-3}d^{-1}$ was attained, but a decrease of the pH value and a formation of gas containing sludge, which was blown out of the gas outlet, made a reduction of the loading rate to 0.73 kg COD $m^{-3}d^{-1}$ necessary. After a period of 60 d, a maximal loading rate of 2.93 kg COD $m^{-3}d^{-1}$ was achieved. A further increase of the loading rate failed because of the instability of the process.

However, in comparison to the UASB-reactor, the formation of gas-containing sludge in the contact process could be reduced by mixing the reactor from top to bottom. Washed out micro-organisms were recirculated daily. Within a period of 47 d, the loading rate could be increased to 2.65 kg COD $m^{-3}d^{-1}$ and until the sixty-eighth day to 6.80 kg COD $m^{-3}d^{-1}$. The further increase of the loading rate was not possible because of formation of gas-containing sludge and the instability of the process.

Analytical Methods

In order to enhance the performance of a biogas fermentation process and to avoid process failure, certain operating parameters must be controlled (14). Therefore, the pH value and gas production were measured daily, Kjeldahl nitrogen, ammonia, total solids, and the organic dry material was measured weekly in samples taken from different heights of the reactor. Sampling was done at the bottom of the reactor and at a height of 0.80 m, 1.4 m, 2.0 m, and 2.6 m. The pH value was also measured daily with a pH meter at different heights of the fermenter in order to determine the pH distribution over the length of the reactor. The COD was analyzed according to Deutsche Einheitsverfahren (DEV) (15). Removal efficiency was calculated from COD difference between the inflow and the effluent of the reactors. Kjeldahl nitrogen was determined after dissolving all samples with sulfuric acid and determining ammonium nitrogen using the distillation method (15). Total and volatile solids, fixed and suspended solids were analyzed according to DEV (15). Total phosphate was analyzed photometrically as vanadate-molybdate-complex according to DEV (15). Total volatile acids were determined with gaschromatography (HP 4890 II) using a FID. Daily gas production was measured by a volumetric gas meter. The gas volume was corrected to standard temperature and pressure. The methane and carbon dioxide concentrations were analyzed by displacement of NaOH in a measuring pipe and absorption of the carbon dioxide by NaOH with an accuracy of 0.1%. H_2S was determined by the reaction of H_2S and lead acetate (colorless) to lead sulfide (yellow), in test tubes (Fa. Dräger, Lübeck,Germany).

RESULTS AND DISCUSSION

Batch Fermentation

Preliminary batch experiments showed the following results: The volcanic stones showed no inhibition on methane formation. One kilogram seed cake residue (organic dry material) gave 446 L of biogas. The theoretical total yield of gas as calculated from different substrate components would be 649 L kg^{-1} organic dry material. The nut shells from the seeds are only slightly degradable and gave only 37 L biogas per kg dry weight. This might be a result of their high lignin content (16). Eighty percent of the substrate was converted during the initial 4 d 90% within 7 d.

Semicontinuous Fermentation

The UASB-reactor showed a high instability during the process as judged by the formation of gas-containing sludge. Variation of the loading rate always caused a decrease of the pH-value and an increase of the amount of volatile acids. The maximal loading was only 2.4 kg COD $m^{-3}d^{-1}$, the biogas yield was 0.31 m^3kg^{-1} CODd^{-1}, and the maximal production rate 1.4 m^3 per m^3 reactor volume and day. One reason for the instability in this case could be that the formation of granular sludge that is necessary for the successfull operation of this type of reactor, could not be observed during the whole experimental period. The presumable reason for this is the high content of solids in the substrate. In comparison with anaerobic sewage sludge as an inoculum, an UASB-reactor showed a 71% removal efficiency and produced 2.94 m^3 biogas per m^3 wastewater with an organic loading rate of 4.5 kg COD $m^{-3}d^{-1}$ *(17)*.

The determining factor for the fermentability of a substrate according to the UASB-principle is also the COD concentration. Highly loaded waste water, i.e., from either pectin or potato starch production, from baker's yeast production or from cellulose pulp plants having a COD higher than 8 kg m^{-3} is normally treated in fixed-bed reactors *(18)*. In accordance with these results it was not surprising that the anaerobic filter was more highly suitable for the treatment of the investigated substrate with a COD of about 29 kg m^{-3} than a sludge-bed reactor.

Bioreactors with a biomass recycle can be operated at higher reactor volume loading rates than UASB-reactors *(19)*. In the contact process used here a production rate of 1.9 $m^3m^{-3}d^{-1}$ and a biogas yield of 0.25 m^3kg^{-1} COD d^{-1} was obtained at a maximum loading rate of 6.8 kg COD $m^{-3}d^{-1}$.

The anaerobic filter allowed the highest loading rates to be attained in comparison to the other two types of bioreactors. The anaerobic filter system was most tolerant to variations of the loading rates, and the formation of gas-containing sludge could be reduced with recirculation of the rising solids to the bottom of the reactor. The influence of the loading rate on biogas yield and production rate of the anaerobic filter is shown in Fig. 2. Many different support materials have been described in the literature such as plastic tubes and organic supports *(20)*. Volcanic stones were chosen because of their availability in Nicaragua and the very low costs. They turned out be highly suitable for the retention of biomass in the reactor.

The highest yield of biogas could be obtained in the anaerobic filter. The yield of gas decreased with increasing loading rates to 0.35 m^3 kg^{-1}COD d^{-1} at a loading rate of 13 kg COD $m^{-3}d^{-1}$. This value is equivalent to 71% of the calculated theoretical yield of gas. These results are in agreement with the values obtained for the COD removal. At a loading of 13 kg COD $m^{-3}d^{-1}$ a conversion of about 75% was measured. A comparison of the three reactor types is given in Table 3.

Compared to the results obtained by other authors *(21)*, treating cotton waste water (COD 5000 mg l^{-1}), the specific biogas yield was observed

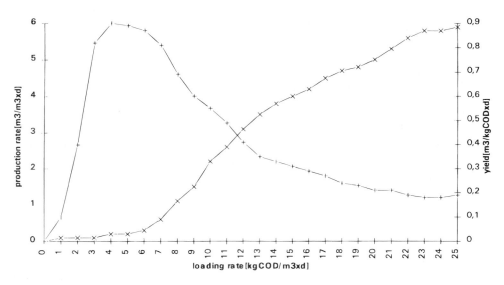

Fig. 2. Biogasproduction in an anaerobic filter with press cake residue as substrate: + biogas yield, x production rate.

Table 3
Comparison of the Results of an UASB-Reactor, a Contact Process and an Anaerobic Filter Using *Jatropha curcas* Seed Cake as a Substrate

	UASB-reactor	Contact process	Anaerobic filter
max. stable loading rate [kgCODm^{-3}d^{-1}]	2.4	6.8	13
biogas yield [m^{3}kg^{-1}CODd^{-1}]	0.31	0.25	0.35
production rate [m^{3}m^{-3}d^{-1}]	1.4	1.9	3.5
removal efficiency [%]	-	-	75

to be 0.44 to 0.48 m3 per kg COD removed for 2.0 d and 1.5 d retention time, respectively. Similary an upflow anaerobic reactor used for the treatment of palm oil mill effluent was operated with organic loads ranging from 1.2 to 11.4 kg COD m^{-3}d^{-1} and hydraulic retention times from 15 to 6 d. The overall substrate removal efficiency was up to 90%. Daily gas production varied in the range 0.69 to 0.79 m^3 kg^{-1} COD d^{-1} (22).

Although the active volume of the anaerobic filter was one-third less than that of the UASB-reactor and the contact system because of the volume of the volcanic stones, production rate was higher than the production rate in the UASB-reactor with a loading rate of 0.9 kg COD m^{-3} d^{-1} and higher than that of the contact process with a loading rate of 2.4 kg COD m^{-3} d^{-1}. A comparison of the maximum production rates between the three reactor types is given in Fig. 3.

When doubling the hydraulic retention time to 3.6 d using a stable loading rate, the anaerobic filter had a 20% higher production rate than before. The maximal production rate of 3.5 m^{3}m^{-3}d^{-1} was obtained at a maximum loading rate for a stable process of 13 kg COD m^{-3}d^{-1}. A further increase of

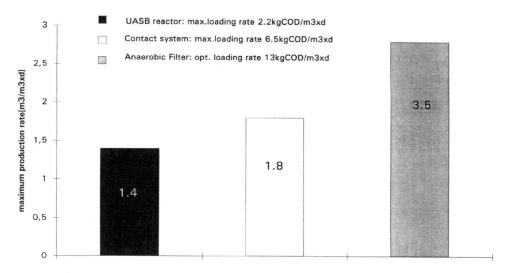

Fig. 3. Biogasproduction in three different bioreactor types: UASB, contact system, anaerobic filter. Comparison of production rates with press cake residue from *Jatropha curcas* as substrate.

production rate should be possible with a higher hydraulic retention time. The substrate concentration could not be increased in order to keep the sedimentability of the shells for their removal from the suspension. Appropriate removal of the shells before fermentation would allow higher substrate concentrations and this would lead to larger hydraulic retention times.

The highest concentration of ammonia measured in the anaerobic filter was 1630 mg L^{-1} at a loading rate of 13 kg COD m^{-3} d^{-1}. The major part of the nitrogen present in the substrate was converted to ammonia, but a concentration till 3000 mg L^{-1} shows no inhibition on methane fermentation according to Henze (23). Micro-organisms can be adapted after a certain time to concentrations higher than 8000 mg^{-1} NH$_4$-N.

In judging the quality of biogas, the content of methane and H$_2$S is the determining factor. The content of methane will depend on the composition of the substrate and on the fermentation conditions. The percentage of CO$_2$ in biogas increases with decreasing pH values and with excessive activity of the hydrolytic and acetogenic micro-organisms since CO$_2$ cannot be completely used by methanogenic bacteria. In case of the anaerobic filter, the methane content varied between 54.2 and 72.4 V% during a 24 h period. A high correlation between pH-values during this interval and a certain effect of the time of substrate addition is observed. The methane content was always measured before substrate addition and was on an average 70.0 V%. In comparison, in the biogas obtained from cotton waste water a methane content of 67% was measured (21).

The average content of H$_2$S in biogas was about 0.39% (v/v). Because of the corrosive effect of H$_2$S a reduction to 0.2% (v/v) before combustion

Table 4
Results of the Anaerobic Filter at a Loading Rate of 13 kg
$CODm^{-3}d^{-1}$ Using *Jatropha curas* Seed Cake as a Substrate

Production rate $[m^3 m^{-3} d^{-1}]$	3.5
Biogas yield $[m^3 kg^{-1} COD d^{-1}]$	0.35
COD Removal efficiency [%]	75
methane content [%]	70
H_2S content [%]	0.39

would be necessary (24). A content of 0.39% (v/v) H_2S in biogas corresponds to a concentration of less than 20 mg L^{-1} H_2S in solution according to Braun (12). A toxic effect of H_2S on is not assumed since the methane fermentation is only inhibited by concentrations of H_2S above 50 mg L^{-1} (25). A summary of all the results from the anaerobic filter is given in Table 4.

In many developing countries high population growth rates, the low economic growth, environmental issues such as deforestation, atmospheric pollution, and water depletion cause serious problems (26,27). The provision of adequate and decentralized produced energy is crucial for the overall development prospects. A number of options for moving towards sustainable development are put forward. These options encompass, among others, biomass and biogas. Biogenic fuels supply at the moment approx 15% of the world's energy (28), but in the future energy scenario these alternative gaseous fuels will most likely be the most important fuels.

ACKNOWLEDGMENTS

This project was supported by Sucher and Holzer and the Austrian Government.

REFERENCES

1. Quisumbing, E., *Medicinal Plants of the Philippines*, Katha Publishing Co., Quezon City, Philippines.
2. Amaral Liberalino, A. A., Alves Bambirra, E., Moraes-Santos, T., and Cardillo Vieira, E. (1988), *Arq. Biol. Technol.* **31**, 539–550.
3. Tentscher, W. A. K. (1995), *Food Technol.* **49**, 80–85.
4. Brown, K. A. and Maunder, D. H. (1994), *Water Sci. Technol.* **30**, 143–151.
5. Tafdrup, S. (1994), *Water Sci. Technol.* **30**, 133–141.
6. Sinclair, R. and Kelleher, M. (1995), *BioCycle* **36**, 50–53.
7. Navarro, L. B. and Bernardo, J. Y. (1994), *Renewable Energy* **5**, 1382–1386.
8. Stassen, I. H. E. M. (1994), *Renewable Energy* **5**, 819–823.
9. Rubab, S. and Kandpal, T. C. (1995), *Int. J. of Ambient Energy* **16**, 49–53.

10. Tomar, S. S. (1994), *Renewable Energy* **5**, 829–831.
11. Thom, C. and Banks, D. I. (1994), *J. Energy Southern Africa* **5**, 121–125.
12. Braun, R. (1982), *Biogas-Methangärung organischer Stoffe*, Springer Verlag Wien.
13. Anderson, B. C., Mavinic, D. S., and Oleszkiewicz, J. A. (1995), *Can. J. Civil Eng.* **22**, 223–234.
14. Sambo, A. S., Garba, B., and Danshehu, B. G. (1995), *Renewable Energy* **6**, 343–344.
15. Fachgr. f. Wasserchem. i. d. Gesellschaft f. deutsche Chem. (1981), *Deutsche Einheitsverfahren zur Wasser- Abwasser- und Schlammuntersuchung*, VCH Verlagsgesellschaft, Weinheim, FRG.
16. Kivaisi, A. K. and Eliapenda, S. (1995), *Biomass Bioenergy* **8**, 45–50.
17. Peng, D., Zhang, X., Jin, Q., Xiang, L., and Zhang, D. (1994), *J. Chem. Techn. Biotechnol.* **60**, 171–176.
18. Temper, U., Pfeiffer, W., and Bischofsberger, W. (1986), *Stand und Entwicklungspotentiale der anaeroben Abwasserreinigung*, Techn. University Munich/FRG.
19. Pereboom, J. H. F. and Vereijken, T. L. F. (1994), *Water Sci. Technol.* **30**, 9–21.
20. Guitonas, A., Paschalidis, G., and Zouboulis, A. (1994), *Water Sci. and Technol.* **29**, 257–263.
21. Chakradhar, B., Kaul, S. N., and Nageswar, G. G. (1995), J. Envir. Sci. Health, Part A: Envir. Sci. Engin. Tox. Haz. Substance Contr. **30**, 971–979.
22. Borja, R. and Banks, C. J. (1994), *J. Chem. Techn. and Biotechnol.* **61**, 103–109.
23. Henze, M. (1983), *Water Sci. Technol.* **15**, 112–121.
24. Wittrup, L. (1995), *BioCycle* **36**, 48–49.
25. Krois, H. and Wabenegg, F. (1982), *Wiener Mitteilungen* **49**, 3–15.
26. Stassen, G. and Kotze, I. A. (1995), *J. Energy Southern Africa* **6**, 35–39.
27. Hyman, E. L. (1994), *Environ. Manag.* **18(1)**, 23–32.
28. Hall, D. O. and House, J. (1995), *J. Power Energy* **209**, 203–213.

Evaluation of PTMSP Membranes in Achieving Enhanced Ethanol Removal from Fermentations by Pervaporation

SHERRY L. SCHMIDT, MICHELE D. MYERS, STEPHEN S. KELLEY, JAMES D. MCMILLAN, AND NANDAN PADUKONE*

National Renewable Energy Laboratory, 1617 Cole Blvd., Golden, CO 80401-3393

ABSTRACT

The use of membrane processes for the recovery of fermentation products has been gaining increased acceptance in recent years. Pervaporation has been studied in the past as a process for simultaneous fermentation and recovery of volatile products such as ethanol and butanol. However, membrane fouling and low permeate fluxes have imposed limitations on the effectiveness of the process. In this study, we characterize the performance of a substituted polyacetylene membrane, poly[(1-trimethylsilyl)-1-propyne] (PTMSP), in the recovery of ethanol from aqueous mixtures and fermentation broths. Pervaporation using PTMSP membranes shows a distinct advantage over conventional poly(dimethyl siloxane) (PDMS) membranes in ethanol removal. The flux with PTMSP is about threefold higher and the concentration factor is about twofold higher than the corresponding performance achieved with PDMS under similar conditions. The performance of PTMSP with fermentation broths shows a reduction in both flux and concentration factor relative to ethanol–water mixtures. However, the PTMSP membranes indicate initial promise of increased fouling resistance in operation with cell-containing fermentation broths.

Index Entries: Pervaporation; PTMSP; PDMS; ethanol recovery; membrane separations.

---

*Author to whom all correspondence and reprint requests should be addressed.

INTRODUCTION

Importance of *In Situ* Product Removal

Fermentations often exhibit strong product inhibition, especially fermentations for the production of alcohols and organic acids *(1–3)*. In batch or fed-batch fermentations, product inhibition limits the amount of substrate that can be efficiently fermented. If high substrate loadings are used, the product accumulates to sufficiently high levels so that the process becomes inhibited. Product inhibition is typically manifested by incomplete substrate utilization, reduced fermentor productivity and low process yield. In continuous fermentations, the impact of product inhibition is more severe than in batch processes because the fermentative micro-organisms are constantly exposed to the final (effluent) product levels. In cases where fermentation processes are limited by product inhibition, simultaneous product removal to maintain product concentrations at lower, less inhibitory conditions can improve the process. For example, continuous product removal has been shown to increase fermentation productivity in ethanol and butanol fermentations *(3–10)* and in lactic acid fermentations *(3,11)*.

Simultaneous product removal is increasingly being pursued as a promising technology in cases where product inhibition is manifest *(12–14)*. In a fermentation process, *in situ* product removal involves combining the fermentation step with a compatible product recovery method such as solvent extraction, membrane separation, or pervaporation *(8)*. Simultaneous fermentation and product removal also offers the opportunity to concentrate the product stream prior to downstream purification *(14)* and to remove other components such as diacetyl *(15)* or inhibitory fermentation by-products like acetaldehyde, ethyl acetate, and acetic acid *(7)*. Recovery of alcohols and organic acids from fermentation broths is one promising application of this technology, since conventional recovery techniques such as distillation are generally not cost-effective when carried out at low feed concentrations. The potential application of simultaneous fermentation and product recovery to a commodity chemical such as ethanol would require not only enhancing fermentation yield and productivity, but also developing a recovery process with low operating and capital costs.

There are a variety of simultaneous fermentation and product recovery schemes that have been investigated for recovering alcohols *(8,12,13)*. The Biostill process combines biomass retention by centrifugation with *in situ* product recovery by stripping *(16)*. The vacuferm process similarly operates a fermentation under vacuum conditions to recover ethanol by evaporation *(17–19)*. These methods suffer from low product selectivity that results in a high downstream product purification cost. *In situ* liquid extraction is another technique that has been demonstrated for a number of fermentation products *(20–22)*; however, the cost of product recovery from the extract can be energy-consuming *(21)*, or carryover of the solvent into the fermentation can cause the micro-organism to be inhibited *(23,24)*.

Ethanol Removal by Pervaporation

Membrane systems have several advantages over conventional separation processes such as distillation, adsorption, and extraction *(7)*. They often offer low operating temperatures, simplicity of design, and favorable economics, thus complementing the attributes of biotechnological processes. Membrane separations such as microfiltration and pervaporation are gaining attention for *in situ* recovery of products like ethanol *(4,25)*. The drawbacks of microfiltration are severe membrane fouling, low product selectivity in the permeate, large membrane area requirement, and relatively complicated operation. Pervaporation offers a higher product selectivity and simpler operation; however, the permeate flux can be an order of magnitude lower than that in microfiltration. Groot et al. *(4,25)* compared the integration of ethanol production with recovery by either microfiltration or pervaporation. The use of pervaporation resulted in a sixfold increase in volumetric productivity over conventional continuous operation, whereas a combination of microfiltration and pervaporation yielded a 16-fold higher productivity. Although a detailed cost estimate is not available, a process coupling microfiltration to the fermentation and pervaporation to the cell-free broth merits further investigation.

In pervaporation membrane systems, a liquid feed mixture, which can be a recycle loop with the fermentor, is contacted with the membrane, and the membrane permeate is removed from the other side of the membrane as a vapor *(6,13)*. A low concentration (vapor pressure) of the product is maintained on the permeate side to provide the driving force for diffusion through the membrane *(8,13)*. Thus, pervaporation can be particularly effective in the separation of a volatile product such as ethanol. In the case of ethanol, three distinct approaches can be used, individually or in combination, to control the permeation rate and selectivity of a particular membrane:

1. Decrease the vapor pressure on the permeate side;
2. Introduce a sweep gas stream to carry away the vapor; and
3. Use a temperature differential to increase the driving force for permeation *(6,26)*.

Previous research has reported low flux rates (membrane flux is a measure of the amount of material that passes through the membrane [$g\ m^{-2}\ hr^{-1}$]), low membrane selectivity (membrane selectivity is a measure of the separation efficiency of the molecule isolated by the process), and membrane fouling in fermentations *(7,25–27)*.

The development of new membranes based on poly[(1-trimethylsilyl)-1-propyne] (PTMSP) with improved permeability and/or selectivity relative to conventional systems of poly(dimethyl siloxane) (PDMS) will increase the potential of realizing efficient product removal *(6,28,29)*. Specifically, the selectivity of PTMSP for ethanol over water is more than four times higher than that of PDMS, whereas the permeability is 30 times higher *(28,29)*. This higher selectivity allows ethanol to be concentrated from 5 to over 50% in process streams with the PTMSP membrane. A traditional PDMS membrane would yield only 30% ethanol as the concentrate, and it would require 30 times more membrane area. Substituted polyacetylenes, such as PTMSP, are rigid rod polymers with bulky substituents that restrict rotational mobility and limit the polymer ability to pack together *(30,31)* (*see* Fig. 1). This limited molecular packing results in an unusually high free volume for these polymers, yielding the highest air permeability of any organic polymer and a high organic vapor permeability. PTMSP also has high ethanol permeability and selectivity over water, which makes it an ideal candidate for recovery of ethanol from fermentation broths *(28,29)*. The selectivity of PTMSP can be further increased by modifying its structure, for example by incorporating copolymers or blends *(32–34)*. Additional advantages of glassy PTMSP over rubbery PDMS include higher possible transmembrane operating pressures (the modulus of PTMSP is more than three orders of magnitude higher than that of PDMS), greater chemical stability, limited swelling of the glassy polymer resulting in greater durability, and limited fouling of the membrane surface by using copolymers or surface fluorination *(30)*.

National Renewable Energy Laboratory (NREL) researchers have recently developed the ability to tailor the permselective properties and molecular morphology of PTMSP through the preparation of copolymers and blends. It is reported here on studies carried out to evaluate the characteristics of NREL-produced PTMSP membranes for achieving enhanced pervaporative ethanol removal from fermentation broths.

MATERIALS AND METHODS

Membranes

PDMS was a standard membrane obtained from Membrane Technology and Research, Inc. (Menlo Park, CA). It is a rubbery dense film (20 µ) on a microporous support. It was used as received.

PTMSP was synthesized by the method of Masuda et al. *(35)*. $TaCl_5$ (1.20 g, 3.35 mmol) was dissolved in 150 mL toluene by heating at 80°C for 15 min. A solution of 22.44g (0.202 mol) 1-(trimethylsilyl)-1-propyne in 50 mL of toluene was cannulated into the $TaCl_5$ solution and heated at 80°C for 24 h. After cooling to room temperature, the solid was stirred with 400 mL methanol and filtered to give a brown solid. The polymer was redis-

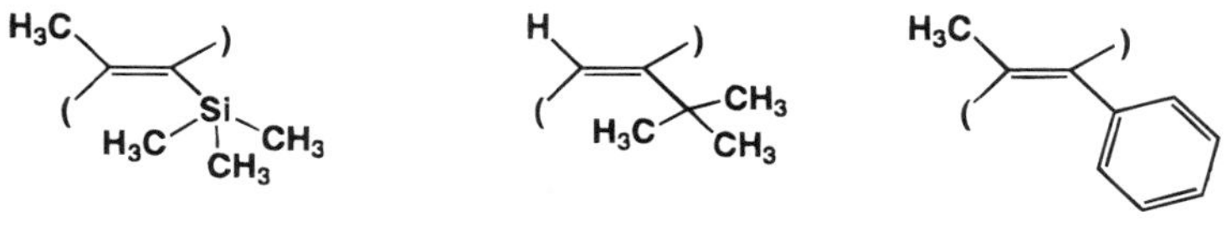

Fig. 1. Chemical structures of substituted polyacetylenes of interest for this project; **(A)** poly[(1-trimethylsilyl)-1-propyne] (PTMSP), **(B)** poly(t-butylacetylene) (PTBA), and **(C)** poly(1-phenyl-1-propyene) (PPP).

solved in 500 mL THF, precipitated by slow addition of methanol (500 mL) and filtered to give 22.20g (99%) white powder. This sample had a number average molecular weight of 300,000 Dalton and a glass transition temperature of greater than 150°C. PTMSP is a rigid polymer with good chemical stability and very good mechanical strength. The PTMSP polymer was dissolved to form a 5% w/v solution in toluene, and filtered to 7 µ. Films were hand-cast using a six-inch BYK Gardner casting knife, at 750 microns onto Teflon-taped glass. PTMSP dense films were 20–70 µ in thickness.

Experimental Set-Up

A Minitan S Ultrafiltration unit (Millipore) was used as the pervaporation cell. Membranes were installed between the two acrylic plates of the Minitan S using two silicone rubber gaskets (one above and one below the membrane) that sealed the perimeter of the filtration surface of the Minitan S. The effective membrane surface area in the unit was 0.0055 m². The inlet and retentate lines were connected to the feed vessel by silicone tubing. Feed was circulated at 100 mL/min by means of a Cole-Parmer pump. The filtrate lines of the unit were piped with stainless steel tubing to a simple on/off valve. Downstream of the valve, the collection vessel, a vacuum flask, was connected by stainless steel tubing inserted through a rubber stopper in the top of the flask. The side-arm of the collection vessel was then connected via vacuum tubing and stainless steel tubing through a rubber stopper into the top of the trap (a larger vacuum flask). Vacuum was applied to the system with a Welch 1400 Duoseal vacuum pump. A McCleod gauge in-line between the vacuum and trap monitored the vacuum pressure and a second on/off valve between the vacuum and the gauge controlled when vacuum was applied to the system. Vacuum was between 1 and 3 mm Hg. Both the collection vessel and trap were cooled by a dry ice/ethanol bath (about –40°C). Samples from the collection vessel were taken by closing the valve connected to the filtrate lines and then closing the valve controlling the vacuum. The stainless steel tubing between the filtrate lines and the collection vessel was then disconnected, and the vacuum tubing on the collection flask removed, and the openings covered with Parafilm to prevent evaporation while the flask was warmed to thaw the frozen collected sample. The collection vessel was weighed, and the increase in weight recorded as the weight

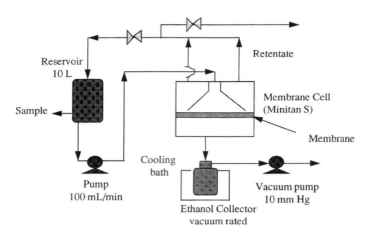

Fig. 2. Pervaporation system schematic.

of sample collected for that time period. Feed samples were also collected at each sampling time. To resume collection, the collection vessel was reconnected to the filtrate tubing, the vacuum valve opened, and then the filtrate valve opened. At the end of a run, the trap was weighed in the same manner as the collection vessel. The pervaporation experiments were carried out at an ambient temperature of about 25°C. Figure 2 shows a schematic of the experimental set-up. The concentration of the feed in the pervaporation runs showed only a small change (about 4%) during these experiments.

Fermentations

Saccharomyces cerevisiae D_5A was used in the yeast fermentations. Inoculum was grown for 12 h at 37°C, 150 rpm, in a baffled shake flask with Morton closure using 2% w/v yeast extract, 1% w/v peptone, 5% w/v glucose It was then used 10% v/v to inoculate a baffled shake flask with Morton closure containing 2% w/v yeast extract, 1% w/v peptone, 12% w/v glucose. The flask was incubated at 37°C, 150 rpm overnight. For cell-free broth, the material was then centrifuged at 4000 rpm for 10 min (1800 g), and the supernatant passed through a 0.2 mm filter.

Zymomonas mobilis 39676 (pZB4L) (NREL's proprietary strain) was used for the bacterial fermentations. Inoculum was grown in a baffled shake flask with Morton closure at 30°C, 150 rpm for 12 h using RM medium (1% w/v yeast extract, 0.02% w/v KH_2PO_4, 2.5% w/v glucose, 2.5% w/v xylose, plus 12.5 mg/L tetracycline). The inoculum was concentrated by centrifugation at 4000 rpm for 10 min (1800 g), and then used to inoculate a 2.5 L Bioflo fermentor (New Brunswick, NJ) containing RM medium with 5% w/v glucose, 3% w/v xylose, and 12.5 mg/L tetracycline. Temperature was controlled at 37°C, and pH was controlled at 6.0 with $3M$ KOH. The broth was harvested at 48 h. For cell-free broth, the same centrifugation/filtration procedure was used as was used for the yeast fermentations.

PTMSP Membranes

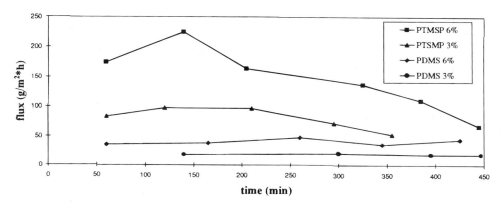

Fig. 3. Effect of feed concentration on membrane flux.

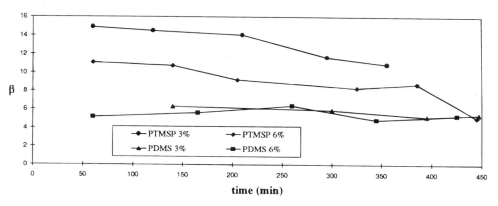

Fig. 4. Effect of feed concentration on β.

Analytical Techniques

Samples from the feed, collection vessel and trap were analyzed by HPLC using a Hewlett Packard Series II 1090, with an Aminex 87H organic acid column from Biorad installed and mobile phase of 0.01N sulfuric acid. The temperature of the column was 65°C.

RESULTS AND DISCUSSION

Comparison of PDMS and PTMSP in Ethanol–Water Mixtures

Effect of Concentration on Membrane Performance

Ethanol mixtures with water were tested as feed to the pervaporation cell in concentrations of 3% w/v and 6% w/v to represent the ethanol range achieved in fermentations. Figures 3 and 4 describe the performance of the two membranes by the ethanol flux and the concentration factor, β, respectively. Flux is expressed as rate of ethanol removal in g ethanol per square meter of membrane surface per hour. The factor, β, is the ratio of ethanol concentration in the permeate to that in the feed. With PDMS, the ethanol in the permeate was concentrated by a factor of about 5.5 for both feed concentrations. The flux of ethanol showed a marked dependence on the feed concentration of ethanol; the flux at 6% w/v concentration of 39 g/m^2-h was twofold higher than that at 3% w/v ethanol feed. The feed concentration provides the key driving force for ethanol permeation through the membrane.

In contrast to PDMS, PTMSP showed a dependence of both β and flux on the feed concentration. The flux at 6% w/v feed concentration was 149 g/m^2-h compared to 79 g/m^2-h at 3% w/v. The concentration factor showed the reverse trend; at 6% w/v feed, the permeate was about 10-fold more concentrated than the feed whereas, at 3% w/v feed, the β was about 13. Previous work with PDMS and PTMSP has been carried out at operating conditions different from those in our experiments. However, the flux and concentration factors in our studies of PTMSP carried out at ambient conditions correspond well with previous reports by Nagase et al. (32,33). Nagase et al. report a membrane permeability of 1.91×10^{-2} g-m/m^2-h compared to about 6.0×10^{-3} g-m/m^2-h in our experiments. The concentration factor is about 6.8 from earlier reports compared to 8.0 in the current study.

Comparison of Membranes in Ethanol Removal

PTMSP showed a distinct improvement over PDMS in both flux and β of ethanol. At both feed concentrations, the flux with PTMSP was at least fourfold higher than that obtained with PDMS. The concentration factor achieved with PTMSP at 6% w/v ethanol feed was about twofold higher and that at 3% w/v, feed was threefold higher than the corresponding values shown by PDMS membranes. The PTMSP performance seemed to indicate a gradual declining trend with time against the steady results obtained with PDMS. The PTMSP membranes that were prepared in-house were used unsupported in the pervaporation cell. The slight deterioration of performance may be a result of gradual physical degradation of the membrane during the experiment. In a commercial set-up, use of supported PTMSP membranes is not likely to show the same deterioration in performance.

Effect of Acetic Acid on Ethanol Separation

Organic acids such as acetic, lactic, and succinic acids are common by-products in an ethanol fermentation. Acetic acid at 1.5% w/v concentration in the feed was selected as a representative by-product. Mixtures of ethanol with acetic acid were used as feed to study the effect of an acidic fermentation by-product on ethanol recovery with PTMSP membranes. Figure 5 describes the performance of PTMSP on a 6% w/v ethanol, 1.5% w/v acetic acid mixture. Both the flux and ethanol concentration factor declined in comparison with the 6% w/v ethanol feed. In the acetic acid runs, the flux of 96 g/m^2-h was a 33% decrease and the β of 8.4 represented a 17% decrease over the corresponding parameters in runs with no feed acetic

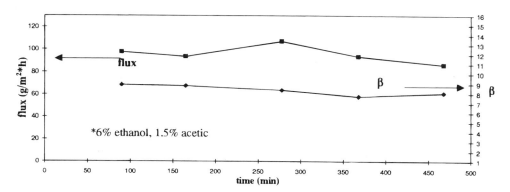

Fig. 5. Effect of acetic acid on ethanol pervaporation.

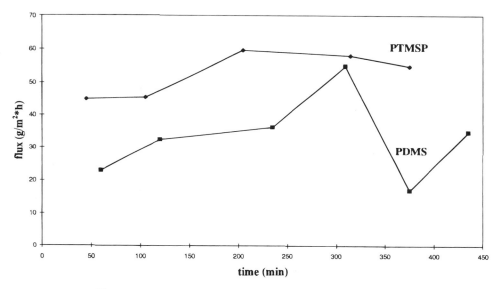

Fig. 6. Pervaporation of cell-free yeast fermentation broth.

acid. Acetic acid was detected in the permeate at concentrations of about 4.5 g/L, a threefold decrease compared to the feed concentration. The flux of acetic acid was two orders of magnitude lower than that of ethanol. These results indicate that the PTMSP selectivity for ethanol is much greater than that for acetic acid. Thus, PTMSP can be used effectively for ethanol recovery in the presence of an organic acid; however, the presence of the impurity does impact the performance to some extent.

Membrane Performance with Fermentation Broths

Figures 6 through 8 show a comparison of flux and β for ethanol-containing broths obtained from glucose fermentations by two organisms: *Saccharomyces cerevisiae* and *Zymomonas mobilis*. The final ethanol concen-

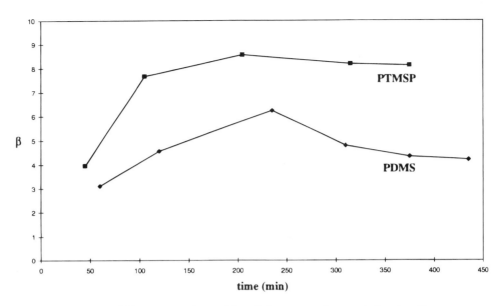

Fig. 7. Pervaporation with cell-free yeast fermentation broth.

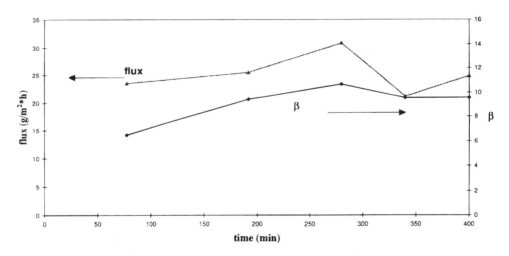

Fig. 8. Pervaporation of cell-free bacterial fermentation broth.

trations for the yeast and bacterial fermentations were about 6% w/v and 3% w/v, respectively, and were, therefore, compared with previous results from corresponding ethanol feed concentrations.

Pervaporation with Cell-Free Broths

The flux of ethanol from fermentation broths achieved with PTMSP were 25 and 52 g/m² for *Zymomonas* and *Saccharomyces*, respectively. These were about threefold lower than those achieved previously with corresponding ethanol–water mixtures. The concentration factor declined by

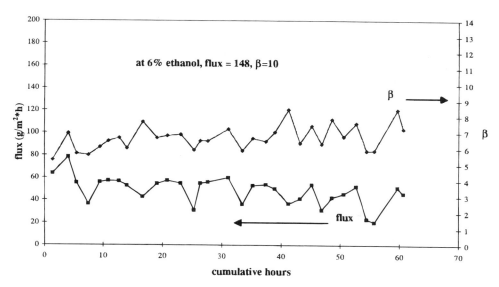

Fig. 9. PTMSP performance in extended operation with cell-free fermentation broth.

about 25% for both fermentation broths relative to the ethanol mixtures. A decline in performance was also observed with PDMS membranes; however, the relative decrease in flux and concentration factors was not as high as that observed with the PTMSP membranes. This may be attributed to the fact that the PDMS membranes were obtained from a commercial source, whereas PTMSP were developed in-house and were used unsupported in the experiments. The spent fermentation broth contains organic acids, proteins, ions, and unused nutrients in addition to product ethanol. The decrease in performance of both membranes with the fermentation broths is likely because of the mixture of components present with ethanol.

An experiment was conducted with fermentation media supplemented with 6% w/v ethanol to study the possible impact of media components on membrane performance. Results with PTMSP were inconclusive (not shown); the flux was comparable to the ethanol–water mixture, but the concentration factor was closer to that described in the previous section with fermentation broth. Our experiments so far appear to indicate a distinct loss in PTMSP performance with fermentation broths relative to ethanol–water mixtures of comparable concentrations. However, comparison of the two membranes indicates that both the flux and β with PTMSP are about 1.5-fold higher than those for similar runs with PDMS.

Analysis of prolonged operation of PTMSP with cell-free fermentation broths is shown in Fig. 9. An 8-h run was conducted each day and the system was shut off at the end of the day. The flux and concentration factor were steady overall at average values of 55 g/m^2-h and 7.0, respectively over a period of 11 d (8-h runs each day). The variations seen in the performance may be attributed to changes in ambient temperature that varied

Table 1
Summary of Pervaporation Performance

No.	Feed	Membrane	Flux, g/sq. m-hour	Average β
1	3% pure ethanol	PDMS	18	5.7
2	3% pure ethanol	PTMSP	79	13.2
3	cell-free bacterial fermentation broth (3% ethanol)	PTMSP	25	9.1
4	6% pure ethanol	PDMS	39	5.5
5	cell-free yeast fermentation broth (6% ethanol)	PDMS	33	4.5
6	6% pure ethanol	PTMSP	149	9.9
7	6% pure ethanol with 1.5% acetic acid	PTMSP	96	8.4
8	cell-free yeast fermentation broth (6% ethanol)	PTMSP	52	7.3
9	whole yeast fermentation broth (6% ethanol)	PTMSP	75	7.6

between 22 and 27°C. The temperature of operation can affect membrane performance significantly (32,33). Membrane performance was observed to increase gradually to its steady level over the first hour of operation as the system reached equilibrium (not shown).

Pervaporation with Cell-Containing Broths

A single 8-h run was conducted on PTMSP with spent fermentation broth containing yeast at about 4 g/L dry cell weight. Results obtained were comparable with those of the cell-free broth. Although some cell accumulation was observed visually on the membrane at the end of the run, this did not affect the pervaporation performance. Fouling resistance of membranes is a crucial factor in the sustained operation of such units. Our results with PTMSP indicate an early promise in providing a fouling-resistant operation of pervaporation although detailed studies of fouling in sustained operation will need to be done in future.

CONCLUSION

Table 1 summarizes the results of experiments conducted with PDMS and PTMSP membranes. Pervaporation using PTMSP membranes shows a distinct advantage over conventional PDMS membranes in ethanol removal. The flux with PTMSP is about three-fold higher and the concentration factor is about twofold higher than the corresponding performance achieved with PDMS under similar conditions. The performance of PTMSP with fermentation broths shows a reduction in both flux and concentration factor relative to ethanol-water mixtures. However, the PTMSP membranes promise increased resistance to fouling in operation with cell-containing fermentation broths.

Future work will be conducted on a 4-inch pervaporation cell connected to a continuous fermentor. Modified PTMSP membranes will be studied for improved performance in pervaporation. Performance of PTMSP membranes will be studied with cell-containing streams from bioreactors for continuous removal of ethanol. The effect of operating conditions such as temperature and vacuum level will be characterized to find optimal operation. Fouling resistance of PTMSP will be investigated in more detail by conducting prolonged runs with continuous fermentation. The benefits of pervaporation in maintaining fermentor concentrations below inhibitory levels will be illustrated. The economic impacts of using pervaporation to increase fermentation productivity and to reduce energy consumption in product purifications will also be assessed.

REFERENCES

1. van Uden, N. (1989), *Alcohol Toxicity in Yeasts and Bacteria*, CRC Press, Boca Raton, FL.
2. Herrero, A. A. (1983), *Trends Biotechnol.* **1**, 49–53.
3. Erickson, L. E., Fung, D. Y. C., and Tuitemwong, P. (1993), in *Biotechnology*, second edition, Volume 3: Bioprocessing, G. Stephanopoulos, ed., VCH, New York, NY, pp. 295–318.
4. Groot, W. J., van der Lans, R. G. J. M., and Luyben, K. Ch. A. M. (1991), *Appl. Biochem. Biotechnol.* **28/29**, 539–547.
5. Kim, Y.-J. and Weigand, W. A. (1992), *Appl. Biochem. Biotechnol.* **34/35**, 419–430.
6. Strathman, H. and Gudematsch, W. (1991), in *Extractive Bioconversions*, Mattiasson, B. and Holst, O., eds., Marcel Dekker, New York, NY, pp. 67–89.
7. Boddeker, K. W. and Bengtson, G. (1991), in *Pervaporation Membrane Separation Processes*, Huang, R. Y. M., ed., Elsevier Science Publishers B. V., Amsterdam.
8. Park, C.-H. and Geng, Q. (1992), *Separation Purif. Methods* **21**, 127–174.
9. Daugulis, A. J., Axford, D. B., Ciszek, B., and Malinowski, J. J. (1994), *Biotechnol. Lett.* **16**, 637–642.
10. Geng, Q. and Park, C.-H. (1994), *Biotech. Bioeng.* **43**, 978–986.
11. Davison, B. H. and Thompson, J. E. (1992), *Appl. Biochem. Biotechnol.* **34/35**, 431–439.
12. Mattiasson, B. and Holst, O. (1991), in *Extractive Bioconversions*, Mattiasson, B. and Holst, O. eds., Marcel Dekker, New York, NY, pp. 1–9.
13. Fleming, H. L. and Slater, C. S. (1992), in *Membrane Handbook*, Ho, W. S. W. and K. K. eds., van Nostrand Reinhold, New York, NY, pp. 105–116.
14. Freeman, A., Woodley, J. M., and Lilly, M. D. (1993), *Bio/Technology* **11**, 1007–1012.
15. Rajagopalan, N., Cheryan, M., and Matsuura, T. (1994), *Biotech. Techniques* **8**, 869–872.
16. Ehnström, L., Frisenfelt, J., and Danielsson, M. (1991), In: *Extractive Bioconversions*, Mattiasson, B. and Holst, O. eds., Marcel Dekker, New York, NY, pp. 303–321.
17. Ramalingham, A. and Finn, R. K. (1977), *Biotech. Bioeng.* **19**, 583–589.
18. Cysewski, G. R. and Wilke, C. R. (1977), *Biotechnol. Bioeng.* **19**, 1125–1143.
19. Sundquist, J., Blanch, H. W., and Wilke, C. R. (1991), in *Extractive Bioconversions*, Mattiasson, B. and Holst, O. eds., Marcel Dekker, New York, NY, pp. 237–258.
20. Daugulis, A. J., Swaine, D. E., Kollerup, F., and Groom, C. A. (1987), *Biotechnol. Lett.* **9**, 425–430.
21. Ruiz, F., Gomis, V., and Botella, R. F. (1987), *Ind. Eng. Chem.* **26**, 696–699.
22. Busche, R. M. (1991), *Appl. Biochem. Biotechnol.* **28/29**, 605–621.
23. Wang, D. I. C., Fleischaker, R. J., and Wang, G. Y. (1978), in *Biochemical Engineering: Renewable Sources of Energy and Chemical Feedstocks*, Nystrom, J. M. and Barnett, S. M. eds., *AIChE Symp.* **74**, 81–88.

24. Bar, R. and Gainer, J. L. (1987), *Biotech. Prog.* **3,** 109–114.
25. Groot, W. J., Kraayenbrink, M. R., Waldram, R. H., van der Lans, R. G. J. M., and Luyben, K. Ch. A. M. (1992), *Bioproc. Eng.* **8,** 99–111.
26. Park, C.-H., and Janni, K. (1994), In: *Environmentally Responsible Food Processing*, AIChE Symposium Series No. 300, **90,** 63–79.
27. Shabtai, Chaimovitz, Y. S., Freeman, A., Katchalski-Katzir, E., Linder, C., Nemas, M., Perry, M., and Kedem, O. (1991), *Biotech. Bioeng.* **38,** 869–876.
28. Ishihara, K., Nagase, Y., and Matsui, K. (1986), *Makromol. Chem., Rapid Commun.* **7,** 43–46.
29. Higashimura, T. and Masuda, T. (1986), U.S. Patent No. 4,591,440.
30. Masuda, T. and Higashimura, T. (1986), *Adv. Polymer Sci.* **81,** 121–165.
31. Masuda, T., Iguchi, Y., Tang, B.-Z., and Higashimura, T. (1988), *Polymer* **29,** 2041–2049.
32. Nagase, Y., Takamura, Y., Matsui, K. (1991), *J. Appl. Polymer Sci.* **42,** 185–190.
33. Nagase, Y., Sugimoto, Y., Takamura, K., Matsui, K. (1991), *J. Appl. Polymer Sci.* **43,** 1227–1232.
34. Hamono, T., Masuda, T., and Higashimura, T. (1988), *J. Appl. Polymer Sci.: Part A: Polymer Chem.* **26,** 2603–2612.
35. Masuda, T., Isobe, E. and Higashimura, T. (1985), *Macromolecules* **18,** 841–848.

Performance of Coimmobilized Yeast and Amyloglucosidase in a Fluidized Bed Reactor for Fuel Ethanol Production

MAY Y. SUN,[1,2] PAUL R. BIENKOWSKI,[1,2]
BRIAN H. DAVISON,[1,2] MERRY A. SPURRIER,[1]
AND OREN F. WEBB*,[1]

[1]*Oak Ridge National Laboratory, P.O. Box 2008, MS-6226, Chemical Technology Division, Oak Ridge, TN 37831-6226; and* [2]*University of Tennessee, Department of Chemical Engineering, Knoxville, TN 37916*

ABSTRACT

The performance of coimmobilized *Saccharomyces cerevisiae* and amyloglucosidase (AG) was evaluated in a fluidized-bed reactor. Soluble starch and yeast extracts were used as feed stocks. Conversion of soluble starch streams to ethanol has potential practical applications in corn dry and wet milling and in developmental lignocellulosic processes. The biocatalyst performed well, and demonstrated no significant loss of activity or physical integrity during 10 wk of continuous operation. The reactor was easily operated and required no pH control. No operational problems were encountered from bacterial contaminants even though the reactor was operated under nonsterile conditions over the entire course of experiments. Productivities ranged between 25 and 44 g ethanol/L/h. The experiments demonstrated that ethanol inhibition and bed loading had significant effects on reactor performance.

Index Entries: Ethanol; glucose; starch; simultaneous saccharification and fermentation; fluidized-bed reactor.

INTRODUCTION

Domestic ethanol use and production are presently undergoing significant increases along with planning and construction of new production facilities. Raw material costs typically make up 55–75% of the final alcohol

*Author to whom all correspondence and reprint requests should be addressed.

selling price *(1)*. Significant efforts are ongoing to reduce ethanol production costs by investigating new inexpensive feedstocks (woody biomass) and by process improvements in the fermentation and separation steps. Increasing reactor productivity is a potent method for reducing capital costs associated with new construction and expansion of existing facilities. Selection of fermentative organism and the reactor configuration also affect operating cost factors, such as yield, energy demand, and control of bacterial infections.

A key element in the development of advanced bioreactor systems capable of very high conversion rates is the retention of high biocatalyst concentrations within the bioreactor and a reaction environment that ensures intimate contact between substrate and biocatalyst. Such strategies include cell recycle by filtration, sedimentation, entrapment by membranes, and immobilization in gel beads *(2,3)*. These retention schemes can then be used with various reactor configurations, including continuous stirred-tank (CSTR), packed-bed (PBR), and fluidized-bed reactors (FBR). Typical batch reactors, commonly used in industry, have volumetric ethanol (EtOH) productivities between 2 and 5 g EtOH/L/h *(4,5)*. On a total reactor volume basis, volumetric productivity for continuous systems with high conversion is reported as approx 6–8 g EtOH/L/h for a free-cell CSTR, 10–16 g/L/h for an immobilized-cell CSTR, 10–30 g/L/h for a hollow-fiber reactor, 16–40 g/L/h for a vertical PBR, and 50–120 g/L/h for an immobilized-cell FBR *(6)*. One very effective method is to use an immobilized biocatalyst that can be placed into a reaction environment that provides effective mass transport, such as a fluidized bed. Previous studies have shown that such systems may be more than 10–50 times as productive as industrial benchmarks *(6,7)*. Economic impacts of the FBR for ethanol production may be significant *(8)*.

In this article, we describe FBR experiments for simultaneous saccharification and fermentation of starch, which employ entrapped yeast in a covalently crosslinked gelatin, chytosan, amyloglucosidase (AG) matrix. This study attempts to provide a comparison between yeast and previous FBR investigations. Previous studies with *Zymomonas mobilis* demonstrated significant performance advantages, such as very high productivities of 50–200 g ethanol/L/h and high yields around 97% of theoretical *(6,7)*. Yeasts have operational advantages, such as excellent pH tolerance. Such a simultaneous saccharification and fermentative approach has the advantage that AG and yeast are easily retained in the reactor for continuous use. Starch was used as raw material instead of glucose. Combining saccharification and fermentation in one vessel could reduce capital costs.

MATERIAL AND METHODS

The *Saccharomyces cerevisiae* cells were immobilized in covalently crosslinked gelatin (6 wt%) and chytosan (0.25 wt%) with AG. Crosslinking was accomplished by glutaraldehyde. Biocatalyst diameters

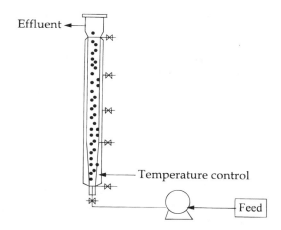

Fig. 1. FBR schematic. The reactor consisted of an expanded 30-cm inlet section (1.27–2.5 cm in id), three 30-cm sections of 2.5-cm ID jacketed glass pipe, and a 10-cm disengagement section of 9-cm ID with a screened sidearm for disengagement of beads from the reactor effluent.

ranged between 1.2 and 2.8 mm. The developmental biocatalyst was supplied by Genencor International.

Feed solutions consisted of various concentrations of StarDri 100 starch (A. E. Staley, Decatur, IL), 5 g/L Tastetone 900 AG yeast extract (Red Star, Juneau, WI), and 0.1 % w/v Antifoam B (Dow Corning, Midland, MI) in tap water.

The reactor depicted in Fig. 1 was constructed of a 30-cm inlet section, which expanded from 1.27–2.5 cm in ID, three 30-cm sections of 2.5-cm id jacketed glass pipe, and a 10 cm disengagement section of 9 cm id with a screened sidearm for disengagement of beads from the reactor effluent. Temperature was controlled at 34°C by a Haake A82 (Berlin, Germany) recirculating water bath. The feed was introduced at the reactor bottom using a model 7550-60 Masterflex peristaltic pump (Cole Parmer, Niles, IL). The pump was calibrated daily. The reactor was open to the atmosphere for gas-liquid biocatalyst disengagement at the outlet.

Starch, glucose, and ethanol concentrations were measured using a Shimadzu high-performance liquid chromatograph (HPLC) consisting of an RID6A refractive index detector, an SIL 10A autoinjector, an LC10AD pump, an SCL10A system controller, CTD10A column oven, and a CR501 integrator. An Aminex HPX-87H (Bio-Rad Laboratories, Hercules, CA) column with a 5-mM H_2SO_4 mobile phase separated analytes.

Minimal procedures were used for mitigation of contaminant growth. These included:

1. Changing feed lines with each new charge of feedstocks;
2. Replacement of feed containers with each charge of fresh feed; and

3. Medium was autoclaved prior to use, because significant amounts of contaminants existed in the yeast extract.

Previous investigations demonstrated that sterile operation is not necessary: however, the feed must be kept free of high levels of contamination (e.g., contamination >10^8 cells/mL). The reactor was not sterilized or cleaned after the 10-wk experiment began.

RESULTS AND DISCUSSION

The operability of the reactor was good throughout the experiment. The reactor system generally operated without operator intervention or attendance. The pH within the reactor was not controlled, and ranged from approx 6.5 at the inlet to about 3.5 at the column outlet. The temperature was maintained at 34°C throughout the experiment. The biocatalyst was used continuously for 10 wk in the FBR without recharging. There was no noticeable loss of biocatalyst from the bed. There were no obvious signs of physical degradation of the biocatalyst, except for very few beads that demonstrated significant diameter increases. The average AG activity was approx 8 µmol of glucose produced/min/mL of biocatalyst.

The fluid dynamics and reaction kinetics are coupled, and are complex functions of reaction rate, starch and glucose concentration, solids loading, and gas-liquid-solid properties *(9,10)*. The fluidization of the bed changes rapidly with axial position owing to significant changes in fluid flow rates and physical properties. Fluidization of the bed can be thought of as occurring in three zones that may be distinguished visually. The first zone, located at the bed entrance, is fluidized by liquid. The second zone, fluidized by gas product, starts within the expansion section and encompasses most of the bed. The third zone, termed the disengagement section, is characterized by high gas holdup (e.g., 5–20%) and significant mixing. The axial reaction rate in the FBR is a strong function of reactor biocatalyst concentration. Gas and liquid holdup, and dispersion, which become larger in the upper parts of the reactor, reduce the biocatalyst concentration. Inside the biocatalyst matrix, starch is hydrolyzed by AG to form glucose and then converted to ethanol and carbon dioxide by yeast:

$$(C_6H_{10}O_5)_n + (n-1)H_2O \xrightarrow{AG} nC_6H_{12}O_6 \quad (1)$$

$$nC_6H_{12}O_6 \xrightarrow{Yeast} 2nCH_3CH_2OH + 2nCO_2 \quad (2)$$

The biocatalyst was loaded with sufficient AG to convert low concentrations of starch at significant rates. The AG reaction rate was thus expected to be rapid at higher starch concentrations.

Figure 2 depicts an example concentration profile for the FBR. The flow rate and mass flow rate were 10 mL/min and 60 g starch/h, respectively. Bed loading is moderate, since the range of starch mass loading for

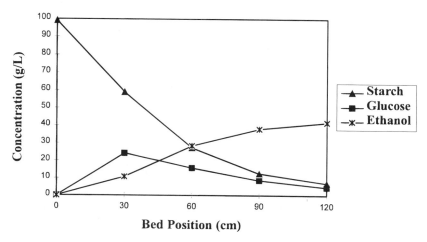

Fig. 2. Example of concentration profile. Starch concentration decreased with reactor position as AG conversion proceeded. Glucose is an intermediate between the AG and yeast reactions initially increased and then decreased. Ethanol concentration increased as the dextrose was converted.

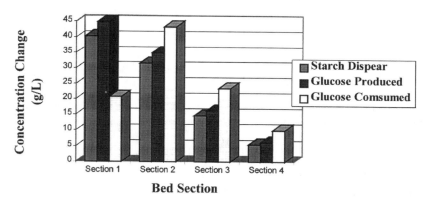

Fig. 3. The conversion and/or accumulation of starch, glucose, and ethanol as change within individual reactor sections.

the 10-wk experiment ranged from 20–120 g/h^{-1}. Starch concentration decreased with reactor position as AG conversion proceeded. Glucose is an intermediate between the AG and yeast reactions. Therefore, axial glucose concentration will be a function of both AG and yeast reaction rates. Glucose accumulates in the first reactor section because of high AG conversion rate (high starch concentration). The ethanol concentration increases as glucose is made available by the AG and then converted by the yeast. Figure 3 illustrates the conversion and/or accumulation of starch, glucose, and ethanol within individual reactor sections. Glucose accumulation in the first section indicates that the yeast reaction is slower than the

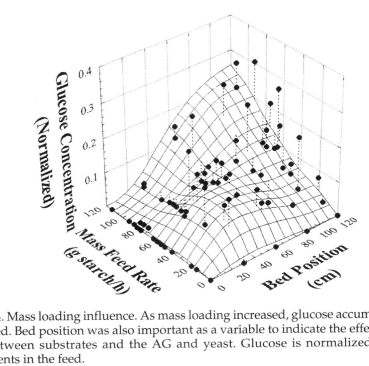

Fig. 4. Mass loading influence. As mass loading increased, glucose accumulation also increased. Bed position was also important as a variable to indicate the effect of contact time between substrates and the AG and yeast. Glucose is normalized to glucose equivalents in the feed.

AG reaction in this section. Yeast reaction rate is a function of glucose concentration, but is also affected by substrate and product inhibition. The conversion of starch to glucose drops in later sections as the available starch is reduced by prior reaction. Significant glucose conversion occurred in the second section owing to high glucose availability. Product may inhibit yeast in later sections of the reactor.

The surface in Fig. 4 allows visualization of the effect of mass loading on reactor performance. The data clearly demonstrate that as mass loading increased, glucose accumulation also increased. Bed position was also important as a variable to indicate the effect of contact time between substrates and the AG and yeast. Thus, as the contact time increased, the total conversion of starch and glucose also increased. The case depicted in Fig. 2 (previously discussed) is shown in Fig. 4 as the first data series closest to the Bed Position axis. In this case, glucose initially accumulated; however, as the starch supply rapidly decreased via AG conversion, almost complete conversion of substrates was achieved. The AG reaction is influenced greatly by starch concentration and to a minor extent by glucose concentration. The yeasts, on the other hand, are limited at the reactor inlet by a lack of glucose. As ethanol is formed, the yeasts become inhibited. The onset of ethanol inhibition begins at concentrations of approx 20 g/L^{-1} (1). Figure 4 also indicates that the reaction was limited by yeast and not the

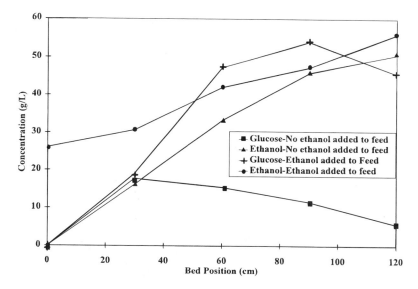

Fig. 5. Ethanol inhibition example. The concentration profile of glucose in the reactor was a function of ethanol inhibition demonstrated by addition of ethanol to the feed.

AG. The data clearly demonstrate that as mass loading increased, the accumulation of glucose increased. Under actual practice, the AG activity would ideally be balanced against the yeast activity to maintain a low glucose concentration throughout the reactor. Yeasts are primarily the greatest unknown in this system, because their performance not only depends on glucose and ethanol concentration, but also on nutrition, yeast, and biocatalyst age, and other factors. The surface was produced by fitting the data using the distance-weighted least-squares method in Statistica (StatSoft, Inc., Tulsa, OK) similar to the method of McLain (11). The surface in Fig. 4 is qualitative because of the complex coupling of reactions (AG and yeast) and three-phase hydrodynamics. Figure 4 demonstrates that care must be taken to balance the activities of the yeast and the AG.

In a series of experiments, ethanol was added to the reactor feed solution for verifying model kinetics and for reactor scale-up. Ethanol inhibition at the top of the reactor is masked by three-phase hydrodynamic effects, gas holdup in particular. Figure 5 depicts an example of ethanol inhibition. Ethanol concentration increased slowly relative to the case without ethanol in the feed. As yeast ethanol inhibition increases, the difference in yeast rates and AG will increase. Ethanol is not expected to inhibit AG activity under fermentation conditions (12,13,14). Thus, the ratio of glucose to starch concentrations at different points within the reactor was used as a general indicator of ethanol inhibition. Figure 6 allows visualization of the effects of ethanol inhibition on reactor performance. The data clearly show that as the ethanol in the feed increased, that rate differences between AG

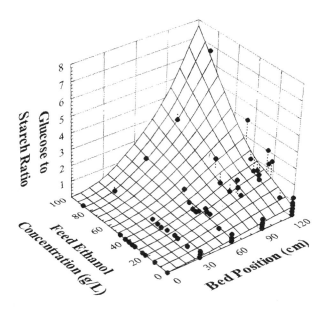

Fig. 6. Visualization of ethanol inhibition effects. As yeast ethanol inhibition increased, the difference in glucose and starch conversion rates increased, because AG was not inhibited. Under reactor conditions, the AG-catalyzed, starch-to-glucose equilibrium is greatly shifted to the dextrose product. The ratio of glucose to starch concentrations at different points within the reactor was used as a general indicator of ethanol inhibition.

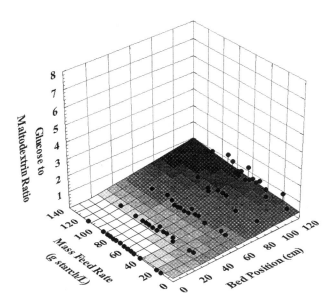

Fig. 7. Influence of mass loading on dextrose conversion and accumulation. The ratio of glucose to starch is a function of yeast and AG reaction rates. There is a small affect on the ratio of glucose to starch within the bed with bed position and mass feed rate.

Coimmobilized Yeast and Amyloglucosidase

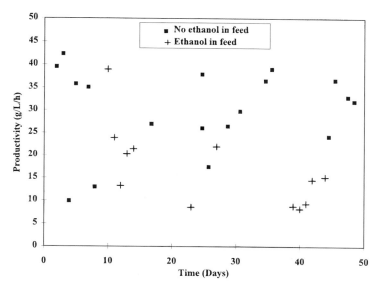

Fig. 8. Productivity profile. The data are grouped into two subsets: one subset containing feeds with ethanol and one subset containing feeds without ethanol. Most of the data with no ethanol in the feed were grouped in the range of 25–44 g ethanol/L/h. Low values were owing to very low loading of the reactor. When ethanol was included in the feed, the range of productivities generally ranged between 5 and 25 g ethanol/L/h. The differences in these productivities further demonstrate that ethanol inhibition had a significant effect on reactor performance. From both set of data, there were no obvious declines in productivity.

and yeast became more pronounced. The ratios also depend somewhat on feed concentrations and liquid feed rates. Figure 7 demonstrates that the mass loading effect was small compared to ethanol inhibition. No ethanol was added to the feed in Fig. 7. There was a small increase in the glucose-to-starch mass ratio with bed position and with mass loading when ethanol was not added to the feed. The data for Figs. 6 and 7 were generated using the same ranges of starch feed concentrations and flow rates. Statistica generated the surfaces for visualization in Figs 6. and 7.

Figure 8 depicts productivity over the course of the last 8 wk of the 10-wk experiment. Two weeks were allowed for stabilization of yeast activity. Mass loading varied throughout the experiment by varying feed concentrations and slow rates. The data are grouped into two subsets: one subset containing feeds with ethanol and one subset containing feeds without ethanol. Most of the data with no ethanol in the feed were grouped in the range of 25–44 g ethanol/L^{-1}/h^{-1}. Three points not falling in this group were owing to very low mass loading of the reactor. When ethanol was included in the feed, the range of productivities generally was between 5 and 25 g ethanol/L/h. The differences in these productivities further demonstrate that ethanol inhibition had a significant effect on reactor per-

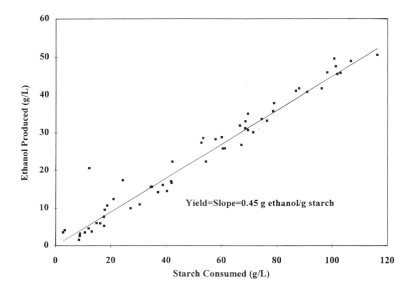

Fig. 9. Average yield. The yield for these experiments ranged around 80% of theoretical as measured by the slope of substrate converted to ethanol produced. Possibly, yeast nutrition limited yield.

formance. Also, there were no obvious declines in productivity in either set of data over time.

The average yield was calculated by the slope of substrate conversion to ethanol production (Fig. 9). The yield for these experiments ranged around 80% of theoretical. A yield of >90% of theoretical has been demonstrated with stirred tank reactors (STRs) and similar biocatalysts (15). Possibly yeast nutrition played a significant role in this lower yield. In a batch reactor system, Cysewski (16) demonstrated that ethanol yield was severely restricted if the concentration of yeast extract was 4.0 g/L^{-1} or 6.0 g/L^{-1}. He suggested a yeast extract requirement of 8.5g L^{-1}. The physiology of immobilized yeast may differ from free-cell physiology. Yeast extract was used at 5 g/L^{-1} concentration levels for direct comparison with previous work with Z. mobilis. Under these conditions, Z. mobilis demonstrated yields of 97 and 96% of theoretical for experiments at the batch and pilot scales (6,7). These data suggest that yeasts have higher nutritional requirement than Z. mobilis when immobilized. This may not be a significant issue in industrial processes where very rich nutrient streams, such as steep water, are routinely used in the fermentation.

SUMMARY

Coimmobilized yeast and AG were used to convert soluble starch to ethanol effectively in one reactor. This research demonstrates that yeast is very hardy and produces ethanol over a very long period of time. The biocatalyst performed well, and demonstrated no significant loss of activity or physical integrity during 10 wk of continuous operation. The reactor was easily operated and required no pH control. No operational problems were encountered from bacterial contaminants, even though the reactor was operated under nonsterile conditions over the entire course of experiments. Ethanol inhibition is an important factor. Productivities ranged significantly above industrial benchmarks. Coimmobilized yeast-AG biocatalyst continues to demonstrate the potential for FBR use.

ACKNOWLEDGMENT

Research was supported by the Office of Transportation Technologies of the US Department of Energy under contract DE-AC05-96OR22464 with Lockheed Martin Energy Research Corp.

The submitted manuscript has been authored by a contractor of the US government under contract DE-AC05-96OR22464. Accordingly, the US government retains a nonexclusive, royalty-free license to published form of the contribution, or allow others to do so, for US government purposes.

REFERENCES

1. Maiorella, B. L., Blanch, H. W., Willee, C. R. (1985), *Biotech. Bio.* **26**, 1003–1025.
2. Crueger, W. and Crueger, A. (1982), *Biotechnology: A Textbook of Industrial Microbiology*, Science Tech, Madison, WI.
3. Inloes, D. S., Michaels, A. S., Robertson, C. R., and Matin, A. (1985), *Appl. Microbiol. Biotechnol.* **23**, 85–91.
4. Silman, R. W. (1984), *Biotechnol. Bioeng.* **26**, 247–51.
5. Bajpai, P. K. and Margaritis, A. (1985), *Enzyme Microb. Technol.* **7**, 462–464.
6. Davison, B. H. and Scott, C. D. (1988), *Appl. Biochem. Biotechnol.* **18**, 19.
7. Webb, O. F., Scott, T. C., Davison, B. H., and Scott, C. D. (1995), *Appl. Biochem. Biotechnol.* **51/52**, 559.
8. Harshbarger, D., Bautz, M., Davison, B. H., Scott, T. C., and Scott, C. D. (1995), *Appl. Biochem. Biotechnol.* **51/52**, 593.
9. Petersen, J. N. and Davison, B. H. (1995), *Biotechnol. Bioeng.* **46**, 139.
10. Webb, O. F., Davison, B. H., and Scott, T. C. (1996), *Appl. Biochem. Biotechnol.* **57/58**, 639.
11. McLain, D. H. (1974), *Comput. J.* **17**, 318.
12. Lee, J. H., Pagan, R. J., and Rogers, P. L. (1983), *Biotechnol. Bioeng.* **25**, 659.
13. Kim, C. H., Lee, G. M., Zanial, A., Han, M. H., and Rhee, S. K. (1988), *Enzyme Microb. Technol.* **10**, 426.
14. Lee, C. G. Kim, C. H., and Rhee, S. K. (1992), *Bioprocess Eng.* **7**, 335.
15. Wallace, T. C. (1993), Personal communication.
16. Cysewski, G. R. (1976), Fermentation kinetics and process economics for the production of ethanol, Ph.D. dissertation, University of California, Berkeley.

A Mathematical Model of Ethanol Fermentation from Cheese Whey

I: Model Development and Parameter Estimation

CHEN-JEN WANG AND RAKESH K. BAJPAI*

Department of Chemical Engineering, University of Missouri-Columbia Columbia, MO 65211

ABSTRACT

The cybernetic approach to modeling of microbial kinetics in a mixed-substrate environment has been used in this work for the fermentative production of ethanol from cheese whey. In this system, the cells grow on multiple substrates and generate metabolic energy during product formation. This article deals with the development of a mathematical model in which the concept of cell maintenance was modified in light of the specific nature of product formation. Continuous culture data for anaerobic production of ethanol by *Kluyveromyces marxianus* CBS 397 on glucose and lactose were used to estimate the kinetic parameters for subsequent use in predicting the behavior of microbial growth and product formation in new situations.

Index Entries: Cheese whey; cybernetic model; *Kluyveromyces marxianus*; kinetics; chemostat.

INTRODUCTION

Glycosides are utilized in many anaerobic fermentations as a carbon source for ethanol production. Commonly encountered examples are those involving sucrose in molasses, sucrose, maltose, and maltotriose in brewer's wort, and lactose in cheese whey. Oligo- and polysaccharides in cellulose and starch are often used, too. Although fermentations start with glycosides, they may result in emergence of monosaccharides in broth to hydrolysis of glycosides. Cofermentations with other sugars to enhance ethanol production are also common. For example, fermentations from reconstituted cheese whey powder (CWP) may result in extracellular accu-

*Author to whom all correspondence and reprint requests should be addressed.

mulation of glucose and galactose in addition to the lactose, depending on the source of whey, the initial concentration of CWP, and the operating conditions (1,2). Cheese whey is often either spiked with additional sugars (3) or cofermented with agricultural products (4–6). As a result, fermentation from whey may involve the presence of multiple substrates in broth.

When multiple sugars are present in a pure culture system, sugars may be taken up by the cells either sequentially and/or simultaneously, depending on environmental conditions, microorganism, and sugars. In our previous study (2) on fermentations of multiple sugars in cheese whey by yeast *Kluyveromyces marxianus* CBS 397, it was shown that glucose and lactose are consumed simultaneously in batch fermentations under conditions of growth limitation by the nitrogen source. Under nitrogen-sufficient situations, however, a diauxic growth on both sugars was observed.

Although multiple-substrate systems are very common, only sporadic attempts have been made to study quantitatively the interactions between mono- and disaccharides and their impact on the overall kinetics observed in product-forming fermenters. A systematic study dealing with the effect of monosaccharides formed by the hydrolysis of sucrose in ethanol fermentations from molasses has been conducted by Jayanata (7). In this case, it was found that fast invertase-mediated hydrolysis of sucrose causes rapid increase in concentrations of glucose and fructose in the broth immediately after the start of the fermentation. The rates of cell growth and product formation from each of these monosaccharides were inhibited by the presence of the other sugar. Since the hydrolysis rate was fast and the yeast exhibited very similar uptake behavior of these monosugars, it was possible to consider the system as one consisting of a single sugar with its concentration equal to the sum total of those of glucose and fructose.

In another system involving five sugars in brewer's wort (8), a complex interaction between the sugars as a result of both simultaneous and sequential sugar uptake behaviors has been shown. Although glucose, fructose, and sucrose were simultaneously taken up by the cells, maltose and maltotriose were consumed later sequentially. This was because of the glucose-mediated repression of maltose uptake; utilization of maltotriose was shut off by both glucose and maltose. Given the critical levels of glucose and maltose in broth, the model ably described sugar consumption behavior and cell growth profile; kinetics of ethanol formation was, however, not presented.

The scarcity of a robust mathematical model for the product-forming systems is primarily because most microbial genomic as well as physiological characteristics, such as intracellular regulatory processes, are usually not well understood. This difficulty may be resolved by cybernetic modeling. The cybernetic approach introduced by Ramkrishna and coworkers (9–13) has found great success in modeling the diverse behavior of bacterial growth on multiple substrates in batch, fed-batch, and chemostat operations. The underlying assumption is that the complex cellular metabolism is driven by an optimal policy that attempts to maximize either cell growth or the efficiency of resource utilization under given environmental conditions. These models are based on a general observation that in an environment containing more than one nonessential substrate, the order of utilization of the substrates follows the decreasing pattern of growth rates. Formulation of the cybernetic models does not require *a priori* knowledge of actual mechanisms of intracellular regulatory processes, such as the order of preferential uptake of the substrate. As a result, the same cybernetic approach may find a wide range of applicability in various microbial systems with different regulatory mechanisms under varied environmental conditions.

A straightforward application of the cybernetic model to the product-forming system, however, requires consideration of product formation, which represents an important and major metabolic flux for carbon substrates. In light of the energetic coupling and net energy production involved in the formation of ethanol, it also requires a reconsideration of maintenance terms. In this article, a cybernetic model has been developed with special attention to anaerobic growth of yeasts on multiple substrates. Kinetic parameters have been estimated using continuous-culture data. Applications of this model have been considered in a companion article (Part II).

MATHEMATICAL MODEL

The cybernetic models *(9,10)* consider that cellular metabolism of a given substrate is controlled by a single key enzyme whose synthesis and activity are modulated by appropriate cybernetic variables. Maintenance consumption of substrate is interpreted as an outcome of uncoupling between catabolism and anabolism and, hence, as a linear function of specific growth rates of cells *(11)*. Since the control of substrate utilization in yeasts is also achieved via induction/repression of enzyme synthesis coupled with inhibition/activation of those already synthesized, fermentations involving yeasts can be modeled in a similar way.

Unlike the aerobic growth of bacteria, anaerobic growth of yeasts involves both cell growth and ethanol production. In a cybernetic framework where synthesis and activity of key enzyme are considered to be governed by appropriate allocations of critical resources, a question arises regarding whether product formation and cellular growth involve a single common enzyme or two separate key enzymes. In order to resolve this question, it is necessary to understand the interactions between growth and ethanol production in yeasts.

Under unsteered conditions (i.e., in the absence of bisulfite or salt stress) of anaerobic growth of yeasts, the nicotinamide adenine dinucleotide (NAD) reduced by metabolism of sugar is regenerated during reduction of pyruvate to ethanol *(14)*. Thus, ethanol is the major product formed under these conditions. The adenosine 5'-triphosphate (ATP) produced during the metabolism is utilized to support growth of cells, the extent of which depends on the coupling between catabolic and anabolic processes. This is also manifested in the observation that ethanol production rate is often growth-related in batch cultures *(15)*. Therefore, it should be possible to use a single common key enzyme for growth as well as the ethanol production process. Under this circumstance, each substrate will be associated with a single set of cybernetic variables (ε_i, δ_i) representing the control of the synthesis and inhibition of the key enzyme, respectively. Another cybernetic variable, δ_M, is associated with maintenance-related activities as proposed by Turner and Ramkrishna *(11)*. It indicates the uncoupling of anabolic and catabolic processes, and is representative of the state of cells rather than of specific physiological state of broth. In other words, it has the same value for all carbon substrates in the system.

As a result, the model equations, also considering substrate and product inhibitions, for fermentation of multiple substrates by yeasts can be expressed as:

Specific cell growth rate:

$$r_X = \sum_{i=1}^{n} r_{x,i}\delta_i = \sum_{i=1}^{n} [\mu_i^e e_i S_i / (K_{SX,i} + S_i + S_i^2 / K_{IX,i})] \quad (1)$$
$$[1/(1 + P/K_{P1,i} + P^2/K_{P2,i})]\delta_i$$

Specific product formation rate:

$$r_P = \sum_{i=1}^{n} r_{P,i}\delta_1 = \sum_{i=1}^{n} [v_i^e e_i S_i / (K_{SP,i} + S_i + S_i^2 / K_{IP,i})][K_{P,i}/(K_{P,i} + P)]\delta_i \quad (2)$$

Specific formation rate of enzyme e_i:

$$r_{e,i} = [\alpha_i S_i / (K_{SX,i} + S_i + S_i^2/K_{IX,i})][1/(1 + P_i/K_{P1,i} + P_i^2/K_{P2,i})]\varepsilon_i \quad (3)$$
$$- e_i[\beta_i + (1/X)(dX/dt)] + \alpha_i$$

Specific rate of consumption of substrate *i* for maintenance:

$$r_{M,i} = \phi_{M,i} e_i S_i / (K_{SM,i} + S_i) \quad (4)$$

Specific substrate consumption rate:

$$r_{S,i} = (r_{X,i}\delta_i / Y_{X/S_i}) + (r_{P,i}\delta_i / Y_{P/S_i}) + r_{M,i}\delta_M + \phi_{MO,i} \quad (5)$$

Here, the specific rate of substrate consumption for maintenance purposes is split into a growth-independent value ($\phi_{MO,i}$) and a growth-dependent value ($r_{M,i}$). This is in accordance with the observation of Pirt *(16)*. The maintenance process is characterized by uncoupling of catabolism from anabolism, which becomes more pronounced as growth rate decreases. Under conditions of fast growth, the efficiency of coupling appears to be at a maximum. One major cause of lowered cell yields at low specific growth rates has been suggested to be the energy uncoupling under these conditions *(17)*. The parameters μ_i^e, v_i^e, and $\phi_{M,i}^e$ are related to μ_i^{max}, v_i^{max}, and $\phi_{M,i}^{max}$, respectively, as *(9)*:

$$\mu_i^e = \mu_i^{max}[(\mu_i^{max} + \beta_i)/(\alpha_i + a_i)] \quad (6a)$$

$$v_i^e = v_i^{max}[(\mu_i^{max} + \beta_i)/(\alpha_i + a_i)] \quad (6b)$$

$$\phi_{M,i}^e = \phi_{M,i}^{max}[(\mu_i^{max} + \beta_i)/(\alpha_i + a_i)] \quad (6c)$$

The cybernetic variable δ_i *(10)* represents the regulation of enzyme activity by inhibition/activation and is given by:

$$\delta_i = [r_{X,i}/\max(r_{X,i})] \quad (7)$$

The cybernetic variable ε_i *(10)* in Eq. (3) represents induction/repression of synthesis of key enzyme e_i, and is given by:

$$\varepsilon_i = \left(r_{X,i} \bigg/ \sum_{i=1}^n r_{X,i}\right) \quad (8)$$

In cybernetic models, δ_M is considered as a measure of cellular control of activity of a key enzyme for maintenance functions. Turner et al. *(13)* have suggested a linear functional dependence of δ_M on specific growth rate in their model. Similar dependence has been suggested by the experimental data of Neijssel and Tempest *(18)* and further modeled by Pirt *(16)*. However, all of these studies have involved growth-only processes. For a product formation system, especially one that involves ethanol production, which results in net ATP generation, product formation rate should also affect the maintenance metabolism. This is in agreement with the cybernetic perspective according to which the cells aspire to achieve maximum return (growth rate) by efficiently allocating the resources to the metabolic function. Under anaerobic conditions, the cells cannot achieve such a cybernetic goal without a continuous regeneration of NAD via ethanol production. As a result, cell growth and ethanol production are interrelated in the sense of cybernetics. This dependence of δ_M on specific growth and product formation rates has been derived as follows.

For a single-substrate environment with adapted cells, the cybernetic variables δ and ε can be taken to be equal to unity. At any specific rate of growth and product formation, Eq. (5) may be written as

$$r_S = (1/Y_{X/S})r_X + (1/Y_{P/S})r_P + r_M\delta_M + \phi_{MO} \quad (9)$$

For this problem, r_S may reach its maximum value when r_X and r_P reach their maxima. These maxima of r_X and r_P appear together for ethanol fermentations *(15)*. Under these conditions, cells are very active, and therefore, the maintenance consumption of substrate is minimal; hence, δ_M may be equated to zero, and Eq. (9) becomes:

$$r_S^{max} = (1/Y_{X/S})\mu^{max} + (1/Y_{P/S})v^{max} + \phi_{MO} \tag{10}$$

Since $r_S \leq r_S^{max}$:

$$(r_X/Y_{X/S}) + (r_P/Y_{P/S}) + r_M\delta_M + \phi_{MO} \leq (\mu^{max}/Y_{X/S}) + (v^{max}/Y_{P/S}) + \phi_{MO} \tag{11}$$

Therefore:

$$\delta_M \leq (1/r_M)[(1/Y_{X/S})(\mu^{max} - r_X) + (1/Y_{P/S})(v^{max} - r_P)] \quad \text{for } r_M > 0 \tag{12}$$

Based on the above arguments, this equation is subject to the following equality constraints:

$$\delta_M = 0 \text{ at } r_X = \mu^{max} \tag{13a}$$

$$\delta_M = 0 \text{ at } r_P = v^{max} \tag{13b}$$

$$\delta_M = 1 \text{ at } r_X = 0 \text{ and } r_P = 0 \tag{13c}$$

Since this is true for all conditions of fermentations (even when $r_M = \phi_M^{max}$, its maximum value):

$$\delta_M \leq (1/\phi_M^{max})[(1/Y_{X/S})(\mu^{max} - r_X) + (1/Y_{P/S})(v^{max} - r_P)] \tag{14}$$

Therefore:
$$\delta_M \leq [c_1(\mu^{max} - r_X) + c_2(v^{max} - r_P)] \tag{15}$$

where $c_1 = (1/\phi_M^{max}) \cdot (1/Y_{X/S})$ and $c_2 = (1/\phi_M^{max}) \cdot (1/Y_{P/S})$, and $c_1 \neq 0$ and $c_2 \neq 0$. Equation (15) implies that δ_M is a function of both r_X and r_P. The inequality in Eq. (15) can be rewritten in a form of equality as:

$$\delta_M = -w + c_1(\mu^{max} - r_X) + c_2(v^{max} - r_P) \tag{16}$$

where $w \geq 0$.

The term w in Eq. (15) may be a function of r_X, or r_P, or both. Its form may be chosen to satisfy the equality constraints (Eqs. [13a–c]). Assuming a function of the following form for w: $w = w_1 + w_2 r_X + w_3 r_P + w_4 r_X r_P$, the following equation can be derived with equality constraints (Eqs. [13a–c]):

$$\delta_M = [1 - (r_X/\mu^{max})][1 - (r_P/v^{max})] \tag{17}$$

Table 1
Composition of Semisynthetic Medium

Nutrient	Composition g/L
Sugar	5.20
Yeast Extract	4.0
$(NH_4)_2SO_4$	2.038
KH_2PO_4	0.334
$MgSO_4 \cdot 7H_2O$	0.122
$FeSO_4 \cdot 7H_2O$	0.012
$CaCl_2 \cdot 2H_2O$	0.007

According to Eq. (17), δ_M is a joint function of deviations from maximum growth and product formation rates of the cells. For multiple substrate systems, an analogous function for δ_M can be suggested as a joint function of these deviations

$$\delta_M = \left[1 - \left(\sum_{j=1}^n r_{X,j}\delta_j \bigg/ \sum_{j=1}^n \mu_j^{max}\varepsilon_j\right)\right]\left[1 - \left(\sum_{j=1}^n r_{P,j}\delta_j \bigg/ \sum_{j=1}^n v_j^{max}\varepsilon_j\right)\right] \quad (18)$$

Equations (18) and (1)–(5) constitute the cybernetic model for a multiple-substrate, product-forming system.

MATERIALS AND METHODS

Experimental conditions for fermentations of semisynthetic media were kept the same as those reported before (2), except for the composition of supplemental nutrients as shown in Table 1. This composition was made to ensure the growth conditions without limitation of nitrogen source.

Analytical methods were also the same as previously described (2). Samples containing <0.3 g/L glucose or lactose were analyzed by the Nelson-Somagyi method (19). In a mixture of glucose and lactose at low concentrations, glucose content was analyzed by a glucose analyzer (Yellow Springs Instrument, Yellow Springs, OH), and lactose concentration was determined by subtracting glucose concentration from the total sugar concentration obtained from the Nelson-Somagyi method.

For determination of cell concentration, 5-mL samples were vacuum-filtered on preweighed 0.2-μm cellulose-acetate membrane filter (Gelman Sciences, Ann Arbor, Michigan). The solids were washed once with 5 mL distilled water and dried along with the filter in a microwave oven for 5 min at full power (700 W). The solids were cooled in a desiccator for 10 min before weighing.

ESTIMATION OF KINETIC PARAMETERS

The distinctive advantage of the cybernetic approach over the other models describing microbial metabolism in the presence of multiple substrates is its sole use of single-substrate kinetic parameters as the model inputs to simulate the multiple-substrate system. Single-substrate kinetic parameters can be estimated from either batch or continuous-culture data. With batch data, the dynamic equations for cell mass (X), product (P), and sugar (S) need to be solved together. These solutions depend on initial conditions, including the initial concentration of "key enzyme," which is unknown. Therefore, the solutions always involve an unknown initial condition, which may change from experiment to experiment. Use of steady-state continuous-culture data presents no such problem and was made in estimation of the kinetic parameters for sugars in semisynthetic media.

For continuous cultures of *Kluyveromyces marxianus* on two substrates, the governing equations for the rates of cell growth, ethanol production, substrate utilization, and enzyme production can be written as follows

$$(dX/dt) = \sum_{j=1}^{2} r_{X,j} \delta_j X - DX \qquad (19)$$

$$(dP/dt) = \sum_{j=1}^{2} r_{P,j} \delta_j X - DP \qquad (20)$$

$$(dS_i/dt) = D(S_{i,o} - S_i) - X\{[(r_{x,i}/Y_{X/S_i}) + (r_{P,i}/Y_{P/S_i})]\delta_i + r_{M,i}\delta_M\} \quad i=1,2 \qquad (21)$$

$$(de_i/dt) = [\alpha_i S_i / (K_{SX,i} + S_i + S_i^2/K_{IX,i})] [1/(1 + S_i/K_{P_1,i} + S_i^2/K_{P2,i})]\varepsilon_i \\ - e_i\left(\beta_i + \sum_{j=1}^{2} r_{X,i}\delta_i\right) + a_i - De_i \quad i=1,2 \qquad (22)$$

δ_i, ε_i ($i = 1,2$), and δ_M are the cybernetic variables defined in Eq. (7), (8), and (18), respectively. $S_{i,0}$ is the concentration of substrate i ($i = 1,2$) in the feed medium.

For a single substrate, the cybernetic parameters (ε and δ) are equal to 1. The constant specific maintenance rate, ϕ_{M0}, is small compared to the rest of maintenance contribution and has been neglected. Under steady-state conditions, Eqs. (19)–(22) yield:

$$D = r_X = [\mu^e eS/(K_{SX} + S + S^2/K_{IX})] [1/(1 + P^2/K_{P1} + P^2/K_{P2})] \qquad (23)$$

$$D = r_P[X/P] = [(v^e eS)/(K_{SP} + S + S^2/K_{1P})] [(K_P + P)](X/P) \qquad (24)$$

$$D = X/(S_0 - S)[(r_X/Y_{X/S}) + (r_P/Y_{P/S}) + (\phi_M^e eS)/(K_{SM} + S)\delta_M]$$
$$= [X/(S_0 - S)][(r_X/Y_{X/S}) + (r_X/Y_{X/S}) + \phi_M^e e\delta_M] \quad \text{for } S \gg K_{SM} \quad (25)$$

$$e = \{[\alpha S/(K_{SX} + S + S^2/K_{IX})][1/(1 + S/K_{P1} + S^2/K_{P2})] + a\}/(\beta + D) \quad (26)$$

$$\delta_M = [1 - (r_X/\mu^{max})][1 - (r_P/\nu^{max})] \quad (27)$$

Substitution of e from Eq. (26) into Eq. (23) yields:

$$Z = \beta D + D^2$$
$$= \{[\alpha S/(K_{SX} + S + S^2/K_{IX})][1/(1 + P/(K_{P1} + P^2/K_{P2}))] + a\} \cdot \quad (28)$$
$$[\mu^e S/(K_{SX} + S + S^2/K_{IX})][1/(1 + P/K_{P1} + P^2/K_{P2})]$$

The values of α, β, and a were assumed to be the same as those suggested by Kompala et al. (10) and Turner and Ramkrishna (11). These values are 0.001, 0.05, and 10^{-5}, respectively. The parameters μ^e, K_{SX}, K_{IX}, K_{P1} and K_{P2} were estimated using Eq. (28) with the steady-state measurements of sugar and ethanol concentrations at different dilution rates.

The estimation procedure involves minimization of an objective function, which is usually formulated as the sum of the squares of the weighted differences between experimental data ($A_{i,j}$) and model predictions ($\hat{A}_{i,j}$). The subscript i refers to the measurements (1, ..., n) and j refers to the component (1, ..., m). A minimization with equal weights may be unsatisfactory where magnitudes of data for various components differ significantly. For example, in most fermentation experiments, the magnitudes of the concentrations of cell mass are considerably smaller than those of substrate and product. At the same time, the choice of weighting factors affects the calculated values of kinetic parameters and leads to different explanations of experimental observations.

Himmelblau et al. (20) have suggested the following weighting factors for different components.

$$W_j = \left\{\sum_{k=1}^{n}\left[A_{k,j} - (1/n)\sum_{i=1}^{n}A_{i,j}\right]^2\right\}^{1/2} \quad (29)$$

These weighting factors were used in this study. The objective function to be minimized is then:

$$\sum_{i}^{n}\sum_{j}^{m}\{(A_{i,j} - \hat{A}_{i,j}) \cdot W_j\}^2 \quad (30)$$

In estimation of the parameters μ_e, K_{SX}, K_{IX}, K_{P1}, and K_{P2}, for example, $A_{i,j}$ has been substituted by the corresponding measured variable in Eq. (28).

Table 2
Kinetic Parameters

Parameter		Glucose[a]	Lactose[b]
μ^e		425	293
μ^{max}	(hr^{-1})	0.628	0.517
K_{SX}	(g/L)	0.055	0.094
K_{IX}	(g/L)	10,000	370
K_{P1}	(g/L)	7.8	12.1
K_{P2}	(g^2/L^2)	330	330
v^e		1037	867
v^{max}	(hr^{-1})	1.531	1.531
K_{SP}	(g/L)	0.050	0.088
K_{IP}	(g/L)	10,000	250
K_P	(g/L)	24.9	33.0
ϕ_M^e		1.7	4.0
ϕ_M^{max}	(hr^{-1})	0.0025	0.007
K_{SM}	(g/L)	10^{-6}(13)	10^{-6}(13)
$Y_{X/S}$	(g/g)	0.957	0.839
$Y_{P/S}$	(g/g)	0.510 (theor.)	0.538 (theor.)

μ^e, v^e, and ϕ_M^e are in unit of (g dry wt/[h of unit enzyme activity]).

μ^{max}, v^{max}, and ϕ_M^{max} were calculated from the estimated values of μ^e, v^e, and ϕ_M^e, respectively.

K_{SM} was fixed at the value of 10^{-6} as reported by Turner et al. (13).

[a] Estimated from continuous-culture data using glucose feed (5 and 20 g/L).

[b] Estimated from continuous-culture data using lactose feed (6.4 and 20 g/L).

Minimization of the objective function was conducted using a simplex algorithm proposed by Nelder and Mead (21). Computations were conducted on an IBM mainframe computer, and the estimated parameters are listed in Table 2.

The same procedures were also applied to Eqs. (24) and (25) to compute the rest of the kinetic parameters (K_{SP}, K_{IP}, v^e, K_P, and ϕ_M^e). The yield coefficients, $Y_{P/S}$, were set at their theoretical values, i.e., 0.510 for glucose and 0.538 for lactose. The value of K_{SM} has also been fixed at 10^{-6} g/L, as suggested by Turner and Ramkrishna (11). All of the kinetic parameters estimated from pure culture data of glucose and lactose are presented in Table 2. The table shows, additionally, the values of μ^{max}, v^{max}, and ϕ_M^{max}, calculated by rearrangement of Eqs. (6a)–(6c). The parameters listed in Table 2 were used to simulate batch and continuous-culture experiments involving fermentation of pure and mixed sugars. Results of these simulations and their comparisons with experimental data are presented in the accompanying article (Part II).

Model Development and Parameter Estimation

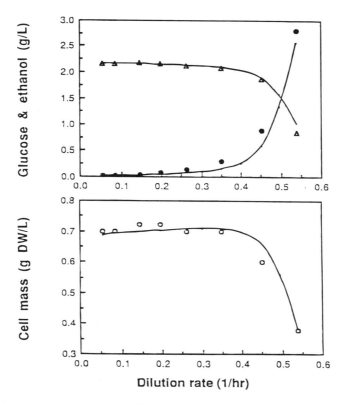

Fig. 1. Chemostat fermentation of 5 g/L glucose. Symbols: ●, glucose; ■, cell mass; Δ, ethanol. Solid lines represent simulation results.

The values of μ^{max} and v^{max} are in the same range as those in ethanol fermentations *(22–25)*, whereas K_{SX} and K_{SP} are relatively low. Lactose has been observed to be inhibitory to cell growth as well as product formation processes, as suggested by finite low values of constants K_{IX} and K_{IP}. This too is in accordance with the experimental observations of several researchers *(25–28)*. For the concentration range used in this work, glucose was not inhibitory, and K_{IX} and K_{IP} were given relatively large values (10,000). Ethanol is an inhibitory product in fermentations of glucose as well as lactose *(2,25–28)* as shown by the low values of K_{PI} and K_P. The values of cell yield parameter $Y_{X/S}$ is rather high, but not surprising in the presence of the complex nature of medium.

RESULTS

In order to show the goodness of fit between the experimental data and simulation results using estimated parameters, Eqs. (19–22) were solved in a dynamic manner for single-substrate situations ($i = j = 1$) until steady state was achieved. These simulation results have been presented in Figs. 1–4 for fermentations of glucose and lactose at two different feed

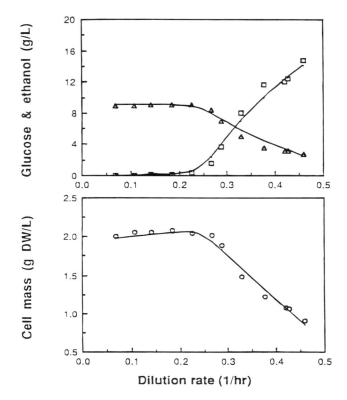

Fig. 2. Chemostat fermentation of 20 g/L glucose. Symbols: ●, glucose; ■, cell mass; Δ, ethanol. Solid lines represent simulation results.

concentrations levels. The feed concentrations were chosen to be the same as the values used in experimental work. The steady-state experimental data points for the concentrations of sugar (glucose or lactose), ethanol, and cell mass have been shown in these figures as discrete points. Simulation results have been presented as solid lines. Although the experimental measurements were used to calculate the parameters, it is reassuring to see that the predictions match the experimental observations quite closely in all the cases.

The simulation results with other conditions involving mixtures of sugars in batch and continuous cultures have been presented in detail in the accompanying article (Part II).

CONCLUSIONS

Product formation has been interpreted as the outcome of the cybernetic goal for yeasts to grow in ethanol fermentation and incorporated into the cybernetic framework. As a result, the cybernetic variable for cell maintenance was modified to consider the effect of product formation activity.

Model Development and Parameter Estimation

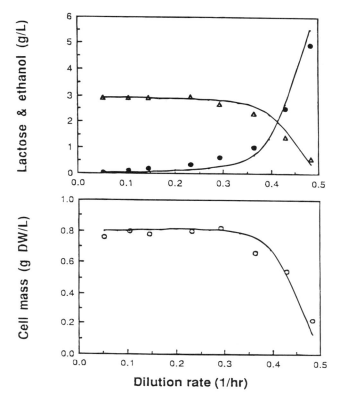

Fig. 3. Chemostat fermentation of 6.4 g/L lactose. Symbols:, ● lactose; ■, cell mass; Δ, ethanol. Solid lines represent simulation results.

The kinetic parameters of the proposed model were estimated using the continuous data. These parameters have been used with the model to simulate fermentations under various conditions as described in the accompanying article (Part II).

NOMENCLATURE

a   basal enzyme production, unit of enzyme activity/g
D   dilution rate, 1/h
e   key enzyme level, unit of enzyme activity/g
$K_{IX,i}$ substrate inhibition constant for cell growth, g/l
$K_{IP,i}$ substrate inhibition constant for product formation, g/l
K_P product inhibition constant for product formation, g/l
$K_{P1,j}$ product inhibition constants for cell growth, g/l
$K_{P2,j}$ product inhibition constants for cell growth, g^2/l^2
$K_{SX,i}$ substrate saturation constant for cell growth, g/l
$K_{SP,i}$ substrate saturation constant for product formation, g/l
$K_{SM,i}$ substrate saturation constant for maintenance, g/l

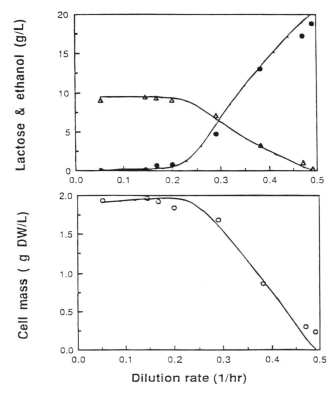

Fig. 4. Chemostat fermentation of 20 g/L lactose. Symbols:, ● lactose; ○ cell mass; Δ, ethanol. Solid lines represent simulation results.

P	ethanol concentration in broth, g/l
S	sugar concentration in broth, g/l
t	time, h
X	cell mass concentration in broth, g/l
α	enzyme synthesis constant
β	enzyme degradation constant
δ	cybernetic variable for inhibition/activation of enzyme activity
δ_M	cybernetic variable for uncoupling between catabolism and anabolism
ε	cybernetic variable for induction/repression of enzyme synthesis
μ^{max}	maximum specific rate of cell growth, 1/h
v^{max}	maximum specific rate of product formation, 1/h
ϕ_M^{max}	maximum specific substrate consumption rate for maintenance, 1/h

REFERENCES

1. Tu, C. W., Jayanata, Y., and Bajpai, R. K. (1985), *Biotechnol. Bioeng. Symp.* **15,** 295–305.
2. Wang, C. J., Jayanata, Y., and Bajpai, R. K. (1987), *J. Ferment. Technol.* **65,** 249–253.
3. Yoo, B. W. and Mattick, J. F. (1969), *J. Dairy Sci.* **52,** 900.

4. Whalen, P. J., Shahani, K. M., and Lowry, S. R. (1985), *Biotechnol. Bioeng. Symp.* **15**, 117–128.
5. Friend, B. A. and Shahani, K. M. (1981), *Fuels from biomass and wastes*, Klass, D. L. and Emert, G. H., eds., Ann Arbor Science Publishers, Michigan, pp. 343–355.
6. Friend, B. A., Cunningham, M. L., and Shahani, K. M. (1982), *Agricultural Wastes* **4**, 55–63.
7. Jayanata, Y. (1983), Einsatz Reaktionstechnischer Modelle zur Kinnzeichung und Optimierung der Kontinuierlichen Äthanolproduktion mit Biomasserückführung, Ph.D. thesis, TU Berlin, D83/FB 13 Nr. p. 148.
8. Fidgett, M. and Smith, E. L. (1975), *Appl. Chem. Biotechnol.* **25**, 355–366.
9. Kompala, D. S., Ramkrishna, D., and Tsao, G. T. (1984), *Biotechnol. Bioeng.* **26**, 1272–1281.
10. Kompala, D. S., Ramkrishna, D., Jansen, N. B., and Tsao, G. T. (1986), *Biotechnol. Bioeng.* **28**, 1044–1055.
11. Turner, B. G. and Ramkrishna, D. (1988), *Biotechnol. Bioeng.* **31**, 41–43.
12. Turner, B. G., Ramkrishna, D., and Jansen, N. B. (1989), *Biotechnol. Bioeng.* **34**, 252–261.
13. Turner, B. G., Ramkrishna, D., and Jansen, N. B. (1988), *Biotechnol. Bioeng.* **32**, 46–54.
14. Rose, A. H. (1976), *Chemical Microbiology*, 3rd ed., Plenum, New York, p. 209.
15. Gaden, E. L., Jr. (1959), *J. Biochem. Microbiol. Technol. Eng.* **1**, 413–420.
16. Pirt, S. J. (1982), *Arch. Microbiol.* **133**, 300–302.
17. Senez, J. C. (1962), *Bacteriol. Rev.* **26**, 95–107.
18. Neijssel, O. M. and Tempest, D. W. (1976), *Arch. Microbiol.* **107**, 215–221.
19. Somagyi, M. (1952), *J. Biol. Chem.* **195**, 19–23.
20. Himmelblau, D. M., Jones, C. R., and Bischoff, K. B. (1967), *Ind. Eng. Chem. Fundam.* **6**, 539–543.
21. Nelder, J. A. and Mead, R. (1965), *Comp. J.* **7**, 308–313.
22. Bazua, C. D. and Wilke, C. R. (1977), *Biotechnol. Bioeng. Symp.* **7**, 105–118.
23. Ghose, T. K. and Tyagi, R. D. (1979), *Biotechnol. Bioeng.* **21**, 1401–1420.
24. Hoppe, G. K. and Hansford, G. S. (1982), *Biotechnol. Lett.* **4**, 39–44.
25. Ruggeri, R., Specchia, V., and Gianetto, A. (1988), *Chem. Eng. J.* **37**, B23–B30.
26. Moulin, G., Boze, H., and Galzy, P. (1980), *Biotechnol. Bioeng.* **22**, 2375–2381.
27. Harbison, A. M., Kempton, A. G., and Stewart, G. G. (1984), *Dev. Ind. Microbiol.* **25**, 467–473.
28. Vienne, P. and von Stockar, U. (1985), *Enzyme Microbiol. Technol.* **7**, 287–294.

A Mathematical Model of Ethanol Fermentation from Cheese Whey

II. Simulation and Comparison with Experimental Data

CHEN-JEN WANG AND RAKESH K. BAJPAI*

Department of Chemical Engineering, University of Missouri-Columbia, Columbia, MO 65211

ABSTRACT

A cybernetic model for microbial growth on mixed substrates, was used to simulate the anaerobic fermentation of cheese whey and multiple sugars in semisynthetic media by *Kluyveromyces marxianus* CBS 397. The model simulations quite successfully predicted the observed behavior in batch and during transients in continuous operation, in single-substrate systems as well as in media involving multiple substrates, and in semisynthetic and reconstituted cheese whey solutions. The results of simulations and their comparison with the experimental data are presented.

Index Entries: Cybernetic model; batch culture; chemostat; transient; simulations.

INTRODUCTION

In Part I of this two-part article, the cybernetic approach of microbial metabolism in presence of multiple substrates was extended to include the product formation observed in anaerobic fermentations. Therein, the cybernetic variable controlling the cell-maintenance function was also modified considering the crucial energy-supplying function of product formation in achieving the cybernetic goal. In this article, the applicability of the extended model has been tested with experimental results under diverse operating conditions.

Some of the experimental data dealing with fermentations of cheese whey by *Kluyveromyces marxianus* CBS 397 have already been presented elsewhere [1], Additional experimental data with single or mixed carbon

---

*Author to whom all correspondence and reprint requests should be addressed.

Table 1
Composition of semisynthetic medium

Nutrient	Composition, g/L
Sugar	5–20
Yeast extract	4.0
$(NH_4)_2SO_4$	2.038
KH_2PO_4	0.334
$MgSO_4 \cdot 7H_2O$	0.122
$FeSO_4 \cdot 7H_2O$	0.012
$CaCl_2 \cdot 2H_2O$	0.007

substrates (glucose and/or lactose) were collected through specifically designed fermentations in order to identify the parameters of the model. These experiments were conducted in a semisynthetic medium. In this work, the influence of galactose on ethanol fermentations from cheese whey has been disregarded, since it is not significant compared to that of glucose (1).

MATERIALS AND METHODS

Experimental conditions for fermentations of semisynthetic media were kept the same as those reported before (1) except for the composition of supplemental nutrients as shown in Table 1. This composition was made to ensure the growth conditions without limitation of nitrogen source.

Analytical methods were also the same as previously described (1). Samples containing <0.3 g/L glucose or lactose were analyzed by the Nelson-Somagyi method (2). In a mixture of glucose and lactose at low concentrations, glucose content was analyzed by a glucose analyzer (Yellow Springs Instrument, Yellow Springs, OH), and lactose concentration was determined by subtracting glucose concentration from the total sugar concentration obtained from the Nelson-Somagyi method.

For determination of cell concentration, 5-mL samples were vacuum-filtered on preweighed 0.2-µm cellulose-acetate membrane filter (Gelman Sciences, Ann Arbor, MI). The solids were washed once with 5 mL distilled water and dried along with the filter in a microwave oven for 5 min at full power (700 W). The solids were cooled in a desiccator for 10 min before weighing.

MATHEMATICAL MODEL

The mathematical model has been described in detail in the previous article, Part I.

RESULTS AND DISCUSSIONS

Simulation of Single-Substrate System

Several batch fermentations were conducted with low concentrations of glucose or lactose in the medium. In lactose media, glucose and galactose were not detected at any time. These batch fermentations were simulated by the extended cybernetic model (Eqs. [19–22] with D = 0 and $i = j = 1$ in Part I) using kinetic parameters obtained from continuous-culture data. The kinetic parameters have been listed in Table 2 of Part I (previous article). Michelson's (3) method was used for the solution of differential equations.

Since the initial concentration of the "key enzyme" significantly influences the duration of lag phase in batch cultures, a successful simulation is determined by the consistent use of the same enzyme level. A low specific enzyme level leads to a longer lag period, whereas a high level reduces it. The specific enzyme concentration depends on culture history. Since the inocula for all the semisynthetic media were prepared with a glucose medium, it is reasonable to set the initial levels at 90% of the maximum specific enzyme activity for fermentations on glucose and at 10% for those on lactose. These hypothetical level have been found to give the best simulation results in this work and in the other study (4). The maximum specific activity of enzyme, e^{max}, was calculated as

$$e^{max} = [(\alpha + a)/(\mu^{max} + \beta)] \qquad (1)$$

Figures 1A–1D show the experimental and the computed results of simulation. Profiles of the concentration of key enzymes are not shown, since no corresponding experimental measurements were obtained. Generally, the batch experimental data are in good agreement with model predictions with parameters estimated from continuous-culture data. Evidently, the cybernetic model can be used to simulate batch fermentations satisfactorily as well as the continuous data (see Part I). Experimental observations of batch fermentations with lactose generally lasted longer than those with glucose (see Fig. 1A–D). The simulations also predicted a similar behavior. When the inoculum was initially grown on glucose and the cells were transferred to a medium containing lactose only, these require a longer period to synthesize enough enzymes to utilize lactose effectively. This lag period is marked by a very slow growth at the beginning of batch cultures. As shown in the Figs. 1A-D, ethanol production profiles follow the same trend as those of cell growth.

Transients in chemostat were studied by changing feed sugar concentrations at constant dilution rate. Both shift-up and shift-down experiments were conducted, and the concentrations of cell mass, sugar, and product (ethanol) were monitored during shift from one steady state to another. These transient results are plotted in Figs. 2A–D as discrete points. The time in abscissa with negative values indicates the original steady state before a transient experiment was started. For fermentations of glucose,

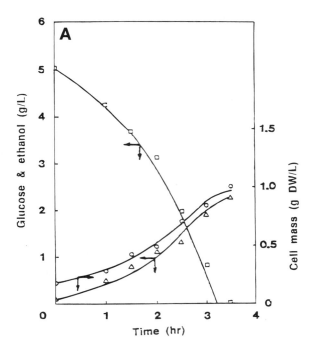

Fig. 1A. Batch fermentation of 5 g/L glucose. Symbols: □, glucose; ○, cell mass; △, ethanol. Solid lines represent simulation results.

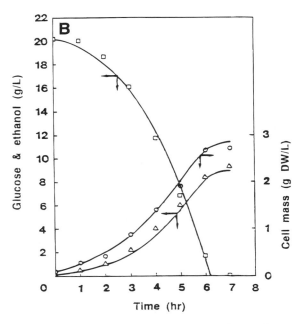

Fig. 1B. Batch fermentation of 20 g/L glucose. Symbols: □, glucose; ○, cell mass; △, ethanol. Solid lines represent simulation results.

Simulation and Comparison

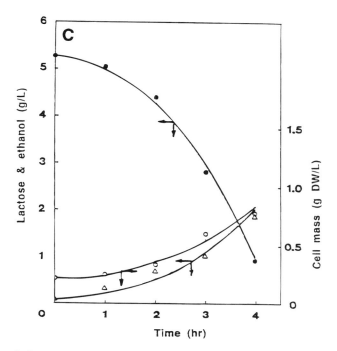

Fig. 1C. Batch fermentation of 5.3 g/L lactose. Symbols: ●, lactose; ○, cell mass; △, ethanol. Solid lines represent simulation results.

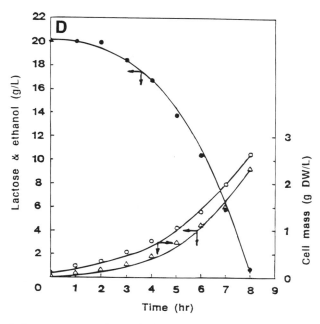

Fig. 1D. Batch fermentation of 20 g/L lactose. Symbols: ●, lactose; ○, cell mass; △, ethanol. Solid lines represent simulation results.

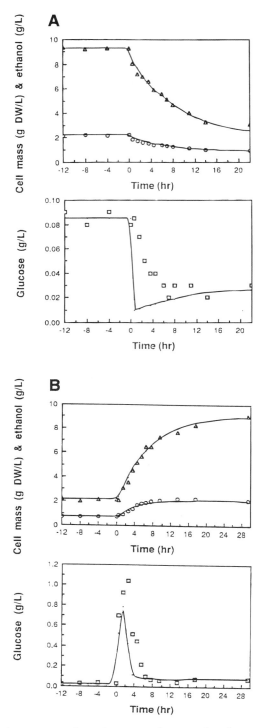

Fig. 2A. Transient response of continuous culture to the change in feed glucose concentration from 20–5 g/L at a dilution rate of 0.144 h^{-1}. B. Transient response of continuous culture to the change in feed glucose concentration from 5–20 g/L at a dilution rate of 0.144 h^{-1}. Symbols: □, glucose; ○, cell mass; △, ethanol. Solid lines represent simulation results.

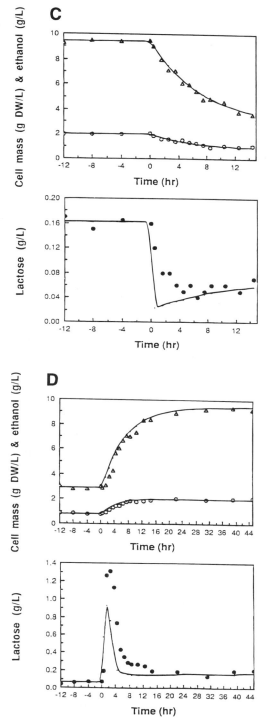

Fig. 2C. Transient response of continuous culture to the change in feed lactose concentration from 20–6.4 g/L at a dilution rate of 0.146 h^{-1}. **D.** Transient response of continuous culture to the change in feed lactose concentration from 6.4–20 g/L at a dilution rate of 0.146 h^{-1}. Symbols: ●, lactose; ○, cell mass; △, ethanol. Solid lines represent simulation results.

the employed concentrations of sugar in feed media were 5 and 20 g/L, and the constant dilution rate was set at 0.144 h^{-1}. For lactose, these were 6.4 and 20 g/L along with a constant dilution rate of 0.146 h^{-1}.

Simulation of the transient response was conducted with the help of dynamic Eqs. (19)–(22) with $i = j = 1$ in the companion article (Part I) to yield a steady state at the beginning of the transient. The simulation results of cell mass, sugar, ethanol, and enzyme at steady state were used as initial conditions to simulate the transients, and the results are shown in Figs. 2A–D as solid lines.

For both types of transient experiments, the observed concentrations of cell mass and ethanol were in good agreement with the model predictions. The model predicted net changes in sugar concentration rather remarkably; there were, however, some discrepancies between model predictions and the experimental observations of this variable. The model predicted sharp minima in sugar concentrations during shift-downs in sugar concentration; the observed changes in sugar concentration were smooth and without a minimum. During shift-up, both (the experimental observations and the model predictions) showed sharp maximum, but the magnitude of increase in sugar concentration was not predicted to be as large as observed in the experiments. Both the discrepancies suggest that the activities of "key enzyme" in model predictions were higher than the real levels. The concentrations of cell mass and ethanol in the model are not very sensitive to the precise enzyme levels and therefore, show a good agreement with the experimental observations. No effort was made to achieve a better fit. It should be noted that the parameters used in these simulations were estimated from steady-state experiments, and we aimed at using the same parameters for new experiments.

These delays and differences will not be observed in batch cultures and in steady-state continuous cultures. In batch cultures, the initial concentrations of key enzyme were assumed in order to obtain the best simulation results. The real process (or the time period) to reach these levels was ignored, although they are dependent on culture history. In steady-state continuous cultures, cells have ample time to synthesize enough enzymes to utilize sugar effectively for growth. The present model, involving enzyme synthesis, essentially introduces an inertia in cellular response to environmental changes. This structure improves the model performance during transients in continuous culture.

There have been few other attempts to model enzyme production under transient conditions. Imanaka et al. (5) have modeled α-galactosidase production during transient response using a model similar to ours. Small changes in dilution rate (0.140–0.142 h^{-1}) were used in shift-up experiments by these workers, and the response of cells to such a small change could be well described by the model. Incorporation of "more structures" in the form of the amount of RNA polymerase, its affinity to promoter (6), synthesis and activation of ribosomes, and so forth (7), may be required to predict the effect of large changes successfully in system behavior during transients.

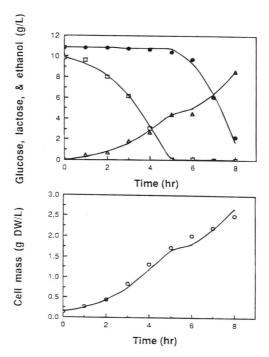

Fig. 3. Batch fermentation of a mixture of glucose (10 g/L) and lactose (10 g/L). Symbols: ●, lactose; □, glucose; ○, cell mass; △, ethanol. Solid lines represent simulation results.

Simulation of Two-Substrate System

The usefulness of the cybernetic model is best realized in predicting the behavior of substrate utilization in a mixed-substrate environment. The kinetic parameters obtained from fermentations on single substrates can be used to simulate ethanol fermentation from a mixture of substrates. Therefore, batch and continuous cultures were conducted with a mixture of glucose and lactose.

A batch fermentation was conducted where glucose-grown cells were used to inoculate a nutrient solution containing 10 g/L glucose and 10 g/L lactose. The experimental results are plotted in Fig. 3. A sequential consumption of the two sugars was observed. Cell growth and ethanol production profiles showed an intermediate lag between two distinct phases. Presence of glucose catabolically repressed utilization of lactose. As glucose in the broth was nearly exhausted, consumption of lactose began. Equations (19)–(22) in the companion article (Part I) with $D = 0$ were solved to simulate the results of this batch experiment. The parameter values used were the ones established from pure substrate experiments (Table 2 of Part I). Since the inoculum was grown on glucose, initial concentrations of the key enzymes were set at 90% of the maximum for glucose and 10% for lac-

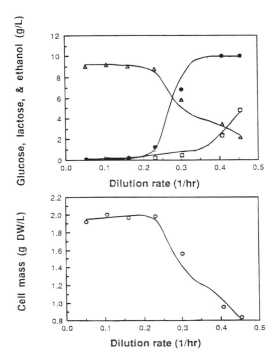

Fig. 4. Chemostat fermentation of a mixture of glucose (10 g/L) and lactose (10 g/L). Symbols: ●, lactose; □, glucose; ○, cell mass; △, ethanol. Solid lines represent simulation results.

tose as described before. The results of simulation are also shown in Fig. 3 as solid lines. The model also predicted a diauxic growth on glucose and lactose. Glucose, which supported faster growth rate, as expected, was consumed first, followed by the utilization of lactose.

Continuous-culture studies involving ethanol production from mixed substrates were also conducted. The feed solution consisted of a mixture of glucose (10 g/L) and lactose (10 g/L). Steady-state values of X, P, and S were measured at different dilution rates, and the data are presented in Fig. 4. Simulation of the continuous culture on mixed substrates was conducted with the help of model Eqs. (19)–(22) presented in Part I. For simulation of continuous cultures, the initial concentrations of variables are not critical, although they influence the time to reach steady states. Parameter values used in the simulation were again those for single sugars listed in Table 2 of Part I. The steady-state results were obtained by simulating the dynamic equations for long periods (about 500 h) until steady state was achieved.

The model satisfactorily predicted fermenter performance in the presence of a mixture of sugars. In contrast to the sequential uptake in batch fermentation, a simultaneous uptake of both sugars was observed at dilution rates below 0.4 h^{-1}. Above this value, only glucose was consumed.

This switch in sugar uptake pattern was also predicted by the cybernetic model utilizing the information obtained from single sugar experiments.

Simulation of Fermentations from Sugars in Cheese Whey

In fermentations of cheese whey, monosaccharides may be present as initial contaminants or as additives to enhance supply of carbon substrates. The presence of mixed sugars can affect the fermentation patterns. Cybernetic models offer a potential tool to simulate and predict these effects. The applicability of the cybernetic approach to such mixed-substrate fermentations was demonstrated through the following experiments and simulations.

In these fermentations, no nutrient supplementation was done. Therefore, the growth characteristics may be different from those on glucose and lactose presented earlier. Therefore, new estimates of certain parameters (μ^e, $Y_{X/S}$, $Y_{P/S}$, and ϕ^e_M) were made from experimental data. The rest of the parameters were kept the same as those identified previously in Table 2 (accompanying article—Part I).

Figures 5A and B show batch fermentation data for two different concentrations of cheese whey powder (CWP). In these experiments, monosaccharides were either not observed or were very low in concentrations (not shown). Therefore, the fermentation may be considered as one having only single substrate (lactose). The experimental results, reported elsewhere (1), for 100 g/L CWP were used to obtain new estimates of growth parameters on lactose in CWP. The inoculum media for fermentations involving cheese whey were prepared from cheese whey (1). Therefore, the initial enzyme levels for glucose and lactose were set at 10 and 90%, respectively, of the maximum enzyme level in parameter estimation and in the following simulation. The estimated parameters along with all the others are tabulated in the second column of Table 2.

Any nutritional variation appears to affect only the maximum specific growth rate, maintenance, and yield parameters, as shown by a good match between predictions (solid lines) and experimental observations, shown in Fig. 5A. The same set of parameters were used to simulate the fermentations with 25 g/L CWP. Results of this simulation are also presented in Fig. 5B. Again, an excellent match is observed for 25 g/L CWP experiment.

To verify the applicability of cybernetic models to the complex system of glucose-supplemented cheese whey, batch fermentations were carried out with a mixture of cheese whey and glucose. In these experiments, 100 g/L CWP solutions were spiked with 10 and 20 g/L glucose, respectively. No additional nutrient supplement was provided. The experimental results are shown in Figs. 6A and B.

Since these fermentations were also not supplemented with nutrient, the growth parameters for glucose can be different from those in Table 2 of Part I. Therefore, one of the experiments (Fig. 6A) was used to estimate μ^e, $Y_{X/S}$, and ϕ^e_M on glucose. The other parameters for glucose and the

Table 2
Kinetic Parameters

Parameter		Glucose[a]	CWP[b]
μ^e		180	93
μ^{max}	(h^{-1})	0.414	0.281
K_{SX}	(g/L)	0.055	0.094
K_{IX}	(g/L)	10,000	370
K_{P1}	(g/L)	7.8	12.1
K_{P2}	(g^2/L^2)	330	330
v^e		688	506
v^{max}	(h^{-1})	1.531	1.531
K_{SP}	(g/L)	0.05	0.088
K_{IP}	(g/L)	10,000	250
K_P	(g/L)	24.9	33.0
ϕ_M^e		2.3	20.0
ϕ_M^{max}	(h^{-1})	0.005	0.06
K_{SM}	(g/L)	10^{-6}	10^{-6}
$Y_{X/S}$	(g/g)	0.532	0.433
$Y_{P/S}$	(g/g)	0.510	0.445

μ^e, v^e, and ϕ_M^e are in units of (g dry wt/[hr./u enzyme activity]).

[a]estimated using data from batch culture on 100 g/L CWP plus 10 g/L glucose.

[b]Estimated using data from batch culture on 100 g/L CWP.

parameters for CWP were kept the same as those in Table 2 of Part I. These estimated parameters are also listed in Table 2 of Part I and used to simulate the experiment of a mixture of 100 g/L CWP and 20 g/L glucose. The predictions are presented in Fig. 6B as solid lines, and an excellent agreement is obtained with experimental data.

Since the maximum rate of substrate consumption for maintenance is relatively low compared to the values of μ^{max} and v^{max}, it is difficult to verify the applicability of Eq. (18) in Part I solely from the results shown in figures. Verification of this equation requires further experimentation under conditions of low growth rates.

CONCLUSIONS

Experimental results from anaerobic fermentations of sugars in semi-synthetic media as well as those from cheese whey (2) have been used in simulation. The extended cybernetic approach can satisfactorily model the

Simulation and Comparison

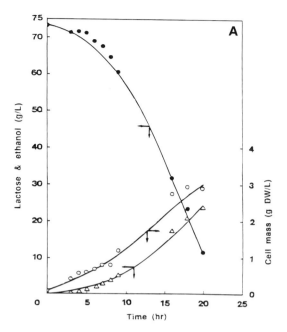

Fig. 5A. Batch fermentation of 100 g/L CWP. Symbols: ●, lactose; ○, cell mass; △, ethanol. Solid lines represent simulation results.

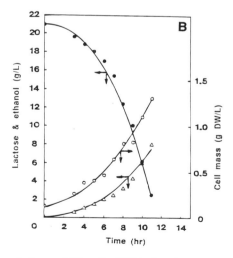

Fig. 5B. Batch fermentation of 25 g/L CWP. Symbols: ●, lactose; ○, cell mass; △, ethanol. Solid lines represent simulation results.

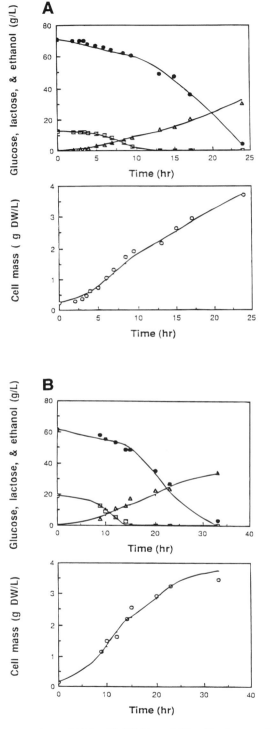

Fig. 6A. Batch fermentation of 100 g/L CWP and 10 g/L glucose. B. Batch fermentation of 100 g/L CWP and 20 g/L glucose. Symbols: ●, lactose; □, glucose; ○, cell mass; △, ethanol. Solid lines represent simulation results.

fermentations of cheese whey/semisynthetic media to ethanol under diverse experimental conditions. These include systems containing single substrates and multiple substrates.

NOMENCLATURE

a	basal enzyme production, unit of enzyme activity/g
D	dilution rate, L/h
e	key enzyme level, unit of enzyme activity/g
$K_{IX,i}$	substrate inhibition constant for cell growth, g/L
$K_{IP,i}$	substrate inhibition constant for product formation, g/L
K_P	product inhibition constant for product formation, g/L
$K_{P1,i}$	product inhibition constants for cell growth, g/L
$K_{P2,i}$	product inhibition constants for cell growth, g^2/L^2
$K_{SX,i}$	substrate saturation constant for cell growth, g/L
$K_{SP,i}$	substrate saturation constant for product formation, g/L
$K_{SM,i}$	substrate saturation constant for maintenance, g/L
P	ethanol concentration in broth, g/L
S	sugar concentration in broth, g/L
t	time, h
X	cell mass concentration in broth, g/L
α	enzyme synthesis constant
β	enzyme degradation constant
δ	cybernetic variable for inhibition/activation of enzyme activity
δ_M	cybernetic variable for uncoupling between catabolism and anabolism
ε	cybernetic variable for induction/repression of enzyme synthesis
μ^{max}	maximum specific rate of cell growth, 1/h
ν^{max}	maximum specific rate of product formation, 1/h
ϕ_M^{max}	maximum specific substrate consumption rate for maintenance, 1/h

REFERENCES

1. Wang, C. J., Jayanata, Y., and Bajpai, R. K. (1987), *J. Fermentation Technology* **65**, 249–253.
2. Somagyi, M. (1952), *J. Biol. Chem.* **195**, 19–23.
3. Michelson, M. L. (1976), *AIChE J.* **22**, 594–597.
4. Kompala, D. S., Ramkrishna, D., Jansen, N. B., and Tsao, G. T. (1986), *Biotechnol. Bioeng.* **28**, 1044–1055.
5. Imanaka, T., Kaieda, T., and Taguchi, H. (1973), *J. Fermentation Technology* **51**, 423–430.
6. Maaloe, O. (1969), *Dev. Biol. Suppl.* **3**, 33–58.
7. Wanner, B. L., Kodaira, R., and Neidhardt, F. C. (1977), *J. Bacteriol.* **130**, 212–222.

Modeling Fixed and Fluidized Reactors for Cassava Starch Saccharification with Immobilized Enzyme

GISELLA M. ZANIN* AND FLÁVIO F. DE MORAES

State University of Maringá, Chemical Engineering Department, Av. Colombo, 5790-BL. E46-09, 87020-900, Maringá, PR, Brazil

ABSTRACT

Cassava starch saccharification in fixed-and fluidized-bed reactors using immobilized enzyme was modeled in a previous paper using a simple model in which all dextrins were grouped in a single substrate. In that case, although good fit of the model to experimental data was obtained, physical inconsistency appeared as negative kinetic constants. In this work, a multisubstrate model, developed earlier for saccharification with free enzyme, is adapted for immobilized enzyme. This latter model takes into account the formation of intermediate substrates, which are dextrins competing for the catalytic site of the enzyme, reversibility of some reactions, inhibition by substrate and product, and the formation of isomaltose. Kinetic parameters to be used with this model were obtained from initial velocity saccharification tests using the immobilized enzyme and different liquefied starch concentrations. The new model was found to be valid for modeling both fixed- and fluidized-bed reactors. It did not present inconsistencies as the earlier one had and has shown that apparent glucose inhibition is about seven times higher in the fixed-bed than in fluidized-bed reactor.

Index Entries: Cassava starch; amyloglucosidase; immobilized enzyme; fluidized bed; controlled pore silica.

INTRODUCTION

In a previous paper *(1)*, the performance of fixed- and fluidized-bed reactors with immobilized amyloglucosidase was examined for the purpose of saccharifying liquefied cassava starch. Data showed that for equal

*Author to whom all correspondence and reprint requests should be addressed.

normalized residence time (enzyme activity per volume multiplied by real residence time) and lower fluid-bed porosities, the fluidized-bed reactor led to higher conversion than fixed bed. The immobilized-bed enzyme (IE) support particle was controlled pore silica (CPS) with 0.5-mm particle diameter and an average pore diameter of 37.5 nm. Interparticle mass transfer experiments showed absence of diffusion limitations, and reactor conversion was modeled by grouping all intermediate dextrins of starch hydrolysis in a single substrate model, which allowed for substrate and product inhibition. Taking the fixed- and fluidized-bed systems as plug flow reactors, this model yielded the simple integrated Eq. (1) of ref. *(1)*.

Conversion-residence time data for both fixed- and fluidized-bed reactors were satisfactorily fitted by this equation. Analysis of the adjusted parameters, however, revealed a negative value for the reaction velocity constant that is not physically meaningful. This result suggested the need for improvements in reactor modeling.

A more realistic cassava starch hydrolysis model was developed and applied initially to a batch reactor containing free enzyme *(2)*. This is a multisubstrate model that considers intermediate dextrins, reversibility of some reactions, substrate and product inhibition, in addition to allowing for isomaltose synthesis from glucose, and competition among dextrins for the active site of amyloglucosidase. This model does not present the aforementioned inconsistency of negative kinetic constants, and showed good agreement with the batch reactor data in the presence or absence of added glucose. Although still relatively easy to use, this model is mathematically more complex than the earlier one, and requires the solution of six ordinary differential equations coupled to five algebrical relations *(2,3)*.

In this article, the later model *(2)* will be adapted for saccharification of cassava starch in fixed- and fluidized-bed reactors of immobilized amyloglucosidase. The earlier data *(1)* will be compared to modeling.

MATERIALS AND METHODS

To avoid extensive repetition, reference is made to information contained in the previous paper *(1)*. There are descriptions to be found of the Enzyme, Substrate, Reactors, Assay Methods, and Conversion Tests.

In addition, described here are the tests used to determine:

1. The influence of immobilized enzyme mass on initial starch hydrolysis rate;
2. Intraparticle diffusion limitations; and
3. Kinetic parameters observed with the immobilized enzyme.

The IE used inside the reactors and in test 2 had a higher amyloglucosidase load (17.3 mg/g_{IE}) and activity (676.8 U/g_{IE}) *(1)*, whereas the IE used with tests 1 and 3 had a load of 4.12 mg/g_{IE} and activity equal to 188.5 U/g_{IE}. Each enzyme unit corresponds to the quantity of enzyme that produces 1 µmol of glucose/min at 45°C, pH 4.5, and with 30% (w/v) liquefied cassava starch as substrate.

Influence of Immobilized Enzyme Mass on Initial Starch Hydrolysis Rate

Since the kinetic parameters to be found for the immobilized enzyme will be determined in a batch microreactor containing a small basket where the IE is retained, it was necessary to know up to what mass of IE could be used inside the basket before reaction limitations would cause the initial hydrolysis rate to deviate from being linearly proportional to the quantity of enzyme used.

This test was conducted with 50 mL of liquefied starch solution, 1% (w/v) at 45°C, and pH 4.5. Immobilized enzyme mass loaded into the basket varied from 0.08–1.0 g, dry wt. Starting from time zero when the loaded basket was dipped into the starch solution, samples were taken at regular intervals of 2 mins. These samples were boiled for 10 min and kept at 4°C until assayed for glucose produced. A linear fit of micromoles of glucose produced vs time gave the initial rate of liquefied starch hydrolysis.

Intraparticle Diffusion Limitations

Arrhenius plots (log of initial rate of hydrolysis reaction vs the inverse of absolute temperature) obtained for the soluble and immobilized enzyme, at the same substrate concentration, allow inference of the presence of intraparticle diffusion limitation from differences that might result in the comparison of the activation energy (E_a) of both cases (4,5).

The experimental test was conducted with 0.5 g (dry wt) of immobilized enzyme and the same procedure described above for the influence of IE mass on initial starch hydrolysis rate, whereas for the soluble enzyme, the same batch reactor was used with 0.5 mL of the stock enzyme, diluted 1 to 500 (2). Two substrate concentrations were tested, namely 1 and 30% (w/v), and temperatures ranged from 35–70°C.

Initial Rate of Saccharification and Substrate Inhibition Test

This test was carried out also with the same procedure described above, but the immobilized enzyme mass was fixed at 0.57 g (dry wt), and starch concentration varied from 2–300 g/L (total of 24 determinations, with new enzyme for each case). Total reaction time decreased from 30 min used for more concentrated solutions down to 16 min for more diluted ones, so that the rate of hydrolysis in each case could be considered constant and equal to the initial rate of hydrolysis at that starch concentration.

Plotting the initial rate of hydrolysis vs starch concentration shows the influence of substrate inhibition. The same data treated as described in ref. (2), Eq. (23), and the fitting description below it yield the Haldane parameters: K_{m4} (Michaelis-Menten constant for the susceptible oligosaccharides), V_{m4} (maximum velocity constant for the hydrolysis of the susceptible oligosaccharides), and K_S (substrate inhibition constant).

Fixed- and Fluidized-Bed Reactor Modeling

The multisubstrate model that was developed for batch saccharification of liquefied cassava starch with soluble amyloglucosidase (2) is here adapted for immobilized enzyme and continuous fixed- or fluidized-bed reactors. The main features of the model are preserved, and only minor modifications are necessary.

Earlier work (6) has shown that for the liquid fluidized beds of these systems, liquid axial dispersion is of intermediate level (dispersion number < 0.03), and therefore when used as a reactor, the performance of these fluidized beds will not deviate significantly from a plug flow reactor (PFR) (7a). Fixed-bed reactors with the same range of variables should have even smaller dispersion. Consequently, both fixed- and fluidized-bed reactors with immobilized amyloglucosidase have been modeled as a PFR.

For constant density systems, as in the case of starch saccharification, both batch and PFR reactors lead to equivalent performance equations, in which the only difference is that where reaction time appears in the batch equation, residence time appears in the case of PFR (7b). For this reason, the same computer program developed for saccharification of cassava starch in a batch reactor (2) has been used here with the fixed- and fluidized-bed reactors. To calculate the real residence time (t_R) in these reactors, the interstitial liquid volume (V_L) was considered the relevant volume.

$$t_R = V_L/v = (V_T\varepsilon)/v = M_e\varepsilon/[\rho_p(1-\varepsilon)v] \qquad (1)$$

The parameters of the model related to maximum initial hydrolysis velocities now are given in units of mol/(h g_{IE}), and therefore the ratio of immobilized enzyme to liquid volume inside the reactor (E) is an important experimental datum. Interstitial liquid volume was also considered the relevant liquid volume to calculate this ratio. It is constant for fixed bed, since a fixed bed has a constant volume, but is different for each bed porosity (ε) in the fluidized bed, because this bed has higher expansions for greater flow rates.

$$E = M_e/V_L = M_e/(V_T\varepsilon) = \rho_p(1-\varepsilon)/\varepsilon \qquad (2)$$

The apparent glucose inhibition parameter remains as the only adjustable parameter to make the model fit the reactor data.

Model Hypothesis

1. Since the substrate is starch that has been extensively hydrolyzed with α-amylase, the average degree of polymerization (n) of the remaining oligosaccharides with degree of polymerization > 3, was assumed to be 5, based on Reilly's description of α-amylase action (8).
2. These oligosaccharides are considered to be formed of two fractions called G_4 and G_6. G_4 is more susceptible to hydrolysis, comprises approx 77% of the molecules, and contains the α-1,4 chemical bonds that are rapidly hydrolyzed. The second fraction, G_6, is resistant to hydrolysis, comprises the remaining 23% of the molecules, and contains the α-1,6 chemical bonds associated with branching that are hydrolyzed at a slower rate than the α-1,4 bonds (9).
3. Saccharification proceeds through multiple reactions that occur simultaneously and are divided into three classes:

    a. Hydrolysis reactions of oligosaccharides with degree of polymerization (n) > 3 are lumped together as a single class that is divided in the two aforementioned fractions—susceptible (G_4) and resistant (G_6):

    $$\text{Susceptible oligos. } (G_4) + (n-3)\, H_2O \rightarrow \text{maltotriose } (G_3) + (n-3) \text{ glucose } (G) \quad (3)$$

    $$\text{Resistant oligos. } (G_6) + (n-3)\, H_2O \rightarrow \text{maltotriose } (G_3) + (n-3) \text{ glucose } (G) \quad (4)$$

    b. Hydrolysis of maltotriose (G_3) produces maltose (G_2) and is reversible:

    $$\text{Maltotriose } (G_3) + H_2O \rightleftarrows \text{maltose } (G_2) + \text{glucose } (G) \quad (5)$$

    c. Hydrolysis of maltose produces glucose (G) and is reversible:

    $$\text{Maltose } (G_2) + H_2O \rightleftarrows 2 \text{ glucose } (G) \quad (6)$$

4. Glucose can undergo condensation to isomaltose (G_I). This reaction is reversible:

    $$2 \text{ glucose } (G) \rightleftarrows \text{isomaltose } (G_I) \quad (7)$$

5. There is product inhibition (K_i) in the case of glucose (G), for reactions given by Eqs. (3)–(6).
6. There is substrate inhibition (K_S) by oligosaccharides with degree of polymerization >3, but not for maltotriose and maltose.

7. The various substrates (G_4, G_6, G_3, G_2) compete for the amyloglucosidase active site.
8. During saccharification, water in the reaction medium is in excess, and therefore, its concentration is assumed constant.
9. Thermal deactivation of the enzyme was not considered in this work, since at 45°C and 30% substrate concentration, it was observed that amyloglucosidase stability is high (3).
10. The reactors are modeled by the isothermal, plug flow, one-dimensional pseudohomogeneous model.

Hydrolysis Rate and Mass Balance Equations

Given the aforementioned considerations, Eqs. (8)–(21) apply:

1. Hydrolysis rate for the susceptible oligosaccharides (r_4):

$$r_4 = V_{m4} E\, G_4 / [K_{m4}(1 + G/K_i) + G_4 \\ + G_2 K_{m4}/K_{m2} + G_3 K_{m4}/K_{m3} + G_6 K_{m4}/K_{m6} + G_4^2/K_S] \tag{8}$$

2. Hydrolysis rate for the resistant oligosaccharides (r_6):

$$r_6 = V_{m6} E\, G_6 / [K_{m6}(1 + G/K_i) + G_6 \\ + G_2 K_{m6}/K_{m2} + G_3 K_{m6}/K_{m3} + G_4 K_{m6}/K_{m4} + G_6^2/K_S] \tag{9}$$

3. Rate of consumption of susceptible (G_4) and resistant (G_6) oligosaccharides:

$$dG_4/dt_R = -r_4 \tag{10}$$

$$dG_6/dt_R = -r_6 \tag{11}$$

4. Rate of hydrolysis of maltotriose (r_3):

$$r_3 = V_{m3}\, E\, (G_3 - G_2 G/K_{eq3})/[K_{m3}(1 + G/K_i) + G_3 + G_2 K_{m3}/K_{m2} + G_4 K_{m3}/K_{m4} + G_6 K_{m3}/K_{m6}] \tag{12}$$

At equilibrium:

$$K_{eq3} = (G_{2eq} G_{eq})/G_{3eq} \tag{13}$$

5. Rate of hydrolysis of maltose (r_2):

$$r_2 = V_{m2} E(G_2 - G^2/K_{eq2})/[K_{m2}(1 + G/K_i) + G_2 + G_3 K_{m2}/K_{m3} + G_4 K_{m2}/K_{m4} + G_6 K_{m2}/K_{m6}] \tag{14}$$

At equilibrium:

$$K_{eq2} = G_{eq}^2 / G_{2eq} \tag{15}$$

6. Net rate of formation of maltotriose and maltose:

$$dG_3/dt_R = r_4 + r_6 - r_3 \tag{16}$$

$$dG_2/dt_R = r_3 - r_2 \tag{17}$$

7. Rate of formation of isomaltose (r_I) resulting from the α-1,6 condensation of two molecules of glucose (Eq. [7]):

$$r_I = V_{IM}E(G^2 - G_I/K_{eqI}) \tag{18}$$

At equilibrium:

$$K_{eqI} = G_{Ieq}/G_{eq}^2 \tag{19}$$

$$dG_I/dt_R = r_I \tag{20}$$

8. Net rate of glucose formation:

$$dG/dt_R = (n-3)(r_4 + r_6) + r_3 + 2(r_2 - r_I) \tag{21}$$

Conversion of the multiple substrates present in the liquefied starch solution (G_2, G_3, G_4, G_6) was calculated on a mass basis by the following equation:

$$X_A = 100f(C_g - C_{gi})/[C_{40} - f(C_{gi} - C_g)] \tag{22}$$

RESULTS AND DISCUSSION

Influence of Immobilized Enzyme Mass on Initial Starch Hydrolysis Rate

Figure 1 shows the initial starch hydrolysis rate as a function of the mass of immobilized enzyme used inside the basket. It can be observed that a linear relation holds up to 0.6 g of IE, equivalent to approx 50 mg/L of immobilized enzyme protein. Although starch concentration is relatively low (1% w/v) it is sufficient to ensure complete saturation of the enzyme's active site (200 mg of liquefied starch/mg of protein). These results allow one to select 0.57 g (dry weight) as the IE mass used to determine the initial rate of saccharification and substrate inhibition. Since starch concentration varied 2 to 300 g/L, an IE mass near to the upper limit is a good choice so that high initial velocities would be obtained also for the smaller starch concentrations.

Intraparticle Diffusion Limitations

Figure 2A show Arrhenius plots for the soluble and immobilized enzyme with 1% (w/v) starch concentration, whereas Fig. 2B is for 30% (w/v). The energy of activation (E_a) for each concentration is nearly coincident for both free and IE, and that indicates absence of intraparticle diffu-

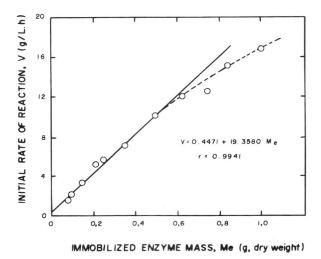

Fig. 1. Initial rate of saccharification of liquefied cassava starch by immobilized amyloglucosidase, 45°C, pH 4.5, 50 mL of substrate with starch concentration 1% (w/v), reaction time 16 min. Enzyme activity = 188.5 U/g IE.

sion limitation, but nevertheless, E_a is different for each concentration, namely about 9 and 13 kcal/mol respectively for 1 and 30% (w/v) starch. Since Arrhenius equation is only strictly valid for single elementar reactions (10), the energy of activation derived from Arrhenius plot for complex kinetics may be a function of temperature and concentration. This has also been shown to be true for some reactions believed to be elementar (11,12). The absence of intraparticle diffusion limitation is attributed to the large and regular pores of the CPS particle used to immobilize the enzyme and also to the slowness of the liquefied starch hydrolysis reaction at 45°C.

Initial Rate of Saccharification and Substrate Inhibition Test

Figure 3 is a plot of the initial rate of saccharification with IE vs concentration of the liquefied cassava starch. It shows clearly the presence of substrate inhibition for starch concentrations above 30 g/L. Maximal reaction velocity is observed around 50 g/L, whereas at 300 g/L (typical industrial starch concentration), the rate of saccharification is 23% lower. For the soluble enzyme (2), maximum rate occurred at 90 g/L, and at 300 g/L, the rate was 25% lower. As mentioned in the methodology and described in ref. (2), the Haldane parameters: Michaelis-Menten constant for the susceptible oligosaccharides (K_{m4}), maximum velocity constant for the hydrolysis of susceptible oligosaccharides (V_{m4}), and the substrate inhibition constant (K_S) were calculated from this data. Since the enzyme used in the reactors is 3.59 times more active than the enzyme used in this test, V_{m4} was corrected accordingly. The value of maximum rate of hydrolysis for resistant oligosaccharides (V_{m6}) was then taken as half V_{m4}. Maximum

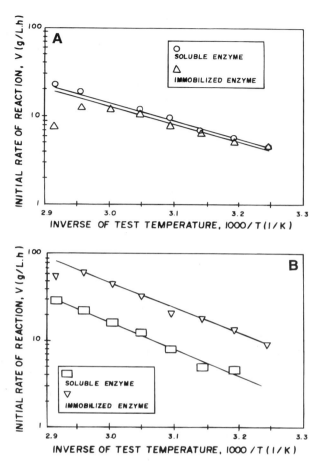

Fig. 2. Arrhenius plot for the saccharification reaction of liquefied cassava starch with soluble and immobilized amyloglucosidase. **(A)** Starch concentration, 1% (w/v); **(B)** starch concentration, 30% (w/v). For both cases, the pH was 4.5. Enzyme loading: soluble—0.5 mL of stock enzyme diluted 1 to 500; immobilized enzyme 0.5 g (dry wt).

velocities for maltose (V_{m2}) and isomaltose (V_{IM}) were taken from ref. (2) and expressed in terms of g_{IE} by multiplying them by the ratio of enzyme activities between the immobilized and soluble enzymes (676.8 U/g_{IE}/5358.4 U/mL). Maltotriose maximum velocity (V_{m3}) was then taken as twice the value of V_{m2}. The ratios between V_{m6} and V_{m4}, and V_{m3} and V_{m2} were taken from the literature (9). Table 1 shows the complete set of parameters used to model the reactor data.

Fixed- and Fluidized-Bed Reactor Modeling

Figures 4 and 5 compare reactor performance data with the adapted multisubstrate model and show a very satisfactory agreement. The fluidized-bed reactor data (Fig. 5) are fitted with the same product inhibition

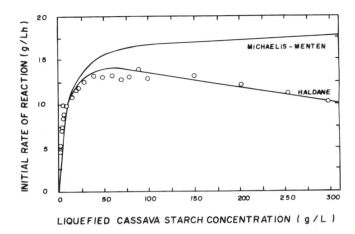

Fig. 3. Initial rate of liquefied cassava starch saccharification with immobilized amyloglucosidase, 45°C, pH 4.5. Reactor of basket type, 50-mL volume. Immobilized enzyme mass 0.57 g (dry wt); substrate concentration varied from 2–300 g/L.

Table 1
Complete Set of Kinetic Parameters Used with the Multisubstrate Model of Liquefied Cassava Starch Saccharification by Immobilized Amyloglucosidase

	Parameter	Parameter Values
Equilibrium constants	K_{eq2}	200 mol/L
	K_{eq1}	0.0544 mol/L
	K_{eq3}	8 mol/L
Michaelis-Menten constants	K_{m2}	0.0030 mol/L
	K_{m3}	0.004916 mol/L
	K_{m4}	0.004916 mol/L
	K_{m6}	0.004916 mol/L
Inhibition constants	K_S	0.2395 mol/L
	K_i (fixed bed)	0.030 mol/L
	K_i (fluidized bed)	0.200 mol/L
Maximum velocities	V_{m2}	0.009814 mol/h g_{IE}
	V_{m3}	0.019630 mol/h g_{IE}
	V_{m4}	0.01587 mol/h g_{IE}
	V_{m6}	0.007940 mol/h g_{IE}
	V_{IM}	0.00001378 L²/mol g_{IE}

constant (K_i = 0.200 mol/L), whereas for the fixed-bed reactor data (Fig. 4), K_i is 0.030 mol/L. This result reveals a much stronger apparent inhibition in the fixed bed than in the fluidized bed.

Since inherent immobilized enzyme inhibition constants should be independent of the reactor type, diffusion limitations were investigated as

Cassava Starch Saccharification

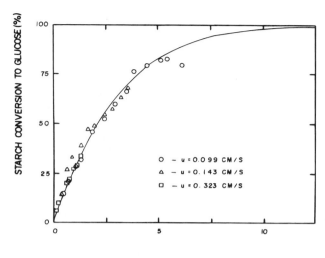

Fig. 4. Comparison of fixed-bed reactor data and reactor modeling for the saccharification of liquefied cassava starch with immobilized amyloglucosidase. Average bed porosity: 0.418. Starch concentration solution 30% (w/v), 45°C, pH 4.5. (u = superficial liquid velocity, ε = bed porosity)

possible causes for the difference in the apparent glucose inhibition values. However, previous work (1) has shown absence of interparticle diffusion limitation, and the data on Energy of Activation obtained in this work have indicated absence of intraparticle diffusion limitation. Therefore, the difference observed in apparent glucose inhibition with both types of reactors incorporates the effects of the macroscopic mixing differences of both reactors. Whereas in the fixed bed, particles are stationary and there is very little liquid axial mixing, in the fluidized bed, particles are free to move and engage in an overall slow recirculation pattern, leading to an axial liquid–solid mixing of intermediate value. For common reactions of positive order, macroscopic mixing in the reactors is detrimental to reactor performance, but for reactions with substrate inhibition, some axial mixing is beneficial because it lowers substrate concentration and, consequently, its inhibition. Channeling may also contribute to inferior performance of fixed bed IE reactors (13).

CONCLUSIONS

The conclusions reached in this work are:

1. In the saccharification of liquefied cassava starch with immobilized amyloglucosidase, a strong substrate inhibition effect is observed for starch concentration above 30 g/L.

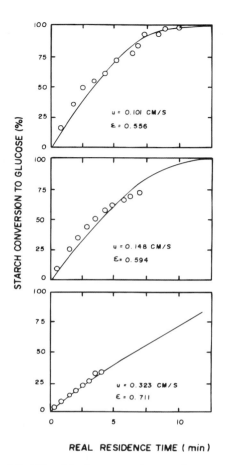

Fig. 5. Comparison of fluidized-bed reactor data and reactor modeling for the saccharification of liquefied cassava starch with immobilized amyloglucosidase. Starch concentration solution 30 % (w/v), 45°C, pH 4.5. (u = superficial liquid velocity, ε = bed porosity).

2. The multisubstrate starch hydrolysis model, previously developed for batch saccharification of liquefied cassava starch with soluble enzyme, was easily adapted for continuous fixed- and fluidized-bed reactors, giving very satisfactory agreement with experimental data.
3. Diffusional limitation, both inter- and intraparticle, was not observed for the reactor flow conditions used in this work and for the type of controlled pore silica of large pores used as support for enzyme immobilization.
4. The only parameter that had to be adjusted in the multisubstrate model to cope with experimental differences in fixed and fluidized bed results was apparent glucose inhibition.

5. For good modeling, fixed bed required a much lower apparent glucose inhibition constant than fluidized bed, and that means a much higher apparent glucose inhibition (about seven times). The difference should not result from differences in the intrinsic kinetic parameter in each bed, but would more likely be a consequence of the different macroscopic liquid mixing of each reactor type.

ACKNOWLEDGMENTS

The authors thank the financial support received of FINEP and the State University of Maringá. The companies that supplied materials (COPAGRA, NOVO, Corning Glass Works) are also acknowledged.

NOMENCLATURE

C_{A0} initial starch concentration, 300 g/L
C_g glucose concentration, g/L
C_{ga} concentration of added glucose, g/L
C_{gi} glucose concentration at the start of saccharification, g/L
E   mass of immobilized enzyme per volume of interstitial liquid, g_{IE}/L
f   ratio of molecular weights for the anhydroglucose unit in starch and glucose, f = 162/180 = 0.9
G   glucose molar concentration, mol/L
G_2 maltose molar concentration, mol/L
G_3 maltotriose molar concentration, mol/L
G_4 susceptible oligosaccharides molar concentration, mol/L
G_6 resistant oligosaccharides molar concentration, mol/L
G_I isomaltose molar concentration, mol/L
G_{eq} G_{2eq}, G_{3eq}, G_{Ieq} equilibrium molar concentration for glucose, maltose, maltotriose, and isomaltose, respectively, mol/L
K_{eq2} K_{eq3}, K_{eqI}, equilibrium constants for maltose (mol/L), maltotriose (mol/L), and isomaltose (L/mol), respectively
K_i apparent glucose inhibition constant, mol/L
K_{m2} K_{m3}, K_{m4}, K_{m6}, Michaelis-Menten constants for maltose, maltotriose, susceptible oligosaccharides, and resistant oligosaccharides, respectively, mol/L
K_S substrate inhibition constant, mol/L
M_e immobilized enzyme mass, dry wt, g
n   average degree of polymerization, dimensionless
r_2 r_3, r_4, r_6, r_I, rate of reaction for maltose, maltotriose, susceptible oligosaccharides, resistant oligosaccharides, and isomaltose, respectively, mol/(L h)

t_R fluid real residence time, h
v volumetric liquid flow rate, cm^3/s
V initial rate of glucose production, g/(L h)
V_L interstitial liquid volume, $V_L = V_T (1 - \varepsilon)$, cm^3
V_{IM} second-order rate constant for isomaltose, L$_2$ (mol·h·mL of enzyme)
V_{m2} V_{m3}, V_{m4}, V_{m6}, maximum velocity constants associated with the reaction rate of maltose, maltotriose, susceptible oligosaccharides, and resistant oligosaccharides, respectively, mol/(h gIE)
V_T total reactor bed volume, $V_T = M_e / [\rho_p(1 - \varepsilon)]$, cm^3
X_A conversion of liquefied starch to glucose, %
ε bed porosity
ρ_p particle density, $\rho_p = 0.939$ g/cm^3

REFERENCES

1. Zanin, G. M. and de Moraes, F. F. (1994), *Appl. Biochem. Biotechnol.* **45/46,** 627–640.
2. Zanin, G. M. and de Moraes, F. F. (1996), *Appl. Biochem. Biotechnol.* **57/58,** 617–625.
3. Zanin, G. M. (1989), Sacarificação de amido em reator de leito fluidizado com enzima amiloglicosidase imobilizada. Ph.D. thesis, Universidade Estadual de Campinas, Campinas-SP, Brasil.
4. Engasser, J. M. and Horvath, C. (1976), in *Applied Biochemistry and Bioengineering*, vol. 1, Wingard, L. B., Jr., Katchalski-Katzir, E., and Goldstein, L., eds., Academic, New York, p. 182.
5. Pitcher W. H., Jr., (1975), in *Immobilized Enzyme for Industrial Reactors*, Messing, R. A, ed., Academic, New York, p. 170.
6. Zanin, G. M., Neitzel, I., and de Moraes, F. F. (1993), *Appl. Biochem. Biotechnol.* **39/40,** 477–489.
7. Levenspiel, O. (1972), in *Chemical Reaction Engineering*, 2nd ed., John Wiley, New York, (a) p. 287; (b) p. 111.
8. Reilly, P. J. (1985), in *Starch Conversion Technology*, Van Beynum, G. M. A. and Roels, J. A., eds., Marcel Dekker, New York, pp. 101–114.
9. Marc, A. (1985), Cinetique et modelisation de reacteurs a glucoamylase soluble et immobilisee. Ph.D. thesis, Inst. Nat. Polytechnique de Lorraine, France.
10. Froment, G. F. and Bischoff, K. B. (1990), in *Chemical Reactor Analysis and Design*, John Wiley, New York, p. 36.
11. Nauman, E. B. (1992), in *Chemical Reactor Design*, Krieger Publishing, Malabar, pp. 89–93.
12. Eyring, H., Lin, S. H., and Lin, S. M. (1980), in *Basic Chemical Kinetics*, John Wiley, New York, pp. 196–197.
13. O'Neill, S. P., Dunnill, P., and Lilly, M. D. (1971), *Biotechnol. Bioeng.* **13,** 337–352.

Fumaric Acid Production in Airlift Loop Reactor with Porous Sparger

JIANXIN DU,*,[1] NINGJUN CAO,[1] CHENG S. GONG,[1] GEORGE T. TSAO,[1] AND NAIJU YUAN[2]

[1]*Laboratory of Renewable Resources Engineering, 1295 Potter Center, Purdue University, West Lafayette, IN 47907; and [2]Department of Chemical Engineering, Tsinghua University, Beijing, P. R. China, 100084*

ABSTRACT

Airlift loop reactors with porous spargers were investigated and used in the process of fumaric acid production by *Rhizopus oryzae* ATCC 20344. In order to enhance oxygen mass transfer, which is very important for organic acid production, two kinds of porous spargers (stainless steel membrane tube and porcelain tube) were examined. Gas holdup, liquid circulation velocity, mixing time, bubble size, and bubble rise velocities were measured in a 50 L rectangular airlift loop reactor with different ratios of the cross-sectional area of the riser and downcomer. The local volumetric mass transfer coefficient ($K_L a$) was also measured in the gas sparger zone. The results indicated that high $K_L a$ and excellent hydrodynamics can be obtained in the airlift loop reactor with a porous sparger. A 10 L laboratory airlift loop reactor was employed for the fumaric acid fermentation. Results showed that the turbulence of two-phase flow in the airlift loop reactor not only produced favorable conditions for mass transfer, but was also useful for forming and suspending small, well-distributed mycelial pellets (1~2 mm). A production rate of up to 0.814 g/L/h and efficiency yield of 50.1% (w/w) was obtained in the airlift loop reactor. The performance was compared with the typical stirred tank fermentor fermentation results.

Index Entries: Airlift loop reactor; fumaric acid; bubble characteristic; mass transfer coefficient; porous sparger, mycelial pellet; *Rhizopus*.

*Author to whom all correspondence and reprint requests should be addressed.

INTRODUCTION

Fumaric acid is a naturally occurring four-carbon dicarboxylic acid that has many potential applications, such as in the manufacture of unsaturated polyester resins in furniture lacquers, as a food acidulent, and in quick-setting inks (1). Fumaric acid can be produced either chemically or biologically. Many strains of mycelial fungi, especially those belonging to the genus *Rhizopus*, are known to produce appreciable quantities of fumaric acid by utilizing either glucose or xylose (2,3). Because of the ability of *Rhizopus* to fix carbon dioxide and combine it with the metabolic intermediate, pyruvic acid, the high weight yield of up to 93% of fumaric acid from glucose can be achieved.

Fumaric acid is known to be a strong inhibitor of its own production. Therefore, in a typical fumaric acid fermentation, calcium carbonate is usually added to neutralize the acid produced and maintain the fermentation broth at a controlled pH. Calcium fumarate, instead of fumaric acid, is obtained. When fumaric acid production is developed in the conventional stirred tank reactor, the formation of calcium fumarate can cause a drastic change of hydrodynamic conditions of the fermentor. This can result in the premature termination of operation caused by the failure of agitators. On the other hand, the cells of *Rhizopus* tend to grow into large mycelial pellets or clumps, which result in the limitation of oxygen and substrate mass transfer, and encouraged ethanol instead of fumaric acid formation. Fumaric acid production is essentially an aerobic process; it is recognized that the metabolic pathway may shift more toward ethanol formation under oxygen-starved conditions and result in low fumaric acid yield. Jiang (4) found that with the increasing air flow rate, the fumaric acid production rate increased, and corresponding ethanol production decreased considerably in a 1 L stirred tank reactor. It is important that the choice and/or design of a suitable bioreactor for fumaric acid production should place emphasis on better hydrodynamics and high oxygen mass transfer rate in order to obtain a high yield and productivity.

The airlift loop reactor has been widely investigated in recent years. It offers several advantages over the traditional bubble column and stirred tank reactor, including the ease of construction and operation, lower energy consumption, lower shear rates, as well as its good hydrodynamics and mass transfer characteristics. Therefore, this type of reactor has been widely employed in industrial applications that require high mass transfer and good dispersion; it is more notable in large scale fermentation industries (5–7). Previously, it has been shown that it is possible to suspend solid particles with a density as high as 2800 kg/m^3 at a solid content of 300 g/L in an airlift loop reactor (8). This indicates that airlift loop reactors are suitable for cultures containing solid particles and mycelial pellets.

Some researchers have studied the cultivation of mycelial fungi in airlift loop reactors. Trager et al. *(9)* used a simple laboratory airlift fermentor to produce gluconic acid with *Aspergillus niger*, and found that smaller pellets were obtained as a result of low stress conditions, as compared to the stirred tank. Okabe et al. *(10)* used an airlift bioreactor in which the top and bottom of the draft tube were covered with stainless steel sieves (four meshes) for the production of itaconic acid by *Aspergillus terreus*. The results showed that the itaconic acid production rate was 100% greater than that of a jar fermentor.

In this study, the authors examined two laboratory-constructed airlift reactors equipped with internal loop and porous gas spargers with 10 and 50 L of working volume, respectively. In the 50 L reactor, bubble characteristics of different regions were measured by means of a dual-probe automatic measurement system that was controlled by an on-line computer. The experimental measurements comprised bubble size, bubble rising velocity, and gas holdup in the riser and downcomer. In order to obtain a better understanding of the hydrodynamics, the liquid circulation velocity, the mean circulation time, and the mixing time were systematically measured in a broad operation range, using the conductive tracking approach. The mass transfer coefficients ($K_L a$) in the gas sparger zones were also studied using a dissolved-oxygen probe to demonstrate the advantages of porous spargers. In addition, fumaric acid production was carried out in the 10 L airlift loop reactor with controlled temperature; results obtained were compared to those with stirred tank reactor.

MATERIALS AND METHODS

Description of Airlift Loop Reactors

50 L Airlift Loop Reactor

The configuration of an airlift loop reactor with a 50 L working volume for the hydrodynamic measurements is shown in Fig. 1. This reactor was constructed with polymethyl methacrylate that has a rectangular cross-section for the installation of conductivity probes at different axial positions. A baffle divides the rectangular part into two sections: the riser and the downcomer. The position of the baffle is designed to be mobile so the ratios of different cross-sectional areas of the riser and downcomer can be obtained in one reactor. The top part (head), where gas–liquid separation occurs, is designed for good circulation and mixing of gas and liquid. The head portion of the reactor, the so-called gas-liquid separator, can be operated in either the closed or opened modes, as shown in Fig. 1. At the bottom of the reactor, a stainless steel membrane tube (pore size, 30–40 µm; diameter, 40 mm; and length, 120 mm) was used as the gas sparger, which can distribute air evenly into the reac-

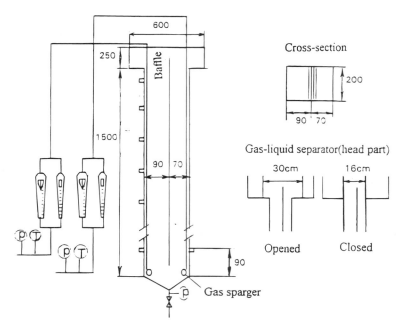

Fig. 1. The configuration of 50 L airlift loop reactor. *See* Table 1 for dimensions.

tor as tiny air bubbles of 1~2 mm in diameter. Dimensions of the reactor are given in Table 1.

10 L Airlift Loop Reactor

The configuration of the 10 L working volume airlift loop reactor with attachments for the fumaric acid production is shown in Fig. 2. This 10 L reactor was constructed of Pyrex glass pipe and equipped with stainless steel fittings. A concentric draught tube inside the reactor was used as the riser section to form an inner gas-liquid loop. This reactor contains several ports for installing the dissolved oxygen, pH probes, and temperature sensor for addition of nutrient medium, antifoam, neutralization agent, and for removing exhaust gas. By circulating water through a heat exchanger inside the reactor, the temperature during fermentation can be controlled. A porcelain tube (pore size, 80 µm; diameter, 40 mm; and length, 50 mm. Fisher Scientific Co.) was inserted into the bottom of the reactor as the air sparger. The dimensions of this 10 L reactor are also given in Table 1.

Dual-Probe Conductivity Sensor

The schematic diagram of a dual-probe conductivity sensor made in the laboratory is shown in Fig. 3. The use of the conductivity probe has many advantages for measuring the bubble characteristics in two-phase flow, including fast response, high sensitivity, ease of operation, and ease

Table 1
Geometrical Characteristics of Airlift Loop Reactors (ALRs)

Geometrical Parameter	ALR for Hydrodynamics Measurement	ALR for Fumaric Acid Production
Working Volume	50 L	10 L
Cross-Sectional Shape	Rectangular	Cylindrical
Total Reactor Length	1750 mm	1165 mm
Cross-Sectional Size	Length: 200 mm	Diameter: 100 mm
	Width: 160 mm	
Length of Baffle	1500 mm	-
Length of Draft Tube	-	900 mm
Diameter of Draft Tube	-	Diameter: 80 mm
$S_r/S_d{}^a$	1.07	1.78
	1.58	
	2.44	
	4.17	
Gas-Liquid Separator	Closed[b]: 200×160×250 mm^3	Diameter: 153 mm
	Opened[c]: 200×300×250 mm^3	Height: 250 mm
Air Sparger	Stainless Steel Membrane Tube	Porcelain Membrane Tube

[a]Ratio of cross-sectional area of riser and downcomer.

[b]Means gas-liquid separator has the same cross-sectional area as the main part of the reactor.

[c]Means gas-liquid separator has larger cross-sectional area than that of the main part of the reactor.

of control by a computer (11). Because of its high hardness, tungsten filament (diameter = 8×10^{-5} mm) was selected as the electrode. Glass capillary (diameter = 0.3 mm) was employed as an insulating sleeve, that made the drying of the probe easier. The measurement results indicate that this dual-probe sensor can measure tiny bubbles as small as 0.6 mm in diameter with satisfactory stability and reproducibility (11). Using this type of dual-probe automatic measurement system controlled by an on-line computer in an airlift loop reactor has not been reported in the literature.

Micro-organism and Inoculum

Rhizopus oryzae ATCC 20344, purchased from American Type Culture Collection (Rockville, MD), was chosen for this study because of its ability to produce fumaric acid from glucose (12). The culture was propagated on YMA (Difco) agar plates. After the spores formed, the agar plates were

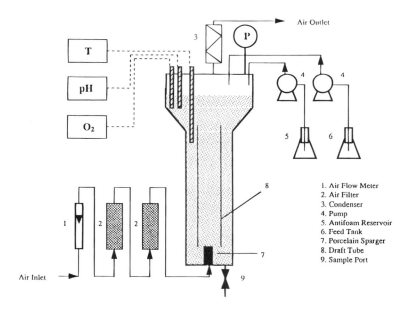

Fig. 2. Schematic diagram of airlift loop reactor (10 L) for fumaric acid production.

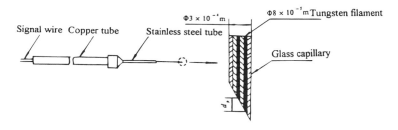

Fig. 3. Schematic diagram of dual-probe conductivity sensor.

maintained at 4°C. For inoculation, the agar plates containing sporangiospores were washed with sterile water to obtain the spore suspension. The spores were then collected by filtration and used as inocula.

Cultivation

Because *Rhizopus oryzae* grows slowly using xylose as the carbon source, small pellets can be formed in the medium containing xylose *(13)*. In order to obtain small- and uniform-sized pellets in the airlift loop reactor, preculture was performed in the flasks on the shaker. Cultivation medium consisted of 50 g xylose, 2 g urea, 0.6 g KH_2PO_4, 0.25 g $MgSO_4 \cdot 7H_2O$, and 0.088 g $ZnSO_4 \cdot 7H_2O$, per liter of distilled water. The spore solution was inoculated into 2500-mL Erlenmeyer flasks containing 1000 mL sterilized cultivation medium. Incubation was carried out at

30°C and 150 rpm in a gyratory shaker for about 2 d. The mixture of filamentous fungi and small fluffy pellets with hairy surfaces were obtained for fermentation studies.

Fermentation

Mycelial pellets from flask cultures were harvested and transferred into the 10 L airlift loop reactor containing fermentation medium that consisted of 100 g glucose, 0.5 g urea, 0.6 g KH_2PO_4, 0.25 g $MgSO_4 \cdot 7H_2O$, and 0.088 g $ZnSO_4 \cdot 7H_2O$, in one L distilled water. The fermentation was operated at 35°C with an air flow rate of 8.5 L/min (about 1 vvm). $CaCO_3$ was added whenever needed, to maintain pH in the broth at around 5.

Analytical methods

High Performance Liquid Chromatography (HPLC)

High Performance Liquid Chromatography (HPLC) with an RI detector, an automatic injector, and an integrator (Hitachi, Tokyo, Japan) was used to determine sugar, fumaric acid, and by-product concentrations. The mobile phase was $0.005M$ H_2SO_4 at a flow rate of 0.8 mL/min through a BioRad HPX-87H Ion-Exclusion column (BioRad Laboratory, Hercules, CA) at 60°C.

Final Fumaric Acid Concentration

Because of the low solubility of calcium fumarate (~25 g/L at 35°C), fumarate may precipitate during fermentation. In order to obtain an accurate total amount of the fumaric acid produced, water was added to the broth after fermentation until the fumarate concentration was lower than its solubility. This was followed by heating the broth for 1 h at 80°C. Samples were collected for analysis.

RESULTS AND DISCUSSION

Bubble Characteristics in Airlift Loop Reactor

Previous studies on the hydrodynamics of the airlift loop reactor have been based mostly on the concept of viewing the reactor as a whole. Very little attention was focused on the local properties that resulted in many problems that hindered further applications of the airlift reactor in the industry. For this reason, local gas holdup, bubble size, and bubble velocity were measured by a dual-probe conductivity sensor in both the riser and the downcomer in the 50 L airlift loop reactor.

In an air-water system, when a single nozzle sparger or perforated ring sparger was employed, large spherical-capped bubbles (5~10 mm in diameter) were predominant in the riser section, and many larger bubbles were trapped in the downcomer as well. As the superficial air velocity increased, bubbles in both the riser and the downcomer coalesced rapidly

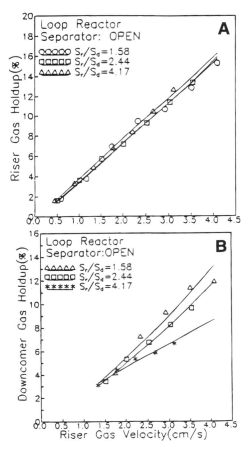

Fig. 4. Effects of superficial air velocity on the mean gas holdup in 50 L airlift loop reactor. **(A)** The riser. **(B)** The downcomer.

to form large bubbles. Some of the bubbles can be as large as the width of the reactor. In this study, very small and uniform bubbles were formed from the porous holes of the stainless steel membrane sparger. Bubble diameters 1–5 mm were observed along the length of the riser.

Gas holdup is one of the most important parameters characterizing the performance of the airlift loop reactor. Effects of superficial air velocity on the mean gas holdup in the riser and the downcomer is shown in Fig. 4a and 4b. It can be seen that gas holdup in the riser displayed the same principle at different riser-downcomer cross-sectional area ratios (S_r/S_d); a linear increase in gas holdup was found with an increasing superficial gas velocity. In the downcomer, gas holdup was a function of S_r/S_d and tended to decline with increasing S_r/S_d. Liquid circulation velocity increased with increasing S_r/S_d because of the rising of circulation drive force. Higher gas holdup was obtained in this study when compared to the data obtained by Kawase and Moo-Young *(14)*.

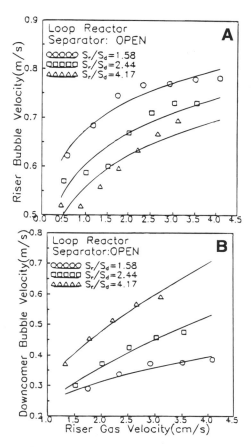

Fig. 5. Effects of superficial air velocity on the mean bubble velocity in 50 L airlift loop reactor. (**A**) The riser. (**B**) The downcomer.

Gas-liquid relative velocity and bubble size are two main parameters influencing gas-liquid mass transfer. Figures 5a and 5b show the effects of the superficial gas velocity in the riser on bubble velocity. The effects of S_r/S_d and riser superficial gas velocities on bubble size in the riser and downcomer are given in Figures 6a and 6b. In the riser, bubble size increased with an increase in both S_r/S_d and riser superficial gas velocity. In the downcomer, bubble size remained almost constant with increasing riser superficial gas velocity at lower values of S_r/S_d, and decreased at higher values of S_r/S_d because bubble breakup frequency increased with higher liquid circulation velocity.

Liquid Circulation Velocity and Mixing

Liquid circulation velocity is another important parameter affecting the behavior of the airlift loop reactor. Based on previous observation, liquid circulation velocity not only depends on the gas flow rate, but also is

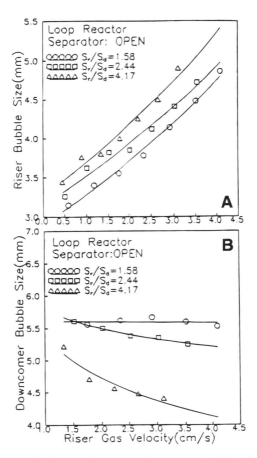

Fig. 6. Effects of superficial air velocity on the mean bubble diameter in 50 L airlift loop reactor. **(A)** The riser. **(B)** The downcomer.

sensitive to the structure of the gas-liquid separator (head part) *(11)*. "Opened mode," which means the cross section of the head is larger than that of the main part of the reactor, accounts for the expected high liquid circulation velocity. This is a result of a sufficient gas-liquid separation region in the head. Figure 7 shows the effect of riser superficial air velocity on the liquid circulation velocity in the riser with the head opened. Higher liquid velocity was obtained at a lower S_r/S_d because of a larger difference of gas holdup in the riser and downcomer. There was no significant difference in liquid circulation time at riser superficial gas velocity above 3.0 cm/s (Fig. 8). However, liquid mixing time decreased with increasing riser superficial air velocity and S_r/S_d, because of the enhancement of gas-liquid turbulence (Fig. 9).

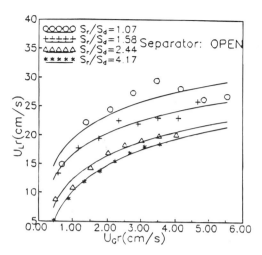

Fig. 7. Effects of riser superficial air velocity on the liquid circulation velocity in the riser of 50 L airlift loop reactor.

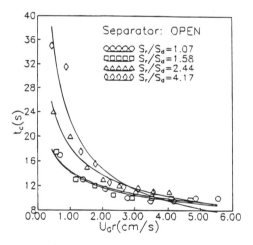

Fig. 8. Effects of riser superficial air velocity on the liquid circulation time of 50 L airlift loop reactor.

Mass Transfer Characteristics in Airlift Loop Reactor

The volumetric mass transfer coefficient ($K_L a$) in the airlift loop reactor is affected by many factors, such as; liquid phase diffusibility, bubble and liquid velocity, bubble size and its distribution, gas holdup, and gas-liquid turbulence. However, $K_L a$ is very sensitive to bubble size that affects mass transfer coefficient (K_L) and gas-liquid interfacial area per unit volume (a). When a stainless steel membrane sparger is used in the airlift loop reactor, small bubbles were obtained, especially in the gas sparger zone. For the air-water system, the initial bubble size formed from the membrane sparger

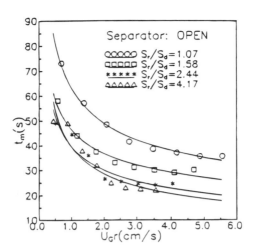

Fig. 9. Effects of riser superficial air velocity on the mixing time of 50 L airlift loop reactor.

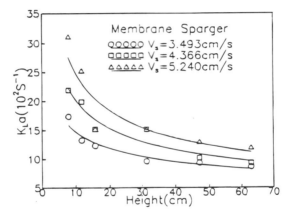

Fig. 10. Effects of height on volumetric mass transfer coefficient in sparger zone of 50 L airlift loop reactor.

was in the range of 1~2 mm. In the gas sparger zone (height of 40 cm in the 50 L airlift loop reactor), the bubble size increased rapidly along the axial positions leading to a decrease of K_La. K_La measured by a dissolved oxygen probe in the gas sparger zone as a function of height, is given in Fig. 10. It can be seen that at superficial air velocity of 5.24 cm/s, K_La decreased from 0.3 to 0.14 S^{-1}. For comparison, the experimental data obtained for air-water system from this work, results reported by Kawase and Moo-Young (15) and by Shah et al. (16), using single nozzle sparger, were plotted and compared and are shown in Fig. 11. It was obvious that because of the use of a membrane sparger, K_La increased 3~4 times, compared with K_La found using a single nozzle sparger.

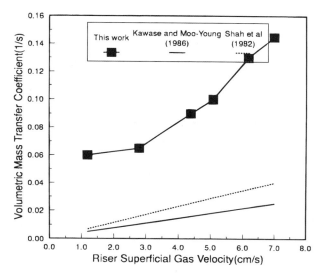

Fig. 11. Comparison between volumetric mass transfer coefficients obtained from published data with single gas sparger and with stainless steel membrane gas sparger in 50 L airlift loop reactor.

Mycelial Pellet Formation in Airlift Loop Reactor

When filamentous fungi were grown in submerged cultures, the type of growth varied from the pelletes to filamentous forms and/or a mixture of both forms. Mycelial pellets consist of compact discrete masses of hyphae, and the filamentous forms consist of homogeneous hyphae suspension were distributed throughout the medium. The filamentous form is preferred for fumaric acid production because of its good properties of oxygen and substrate mass transfer. However, when $CaCO_3$ was added to neutralize the acid produced, the filamentous fungi coagulated to form large clumps of mass that were easily precipitated to the bottom of the reactor. Therefore, the small pelleted form is preferred for fermentation. After cultivation, the mixture of filamentous and small fluffy pellets was transferred to an airlift loop reactor containing fermentation medium containing a limited quantity of nitrogen source (0.5% urea) in order to shift the organism from growth stage to acid production stage. When the air rate was 1 vvm, very small and uniform pellets formed with an average diameter of about 2 mm. Based on the visual observation, the small pellets in the airlift loop reactor were easily suspended and circulated and very well distributed in the entire zone of the reactor. During mechanically agitated fermentation in a stirred tank reactor, some very large mycelial clumps were formed whereas others retained their original mycelial form as a result of shear stress exerted by the agitator. The formation of large mycelial pellets often led to premature termination of fermentation (2). Another advantage of operating in small pelleted form was a low viscosity of broth that not only benefited mass transfer, but also avoided foaming as compared to filamentous form.

Fumaric Acid Production in an Airlift Loop Reactor

Two pathways, oxidative and reductive, are involved in fumaric acid metabolism in fungi. The oxidative pathway (TCA cycle) will generate one mole of fumarate per mole of glucose consumed. During active cell growth, however, this pathway cannot lead to the accumulation of fumarate. The generated fumarate was utilized for biosynthesis of cell constituents (2). On the other hand, the accumulation of fumaric acid is dependent on the operation of the reductive branch of the TCA cycle (17,18). The carbon dioxide-fixing reductive branch is capable of producing two moles of fumarate per mole of glucose consumed. The enzyme responsible for fumarate accumulation in *Rhizopus* is pyruvate carboxylase (EC 6.4.1.1) (19). This enzyme catalyzed the ATP-dependent condensation of pyruvate and CO_2 to form oxaloacetic acid, the key intermediate of TCA cycle. The ability of fungi, such as *Rhizopus*, to incorporate CO_2 into the reductive branch of oxidative pathway render them the ability to produce organic acids in very high yield with fumaric acid as the major acid product (20).

By limiting the nitrogen source, *Rhizopus* cell growth can be kept minimal. During this non-growth stage, fumaric acid can be accumulated with a maximum yield of 2 moles per mole glucose consumed or 1.29 g fumarate per g of glucose consumed on the basis of weight (17). In reality, the reductive bypass requires the supply of NADH from TCA cycle. Therefore, the obtainable yield is about 1.45 moles of fumarate from each mole of glucose (as much as 0.93 g per g of glucose consumed) (2).

Figure 12 shows the time course of fumaric acid fermentation in 10 L airlift loop reactor. It can be seen that the final fumaric acid concentration reached 37.8 g/L after 46 h, and the productivity and efficiency yield were 0.814 g/L/h and 50.1%, respectively. The concentration of ethanol produced in the broth was under 10 g/L. The comparison of the results in the airlift loop reactor and stirred tank was summarized in Table 2. Noticeably, the productivity of fumaric acid in the airlift loop reactor was higher than in the stirred tank reactor. Furthermore, the maximum ethanol concentration in the stirred tank reactor reached 20g/L, which was twice as as that in the airlift loop reactor. The measured results of dissolved oxygen show that nearly 90% of saturation was obtained in the airlift loop reactor throughout the entire course of fermentation at an air supply of 1 vvm.

CONCLUSION

A dual-probe automatic measurement system controlled by an on-line computer was used to measure the bubble characteristics in the airlift loop reactor. The results indicated that K_La in the gas sparger zone

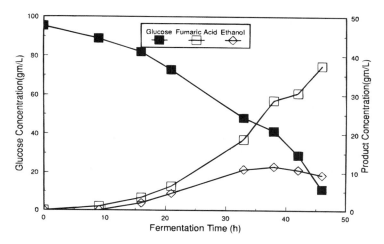

Fig. 12. Time course of fumaric acid fermentation in 10 L airlift loop reactor with porous gas sparger. Air flow rate: 1 vvm; Temperature: 35°C.

Table 2
Comparison of Fumaric Acid Production by *R. Oryzae* in Different Fermentors

Process	Airlift Loop Reactor	Stirred Tank Fermentor
Volume(L)	10	6.5
Neutralizing Agent	$CaCO_3$	$CaCO_3$
Initial Glucose(g/L)	95	100
pH	4.0~5.0	5.0
Fermentation Time(h)	46	89
Air Flow Rate(L/L/min)	0.9	2.0
Stirring Speed(rpm)	-	600
Yield(%)[a]	75.4	60
Productivity(g/L/h)	0.814	0.66

[a]Yield = (maximum g/L of fumaric acid)/(g/L of glucose consumed)

decreased rapidly along the axial position because of the increase in bubble size. By employing the porous gas sparger, K_La in the airlift loop reactor increased significantly compared with K_La found using single nozzle spargers. The results of fumaric acid fermentation show that the hydrodynamics in the laboratory airlift loop reactor with porous sparger produced favorable conditions for mass transfer. In addition, it is useful for forming small uniform mycelial pellets. Higher production rates and product yields were obtained in the airlift loop reactor than in the typical stirred tank fermentor. Airlift loop reactors have the potential for application in a larger scale fungus fermentation system.

ACKNOWLEDGMENTS

This research was supported in part by Division of Biological and Critical System, National Science Foundation grant BES-9412582.

NOMENCLATURE

a	gas-liquid interfacial area per unit volume, m^2/m^3
K_L	mass transfer coefficient, m/s
$K_L a$	volumetric mass transfer coefficient, 1/s
S_r/S_d	ratio of cross-sectional area of riser and downcomer
t_c	liquid circulation time, s
t_m	liquid mixing time, s
$U_L r$	liquid circulation velocity, cm/s
$U_G r$	riser gas velocity, cm/s
V_s	superficial gas velocity, cm/s.

REFERENCES

1. Robinson, W. D. and Mount, R. A. (1981), in *Kirk-Othmer Encyclopedia of Chemical Technology*, Vol. 14, Grayson, M. and Eckroth, D., eds., Wiley, New York, p. 770.
2. Rhodes, R. A., Lagoda, A. A., Misenheimer, T. J., Smith, M. L., Anderson, R. F., and Jackson, R. W. (1962), *Appl. Microbiol.* **10**, 9–15.
3. Kautola, H. and Linko, Y. (1989), *Appl. Microbiol. Biotechnol.* **31**, 448–454.
4. Jiang, Y. H. (1995), M. S. thesis, Purdue University.
5. Bayer, T., Zhou, W., Holzhaner, K., and Schugerl, K. (1989), *Appl. Microbiol. Biotechnol.* **30**, 26–33.
6. Siegel, M. H., Hallaile, M., and Merchuk, J. C. (1988), Upstream Process Equipment and Techniques, Alan R. Lisss Inc., New York, pp. 79–124.
7. Konig, B., Schugerl, K., and Seewald, C. (1982), *Biotechnol. Bioeng.* **24**, 259–280.
8. Heck, J. and Onken, U. (1982), *Chem. Eng. Sci.* **42**, 1211–1212.
9. Trager, M., Qazi, G. N., Onfen, U. and Chopra, C. L. (1989), *J. Fermen. Bioeng.* **68**, 112–116.
10. Okabe, M., Ohta, N., and Park, Y. (1993), *J. Fermen. Bioeng.* **76**, 117–122.
11. Du, J. X. (1995), Ph. D. thesis, Tsinghua University, P. R. China.
12. Yang, C. W. (1994), Ph. D. thesis, Purdue University.
13. Yang, C. W., Lu, Z. J., and Tsao, G. T. (1995), *Appl. Biochem. Biotechnol.* **51/52**, 57–71.
14. Kawase, Y. and Moo-Young, M. (1986), *Chem. Eng. Commun.* **40**, 67–83.
15. Kawase, Y. and Moo-Young, M. (1988), *Chem. Eng. Res. Des.* **66**, 284–288.
16. Shah, Y. T., Kelkar, B. G., Godbole, S. P., and Deckwer, W. D. (1982), *AIChE J.* **28**, 353–379.
17. Gangl, I. C., Weigand, W. A., and Keller, F. A. (1990), *Appl. Biochem. Biotechnol.* **24/25**, 663–677.
18. Peleg, Y., Battat, E., Scrutton, M. C., and Goldberg, I. (1989), *Appl. Microbiol. Biotechnol.* **32**, 334–339.
19. Kenealy, W., Zaady, E., Du Preez, J. C., Stieglitz, B. and Goldberg, I. (1986), *Appl. Environ. Microbiol.* **52**, 128–133.
20. Overman, S. A. and Romano, A. H. (1969), *Biochem. Biophys. Research Commun.* **37**, 457–463.

Maximizing the Xylitol Production from Sugar Cane Bagasse Hydrolysate by Controlling the Aeration Rate

SILVIO S. SILVA,*,[1] JOÃO D. RIBEIRO,[1] MARIA G. A. FELIPE,[1] AND MICHELLE VITOLO[2]

[1]Department of Biotechnology/Faculty of Chemical Engineering of Lorena/Rod. Itajubá/Lorena Km 74,5, 12600000, Lorena, SP, Brazil; and [2]Department of Biochemical and Pharmaceutical Technology/Faculty of Pharmaceutical Sciences/University of São Paulo, P.Box 66083 São Paulo, SP, Brazil

ABSTRACT

Batch fermentations of sugar cane bagasse hemicellulosic hydrolysate treated for removing the inhibitors of the fermentation were performed by *Candida guilliermondii* FTI 20037 for xylitol production. The fermentative parameters agitation and aeration rate were studied aiming the maximization of xylitol production from this agroindustrial residue. The maximal xylitol volumetric productivity (0.87 g/L · h) and yield (0.67 g/g) were attained at 400/min and 0.45 v.v.m. (K_La 27/h). According to the results, a suitable control of the oxygen input permitting the xylitol formation from sugar cane bagasse hydrolysate is required for the development of an efficient fermentation process for large-scale applications.

Index Entries: Sugar cane bagasse hemicellulose hydrolysate; xylitol; xylose; aeration; *Candida guilliermondii*.

INTRODUCTION

Lignocellulosic materials from forestry and agriculture residues such as rice husk, eucalyptus, and sugar cane bagasse are inexpensive and abundant sources of energy that can be used in several biotechno-

*Author to whom all correspondence and reprint requests should be addressed.

logical processes for the obtention of products of high economic value. Sugar cane bagasse is the most important residue in Brazil and an amount of 5–12 millions t/yr of this biomass is generated by the Brazilian sugar-alcohol industries *(1,2)*. Every year large amounts of waste biomass are accumulated in nature, causing serious environmental pollution problems. It is necessary to find new technologies to use this renewable biomass in different processes to produce economically valuable products. The biotechnological approach is one way to use this biomass as a micro-organism substrate for production of several useful feedstocks.

Xylitol, a valuable product with high sweetening power, anticariogenic properties, and several clinical applications is a substance that can be produced by fermentation processes using sugar cane bagasse as the substrate *(3,4)*.

Currently, xylitol is produced by chemical hydrogenation using Nickel as the catalyst. However, this process is expensive since it requires several steps of xylose purification before the chemical reaction *(5,6)*. Microbial production of xylitol from agroindustrial residues is a simpler and more ecomomic process since it occurs at lower temperatures and pressures and does not require pure xylose.

The use of sugar cane bagasse as the substrate in fermentative processes for xylitol production consists initially in releasing sugars from the hemicellulose portion through a mild acid hydrolysis process. This process is accompanied by the formation of considerable amounts of hemicellulose decomposition products, such as furfural, hydroxymethylfurfural, acetic acid, and other products derived from lignin degradation. These chemical compounds interfere negatively with the yeast cell growth *(7)* and the additional xylitol fermentation *(8)*. Thus, the use of this biomass hydrolysate as a fermentation medium for micro-organism growth is critical, and several treatments are necessary for removing these products. The cell growth in this hydrolysate and the xylitol formation depend on the treatment and the fermentation conditions employed *(3,4)*. The oxygen transfer rate is the most significant of all parameters that affect the biological synthesis of xylitol by xylose-fermenting yeasts *(9–11)*. According to the literature the effect of oxygen on xylitol production is not fully understood and appears to be related to the initial steps of the xylose metabolism and to the NAD/NADH pool *(12)*.

In this communication the authors present a simple method of treatment of sugar cane bagasse hemicellulosic hydrolysate for removing toxic compounds and the xylitol production by *Candida guilliermondii* FTI 20037 from this biomass under different O_2 conditions. It is of fundamental importance to understand the influence of this factor on xylitol formation for the development of an efficient technology for large-scale xylitol production from sugar cane bagasse by the biotechnological process.

MATERIALS AND METHODS

Micro-organism

Candida guilliermondii FTI 20037 from the Biotechnology Department of the Faculty of Chemical Engineering of Lorena, FAENQUIL, Lorena, S.P.–Brazil, was used. The culture was maintained in malt extract agar slants at 4°C.

Preparation and Clarification of the Acid Sugar Cane Bagasse Hemicellulosic Hydrolysate

The hemicellulosic hydrolysate was obtained by acid hydrolysis of sugar cane bagasse in a 360 L stainless steel reactor. The sugar cane bagasse was percolated with 10% H_2SO_4 per dry weight of bagasse for 20 min at 120°C. To reach a higher sugar concentration the hydrolysate was concentrated under reduced pressure in a lab-scale evaporator at 70°C. The concentrated hydrolysate was treated prior to the fermentations for removing the toxic components formed by acid hydrolysis. The hydrolysate was treated according to Felipe et al. (3) and clarified with active charcoal (30 g/L) under 200 rpm stirring for 1 h at room temperature. In this step CaO and H_2SO_4 were used, mainly because of beneficial effects of the Ca^{++} ions on the possible removal of unknown compounds during precipitation. The precipitate formed was removed by centrifugation at $1000g$ for 20 min. The treated hydrolysate was then autoclaved with steam at 100°C for 20 min, and asseptically supplemented with rice bran (10 g/L) to provide vitamins and $(NH_4)_2SO_4$ (2 g/L). This treated hydrolysate was then used as a fermentation medium to evaluate the xylitol production by *C. guilliermondii* FTI 20037.

Inoculum Preparation

The inoculum was grown in the aforementioned treated hydrolysate containing the same nutrients that were used in the fermentation medium. A loopful of cells from stock culture was inoculated in 50 mL of this medium in 125 mL Erlenmeyer flasks and incubated for 24 h at 30°C and 200 rpm in a rotary shaker. The cells were harvested by centrifugation at $1000g$ for 20 min, washed twice with distilled water, and resuspended in 10 mL distilled water. The cell concentration was determined and a volume sufficient to give 0.5 g/L cell dry weight was used to inoculate the fermenter.

Fermentation Conditions

Batch fermentation runs were performed in a 1 L fermenter (MULTI-GEN-New Brunswick Scientific, Edison, NJ) containing baffles and two sets of disk Rushton turbines with six flat-blades and a working volume of 0.55 L of medium prepared as described according to the preparation and clarification of the hydrolysate. The fermentation system was equipped with

Table 1
Concentration of Some Components in the Sugar Cane
Bagasse Hemicelulosic Hydrolysate

Components	Hydrolysate Concentrations (g/L)	
	original	concentrated
Glucose	5.54	8.04
Xylose	26.38	62.13
Arabinose	2.07	5.11
Acetic acid	4.52	7.00
Furfural	< 0.5	< 0.1
Hydroximethylfurfural	< 0.1	< 0.1
Lignin degradation products	n.d	n.d

n.d = not determined.

pH, pO_2, temperature and aeration rate controllers. The temperature was maintained at 30°C and the agitation/aeration rates were set at different values according to a previous statistical factorial design. The agitation was set at 200, 300, and 400 min^{-1} and the oxygen supply was varied from 0.10 to 0.80 vvm (volume of air per volume of medium per minute). The oxygen volumetric transfer coefficient (K_La) in all conditions was determined.

Analytical Methods

Batch fermentation runs were monitored by periodic sampling to determine the sugar consumption and xylitol formation. Samples of appropriate dilutions were prepared by filtration through a 0.22 micron filter (Waters Set-pak Cartridge, Millipore, Bedford, MA)

Xylose, glucose, arabinose, acetic acid, and xylitol were analyzed in a Shimadzu high performance liquid chromatograph (HPLC), using a Bio-Rad Aminex HPX-87 H column at 45°C and 0.02 NH_2SO_4 as the eluent at a flow rate of 0.6 mL/min.

Growth was monitored by measuring the culture turbidity at 600 nm. The cell mass was estimated using a relationship between optical density and dry cell weight.

The volumetric oxygen transfer coefficient (K_La) was determined under standard fermentation conditions by the gassing-out method as described by Pirt *(13)*.

RESULTS AND DISCUSSION

The basic composition of the sugar cane bagasse hydrolysate used in this work is shown in Table 1. Under the conditions used, a mixture of sugars (pentoses and hexoses) was obtained. Xylose was the major pentose present in this hydrolysate representing about 70% of the total monosaccharides. The presence of acetic acid was a result of the de-*o*-acetylation of acetylated

Table 2
Xylitol Production Rates from Sugar Cane Bagasse Hydrolysate
by *C. guilliermondii* FTI 20037 under Different Oxygen Conditions

Physical Parameters										
Agitation (min^{-1})	Aeration (v.v.m)	$K_L a$ (h^{-1})	ΔS (%)	XOH (g/L)	Final biomass (g/L)	Final pH	Q_P (g/L.h)	Q_x (g/L.h)	q_P (g/g.h)	$Y_{P/S}$ (g/g)
200	0.10	2	7	1.44	0.71	5.85	0.02	0.06	0.09	0.36
200	0.45	7	19	5.76	0.92	5.35	0.08	0.15	0.22	0.52
200	0.80	11	35	4.32	0.80	5.33	0.06	0.30	0.19	0.20
300	0.10	2	15	5.76	0.56	5.19	0.08	0.13	0.80	0.61
300	0.45	9	82	34.75	2.00	5.62	0.58	0.82	0.38	0.62
300	0.80	18	100	40.20	2.19	6.61	0.67	1.00	0.39	0.67
400	0.10	6	12	2.88	0.75	5.70	0.04	0.10	0.17	0.45
400	0.45	27	100	41.76	5.19	7.70	**0.87**	1.30	0.18	**0.67**
400	0.80	38	80	25.20	10.29	6.83	0.70	1.43	0.07	0.49

ΔS = xylose consumed; XOH = final xylitol concentration; Qp = xylitol volumetric productivy, Qx = xylose uptake rate; qp = specific rate of xylitol production; Yp/s = xylitol produced/substrate consumed.

sugars from the hemicellulosic fraction. This acid has been described as toxic for *C. guilliermondii* at concentrations up to 6 g/L. Consequently it interferes with xylitol production (3). After the hydrolysate concentration step the acetic acid concentration increased from 4.52 to 7.0. A parallel increase in the xylose:glucose ratio was also observed from 4.8 to 7.7. Hence, the concentration of the original hydrolysate corresponded to a second and convenient hydrolytic step (xylo-oligosaccharides to free xylose with simultaneous de-*o*-acetylation). Other compounds like furfural and hydroxymethyl furfural were also present in this hydrolysate at low concentrations (Table 1).

The results of Table 2 demonstrated that the yeast *Candida guilliermondii* FTI 20037 was able to produce xylitol at different rates during the fermentation of pretreated sugar cane bagasse hemicelullosic hydrolysate, under all aeration conditions employed. It is known that the hemicellulosic hydrolysate fermentation is complex and critical since this hydrolysate contains several chemical compounds that are toxic to the micro-organisms. By treating the sugar cane bagasse hydrolysate the toxicity of these compounds was significantly reduced and therefore cell growth, substrate uptake, and xylitol formation could be observed (Table 2). By increasing the original pH of the hydrolysate from 1.1 to 10 and in the presence of active charcoal, the Ca^{++} ions may bind and precipitate some of these toxic compounds and improve further fermentations. The real beneficial effect of Ca^{++} ions on the treatment of the hydrolysate is difficult to detect. According to Van Zyl et al. (14), this may be attributed to the poor solubility of calcium salts formed during the neutralization step and possible removal of unknown toxic substances during precipitation.

According to Table 2 the highest level of final biomass (10.29 g/L) was attained with the highest O_2 levels, whereas the lowest growth occurred at a substantially low aeration level. This fact reflects the importance of oxygen in the utilization of xylose by xylose-fermenting yeasts. *C. guilliermondii* FTI 20037 is considered a good xylose-fermenting yeast with a great potential for large-scale xylitol production *(15,16)* since this strain contains xylose reductase and xylitol dehydrogenase, which are the key enzymes for the xylose metabolism *(17,18)*. It is known that yeast growth depends on the formation of xylulose in the initial steps of the xylose metabolism. Xylose is first reduced to xylitol through a NADPH-linked xylose reductase. Then the xylitol is oxidized to xylulose, through a NAD-linked xylulose dehydrogenase. This is followed by the formation of xylulose-5-phosphate through an ATP-dependent xylulose kinase and by the further entrance of xylulose-5-phosphate into the pentose phosphate pathway. The formation of ATP and the oxidation of NADH are carried out at the respiratory chain in the mitochondria and are strongly dependent on the oxygen availability. The oxygen transfer rate is fundamental for the regeneration of cofactors, which are essential for xylose metabolism, biomass formation, and xylitol excretion.

According to Shook and Hahn-Hägerdahl *(19)*, the available oxygen has a great influence on xylose fermentation. Under aerobic conditions the organism mainly produces cell mass, and under semiaerobic or anaerobic conditions by-products are formed. In our experiments, the xylitol formation was strongly dependent on the oxygen supply. By increasing the aeration rate and consequently the oxygen input, the xylitol formation increased until a suitable agitation/aeration rate relationship was reached. Similar results were found by Silva et al *(11)* using synthetic medium containing xylose as the major carbon source. Thus, for the maximum xylitol production, it is fundamental to control the oxygen transfer rate. In our experiments, the maximum xylitol production (41.76 g/L) from sugar cane bagasse and the maximum xylitol volumetric productivity (0.87 g/L.h) were attained under agitation set at 400 min^{-1} and aeration rate of 0.45 vvm (Table 2).

The K_La is an important parameter since it describes the aeration capacity of the fermentation system and supplies information for the process scale-up. Under the experimental conditions used, the K_La varied from 2 to 38 h^{-1} (Table 2). Few reports describe the influence of this parameter on xylitol production and many published data are contractitory. Under our conditions, by controlling the aeration rate, the K_La for maximum xylitol production is near 27 h^{-1}.

The fermentation pH was also affected by the aeration rate (Table 2). At low aeration rates its increase was not pronounced. However, increasing the oxygen the pH increased from 5.5 to near 7.7. These results are in agreement with those found by Van Zyl et. al. *(20)*. According to these authors, under aerobic conditions, the xylose and the acetic acid present in sugar cane bagasse hydrolysate, were consumed simultaneously, whereas under anaerobic conditions there was no acid consumption. The consumption of acetic acid, as a carbon source, by *Candida guilliermondii* is a fact observed in previous work *(8)*, and the effect of this acid on cell growth is highly dependent on the fermentation pH. It is known that the acetic acid toxicity depends on its concentration and strongly interferes with the yeast energy metabolism by reducing the H^+ gradient across the mitochondrial membrane.

CONCLUSIONS

The demands for xylitol by the food and pharmaceutical industries have aroused great interest in the development of a low cost techonology for xylitol production. Biotechnological processes using agroindustrial residues as the substrate for xylitol production appear to be more efficient and more economically advantageous when compared to chemical processes. It can be inferred from our results that the pretreated sugar cane bagasse hemicellulosic hydrolysate is a valuable substrate for xylitol fermentation by *Candida guilliermondii* FTI 20037.

The maximum xylitol production rates can be attained at an adequate agitation/aeration rate relationship. A suitable control of the oxygen input permitting the xylitol formation from sugar cane bagasse hydrolysate is required for the development of an efficient fermentation process for large-scale applications.

ACKNOWLEDGMENTS

The authors gratefully acknowledge for the financial support of this research provided by Fundação de Amparo à Pesquisa do Estado de São Paulo (FAPESP).

REFERENCES

1. Burgi, R. (1988), *A Granja*, **44,** 16–26.
2. Molina Junior, W. F., Ripoli, T. C., Geraldi, R. N., and Amaral, J. R. (1995), *Açúcar, Álcool e Subprodutos*, **13,** 28–31.
3. Felipe, M. G. A., Mancilha, I. M., Vitolo, M., Roberto, I. C., Silva, S. S., and Rosa, S. A. M. (1993), *Arquivos de Biol. e Tecnol.* **36,** 103–114.
4. Pfeifer, M. J., Silva, S. S., Felipe, M. G. A., Roberto, I. C., and Mancilha, I. M. (1996), *Appl. Biochem. Biotechnol.* **57/58,** 423–430.
5. Melaja, A. J. and Hämäläinen, L. (1977), US Patent no. 4.008.285.
6. Hyvönen, L., Koivistoinen, P., and Voirol, F. (1982), *Adv. Food Res.* **28,** 373–403.
7. Sanchez, B. and Bautista, J. (1988), *Enzyme Microb. Technol.* **10,** 315–318.
8. Felipe, M. G. A., Vieira, D. C., Vitolo, M., Silva, S. S., Roberto, I. C., and Mancilha, I. M. (1995), *J. Basic Microbiol.* **35,** 171–177.
9. Furlan, S. A., Bouilloud, P., and Castro, H. F. (1994), *Process Biochemistry*, **29,** 657–662.
10. Nolleau, V., Preziosi-Belloy, L., and Navarro, J. M. (1995), *Biotechnol. Lett.* **17,** 417–422.
11. Silva, S. S., Roberto, I. C., Felipe, M. G. A., and Mancilha, I. M. (1996), *Process Biochem.* **31,** 549–553.
12. Thornat, P., Guerreiro, J. G., Foucart, M., and Paquot, M. (1987), *Mededelingen. Facultait. Landbouwweteschapp Rijksuniversitat*, **52,** 1517–1528.

13. Pirt, S. J. (1975), in *Principles Microbe Cell Cultivation*, Blackwell, Oxford.
14. Van Zyl, C., Prior, B. A., and Du Preez, J. C. (1988), *Appl. Biochem. Biotechnol.* **17,** 357–369.
15. Barbosa, M. F. S., Medeiros, M. B., Mancilha, I. M., Schneider, H., and Lee, H. (1988), *J. Industrial Microbiol.* **3,** 241–251.
16. Roberto, I. C., Felipe, M. G. A., Mancilha, I. M., Vitolo, M., Sato, S., and Silva, S. S. (1995), *Biores. Technol.* **51,** 255–257.
17. Silva, S. S., Vitolo, M., Pessoa-Junior, A., and Felipe, M. G. A. (1996), *J. Basic Microbiol.* **36,** 187–191.
18. Lee, H., Sopher, C. R., and Yau, K. Y. F. (1996), *J. Chem. Technol. Biotechnol.* **66,** 375–379.
19. Shook, K. and Hahn-Hägerdahl, B. (1988), *Enzyme Microb. Technol.* **10,** 66–80.
20. Van Zyl, C., Prior, B. A., and du Preez, J. C. (1991), *Enzyme Microb. Technol.* **13,** 82–86.

Production of Succinic Acid by *Anaerobiospirillum succiniciproducens*

NHUAN P. NGHIEM,* BRIAN H. DAVISON, BRUCE E. SUTTLE, AND GERALD R. RICHARDSON

Chemical Technology Division Oak Ridge National Laboratory P.O. Box 2008, MS-6226 Oak Ridge, TN 37831-6226

ABSTRACT

The effect of an external supply of carbon dioxide and pH on the production of succinic acid by *Anaerobiospirillum succiniciproducens* was studied. In a rich medium containing yeast extract and peptone, when the external carbon dioxide supply was provided by a 1.5M Na_2CO_3 solution that also was used to maintain the pH at 6.0, no additional carbon dioxide supply was needed. In fact, sparging CO_2 gas into the fermenter at 0.025 L/L-min or higher rates resulted in significant decreases in both production rate and yield of succinate. Under the same conditions, the production of the main by-product acetate was not affected by sparging CO_2 gas into the fermenter. The optimum pH (pH 6.0) for the production of succinic acid was found to be in agreement with results previously reported in the literature. Succinic acid production also was studied in an industrial-type inexpensive medium in which light steep water was the only source of organic nutrients. At pH 6.0 and with a CO_2 gas sparge rate of 0.08 L/L-min, succinate concentration reached a maximum of 32 g/L in 27 h with a yield of 0.99 g succinate/g glucose consumed.

Index Entries: Succinic acid; fermentation; renewable resources; corn sugars; *Anaerobiospirillum succiniciproducens*.

INTRODUCTION

Succinic acid has been used for applications in many areas, including agriculture, food, medicine, plastics, cosmetics, textiles, plating, and waste-gas scrubbing *(1)*. Catalytic processes have recently been developed for the conversion of succinic acid to a number of industrially important chemicals

*Author to whom all correspondence and reprint requests should be addressed.

that include 1,4-butanediol, tetrahydrofuran and gamma-butyrolactone *(2,3)*. This new development has rendered the market for succinic acid much larger.

Succinic acid currently is produced commercially by chemical processes *(1)*. A fermentation process for its production is of great interest because in such processes, renewable resources such as corn-derived glucose can be used as starting material. There is not a current biological process for the commercial production of succinic acid, although a number of patents have been issued on the production of succinic acid by micro-organisms *(4–8)*.

Succinic acid is an intermediate of the tricarboxylic acid cycle and also is a product of anaerobic metabolism *(9)*. As such, its accumulation in fermentation broth has been observed with a number of micro-organisms, which were both aerobes and anaerobes *(4–8,10–12)*. The anaerobic bacterium, *Anaerobiospirillum succiniciproducens*, is considered among the best succinic acid producers. It has been observed that the main products of fermentation of this organism included succinic and acetic acids; other products included lactic acid and ethanol. Based on these observations, a biochemical pathway for the synthesis of succinic acid by *A. succiniciproducens* was proposed. The proposed pathway involved the conversion of phosphoenolpyruvate (PEP) to oxaloacetate by a carbon dioxide-fixing enzyme, PEP-carboxykinase *(13)*. It has been shown that extracellular supply of carbon dioxide was needed for succinic acid synthesis *(7,13)*.

Our efforts to develop a biological process for the production of succinic acid by *A. succiniciproducens* have focused on the establishment of process conditions and the development of an inexpensive fermentation medium. Some of the conditions for the production of succinic acid by *A. succiniciproducens* have been reported by Datta *(7)*. The author studied succinic acid production in a one-L fermenter and reported on the effects of pH and CO_2 gas sparge. However, only end-point results were reported. In addition, only one CO_2 gas sparge rate of 0.01 mL/min was examined. Therefore, it was decided to re-examine the effects of pH and CO_2 gas sparge rates during the course of the fermentations. In this investigation, wider CO_2 gas sparge rates were used. The use of light steep water, which is an inexpensive source of organic nutrients, in the fermentation medium was also investigated. The development of the light steep water medium was still in its very early stage, and therefore, only preliminary results are reported.

MATERIALS AND METHODS

The culture of *A. succiniciproducens* (ATCC 53488) was provided by Michigan Biotechnology Institute. The stock culure was prepared and stored in 25% glycerol at –70°C as described earlier *(14)*.

To prepare inoculum for fermentation experiments, one glycerol vial was used to inoculate a serum bottle containing 100 mL medium. The inoculum medium that has been described by Datta *(7)* contained 20 g/L glucose, 10 g/L polypeptone(Difco), 5 g/L yeast extract (Difco), 3 g/L K_2HPO_4, 1 g/L NaCl, 1 g/L $(NH_4)_2SO_4$, 0.2 g/L $CaCl_2\cdot 2H_2O$, and 0.2 g/L $MgCl_2\cdot 6H_2O$. The glucose-free medium was heat-sterilized and allowed to cool to ambient temperature before 1 mL of 0.03M Na_2CO_3 and 0.15 mL of 0.18M H_2SO_4 were added. Glucose then was added as a 20% solution to bring its concentration to 20 g/L. Finally, 0.5 mL of a solution containing 0.25 g/L cystein.HCl and 0.25 g/L $Na_2S\cdot 9H_2O$ was added and 20 min was allowed for the reduction of the medium before the serum bottle was inoculated with the entire contents of the glycerol vial. The glucose, sodium carbonate, sulfuric acid, and cystein-sodium sulfide solutions were all heat-sterilized. The serum bottle was incubated with gentle shaking at 39°C. The contents of the serum bottle were used to inoculate the fermenter when the residual glucose dropped to about 10 g/L; this normally took about 14 to 16 h. In each experiment, 45 mL broth from the serum bottle was used for inoculation.

All fermentations were batch and performed in 1-L Virtis Omni fermenters. Two fermentation media were used. The composition of the medium that was used to study the effect of pH and the level of external carbon dioxide supply was a slight modification of the one described by Datta *(7)*. With the exception of 50 g/L glucose and 5 g/L $(NH_4)_2SO_4$, and the addition of 5 mg/L $FeSO_4\cdot 7H_2O$, other components were the same as described in the previous paragraph for the inoculum medium. All the ingredients, except glucose and the iron salt, were dissolved in 875 mL deionized water, transferred to the fermenter, autoclaved, and allowed to cool to ambient temperature. To the fermenter then were added 100 mL of 50% glucose, 1 mL of 0.5 g/L $FeSO_4\cdot 7H_2O$, 20 mL of 1.5M Na_2CO_3, 1.5 mL concentrated H_2SO_4, and 5 mL of 0.25 g/L cystein HCl, and 0.25 g/L $Na_2S\cdot 9H_2O$. All these solutions were heat-sterilized prior to being added to the fermenter. In the light steep water medium, both yeast extract and peptone were replaced by 100 mL of light steep water that was obtained from the A. E. Staley corn processing plant in Loudon, TN.

The temperature was maintained at 39°C, which was the temperature used by Datta *(7)*. The pH was controlled by adding a 1.5M Na_2CO_3 solution on demand. This solution also served as an external source of carbon dioxide. In the experiments performed to study the effect of additional external carbon dioxide supply, pure CO_2 gas was sparged into the fermenter at 0.025, 0.05, and 0.1 L/min. The pH in these experiments was maintained at 6.0. The effect of pH on succinic acid production was studied at five pH values, which were 5.0, 5.5, 6.0, 6.5, and 7.0. CO_2 gas was not sparged into the fermenter in these experiments, therefore, the 1.5M Na_2CO_3 solution added for pH control was the only source of external carbon dioxide. Samples were withdrawn at intervals and analyzed for cell growth, residual glucose, succinate and acetate concentrations.

Growth was monitored by measuring optical density at 660 nm with a Milton Roy Spectronic 21D. Glucose was measured with a Yellow

Springs Instrument 2700 Select glucose analyzer. Succinic and acetic acids were determined by gas chromatography using the method developed by Playne (15). A Varian 3700 gas chromatograph equiped with a flame ionization detector and a Chromosorb 101 column maintained at 200°C was used. The carrier gas was helium flowing at 50 mL/min. The injector and detector were maintained at 250°C. Sample was prepared by mixing 500 µL fermentation broth with 100 µL 25% metaphosphoric acid; the mixture then was centrifuged on an Eppendorf microcentrifuge at 12000 rpm for 2 min and the supernatant used for analysis. The injection volume was 2 µL. The integrator was a Hewlett Packard 3396 Series II.

RESULTS AND DISCUSSION

In a previous report on the preliminary results of the effect of biotin on the production of succinic acid by *A. succiniciproducens* (14), it was shown that the final concentration of succinic acid in a fermentation medium containing polypeptone and yeast extract as organic nitrogen sources could be improved by 17% by adding 50 mg/L biotin. It was also pointed out that this was not the optimal concentration. Since the optimal biotin concentration had not been determined yet, it was decided to omit biotin from the polypeptone-yeast extract fermentation medium used in the present investigation.

During the course of all fermentations, the broth volumes in the fermenters increased owing to Na_2CO_3 addition for pH control. The final volumes were from 1.2 to 1.3 L. These volumes were used in the calculations of the yield of succinic acid and acetic acid.

The effect of CO_2 sparge rates is shown in Fig. 1 and summarized in Table 1. The rate of glucose consumption did not seem to be affected by sparging CO_2 gas into the fermenter at 0.025 L/min (Fig. 1A). However, at higher CO_2 sparge rates, the rate of glucose consumption was significantly decreased. In the experiment having the CO_2 sparge rate set at 0.025 L/min, all of the initial glucose was depleted in 23 h; in the control experiment (no CO_2 sparge), although the consumption of glucose was not complete, the glucose concentration remaining at 23 h was only 2.3 g/L. When the CO_2 sparge rates were increased to 0.05 and 0.1 L/min, the glucose concentrations at 23 h were 10.9 and 27.5 g/L, respectively. The authors' results were not in agreement with Datta's results (7). Datta found that when CO_2 gas was sparged into the fermenter at 0.01 L/min, complete utilization of glucose occured at 38 h, whereas in the control experiment (no CO_2 gas sparge), only 71% of the initial glucose was consumed at 40 h. However, in his control experiment, the initial glucose concentration was 54 g/L, whereas in the other experiment it was 47.5 g/L. The high glucose concentration used initially might have caused a long lag before the fermentation took off.

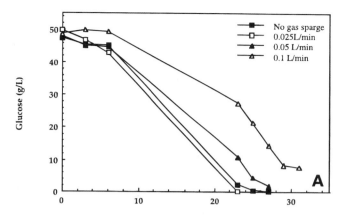

Fig. 1A. The effect of CO_2 gas sparge rates on glucose consumption at pH 6.0 in yeast extract-polypeptone medium.

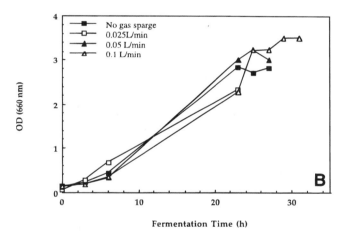

Fig. 1B. The effect of CO_2 gas sparge rates on cell growth at pH 6.0 in yeast extract-polypeptone medium.

The sparge of CO_2 gas into the fermenter did not affect the cell growth rate (Fig. 1B). The cell yield also was not affected by the CO_2 sparge rates up to 0.05 L/min. However, an increase in cell yield was observed when the CO_2 sparge rate was increased to 0.1 L/min.

The sparge of CO_2 gas into the fermenter adversely affected the rate of succinic acid production (Fig. 1C). The succinic acid concentration obtained in the control experiment at 23 h was 34.9 g/L. When CO_2 gas was sparged into the fermenter at 0.025, 0.05, and 0.1 L/min, the succinic acid concentrations at 23 h were 28.0, 22.1, and 12.5 g/L, respectively. The sparge of CO_2 had a slightly different effect on the yield of succinic acid. The succinic acid yield calculated at the exhaustion of glucose in the control experiment was 0.93 g/g glucose consumed. When CO_2 was

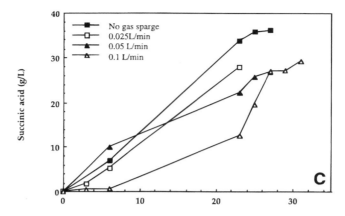

Fig. 1C. The effect of CO_2 gas sparge rates on succinic acid production at pH 6.0 in yeast extract-polypeptone medium.

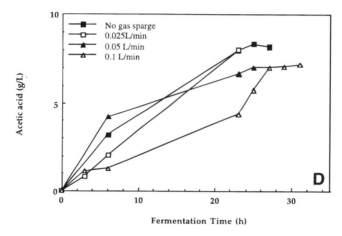

Fig. 1D. The effect of CO_2 gas sparge rates on acetic acid production at pH 6.0 in yeast extract-polypeptone medium.

sparged into the fermenter at 0.025 and 0.05 L/min, the succinic acid yields at glucose exhaustion dropped to 0.69 and 0.71 g/g glucose consumed, respectively. However, the succinic acid yield when the CO_2 sparge rate was increased to 0.1 L/min increased to 0.88 g/g glucose consumed. It should be pointed out that this yield result was calculated at the end of the experiment when the residual glucose concentration was still 7.7 g/L.

The effect of CO_2 sparge on the rate of acetic acid production followed a pattern similar to the one observed for the rate of glucose consumption, i.e., the gas sparge rates had to be increased above 0.025 L/min before a significant decrease in the production rate of acetic acid could be seen (Fig. 1D). The sparge of CO_2 gas into the fermenter, how-

Table 1
Effect of CO_2 Gas Sparge Rates on Succinic Acid and Acetic Acid Production at pH 6.0

CO_2 gas sparge rate (L/min)	0	0.025	0.05	0.1
Succinic acid yield (g/g glucose consumed)	0.93	0.69	0.71	0.88
Acetic acid yield (g/g glucose consumed)	0.21	0.20	0.19	0.21
Succinic acid:acetic acid (mole:mole)	2.23	1.77	1.94	2.08

Note: The results for the CO_2 gas sparge rate of 0.1 L/min were calculated at the end of the experiment when the residual glucose concentration was 7.7 g/L. All other results were calculated at the exhaustion of glucose.

ever, did not affect the acetic acid yield; in all four cases, the yield was unchanged at about 0.2 g/g glucose consumed. The net result of the negative effect on the succinic acid yield and the no-effect on the acetic acid yield was the highest molar ratio of succinic acid:acetic acid of 2.2 obtained in the control experiment.

The effect of pH is shown in Fig. 2 and summarized in Table 2. Highest glucose consumption rate was obtained at pH 6.0. At pH values both above and below 6.0, the rate of glucose consumption was significantly decreased. At pH 5.0, glucose was not consumed at all (Fig. 2A). The growth of cells followed the same pattern (Fig. 2B). Similar results were obtained for the production of succinic and acetica acids (Figs. 2C, D). For both products, the highest production rate was observed at pH 6.0. Maximum succinic acid yield also was obtained at this pH. At pH values above and below 6.0, the succinic acid yield was significantly decreased. Our results are in agreement with those of Datta's patent (7), in which it was reported that the succinic acid yield was significantly lowered at pH below and above 6.0. Samuelov et al. (13) also observed that the production of succinic acid was significantly lower at pH 7.2 than at pH 6.2. However, Datta (7) reported significant production of lactic acid (more than 20 g/L) at pH values above 6. In our study, very little lactic acid (less than 5 g/L) was produced. pH did not seem to affect the acetic acid yield. At all pH values studied, except pH 5.0 at which no glucose consumption

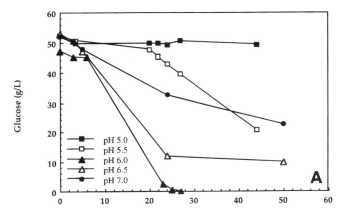

Fig. 2A. The effect of pH on glucose consumption in yeast extract-polypeptone medium without CO_2 gas sparge.

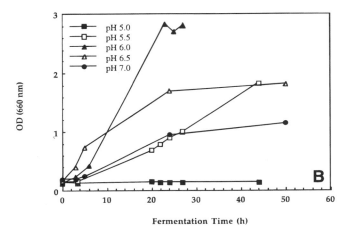

Fig. 2B. The effect of pH on cell growth in yeast extract-polypeptone medium without CO_2 gas sparge.

was observed, the acetic acid yield was unchanged at about 0.2 g acetic acid/g glucose consumed.

The results of succinic acid fermentation in the light steep water medium are shown in Fig. 3. In this study, a CO_2 sparge rate of 0.08 L/min, which was an intermediate value of the two highest CO_2 sparge rates used in the study of succinic acid production in the polypeptone-yeast extract medium, was used. Good production of succinic acid was observed. At 23 h, 24.5 g/L succinic acid was produced. This compared favorably with the succinic acid concentration of 22.1 and 12.5 g/L obtained in the polypeptone-yeast extract medium when the CO_2 sparge rates were 0.025 and 0.5 L/min, respectively. A maximum succinic acid concentration of 32.2 g/L was obtained at 27 h. This was equivalent to a productivity of 1.2 g/L-h.

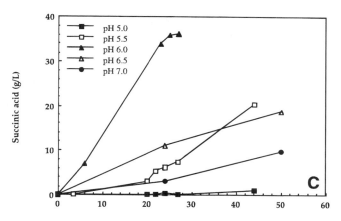

Fig. 2C. The effect of pH on succinic acid production in yeast extract-polypeptone medium without CO_2 gas sparge.

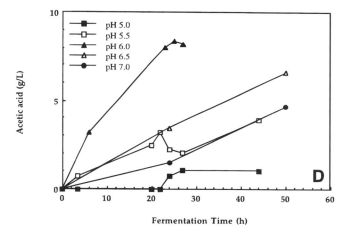

Fig. 2D. The effect of pH on acetic acid production in yeast extract-polypeptone medium without CO_2 gas sparge.

The yield calculated at maximum succinic acid concentration was 0.99 g succinic acid/g glucose consumed and the molar ratio of succinic acid:acetic acid was 1.9. Since the inoculum was raised in the polypeptone-yeast extract medium, there obviously were questions on the effect of nutrient carry-over. The concentrations of polypeptone and yeast extract carried over from the inoculum into the light steep water medium in the fermenter were 0.5 and 0.25 g/L, respectively. In our early study of *A. succiniciproducens* fermentation, it was observed that even with 20 g/L glucose, succinic acid could not be produced beyond 5 g/L at those levels of polypeptone and yeast extract. Therefore, the contribution of nutrients by these two organic nitrogen sources toward succinic acid production in the light steep water medium was insignificant.

Table 2
Effect of pH on Succinic Acid and Acetic Acid
Production Without CO_2 Gas Sparge

pH	5.5	6.0	6.5	7.0
Succinic acid yield (g/g glucose consumed)	0.74	0.93	0.53	0.41
Acetic acid yield (g/g glucose consumed)	0.15	0.21	0.19	0.20
Succinic acid:acetic acid (mole:mole)	2.65	2.23	1.45	1.05

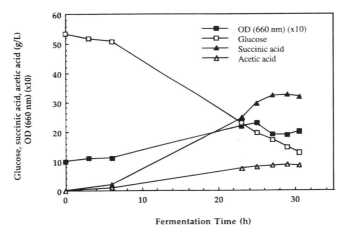

Fig. 3. Concentration profiles of OD(660), glucose, succinic acid, and acetic acid in light steep water medium at pH 6.0 and CO_2 gas sparge rate of 0.08 L/min.

The results showed that light steep water could replace yeast extract and polypeptone in the production of succinic acid by *A. succiniciproducens*. Light steep water is a waste product in a corn processing plant and normally is available at no cost. However, its use in a succinic acid fermentation process will be beneficial only if the succinic acid manufacturing plant is located next to the corn processing plant. The transportation costs of light steep water to a distant location will add significantly to the manufacturing costs of succinic acid. Its use for the production of succinic acid in this case will not be economical.

CONCLUSION

The effect of external supply of carbon dioxide and pH on the production of succinic acid by *A. succiniciproducens* have been studied. The following conclusions can be made:

1. In the yeast extract-peptone medium, when the external carbon dioxide suppy was provided by a $1.5M$ Na_2CO_3 solution that also was used to maintain the pH at 6.0, no additional carbon dioxide supply was needed. In fact, sparging CO_2 gas into the fermenter at 0.025 L/min or higher rates resulted in significant decreases in both production rate and yield of succinate.
2. Under the same conditions, the production of acetate was not affected by sparging CO_2 gas into the fermenter.
3. The optimum pH (pH 6) for the production of succinic acid was found to be in agreement with previously reported results.
4. Succinic acid could be produced in an industrial-type inexpensive medium in which light steep water was the only source of organic nutrients. Under the conditions studied, succinate concentration reached a maximum of 32.2 g/L in 27 h with a yield of 0.99 g succinate/g glucose consumed.

ACKNOWLEDGMENT

This research was supported by the US Department of Energy at Oak Ridge National Laboratory, which is managed by Lockheed Martin Energy Research Corporation, under contract DE-AC05-96OR22464.

Michigan Biotechnology Institute provided the culture of *A. succiniciproducens* and assisted in the early stage of the fermentation study.

The light steep water was a gift from the A. E. Staley corn processing plant in Loudon, TN.

The submitted manuscript has been authored by a contractor of the US government under contract DE-AC05-96OR22464. Accordingly, the US government retains a nonexclusive, royalty-free license to publish or reproduce the published form of the contribution, or allow others to do so, for US government purposes

REFERENCES

1. Winstrom, L. O. (1978), *Kirk-Othmer Encyclopedia of Chemical Technology*, vol. 21, Wiley, New York, pp. 848–864.
2. Rao, V. N. M. (1988), US Patent 4782167.
3. Mabry, M. A., Prichard, W. W., and Zlemecki, S. B. (1985), US Patent 4550185.
4. Ling, L. B. and Ng, T. K. (1989), US Patent 4877731.
5. Lemme, C. and Datta, R. (1987), Eur. Pat. Appl. 249–773.
6. Glassner, D. A. and Datta, R. (1990), Eur. Pat. Appl. 389–103.

7. Datta, R. (1992), US Pat. 5143833.
8. Glassner, D. A. and Datta, R. (1992), US Pat. 5143834.
9. Gottschalk, G. (1986), *Bacterial Metabolism*, Springer-Verlag, New York, p. 21 and p. 244.
10. Sato, M., Nakahara, T., and Yamada, K. (1972), *Agr. Biol. Chem.* **36,** 1969–1974.
11. Hopgood, M. F. and Walker, D. J. (1966), *Aust. J. Biol. Sci.* **20,** 165–182.
12. Weimer, P. J. (1993), *Arch. Microbiol.* **160,** 288–294.
13. Samuelov, N. S., Lamed, R., Lowe, S., and Zeikus, J. G. (1991), *Appl. Environ. Microbiol.* **57,** 3013–3019.
14. Nghiem, N. P., Davison, B. H., Suttle, B. E., and Richardson, G. R. (1996), *Appl. Biochem. Biotechnol.* **57/58,** 633–638.
15. Playne, M. J. (1985), *J. Sci. Food Agriculture* **36,** 638–644.

Spiral Tubular Bioreactors for Hydrogen Production by Photosynthetic Microorganisms
Design and Operation

SERGEI A. MARKOV,* PAUL F. WEAVER, AND MICHAEL SEIBERT

National Renewable Energy Laboratory, 1617 Cole Boulevard, Golden, CO 80401

ABSTRACT

Spiral tubular bioreactors were constructed out of transparent PVC tubing for H_2 production applications. Both a cyanobacterial *Anabaena variabilis* mutant that lacks uptake hydrogenase activity and the photosynthetic bacterium *Rhodobacter* sp. CBS were tested in the bioreactors. Continuous H_2 photoproduction at an average rate of 19 mL · min^{-2} · h^{-1} was observed using the *A. variabilis* mutant under an air atmosphere (without argon sparging or application of a partial vacuum). The cyanobacterial photobioreactor was run continuously for over one month with an average efficiency of light energy conversion to H_2 of 1.4%. Another H_2-producing approach employed a unique type of activity found in a strain of photosynthetic bacteria that shifts CO (and H_2O) into H_2 (and CO_2) in darkness. Continuous dark H_2 production by *Rhodobacter* sp. CBS from CO (in anticipation of using synthesis gas as the future substrate) at rates up to 140 mL · g cdw^{-1} · h^{-1} was observed in a bubble-train bioreactor for more than 10 d.

Index Entries: Hydrogen; bioreactors; *Anabaena variabilis*; water-gas shift reaction; *Rhodobacter*.

INTRODUCTION

Hydrogen is considered to be an environmentally desirable fuel because it can be produced from renewable resources, and its combustion product (water) is nonpolluting. Several biological approaches are being used to produce H_2 either from water and solar energy or from biomass [1,2]. The main challenge here is to design simple, efficient bioreactors that

*Author to whom all correspondence and reprint requests should be addressed.

consume as little energy as possible. In recent years, several groups have studied the efficacy of tubular bioreactors for cultivation of photosynthetic micro-organisms (3–5). In comparison to open ponds or tank reactors, these tubular bioreactors have the following advantages:

1. A high surface area to culture volume ratio, allowing photosynthetic micro-organisms to absorb light energy more effectively;
2. Better gas mass transfer rates into liquid media; and
3. Low mixing energy requirements.

In the present study, two simple polyvinyl chloride (PVC) tubular bioreactors vertically spiraled to facilitate continuous H_2 production by photosynthetic micro-organisms were analyzed.

The first type of bioreactor incorporated a cyanobacterial mutant of *Anabaena variabilis*. Hydrogen was photoevolved from water and released from solution at atmospheric pressure, under conditions where the organisms were continuously exposed to ambient levels of O_2. This was possible because of the use of the *A. variabilis* PK84 mutant obtained from Prof. S. V. Shestakov (Moscow State University) that lacks uptake hydrogenase activity (6). The second type of bioreactor utilized a unique type of H_2-producing activity originally found in a strain of photosynthetic bacteria by Uffen (7). Fermentative dark cultures of this strain in complex media with CO carried out a water-gas shift reaction to produce H_2 according to the reaction, $CO + H_2O \rightarrow CO_2 + H_2$. Numerous strains of photosynthetic bacteria, including *Rhodobacter* sp. CBS, have been isolated at the National Renewable Energy Laboratory that utilize CO in the light as well as in darkness and do not require complex organic substrates (8). These strains quantitatively shift the CO component of synthesis gas (e.g., from thermally gasified biomass) into H_2. However, mass transport of gaseous CO into an aqueous bacterial suspension is the rate-limiting step in the process and was the main concern in bioreactor design for the current study.

MATERIALS AND METHODS

Microbial Cultures

Prior to inoculation into bioreactors, *A. variabilis* PK84 was grown with shaking as a batch culture in the medium of Allen and Arnon (9) without combined nitrogen. Continuous light was provided by cool white fluorescent lamps ($3.0 \text{ W} \cdot \text{m}^{-2}$). Before inoculation into the bioreactor, *Rhodobacter* sp. CBS was cultivated as a batch culture in closed bottles on basal medium (10) plus 10% CO with shaking and illuminated with incandescent lamps ($35 \text{ W} \cdot \text{m}^{-2}$). Cell dry weights were determined by trapping the cyanobacteria or bacteria on Whatman #114 filter paper and drying the cell suspensions at 90°C to constant weight.

Spiral Tubular Bioreactors

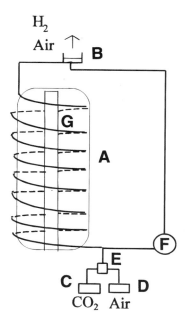

Fig. 1. Schematic diagram of a helical PVC tubular photobioreactor for H_2 production by an *A. variabilis* mutant. **A**, PVC tubing; **B**, H_2 measurement port; **C**, CO_2 gas cylinder; **D**, an air input line; **E**, rotameter; **F**, pump; **G**, lamp.

Bioreactors

Two types of bioreactors were designed and constructed. The first was a 2 L (total volume), 0.4 m high photobioreactor for cyanobacteria as shown in Fig. 1. The bioreactor consisted of:

1. A 42 m transparent PVC (Nalgene, Rochester, NY), 7.9 mm internal diameter tube wound helically on a vertical transparent cylindrical supporting structure;
2. A H_2 measurement port, where H_2 and air are vented from the cyanobacterial suspension (gas flow rate 46 mL· min^{-1});
3. A CO_2 gas cylinder;
4. An air input line;
5. A gas proportioning rotameter (Omega, Stamford, CT), which was used to measure and mix CO_2 and air;
6. A peristaltic pump (Masterflex, Cole-Palmer Instrument, Niles, IL) for circulating (35 mL · min^{-1}) the cyanobacterial suspension. Cell concentration for the cyanobacterial bioreactor during inoculation was 0.65 mg cdw · mL^{-1} and increased because of cyanobacterial growth during the bioreactor operation. It was difficult to measure

accurately the cell biomass during the bioreactor operation because of cell adhesion to the bioreactor walls; and
7. A 33 W cool white fluorescent lamp.

The bioreactor suspension was bubbled with a mixture of CO_2 (about 5%) and air through a needle/septum connection at the base of the photobioreactor to supply the cells with a carbon source and remove H_2. The inner cylindrical surface of the bioreactor (0.22 m^2) was illuminated continuously with fluorescent light (average irradiance 3.0 $W \cdot m^{-2}$). Light irradiance was measured using a radiometer (Model 65A, Yellow Spring Instruments, Yellow Springs, OH) at different points on the inner surface of the cyanobacterial photobioreactor.

The second bioreactor was a 0.5 L (liquid volume), 0.8 m high device for the dark bacterial production of H_2 from CO (and H_2O) as diagrammed in Fig. 2. The bioreactor was constructed from:

1. A 9.8 m transparent PVC (Tygon, Akron, OH), 6.3 mm inner diameter, tube wound helically on a vertical cylindrical supporting structure;
2. A pump (Masterflex, Cole-Palmer Instrument) for circulating (pumping speed 15 $mL \cdot min^{-1}$) the bacterial suspension (0.36 $mg\ cdw \cdot mL^{-1}$);
3. A port for injection of the bacterial suspension into the PVC tubing;
4. A needle injector for 20% CO in N_2 (2 $mL \cdot min^{-1}$); and
5. A 300 mL gas reservoir.

The bioreactor was designed so that small bubbles containing CO were injected continuously through a needle/septum connection from the gas reservoir (initially 20% CO in N_2). The bubbles rose with the pumped medium from the bottom of the bioreactor to the top (3.5 min transit time). The high surface area of the bubble train promoted enhanced mass transport of gaseous CO into the aqueous bacterial suspension. In order to keep the medium pH from dropping due to bicarbonate build-up, the gas phase of the reservoir was degassed with N_2 once every day, and then CO (20% in a N_2 balance) was reinjected into the system. The bioreactor was covered with a black cloth to prevent photosynthetic H_2 consumption from exposure to ambient light according to the reaction: $2H_2 + CO_2 \rightarrow (CH_2O)_n + H_2O$.

The cyanobacterial bioreactor was sterilized with a 5% sodium hypochlorite solution and washed with sterile distilled water several times before inoculation. The photosynthetic bacterial bioreactor did not require sterilization because CO is either toxic to, or will not support growth of, most potential invading organisms. Both bioreactors were maintained at room temperature (23–24°C).

Hydrogen Production

Hydrogen production rates were measured using a Varian Model 3700 gas chromatograph (Walnut Creek, CA) equipped with a molecular

Spiral Tubular Bioreactors

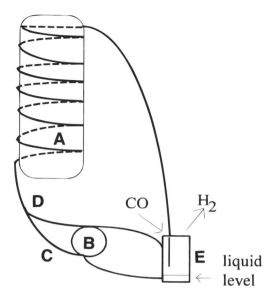

Fig. 2. Schematic diagram of a helical PVC tubular bioreactor to shift CO into H_2 by *Rhodobacter* sp. CBS. **A**, PVC tubing; **B**, pump; **C**, bacterial suspension entrance; **D**, CO injector; **E**, gas reservoir.

sieve 5A column and a thermal conductivity detector. Argon was used as the carrier gas. Light energy conversion efficiencies (to H_2) in the cyanobacterial photobioreactor were calculated as follows:

$$\text{Efficiency (\%)} = \frac{H_2 \text{ production rate} \times H_2 \text{ energy content}}{\text{Incident Light Irradiance}} \times 100\% \quad (1)$$

The heat of H_2O formation, (241,000 J · mol^{-1}) was used as the energy content of the H_2 produced.

RESULTS

Operation of the Photobioreactor for H_2 Production by *A. variabilis*

H_2 production by the *A. variabilis* mutant is shown in Fig. 3. Initially, H_2 production increased as the cyanobacterial culture grew and then decreased as the cyanobacteria aged. H_2 production was observed for about one month during the period that the culture was most active. The percentage of H_2 in the effluent gas varied from 0.03 to 1%. After 25 d under H_2-producing conditions, the cyanobacterial culture appeared to turn more green in color from its natural blue-green appearance. This change in color coincided with the loss of H_2 evolution activity, which in turn was probably a result of nutrient limitation. The CO_2 was consumed during the bioreactor run (<1% at the exit port). Cell adhesion to the bioreactor walls was also observed. Control exper-

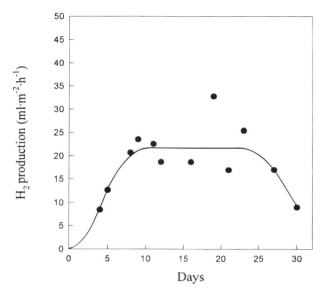

Fig. 3. H_2 production by an *A. variabilis* mutant in the PVC tubular photobioreactor.

iments, after removal of all suspended nonbound cells, indicated that the adsorbed cells produce H_2. These adsorbed cells could be removed easily by scouring the PVC tubing with pressurized air from the bottom end *(11)*.

The efficiency of light energy conversion to H_2 in the photobioreactor was calculated using Equation 1. The average rate of H_2 production over 30 d of the run (Fig. 3) was 18.9 mL · m^{-2} · h^{-1}, which is equal to 6.29 · 10^{-4} mol · m^{-2} · h^{-1} or 1.74 · 10^{-7} mol · m^{-2} · s^{-1}. Using the aforementioned information, we calculate an average efficiency of

$$\frac{1.74 \cdot 10^{-7} \text{ mol} \cdot \text{m}^{-2} \cdot \text{s}^{-1} \times 24 \times 10^5 \text{ J} \cdot \text{mol}^{-1}}{3 \text{ J} \cdot \text{m}^{-2} \cdot \text{s}^{-1}} \times 100\% = 1.4\% \qquad (2)$$

Operation of the Bioreactor for H_2 Production by *Rhodobacter* sp. CBS

Continuous H_2 production from CO at rates up to 140 mL H_2 · g cdw^{-1} · h^{-1} was observed in a bubble-train bioreactor for more than 10 d (Fig. 4). Rates of H_2 production were low at first, probably because of the exposure of the bacterial culture to O_2 during bacterial transfer to the bioreactor. Then, under more favorable anaerobic conditions for the bacteria in the bioreactor, rates of H_2 production started to increase. At the higher rates, 2 h was sufficient to shift all of the added CO in the reservoir gas phase into H_2. No detectable level of CO remained in the gas phase (less than 18 ppm). The bulk of the added CO was shifted during the first hour after feeding. Repetitive batch feeding of CO (the gas phase was changed once a day and reestablished with 20% CO in N_2) maintained the culture in a highly active

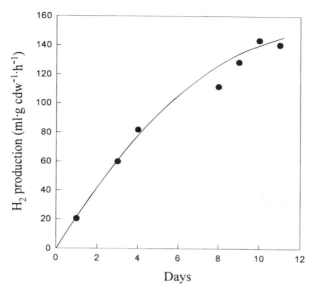

Fig. 4. Shifting of CO into H_2 by *Rhodobacter* sp. CBS in the PVC tubular dark bioreactor.

state. The product gas, containing up to 20% H_2 in N_2 and devoid of any remaining CO, was sufficiently clean for direct injection into a H_2 fuel cell.

DISCUSSION

In this preliminary study, helical bioreactors made of transparent PVC tubing were employed for the first time to examine H_2 production by two distinct types of micro-organisms. It is shown that it is possible to produce H_2 using this type of simple, low-cost bioreactor. In the case of the photobioreactor employing the *A. variabilis* mutant, the important new result was that H_2 could be produced from water under ambient conditions. Previous work employed an argon sparging system (90% argon with CO_2 and N_2) to produce H_2 from a tubular (glass) bioreactor (12), a system inherently more complex. The second bioreactor system (bubble-train bioreactor) used *Rhodobacter* sp. CBS to produce H_2 from CO (in anticipation of using synthesis gas as substrate). Prolonged movement of small bubbles of CO through this bioreactor increased the contact time between the bacterial suspension and CO, which enhanced mass transport of the gas into the aqueous bacterial suspension. The bioreactor is now being modified by adding more PVC tubing so that the bulk of the CO will be shifted in a single pass of entrained bubbles.

Our results with cyanobacterial photobioreactors suggest that to achieve long-term operation and steady-state H_2 production levels, it may be necessary to supply fresh medium periodically and remove the old cells. By doing this, the authors expect to achieve H_2 production for periods as long as 9 mo. They have operated continuous, H_2-producing bioreactors for such periods of time with the same microbial cells immobilized on hollow-fiber arrays *(13)*. However, the cost of tubular PVC bioreactors is significantly less than that of hollow-fiber bioreactors. Development of a computerized bioreactor system will also help to optimize H_2 production by careful regulation of the gas supply and control of pH *(14)*. To improve the energy balance of the bioreactor system, air-lift designs will save on energy consumption by pumps. Also, to further improve the economics, H_2 production can be combined with the synthesis of secondary products, such as commodity chemicals or animal feed.

ACKNOWLEDGMENTS

This work was supported by the U.S. Department of Energy Hydrogen program.

REFERENCES

1. Weaver, P., Lien, S., and Seibert, M. (1980), *Solar Energy* **24,** 3–45.
2. Markov, S. A., Bazin, M. J., and Hall, D. O. (1995), *Adv. Biochem. Engineer. Biotechnol.* **52,** 59–86.
3. Lee, E. T.-Y. and Bazin, M. J. (1990), New Phytol. **116,** 331–335.
4. Watanabe, Y., de la Noüe J., and Hall, D. O. (1995), *Biotechnol. Bioeng.* **47,** 261–269.
5. Tredici, M. and Materassi, R. (1992), *J. Appl. Phycol.* **4,** 221–231.
6. Mikheeva, L. E., Schmitz, O, Shestakov S. V., and Bothe, H. (1995), *Zeitschrift für Naturforschung* **50,** 505–510.
7. Uffen, R. L. (1976), *Proc. Nat. Acad. Sci. USA* **73,** 103–119.
8. Maness, P.-C. and Weaver, P. F. (1994), *Appl. Biochem. Biotechnol.* **45/46,** 395–406.
9. Allen, M. B. and Arnon, D. I. (1955), *Plant Physiol.* **30,** 366–372.
10. Schultz, J. and Weaver, P. F. (1981), *J. Bacteriol.* **149,** 181–190.
11. Lee, Y.-K., Ding, S.-Y., Low, C.-S., Chang Y.-C., Forday, W. L. and Chew, P.-C. (1995), *Appl. Phycol.* **7,** 47–51.
12. Miyamoto, K., Hallenbeck, P. C., and Benemann, J. R. (1979), *J. Ferment. Technol.* **57,** 287–293.
13. Markov, S. A., Weaver, P., and Seibert, M. (1996), in *Hydrogen Energy Progress XI, Proceeding of the 11th World Hydrogen Energy Conference*, Stuttgart, Germany, June 23–28, 1996, vol. 3, Veziroğlu, T. N., Winter, C.-J., Baselt, J. P., and Kreysa, G., eds., pp. 2619–2624.
14. Markov, S. A., Thomas A., and Bazin M. J. (1994), in *Abstracts of VIII International Symposium on Phototrophic Prokaryotes*, Urbino, Italy, September 10–15, 1994. p. 106.

Use of a New Membrane-Reactor Saccharification Assay to Evaluate the Performance of Cellulases Under Simulated SSF Conditions

Effect on Enzyme Quality of Growing *Trichoderma reesei* in the Presence of Targeted Lignocellulosic Substrate

JOHN O. BAKER,* TODD B. VINZANT, CHRISTINE I. EHRMAN, WILLIAM S. ADNEY, AND MICHAEL E. HIMMEL

Enzyme Technology Team, Biotechnology Center for Fuels and Chemicals, National Renewable Energy Laboratory, Golden, CO

ABSTRACT

A new saccharification assay has been devised, in which a continuously buffer-swept membrane reactor is used to remove the solubilized saccharification products, thus allowing high extents of substrate conversion without significant inhibitory effects from the buildup of either cellobiose or glucose. This diafiltration saccharification assay (DSA) can, therefore, be used to obtain direct measurements of the performance of combinations of cellulase and substrate under simulated SSF conditions, without the saccharification results being complicated by factors that may influence the subsequent fermentation step. This assay has been used to compare the effectiveness of commercial and special in-house-produced *Trichoderma reesei* cellulase preparations in the saccharification of a standardized microcrystalline (Sigmacell) substrate and a dilute-acid pretreated lignocellulosic substrate. Initial results strongly suggest that enzyme preparations produced in the presence of the targeted lignocellulosic substrate will saccharify that substrate more effectively. These results call into question the widespread use of the "filter paper assay" as a reliable predictor of enzyme performance in the extensive hydrolysis of substrates that are quite different from filter paper in both physical properties and chemical composition.

*Author to whom all correspondence and reprint requests should be addressed.

Index Entries: Cellulase digestion; *Trichoderma reesei*; pretreated hardwood; diafiltration saccharification; membrane reactor.

INTRODUCTION

Cellulase enzymes are widely sold, and their industrial utilization projected, on the basis of the "filter-paper unit" of activity. The "filter-paper assay" *(1)* is, however, severely limited as a predictor of cellulase performance in the extensive (80–90%-plus) saccharification of actual industrial lignocellulosic substrates. These limitations are traceable both to the chemical and physical differences between filter paper and the industrial substrates, and to the nonhomogeneous nature of most cellulosic substrates (filter paper included). As a result of the nonhomogeneity of the substrates, assays that are run to very limited extents of conversion (such as the 4% conversion target in the filter-paper assay) *(1)* measure the digestibility of only the most easily digestible fraction of the substrate, and reveal little about the convertibility of the bulk of the substrate.

Actual performance of cellulases is estimated better by assays that utilize the actual application substrate *and* are run to the extents of conversion required in the process. Because of the inhibitory nature of the products of cellulase action (primarily glucose and cellobiose), such high-conversion assays encounter the problem of significant product inhibition, if run as simple saccharifications in "closed" systems *(2–5)*. Assays in which product inhibition is a significant factor cannot reliably predict the performance of a cellulase/substrate combination under conditions of simultaneous saccharification and fermentation (SSF), where consumption of the solubilized sugars by the fermentative organism holds the concentrations of saccharification products to very low pseudo-steady-state levels. One obvious strategy for getting around the problem of cellulase inhibition by saccharification products is simply to use actual small-scale SSF as the assay of cellulase effectiveness. This, however, provides a very indirect measure of cellulase performance, because the final results depend not only on cellulase performance, but also on the vagaries of microbial metabolism. Such "SSF assays" cannot, for instance, readily distinguish poor overall performance because of cellulase ineffectiveness from that caused by the toxicity to the fermenting organism of substances that may be found in chemically pretreated biomass. To deal effectively with such problems during process development, one must know which class of problem one is encountering. In addition, the quantitation of volatile products (such as ethanol) under the conditions of such small-scale, improvised SSF assays may present more severe analytical challenges than are posed by sugar determination in liquid samples using standard HPLC methods.

The membrane-diafiltration assay described herein achieves both high extents of saccharification of solid cellulosic substrates and the maintenance of low concentrations of inhibitory soluble products, thus

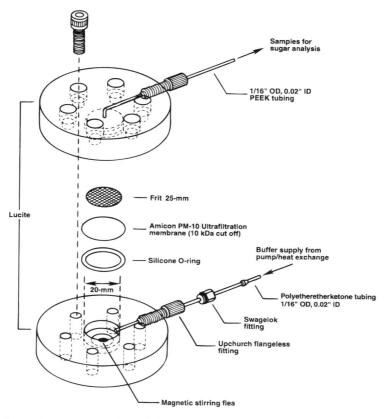

Fig. 1. Schematic representation of the diafiltration saccharification assay (DSA) cell.

effectively mimicking SSF conditions. In the DSA approach, a magnetically-stirred membrane reactor vessel (Fig. 1) is constantly swept by a buffer flux. Cellulase enzymes and solid cellulosic substrate particles are confined to the reaction vessel by an ultrafiltration membrane (10-kDa cutoff) at the exit side of the cell and, at the entrance side of the vessel by the relatively high linear velocity of incoming buffer passing through the small-diameter (0.02-in) entrance port. Soluble saccharification products are swept through the ultrafiltration membrane and out of the cell for detection by HPLC.

Figure 2 presents a schematic diagram of the overall apparatus used in collecting diafiltration-saccharification data. Essential components are an HPLC pump used to deliver buffer, at highly consistent flow rates, through a heat-exchanger to the "high-pressure" side of a custom-built membrane reactor that is temperature-controlled in an oven, and ultimately to a fraction collector set to collect timed fractions. The saccharification progress curve is monitored by weighing the tared fraction-collector tubes (which, for these dilute solutions, gives a satisfactorily accurate estimate of the fraction volume) and then determining the sugar concentra-

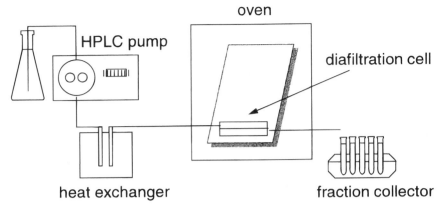

Fig. 2. Schematic of the overall DSA system.

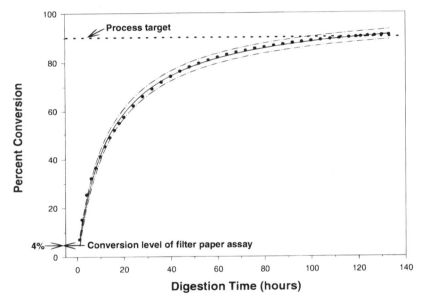

Fig. 3. Illustrative example of DSA data, comparing the relative magnitudes of the cumulative substrate conversion in the DSA with those of the target conversions in the filter-paper assay and in the bioethanol process. Total protein loading (*T. reesei* EG I + CBH I in a 40:60 molar ratio) was equivalent to the protein loading in a 20-FPU/g cellulose loading of native *T. reesei* cellulase. Substrate was 5% (w/v) Sigmacell type 20, in pH 5.0, 20 mM acetate; reaction temperature was 50°C. The solid curve is a nonlinear best fit to the data; the dotted lines delineate the 95% confidence interval.

tions of the fractions by HPLC. The cumulative sugar production is then compared with the quantity of cellulose originally loaded into the reactor, as illustrated in Fig. 3, which also graphically compares the extents of conversion achieved in typical diafiltration assays with the target conversion for the filter paper assay.

The basic idea behind the diafiltration assay was presented in 1975 by Howell and Stuck (6), who used a standard membrane concentrator with periodic manual buffer replenishment to accomplish the same result. The advantages of our custom-built apparatus are that

1. It has a substantially smaller volume, which reduces the required quantities of purified enzymes when such are being assayed;
2. It maintains a constant reactor volume and well-controlled buffer flux, thus allowing accurate description of the product concentrations seen by the enzymes at different times during the reaction; and
3. It can operate unattended for days, if need be.

MATERIALS AND METHODS

Diafiltration Assay Apparatus Construction

The diafiltration reactor cell was milled from Lucite to have a "high-pressure" chamber 2.2 mL in volume, and utilizes standard 25-mm diameter Amicon PM-10 membranes (10,000 MW cutoff) and the corresponding Amicon O-rings. Inlet and outlet ports are 11-cm lengths of polyetheretherketone (PEEK) tubing with 0.02-in inside diameter, connected to milled, threaded ports on the cell and connected to input and output lines with Swagelok fittings. The precell heat-exchanger shown in Fig. 2 is formed from a coiled, 10-ft length of 1/8 in. stainless-steel tubing.

Sugar concentrations in collected effluent fractions were determined by chromatography on a Bio-Rad HPX-87P (lead-form) carbohydrate analysis column (Bio-Rad deashing precolumns), installed in a Hewlett-Packard 1090 chromatograph and operated at 85°C with deionized water as mobile phase at a flow rate of 0.6 mL/min.

Enzyme Sources

As an example of commercial cellulase produced using soluble sugar as both inducer and carbon source (7), Iogen Cellulase, lot PRC-191095, was obtained from Iogen, Inc., Ottawa, Ontario, Canada. Spezyme CP, lot GC 30952.3E1P1Z1, was obtained from Genencor, South San Francisco, CA. Substrate-induced *T. reesei* cellulase (hereinafter referred to as "PYP-grown cellulase") was produced in our laboratory by growing the organism (NREL strain MTCA-13, which was derived from MG-80) in the presence of 1% pretreated yellow poplar and 6% lactose.

The growth medium used in this study was based on the original Mandels medium (8), substituting corn steep with a 50/50 mixture of peptone and yeast extract. The exact composition was: $CaCl_2$, 0.4 g/L; $MgSO_4$, 0.3 g/L; KH_2PO_4, 2 g/L; $(NH_4)_2SO_4$, 1.4 g/L; peptone, 5 g/L; yeast extract, 5 g/L; Tween 80, 0.2 mL/L; and trace mineral solution.

All the fermentations were carried out using 10 L of the above medium in New Brunswick Scientific (NBS) 14 L Microferm bioreactors (series MGF114) with agitation and temperature control. The pH was controlled using a model 7600 Cole-Parmer controller with an Ingold probe. Dissolved oxygen level was maintained using an NBS dissolved oxygen controller model DO-81 and an Ingold galvanic probe. The fermentation vessel was modified by adding a port to allow for homogenous sampling from the bottom of the reactor.

The media components described previously (9) were sterilized in the presence of the pretreated poplar substrate in the bioreactor. Lactose was added aseptically post sterilization. The reactors were allowed to equilibrate to temperature (26°C), at which time the dissolved oxygen and pH probes were calibrated. Adjustments to pH were automatically controlled with the additions of $3M$ H_3PO_4 and $3M$ NH_4OH. The concentration of dissolved oxygen was maintained at 20% by the automatic adjustment to the flow rate of a 50:50 mixture of compressed air and pure oxygen. The fermentation mixture was constantly mixed at 300 rpm for maximum oxygen mass transfer and minimum shear.

The fermentations were inoculated with a 10% culture produced in two shake flasks; this constituted time zero. Fifty-mL samples were taken at 24-h intervals until the cellulase production leveled. The activity was determined using the FPU assay (1) or general glucose release from filter paper and recorded as a function of fermentation time.

Cellulosic Substrates

Microcrystalline cellulose (Sigmacell, Type 20) was purchased from Sigma Chemical (St. Louis, MO), and was weighed, dry, into the cell. Pretreated yellow poplar (PYP) sawdust used in SSF studies was subjected to dilute-acid pretreatment as described previously (9). PYP used in DSAs was further subjected to wet milling using an Ultra-Turrax (Tekmar, Cincinnati, OH) to reduce the particle size to a distribution in the same general range as that of Sigmacell, type 20. This finely divided material was then exchanged into 20 mM acetate buffer, pH 5.0, containing 0.02% (w/v) sodium azide. A procedure was developed for pipetting from a well-stirred 3.54% (w/v) slurry of this material to produce a series of standardized substrate aliquots that contained an average of 0.0520 g (dry weight) of PYP, with a standard deviation (n = 4) of ±0.7%. Two of these aliquots were quantitatively transferred to the diafiltration cell as PYP loadings for each assay.

Compositional analysis, carried out as described in reference 9, revealed that the PYP consisted of 67.7% glucan, 1.68% xylan, 32.42% Kjehldal lignin, 1.45% acid-soluble lignin, and 0.39% total ash. Particle-size analysis of both substrates was carried out using a Coulter LS-130 light-scattering analyzer with a fluid module.

Simultaneous Saccharification and Fermentation

Effectiveness of the two enzyme preparations was compared following the standard (shake-flask, quadruplicate) procedure described earlier *(9)*, except that the enzyme loadings were different from those used in the previous study.

Diafiltration Assay Conditions

All DSAs were carried out at pH 5.0 in 20 mM sodium acetate buffer containing 0.02% (w/v) sodium azide to prevent microbial growth, at either 37 or 50°C. All substrate loadings were equal in terms of cellulose content, being set at 70.4 mg cellulose in each 2.2-mL digestion mixture. In the case of PYP, a loading of 104 mg (dry wt) of biomass was required to achieve this cellulose loading. Solids loadings (as opposed to cellulose loadings) were thus approx 3.2% (w/v) for the microcrystalline cellulose, and approx 4.7% for PYP. Cellulase enzyme loadings were based on cellulose content of the substrate and were fixed at 20 FPU/g cellulose, using filter-paper activities determined in our laboratory for both enzyme preparations. Buffer flux through the 2.2-mL reaction cell was 0.072 mL/min.

RESULTS AND DISCUSSION

Substrate Particle-Size Analysis

Both substrates used in the DSA cell were finely divided material. As shown in Fig. 4, the two size-distributions are significantly different, but the two substrates, Sigmacell Type 20 and PYP, can be described as roughly comparable in terms of particle size. If simple particle size should affect access by the enzymes to potential cleavage sites, the "smaller" Sigmacell substrate would be favored.

SSF Comparisons

Figure 5 shows the performance of the two enzyme preparations in the SSF of PYP. At both the 5 and 10 FPU/g enzyme loadings, the SSF mixtures containing the enzyme produced by *T. reesei* in the presence of PYP outperformed the commercial preparation grown on soluble sugars alone, although the preparations were used at equal loadings in terms of "filter-paper activity." This is not an especially surprising result, the suggestion having been made earlier that the presence of the actual intended "target" biomass during enzyme production may induce the production of more appropriate ratios of major cellulase components, or of minor but important ancillary enzymes important for digesting complex substrates *(10–12)*. We believe, however, that until now there has been little if any published systematic experimental testing of this idea.

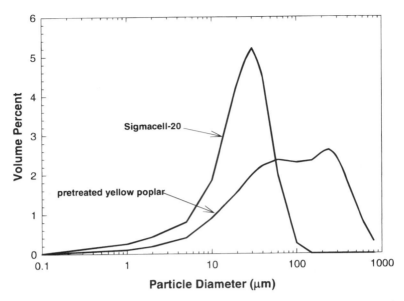

Fig. 4. Comparative particle size distribution for the two insoluble cellulosic substrates used in this study, Sigmacell type 20 and pretreated yellow poplar (PYP).

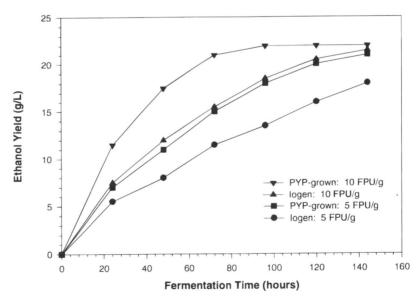

Fig. 5. Use of actual SSF assay to evaluate the relative effectiveness of PYP-grown *T. reesei* cellulase and commercial cellulase grown on soluble sugars alone, in the fermentation of PYP. Fermentations by *S. cerevisiae* D_5A were carried out at 37°C at pH 5.0 in a total volume of 100 mL in 250-mL Erlenmeyer flasks shaken at 15/s. Solids loading was 10%, with enzyme loadings of either 5 FPU or 10 FPU of the tested cellulase per g cellulose, plus 25 βGU (NOVO SP188) per g cellulose.

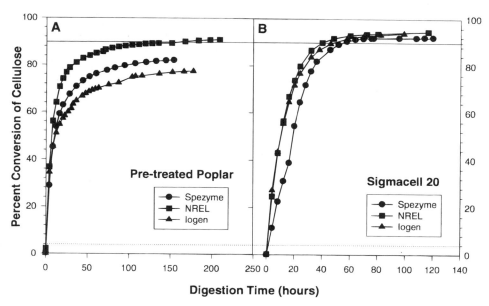

Fig. 6. Use of the direct DSA to evaluate the effectiveness of three different *T. reesei* enzyme preparations in saccharifying (**A**) PYP and (**B**) microcrystalline cellulose (Sigmacell, type 20) at pH 5.0 and 37°C. Substrate concentrations were 70.4 mg cellulose in each 2.2-mL reactor volume; in the case of PYP, this cellulose loading was achieved by loading 104 mg of pretreated biomass in each cell. Enzyme loadings were fixed at 20 FPU/g cellulose, based on filter-paper activities determined in our laboratory.

Diafiltration-Saccharification Comparisons

In addition to being both interesting and significant in their own right, the SSF results provided an opportunity to "calibrate" the DSA as a method of predicting SSF performance by combinations of cellulase enzymes and cellulosic substrates. As shown in Fig. 6A, the directly-measured saccharification results correlate well with the SSF results, in that the enzyme produced in the presence of PYP is markedly superior, both in terms of kinetics and apparent ultimate yields, in the saccharification of PYP. The PYP-grown enzyme reaches 80% conversion in less than 33 h digestion; at this time a equivalent loading (in terms of filter-paper activity) of the commercial preparation, produced on soluble sugars alone, has reached only 63% conversion, and does not in fact reach the 80% level even by 176 h. The PYP-grown enzyme goes on to achieve 90% conversion of the cellulose content by 182 h. The performance of a second commercial cellulase preparation, Spezyme (Genencor International) is intermediate between the performances of the NREL and Iogen enzymes.

The saccharification progress curves shown in Fig. 6B strongly suggest that the differences seen in Fig. 6A are substrate-dependent. In the digestion of microcrystalline cellulose (Sigmacell, type 20), equivalent filter-paper-activity loadings of the two enzymes perform equivalently (Fig. 6B). The major compositional difference between the two substrates is that Sigmacell is essentially all cellulose, whereas PYP (as noted under MATERIALS AND METHODS) is more than 32% Kjehldal lignin. It may be that the breaking of tight lignin carbohydrate complexes is key to the enhanced kinetics and 11%+ better ultimate yield (relative to that of the Iogen preparation) shown by the PYP-grown enzyme in the digestion of the PYP as a substrate. If so, either breakdown products from the substrate, or the simple presence of the macromolecular matrix during production of the enzyme mixture, may serve as an inducer of an enzyme (or enzymes) that may be minor in quantity of protein, but very important in terms of activity on actual process substrates.

CONCLUSIONS

The ability to obtain 80–90% conversion of biomass cellulose by SSF in 5–7 d at reasonable cellulase enzyme cost is considered to be an important requirement for economic viability of an SSF-based bioethanol process *(13,14)*.

The results of this study strongly suggest that production of cellulase enzyme mixtures in the presence of the actual target substrate is a useful approach to optimizing enzyme mixtures for reaction time and yield criteria. Furthermore, it has been demonstrated that the DSA assay is an effective and enzyme-efficient means of predicting the SSF performance of enzymes according to both of these criteria.

ACKNOWLEDGMENTS

This work was funded by the Biochemical Conversion Element within the Biofuels Systems Program of the Office of Fuels Development of the US Department of Energy.

REFERENCES

1. Ghose, T. K. (1987), *Pure Appl. Chem.* **59,** 257–268.
2. Ladisch, M. R., Lin, K. W., Voloch, M., and Tsao, G. T. (1983), *Enz. Microb. Technol.* **5,** 82–102.
3. Lee, Y.-H. and Fan, L. T. (1983), *Biotechnol. Bioeng.* **25,** 939–966.
4. Holtzapple, M., Cognata, M., Shu, Y., and Hendrickson, C. (1990), *Biotechnol. Bioeng.* **36,** 275–287.
5. Gusakov, A. V. and Sinitsyn, A. P. (1992), *Biotechnol. Bioeng.* **40,** 663–671.
6. Howell, J. A. and Stuck, J. D. (1975), *Biotechnol. Bioeng.* **17,** 873–893.
7. Douglas, L. J. (1989), *A Technical and Economic Evaluation of Wood Conversion Processes; Entropy Associates Report of Contract File 051SZ.23283-8-6091*, Efficiency and Alternative Energy Technology Branch, Energy, Mines and Resources, Ottawa, Ontario, Canada.
8. Mandels, M., Weber, J., and Parizek, R. (1971), *Appl. Microbiol.* **21,** 152–154.
9. Vinzant, T. B., Ponfick, L., Nagle, N. J., Ehrman, C. I., Reynolds, J. B., and Himmel, M. E. (1994), *Appl. Biochem. Biotechnol.* **45/46,** 611–626.

10. Esterbauer, J., Steiner, W., Labudova, I., Hermann, A., and Hayn, M. (1991), *Bioresource Technol.* **36,** 51–65.
11. Pourquie, J. and Warzywoda, M. (1993), in *Bioconversion of Forest and Agricultural Plant Residues,* Saddler, J. N., ed., CAB International, pp. 107–116.
12. Hayn, M., Steiner, W., Klinger, R., Steinmuller, H., Sinner, M., and Esterbauer, H. (1993), in *Bioconversion of Forest and Agricultural Plant Residues,* Saddler, J. N., ed., CAB International, pp. 33–72.
13. Lynd, L. R., Cushman, J. H., Nichols, R. J., and Wyman, C. E. (1991), *Science* **251,** 1318–1323.
14. Hinman, N. D., Schell, D. J., Riley, C. J., Bergeron, P. W., and Walter, P. J. (1992), *Appl. Biochem. Biotechnol.* **34/35,** 639–649.

Copyright © 1997 by Humana Press Inc.
All rights of any nature whatsoever reserved.
0273-2289/97/63-65—0597$8.25

Session 4

Industrial Needs for Commercialization

DALE MONCEAUX[1] AND JAMES L. GADDY[2]

*[1]Raphael Katzen Associates International, Cincinatti, OH;
and [2]Bioengineering Resources, Fayetteville, AR*

This technical discussion session focused on the obstacles to commercial applications of bioprocesses for production of fuels, chemicals, and materials. Commercial successes in these technology areas have been slow to emerge, not because of a lack of innovative or cogent technology, but for other reasons. The presentations and discussions in this session sought to identify these reasons and to suggest solutions.

The panel of distinguished industrialists that addressed these issues included:

Dr. Charles Abbas, Group Leader, Research Division,
　Archer Daniels Midland
Dr. Robert Dorsch, Director Biotechnology Development,
　Central R&D, DuPont
Dr. Raphael Katzen, Chairman, Raphael Katzen Associates
Dr. Robert E. Lumpkin, Vice-President Technology, Swan Biomass

The model for industrial development in the commodity biotech industry is ethanol, produced in the United States in quantities of 1.5 billion gal annually. Different perspective in ethanol/biotech commercialization were provided; from the leading ethanol producer from grain; from the emerging cellulose to ethanol industry; from the leading chemical producer in the United States; and from an historical engineering viewpoint.

A summary of potential problems and industrial needs for commercialization that were identified included:

- Nontechnical forces (externalities)
- Lack of business focus (marketing, and so forth)
- Financing for demonstrations
- Reduced risk for investors
- Increased government support (subsidies?)
- Coordination of national strategy
- Environmental issues (all streams considered)
- Developed product markets
- Confirmed economic projections

Net Present Value Analysis to Select Public R&D Programs and Valuate Expected Private Sector Participation

NORMAN D. HINMAN* AND MARK A. YANCEY

Center for Renewable Fuels and Biotechnology, National Renewable Energy Laboratory, Golden, CO 80401-3393

ABSTRACT

One of the main functions of government is to invest taxpayer dollars in projects, programs, and properties that will result in social benefit. Public programs focused on the development of technology are examples of such opportunities. Selecting these programs requires the same investment analysis approaches that private companies and individuals use. Good use of investment analysis approaches to these programs will minimize our tax costs and maximize public benefit from tax dollars invested. This article describes the use of the net present value (NPV) analysis approach to select public R&D programs and valuate expected private sector participation in the programs.

Index Entries: Investment analysis; net present value; R&D; public/private partnerships.

INTRODUCTION

One of the main functions of government is to invest tax dollars in programs, projects, and properties that will result in greater social benefit than would have resulted from leaving those tax dollars in the private sector or using them to pay off the public debt. One traditional area for investment by government is Research and Development (R&D). According to Battelle, US R&D expenditures reached $164.5 billion in 1994, and federal support represented $69.8 billion (42.4%) of the total *(1)*. If invested wisely, these tax dollars can lead to greater social benefit than would be obtained by leaving them in the private sector or using the money to pay off the federal debt. However, if

---

*Author to whom all correspondence and reprint requests should be addressed.

not invested wisely, this could result in less-than-optimal social benefit or, even worse, in less social benefit than could be obtained from the other two options. The purpose of this article is to describe an approach to analyzing and selecting investment opportunities for federal money in public R&D programs and valuating expected private sector participation in the programs.

BASICS OF INVESTMENT ANALYSIS

For all investment situations, there are five basic variables:

1. Costs;
2. Profits or benefits;
3. Time;
4. The discount rate; and
5. Risk.

In the analysis of investment alternatives for a given situation, the alternatives under consideration may have differences with respect to costs and profits or benefits, project lives, and uncertainties. If the effects of these factors are not quantified systematically, correctly assessing which alternatives have the best potential is very difficult.

Many methods are available to decision makers to evaluate investment options systematically. These methods, described in detail in a variety of books and articles (2), include present, annual, and future value; rate of return; and break-even analysis. The application of each method depends on whether the analysis is for a single opportunity, two mutually exclusive opportunities, or several nonmutually exclusive opportunities. For the single-opportunity situation, the decision maker is simply trying to decide if the single investment option meets a minimum expected financial return. For the mutually exclusive situation, the decision maker has two investment options and is trying to decide whether the options meet the minimum expected financial return and, if both do, which is the best choice. For the nonmutually exclusive situation, the decision maker has several investment options and is trying to decide which of these meets the minimum expected financial return and, of those that do, which combination of these will provide the maximum return on total investment dollars available.

One must be careful in applying rate of return analysis to mutually exclusive and nonmutually exclusive situations. If one simply calculates the rate of return for each alternative and then chooses the alternative or alternatives with the largest rates of return, this can, and often does, lead to the wrong choice. The correct application of rate-of-return analysis to either situation is known as incremental rate of return, and can be very tedious and time-consuming. One must take extra steps to account for differences in project lives. Net present value (NPV) is the tool of choice for evaluating mutually exclusive or nonmutually exclusive investment options, because it is much less time-consuming, is straightforward, does

Net Present Value Analysis

not require additional steps or considerations for projects with different lives, allows direct comparison between projects of widely differing objectives and scopes, and allows a rational approach to valuating private sector participation in public programs.

NPV APPROACH TO NONMUTUALLY EXCLUSIVE INVESTMENTS

A nonmutually exclusive investment situation is one where more than one investment option can be selected, depending on available capital or budget restrictions. The objective is to select those projects that maximize the cumulative profitability or benefit from the available investment dollars. To maximize the cumulative profitability or benefit, the decision maker selects the combination of projects that maximize the cumulative NPV.

To apply NPV to nonmutually exclusive alternatives, the NPV for each alternative is calculated by determining the present value of the profit/benefit stream calculated at the minimum rate of return (hurdle rate) and subtracting the present value of investment dollars and other costs, also calculated at the minimum rate of return.

$$NPV = \text{present value revenues @ } i^* - \text{present value costs @ } i^*$$

where i^* = minimum rate of return.

If the project NPV is zero, there is enough revenue or benefit to cover the costs at a rate of return that is equal to the minimum rate of return required by the investor. Projects with an NPV less than zero are dropped from further consideration, because their rate of return is less than the minimum required return. If the NPV is greater than zero, the NPV represents how many present value dollars will be returned to the investor above and beyond those that will be returned at the minimum rate of return. Once the NPV for each project is calculated, the decision maker looks at all possible combinations of projects to determine which combination (whose total investment does not exceed the amount of money available) has the largest cumulative NPV. This is the best possible investment portfolio. Often, selecting the best portfolio does not involve selecting projects with the largest individual project NPV and, as will be seen in the following example, does not necessarily involve selecting projects with the highest rates of return.

For example, consider the following three investment alternatives. Each has a different life. Assume that the investor's minimum rate of return is 10% and the investor has $50,000 to invest.

Alternative	Investment	Profit/benefit at the end of each year	Project life
1	$50,000	$26,000	3 yr
2	$30,000	$10,000	5 yr
3	$20,000	$6000	7 yr

The NPVs and rates of return for each alternative follow:

Alternative	NPV	Rate of return
1	+$14,658	26%
2	+$7908	20%
3	+$9211	23%

Because all these projects have positive NPVs, they will all provide a rate of return >10% to the investor. Thus, all are to be considered for possible inclusion in the optimum portfolio. The two possible portfolios that do not exceed the $50,000 available to invest are: (1) Alternative 1 and (2) the combination of Alternatives 2 and 3. The combination of Alternatives 2 and 3 has a cumulative NPV of $17,119, whereas Alternative 1 has an NPV of $14,658. Thus, the best portfolio is the combination of Alternatives 2 and 3. In fact, this combination of investments will return $2,460 to the investor above and beyond what Alternative 1 by itself will do. Also, ranking the alternatives by regular rate of return does not give the correct answer, because Alternative 1 has a short project life, and when it is finished, the investor would have to reinvest at the minimum rate of return of only 10%, giving an overall lower rate of return to the investor compared to investing in Alternatives 2 and 3, which have longer lives at a return much better than the minimum rate of return, providing an overall greater rate of return for the combination of Alternatives 2 and 3.

If one is faced with the daunting task of selecting an investment portfolio when there are dozens of investment options, an alternate method may be used to simplify the process. Growth rate of return or ratio analysis may be used to rank nonmutually exclusive alternatives rather than cumulative NPV analysis *(2)*. Large companies and government programs are often faced with the task of evaluating literally hundreds of potential projects. Many combinations of projects must be analyzed to determine the optimum group of projects that will maximize the cumulative NPV for a given budget. The use of growth rate of return or ratio analysis only requires the calculation of the respective values for each project and then ranking the projects in the order of decreasing values. The illustration of these concepts will not be demonstrated here, but the reader should be aware of these methods to evaluate a complex investment portfolio.

SPECIAL CONSIDERATIONS FOR NONMUTUALLY EXCLUSIVE GOVERNMENT INVESTMENTS

Converting Intangible Benefits and Costs into Dollar Values

A basic tenet of this article is that to make rational investments of public dollars, one must have some approximate, quantitative idea of the value of critical costs and benefits. Moreover, as a practical matter, it is essential that the measure of value be the same for both costs and benefits, so that direct comparisons between costs and benefits can be made. The most universal measure of value is the dollar. In the private sector, this is the measure of cost and benefit. In the public sector, particularly with respect to R&D programs, it is the established measure of cost. However, on the benefit side, there is no established measure of value. The authors contend that the dollar should be the measure of benefit, so that direct comparisons can be made with costs and so that the established and the well-recognized investments analysis methodology described above can be employed in the public sector.

In many cases, converting benefits to dollars is fairly straightforward. For example, a key benefit that the US Department of Energy (DOE) is interested in is reducing imported petroleum. The dollar value of the yearly benefit can easily be calculated from the present and projected price of petroleum (3). As another example, it is possible to estimate the net annual increase or decrease in jobs that results from introducing new technology. In addition, it is fairly straightforward to place a dollar value on these jobs (4). Other possible costs and benefits are environmental and social, which are more difficult to quantify. Nevertheless, the US Environmental Protection Agency has studied these issues carefully, and has given dollar estimates of health costs associated with various types and levels of pollution. Clearly, more effort needs to be made to develop methodologies to convert public benefit to dollar value, so that time-tested investment decision-making methodologies can be employed.

Minimal Rate of Return for Public Projects

Establishing a minimal rate of return for public projects requires some special considerations, which have been reviewed extensively by Heaps and Pratt (5) for Canadian public projects. They concluded that the correct social discount rate for Canada was 3–7%. In another study performed by Wilson Hill Associates (3), a discount rate of 7% was used for projects evaluated for the Office of Transportation Programs in DOE.

SELECTING PUBLIC R&D PROGRAMS AND VALUATING EXPECTED PARTICIPATION BY THE IMPLEMENTING INDUSTRY

Commonly, a government R&D program is initiated without the private sector, but the private sector is expected to "come on board" at some point to carry, the ball forward into the commercial arena. For these situations, the government and the private sector make investments in R&D and technology commercialization in order to obtain what each desires—social benefit in the case of government, and profit in the case of the private companies.

Analysis of the value of these programs demands answers to three questions:

1. What portion of the R&D cost can the private sector incur and still obtain its minimum return from implementing the technology?
2. When this private sector cost allowance is subtracted from the total estimated cost to carry out R&D to obtain an estimate of the R&D cost that must be borne by government, is the estimated government R&D cost justified given the expected social benefit from implementing the technology?
3. If the answer to questions 2 is positive, does the program represent one of government's best opportunities for its limited investment dollars?

The NPV approach to investments provides the answer to all three questions. For example, to answer the first question, one calculates the industry NPV. To do this, one estimates over time the capital and operating costs the industry at large will incur to implement a new technology and, using the average minimum interest rate for the industry, calculates the present value of these costs to industry at the initial time of commercialization. One also estimates over time the present value at the time of commercialization of the expected increased revenues or savings the industry should experience from implementing the technology. Subtracting the present value costs from the present value revenues gives the industry NPV at the time when commercialization is expected to begin. If the NPV is negative, the industry cannot afford to contribute to the R&D effort, and cannot afford the capital and/or operating costs of commercialization. As a result, it will not "come on board," and the government should drop consideration of the program. If the industry NPV is zero, industry cannot afford to contribute to the R&D costs, but can afford the capital and operating costs to implement the technology. In this situation, the government will have to incur all the R&D costs in order for industry to adopt the technology. If the industry NPV is positive, the government can expect the industry to participate in the R&D costs at a level equivalent to the NPV. This participation may be provided in the form of cost-sharing or through licensing arrangements.

To answer the second question, one calculates the government NPV. To do this, the expected social benefits are estimated over time, and dollar values are assigned. Then the present value of these benefits is calculated at the time the program was initiated using the social discount factor. Next, the entire R&D costs over time are estimated and discounted to the time the program began using the social discount factor. Next, the expected R&D contribution from industry, calculated above as industry NPV, is discounted to the time of initiating the program using the industry discount factor. This industry R&D contribution, discounted to when the program began, is then subtracted from the entire R&D costs, also discounted to when the program began, to obtain the governments expected R&D costs discounted to the time the program began. These discounted government R&D costs are then subtracted from the discounted benefits to obtain the government NPV for the program at the time the program was initiated. If the government NPV is less than zero, the program should not be considered for investment of tax dollars. If the government NPV is zero or greater, it should be thrown in the pot of possible government investments.

To answer the third question, government should list all investment options with an NPV greater than zero and select that combination of projects that will maximize the governments cumulative net present value.

VALUATING EXPECTED PARTICIPATION BY INDIVIDUAL COMPANIES

If, from the above analysis, the industry NPV is positive, individual companies that are members of the industry can be expected to cost share in the R&D phase of a program or purchase licensing arrangements. However, the level of cost-sharing or license fees will depend on each company's circumstances. The expected level of cost-sharing or the licensing fee for a given company can be calculated using NPV analysis. For example, for a given company, one can estimate over time the capital and operating costs the company will incur to implement a new technology and, using the company's minimum interest rate, calculate the present value of these costs to the company at the time of expected commercialization. One can also estimate over time the present value of the expected increased revenues or savings a company should experience from implementing the new technology. Subtracting the present value costs from the present value revenues gives the NPV to the company when it is expected to begin its commercialization effort. If the company NPV is negative, the particular company cannot afford to implement the technology, even if the technology is provided free. Such a company is not a viable partner to the government program. If the company NPV is zero, the company may be a partner only in the sense that it will implement the government—developed technology if it is free to the company. If the company NPV is positive, the company can afford to cost-share the R&D effort or purchase a licensing arrangement at a level equal to the company NPV. Such companies are potentially the most valuable partners to the program.

RISK ANALYSIS, INFLATION, AND ESCALATION

All the above analyses presumed that no risk was involved. To account for the risk associated with a program or project, one can use a higher discount rate, or modify the various expected benefits, profits, or costs accordingly. In addition, when conducting NPV analysis, one must account for anticipated inflation and escalation of costs, revenues, and benefits.

CONCLUSIONS

With federal R&D dollars dwindling, wise choices must be made with regard to investing in projects so that maximum social benefit results. In addition, the government must work closely in partnership with the private sector to leverage public dollars.

To accomplish this, an accepted methodology must be developed to evaluate the "goodness" of government programs and the expected contributions from the private sector. The authors contend that NPV analysis is the most appropriate methodology. The major advantages of this methodology is that it allows direct comparison between public projects of widely differing purposes and scopes, and it allows direct comparison of the value of public endeavors vs public endeavors. In addition, through NPV analysis, it is possible to separate clearly the costs and benefits of a joint project between government and industry. The description of this last application is a unique aspect of this article.

We recognize that a significant hurdle to using NPV analysis is that methodologies for valuating public benefits in terms of dollars are not always available. However, because the value of NPV analysis is so great, particularly when analyzing joint public/private ventures, it behooves public agencies to work to establish methodologies to put dollar values on pubic benefits.

The authors hope that this article will stimulate discussion toward a model that attempts to compare competing programs for limited resources based on parameters that are both relevant and measurable. Such discussion will certainly lead to refinement of the ideas in this article, to detailed examples of their use, and hopefully, to the development of improved methodologies to valuate public benefits in terms of dollars.

ACKNOWLEDGMENT

This work was funded by the Biochemical Conversion Element of the Office of Fuels Development, US Department of Energy.

REFERENCES

1. *Manufacturing Engineering*, vol. 112, no. 2, 1994. 2.
2. Stermole, F. J. (1984), *Economic Evaluation and Investment Decision Methods*, 5th ed., Golden, CO.

3. Santone, L. C. (1981), *Methods for Evaluating and Ranking Transportation Energy Conservation Programs Final Report*, Washington, DC, April 30.
4. Tyson, K. S., Putsche, V., and Bergeron, P. (1996), *Modeling the Penetration of the Biomass-Ethanol Industry and Its Future Benefits*, Golden, CO, March 15.
5. Heaps, T., and Pratt, B. (1989), FRDA Report 071, *The Social Discount Rate for Silvicultural Investments*, Victoria, BC, March.

A Review of Techno-Economic Modeling Methodology for a Wood-to-Ethanol Process

DAVID J. GREGG AND JOHN N. SADDLER*

Chair of Forest Products Biotechnology, Faculty of Forestry, University of British Columbia, #270-2357 Main Mall, Vancouver, British Columbia, Canada V6T 1Z4

ABSTRACT

Techno-economic modeling has been a valuable tool in directing and assessing the research and development efforts for biomass-to-ethanol processes. In developing a techno-economic model of a "generic" wood-to-ethanol process, we decided to follow a three-pronged design approach. This initially consisted of a detailed review of the current definition and technical maturity of the process, which concluded that the process remains complex and immature. More recently, we have critically assessed/compared two inherited models, and examined the historical and current trends in modeling design. We confirmed that process complexity and immaturity, in association with the capabilities of the available modeling tools and the ease with which they can be used, influenced the design and implementation of past models. We have discussed these influences with reference to our own model development decisions. For example, on review of two inherited techno-economic models, we decided that our new model would require a greater degree of flexibility in its structure and user interface.

Index Entries: Techno-economic modeling; biomass-to-ethanol process; lignocellulose-to-ethanol process; spreadsheet models; flowsheet simulators.

INTRODUCTION

Past techno-economic modeling work has been used for a variety of reasons. Although many groups have used this approach to estimate the likely cost of producing ethanol from various lignocellulosic substrates,

---

*Author to whom all correspondence and reprint requests should be addressed.

past models have tended to be most useful in establishing the level of maturity of each of the component steps. For example, we have used our past techno-economic model (1) to assess various biomass-to-ethanol options, and found that the front-end (pretreatment, fractionation, enzymatic hydrolysis) steps were both technically immature and represented a large component of the total product cost. Consequently, our more recent work has looked at various process options (SO_2 pretreatment, alkali and peroxide-enhanced fractionation, and enzyme recycle hydrolysis) in an attempt to provide a more defined overall process and reduce the proportion of the total product costs attributed to the front-end steps. We have previously described both the technical results of our recent research efforts and our current process development philosophy (2–4). These process options along with our previous modeling experience suggested that modification or development of a new model must address the need for greater flexibility, ease of development, or modification and ease of use.

HISTORICAL BIOMASS-TO-ETHANOL MODELING

Process modeling of the biomass-to-ethanol process has been ongoing since the early 1980s, with most earlier efforts (5,6) concentrating on the enzymatic hydrolysis portion of the process and with the overall economics based on the price of producing glucose. At this time, most computationally intensive tasks, such as techno-economic modeling, were restricted to large, expensive mainframe computers. Microcomputers were only beginning to enter the market, and the process modeling tools were restricted to high-level programming languages and procedural structuring. This often meant that this type of modeling was viewed as being too complex and expensive. Spreadsheets were severely limited in their computational and storage capabilities, and flowsheet simulators were not yet available. At this time, modeling efforts did not simulate the entire process. Consequently, they did not imitate the complex interrelated nature of the biomass-to-ethanol process steps, or provide the means to determine the influence of both the individual and combined process steps on the production cost of ethanol. Thus, these models effectively represented the mathematical equivalent of a nonintegrated physical model, such as a lab-scale process development unit (PDU) or a small pilot plant.

Integrated biomass-to-ethanol process models (7,8) began to appear around the mid-1980s, and continued, along with most engineering and computationally intensive applications, to utilize high-level programming languages and procedural structuring for their development. The graphical user interface and object-oriented programming concepts were only beginning to influence the microcomputer market for personal and business applications. Although the complexity of the biomass-to-ethanol process had always been recognized, it was only with the development of integrated models that the relative importance of the various process steps or parameters and the influence of byproduct production could be readily determined through sensitivity or parametric analysis. These models represented the mathematical equivalent of an integrated physical model, such as a development- or commercial-scale plant, and they provided, for example, scale-up influences on both the technology and economics of the overall process being evaluated. Although linear programming (9) was also attempted, it was found to be an inappropriate modeling method, since much of the bioconversion process is not linear, thus requiring linear approximations and the determination of a large number of coefficients that may be difficult to measure or estimate. A representative example of the models that were developed over this period is the SERI-Chem Systems-Lawrence Berkeley Laboratories Study (7). This model used a modular structure, was written in a high-level programming language (APL), and accessed the ICARUS equipment capital cost data base. It was based on an aspen feedstock and contained the following subprocesses: steam-exploded pretreatment, fed-batch enzyme production using RUT C30 strain of *Trichoderma viride*, separate enzyme hydrolysis and fermentation, and vapor reuse (benzene) distillation. An enzyme recycle option using the Lawrence Berkeley Laboratories countercurrent adsorption research was also included. Although the simulation program modeled the back end of the process (fermentation, distillation, waste treatment, and heat generation), there was little detail provided. It was also proposed by researchers, and later shown by these models, that the immaturity of the process front end (pretreatment, enzyme production, and hydrolysis) resulted in a major contribution of this component to the overall cost of producing ethanol. A comprehensive kinetic model to describe cellulose conversion for various pretreatments, feedstocks, and hydrolysis conditions was not available, and consequently, the parametric analysis did not account for these interactions.

Spreadsheet modeling (10,11) began to appear around the end of the 1980s. Microcomputers and spreadsheet software had become well established in the personal and business markets. The middle to late 1980s was a time period in which there were tremendous growth and development of both the hardware and software capabilities of microcomputers. Models and other computationally intensive applications that had previously required the computational and storage capabilities of a larger computer were now being developed or ported to these smaller stand-alone microcomputers. Spreadsheets became popular as a way to develop the calculational relationships rapidly between process variables, via cell addressing or cell naming conventions, without having to learn a programming language. Model development using spreadsheet software acquired many capabilities that previously had to be programmed, such as the ability to format and generate reports and/or charts, the ability to read, illustrate, and write the contents of storage files, and in a small way, internal error checking and debugging by showing intermediate results. The modeling also acquired capabilities that were unique to spreadsheets, such as the easy incorporation of supplied spreadsheet functions (interest calculations, look-up tables, random number generators, and so forth), the ability to do sensitivity analysis rapidly by changing the value of one cell or variable and noting the change in the final ethanol price, development of simple flat-file data bases, and the ability to automate common tasks through the development of small programs using macro languages. The IOGEN study *(11)* is representative of the type of modeling carried out at this time. The model was built using Lotus 1-2-3 spreadsheet software and based on an aspen feedstock. Residual solids from the process were assumed to supply all the energy requirements for the plan, and surplus electricity was projected to be sold: Very high enzyme production yields from lactose were also assumed.

Although flowsheet simulators were available in the mid-1980s, they were only generally accepted as a means to model processes or unit operations, and were used on a regular basis by process and research engineers in the latter part of the 1980s and early 1990s. Flowsheet simulators, like spreadsheets, provided the user with commands to format and generate reports, plus read, illustrate, and write the contents of storage files. However, unlike spreadsheets, they were somewhat limited in their capability to produce charts. They utilized a graphical interface and ready-made unit processes and data bases/estimation routines for estimating flow-stream component properties to build a model. However, because of their computationally intensive nature, flowsheet simulators often required long periods of time to provide modeling results. The latter feature often makes flowsheet simulators an inappropriate application for sensitivity analyses or Monte Carlo simulations. This type of software, owing in large part to its size and computational requirements, was at this time restricted to mainframe or minicomputers. The hardware constraint along with the cost of the software, lack of economics modeling, and steep learning curve (generally associated with the requirement to learn which of many options to select for the determination of thermodynamic and other chemical properties) tended to limit the use of flowsheet simulators for techno-economic modeling. Furthermore, the commercially available-simulators were developed primarily for the design and evaluation of physicochemical processes, and as a result, did not provide ready-made unit operations for biological processes or readily characterize biologically derived materials. Users were forced to develop modules in a high-level programming language for the biological processes, which were subsequently linked to the flowsheet simulator kernel and users were also forced to assign properties directly to the biologically derived materials. Even with these constraints, flowsheet simulators were primarily used for techno-economic modeling during this time.

Techno-economic modeling since the late 1980s has been carried out by a few groups using object-oriented high-level languages to build whole applications that include some of the features of flowsheet simulators *(12)*. However, most modeling has used flowsheet simulators *(13,14)* or enhanced spreadsheets *(1,15)*. There has generally been a convergence of features within these three previously distinct modeling methods, primarily as a result of the proliferation of graphical user interfaces in microcomputers. The graphical user interface has spawned the use of graphical controls and dialog boxes, object-oriented programming, and dynamic linkage of applications and their files.

Object-oriented high-level languages, such as C++ and Visual Basic, have become popular over the period since the late 1980s primarily as a result of their recognized productivity gain for developing graphical user interfaces. At least one research group *(12)* interface and modular structuring similar to flowsheet simulators. However, it also includes economic evaluation, sensitivity, and Monte Carlo simulation features. Furthermore, because it is strictly limited to the biomass-to-ethanol processes, the training time should be significantly shorter than an equivalent commercial flowsheet simulator.

Flowsheet simulators since the late 1980s, as mentioned previously, have become a valuable design and assessment tool for process engineers. With the growing computational and storage capabilities of microcomputers, many of the previously developed commercially available simulators have been ported to the desktop. It has also been recognized by the developers that simpler, less-full-featured programs can reduce the training time and also the computer hardware demands. Some of the developers are also beginning to recognize the need for economic analysis modules, modules that model biological unit operations, and data base/calculational routines that estimate biological material properties. The basic principles associated with flowsheet simulators are generally recognized as being good for the rapid development of process scenarios, in that the user only has to select the required unit operations and specify the values for the required number of variables to define the solution fully. However, the current flowsheet simulators are not capable of providing a user who is both knowledgeable in modeling this particular type of process and reasonably computer-literate with the ability to produce lignocellulosic-to-ethanol process scenarios easily and assess their economic feasibility. An example of a flowsheet simulation over this time period is the University of Lund Study *(14)*, which compared three wood-to-ethanol processes (concentrated hydrochloric acid process [CHAP]; Canada, America, Sweden hydrolysis [CASH], a two-step weak acid hydrolysis; and an enzymatic process using a dilute sulfur dioxide pretreatment). The modeling was done using ASPEN PLUS (Aspen Tech Inc., Cambridge, MA), a commercial flowsheeting program that is normally used for thermochemical process design. All three processes used pine as a feedstock and were based on the same plant capacity. The main product was 95% ethanol (w/v) with the residual solids and methane used to produce process steam while the carbon dioxide was sold on the open market.

Since the late 1980s spreadsheets have become more graphically oriented with the incorporation of more drawing capabilities. For example, three-dimensional spreadsheet capabilities can be accessed through tabbing of worksheets within a workbook; the inclusion of graphical controls (including buttons, check boxes, menus, drop-down list menuing, scroll bars, spinners, and dialog boxes) is possible, and more sophisticated macro languages or full programming languages that allow the development of graphical user interfaces similar to those in flowsheet simulators can be used. Spreadsheets have also become more flexible in that they now have the ability to link dynamically to various other entities (i.e., objects, dialogs, worksheets, workbooks, and applications). Two spreadsheet models (the Virginia Polytechnical Institute [VPI] Study *(15)* and the Forintek Study *(16)*) representative of this time period were inherited by our research group.

Although there have been a number of techno-economic models built over the last 10 years for a hardwood-to-ethanol process, most of the details associated with their calculational logic have not been published or made publicly available. This is not altogether unreasonable, since they require a considerable amount of time to develop and would require a sizable document to describe their logic fully. Furthermore, it is difficult to assess the effect of changes in one particular portion of the whole process without building a complete process model, because the individual subprocesses are strongly interdependent and the measure of progress in the development of the process is the production price of the final product. Consequently, it was recognized that although a full model would be required to provide a reasonable assessment of new developments in any of the subprocesses, the amount of time required to develop a completely new model would be substantial. Fortunately, we recently acquired the complete description of the two hardwood modeling efforts mentioned previously, a model of the whole hardwood-to-ethanol process from Forintek Canada completed in 1991 and a model of the steam pretreatment and fractionation subprocesses from VPI completed in 1994. Through a process of integration and refinement of these two models (discussed in more detail below), the time and effort required to build a complete hardwood-to-ethanol model were shortened. However, it must be recognized that these past models did not, for the most part, reflect the results of research completed since the early 1990s. Therefore, some new subprocess modules based on more recent data, particularly in the pretreatment and hydrolysis areas, had to be developed.

ASSESSMENT AND MODIFICATION OF INHERITED MODELS

Model Structure

The biomass-to-ethanol process, with its multiple components and processing steps, is complex. Furthermore, most of the front-end components of the process have not been tested in an integrated large-scale facility. A model of this process should therefore be capable of easily including

Techno-Economic Modeling

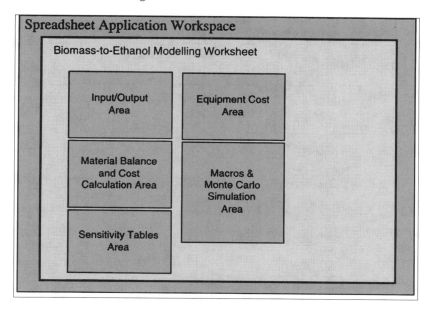

Fig. 1. Forintek model structure.

and evaluating a number of process and equipment options. Two structural concepts, modularization and encapsulation (discussed in more detail below), have been shown to enhance both software development in general and the ability to build on past techno-economic modeling efforts. The incorporation of these structural concepts should provide the framework for a flexible and long-lived model.

Modularization

A review of the Forintek model indicated that it was organized on the type of activity that was carried out (Fig. 1), i.e., Input/Output, Material Balance Calculation, Energy Balance Calculation, Capital Cost Estimation, Operating Cost Estimation, Sensitivity Analysis, and Monte Carlo Simulation. As a result, the subprocess input/output and calculational routines were intertwined. This structure did not allow the user to change the process or the process options rapidly, since the calculations for each process step were spatially dispersed over the spreadsheet area.

The VPI model was organized into modules (Fig. 2) that represented subprocess or unit operational options. This structuring allowed more rapid changes to the process by the modeler than was provided by the Forintek model, since the calculations for each subprocess or unit operation were located in the same general area on the spreadsheet. The modules appear to have been developed in one spreadsheet, and then the appropriate subprocesses for a particular process scenario were copied to another spreadsheet and linked. Although this structure is superior to the

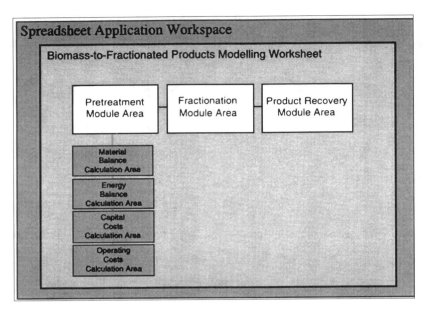

Fig. 2. Virginia Polytechnical Institute model structure.

Forintek model, it still lacks the potential for rapid process modification found in a flowsheet simulator-type of structure.

The latest version (5.0) of Microsoft Excel (used for the current STeam Explosion Assessment Model [STEAM]) provides the means to develop easily a modular structure that represents various subprocesses and types of calculations within each subprocess (Fig. 3). Each of the files is now referred to as a "workbook," since each layer of the three-dimensional file is now called a "worksheet" and can be accessed through a graphical tabbing system analogous to a physical filing cabinet or card file. The graphical structuring allows the designer to separate the various calculational or storage requirements easily, for a particular subprocess, into separate graphical layers. As in previous three-dimensional spreadsheets, the "worksheets" within the "workbooks" can be linked either through worksheet-row-cell referencing or through cell naming.

A further modular structuring feature of Excel is the ability to link various workbooks. This provides the developer with the ability to link the various subprocesses into a complete process scenario. Linkages can be fixed through methods, such as worksheet-row-cell referencing or worksheet-cell naming conventions. Alternatively, they can be more flexible and dynamic, through development of an "executive workbook" and/or an executive program. Consequently, there is a substantial level of flexibility now available to the model developer for designing and implementing a structure that is modularized, and reflects a structure similar to flowsheet simulators or previous high-level language models.

Techno-Economic Modeling

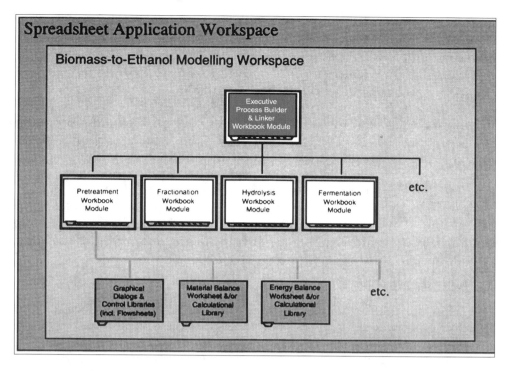

Fig. 3. Forest Products Biotechnology STEAM model structure.

Object-Oriented Encapsulation (Flowsheet, Input/Output, Calculation)

Encapsulation is a term used in computing science to refer to the capability of combining routines that operate on a data structure together with the data structure itself. Spreadsheets such as those used for the Forintek and VPI models incorporate this concept by keeping together the data (cell values), data structure (array represented visually by the row-column addresses), and the routines that operate on the data structure (cell formulae). However, spreadsheets at the time that the Forintek and VPI models were built did not include the encapsulation of the flowsheet that was used to develop or design the overall data or process structure. Object-oriented programming (OOP), which has become popular since the introduction of graphical user interfaces for microcomputers, incorporates and extends the encapsulation concept through including both the calculational routines that operate on a data structure and methods or routines that draw a graphical representation of the object for the user. This extra graphical abstraction allows the user to manipulate the relationships between data structures more easily and/or to create rapidly new objects that inherit the characteristics of the parent. Flowsheet simulators, e.g., ASPEN Plus, are an example of a type of program that uses the extended graphical characteristics of OOP programming for the user interface. Flowsheet simulators map a graphical representation of process objects (subprocesses and/or unit operations) and process interrelationships (flowstream) to the associated data structures and routines (calculational and graphical). This graphical flowsheeting feature is often recognized as being a key to the popularity and flexibility of flowsheet simulators.

Through the use of current spreadsheets, such as Excel, with its graphical objects (controls, dialogs, toolbars, and menus) and a programming language (Visual Basic), it is now possible to include the flowsheet, input/output, and calculational relationships of the various subprocesses and unit operations within a spreadsheet format. The graphical objects can now be mapped directly to the data structures and calculational routines associated with the subprocess and/or unit operations. This allows for rapid manipulation of the relationships between these operations, and the potential for graphical vs manual linkage of the subprocesses and unit operations to create new scenarios.

Model User Interface

The user interface, in this study, refers to both the interface used to develop and evaluate various process scenarios, as well as the interface for editing the input variables, calculational routines and accessing, or reporting the output variables. To develop techno-economic models, the first step is generally the development of a series of flowsheets for the overall process and for each of the process steps. This flowsheeting process, as mentioned previously, is an integral part of flowsheet simulators, and this feature is often recognized as being a key to their popularity and flexibility.

The model user interface of the two inherited models was essentially a standard spreadsheet interface (Fig. 4) with the user entering the values and formulae for the various variables via the formulae entry bar in spreadsheet row/column format. Both models did not include a specific error checking routine or formulae security. However, there was a slight difference in the overall interface of the two models. The Forintek designers expected the user or developer to find the appropriate areas for input/output and modification of the existing calculational routines, whereas the VPI model provided, through a structuring that emphasized process steps rather than calculational types, an easier environment to modify.

In both the inherited models, the user was forced to refer to process flowsheets that were not part of the modeling application in order to understand the process details, e.g., flowstream numbering. The process flowsheets were either hand-drawn (Forintek model) or computer-drawn (VPI model) in another computer application.

Current spreadsheet capabilities now allow the model developer to include features previously found only in flowsheet simulator user interfaces. Within the STEAM model, each process/subprocess is built from

Techno-Economic Modeling

	A	B	C	D	E	F	G	H	I	J	K
						Forintek Model					
1	Input		500	OD tonnes aspen/day		(Range= 100 to 1000 OD tonnes/day)					
2											
3	Output		5.5E+07	L ethanol/year		Total capital cost	5.3E+07	dollars			
4			158504	L ethanol/day			0.97677	$/L			
5			317.008	L ethanol/OD tonne wood		PEtOH =	0.86717	TOTOC =			
6											
7			PROCESS PARAMETERS								
8											
9	Feedstock Handling	Wood	20833.3	Throughput of wood chips, dry kg/hr							
10	Steam Pretreatment	MC	50	Moisture content of wood chips, wt% (wet basis)							
11	Steam Pretreatment	WD	433	Density of wood, kg/m3							
12	Steam Pretreatment	BDW	112	Bulk density of wood chips, kg/m3 of reactor volume							
13	Steam Pretreatment	TEMP	240	Temperature of high pressure steam, deg C							
14	Steam Pretreatment	TWOOD	20	Temperature of input wood chips, deg C							
15	Steam Pretreatment	SHW	0.5	Specific heat of wood, cal/g.deg C							
16		MCLCl	0.4	Moisture content of wet lignin cake, wt fraction							
17	Enzyme Production	cEP	0.1	Substrate consistency in enzyme production fermentors, w/v fraction							
18	Enzymatic Hydrolysis	cCH	0.1	Substrate consistency in cellulose hydrolysis fermentors, w/v fraction							
19	Pentose Fermentation	cXH	0.1	Dissolved solid concentration in xylose-to-ethanol fermentors, w/v fraction							
20	Fractionation - Water Wash	cSX	0.08	Dissolved solid concentration in hemicellulose extract, w/v fraction							
21	Fractionation - Alkali Wash	cLX	0.04	Lignin concentration in lignin extract, w/v fraction							
22	Fractionation - Alkali Wash	cNaOH	0.004	Concentration of NaOH in caustic solution feed to lignin extractors, w/v fraction							
23	Enzymatic Hydrolysis	eCL	12.5601	Cellulase loading in cellulose hydrolysis fermentors, FPU/g substrate							
24	Pentose Fermentation	eXL	0	Xylanase loading in xylan hydrolysis fermentors, IU/g substrate							
25	Fractionation - Alkali Wash	fNaOHr	0	Caustic recovery, fraction of NaOH input							
26	Fractionation - Water Wash	fC5WS	0.75	Fermentable sugars in water solubles, wt fraction of total dissolved solids							
27	Fractionation - Alkali Wash	fRH	0.888	Fraction of caustic insolubles (SEW-WIA) digested in cellulose hydrolysis fermentors							

Fig. 4. Spreadsheet input/output interface of Forintek model.

graphical objects representing the various unit processes/major pieces of equipment and their associated flowstreams. Both the graphical elements representing the unit processes/pieces of equipment and the flowstreams have associated properties that are now accessed through input/output dialog boxes (Fig. 5). In this case, the chemical, physical, process-related and economic attributes of the feedstock flowstream are shown. Each graphical object is linked to the next lower level of detail through simply "clicking" with the mouse on the graphical representation or icon of the object. For example, "clicking on" the pretreatment subprocess icon in the current process (Fig. 6) will expose or bring up the pretreatment subprocess flowsheet (Fig. 7). Each unit process/piece of equipment within a subprocess has associated properties, including calculational routines. For example, details for the steam reactor (object with steam shown entering) contained within the steam-pretreatment subprocess (Fig. 7), can be accessed through clicking on the icon of the reactor. This brings up the equipment option dialogs (Fig. 8) with the currently calculated values for the balances (material and energy) and cost estimations (operating and capital) for this particular piece of equipment. This dialog also provides access to the programming modules associated with the various calculations (Fig. 9) through clicking on the appropriate button within the Calculation Libraries area of the dialog.

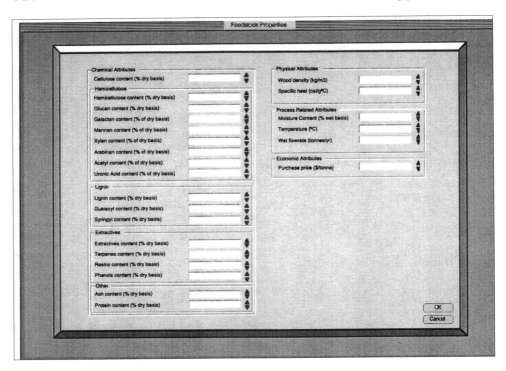

Fig. 5. An example of an input/output dialog box from the STEAM model.

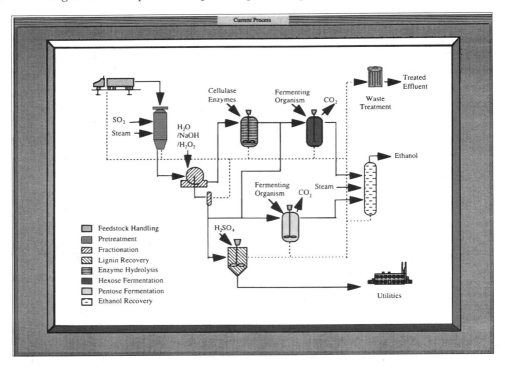

Fig. 6. STEAM model generic wood-to-ethanol process flowsheet.

Techno-Economic Modeling

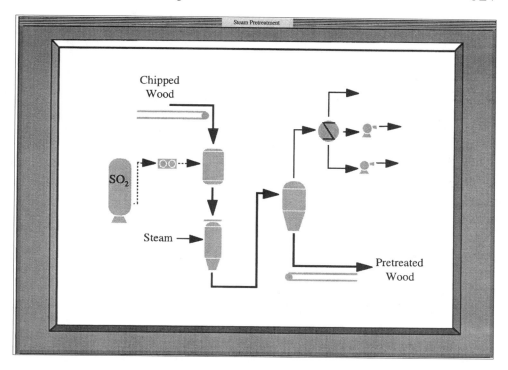

Fig. 7. STEAM model pretreatment subprocess flowsheet.

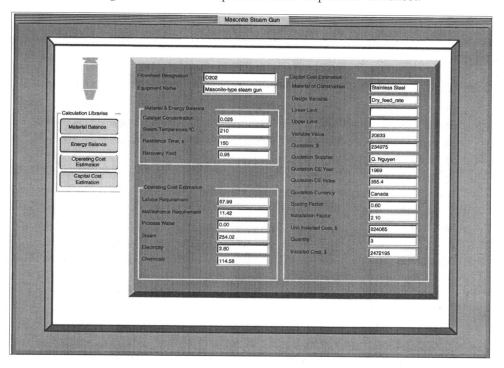

Fig. 8. STEAM model equipment option and calculational routine access.

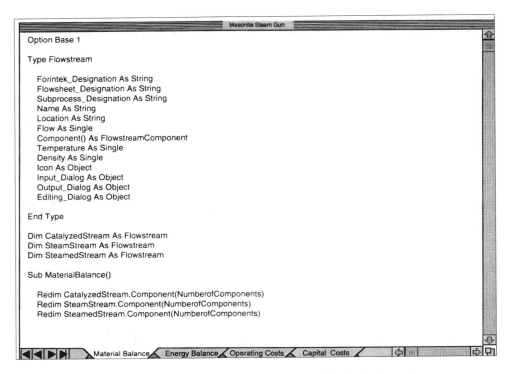

Fig. 9. An example of a STEAM model calculational routine.

CONCLUSIONS

The biomass-to-ethanol process, with its multiple components and processing steps, is complex. Furthermore, most of the front end of the process has not been tested in an integrated large-scale facility. The structure of the model dictates to a large extent the flexibility of the model as a whole. Any techno-economic model of this process should therefore be capable of easily including and evaluating a number of process and equipment options. Two structural concepts, modularization and encapsulation, have been shown to enhance both software development in general and the ability to build on past techno-economic modeling efforts. The STEAM model structure includes elements of modularization and encapsulation to ensure the adequate calculational routine and user interface flexibility generally associated with good models. The incorporation of these structural concepts has provided the framework required for a flexible and long-lived model that currently forms the basis for regular interaction among the International Energy Agency (IEA) network on bioconversion.

REFERENCES

1. Nguyen, Q. A. and Saddler, J. N. (1991), *Bioresource Technol.* **85,** 275–282.
2. Gregg, D. J. and Saddler, J. N. (1995), *Biomass and Bioenergy* **9(1–5),** 287–302.
3. Gregg, D. J. and Saddler, J. N. (1995), *Appl. Biochem. Biotechnol.* **57/58,** 711–727.
4. Gregg, D. J. and Saddler, J. N. (1996), *Biotechnol. Bioeng.* **51,** 375–383.
5. Perez, J., Wilke, C. R., and Blanch, H. W. (1981), *Enzymatic Hydrolysis of Corn Stover Process Development and Evaluation*, Lawrence Berkeley Laboratory, Berkeley, CA Report 14223.
6. Wald, S. A. (1981), *Enzymatic Hydrolysis of Rice Straw for Ethanol Production*, M.S. Thesis, Department of Chemical Engineering, University of California at Berkeley, Berkeley, CA.
7. Isaacs, S. H. (1984), *Ethanol Production by Enzymatic Hydrolysis—Parametric Analysis of a Base-Case Process*, SERI-Chem Systems-Lawrence Berkeley Laboratory, Berkeley, CA, Report SERI/TR-231-2093.
8. Arthur D. Little (1985), *Technical and Economic Feasibility of Enzymatic Hydrolysis for Ethanol Production from Wood*, Solar Energy Research Institute, Golden, CO, Report 625-RIER-BEA-84.
9. Ung, R. (1986), *The Economics of Forest Chemicals: Process Development and Economic Analysis*, Forintek Canada Corp., Ottawa, ON, Report 53-43-393.
10. Wright, J. D. (1986), *Fuel Alcohol Technical and Economic Evaluation*, NREL for US Department of Energy, Golden, CO., Report SP-231-2904.
11. Douglas, L. J. (1989), *A Technical and Economic Evaluation of Wood Conversion Processes*, Energy, Mines and Resources Canada, Ottawa, ON, Report DSS Contract File 23283-8-6091.
12. von Sivers, M. (1995), *Ethanol from Wood–A Technical and Economic Evaluation of Ethanol Production Processes*, Licentiate Dissertation, Department of Chemical Engineering I, Lund Institute of Technology, Lund, Sweden.
13. Hinman, N. D., Schell, D. J., Riley, C. J., Bergeron, P. W., and Walter, P. J., (1992), *Appl. Biochem. Biotech.,* **34/35,** 639–649.
14. von Sivers, M. and Zacchi, G. (1993), *A Techno-Economical Comparison of Three Processes for the Production of Ethanol from Wood*, Lund Institute of Technology, Department of Chemical Engineering I, Lund, Sweden, Report LUTKDH/(TKKA-7006)/1-27/(1993).
15. Avellar, B. K. (1994), *Engineering and Economic Considerations for Fractionation of Steam-Exploded Biomass*, M.Sc. Thesis, Department of Chemical Engineering, Virginia Polytechnical Institute.
16. Nguyen, Q. (1990), *Computer Simulation Model of an Enzymatic Biomass Conversion Process*, Forintek Canada Corp., Ottawa, ON, Report 53-43-B-407.

Copyright © 1997 by Human Press Inc.
All rights of any nature whatsoever reserved.
0273-2289/97/63-65—0625$8.50

Session 5

Emerging Topics in Industrial Biotechnology

BRUCE E. DALE[1] AND ERIC N. KAUFMAN[2]

*[1]Michigan State University, East Lansing, MI;
and [2]Oak Ridge National Laboratory, Oak Ridge, TN*

Conversion of cellulosic materials and sugars is but one area in which biotechnology can contribute to the production of fuels and chemicals. Emerging areas of research hold great promise for utilizing a wider array of potential substrates for fuel and chemical production as well as complementing conventional catalysis in fuel and chemical processing. The papers presented at the Eighteenth Symposium on Biotechnology for Fuels and Chemicals address several areas in which emerging technologies will impact the future of bioprocessing.

Advancements in our fundamental understanding of photosynthetic pathways have led to the discovery of algal mutants that lack one of the two photosystems previously thought necessary for conversion of light into chemical energy. The discovery of photosynthetic conversion using only one photosystem has the potential to double the thermodynamic conversion efficiency of solar energy into stored chemical energy. Developments in the area of reactor design for photoautotrophs enable higher cell densities and a higher frequency of light–dark cycles needed for greater light utilization and hence more efficient photoconversion processes.

Synthesis gas (CO, H_2 and CO_2) represents a widely available and more easily handled substrate for chemical and fuel production than raw biomass. Gasification of biomass, waste paper, and coal can provide this substrate without pretreatment and digestion. Recent research has developed processes for the conversion of synthesis gas into acetic acid and ethanol and has demonstrated that sulfate-reducing bacteria can utilize synthesis gas as carbon and energy sources; an alternative to sugars and organic acids. Advancements in reactor design for gaseous substrates are leading to reactors better able to deliver sparingly soluble substrates to the biocatalyst.

Advances in biocatalyst research and development are enabling the use of enzymes in increasingly hostile environments and promise to broaden the spectrum of chemical processes in which biotechnology may be utilized. Native phosphotriesterases have been demonstrated to be capable of transesterification and optical resolution of racemic alcohol, phosphoric acids in a variety of organic solvents. Hyperthermophilic hydrogenase has shown to be capable of performing its native biocatalytic function after it was chemically modified so as to be soluble and active in neat toluene, an environment in which unmodified enzymes are both insoluble and inactive.

Finally, novel electric field bioreactors are being developed to efficiently contact a biocatalyst-containing aqueous phase with organic substrates. These reactors create tremendous surface area between the two phases by focusing electrically induced shear forces at liquid/liquid interfaces rather than by supplying energy to the entire bulk solution as is done in impeller-based reactors. Such systems are expected to find utility in a number of biological processes including biodesulfurization and upgrading of crude oil.

Coupling of Waste Water Treatment with Storage Polymer Production

H. Chua,*,[1] P. H. F. Yu,[2] and L. Y. Ho[1]

Departments of [1]Civil and Structural Engineering; and [2]Applied Biology and Chemical Technology, The Hong Kong Polytechnic University, Kowloon, Hong Kong

ABSTRACT

Storage polymers in bacterial cells can be extracted and used as biodegradable thermoplastics. However, widespread applications have been limited by high production costs. In this study, activated sludge bacteria in a conventional waste water treatment system were induced, by controlling the carbon–nitrogen (C:N) ratio in the reactor liquor, to accumulate storage polymers. Specific polymer yield increased to a maximum of 0.374 g polymer/g cell when the C:N ratio was increased from from 24 to 144, whereas specific growth yield decreased with increasing C:N ratio. An optimum C:N ratio of 96 provided the highest overall polymer production yield of 0.093 g of polymer/g of carbonaceous substrate consumed, without significantly affecting the organic treatment efficiency in the waste water treatment system.

Index Entries: Activated sludge; carbon–nitrogen ratio; poly-hydroxy-alkanoates; storage polymer accumulation.

INTRODUCTION

In Hong Kong, 9500 t of municipal solid wastes are disposed of each day *(1,2)*. A very high proportion, 11 wt%, of these wastes are plastics packaging materials and disposable products. Plastics usage and plastics-waste generation are forecast to increase at 15% a year over the next decade *(3–5)*. These conventional plastics, which are synthetically derived from petroleum, are not easily decomposed in nature by microorganisms *(6,7)*, and are among the most environmentally harmful wastes *(1)*.

---

*Author to whom all correspondence and reprint requests should be addressed.

In the past decades, there has been much interest in the development and production of biodegradable plastics as an environmentally friendly substitute for conventional plastics, particularly for packaging materials and disposable products. Various biodegradable plastics have been produced either by incorporating natural polymers into conventional plastics formulations, by chemical synthesis, or by microbial fermentation *(8)*. Among these biodegradable plastics, a family of more than 40 polyhydroxyalkanoates (PHAs) and their copolymeric derivatives have emerged as very attractive materials for their complete biodegradability *(9)*, wideranging physical properties by copolymerization, and biocompatibility to human tissue in surgical applications *(10,11)*. A number of bacteria, including *Alcaligenes* spp., *Pseudonomas* spp., recombinant *Escherichia coli*, and a number of filamentous genera accumulate these polymers or copolymers as an intracellular carbon reserve when unfavorable environmental conditions are encountered *(12)*. A nutrient-deficient condition could result in accumulated polymers of up to 75% of the dry cell mass *(13)*. These extracted and processed polymers have a number of properties that are comparable to commonly used plastics, namely thermoplastic processability and 100% water resistance. However, widespread application of PHAs is hampered by high costs of production.

Much effort has been spent in optimizing the PHA production process and reducing costs. Lee et al. investigated various recombinant *E. coli* using different complex culture media *(14)*. *E coli* strain XL1-blue in LB plus 20 g glucose/L could accumulate up to 0.369 g PHA/g glucose, or equivalent to 7 g PHA/L. Shirai et al. used a photosynthetic bacteria, *Rhodobacter spheroides*, in a fed-batch culture with glucose as the sole carbon to achieve a PHA production of 6 g/L *(15)*. Shimizu et al. used a cell-growth phase followed by a separately optimized nutrient-deficient PHA accumulation phase to improve the specific production yield to as high as 0.70 g PHA/g cell mass *(16)*. Despite these efforts, the current cost of PHA is still around 10 times higher than that of conventional plastics *(8)*.

This article reports a novel technique that induced the activated sludge bacteria in a conventional waste water treatment process to produce PHAs. This technique could significantly reduce the cost of PHA production and, at the same time, reduce the quantity of excess sludge that required further treatment.

METHODS

A reactor vessel with 10-L effective volume was seeded with returned activated sludge collected from municipal sewage treatment works. The reactor was fed with a synthetic waste water containing glucose at 4 g/L and NH_4Cl at 0.252 g/L, resulting in a C:N mass ratio of 24. The C:N ratio of 24 is widely accepted as required for normal bacterial synthesis in activated sludge processes (17). The synthetic waste water was also supplemented with phosphorus, trace minerals, and a growth factor with the following formulation in g/L: KH_2PO_4, 0.0037; $MgSO_4 \cdot 7H_2O$, 0.0200; $FeCl_3$, 0.0284; $MnCl_2 \cdot 2H_2O$, 0.0003; $Al_2(SO_4)_3 \cdot 18H_2O$, 0.0022; $CaCl_2$ 0.0400; $CoCl_2 \cdot 6H_2O$, 0.0080; $NaSiO_3 \cdot 5H_2O$, 0.0040; H_3BO_3, 0.0040; $ZnSO_4 \cdot 7H_2O$, 0.0020; $CuSO_4 \cdot 5H_2O$, 0.0020; $(NH_4)2MoO_4$, 0.0020; thiamine hydrogen chloride, 0.0080.

The reactor was operated in a sequencing batch mode with a batch loading rate of 0.4 mg COD/mg MLVSS-d. The react-to-contact time ratio was 0.6, and the average organic reduction efficiency was 98.1%. The detailed operation and performance of the sequencing batch reactor (SBR) system were described by Ho (18). When the reactor was operating under stable conditions, the nitrogen concentration in the synthetic waste water was sporadically reduced to result in C:N mass ratios of 48, 96, and 144, creating different degrees of nutrient deficiency.

When stable operation was attained in the SBR under each C:N ratio, samples were periodically collected and analyzed during the 2-h reaction time in one randomly selected SBR operation cycle. The samples were analyzed for total organic carbon (TOC), total kjeldahl nitrogen (TKN), dissolved oxygen, pH, and dry cell mass. The analytical techniques were carried out according to the standard methods (19). The mass of PHA extracted by 1,1,2-trifluoro-1,2,2-trichloroethane from the cell mass was also measured. The organic solvent extraction and precipitation procedure for PHA was in accordance with that described by Suzuki et al. (20).

RESULTS AND DISCUSSION

The SBR was in operation for 180 d and was consistently treating the synthetic waste water with an organic reduction efficiency around 98.1%. Residual organic and nutrient concentrations, cell growth, and polymer accumulation, with a C:N ratio of 24, are summarized in Table 1. The profiles of depletion of carbon, measured as TOC, and nitrogen, measured as TKN, in the reactor liquor are shown in Fig. 1. Both the TOC and TKN maintained a consistent depletion rate throughout the 2-h reaction time. The dissolved oxygen profile corresponded with the microbial activities in the activated sludge during the reaction. The pH of the reactor liquor remained between 6.50 and 6.91. There was a net cell growth of 1.74 g in the reactor, as measured in dry cell mass, and an accumulation of 0.11 g of intracellular storage polymer (Fig. 2).

On the other hand, when the C:N ratio was increased to 96, the TKN was almost depleted within the first 15 min, and the reactor entered into nitrogen-deficient condition (Fig. 3). The characteristics in the reactor are summarized in Table 2. The net cell growth of 0.92 g in the reactor was lower than that when the C:N ratio was 24. However, the accumulation of

Table 1
Reactor Operated Under C:N Ratio of 24

React time, h	Residual TOC, mg/L	Residual TKN, mg/L	D.O., mg/L	pH	Dry cell mass, g	Polymer accumulation, g
0	274	12.6	3.23	6.91	42.85	0.01
0.25	228	8.3	4.81	6.64		
0.50	179	5.9	4.52	6.61	43.58	0.04
0.75	135	5.3	1.33	6.59		
1.00	124	4.7	0.27	6.56	43.79	0.09
1.25	72	4.0	0.08	6.50		
1.50	44	3.0	0.35	6.56	44.59	0.12
1.75	31	2.8	0.42	6.58		

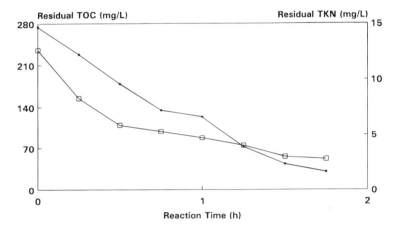

Fig. 1. Carbon and nitrogen profiles under C:N ratio of 24. ■ TOC, □ TKN.

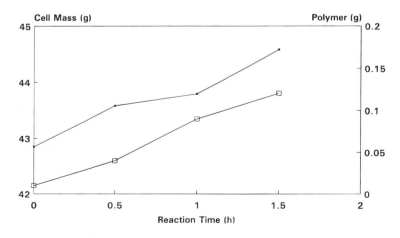

Fig. 2. Growth and polymer accumulation under C:N ratio of 24. ■ Cell mass, □ polymer.

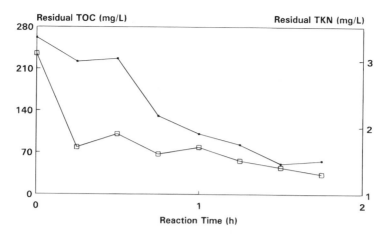

Fig. 3. Carbon and nitrogen profiles under C:N ratio of 96. ■ TOC, ▫ TKN.

Table 2
Reactor Operated Under C:N Ratio of 96

React time, h	Residual TOC, mg/L	Residual TKN, mg/L	D.O., mg/L	pH	Dry cell mass, g	Polymer accumulation, g
0	262	3.1	3.46	6.92	40.68	0.17
0.25	222	1.7	3.06	6.91		
0.50	227	1.9	2.32	6.88	40.68	0.15
0.75	131	1.6	1.67	6.89		
1.00	101	1.7	0.07	6.73	40.91	0.23
1.25	83	1.5	1.25	6.84		
1.50	51	1.4	1.32	6.87	41.60	0.40
1.75	56	1.3	1.18	6.90		

0.23 g of intracellular storage polymer was more than that when the C:N ratio was 24. The rate of polymer accumulation increased significantly after 0.5 h of reaction time (Fig. 4), which was 15 min after the reaction entered into nitrogen deficiency.

Overall organic consumption, cell growth, polymer accumulation, and yields during the 2-h reaction under four different C:N ratios are summarized in Table 3. An increase in C:N ratio from 24–144 resulted in a decline in specific growth yield, $Y_{x/s}$, from 0.579–0.232 g cell mass/g TOC (Fig. 5). This indicated that a nitrogen-deficient condition affected the growth of biomass in the activated sludge. On the other hand, the increased C:N ratio caused an increased specific polymer yield or intracellular polymer fraction, $Y_{p/x}$, from 0.066–0.374 g polymer/g cell mass. This demonstrated that the unfavorable condition resulting from nitrogen deficiency induced the microorganisms in the activated sludge to accumulate

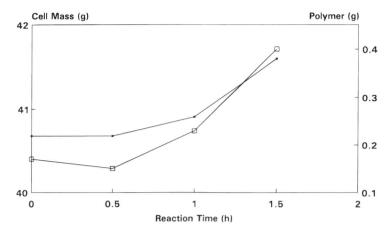

Fig. 4. Growth and polymer accumulation under C:N ratio of 96. —■— Cell mass, —□— polymer.

Table 3
Polymer Productivity Under Different C:N Ratios[a]

C:N ratio	del X, g	del P, g	del S, g	$Y_{X/S}$, g/g	$Y_{P/X}$, g/g	$Y_{P/S}$, g/g	TOC removal efficiency, %
24	1.74	0.11	2.92	0.597	0.066	0.038	98.1
48	1.51	0.18	3.11	0.485	0.122	0.058	98.1
96	0.92	0.23	2.47	0.374	0.246	0.093	96.5
144	0.57	0.21	2.47	0.232	0.374	0.085	96.9

del X = net cell growth as measured in dry cell mass during the 2-h reaction time;
del P = net accumulation of intracellular polymers during the 2-h reaction time;
del S = net consumption of TOC during the 2-h reaction time.

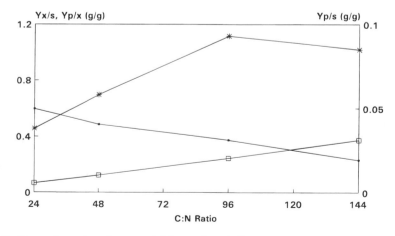

Fig. 5. Growth and polymer yields under different C:N ratios. —■— $Y_{x/y}$, —□— $Y_{p/x}$, —✱— $Y_{P/S}$.

Waste Water Treatment and Storage Polymer

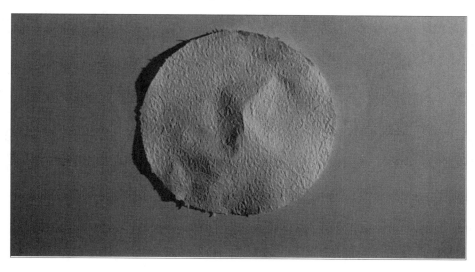

Fig. 6. Polymeric materials extracted from activated sludge.

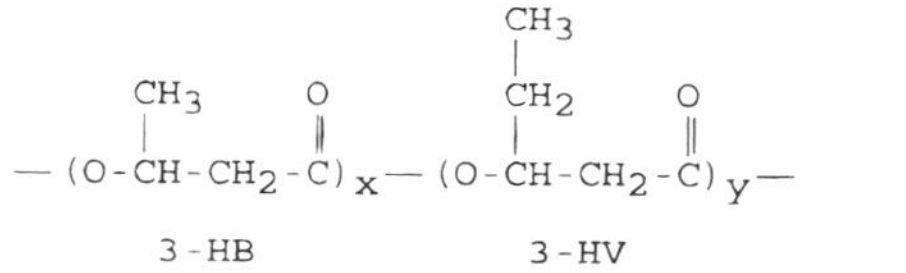

$$-(O-\underset{\underset{CH_3}{|}}{CH}-CH_2-\underset{\underset{}{||}}{\overset{O}{C}})_x-(O-\underset{\underset{CH_2}{|}\atop CH_3}{CH}-CH_2-\underset{\underset{}{||}}{\overset{O}{C}})_y-$$

3-HB 3-HV

where x:y ratio varied from 0.79 to 5.67

Fig. 7. Structural formula of copolymers.

more intracellular storage polymers. The overall polymer production yield, $Y_{p/s}$, which was a product of $Y_{x/s}$ and $Y_{p/x}$, reached a maximum of 0.093 g polymer/g TOC under the C:N ratio of 96.

Accumulated polymeric materials extracted from the microorganisms and precipitated are shown in Fig. 6. These polymeric materials were analyzed by gas chromatographic methods to contain mainly poly-β-hydroxybutyric acid, and copolymers of β-hydroxybutyric (3-HB) and β-hydroxyvaleric (3-HV) acids (18). The structural formula of these copolymers is shown in Fig. 7. The intracellular polymer fraction, $Y_{p/x}$, was 0.374 g polymer/g cell mass, indicating that 37.4 wt% of the activated sludge was composed of the polymers. If this portion was extracted for use, the requirements for treatment and disposal of the excess sludge produced from the waste water treatment process would be reduced by 37.4%.

Sporadic changes of C:N ratio in the synthetic waste water above 24, creating a nitrogen-deficient condition in the activated sludge, did not significantly affect the efficiency of organic reduction by the SBR. The TOC removal efficiency remained above 96.5% for all the C:N ratio investigated

(Table 3). These observations were in contrast with the widely accepted view that C:N ratio in activated sludge processes must be kept around 24 in order to enable normal microbial cell synthesis *(17)*. Nitrogen deficiency would result in a slowdown in microbial growth, which would, in turn, have an adverse effect on the organic treatment performance by the process. However, it must be noted that if the nitrogen-deficient condition in the SBR was prolonged to have a continual polymer production, cell growth and TOC removal efficiency would have been adversely affected. Therefore, an intermittent nitrogen feeding program must be established in order to optimize the polymer production without significantly affecting the normal treatment performance of the activated sludge process.

CONCLUSION

Activated sludge bacteria were induced by controlling the C:N ratio in the SBR reactor liquor to accumulate storage polymers. Specific polymer yield increased with increasing C:N ratio, whereas specific growth yield decreased with increasing C:N ratio. An optimum C:N ratio of 96 provided the highest overall polymer production yield. Sporadic adjustments of the C:N ratio did not significantly affect the treatment efficiency in the SBR. Production and recovery of PHAs from activated sludge could significantly reduce the cost of PHAs and, at the same time, reduce the quantity of excess sludge that required further treatment.

REFERENCES

1. Hong Kong Environmental Protection Department (1994), in *Environment Hong Kong 1994*, Hong Kong Government Press, pp. 51–66.
2. Hong Kong and Kowloon Plastic Product Merchants United Association (1992), Plastics and the Environment, Hong Kong Plastic Industry Bulletin, **33**, December.
3. Hong Kong Government Industry Department (1993), *Hong Kong's Manufacturing Industries 1993*, Hong Kong Government Press.
4. Hong Kong Government Industry Department (1991), in *Techno-Economic and Market Research Study of Hong Kong's Plastic Industry*, 1990–1991, p. 1.
5. Chua, H., Yu, P. H. F., Xing, S., and Ho, L. Y. (1995), A *Plastics Technol.* **18**, 132–148.
6. Huang, T., Zhao, J. Q., and Shen, J. R. (1991), *Plastics Industry* **4**, 23–27.
7. Young, R. J. (1981), in *Introduction to Polymers*, Chapman and Hall, New York, pp. 9–85.
8. Chang, H. N. (1994), in *Better Living Through Innovative Biochemical Engineering*, Teo, W. K., ed., Singapore University Press, pp. 24–30.
9. Kumagai, Y. (1992), *Polymer Degradation and Stability* **37**, 253–256.
10. Industrie-Anzeiger (1987), *Industrie-Anzeiger* **109**, 26.
11. Pelissero, A. (1987), *Imballaggio* **38**, 54.
12. Pfeffer, J. T. (1992), in *Solid Waste Management Eng.* 72–84.
13. Billmeyer, F. W. (1971), in *Polymer Science* Wiley Interscience, New York, pp. 379–490.
14. Lee, S. Y., Chang, H. N., and Chang, Y. K. (1994), in *Better Living Through Innovative Biochem. Engineering*, Teo, W. K., ed., Singapore University Press, pp. 53–55.
15. Shirai, Y., Yamaguchi, M., Kusubayashi, N., Hibi, K., Uemura, T., and Hashimoto, K. (1994), in *Better Living Through Innovative Biochemical Engineering*, Teo, W. K., ed., Singapore University Press, pp. 263–265.

16. Shimizu, H., Sono, S., Shioya, S., and Suga, K. (1992), in *Biochemical Engineering for 2001*, S. Furusaki, I. Endo, and R. Matsuno, eds., Springer-Verlag, Tokyo, pp. 195–197.
17. Metcalf and Eddy, Inc. (1991), in *Wastewater Engineering*, McGraw-Hill, Singapore, pp. 529–662.
18. Ho, L. Y. (1996), *Synthesis of Environmentally Friendly Materials*, Master's Thesis, The Hong Kong Polytechnic University.
19. American Public Health Association (1992), *Standard Methods for the Examination of Water and Wastewater*, 18 ed., APHA, Washington, DC.
20. Suzuki, T., Mori, H., Yamane, T., and Shimizu, H. (1985), *Biotechnol. Bioeng.* **27,** 192–201.

Effect of Surfactants on Carbon Monoxide Fermentations by *Butyribacterium methylotrophicum*

M. D. Bredwell,[1] M. D. Telgenhoff,[1] S. Barnard,[2] and R. M. Worden*,[1]

[1]*Department of Chemical Engineering, Michigan State University, East Lansing, MI 48824; and* [2]*Department of Chemical Engineering, University of Michigan, Ann Arbor, MI 48109*

ABSTRACT

Butyribacterium methylotrophicum has been grown on carbon monoxide as its carbon and energy source in the presence of various surfactants that are capable of forming microbubble dispersions for the screening of surfactants for use in microbubble-sparged synthesis gas fermentations. In the range of 0–3 times the critical micelle concentration, the presence of Tween surfactants was not significantly inhibitory to growth, final cell density, and fermentation stoichiometry, although some of the Brij surfactants caused significant inhibition. As the batch fermentations entered the stationary phase both the pH and the ratio of acetate to butyrate decreased.

Index Entries: *Butyribacterium methylotrophicum*; synthesis gas; microbubbles; aphrons.

INTRODUCTION

The obligate anaerobe *Butyribacterium methylotrophicum* consumes carbon monoxide, CO, to produce acetate, ethanol, butyrate, and butanol *(1)*. Carbon monoxide is a primary component in synthesis gas, which can be obtained from the pyrolysis of biomass *(2)* or the gasification of coal *(3)*. The rate of synthesis gas fermentations is limited by low gas-to-liquid mass-transfer rates arising from the low solubility of CO in aqueous solu-

---

*Author to whom all correspondence and reprint requests should be addressed.

tions (4). One technique to enhance gas-to-liquid mass transfer is to use microbubble dispersions (5). Microbubbles are surfactant-stabilized bubbles with diameters on the order of 50 µm. The surfactant layer surrounding a microbubble generates a diffuse electric double layer that acts to repel bubbles and prevent coalescence (6). A microbubble dispersion exhibits colloidal properties and is stable enough to be pumped (7). To be successfully used in synthesis gas fermentations, the surfactant used must be nontoxic to the cells at levels necessary to generate microbubbles. The surfactant should also have no detrimental effect on formation of the desirable products.

This article examines the growth and product formation of *B. methylotrophicum* during CO fermentations in the presence of surfactants that are suitable for generating microbubble dispersions. This information is needed to select surfactants for use in microbubble-sparged synthesis gas fermentations.

METHODS

Culture Techniques

All chemicals and vitamins were obtained from Sigma Chemical Company (St. Louis, MO), except for sodium dodecyl sulfate, which was obtained from Boehringer Manheim (Mannheim, Germany). The nitrogen (N_2) and CO gases were purchased from AGA Gas and Welding (Lansing, MI). *B. methylotrophicum* was obtained from the Michigan Biotechnology Institute (Lansing, MI) and was grown anaerobically at 37°C in a phosphate-buffered (PB), sulfide-reduced medium prepared with 0.5% yeast extract as previously described (8) on 100% carbon monoxide. Resazaurin was used as an oxygen indicator in maintaining stock cultures, but was not used in the experimental fermentations owing to interference with measuring optical density. The medium was dispersed into 125-mL sera bottles (Wheaton, Millville, NJ) at 50 mL/bottle, and sealed with butyl rubber stoppers and aluminum crimps (Wheaton, Millville, NJ). For the experimental runs with surfactant present in the media, the surfactant was added to the media prior to autoclaving. A 1% inoculum of an actively growing culture of *B. methylotrophicum* was used. The cultures were incubated in an Innova 4000 shaker (New Brunswick Scientific, New Brunswick, NJ) at 100 rpm in the dark.

Culture Analysis

Liquid samples (2 mL) were taken throughout the duration of each experiment. Growth was analyzed using a Lambda 3 spectrophotometer (Perkin Elmer, Norwalk, CT) at 660 nm immediately after withdrawing the sample. The sample was then frozen for later product analysis. In experiments in which the pH was monitored, a 1-mL sample of cells was

A

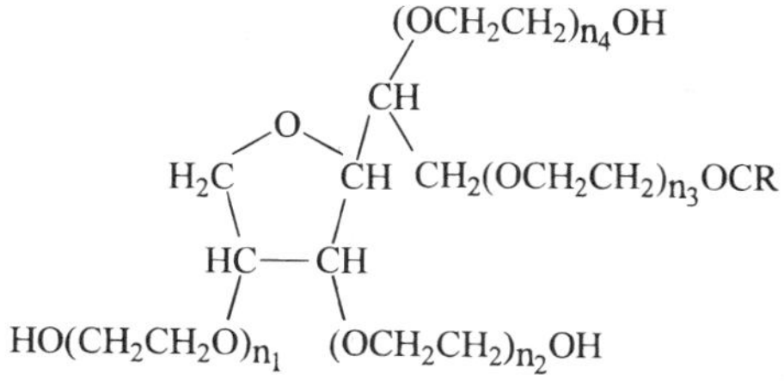

$(OCH_2CH_2)_{n_4}OH$
CH
H_2C O CH $CH_2(OCH_2CH_2)_{n_3}OCR$
HC—CH
$HO(CH_2CH_2O)_{n_1}$ $(OCH_2CH_2)_{n_2}OH$

EO adducts $= 20 = n_1 + n_2 + n_3 + n_4$

Tween 20: $R =$ *monolaurate*
Tween 40: $R =$ *monopalmitate*
Tween 80: $R =$ *monooleate*

B

$CH_3(CH_2)_{15}O[CH_2CH_2O]_nH$

Brij 52: $n = 2$
Brij 56: $n = 10$
Brij 58: $n = 20$

Fig. 1. Chemical structure for Tween and Brij surfactants *(9)*. EO adducts indicates the number of ethylene oxide adducts.

centrifuged in a Fisher Microcentrifuge 235C (Fisher Scientific, Chicago, IL) at 15,000 g for 10 min. Phenol red was then added to the sample to 20 μM, and the optical density was measured at 560 nm to determine the pH of the sample. The phenol red assay was calibrated using larger samples with a pH electrode. To prepare samples for product analysis, 10% (v/v) of $1M$ phosphoric acid was added. After a 10 min incubation at 37°C, the samples were centrifuged as before, and then analyzed by gas chromatography for acetate, ethanol, butyrate, and *n*-butanol. The separation was done with a Perkin-Elmer Autosystems GC (Perkin Elmer) with a Hayesep R, 6' × 1/4" × 2 mm deactiglass column (Alltech, Waukeegen, WI) and a flame-ionization detector.

The primary surfactants used for this study were Tween (polyoxyethylene sorbitans) and Brij (polyoxyethylene alcohols). The structures of these surfactants are shown in Fig. 1. Dimensionless surfactant concentrations (DSC) are defined as the ratio of the surfactant concentration to its critical micelle concentration. DSC values up to 3 were used because previous studies *(5)* indicated that microbubble dispersions needed DSC values of at least 1 for the formation of stable dispersions. Increasing DSC values above 3 had no effect on the dispersion's stability, presumably because the absorption saturation value of the available gas–liquid interface had been reached. The critical micelle concentrations and aggregation numbers for the surfactants used in these experiments are listed in Table 1 *(9)*.

RESULTS

Figure 2 shows growth curves for *B. methylotrophicum* for the Tween surfactants at different DSC values. Figure 3 shows the growth curves for Brij surfactants at different DSC values. Both the Tween and Brij runs had

Table 1
Surfactant Critical Micelle Concentrations (8)

Surfactant	Critical micelle concentration
Brij 52	> 7 µM
Brij 56	7 µM
Brij 58	77 µM
Tween 20	60 mg/L
Tween 40	29 mg/L
Tween 80	13 mg/L

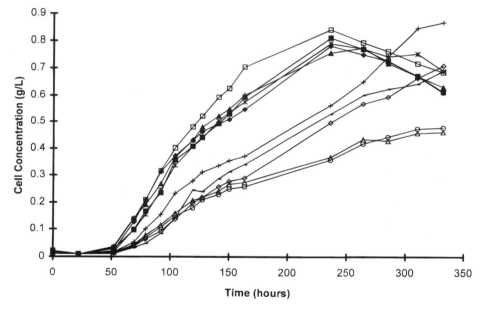

Fig. 2. Growth of *B. methylotrophicum* in batch bottle fermentation with Tween surfactant. ─□─ Control; ─◇─ Tween 20, DSC = 1; ─○─ Tween 20, DSC = 2; ─△─ Tween 20, DSC = 3; ─✶─ Tween 40, DSC= 1; ─+─ Tween 40, DSC = 2; ─·─ Tween 40, DSC = 3; ─■─ Tween 80, DSC = 1; ─◆─ Tween 80, DSC = 2; ─▲─ Tween 80, DSC = 3.

individual controls that were prepared and inoculated at the same time and under the same conditions as the data taken. The data in Figs. 2 and 3 are averages of triplicate runs. Figures 4 and 5 show the specific growth rate as a function of surfactant concentration and surfactant chain length for Tween and Brij, respectively.

Effect of Surfactants

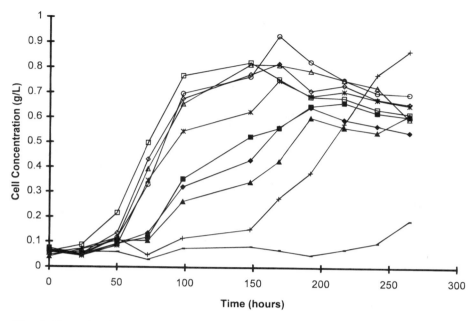

Fig. 3. Growth of *B. methylotrophicum* in batch bottle fermentation with Brij surfactant. ─▣─ Control; ─◇─ Brj 52, DSC = 1; ─○─ Brj 52, DSC = 2; ─△─ Brj 52, DSC = 3; ─✳─ Brj 56, DSC = 1; ─+─ Brj 56, DSC = 2; ─●─ Brj 56, DSC = 3; ─■─ Brj 58, DSC = 1; ─◆─ Brj 58, DSC = 2; ─▲─ Brj 58, DSC = 3;

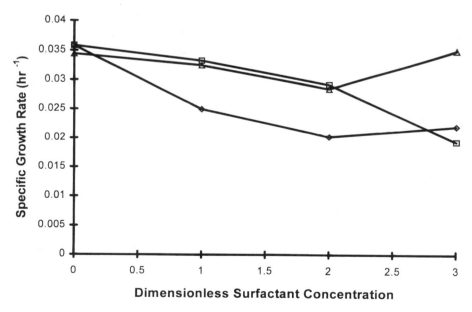

Fig. 4. Specific growth rate of *B. methylotrophicum* determined from data in Fig. 1 for Tween surfactant as a function of surfactant concentration. ─◇─ Tween 20, ─▣─ Tween 40, ─△─ Tween 80.

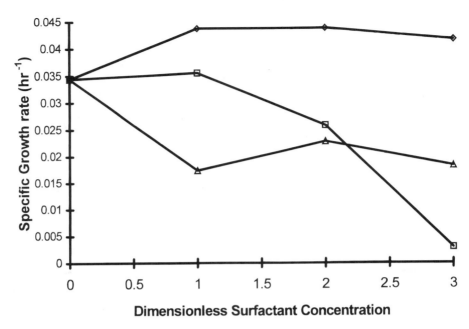

Fig. 5. Specific growth rate of *B. methylotrophicum* determined from data in Fig. 2 for Brij surfactant as a function of surfactant concentration. —◇— Brij 52, —□— Brij 56, —△— Brij 58.

The acetate, ethanol, and butyrate fermentation profiles are shown in Figs. 6, 7, and 8, respectively, as a function of surfactant type and chain length. The remainder of the product concentration data were calculated using carbon and electron balances for each run *(10,11)*. The data shown in Table 2 for Tween surfactants represent the average of triplicate values.

A typical plot of pH as a function of time during the experiment is shown in Fig. 9. The pH profiles for other runs (data not shown) had similar profiles to the ones presented here.

DISCUSSION

None of the Tween surfactants strongly inhibited growth of *B. methylotrophicum* in batch bottle fermentations. However, there did seem to be an effect of the length of the hydrophobic end of the surfactant on the growth of the microorganisms. Tween-20 has a lauric acid group (C12), Tween-40 has a palmitic acid group (C16), and Tween-80 has an oleic acid group (C18-1). In general, the shorter the chain length, the lower the growth rate (*see* Fig. 2). However, for the Tween surfactants, most cultures obtained the same approximate final cell density. It has been suggested that longer chain lengths slow surfactant diffusion into the cellular membrance and thereby result in lower toxicity *(12)*. Tween surfactants have also been found to be nontoxic to other types of culture systems (e.g., soy

Effect of Surfactants

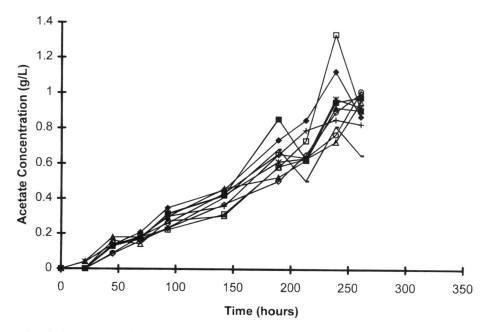

Fig. 6. Acetate production during the Tween surfactant experiments. —□— Control; —◇— Tween 20, DSC = 1; —○— Tween 20, DSC = 2; —△— Tween 20, DSC = 3; —✳— Tween 40, DSC= 1; —+— Tween 40, DSC = 2; —⁃— Tween 40, DSC = 3; —■— Tween 80, DSC = 1; —◆— Tween 80, DSC = 2; —▲— Tween 80, DSC = 3.

and carrot suspension cultures), and in some cases, Tween-20 actually enhanced the growth of the culture (Ames, T. T., unpublished data). Higher surfactant concentrations (10–20 DSC) were found to be toxic to plant suspension cultures.

The Brij surfactants did strongly inhibit growth in some cases, but the inhibitory effect again depended on the chain length. The Brij surfactants are polyoxyethylene (polyethylene glycol—PEG) alcohols (see Fig. 1) and had the following characteristics (in order of increasing chain length): Brij 52-PEG(2) Cetyl alcohol, Brij 56-PEG(10) Cetyl alcohol, and Brij 58-PEG(20) Cetyl alcohol. As shown in Figs. 2 and 5, the specific growth rate and final cell densities are much lower than those for the control for increasing amounts of the two longer chain surfactants, Brij 56 and Brij 58. In general, the higher the surfactant concentration, the lower the specific growth rate and final cell density. The Brij surfactants are apparently more inhibitory to the growth of *B. methylotrophicum* in concentrations necessary to form microbubbles than are the Tween surfactants.

Growth experiments were also done with a few ionic surfactants (e.g., sodium dodecyl sulfate, cetyl pyridinium choride, and sodium do\decyl benzene sulfonate). The growth of *B. methylotrophicum* on these surfactants was strongly inhibited even at the lowest concentrations tested (0.5 DSC)

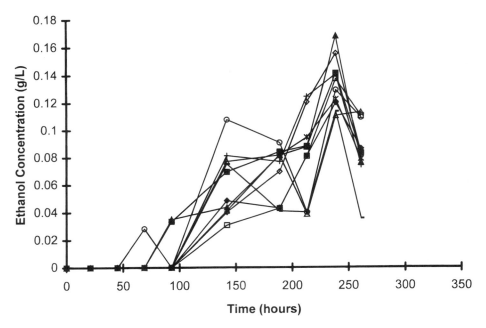

Fig. 7. Ethanol production during the Tween surfactant experiments. —□— Control; —◇— Tween 20, DSC = 1; —○— Tween 20, DSC = 2; —△— Tween 20, DSC = 3; —✻— Tween 40, DSC= 1; —+— Tween 40, DSC = 2; —— Tween 40, DSC = 3; —■— Tween 80, DSC = 1; —◆— Tween 80, DSC = 2; —▲— Tween 80, DSC = 3.

(data not shown). Both anionic and cationic surfactants are prone to bind proteins. The charged groups of the ionic surfactants form relatively strong ionic bonds with charged groups on proteins. Nonionic surfactants, on the other hand, lacking the charged groups bind through much weaker hydrophobic interactions with the protein chain, and thus, are less likely to inactivate or denature enzymes and proteins *(12)*. Nonionic surfactants would thus be expected to have the least influence on cellular metabolism in fermentation.

The specific growth rates were calculated from the exponential portion of the growth curve. There was no inhibition from the products formed at the concentrations seen in this study *(13)*. The deviations from typical exponential growth are because of mass-transfer limitations of the gaseous substrate.

The pH decrease during the fermentation is because of to the production of acetic and butyric acids. The metabolic pathways for acid production have been well characterized *(3)*. A shift in the product ratios for *B. methylotrophicum* has been observed in which less acetate and more butyrate and alcohols are produced at low pH *(13,14)*. Moreover, a pH shift from 6.8 to 6.0 as the cells entered the stationary phase has been shown to induce formation of butyrate as the primary product *(15,16)*.

Effect of Surfactants

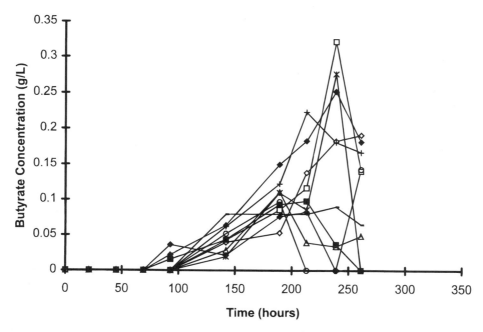

Fig. 8. Butyrate production during the Tween surfactant experiments. ⊟ Control; ◇ Tween 20, DSC = 1; ⊖ Tween 20, DSC = 2; △ Tween 20, DSC = 3; ✳ Tween 40, DSC = 1; ＋ Tween 40, DSC = 2; — Tween 40, DSC = 3; ■ Tween 80, DSC = 1; ◆ Tween 80, DSC = 2; ▲ Tween 80, DSC = 3.

Table 2
Carbon and Electron Balance Results for Tween Surfactants

Control	4 CO →	0.40 Ace +	0.076 Et +	0.065 Bu +	0.61 CM +	2.17 CO_2
Tween 20 (1 DSC)	4 CO →	0.48 Ace +	0.050 Et +	0.028 Bu +	0.70 CM +	2.12 CO_2
Tween 20 (2 DSC)	4 CO →	0.43 Ace +	0.114 Et +	0.039 Bu +	0.58 CM +	2.18 CO_2
Tween 20 (3 DSC)	4 CO →	0.45 Ace +	0.163 Et +	0.012 Bu +	0.52 CM +	2.20 CO_2
Tween 40 (1 DSC)	4 CO →	0.41 Ace +	0.083 Et +	0.057 Bu +	0.61 CM +	2.17 CO_2
Tween 40 (2 DSC)	4 CO →	0.34 Ace +	0.076 Et +	0.079 Bu +	0.66 CM +	2.19 CO_2
Tween 40 (3 DSC)	4 CO →	0.37 Ace +	0.068 Et +	0.082 Bu +	0.62 CM +	2.18 CO_2
Tween 80 (1 DSC)	4 CO →	0.43 Ace +	0.085 Et +	0.047 Bu +	0.62 CM +	2.16 CO_2
Tween 80 (2 DSC)	4 CO →	0.43 Ace +	0.063 Et +	0.031 Bu +	0.77 CM +	2.13 CO_2
Tween 80 (3 DSC)	4 CO →	0.37 Ace +	0.096 Et +	0.078 Bu +	0.55 CM +	2.20 CO_2

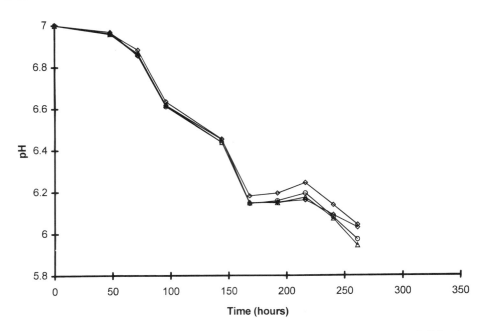

Fig. 9. pH profile of Tween-80 fermentation. -◆- Control; -◆- Tween 20, DSC = 1; -○- Tween 20, DSC = 2; -△- Tween 20, DSC = 3.

Consistent with this trend, Figs. 6 and 8 show that acetate is produced throughout the fermentation, but the butyrate production starts after 70 h. The acetate, ethanol, and butyrate profiles in Figs. 6, 7, and 8 are the averages of triplicate data sets. In the averaged sets of data, the variability was as great as ±25% between bottles. Because of the postautoclave additions done to each bottle and individually inoculating each bottle, these variations are expected.

The carbon and electron balance results for the Tween surfactant experiments given in Table 2 show the ratio of the primary products, acetate, ethanol, and butyrate, was not affected by the presence of the surfactant in the media. This result suggests that the Tween surfactants are a good choice for microbubble-dispersion-sparged CO fermentations.

CONCLUSIONS

The effect of Tween surfactants on growth of *B. methylotrophicum* on CO varied with the chain length. Longer chain lengths had negligible effects on growth of a DSC range of 0–3. Shorter chain lengths slowed growth somewhat, but did not affect the final cell density. None of the Tween surfactants significantly affected the product stoichiometry. Brij surfactants were inhibitory over the same concentration range, in some cases reducing both the growth rate and the final cell density. An increase in the ratio of butyrate to acetate observed late in the fermentations coincided with a decrease in the fermentation pH and the onset of the stationary phase.

ACKNOWLEDGMENTS

This material is based on work supported by the National Science Foundation under grant #BCS 92 20396 and fellowship support was provided from the NIH Biotechnology Training Program Grant and the Biotechnology Research Center at Michigan State University.

REFERENCES

1. Worden, R. M., Grethlein, A. J., Jain, M. K., and Datta, R. (1991), *Fuel* **70,** 615–619.
2. Maschio, G., Lucchesi, A., and Stoppato, G. (1994), *Bioresource Tech.* **48,** 119–126.
3. Grethlein, A. J. and Jain, M. K. (1992), *Trends Biotechnol.* **10(12),** 418–423.
4. Morsdorf, G., Frunzke, K., Gadkari, D., and Meyer, O. (1992), *Biodegradation* **3,** 61–82.
5. Bredwell, M. D., Telgenhoff, M. D., and Worden, R. M. (1995), *Appl. Biochem. Biotechnol.* **51/52,** 501–509.
6. Sebba, F. (1987), in *Foams and Biliquid Foams—Aphrons*, Wiley, New York, pp. 47, 48.
7. Longe, T. A. (1989), Colloidal gas aphrons: Generation, flow characterization and application in soil and groundwater decontamination, Ph.D. dissertation, Virginia Polytechnic Institute and State University, Blacksburg, VA.
8. Kerby, R. and Zeikus, J. G. (1987), *J. Bacteriol.* **169(5),** 2063–2068.
9. Ash, M. and Ash, I. eds., (1981), *Encyclopedia of Surfactants*, Chemical Publishing, New York.
10. Erickson, L. E. and Oner, M. D. (1983), *Ann. NY Acad. Sci.* **413,** 99.
11. Erickson, L. E., Minkevich, I. G., and Eroshin, V. K. (1978), *Biotech. Bioeng.* **20,** 1595.
12. Schwuber, M. J. and Bartnik, F. G. (1980), in *Anionic Surfactants: Biochemistry, Toxicology, Detmatology*, Marcel Dekker, New York, pp. 1–49.
13. Grethlein, A. J. (1991), Metabolic engineering of product formation during carbon monoxide fermentations by *Butyribacterium methylotrophicum*, Ph.D. dissertation, Michigan State University, East Lansing, MI.
14. Grethlein, A. J., Worden, R. M., Jain, M. K., and Datta, R. (1990), *Appl. Biochem. Biotechnol.* **24/25,** 875–884.
15. Grethlein, A. J., Worden, R. M., Jain, M. K., and Datta, R. (1991), *J. Fermentation and Bioeng.* **72,** 58–60.
16. Worden, R.M., Grethlein, A. J., Zeikus, J. G., and Datta, R. (1989), *Appl. Biochem. Biotechnol.* **20/21,** 687–698.

Principles for Efficient Utilization of Light for Mass Production of Photoautotrophic Microorganisms

Amos Richmond* and Hu Qiang

Microalgal Biotechnology Laboratory, The Jacob Blaustein Institute for Desert Research, Ben Gurion University at Sede Boker 84990, Israel

ABSTRACT

Outdoor production of microalgae could be set on a sound industrial basis if solar energy were utilized at a much higher efficiency than presently obtained. Many types of photobioreactors have been developed in the past in an attempt to answer this challenge, but their photosynthetic efficiency has been rather similar to the basically inefficient open raceway commonly used today. Efficient utilization of the oversaturating solar energy flux mandates that reactors should have a narrow lightpath to facilitate ultra-high cell densities, be maximally exposed to sunshine, and have an efficient mixing system to create strong turbulent streaming to affect dark–light cycles of the highest possible frequency.

Index Entries: Photoautotrophs; mass production; photobioreactors; ultra-high cell density; mixing rate.

INTRODUCTION

A central issue in mass cultivation of photoautotrophic microorganisms concerns the mode of cultivation by which to utilize light energy for growth most effectively. The concept of using sunlight energy to mass-produce microalgae for various economic purposes is based on a premise that microalgae may be so cultured as to be limited by light only. Microalgae, however, are usually light-saturated at 5–10% of the solar energy flux available in midday, and therefore, abundant solar light can be only partially used for the production of biomass. Several theoretical approaches have been suggested to address this difficulty, e.g., Kok and

---

*Author to whom all correspondence and reprint requests should be addressed.

Van Dorschot (1), who suggested improving outdoor productivity by using strains of algae with a higher capacity to use strong irradiance for photosynthesis. More recently, Sukenik et al. (2), following this line of thought, suggested algae should be improved to increase their photosynthetic capacity at light saturation by amplification of the carboxylation enzyme relative to the electron transport complexes. Phillips and Myers (3) suggested a different approach. Observing that light of high intensity may be used with higher efficiency if presented in short flashes separated by long dark periods, they proposed that utilization of high-intensity light would be enhanced by inducing turbulent streaming in culture suspension. They showed that a dense algal culture in sunlight exhibited a significant increase in growth when the algal cells were moved in and out of the high light intensity region at the front surface, in rates that produced flash times of 1–100 ms (3).

Thus far, the practical expression of the principles governing the utilization of sunlight for photoproduction of algal mass has been the open raceway, which has become the major reactor available for mass production of microalgae outdoors. This device has both enhanced and impeded the development of algal biotechnology (4), as shall be elucidated in what follows. In this work, we present principles by which to utilize high-intensity light, such as sunlight, in high efficiency.

MATERIALS AND METHODS

Microorganism

The cyanobacteria *Spirulina platensis* was grown in Zarouk medium (5) in which the $NaNO_3$ concentration was raised to 5.0 gL^{-1}. Culture temperature was 35 °C and pH was 9.5. Culture supernatant was replaced with fresh growth medium once a day by filtering the algal suspension through a 300-mesh screen.

Photobioreactor

A flat-plate bioreactor made of glass consisting of 2.6- or 1.3-cm wide flat tanks equipped with a perforated tube extending along the bottom through which a stream of compressed air was passed to affect stirring. The tanks were immersed in a water jacket for temperature regulation. The total illuminated surface of each tank was 0.1 m^2, and mixing and CO_2 were provided by continuous supply of air enriched with 2% CO_2 injected through the perforated tube. Turbulent flow was controlled by adjusting the rate of air passing through a gas flow meter, being expressed in terms of L air/L culture suspension/min ($L L^{-1} min^{-1}$). Illumination was provided by 1500-W halogen lamps, and light intensity was modified by manipulating the distance between the light source and the reactor.

Measurements and Analytical Methods

Output rate of biomass was estimated by measuring changes in biomass concentration at 4-h intervals as reflected in the dry weight, a procedure repeated at least four times. PFD was measured with a Li-Cor model Li-185A, and the oxygen production rate (OPR) was measured according to Guterman et al. (6). Dry weight was determined in duplicates using 5-mL samples (7).

RESULTS AND DISCUSSION

Our early experimentations in outdoor cultures were all carried out in open raceways—shallow pans with a divider in the middle to form a raceway in which the algal suspension is caused to flow at 20–40 cm · s^{-1} by a paddle wheel. This reactor requires simple technology to construct and to maintain, but does not meet the major requirements for efficient utilization of light: its water level must be maintained at a height of 12–15 cm, resulting in very dilute (e.g., a few hundred mg of dry wt/L) algal suspensions. Also, the paddle wheel cannot practically affect in large areas the extent of turbulent streaming required to create a high frequency of the dark/light cycle algal cells undergo when growth is light-limited. It is a particularly poor device for light utilization in winter owing to lack of temperature control (8). As a result of all its drawbacks, the open raceway yields low outputs of algal mass (e.g., 20–40 t · ha^{-1} · y^{-1}) and requires handling very large amounts of algal suspension, which often become readily contaminated as well as increasing the cost of harvesting. The inadequacy of the open raceway prompted research to develop more efficient modes of production with which to facilitate a more effective use of the high photon flux density (PFD) existing outdoors. In continuous cultures, productivity (P) at steady state is the product of uxv, u representing the specific growth rate, x the concentration of algal mass, and v the culture volume. Therefore, the only practical method by which to increase P per unit area is to increase x without lowering u as a result of the enhancement in mutual shading and the ensuing rise in light limitation. We suggest achieving this by employing photobioreactors in which the lightpath has been greatly reduced and optimal cell concentrations (defined as that algal density that would result in the highest output rates under the given environmental circumstances) have been greatly increased, resulting in higher output rates (Fig. 1). Many enclosed photobioreactors facilitate an increase in cell concentration of up to twice one order of magnitude compared with the open raceway. Most of these new devices, however, have not increased the efficiency by which light is utilized for production of chemical energy. We suggest this failure stems from inadequate optimization of the short lightpath in relation to cell density and the extent of mixing. In what follows, these basic relationships will be examined in some detail.

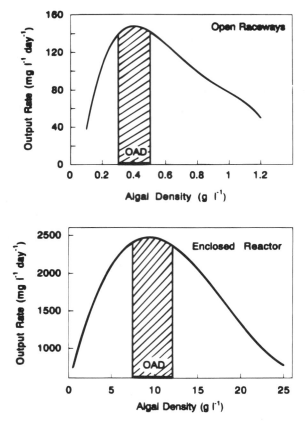

Fig. 1. The optimal algal density and output rates in an open raceway and an enclosed photobioreactor.

The interrelationships between the length of the lightpath and the optimal algal density in a photoautotrophic culture is shown in Fig. 2. As the lightpath is reduced, the optimal density increases because of the exponential increase in the availability of light to the cells in the culture. In addition, Hu et al. (9) discovered recently that the areal output rate also increases with reduction in the lightpath (Fig. 2), indicating that as the lightpath is reduced, photosynthetic efficiency increases, reflecting a more efficient utilization of the light source.

The other important mode by which the efficiency of light utilization is increased concerns the rate of mixing of the culture (10). This is clearly evident in Fig. 3, describing the relationships between the population density and the output rate as affected by varying rates of mixing, under a relatively "low-" and "high-"incident light.

A low mixing rate (0.6 L air/L algal suspension/min) in low light resulted in an optimal algal density (OAD) of ca. 2 g/L^{-1}. As the rate of mixing was increased, the OAD shifted up (to ca. 5 g/L^{-1}), and as appro-

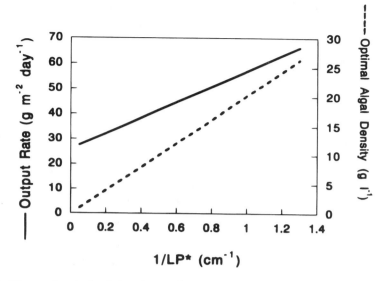

Fig. 2. The optimal algal density and output rate as affected by the length of the lightpath.

priate, the output rate increased significantly from 70 (at the minimal mixing rate) to 100 mg dry wt/L^{-1}/h^{-1} *(10)*. Two aspects concerning the relationship between the output rate and cell density at this low light intensity deserve attention: when cell density was below 2 g/L, there was no difference in output rate in response to a wide range of mixing rates (i.e., 0.6–4.2 $L/L^{-1}/min^{-1}$), indicating the rate of mixing was not limiting productivity under these conditions. Indeed, the magnitude of the effect exerted by mixing was strictly dependent on the strength of the light source as well as on culture density. Thus, as the PFD was increased to 1800 $\mu mol/m^{-2}/s^{-1}$ (the order of magnitude of the energy level existing outdoors at noon), the output rates of biomass obtained at this energy flux indicated a sensitive response to the rate of mixing, an increase in aeration rate from 0.6–4.2 L/L^{-1} resulting in doubling the output rate. A further increase in aeration rate to 6.3 $L/L^{-1}/min^{-1}$ was harmful *(10)*. Clearly, the higher the intensity of the light source, the higher the optimal population density would become and the more significant the degree to which the rate of mixing affects the output rate would be. The role of stirring in affecting maximal light utilization is evidenced from monitoring the photosynthetic efficiency (PE): First, mutual shading becomes ever more severe as the optimal population density is increased in adjustment to the reduction in lightpath. Second, assuming the nutritional requirements are satisfied and the environmental conditions optimized, the light regime to which the individual cells are exposed becomes the predominant parameter affecting productivity in cultures of

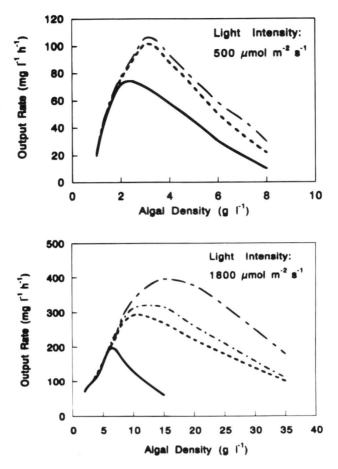

Fig. 3. Interrelationships among light intensity, algal density, and output rate as affected by the rate of mixing.

ultra-high cell densities. The light regime for the average cell in the culture is governed by several factors, among which are the intensity of irradiance at the reactor surface and the average duration of exposure of the cells to the photic as well as the dark volumes of the reactor, i.e., the frequency of fluctuating between these zones, which creates "light–dark cycles" (L–D cycle), well analyzed by Terry *(11)*. The frequency of the L–D cycle to which cells in a photobioreactor are exposed is a function of the lightpath across which axis the algal cells are moved back and forth from the lit to the dark volume in the reactor, as well as the extent of turbulence, which affects the rate of such movement. The estimated average duration of the L–D cycle at a rate of flow of ca. 250 mm/s^{-1} in relation to the length of the lightpath is portrayed in Fig. 4, indicating the smaller the lightpath (i.e., the width of the reactor), the higher the frequency of the L–D cycle.

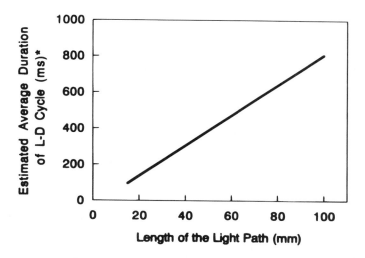

Fig. 4. The relationship between the length of the lightpath and the L–D cycle.

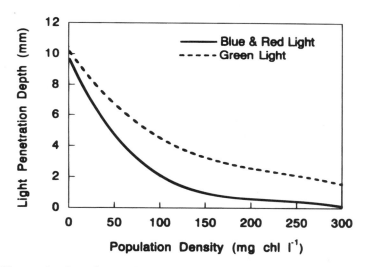

Fig. 5. Blue and red, and green light penetration into the algal suspension as affected by the population density.

The rate of flow of the algal suspension, however, has no effect on the ratio between the duration of the light and that of the dark phase in the L–D cycle. This ratio is determined only by the relative width of the photic zone as determined by the intensity of the light source and the population density. Gitelson et al. (12) have measured the penetration depth of light energy in the range of photosynthetic active radiation (PAR) in relation to algal density (expressed in mg chlorophyll; Fig. 5). Light penetration depth (defined as the depth at which down-welling irradiance is decreased to

10% of incident irradiance) differed for blue and red lights compared with green light (Fig. 5), highlighting the complexity of the photic volume in ultra-high-density cultures. Nevertheless, the general relationship between light penetration depth and population density is such that light penetration (and hence the photic volume) are strictly dependent on the population density. In ultra-high population densities, the photic zone may be ca. 1–4 mm deep in the flat plate reactors used in this study. When light was irradiated on one reactor surface only, the photic volume occupied up to 10% of the total reactor volume. Therefore, the ratio of the light to the dark phase in the L–D cycle would be ca. 10, found by Kok *(13)* as well as by Phillips and Myers *(3)* to be conducive to efficient utilization of light. In reactors with a lightpath of 15–25 mm, the frequency of the L–D cycles under our experimental conditions was roughly estimated to range between 100 and 200 ms. The basic premise is that the higher the frequency in which cells in a severely light-limited culture are exposed to short pulses of light, the more efficient the utilization of a high flux of over saturating light would be. Surprisingly perhaps, the strategy by which to utilize the high PFD existing outdoors at the highest efficiency is by "diluting" it, converting in effect direct beam to diffuse light. Such "conversion" is affected by a method developed by Pulz *(14)*. Accordingly, flat plate panels are placed outdoors in close proximity to one another, resulting in a large algal volume spread, in effect, over a small ground area. Obviously, light available for the culture in such mutually shading plates will be essentially low-energy diffuse light. Therefore, the optimal population density in such plates would necessarily be greatly reduced to ca. 1/5 or less of that maintained in flat plates exposed to direct beam radiation. Since, however, photosynthetic efficiency involved in utilizing low-energy light could be as much as five times higher than that of the high-energy rate provided by direct beam radiation, the output rate per ground area of such a system would be the highest attainable, yielding the highest efficiency possible for strong light utilization (Fig. 6). Record areal output rates reflecting highest PE, however, do not impart economic significance on this system: The volumetric yield (i.e., the amount of cell mass per unit of reactor volume) is necessarily much reduced in such reactor arrangement, and the capital investment involved in stacking expensive photobioreactor hardware to yield low volumetric yields represents a financial burden that seems to counteract the mere benefit of utilizing PAR in a most efficient manner. Indeed, economic constraints in microalgal biotechnology favor high volumetric yields, reflecting the advantage of a high photobioreactor efficiency in terms of output rate per unit reactor volume.

    We suggest that a promising method for effective utilization of solar energy outdoors is to focus on efficient distribution of high light flux to the individual cells in ultra-high-density cultures, rather than on strategies based on spatial dilution of direct beam. We place this assertion on the basis of our results obtained by growing *S. plantensis* in flat plate reactors

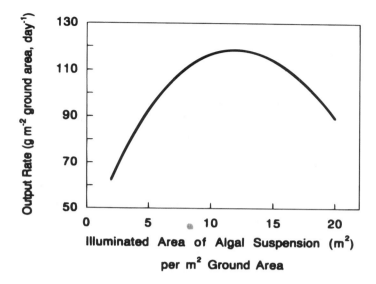

Fig. 6. The areal output rate as related to the total surface area of cultures in vertical flat plate reactors placed varying distances apart.

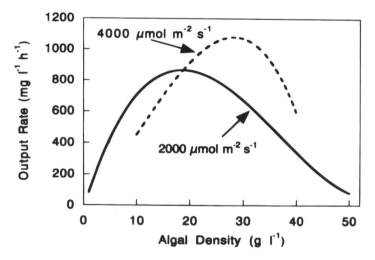

Fig. 7. The effect of the photon flux density on the output rate of cell mass at the optimized population density and mixing rate.

under laboratory conditions, in which light was applied continuously to both reactor surfaces (10). Cultures responded well to an increase in light intensity provided the population density and mixing rates were optimized with added light increment (Fig. 7). At a light flux of 2000 µmol photons/m^{-2}/s^{-1} applied on each side of the flat plate reactor, i.e., ca. twice the light flux available outdoors (at noon), the optimal population density ranged between 18 and 22 g dry wt/L^{-1} and the maximal output rate was

900 mg/L^{-1}/h corresponding to ca. 7.8 g dry wt of algal mass/m^2/h—a record output. This output approaches the value predicted by Raven *(15)*, who calculated the maximal output rate of photosynthetic activity in a hypothetical culture exposed to 2000 µmol photons/m^{-2}/s^{-1} as being ca. 11 g/L^{-1}/h^{-1} assuming the quantum demand was 16 (mole photons per mole CO_2 fixed in cell mass), cell content was 50% carbon and all incident photons absorbed by the algal suspension (Fig. 7). The latter condition seems to be well answered in our ultra-high-cell density cultures, in which measurements of the oxygen production rate (OPR) reveal that no saturation of light takes place, i.e., the OPR in our flat plate reactor is linearly related to light intensity, provided cell concentration and mixing rate are optimized. Indeed, when the culture was exposed to 4000 µmol photons, the range of the optimal population density steadied between 30 and 38 g/L^{-1} and the output rate was increased to 1150 mg dry wt of cell mass/L/h. As expected, this surge in output rate was achieved with a much reduced photosynthetic efficiency, declining from ca. 15–10% (not shown).

In conclusion, we propose that the key for obtaining high output rates of photoautotrophic cell mass at high PE under supersaturating PFD rests in reactors with a narrow lightpath, in which the culture may be maintained at ultra-high, carefully optimized algal density by providing mixing at the maximal rates permissible without cell damage.

REFERENCES

1. Kok, B. and Van Oorschot, J. L. P. (1954), *Acta Botanica Neerlandica* **3**, 533–546.
2. Sukenik, A., Falkowski, P. G., and Bennet, J. (1987), *Biotechnol. Bioeng.* **30**, 970–977.
3. Phillips, J. N. and Myers, J. (1954), *Plant Physiol.* **29**, 152–161.
4. Richmond, A. (1992), *J. Appl. Phycology* **4**, 281–286.
5. Zarouk, C. (1966), Ph.D. thesis, Paris.
6. Guterman, H., Ben-Yaakov, S., and Vonshak, A. (1989), *Biotechnol. Bioeng.* **34**, 143–152.
7. Vonshak, A., Abeliovich, A., Boussiba, S., and Richmond, A. (1982), *Biomass* **2**, 175–186.
8. Richmond, A. (1990), *Prog. Phycological Res.* **7**, 1–62.
9. Hu, Q., Guterman, H., and Richmond, A. (1996), *Biotechnol. Bioeng.* **51(1)**, 15–60.
10. Hu, Q. and Richmond, A. (1996), *J. Appl. Phycology*, **8**, 139–145.
11. Terry, K. L. (1986), *Biotechnol. Bioeng.* **28**, 988–995.
12. Gitelson, A., Hu, Q., and Richmond, A. (1996), *Appl. Environ. Microbiol.* **62(5)**, 1570–1573.
13. Kok, B. (1953), in *Algal Culture from Laboratory to Pilot Plant*, Burlew, J. S., ed., Carnegie Institute, Washington, DC, pp. 63–75.
14. Pulz, O. (1994), in *Algal Biotechnology in the Asia-Pacific Region*, Phang S. M., ed., Kuala Lumpur, University of Malaysia, pp. 113–117.
15. Raven, J. A. (1988), in *Microalgal Biotechnology*, Borowitzka, M. and Borowitzka, L. J., eds., Cambridge University Press, Cambridge, UK, pp. 331–356.

An Optical Resolution of Racemic Organophosphorous Esters by Phosphotriesterase-Catalyzing Hydrolysis

SHOKICHI OHUCHI, HIROYUKI NAKAMURA,
HIROTO SUGIURA, MITSUAKI NARITA, AND KOJI SODE*

Department of Biotechnology, Tokyo University of Agriculture and Technology, 2-24-16 Nakamachi, Koganei, Tokyo 184, Japan

Index Entries: Phosphotriesterase; stereospecific reaction; organophosphorous insecticide; phosphotriesters; kinetic optical resolution.

INTRODUCTION

Commercially available organophosphorous insecticides, such as malathion, vamidothion, profenofos *(1)*, acephate *(2)*, and so forth, often contain chiral carbon, phosphorous, or sulfur atoms. These chiral insecticides are used as racemic mixtures because of the absence of effective resolution procedures. However, the insecticidal activity of the optically pure compound is different from the racemic compound. For example, ethyl 4-nitrophenyl phenylphosphonothioate (EPN), which contains a chiral phosphorous atom, is more toxic as the R-form (Rp-EPN) than as the S-form (Sp-EPN) or as the racemic mixture against mice, hens, and insects *(3)*. Therefore, further systematic analysis of biological activity of chiral insecticides is essential in order to design and develop novel insecticides with specificity. In addition, enantiospecific degradation of organophosphorous insecticides by soil bacterial enzymes should also be studied in detail for further understanding in selectively remaining enantiomers and their toxicity.

Phosphotriesterase (PTE) from native soil bacteria catalyzes the hydrolysis of organophosphate triesters and organophosphonate diesters (Fig. 1). This enzyme is capable of hydrolyzing the various P–X bonds; P–O, P–S, P–F, P–CN, and P–N bonds *(4–6)*, and has been reported to be stereospecific for organophosphorous esters containing chiral phospho-

---

*Author to whom all correspondence and reprint requests should be addressed.

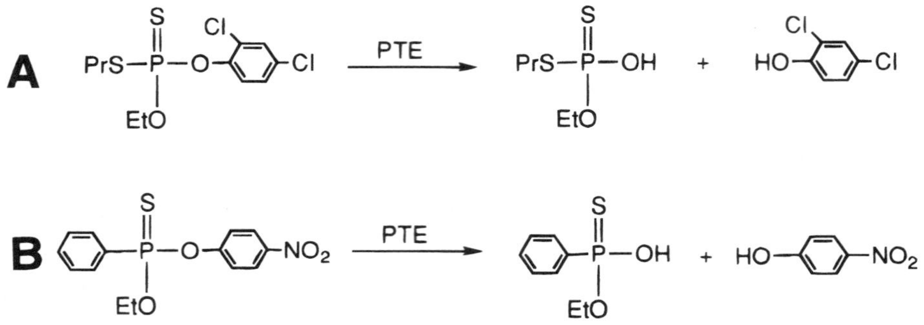

A PrS—P(=S)(OEt)—O—C₆H₃(Cl)₂ —PTE→ PrS—P(=S)(OEt)—OH + HO—C₆H₃(Cl)₂

B Ph—P(=S)(OEt)—O—C₆H₄—NO₂ —PTE→ Ph—P(=S)(OEt)—OH + HO—C₆H₄—NO₂

Fig. 1. The hydrolysis of organophosphorous esters, prothiofos (**A**) and EPN (**B**), by PTE

rous atom *(7,8)* However, the enantiospecificity of PTE toward other chiral centers than chiral phosphorous atom has not yet been reported. The elucidation and application of such enzymatic characteristics will enable us to design a process for the preparation of chiral organophosphate insecticides by kinetic optical resolution of racemic compounds by PTE-catalyzed hydrolysis.

In this article, we investigate the site of chiral recognition by PTE using various type of organophosphorous esters containing chiral phosphorous and carbon atoms (Fig. 2).

MATERIALS AND METHODS

Chemicals

Prothiofos, phenthoate, and vamidothion were purchased from Nanogen Co. (Iza, UK), and ethyl 4-nitrophenyl phenylphosphonothioate (EPN) from Dr. Ehrenstorfer GmbH (Augusburg, Germany). Diethyl α-methylbenzyl phosphate and diethyl α-methylbenzyl phosphamide were prepared from chlorodiethyl phosphate, and the corresponding α-methylbenzyl alcohol and α-methylbenzylamine in the presence of triethylamine. These synthetic derivatives were purified with silica gel column chromatography. All analytical data are as follows: Diethyl α-methylbenzyl phosphate; ^{1}H-NMR (δ, $CDCl_3$) 1.22 (t, CH_3, 6H), 1.75 (d, CH_3, 3H), 3.95–4.05 (m, CH_2, 4H), 4.80–4.90 (m, CH, 1H), 7.10–7.15 (m, ArH, 10H) ppm, ^{31}P-NMR (δ, $CDCl_3$) –1.55 ppm. Diethyl α-methylbenzyl phosphamide, ^{1}H-NMR (δ, $CDCl_3$) 1.20 (t, CH_3, 6H), 1.80 (d, CH_3, 3H), 4.00–4.15 (m, CH_2, 4H), 5.10–5.20 (m, CH, 1H), 6.80–6.85 (m, NH, 1H), 7.10–7.15 (m, ArH, 10H) ppm, ^{31}P-NMR (δ, $CDCl_3$) –1.80 ppm. These racemic esters were analyzed by HPLC with chiral reversed-phase column (chiracel OJ-R, 4.6 × 150 mm, Daicel, Osaka, Japan). All enantiomers were separated in an isocratical condition (methanol:water, 93:7) at 0.7

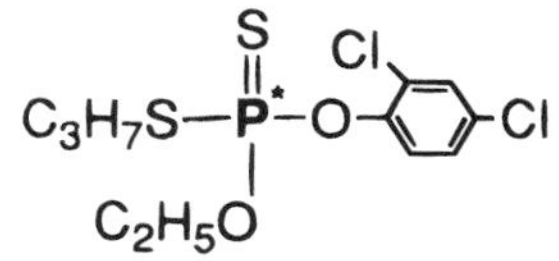

[Chemical structures of: Prothiofos, EPN, Phenthoate, Vamidothion, Diethyl α-methylbenzyl phosphate, Diethyl α-methylbenzyl phosphamide]

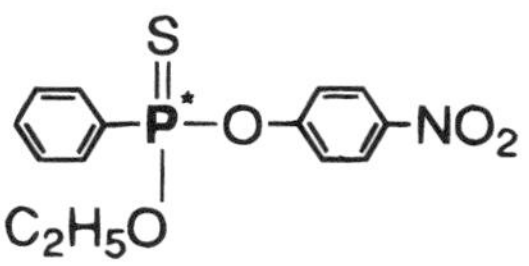

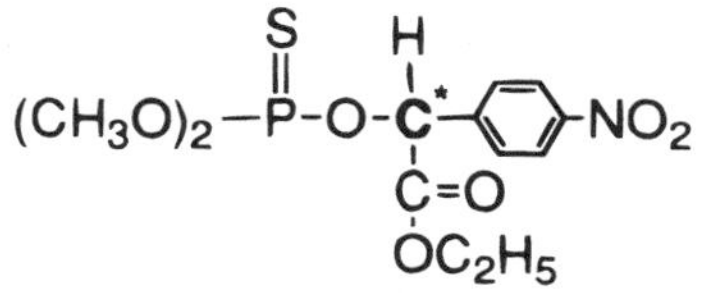

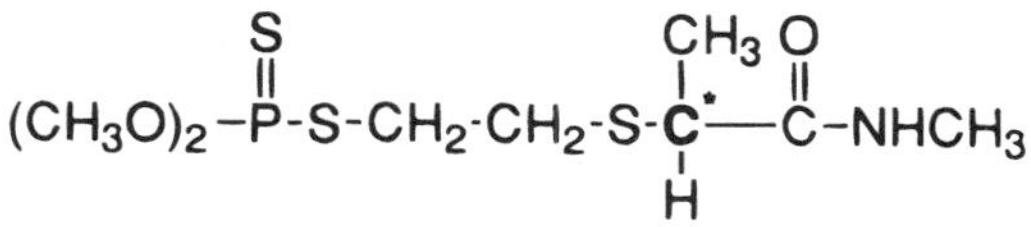

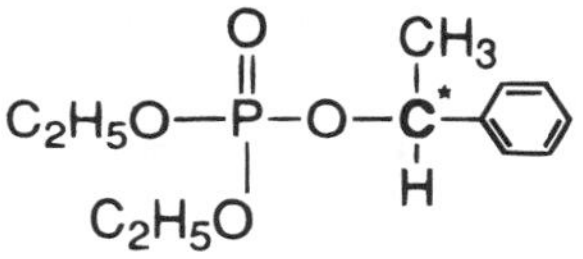

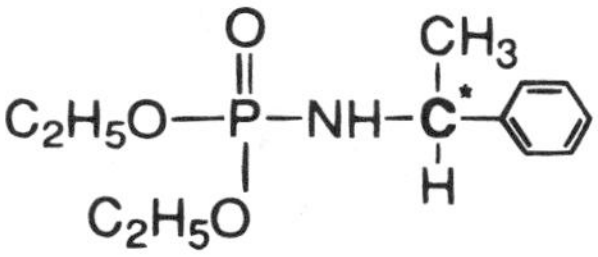

Fig. 2. Various organophosphorous esters containing chiral phosphorous and carbon atoms (asterisk) used in this study.

mL/min. Also, the resolved enantiomers of prothiofos and EPN were estimated from their optical rotations by polarimeter measurement (DIP-1000, Jasco, Tokyo, Japan). The earliest eluting peak in the chromatogram of prothifos by chiral HPLC was confirmed to be the (+)-enantiomer, and the later eluting peak was the (–)-enantiomer. Similarly, the earliest eluting EPN was confirmed to be the (+)-enantiomer, and the later eluting EPN was the (–)-enantiomer. In addition, the configuration of (+)-EPN is corresponded to the (R)-form, and (–)-EPN is the (S)-form, as previously reported (7). The enantiomers of phenthoate, vamidothion, diethyl α-methylbenzyl phosphate, and diethyl α-methylbenzyl phosphamide were confirmed only by the separated two peaks on chromatography.

PTE Preparation

PTE was prepared from *Flavobacterium* sp. ATCC 27551, as follows. Cells were cultivated aerobically in a L-broth medium supplemented with 0.01 mM ZnCl$_2$, at 30°C for 21 h. Cells were then washed twice in 50 mM, pH 7.2 potassium phosphate buffer, and resuspended in 20% sucrose, 0.3M

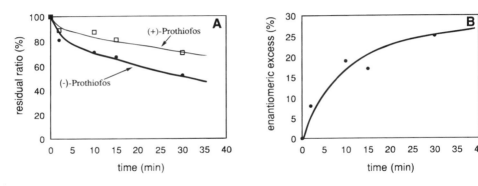

Fig. 3. The time-course of hydrolysis (**A**) and optical resolution (**B**) of prothiofos.

Tris-HCl (pH 8.1), 1 mM EDTA, and 0.5 mM MgCl$_2$. After centrifugation (5000g at 4°C), the cell pellet was decanted and resuspended in the residual liquid. From periplasmic space, PTE was extracted by osmotic shock by adding in ice-cold 0.5 mM MgCl$_2$. After incubation and centrifugation (10,000g at 4°C), the supernatant was dialyzed against 20 mM, pH 8.0, Tris-HCl buffer for 12 h at 4°C. This solution, containing 60 U/mL PTE activity, was then lyophilized and kept at –30°C until usage. PTE activity was determined by measuring the rate of liberation of 4-nitrophenol from tris-4-nitophenyl phosphate in the presence of PTE in 50 mM glycine-NaOH buffer (pH 9.5) at 25°C, by monitoring absorbance increase at 410 nm, by a spectrophotometer (UV-160A, Shimadzu, Japan).

Enzymatic Reaction of Racemic Organophosphorous Ester

Typical kinetic resolution of organophosphorous esters was carried out as follows. To a solution of 9.6 mg of prothiofos (10 mM) in 3 mL of 50 mM glycine-NaOH buffer (pH 9.5) was added 11 mg of lyophilized PTE corresponding to 30 U. The reaction was stirred at room temperature and was monitored by HPLC with a chiral reversed-phase column. The kinetic resolution of esters was estimated with the comparison of the earliest eluting peak and the later eluting peak on chiral HPLC.

RESULTS AND DISCUSSION

Figures 3 and 4 show kinetic resolution of racemic prothiofos and EPN by PTE. These esters have chiral phosphorous atoms that are attacked with nucleophile during hydrolysis. The prothiofos of (–)-enantiomer (later eluting) was degraded faster than (+)-enantiomer (earlier eluting) (Fig. 3A). The highest enantiomeric excess attained was 25% after 30 min (Fig. 3B). Similarly, for EPN (–)-enantiomer (later eluting) was degraded faster than (+)-enantiomer (earlier eluting) (Figs. 4A). The highest enantiomeric excess attained was 42% after 45 min (Fig. 4B). With regard to

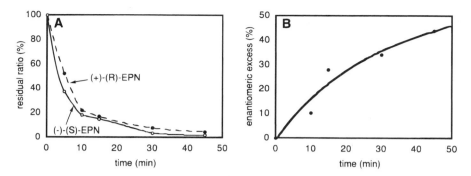

Fig. 4. The time-course of hydrolysis (**A**) and optical resolution (**B**) of EPN.

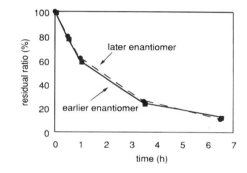

Fig. 5. The time-course of hydrolysis of phenthoate.

EPN, the predominant hydrolysis of (–)-EPN by PTE from *Pseudomonas diminuta* had been demonstrated by Lewis et al. (7). EPN was degraded faster than prothiofos, indicating the difference in substrate specificity of prothiofos and EPN with PTE. These results show that the resolution with PTE has been performed stereospecifically.

Phenthoate, vamidothion, diethyl α-methylbenzyl phosphate, and diethyl α-methylbenzyl phosphamide were tested for PTE hydrolysis as organophosphorous esters having the chiral atom. The chiral center of these esters is located on the carbon atom closest to the phosphorous atom hydrolyzed. These results are shown in Figs. 5, and 6, and Table 1. Every substrate was hydrolyzed with PTE, but the stereospecificity was only found in diethyl α-methylbenzyl phosphamide hydrolysis. The earliest eluting enantiomer of diethyl α-methylbenzyl phosphamide was present in slight excess compared to the later eluting enantiomer (6% enantiomeric excess), whereas no stereospecificity was observed when phenthoate, vamidothion, and diethyl α-methylbenzyl phosphate were used as a substrate.

These results suggest that PTE mainly recognizes the chiral center on phosphorous atom. Recently, Vanhooke et al. reported about the stereo-

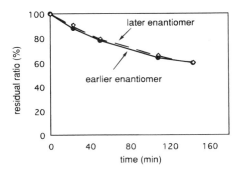

Fig. 6. The time-course of hydrolysis of vamidothion.

Table 1
Hydrolysis of Racemic Phosphorous Esters by Phosphotriesterase[a,b]

Substrate	Residual ratio (%)		e.e. (%)
$(C_2H_5O)_2$-P(=O)-NH-CH(CH$_3$)-C$_6$H$_5$	earliest enantiomer	73	6
	later enantiomer	65	
$(C_2H_5O)_2$-P(=O)-O-CH(CH$_3$)-C$_6$H$_5$	earliest enantiomer	70	0
	later enantiomer	70	

[a] The reaction was performed in glycine-NaOH buffer (pH 8.0) at 40°C at 24 h.
[b] The earliest and the later enantiomers were separated with chiral HPLC.

chemical mechanism of PTE hydrolysis reaction from the binding model of the inhibitor and the enzyme (9). It was suggested that the phosphoryl group of the substrate was positioned within the PTE active site, but the leaving group as a 4-nitrophenyl group of prothiofos and EPN was located in the solvent rather than the interior of PTE. Therefore, phenthoate, vamidothion, and diethyl α-methylbenzyl phosphate containing the chiral center at the leaving group could not be recognized sufficiently by PTE. With diethyl α-methylbenzyl phosphamide, the small value of e.e. is supposed to be the diastereomeric effect by the double-bond character of phosphamide linkage.

CONCLUSION

We have shown that PTE chiral recognition is limited to organophosphorous esters having the chiral center on the phosphorous atoms, which is attacked with nucleophile, not having the chiral center on the other atoms of the leaving group. Therefore, PTE can be utilized for the synthesis of their chiral organophosphorous esters.

REFERENCES

1. Wing K. D., Glickman, A. H., and Casida, J. E. (1984), *Pesticide Biochem. Physicol.* **21,** 22–30.
2. Miyazaki, A., Nakamura, T., Kwawaradani, M., and Marumo, S. (1988), *J. Agric. Food. Chem.* **36,** 835–837.
3. Ohkawa, H., Mikai, N., Okuno, Y., and Miyano, J. (1977), *Bull. Environ. Contam. Toxicol.* **18,** 534–540.
4. Louis, V. E., Donarski, W. J., Wild, J. R., and Raushel, F. M. (1988), *Biochemistry* **27,** 1591–1597.
5. Omburo, G. A., Kuo, J. M., Mullins, L. S., and Raushel, F. M. (1992), *J. Biol. Chem.* **267,** 13,278–13,283.
6. Lai, K., Stolowich, N. J., and Wild, J. R. (1995), *Arch. Biochem. Biophys.* **277,** 155–159.
7. Lewis, V. E., Donarski, W. J., Wild, J. R., and Raushel, F. R. (1988), *Biochemistry* **27,** 1591–1597.
8. Chae, M. Y., Postula, J. F., and Raushel, F. M. (1994), *Bioorg. Med. Chem. Lett.* **4,** 1473–1478.
9. Vanhooke, J. L., Benning, M. W., Raushel, F. M., and Holden, H. M. (1996), *Biochemistry* **35,** 6020–6025.

Session 6

Environmental Biotechnology

MARY J. BECK[1] AND JONI M. BARNES[2]

[1]*Tennessee Valley Authority, Muscle Shoals, AL;* and
[2]*Idaho National Engineering Laboratory, Idaho Falls, ID*

It is a challenge to encompass the varied subjects that deserve consideration under the heading of environmental biotechnology. Even a definition of environmental technology is difficult. The simplest approach is to define it as the use of biological technologies for sustaining or improving the environment. Bioremediation is practically synonymous with environmental biotechnology. Bioremediation deals with past contamination of the environment and the transformation of pollutants into benign substances. Environmental biotechnology has a future role for continued exploration of new, more easily degradable products and for manufacturing processes that prevent disposal problems. An understanding of ecosystems, thorough knowledge of biological pathways, and the means to measure biological activity are necessary to ensure success for cleaning up past problems and addressing future technologies.

Currently, the major focus of environmental biotechnology appears to be the development of bioremediation technologies. This observation is reflected in the compilation of papers in this section. Biological approaches to environmental clean-up are generally viewed in a positive light. Bioremediation is environmentally friendly and potentially less expensive than other remediation technologies. Bioremediation technologies appear to have a strong future because of the number of sites in need of clean-up and the favorable outlook for the bioremediation market.

A variety of bioremediation technologies are covered by the papers in this section. In addition, documentation of intrinsic bioremediation strategies and descriptions of bioreactor designs are provided. A common theme throughout the papers is the interaction of chemical, physical, and biological activities for pollution prevention, maintenance of the environment, and environmental remediation. These studies illustrate the means to harness, fine-tune, and improve reaction rates for biological destruction contaminants.

Hydrodynamic Characteristics in Aerobic Biofilm Reactor Treating High-Strength Trade Effluent

H. CHUA*,[1] AND P. H. F. YU[2]

Departments of [1]Civil and Structural Engineering; and [2]Applied Biology and Chemical Technology, The Hong Kong Polytechnic University, Hung Hom, Kowloon, Hong Kong

ABSTRACT

Four 3-L aerobic biofilm reactors (ABR1, 2, 3, and 4) treating a high-strength food-processing waste water (10 g chemical oxygen demand [COD]/L) were subject to reactor liquor recirculation rates of 1, 3, 15, and 30 L/h, respectively. Treatment performance in terms of COD removal rates of ABR1, 2, and 3 were similar at hydraulic loads of 2.0 g COD/L/d and below. At higher organic loads, ABR3 could achieve a COD removal rate that was over two times higher than that of ABR1 and 2. ABR3 could be operated at a maximum organic load that was two times higher than that of ABR1 and 2. ABR4 experienced a biofilm sloughing from the packing medium at the beginning of operation. Tracer studies showed that recirculation rate of 1 L/h resulted in a plug-flow pattern in the packed bed of the reactor. On the other hand, recirculation rate of 15 L/h, which was equivalent to recirculating the reactor liquor five times per hour, provided effective mixing in the packed bed. Superior performance of ABR3 was attributed to the effective recirculation of reactor liquor, which diluted and distributed the influent, particularly the oil and grease components.

Index Entries: Aerobic biofilm reactor; recirculation rate; high-strength waste water; treatment performance; mixing.

INTRODUCTION

Immobilized biofilm technology has become popular for high-rate aerobic and anaerobic treatments of waste waters with low suspended solids and high organic strengths *(1–4)*. Adequate mixing is essential in

*Author to whom all correspondence and reprint requests should be addressed.

high-rate treatment processes to ensure uniform distribution of substrate, sufficient biomass-substrate contact, and prevention of process instability owing to localized accumulations of toxic matters (5). Adequate mixing is also important to improve oxygen transfer and distribution in aerobic biofilm reactors. In the treatment of food-processing waste water, which contains a high concentration of oil and grease (500–1500 mg/L), adequate mixing is crucial in preventing oil droplets from adhering on and coating the biofilms.

Mixing in the packed bed of a biofilm reactor is only possible through effluent recirculation and aeration. DeWalle and Chian (6) reported that effluent recirculation could dilute the influent and maintain the COD removal efficiency with respect to the diluted influent. On the other hand, Thiramurthi (7) observed that an effluent recirculation beyond a threshold limit could cause process failure. However, very limited information is available on the complex hydrodynamics in the packed bed of a biofilm reactor and the effects of mixing on the treatment performances. This article assesses the effects of mixing on COD removal rate and the process stability of aerobic biofilm reactors treating a high-strength food-processing waste water. Hydrodynamic characteristics in the packed bed of the reactor are also modeled and described.

METHODS

Reactor System

The aerobic biofilm reactor is comprised of a column with a length-to-diameter ratio of 15 (Fig. 1). Four similar columns were packed with fire-expanded clay spheres (FECS) of average diameter 1.5 cm (Fig. 2). The effective volume of each column was 3 L. The reactor liquor was drawn from the top of the four columns, namely ABR1, 2, 3, and 4, and recirculated at 1, 3, 15, and 30 L/h, respectively, through the bottom. Compressed air was supplied, via a distributor, at the bottom of the reactor column. Each reactor was seeded with activated sludge from municipal sewage treatment works and was fed with a food-processing waste water of 10 g chemical oxygen demand (COD)/L for an initial 30-d seeding and acclimatization period. The stabilized reactors were then operated for another 90 d at organic loads of 1.0, 2.0, 5.0, 10.0, and 20.0 g COD/L/d.

Analytical Methods

The COD in the treated effluent was determined by the open reflux method. The volatile suspended solids (VSS) was measured by the volatilization and weighing method. Oil and grease concentrations were measured by the partition-gravimetric method. All the parameters were determined in accordance with the standard methods (8).

Hydrodynamic in Aerobic Biofilm Reactor 671

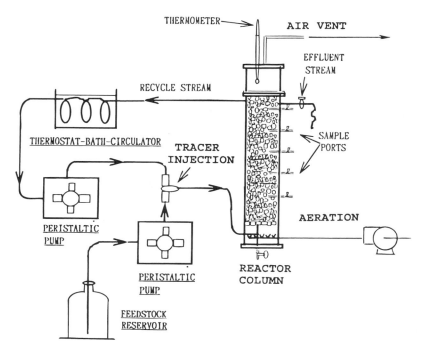

Fig. 1. Schematic diagram of aerobic biofilm reactor system.

Fig. 2. Fire-expanded clay sphere used as packing medium.

Tracer Techniques

Tracer stimulus–response techniques described by Wakao and Kaguei (9) were used to study the residence time distribution (RTD) and flow patterns through the packed bed of the reactor. Experiments were carried out with ABR1 and 3, in which the reactor liquor was recirculated at a rate of 1 and 15 L/h, respectively. Both reactors were operated with an organic load of 10 g COD/L/d. Each tracer input was 30 mL of 500 mg/L sodium chlo-

Table 1
COD Removal Rates and Efficiencies

Organic Loads (gCOD/L-d)	COD Removal Rate (gCOD/L-d)				COD Removal Efficiency (%)			
	ABR1	ABR2	ABR3	ABR4	ABR1	ABR2	ABR3	ABR4
1.0	0.99	0.96	0.98	R.F.	98.8	96.0	97.9	R.F
2.0	1.97	1.94	1.96		98.7	96.9	97.9	
5.0	3.71*	3.99*	4.85		74.1	79.8	97.0	
10.0	R.F.**	R.F.	9.23*		R.F.	R.F.	92.3	
20.0			R.F.				R.F.	

*Optimum performance.
**Reactor failure.

ride solution, injected at the base of the reactor by a high-speed peristaltic dosing pump. Response to the tracer input was taken as a time record of chloride concentrations detected at the top of the reactor by a conductivity meter (LTH Electronics, Type PB5 with a Type CMC5/10/TIK electrode).

RESULTS AND DISCUSSION

Reactor Performances

The COD removal rates and removal efficiencies of the four aerobic biofilm reactors are summarized in Table 1. ABR1, 2, and 3 demonstrated similar performances in terms of COD removal rate and removal efficiency at organic loads of 1.0 and 2.0 g COD/Ld. As the organic load was increased to 5.0 g COD/L/d, the effects of recirculation began to show. COD removal rate and removal efficiency of ABR3 were significantly higher than that of ABR1 and 2. The COD removal rate of ABR3 reached an optimum of 9.23 g COD/L/d at organic load of 10.0 g COD/L/d, whereas that of ABR1 and 2 reached the optima of 3.71 and 3.99 g COD/L/d, respectively, at organic load of 5.0 g COD/L/d. ABR3 could be operated at an organic load that was two times higher than that of ABR1 and 2 before the reactors failed, while achieving a COD removal rate that was over two times higher than that of AR1 and 2. COD removal efficiency of ABR3 remained above 90% throughout the study before reactor failure, whereas that of ABR1 and 2 deteriorated when the organic load was increased to 5.0 g COD/L/d. Oil and grease removal efficiencies in ABR1 and 2 were between 54.6 and 63.2% under different loading rates. ABR3 could achieve an oil and grease removal efficiency as high as 79.7% under an organic loading rate of 10.0 g COD/L/d. Observations on ABR4 will be discussed later.

The optimum range of recirculation rates was around 15 L/h for achieving high COD, oil and grease removal rate, and removal efficiency at organic loads higher than 2.0 g COD/L/d. The recirculation rate of 15 L/h in ABR3 was equivalent to a superficial flow velocity of 480 cm/h. The effects of treated effluent recirculation on dilution and distribution of the organic load, and dispersion of the oil and grease enabled improved treatment performance in ABR3. However, at hydraulic loads of 2.0 g COD/L/d and below, operation with low recirculation rates was a more attractive option, because this could achieve COD removal rates and removal efficiencies that were comparable to that attained with high recirculation rates, while maintaining a lower cost of operation.

Biofilm

During the start-up stage, biomass was retained as suspended bioflocs in the interstices of the packed bed and loosely held biofilm on the surfaces of the packing medium. In ABR1, 2, and 3, the reactor liquor recirculation and the progressively increasing loading rates gradually displaced the excess suspended biomass, and established an uniform and firmly attached biofilm on the surfaces and in the pores of the packing medium (Fig. 3). The VSS concentration in the treated effluent gradually decreased from 617–890 mg/L during the initial 30-d period after seeding, to below 50 mg/L during the entire 90-d operating period of this study. On the other hand, the recirculation rate of 30 L/h in ABR4 resulted in a high superficial flow velocity, 960 cm/h, as experienced by the biofilm. This hydraulic shear caused a sloughing of biofilm from the packing medium at the beginning of operation, which resulted in reactor failure.

Hydrodynamic Characteristics

Tracer studies of the reactors ABR1 and 3 were carried out with recirculation rates of 1 and 15 L/h, respectively, and an organic load of 10.0 g COD/L/d. ABR1 represented reactors with inadequate mixing, and ABR3 represented reactors operated with adequate mixing and produced excellent COD removal rates. The recirculation rates of 1 and 15 L/h were equivalent to recirculating the filter liquor 1/3 and 5 times/h, respectively, and the loading rate was equivalent to a hydraulic retention time of 1 d. The RTDs as indicated by the tracer were plotted in terms of mg/L of chloride vs time in hours (Fig. 4).

The tracer input was approximated by a tall, narrow square wave, represented by the vertical arrow at zero time in Fig. 4. The response signal for ABR1 was a skewed peak that had a sharp, rapidly increasing positive slope and a slow decreasing tail. The lag time was about 23 h, which agreed very closely with the operating hydraulic retention time (HRT) of 1 d. This was a typical response configuration of a plug-flow system with moderate amounts of dispersion. The response signal for ABR3 resembled the typical profile of an ideal mixed-flow system, which could be represented mathematically by an exponential decay function. The response to the

Fig. 3. Biofilm used as packing medium.

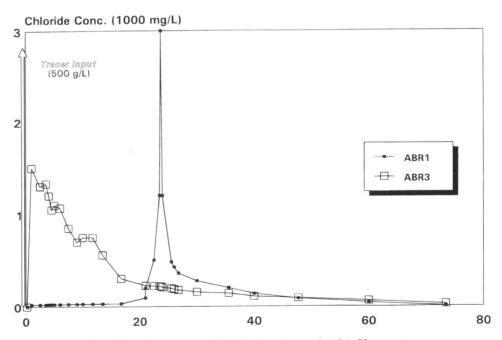

Fig. 4. Residence time distribution in aerobic biofilm reactor.

tracer input was almost instantaneous with a short lag time of about 10 min, which is not visible with the resolution of Fig. 4.

The mean residence time, variance, and dispersion number of the reactors was calculated by the dispersion model presented by Eqs. (1), (2), and (3). The flow pattern in the reactor was described by the single-parameter, dispersion model, which considered the flow in a packed bed to be in a plug-flow pattern with different degrees of dispersion in the axial direction. With varying intensity of dispersion, the predictions of the model ranged from ideal plug flow at one extreme to ideal mixed flow at the other. The model correlated the mean residence time and variance of the RTD curve, obtained through tracer methods, by Eq. (1).

$$S^2/t^2 = 2(D/UL) - 2(D/UL)^2 [1 - \exp(UL/D)] \quad (1)$$

where S^2 = variance of the RTD curve, h^2; t = mean residence time of the RTD curve, h; D/UL = dispersion number, dimensionless; D = axial dispersion coefficient, m^2/h; U = interstitial fluid velocity, m/h; and L = length of reactor column, m. The dimensionless dispersion number (D/UL), the single parameter of the dispersion model, provided a measure of the extent of axial dispersion. The mean residence time (t) located the center of gravity of the RTD curve and was calculated by Eq. (2).

$$t = [SUM(t_i C_i t_i')/SUM(C_i t_i')] \quad (2)$$

where t_i = time after the tracer was introduced, h; C_i = concentration of sodium chloride in the sample taken from the reactor at time t_i, mg/L; and t_i' = interval between successive samples, h; SUM denotes the summation of the argument in the parentheses through each tracer experiment. The variance described the spread of the RTD curve and was given by Eq. (3).

$$S^2 = [SUM(t_i C_i t_i')/SUM(C_i t_i')] - t^2 \quad (3)$$

Table 2 summarizes published values of dispersion numbers for various extents of mixing as predicted by the dispersion model (10). The dispersion number of 0.06 (Table 3) obtained for ABR1 with recirculation rate of 1 L/h indicated that the reactor was essentially a plug-flow column with an intermediate amount of dispersion. This observed dispersion, resulting in the deviation from the truly plug-flow hydrodynamics, which was expected for the packed-bed column with inadequate mixing, was attributed to the slow recirculation, and the diffusion and eddies generated when fluid flowed through the interstitial channels of the packed bed. On the other hand, the dispersion number of 0.65 obtained for ABR3 with recirculation at a rate of 15 L/hr was in the regime where there was a large extent of dispersion (D/UL > 0.2). The recirculation rate of 60 L/h provided effective mixing in the packed bed of the reactor. It was possible that a recirculation rate lower than 60 L/h could also achieve adequate mixing. This could be optimized if necessary.

CONCLUSION

The optimum range of recirculation rates was around 15 L/h for achieving high COD removal rate and removal efficiency at organic loads higher than 2.0 g COD/L/d. However, at hydraulic loads of 2.0 g

Table 2
Values of Dispersion Number at Various Extents of Mixing (10)

Extent of Mixing	Typical Values of Dispersion Number
Ideal plug flow	0.000
Small amount of dispersion	0.002
Intermediate amount of dispersion	0.025
Large amount of dispersion	0.200
Ideal mixed flow	Approaches infinity

Table 3
Results from the Tracer Studies

	Mean Residence Time t (h)	Variance s^2 (h^2)	Dispersion Number
ABR1	62.59	486.92	0.06
ABR3	46.39	1373.46	0.65

COD/L/d and below, operation with low recirculation rates was a more attractive option, because this could achieve COD removal rates and removal efficiencies that were comparable to those attained with high recirculation rates while maintaining a lower cost of operation. Tracer studies confirmed that a recirculation rate of 15 L/h, which was equivalent to recirculating the filter liquor five times per hour, was adequate in achieving effective mixing in the packed bed of the reactor.

REFERENCES

1. Sagy, M. and Kott, Y. (1990), *Water Res.* **9**, 1125–1128.
2. Henze, M. and Harremoes, P. (1983), *Water Sci. Tech.* **15**, 1–101.
3. Yap, M. G. S., Ng, W. J., and Chua, H. (1991), *Bioresource Technol.* **41**, 45–51.
4. Chua, H., Yap, M. G. S., and Ng, W. J. (1993), *Appl. Biochem. Biotechnol.* **34/35**, 789–800.
5. Stafford, D. A. (1982), *Biomass*, **2**, 43–55.
6. DeWalle, F. B. and Chian, E. S. (1976), *Biotechnol. Bioeng.* **18**, 1275–1295.
7. Thirumurthi, D. (1988), *Water Res.* **22:4**, 517–523.
8. APHA (1992), Standard Methods for the Examination of Water and Wastewater, APHA, Washington.
9. Wakao, N. and Kaguei, S. (1982), *Heat and Mass Transfer in Packed Beds*, Gordon and Breach, Science Publishers, New York.
10. Levenspiel, O. (1972), in *Chemical Reaction Engineering*, John Wiley, pp. 253–325.

Copyright © 1997 by Humana Press Inc.
All rights of any nature whatsoever reserved.
0273-2289/97/63-65—0677$12.25

A Biological Process for the Reclamation of Flue Gas Desulfurization Gypsum Using Mixed Sulfate-Reducing Bacteria with Inexpensive Carbon Sources

ERIC N. KAUFMAN,* MARK H. LITTLE, AND PUNJAI T. SELVARAJ

Bioprocessing Research and Development Center, Chemical Technology Division, Oak Ridge National Laboratory, Oak Ridge, TN 37831-6226

ABSTRACT

A combined chemical and biological process for the recycling of flue gas desulfurization (FGD) gypsum into calcium carbonate and elemental sulfur is demonstrated. In this process, a mixed culture of sulfate-reducing bacteria (SRB) utilizes inexpensive carbon sources, such as sewage digest or synthesis gas, to reduce FGD gypsum to hydrogen sulfide. The sulfide is then oxidized to elemental sulfur via reaction with ferric sulfate, and accumulating calcium ions are precipitated as calcium carbonate using carbon dioxide. Employing anaerobically digested municipal sewage sludge (AD-MSS) medium as a carbon source, SRBs in serum bottles demonstrated an FGD gypsum reduction rate of 8 mg/L/h $(10^9 \text{ cells})^{-1}$. A chemostat with continuous addition of both AD-MSS media and gypsum exhibited sulfate reduction rates as high as 1.3 kg FGD gypsum/m^3·d. The increased biocatalyst density afforded by cell immobilization in a columnar reactor allowed a productivity of 152 mg SO$_4^{-2}$/L·h or 6.6 kg FGD gypsum/m^3·d. Both reactors demonstrated 100% conversion of sulfate, with 75–100% recovery of elemental sulfur and chemical oxygen demand utilization as high as 70%. Calcium carbonate was recovered from the reactor effluent on precipitation using carbon dioxide. It was demonstrated that SRBs may also use synthesis gas (CO, H$_2$, and CO$_2$) in the reduction of gypsum, further decreasing process costs. The formation of two marketable products—elemental sulfur and calcium carbonate—from FGD gypsum sludge, combined with the use of a low-cost carbon source and further

*Author to whom all correspondence and reprint requests should be addressed.

improvements in reactor design, promises to offer an attractive alternative to the landfilling of FGD gypsum.

Index Entries: Gypsum sludge; sulfate-reducing bacteria; synthesis gas; sulfur; calcium carbonate; flue gas desulfurization.

INTRODUCTION

The burning of coal at power plants produces sulfur dioxide (SO_2), which causes acid rain. Although regenerable sorbents for SO_2 capture at power plants are gaining popularity, they are frequently economical only at newly constructed facilities. Disposable sorbents, such as limestone, are utilized in many of today's flue gas desulfurization (FGD) systems. As of 1991, 80% of FGD capacity (over 120 GW) consisted of wet limestone scrubbing *(1)*. Limestone processes are also a common choice for retrofitting FGD capacity in pre-existing plants (*see,* for example, Bove et al. *[2]*). The absorption of SO_2 onto limestone produces calcium sulfite and carbon dioxide. Forced oxidation of the resulting sulfite sludge yields a residue rich in calcium sulfate (gypsum). Although other countries utilize gypsum for wallboard or cement, only three plants in the US find this process economical *(3)*. Transportation costs, low purity, and regulations regarding ash constituents in building material do not allow FGD gypsum to compete with the natural gypsum mined in the US. This country utilizes <5% of its FGD gypsum and disposes the remaining 15–20 million t/yr, thus requiring a massive landfill volume *(1,3,4)*. In 1992, the US landfilled more than 12 million t of FGD residue and ponded an additional 8.6 million t *(1)*. The amount of FGD residue a given plant produces is dependent on the size of the plant, the sulfur content of the coal, and the FGD technology employed. A 1000 MW steam plant burning 0.5% sulfur coal (a low-sulfur coal) with a spray dry scrubber removal efficiency of 70% may be expected to produce about 350 kt of FGD residue/yr *(1)*.

It is difficult to determine the actual cost of FGD gypsum disposal, since these numbers are rarely included in the literature, and when published, may not include construction and maintenance costs for the facility. Disposal costs are generally quoted to range between 3 and $12/t *(5,6)*. Some utility companies are able to take advantage of readily available land and lenient regulatory laws that permit them to stack or "rim ditch," their gypsum in an unlined area for much lower operating costs *(5,7)*. Other utility companies are forced to utilize expensive, lower-sulfur-content coal in order to avoid desulfurization entirely. An economically viable process capable of recycling FGD gypsum to reclaim the calcium carbonate sorbent and recover marketable elemental sulfur could dramatically decrease the operating costs for a limestone FGD system. Recycling of the waste product would decrease the land requirements for a limestone facility, and reclamation of calcium carbonate would decrease both the required inventory and the material costs of the sorbent.

The ability of bacteria to reduce sulfates and sulfites to sulfides was first reported in 1864, and a wealth of information has been accumulated regarding sulfate-reducing bacteria (SRB), their environmental and nutritional requirements, and their enzymatic pathways (see Barton [8] and Widdel and Hansen [9] for reviews). Their role in oil well souring has been of particular interest (10,11). SRB oxidize organic acids or hydrogen while reducing sulfates and sulfites into sulfides. Various investigators have suggested that SRB may be utilized to treat FGD gypsum biologically, however, a commercially viable process has not been realized. Apel and Barnes (12) demonstrated that SRB may utilize carbon sources, such as lactate, pyruvate, citrate, alanine, cysteine, glycerol, and ethanol to reduce FGD gypsum to hydrogen sulfide. These investigators concluded that commercial viability of the process could not be achieved until expensive carbon sources could be eliminated and an effective reactor design could be achieved. Uphaus et al. (13) seemingly decreased the cost of a carbon source by utilizing two reactors in series. A photosynthetic green sulfur bacterium, *Chlorobium*, utilized the H_2S and CO_2 evolved from SRB to form elemental sulfur and organic acids. The organic acids were in turn utilized as a "free" carbon source for the SRB. The hidden cost of this process was the 10,000-foot-candles of light required as the energy source for the *Chlorobium*.

Synthesis gas (syn-gas) and sewage digest have been investigated as inexpensive carbon and energy sources for SRB reactors. Syn-gas (40–65%, CO, 25–35% H_2, 1–20% CO_2, and 0–7% CH_4, H_2S, and COS [14]) would be the most desirable feedstock for an SRB process owing to its wide availability particularly at the utility plant and its zero chemical oxygen demand (COD) discharge. Various groups (15–17) have demonstrated that SRB could be supported by carbon dioxide as the sole carbon source and hydrogen as the electron donor. Du Preez et al. (18) operated a sulfate-reducing reactor with a mixed SRB population, which was fed 30% H_2, 59% CO, 8% CO_2, and 3% N_2, demonstrating the feasibility of using whole syn-gas as the feed source for SRB. Recently, van Houten et al. (19) reported the operation of a gas lift sulfate-reducing reactor fed up to 20% CO with the balance being H_2. They reported a slight decrease in productivity owing to the use of CO (12 to 6 kg/m^3·d). Sublette and Gwodzdz (20) assessed the economic viability of a microbial process to reduce the effluent SO_2 from a regenerable FGD process as compared with conventional hydrotreating. This study concluded that although fixed capital investment costs for the two processes were identical, the high cost of the carbon source (corn hydrolysate) rendered the biological process uneconomical. Selvaraj and Sublette (21) demonstrated that sewage digest (anaerobically digested municipal sewage sludge; AD-MSS media) may be utilized as the carbon and energy source for SRB. They defined the targets for necessary organic

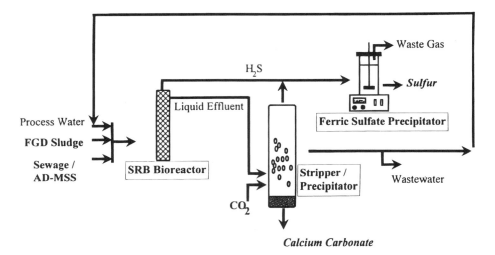

Fig. 1. Reclaiming of sulfur and calcium carbonate from FGD gypsum. A culture of SRB utilizes sewage digest as its carbon and energy source, and reduces the sulfates and sulfites in FGD gypsum sludge to hydrogen sulfide. The sulfide is oxidized by ferric sulfate to form elemental sulfur. The resulting ferrous sulfate may be reoxidized chemically or biologically. Calcium ions accumulating in the liquid of the SRB reactor are precipitated using carbon dioxide to form calcium carbonate, which may be reused as an FGD adsorbent. This process diagram is intended to be generic in nature. Alternative carbon and energy sources, means of sulfide reoxidation, and methods for carbonate precipitation may also be employed.

acid content of the AD-MSS media as well as the biocatalyst density necessary to achieve an economically viable process. Significant progress has been made in achieving higher utilization of COD and organic acids in the AD-MSS, media as well as in improving the biocatalyst density through cell immobilization (22).

Recently, a process has been proposed for recycling FGD gypsum or waste sludge into calcium carbonate and elemental sulfur (23). This process (see Fig. 1) incorporates biological and chemical transformations into an economical method with tremendous capital and environmental impact. Although some of the technologies have been utilized in other applications, a combined system for FGD gypsum recycling has not been addressed. In this process, waste sulfate and sulfite in the form of a gypsum slurry are transformed to hydrogen sulfide by a mixed culture of SRB. The bacteria utilize sewage or syn-gas as their carbon source (20–22,24,25), thus eliminating prohibitive nutrient costs that exist in other reported SRB processes. Hydrogen sulfide is then further processed by ferric sulfate oxidation (26,27) into marketable elemental sulfur. The ferric sulfate may be regenerated using *Thiobacillus* species (28,29). Excess biocatalyst from both the SRB culture and the *Thiobacillus* culture may be removed using a gravity settler in an overall process. Calcium ions that accumulate in the bioreactor owing to the reduction of sulfite and sulfate from the $CaSO_3$ and $CaSO_4$ are combined with the waste gas carbon dioxide under alkaline conditions to re-form $CaCO_3$ see, for example, Wachi and Jones [30], which may then be reused as an SO_2 adsorber at the power plant. Elemental sulfur is a valuable industrial chemical with an annual US market of 13 million t/yr and a selling price of about $50/t (31). The regenerated limestone may be reused as an FGD adsorber at the steam plant, minimizing further purchase of FGD scrubbing materials, which are between 5 and $15/t (5,6).

In this study, many aspects of a biochemical process for the recycling of FGD gypsum to elemental sulfur and calcium carbonate have been demonstrated. The results show that:

1. AD-MSS media or syn-gas can serve as the carbon and electron source for SRBs performing gypsum reduction at rates competitive with cells fed rich media;
2. AD-MSS media can be used in a well-mixed reactor system with continuous, sustained gypsum reduction;
3. An immobilized-cell reactor can be operated with a continuous gypsum feed to achieve improved volumetric productivity;
4. Sulfur can be continuously recovered via ferric sulfate precipitation; and
5. Calcium can be recovered as calcium carbonate.

METHODS

Microbial Culture and Media

The preparation and optimization of AD-MSS media have been discussed in detail previously (22). Municipal sewage solids were obtained from the diffused air flotation (DAF) unit of the Oak Ridge, TN, municipal sewage treatment plant and were fed continuously as a 15% wet wt solid suspension into a 15-L vessel at 37°C. The hydraulic retention time of the sewage solids in this digester was 5.2 d: Chloroform was added at a concentration of 50 ppm to the DAF feed solution during the course of the operation as an inhibiting agent for methanogenesis. Note that unlike previous studies (32), no additional salts or nutrients were added to the sewage solids to promote digestion into organic acids. The supernatant of the effluent from this 15-L digester was used as AD-MSS medium for the gypsum recycling experiments. The pH of this media was 6.9. Effluent samples were analyzed for soluble COD and organic acids as described below.

SRB were isolated from the DAF sewage solids. The cells were incubated at 30°C in serum bottles containing 50 mL AD-MSS media and 0.15 g of gypsum sludge as the terminal electron acceptor. The culture was maintained by weekly repassaging into identical media under aseptic conditions.

FGD Gypsum

FGD gypsum sludge from the Dalman plant of the Springfield Illinois City Water, Light, and Power Company was generously supplied by Y. P. Chugh at the Southern Illinois University at Carbondale. On receipt, the sludge was oven-dried at 100°C to remove free moisture. The resulting powder yielded 551 mg/L sulfate when 0.1 g of sludge was dissolved in 100 mL distilled water at room temperature. Typically, 500 mg/L sulfate resulted when 1 g of sludge was dissolved in 1 L AD-MSS media at room temperature. The maximum solubility of sulfate in AD-MSS media was 1200 mg/L.

Analytical Techniques

Sulfate concentrations were analyzed by either ion chromatography or tubidimetrically. For ion-chromatography analysis, a Dionex 4500i (Dionex, Sunnyvale, CA) with an IonPac AS4A-SC 4-mm column was utilized with a 2 mL/min flow and isocratic elution using HCO_3^-. Turbidimetric sulfate analysis was performed using premeasured Sulfaver IV reagent powder (Hach Chemical, Loveland, CA). Spectrophotometric determination of the analyte was performed at 450 nm using a Spectronic 21 D (Milton-Roy Instruments, Rochester, NY). Organic acids were quantified by gas chromatography using a Hewlett Packard 5890 II gas chromatograph equipped with a flame ionization detector. A 30-m HP-INNOWax column (19091N-133, Hewlett Packard) was used with a helium carrier flow of 1.5 mL/min through a ramped temperature profile from 120°C (1 min) to 240°C (1 min) at 10°C/min. The injector was operated at 250°C, and the detector was operated at 300°C. Hydrogen sulfide, CO, H_2, and CH_4 were determined using a Hewlett-Packard 5890 II gas chromatograph equipped with a thermal conductivity detector. A helium carrier flow of 25 mL/min was passed through a 6 ft × 1/8 in. Teflon column using a Super Q 80/100 mesh stationary phase (Ohio Valley Specialty Chemical, Marietta, OH). The oven temperature was held at 50°C; both the detector and the injector were set at 125°C. The COD was determined spectrophotometrically at 620 nm using Hach premeasured COD reagent vials (0–1500 ppm). Counts of SRB were made using the most probable number (MPN) method (Bioindustrial Technologies, Georgetown, TX).

Serum Bottle Trials

Initial trials to assess the potential use of AD-MSS media in a gypsum reduction process were carried out in 150-mL serum bottles. Three separate series were prepared. The first series contained 55 mL AD-MSS media, 0.15 g dried gypsum sludge, and 5 mL SRB inoculum from the stock culture described above. The second contained the same media and inoculum amounts, but no gypsum was added. Both series were prepared aseptically. The third series (an abiotic control) contained 60 mL of autoclaved media and 0.15 g gypsum sludge. Each series was tested daily for H_2S by withdrawing 500 µL of headspace and injecting onto the gas chromatograph. The headspaces were then evacuated for 1 min by passing nitrogen through each bottle (~500 mL/min) using syringe needles as both injector and vent. Once the gas injector was removed, the vent was left in place for 5 s to ensure that the headspace pressure decreased to ambient. Measured liquid samples were removed every other day, centrifuged at 11,750g for 10 min, and analyzed for sulfate, organic acids, COD, and pH. At d 14, the bottle with SRB and gypsum was sampled for an MPN determination. All other analyses were continued until H_2S evolution ceased.

Mixed Reactor System

Once it was established that the cells could reduce the gypsum sludge to sulfide using the AD-MSS media, a reactor system was constructed the could operate using FGD gypsum in a batch, fed-batch, or continuous mode. The primary reactor consisted of a Virtis OmniCulture chemostat with a 1-L reactor vessel (Virtis, Gardiner, NY). The vessel headplate was fitted for acid and base addition as well as gas and liquid influents and effluents. The pH was regulated at 7.0 with a Chemcadet controller (Cole-Parmer Instrument, Niles, IL); 6N NaOH were used as base, and 6N H_3PO_4 were used for acid addition. Influent and effluent flows (0.2–0.4 mL/min.) were regulated by peristaltic pumps. The reactor was maintained at 30°C and was continuously stirred at 150 rpm. A feed reservoir was constructed that allowed an adjustable nitrogen flow over the AD-MSS media to help prevent contaminant growth. The reactor was continuously sparged with nitrogen to remove the resulting H_2S and carry it to the sulfur-precipitating unit. Sulfate, organic acid, and COD levels were measured daily. The effluent from the reactor was centrifuged daily; the cells were returned to the reactor, and the supernatant was stored for use in calcium precipitation experiments.

To establish biomass in the reactor, a working culture of SRB was grown in a fed-batch mode over a 10-d period with gypsum sludge added to the AD-MSS media. Cells were centrifuged and resuspended in fresh media prior to each gypsum addition, and this cycle continued until a suitable biomass population was established. At this point, the reactor was converted to have continuous AD-MSS feed (0.2 mL/min, 3.5-d residence time) with gypsum manually supplied in a fed-batch mode. Organic acid and COD levels of the influent and effluent AD-MSS media were monitored, as was the sulfate level in the reactor. Gypsum was manually added (typically 1–2 g when the sulfate level fell to near zero).

After the well-mixed reactor had been operated in a fed-batch mode for a period of 34 days, the gypsum feed was made continuous by adding it directly to the AD-MSS feed vessel. FGD gypsum (1–3 g) was added to 1 L of AD-MSS media, which was subsequently autoclaved to ensure that no SRB activity existed outside the reactor. The resulting solution was decanted to remove residual solids, and was then used to feed AD-MSS media and gypsum simultaneously to the reactor at a rate of 0.2–0.4 mL/min.

Immobilized-Cell Reactor

An immobilized-cell reactor was constructed in order to achieve higher reactor productivity owing to the higher biomass density afforded per unit volume. A fully jacketed glass columnar reactor (5821-24, Ace Glass, Vineland, NJ) of internal dimensions 2.5 × 30 cm was filled with BIO-SEP™ beads (Dupont, Glasgow, DE). The resulting fixed-bed reactor had a total volume of 181 mL and a liquid volume in the active portion of the reactor of 81 mL. The porous beads were inoculated using effluent biomass from the above well-mixed reactor, resuspended in fresh AD-MSS media. The liquid stood stagnant over the beads at 30°C for 12 h. At this point, operation was initiated with a continuous and simultaneous gypsum/AD-MSS media feed of 0.17 mL/min (7.9-h liquid residence time). The prolific reduction of sulfate in the columnar reactor resulted in a pH gradient being established within the column. The pH at the column inlet was 6.9, and sulfate reduction increased the pH at the column outlet to 8.3 (above the optimal pH for SRB). To correct this, further trials were performed using a recycle stream (Fig. 2) in which the liquid feed was more rapidly circulated (1 mL/min) through the column, and the effluent was collected in a holding vessel (200 mL) where the pH was adjusted to 7.0 using $6N$ H_3PO_4 before re-entering the reactor. Liquid from the gypsum feed bottle and the holding vessel was introduced into the reactor at a "tee" junction so that the sum of their flow rates equaled 1 mL/min. Liquid was removed from the holding vessel at the same rate as liquid was fed from the gypsum feed bottle. Sulfide was stripped from the reactor and precipitated as sulfur. Analyses of sulfate, organic acids, and COD were performed daily by sampling liquid at the reactor apex.

Sulfur Precipitation and Analysis

In the continuous reactor systems for FGD gypsum reduction, the resulting sulfide was oxidized to elemental sulfur both as an analytical tool to complete the sulfur mass balance and also to demonstrate the formation of a salable product. Hydrogen sulfide was sparged from the reactor units using a continuous flow of nitrogen gas (30 mL/min). The resulting gas stream was contacted with a $0.1M$ ferric sulfate (FX0235-1, EM Science, Gibbstown, NJ) solution with a 0.3-s gas residence time to form elemental sulfur by the reaction.

$$H_2S + Fe_2(SO_4)_3 \rightarrow S^0 \downarrow + 2FeSO_4 + H_2SO_4 \qquad (1)$$

Biological Reclamation of FGD Gypsum

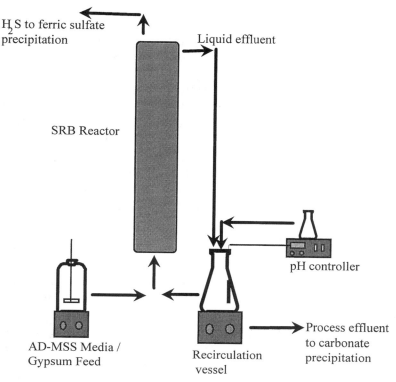

Fig. 2. Immobilized-cell recycle reactor. To regulate the pH within the immobilized-cell reactor, the total liquid flow rate through the reactor was increased to 1 mL min., and the process effluent was adjusted to pH 7.0 before being recirculated to the reactor. The feed rate of fresh AD-MSS media containing FGD gypsum was balanced with the rate of liquid discharge from the recirculation vessel in order to maintain a constant liquid volume. The residence time of the gypsum solution in the reactor was calculated as the liquid volume of the reactor divided by the fresh AD-MSS media/gypsum feed rate.

The method utilized was based on the patented BIO-SR process, *(26–28)* which further reoxidizes the ferrous sulfate back into ferric sulfate using *Thiobacillus* via:

$$2FeSO_4 + H_2SO_4 + 0.5 O_2 \rightarrow Fe(SO_4)_3 + H_2O \qquad (2)$$

The amount of sulfur precipitate was determined gravimetrically by filtration through Whatman 541 grade filter paper and drying at 100°C. When operating reactors in a fed-batch manner, the sulfur was usually collected when a new batch of gypsum was added. During operation of continuous feed reactors, sulfur was collected when the gypsum feed bottles were replenished.

Calcium Carbonate Precipitation and Analysis

To demonstrate the ability to regenerate calcium carbonate, effluent from the well-mixed reactor was brought to pH 11.0 through the addition of 50% w/v NaOH. Carbon dioxide was then bubbled through the vessel resulting in a white precipitate. This precipitate was filtered and dried as above, and was analyzed by X-ray diffraction. The X-ray diffraction sample was prepared by powdering the dried precipitate, dispersing it in toluene, and depositing a thin portion on an Si substrate. A Phillips Model XRG3100 diffractometer with Cu Kα radiation was used to perform the X-ray analysis. The peaks were then identified using search/match software of the International Center for Diffraction Data (ICDD) files for various organic/inorganic phases known in the literature.

Synthesis Gas Experiments

Initial trials to assess the potential use of synthesis gas as the carbon and energy sources in a gypsum reduction process were carried out under anaerobic conditions in a 2-L Schott bottle (Ace Glass Co., Vineland, NJ) fitted with a headplate, which facilitates gaseous influent sparging, and liquid and headspace sampling. A minimal salts media (1.2 g/L Na_2HPO_4, 1.8 g/L KH_2PO_4, 0.7 g/L $MgCl_2$, 0.2 g/L NH_4Cl, 0.2 g/L $FeCl_2$; batch vitamin solution 2 mL/L [24], and heavy metal solution 15 mL/L [24]) containing 1.5 g/L gypsum sludge was prepared anoxically within the bottle by boiling 10 min under a continuous nitrogen sparge. Thirty milliliters of concentrated SRB culture were inoculated into 970 mL media and sparged 10 min with synthesis gas (47% CO, 36% H_2, 10% CO_2, 2% N_2, 0.03% Ar, balance CH_4, Air Liquide, La Porte, TX). Methane and N_2 served as internal standards owing to their inert nature in the reactions under consideration. The pH was readjusted to 6.5 (with 6.5N NaOH), and the bottle incubated at 30°C with gentle shaking.

The bottle headspace was sampled daily by withdrawing 350 µL and injecting onto the gas chromatograph. Liquid samples of 3 mL were removed following gas withdrawal, filtered through 0.2-µM filters, and analyzed for sulfate, organic acids, and pH. Three mL of nonsulfate-containing makeup media was injected immediately after liquid withdrawal so as not to cause a vacuum within the bottle.

RESULTS AND DISCUSSION

Serum Bottle Experiments

Batch experiments conducted in serum bottles demonstrated that a mixed strain of SRB was capable of utilizing organic acids in AD-MSS media to reduce FGD gypsum to hydrogen sulfide (Figs. 3 and 4). The results show that this sulfide production was biological, rather than chem-

Biological Reclamation of FGD Gypsum

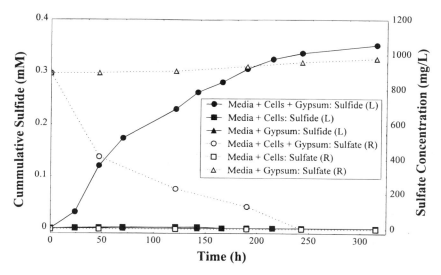

Fig. 3. Sulfate reduction and sulfide production from FGD gypsum sludge by SRBs. Serum bottles were established with (a) media, SRB, and gypsum, (b) media and SRB, and (c) media and gypsum. Sulfate was rapidly depleted in the bottle with media, SRB, and gypsum at the rate of 8 mg/L/h $(10^9 \text{ cells})^{-1}$. No sulfide was detected in either the abiotic control or the control bottle without gypsum, indicating that sulfide production was owing to the biological reduction of FGD gypsum.

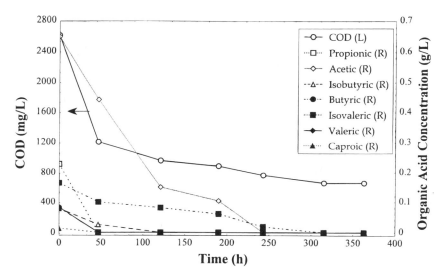

Fig. 4. COD reduction and organic acid utilization by SRB that utilize sewage digest as their carbon and energy sources. Serum bottles were established as described in Fig. 3. SRB utilized short-chain organic acids present in the AD-MSS media as their carbon and energy sources in the reduction of FGD gypsum to hydrogen sulfide. Organic acid utilization reduced the COD of the media from 2617 to 672 mg/L in 13 d.

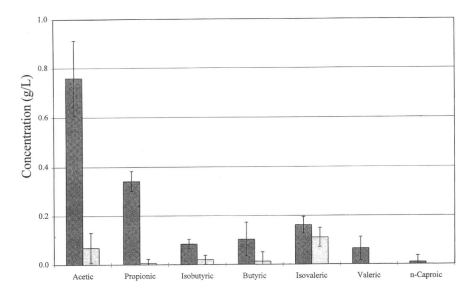

Fig. 5. Utilization of organic acids in the fed-batch well-mixed reactor. Both the AD-MSS media feeding the reactor ($n = 11$) and the reactor effluent ($n = 18$) were monitored during the course of the reactor run to determine which organic acids were utilized by the mixed SRB population. Paired bars denote the influent (dark) and effluent (light) concentrations of the organic acids assayed. Error bars represent the standard deviation of the pooled data. Acetic acid was the most prevalent acid in the feed and was also the product of the utilization of larger organic acids. The influent COD to the reactor was typically 3000 mg/L, whereas the effluent was 1500 mg/L.

ical, in that no sulfide was produced from the abiotic control. The COD was reduced from 2617 to 672 mg/L in 13 d, indicating the utilization of organic acids in the AD-MSS media. Individual organic acids fell below detection limits after 10 d. The sulfate reduction rate in these serum bottle experiments was 8.0 mg/L/h $(10^9 \text{ cells})^{-1}$ when calculated for the first 48 h. This compares favorably with the 18.5 mg/L/h $(10^9 \text{ cells})^{-1}$ reported by Apel and Barnes (12), considering that they utilized pure *Desulfovibrio desulfuricans* with a rich media formulation with lactate as a carbon source, and reported the initial rate for a 20-h period.

Organic Acid Utilization in Reactor Systems

The AD-MSS media contains a variety of short-chain organic acids, including acetic, propionic, butyric, isobutyric, valeric, isovaleric, and caproic, which may be utilized by the mixed culture of SRBs (22). Influent and effluent COD and organic acid concentrations were routinely measured in the continuous reactors to optimize AD-MSS media residence time and demonstrate organic acid utilization by the SRB culture. Such utilization by the well-mixed fed-batch reactor is illustrated in Fig. 5. It is seen that all of the assayed acids were utilized by the culture. In actual process

Table 1
Results of Gypsum Recycling Trials in Continuous Reactors

Inlet SO_4 (mg/L)	SO_4 Conversion (%)	Residence Time (h) /Recycle Ratio[a]	% S Recovered[b]	% COD Reduced	Productivity[c] ($kg/m^3 \cdot d$)
Mixed					
488	100	83/na	100	48	0.25
1,003	100	83/na	78	54	0.52
1,148	100	83/na	74	49	0.59
1,227	100	42/na	80	71	1.27
Column					
578	100	8/na	93	40	1.40
928	100	8/na	98	5	2.25
1107	100	8/na	73	19	2.68
896	100	4/2.8	73	62	4.59
1,000	100	3/2.2	92	64	6.55

[a]Residence times were calculated by dividing the working liquid volume of the reactor by the gypsum feed influent rate. For the well-mixed reactor, the working liquid volume was 1 L. The immobilized-cell reactor had a liquid working volume of 81 mL. When operated in the recycle mode, the recycle ratio for the immobilized-cell reactor was calculated by dividing the total liquid flow to the reactor by the gypsum influent flow rate.

[b]The theoretically possible recoverable sulfur was calculated based on the concentration of sulfate in the feed solution and the duration of the experiment.

[c]The productivity is reported as kg of dried FGD gypsum sludge reduced/m^3 of reactor volume/d. It assumes that the FGD gypsum exists solely as $CaSO_4 \cdot 2H_2O$.

operation, the effluent water would be recycled for preparation of AD-MSS media. Any discharged water would have a reduced COD owing to the biological oxidation of the organic acids present. Allowable COD ranges for discharged water are between 300 and 2000 (33). Although maximum reduction of COD is desired from the point of view of process water treatment, this is not feasible in a well-mixed reactor because such low levels of organic acids would adversely affect culture viability. In the well-mixed reactor, a media residence time of 3.5 d resulted in a typical reduction of COD from 3000 to 1500 mg/L. The columnar reactor with media recycle and an influent flow rate of 0.35 mL/min demonstrated a COD reduction from 3000 to 1100 mg/L.

Reactor Trials

Results from continuous gypsum recycling trials in the well-mixed reactor are summarized in Table 1. The duration for each test was 70–92 h. Trials were first conducted at a liquid residence time of 3.5 d with an increasing gypsum concentration until saturation was reached in the feed solution. At this point, further gypsum loading to the reactor was achieved by increasing the feed rate of saturated solution to the reactor, thus decreasing the liquid residence time. Total conversion of sulfate was realized in each case; however, further increases in gypsum loading were not possible beyond the 42-h residence time trial owing to extensive cell washout in the well-mixed reactor. Between 75 and 100% of the sulfur in the gypsum sludge was recovered as elemental sulfur through ferric sulfate precipitation, and the COD of the process water was typically reduced by 50%. The maximum productivity achieved in the well-mixed reactor was 1.3 kg/m^3/d, being limited by biomass washout. As with the well-mixed reactor, trials in the immobilized-cell reactor were first conducted at a set liquid residence time with increasing gypsum concentration until saturation was reached in the feed solution. With a saturating concentration of gypsum and a one-pass influent flow rate of 0.17 mL/min. however, an appreciable pH gradient was established in the reactor. The pH at the reactor inlet was 7.0, whereas the pH reached 8.3 at the reactor apex, beyond the optimal pH for SRB activity. At this point, further gypsum loading to the reactor was achieved by operating the reactor in a "recycle" mode in which the total flow rate of liquid through the reactor was increased to 1.0 mL/min, and the reactor effluent was returned to a holding vessel where the pH of the liquid was adjusted to 7.0 before re-entering the reactor. Addition of gypsum to the reactor was controlled by altering the influent and effluent flow rates from the holding vessel while keeping the total flow of liquid through the reactor constant at 1 mL/min. In this manner, the gypsum loading to the reactor could be increased, the biomass in the reactor was more effectively utilized, and reactor productivity increased while maintaining 100% conversion of gypsum into sulfide. A maximal reactor productivity of 6.6 kg/m^3/d was achieved with complete sulfate conversion and >60% reduction in the COD of the process liquid. This corresponded to a hydraulic loading rate of 1349 L/m^2/d. To date, the capacity of the immobilized-cell reactor has not been exceeded, and further increases in productivity are expected.

Sulfur and Calcium Carbonate Recovery

Elemental sulfur was recovered from the FGD gypsum by oxidizing the sulfide gas formed through bacterial conversion of sulfate. The theoretically possible amount of recoverable sulfur was calculated based on the known sulfate concentration in the input media and its rate and duration of addition to the reactor. As seen in Table 1, >70% sulfur recovery was achieved in all of the reactor trials. Loss of sulfur could be owing due to the formation of insoluble sulfides with metal constituents of the FGD gypsum or AD-MSS media, the incomplete dissolution of hydrogen sulfide into the ferric sulfate solution, or loss on filtering the resulting sulfur precipitate. Improvements in sulfur recovery efficiency will obviate the need to treat

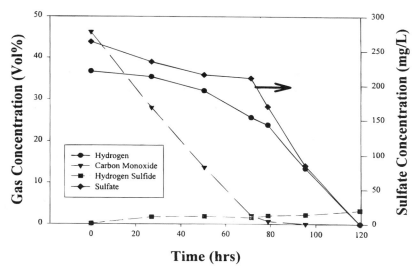

Fig. 6. Synthesis gas as a carbon and energy source for SRB in the reduction of FGD gypsum to hydrogen sulfide. Although reduction rates were not as rapid as those seen in experiments utilizing AD-MSS media, these preliminary studies indicate that synthesis gas may also serve as an inexpensive energy source for gypsum bioconversion.

the liquid effluent further in the overall process. The melting point of the yellow powder resulting from ferric sulfate oxidation was 121°C, consistent with that of elemental sulfur.

Spent process solution was precipitated using carbon dioxide in an alkaline solution in order to recover calcium ions as calcium carbonate—which could be reutilized as an adsorbent for sulfur dioxide in flue gases. X-ray diffraction analysis of the resulting precipitate revealed that major signals were attributable to calcium carbonate and other calcium salts. Additional peaks were attributable to residual organic matter that could be burned off at the coal plant through sorbent injection.

Synthesis Gas Experiments

Batch experiments demonstrated that a mixed strain of SRB was capable of utilizing CO and H_2 to reduce FGD gypsum to H_2S (Fig. 6). Sulfate concentrations were reduced from 263 to 0 mg/L in 120 h, much more slowly than that seen in the AD-MSS media trials. Although results are preliminary and the exact stoichiometry of the reactions are not understood, multiple experiments have demonstrated that sulfide production and sulfate elimination are concomitant with decreases in CO and H_2 levels in the headspace gas. Du Preez et al. *(18)* proposed that the stoichiometry for SO_4^{2-} reduction using H_2 is:

$$4H_2 + SO_4^{2-} \rightarrow H_2S + 2OH^- + 2H_2O \qquad (3)$$

with CO being utilized to produce additional H_2 by:

$$CO + H_2O \rightarrow H_2 + CO_2 \tag{4}$$

van Houten et al. *(19)* suggested that H_2 may also be utilized in the synthesis of organic acids. In our preliminary investigations, without a mass balance closure on CO_2 or the amount of biomass, we are unable to determine the extent to which Eq. (4) enters into the observed sulfate reduction. Like van Houten et al. *(19)*, we observe the production of organic acids (up to 0.5 g/L acetic acid and 0.1 g/L propionic acid) and more H_2 being utilized than is stoichiometrically needed for the reduction of sulfate. Although more research is needed to elucidate the mechanisms of synthesis gas utilization by SRB, it is clear that this inexpensive energy source may be utilized in the bioconversion of FGD gypsum.

CONCLUSIONS

The components of a closed-loop biochemical process for the conversion of FGD gypsum into elemental sulfur and calcium carbonate have been demonstrated at the bench scale, and offer an attractive alternative to the landfilling of waste gypsum. Although an economic assessment of the proposed process has not been completed, this study has established a low-cost carbon source and an efficient reactor design, which improve process economics and move this technology toward economic viability.

ACKNOWLEDGMENTS

This work was supported by the Advanced Research and Technology Development Program of the Office of Fossil Energy, U.S. Department of Energy under contract DE-AC05-96OR22464 with Lockheed Martin Energy Research Corp. Material contributions by Southern Illinois University and Dupont Chemical Company as well as the technical advice of Brian H. Davison are gratefully appreciated.

The submitted manuscript has been authored by a contractor of the U.S. Government under contract DE-AC05-96OR22464. Accordingly, the U.S. Government retains a nonexclusive, royalty-free license to publish or reproduce the published form of this contribution, or allow others to do so, for U.S. Government purposes.

REFERENCES

1. Clarke, L. B. (1993), IEA Coal Research Report # IEACR/62.
2. Bove, H. J., Brodsky, I. S., Nuyt, G. M., and Angelini, E. J. (1995), *Power.* **June,** 33–36.
3. Harben, P. (1991), *Industrial Minerals*, **July 1,** 47–49.
4. Illinois Clean Coal Institute (1994), *Request for Proposals for Research on Utilization and Marketing of Illinois Basin Coal.*

5. Gray, S. M., Puski, S. M., Richard, R., McClellan, T., and Murphy, J. L. (1995), *Results of high efficiency SO2 removal testing at PSI Energy's Gibson station in 1995 SO2 Control Symposium*. Electric Power Research Institute, Miami, FL.
6. Phillips, J. L., Blythe, G. M., and White, J. R. (1995), in *1995 SO2 Control Symposium*. Electric Power Research Institute, Miami, FL.
7. Hedgecoth, M. A. (1995), *Power Eng.* **99**, 24–26.
8. Barton, L. L., ed. (1995), *Sulfate-Reducing Bacteria. Biotechnology Handbooks*, vol. 8, ed. T. Atkinson, and Sherwood, R. F., Plenum: New York, p. 336.
9. Widdel, F. and Hansen, T. A. (1992), in *The Procaryotes*, Balows, A., et al., eds., Springer-Verlag, New York, p. 1027.
10. Montgomery, A. D., McInerney, M. J., and Sublette, K. L. (1990), *Biotechnol. Bioeng.* **35**, 533–539.
11. Chen, C. I., Mueller, R. F., and Griebe, T. (1994), *Biotechnol. Bioeng.* **43**, 267.
12. Apel, W. A. and Barnes, J. M. (1993), in *Biohydrometallurgical Techniques*, Torma, A. E. and Wey, J. E., eds. The Minerals, Metals and Materials Society, Warrendale, PA, pp. 641–651.
13. Uphaus, R. A., Grimm, D., and Cork, D. J. (1983), in *Developments in Industrial Microbiology.* p. 435–442.
14. Ko, C. W., Vega, J. L., Clausen, E. C., and Gaddy, J. L. (1989), *Chem. Engin. Commun.* **77**, 155–169.
15. Brysch, K., Schneider, C., Fuchs, G., and Widdel, F. (1987), *Arch. Microbiol.* **148**, 264–274.
16. Deshmane, V., Lee, C. M., and Sublette, K. L. (1993), *Appl. Biochem. Biotechnol.* **39/40**, 739–752.
17. van Houten, R. T., Pol, L. W. H., and Lettinga, G. (1994), *Biotechnol. Bioeng.* **44**, 586–594.
18. Du Preez, L. A., Odendaal, J. P., Maree, J. P., and Ponsonby, M. (1992), *Environ. Technol.* **13**, 875–882.
19. van Houten, R. T., van der Spoel, H., van Aelst, A. C., Hulshoff Pol, L. W., and Lettinga, G. (1996), *Biotechnol. Bioeng.* **50**, 136–144.
20. Sublette, K. L. and Gwozdz, K. J. (1991), *Appl. Biochem. Biotechnol.* **28**, 635–646.
21. Selvaraj, P. T. and Sublette, K. L. (1996), *Appl. Biochem. Biotechnol.*, **57/58**, 1003–1012.
22. Selvaraj, P. T., Meyer, G. B., and Kaufman, E. N. (1996), *Appl. Biochem. Biotechnol.* **57/58**, 993–1002.
23. Kaufman, E. N. and Selvaraj, P. T. (1995), *Bio-Chemical sulfur reclaiming and limestone regeneration from flue gas desulfurization gypsum using multiple waste streams*. Lockheed Martin Invention Disclosure—ESID No. 1725-X, S-83, 353.
24. Dasu, B. N. and Sublette, K. L. (1989), *Appl. Biochem. Biotechnol.* **20**, 207–220.
25. Plumb, P., Lee, K., and Sublette, K. L. (1990), *Appl. Biochem. Biotechnol.* **24**, 785–797.
26. Asai, S., Konishi, Y., and Yabu, T. (1990), *AIChE J.* **36**, 331–1338.
27. Sonta, H. and Shiratori, T. (1988), US Patent Number 4,931,262.
28. Satoh, H., Yoshizawa, J., and Kametani, S. (1988), *Hydrocarbon Processing* **76**, 76D–76F.
29. Imaizumi, T. (1986), *Biotechnol. Bioeng. Symp. Ser.* 363–371.
30. Wachi, S. and Jones, A. G. (1991), *Chem. Eng. Sci.* **46**, 3289–3293.
31. Chemical Prices (1994), in *Chemical Market Reporter*, Schnell Publishing Company, New York, pp. 28–36.
32. Selvaraj, P. T. and Sublette, K. L. (1995), *Biotech. Prog.* 11, 153–158.
33. Belhateche, D. H. (1995), *Chem. Eng. Prog.* **91**, 32–51.

A Preliminary Cost Analysis of the Biotreatment of Refinery Spent-Sulfidic Caustic

KERRY L. SUBLETTE

Center for Environmental Research & Technology, University of Tulsa, 600 S. College Avenue, Tulsa, OK 74104

ABSTRACT

Caustics are used in petroleum refining to remove hydrogen sulfide from various hydrocarbon streams. Spent-sulfidic caustics from three refineries have been successfully biotreated on the bench and pilot scale, resulting in neutralization and removal of active sulfides. Sulfides were completely oxidized to sulfate by *Thiobacillus denitrificans* strain F. Microbial oxidation of sulfide produced acid, which at least partially neutralized the caustic. A commercial-scale treatment system has been designed that features a bioreactor with a suspended culture of flocculated *T. denitrificans*, a settler and acid and nutrient storage and delivery systems. A cost analysis has been performed for nine cases representing a range of spent caustic sulfide and hydroxide concentrations at a base treatment rate of 10 gpm. This analysis shows that refinery spent-sulfidic caustic can be biotreated for 4–8.3¢/gal.

Index Entries: Sulfidic caustic; sulfide; *Thiobacillus denitrificans*; cost analysis; refinery waste.

INTRODUCTION

Sodium hydroxide (NaOH) solutions are used in petroleum refining to remove hydrogen sulfide (H_2S) from various hydrocarbon streams. Once H_2S reacts with the majority of NaOH in the solution, the solution becomes known as spent-sulfidic caustic. Spent caustics typically have a pH >12.0, sulfide concentrations exceeding 2–3 wt%, and a large amount of residual alkalinity. Depending on the source, spent caustic may also contain phenols, mercaptans, amines, and other organic compounds that are soluble or emulsified in the caustic *(1)*.

Author to whom all correspondence and reprint requests should be addressed.

Currently, most spent-sulfidic caustics generated by refineries are either sent off-site to commercial operations for recovery or reuse (pulp and paper mills, for example) or for disposal by deep-well injection. Biological treatment in the refinery waste water treatment unit is an inexpensive disposal option. However, many refineries do not have the waste water treatment capacity to treat the entire amount of spent caustic generated, and concerns regarding odors and toxicity frequently prohibit this practice.

Recently biotreatment of refinery spent-sulfidic caustic using a microbial culture augmented with the sulfide-oxidizing bacterium *Thiobacillus denitrificans* strain F was shown to be feasible (2,3). It is envisioned that this process could be implemented either by augmenting an existing refinery activated sludge unit so that it could handle higher concentrations of sulfides without toxicity or odor problems, or by using a relatively small bioreactor, which would be specialized for treating spent-sulfidic caustic streams. Reported here are the results of a cost analysis of the biotreatment of refinery spent sulfidic caustic using the sulfide–tolerant strain F of *T. denitrificans* in flocculated, suspended culture.

BIOTREATMENT OF SPENT-SULFIDIC CAUSTIC

Bench-Scale, Stirred-Tank Reactor

The biotreatment of spent-sulfidic caustic was first demonstrated at the bench scale using a B. Braun Biostat M fermenter (2). The reactor was initially charged with 1.5 L of a mineral salts medium containing 1700 mg/L of flocculated *T. denitrificans* strain F (4). The pH and temperatures were maintained at 7.0 and 30°C, respectively. The acid used for pH control was $10N$ HNO_3. This particular acid was used so that acid addition could be monitored by following the nitrate concentration in the culture medium. The culture received a gas feed of 0.3–0.4 L/min of air with 5% CO_2. The outlet gas from the bioreactor was passed to a 500-mL Erlenmeyer flask, where the gas was sparged into 300 mL of 0.3 wt% zinc acetate to trap fugitive H_2S from the bioreactor. A tee connection was located between the bioreactor of zinc acetate trap for gas sampling.

Sample characteristics of all caustics used in these studies are given in Table 1. The spent caustic feed reservoir consisted of a 250-mL graduated cylinder with a cork stopper. Spent-sulfidic caustic samples were diluted 1/5 with deionized water, but not neutralized (pH > 12.0). Diluted caustic was withdrawn through a stainless-steel tube that extended to the bottom of the cylinder. Feed was pumped to the bioreactor at 0.12 mL/min. In this manner, 150 mL of feed (1/10th the culture volume) could be delivered in 21 h. The feed was introduced into the bioreactor at a point about 2.5 cm from the bottom of the vessel and adjacent to one of the agitator impellers.

At start-up 150 mL of the process culture were removed from the bioreactor with the biomass recovered by centrifugation and returned to

Table 1
Characteristics of Spent-Sulfidic Caustic Samples[a]

Sample	Sulfide (M)	COD (mg/L)	MDEA (wt%)	OH⁻(M)
D1	1.06	82100	2.37	2.60
D2	1.05	113800	3.17	1.04
D3	1.06	107000	3.81	1.03
PC1	0.60	73300	2.08	2.46
PC2	0.58	40200		2.91
PC3	0.73	46300		2.80
T1	0.18	26700		2.11

[a]The D, P, and T series caustics were obtained from three different refineries.

the culture. The supernatant from this centrifugation was the initial sample for sulfate, ammonium ion, nitrate, and chemical oxygen demand (COD) determination. At this time, the feed pump was activated, and the 1/5 dilution of spent-sulfidic caustic (sample D1) was delivered to the bioreactor. As noted above, the feed reservoir contained 150 mL of diluted caustic. Therefore, the feed reservoir was emptied in 21 h. The reactor received no feed for the next 3 h. At the end of this time, and each day thereafter, the feeding procedure was repeated with 150 mL removed from the bioreactor, the biomass recovered and returned to the culture. At the end of the 7th day of fed batch operation (after a total of 1050 mL of diluted caustic had been fed to the reactor), the agitation and gas feed were turned off and the biomass allowed to settle under gravity. The biomass settled to 25% of the original volume in <10 min. The clarified liquor was then siphoned off and replaced with fresh medium to a final volume of 1.5 L. The agitation and aeration were then resumed. At this time, 150 mL of the culture were removed, the biomass recovered and returned to the culture, and the feeding schedule resumed as described above using a second sample of caustic (D2). This replenishment of the culture medium was also repeated after 5 d of operation at which time the caustic feed was changed to the D3 sample for the duration of the experiment. The experiment was terminated after 21 full days of operation on the D series feeds.

The PC1 and PC2 samples (Table 1) were biotreated using the same reactor system at a later time using a second flocculated culture of *T. denitrificans* strain F developed for this purpose. The bioreactor was operated as described above for 6 d using PC1 as feed and 10 d using PC2 as feed.

Table 2
Stoichiometry of Sulfide Oxidation by *T. denitrificans* in a Bench-Scale,
Fed-Batch Reactor with a Feed of Spent-Sulfidic Caustic[a]

Sample	SO_4^{-2}/S^{-2} (mole/mole)	HNO_3/S^{-2} (mole/mole)	H^+ prod/S^{-2} (mole/mole)	g MLSS/mole S^{-2}
D1	1.00	1.06	1.39	4.1
D2	1.07	0.048	0.94	6.1
D3	1.10	0.040	0.94	18.3
D3	1.01	0.040	0.94	15.9
PC1	0.98	3.0	1.10	8.5
PC2	1.00	4.0	1.02	11.8
T1	1.02			

[a]HNO_3/S^{-2} is the mol HNO_3/mol sulfide oxidized required to maintain the pH at 7.0–7.1. H^+ prod/S^{-2} is the mol of acid produced/mol of sulfide oxidized.

It is important to note that the spent-sulfidic caustic was introduced into the bioreactor without neutralization. Sulfide oxidation by *T. denitrificans* is acid-producing (5). Therefore, if the reactor is operated on a sulfide-limiting basis, at least partial neutralization of the added caustic can be achieved by the oxidation reaction if the reaction is sufficiently fast. This was indeed the case in this experiment. During the entire course of the experiment, the pH was maintained at 7.0–7.1 with only a small amount of acid addition as shown in Table 2. All of the acid produced by sulfide oxidation by the organism was neutralized by the hydroxide ion in the samples. The amount of HNO_3 required to maintain the pH in the 7.0–7.1 range was dependent on the extent to which the alkalinity in the sample exceeded the acid produced by sulfide oxidation. Greater addition of nitric acid was required when the D1 sample was used as feed because of the much greater alkalinity of the sample (Table 1).

Sulfate accumulated in the culture media as the sulfidic caustics were fed to the bioreactor. A sulfur balance showed complete conversion of sulfide to sulfate (Table 2). No elemental sulfur was detected in the culture medium, and no H_2S was detected in the outlet gas or collected in the zinc acetate trap. It was also observed in these experiments that the process culture acclimated to methyldiethylamine (MDEA), resulting in complete removal of MDEA as well as oxidation of sulfides. The MDEA was metabolized by the mixed heterotrophs of the culture, whereas sulfides were oxidized to sulfate by *T. denitrificans*.

Similar results were obtained with samples PC1 and PC2 (Table 2). Subsequently, bench-scale testing with a similar suspended culture of *T. denitrificans* strain F and a caustic from a third refinery showed that caustic could be fed to the stirred-tank bioreactor undiluted as well as without neutralization (6). The stoichiometry of sulfide oxidation with this caustic (T1, Table 1) is also given in Table 2. With the T1 caustic, the specific activity of flocculated *T. denitrificans* strain F was shown to be 1.1–1.3 mmole sulfide/h/g (MLSS). This agrees well with the specific activity observed in the biotreatment of sour water with this organism in an upflow bubble column (7). Experiments with undiluted T1 caustic also demonstrated that carbonates present in the caustic provided sufficient inorganic carbon to *T. denitrificans* that no other source was necessary to support autotrophic oxidation of caustic sulfide.

Pilot-Scale, Stirred-Tank Reactor

Pilot-scale biotreatment of spent-sulfidic caustic was conducted in a 3.8-m^3 stainless-steel, milk-holding tank manufactured by the Paul Mueller Co. (Springfield, MO). The tank was horizontal and semicylindrical, 170 cm deep, and 660 cm long on the inside. The tank was jacketed with cooling/heating coils running lengthwise in the jacket annular space. A 2-hp variable-speed DC motor and gearbox were mounted on a platform that bridged the center of the vessel. The motor drove a paddle-type stirrer, which was 81 cm in diameter and 12 cm wide. The agitation rate was 50 rpm. On either side of the stirrer platform were stainless-steel lids, which completely closed the top of the vessel. The tank was modified by fitting with stainless-steel baffles, each 1/10 of the major or minor dimensions of the tank, and a sparger. The sparger was fabricated from 2.5-cm stainless-steel tubing in a U-shape. It was fed with air at the bottom of the U through a 2.5-cm stainless-steel tube, which extended through the wall of the vessel at the center and bottom. The sparger was centered under the stirrer with the branches of the U equal in length to the stirrer diameter. The U branches had equally spaced 0.32-cm holes drilled on the bottom, such that the total hole area on each branch was two times the cross-sectional area of the tube.

The 3.8-m^3 stirred-tank reactor was utilized for pilot-scale biotreatment of spent-sulfidic caustic in a fed-batch mode as follows. About 40 L of a concentrated suspension of flocculated *T. denitrificans* strain F were used to inoculate 3.0 m^3 of mineral salts medium. The initial (MLSS) concentration was 630 mg/L. Spent-sulfidic caustic (samples PC2 and PC3, about 180 L of each) was fed (undiluted) from 0.23-m^3 barrels at a feed rate of 30 mL/min. The caustic was conveyed to the reactor through PTFE tubing and introduced below the liquid surface near the impeller tip. The temperature was maintained at 30°C and pH at 7.2 ± 0.1 with 85% H_3PO_4, industrial-grade. The agitation rate was 50 rpm and aeration rate was about 1.1 standard m^3/min (blower + line air). The culture was operated with spent-sulfidic caustic feed during the evenings and on weekends only because of the odor from the caustic. When not receiving a caustic feed (10–12 h at a time), the culture was maintained at temperature with aeration. The total operating time with spent caustic feed was 200 h. A total of 360 L of refinery spent-sulfidic caustic were successfully treated (3).

Sulfate accumulated in the reactor medium as caustic was fed to the reactor. No hydrogen sulfide emissions were detected from the reactor at any time during the 200 h of operation, and no sulfide was detected in the culture medium. The overall sulfate/sulfide ratio observed was 1.3. However, only the soluble sulfide concentrations in these samples were used to calculate this ratio. These caustic samples contained copious amounts of iron sulfides. Iron sulfides in the feed were most notable when feed was initiated from a new barrel (after some agitation in getting the barrel in place) and after about 75% of the caustic in each barrel had been pumped out. In the latter case, the solids had concentrated at the bottom of the barrel. In fact, the sludges were so viscous at the bottom of the barrel that they could not be pumped out.

The elemental sulfur concentration in the reactor medium averaged about 0.3 mg/L, except for one 5-h period when the caustic feed rate was doubled. This increase in feed rate caused an upset condition in which the elemental sulfur concentration became high enough to give the culture a white color. The caustic feed was stopped and the culture aerated overnight. The next day the elemental sulfur was gone (oxidized to sulfate), and the caustic feed was resumed at 30 mL/min with no further difficulties.

COMMERCIAL BIOTREATMENT OF SPENT-SULFIDIC CAUSTIC

The initial MLSS concentration in the 3.8-m^3 reactor was only 620 mg/L. A commercial system would operate with an MLSS of about 4000 mg/L (similar to an activated sludge system). *T. denitrificans* strain F can easily be cultivated to this concentration by growth on thiosulfate (8). As noted above, 1 mmol sulfide/h-g MLSS is a reasonable design figure for the specific activity of flocculated *T. denitrificans* for sulfide oxidation. Based on this specific activity and assuming an MLSS concentration in the bioreactor of 4000 mg/L, a 38 L/min (10 gal/min or gpm) stream with 3 wt% sulfide will require a 535-m^3 bioreactor. A secondary settler and capacity for biomass recycle will also be required for continuous operation. The sulfidic caustic biotreatment system will resemble a small activated sludge treatment system. In fact, the system can be thought of as a specialized activated sludge-treatment system.

A preliminary economic analysis has been conducted for a base case of treating 10 gpm of spent-sulfidic caustic using a suspended culture of flocculated *T. denitrificans* strain F. A commercial-scale treatment system was designed for nine cases: three different sulfide concentrations, 0.2, 0.6, and 1.0 *M*, and three different OH^- alkalinities, 1.0, 2.0, and 3.0*N*. Design

Table 3
Refinery Spent Caustic Treatment System Design Assumptions

Caustic flow rate = 10 gpm

T = 25 C

1 mole H$^+$ per mole sulfide oxidized to sulfate

1.86 moles O$_2$ required per mole of sulfide oxidized to sulfate

Specific activity of biomass = 1.0 mmoles/hr-g MLSS

[MLSS] = 4000 mg/L

[H$_2$SO$_4$] = 18 N

Critical DO concentration = 0.05 mM

Nutrient stock solution = 35 wt% NH$_4$NO$_3$ and 6.3 wt% P$_2$O$_5$

Settler design: overflow rate = 8 m^3/m^2-d

solids loading = 1.0 kg/m^2-hr

depth = 3 m

assumptions are summarized in Table 3. Figure 1 gives a schematic diagram of a commercial-scale system for biotreatment of refinery spent-sulfidic caustic.

The treatment system has three components: the bioreactor/settler, the pH control or acid-delivery system, and the nutrient storage and delivery component. The bioreactor/settler is basically a mixer/settler consisting of a sunken concrete basin. Mixing in the bioreactor is done by aeration. Sulfide oxidation to sulfate occurs in the bioreactor section, and settling of biomass for return to the bioreactor occurs in the settler. The hydraulic retention time and, therefore, the bioreactor effluent flow rate in the bioreactor were based on maintaining a steady-state sulfate concentration in the bioreactor of 0.2M. Sulfate concentrations >0.25M have been shown to be inhibitory to *T. denitrificans* (5). Table 4 gives the sizes of the bioreactor and settler sections for the nine design cases. Table 5 gives calculations of theoretical air requirements for the three sulfide concentrations.

The pH control system consists of a pH meter/controller, a tank to store H$_2$SO$_4$ (chosen for cost), and a pump to deliver the acid as needed to the bioreactor. Acid-delivery rates for the nine cases are given in Table 6, as well as the tank capacity needed for a 30-d supply of acid in each case. The nutrient storage/delivery component consists of two 1000-gal fiberglass tanks, two pumps, and associated piping. A nutrient stock solution (35

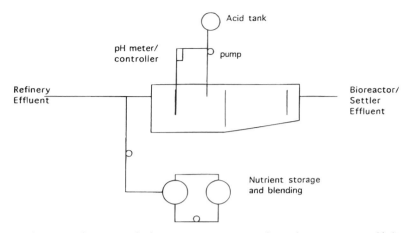

Fig. 1. Schematic diagram of a biotreatment system for refinery spent-sulfidic caustic.

Table 4
Bioreactor and Settler Volumes, 10-gpm Basis

Bioreactor

[sulfide], M	Bioreactor Volume (m^3)
0.2	114
0.6	341
1.0	568

Settler

Settler Volume (m^3)

OH$^-$ (N)	[sulfide], M		
	0.2	0.6	1.0
1.0	82	109	136
2.0	150	177	204
3.0	216	245	272

wt% NH_4NO_3 + 6.3 wt% P_2O_5) can be prepared by adding the appropriate amount of NH_4NO_3 and P_2O_5 and water to one tank, and dissolving by circulating the suspension between the two tanks with a centrifigal pump, transferring the mixture to the second tank, and metering nutrient into the

Table 5
Theoretical Air Requirements, 10-gpm Basis

[sulfide],(M)	molar sulfide feed rate (moles/hr)	Theoretical O_2 (moles/hr)	Theoretical Air (L/min)
0.2	454	844	1630
0.6	1362	2533	4910
1.0	2270	4222	8190

Table 6
Acid-Utilization Rates and Acid Tank Capacity Required for 30-d Supply

Acid Utilization Rate (L/h)

	[sulfide], M		
OH^- (N)	0.2	0.6	1.0
1.0	101	50	0
2.0	227	177	126
3.0	353	303	252

Acid tank size

Tank Volume (m^3)

	[sulfide], M		
OH^- (N)	0.2	0.6	1.0
1.0	73	36	10
2.0	163	127	91
3.0	254	218	181

influent line of the bioreactor. The influent to the bioreactor consists of refinery effluent (assumed to be gravity-fed) plus the nutrient supplement. The nutrient metering rate and the operating capacity (in days)/1000 gal of nutrient tank volume are provided in Table 7.

Estimates of capital costs for a 10-gpm commercial system, including the bioreactor/settler, acid-storage and delivery system, aeration and nutrient storage, and delivery system for each of the nine design cases, were prepared by cost estimators from a major US oil company. These

Table 7
Nutrient Solution Metering Rate
and Tank (1000-gal) Capacity

[sulfide],(M)	L/hr nutrient	Capacity (days)
0.2	15.5	10.1
0.6	46.6	3.4
1.0	77.9	2.0

Table 8
Capital Costs for Biotreatment of Spent Caustic, 10 gal/min Basis ($000)

	[sulfide], M		
OH⁻ (N)	0.2	0.6	1.0
1.0	1452.8	1625.5	1822.9
2.0	1732.8	1907.5	2082.2
3.0	2011.6	2189.5	2361.8

capital costs are given in Table 8. These costs include a 25% contingency and 45% process development allowance. Therefore, these capital costs are heavily burdened with risk factors. Annual costs of acid and nutrients for the nine design cases are given in Tables 9 and 10, respectively. A bulk acid cost of $0.24/gal for 95% H_2SO_4 was assumed. Costs of nutrients were based on $4.13/40 lb bag of NH_4NO_3 and $5.51/40 lb bag of P_2O_5 (9).

Table 11 gives the final cost (¢/gal) for treatment of refinery spent-sulfidic caustic for each of the nine design cases. These calculations were based on 365 d/yr, 24 h/d operation with capital costs annualized over 10 yr. Power, labor, and disposal costs for sulfate (if any) were not included. (These costs cannot be estimated accurately without more extensive pilot testing of the process; however, they are not anticipated to account for a significant fraction of the total operating cost.) Based on a proprietary analysis of alternative caustic treatment technologies by a major US oil company, the costs given in Table 11 show biotreatment of refinery spent-sulfidic caustic to be economically viable.

CONCLUSIONS

Refinery spent-sulfidic caustics have been successfully biotreated on a bench and pilot scales, resulting in the neutralization and complete removal and oxidation of reactive sulfides using suspended cultures of

Table 9
Acid Costs for Biotreatment of Spent Caustic, 10 gal/min Basis ($000)

OH⁻ (N)	[sulfide], M		
	0.2	0.6	1.0
1.0	56.9	28.2	0
2.0	127.9	99.7	71.0
3.0	198.9	170.7	142.0

Table 10
Nutrient Costs for Biotreatment of Spent Caustic, 10 gal/min Basis ($000)

OH⁻ (N)	[sulfide], M		
	0.2	0.6	1.0
1.0	11.2	33.6	56.1
2.0	11.2	33.6	56.1
3.0	11.2	33.6	56.1

Table 11
Cost per gallon for Biotreatment of Spent Caustic, 10 gal/min Basis (¢/gal)[a,b,c]

OH⁻ (N)	[sulfide], M		
	0.2	0.6	1.0
1.0	4.0	4.3	4.5
2.0	5.9	6.2	6.4
3.0	7.8	8.1	8.3

[a]365 d/yr, 24 h/d operation.
[b]Capital cost annualized over 10 yr.
[c]Power and labor not included.

flocculated *T. denitrificans* strain F in stirred-tank reactors. Spent caustic could be fed to the bioreactor without prior neutralization. These observations suggest that biotreatment is a viable process concept for the treatment of these waste streams. An economic analysis shows that caustics can be biotreated for 4–8.3¢/gal.

REFERENCES

1. Bechok, M. R. (1967), *Aqueous Wastes from Petroleum and Petrochemical Plants*, John Wiley, New York.
2. Rajganesh, B. and Sublette, K. L. (1995), *Biotechnol. Prog.* **11,** 228–230.
3. Rajganesh, B., Sublette, K. L., and Camp, C. (1995), *Appl. Biochem. Biotechnol.* **51/52,** 661–671.
4. Ongcharit, C., Dauben, P., and Sublette, K. L. (1989), *Biotechnol. Bioeng.* **33,** 1077–1080.
5. Sublette, K. L. (1987), *Biotechnol. Bioeng.* **29,** 690–695.
6. Kolhatkar, A. and Sublette, K. L. *Appl. Biochem. and Biotechnol.*, in press.
7. Lee, C. and Sublette, K. L. (1993), *Water Res.* **27(5),** 839–846.
8. Hasan, S., Rajganesh, B., and Sublette, K. L. (1994), *Appl. Biochem. and Biotechnol.* **45/46,** 925–934.
9. Raterman, K. T. and Sublette, K. L. (1994), SPE 26396.

The Degradation of L-Tyrosine to Phenol and Benzoate in Pig Manure
The Role of 4-Hydroxy-Benzoate

P. ANTOINE,*,[1] X. TAILLIEU,[1] AND P. THONART[1,2]

[1]Centre Wallon de Biologie Industrielle, Faculté des Sciences Agronomiques, 2, Passage des Déportés B-5030 Gembloux, Belgium; and [2]Université de Liège, Bâtiment 40, Boulevard du Rectorat, B-4000, Sart-Tilman, Belgium

ABSTRACT

The formation of odorous compounds in piggery wastes was investigated. Phenol and *para*-cresol are generally encountered in these typically anaerobic environments. They are produced from L-tyrosine by microbial metabolism. Phenol is further converted to benzoate via *para*-carboxylation.

The biochemical pathways were studied by feeding manure with misclleanous metabolites at concentration between 5 and 20 mM. Metabolites were analyzed by gas chromatography (GC) and high-performance liquid chromatography (HPLC). Experiments were carried out at room temperature.

The degradation of L-tyrosine to phenol, benzoate, and *para*-cresol was confirmed. 4HPPyrA and 4HPAA are not intermediate compounds in phenol production.

It was shown that phenol was converted to benzoate without any production of 4HBA. Other experiments showed that 4HBA was decarboxylated to phenol, but not dehydroxylated to benzoate.

When phenol was added in presence of benzoate (5 mM each) or alone at higher concentrations (10 or 20 mM), transient small amounts of 4HBA were observed (about 0.02 mM).

Our experiments show that 4HBA is not an intermediate metabolite in the conversion of phenol to benzoate. The decarboxylation of 4HBA to phenol is probably the last step of another degradation pathway. This reaction is proposed to have a weakly reversible property, explaining 4HBA production.

*Author to whom all correspondence and reprint requests should be addressed.

Index Entries: L-tyrosine; benzoate; 4-hydroxybenzoate; phenol; degradation; piggery; anaerobic.

Abbreviations: 4HBA, 4-hydroxybenzoate; 4HPPyrA, 4-hydroxyphenylpyruvate; 4HPAA, 4-hydroxyphenylacetate.

INTRODUCTION

The intensive breeding of pigs leads to the accumulation of large amounts of malodorous and polluting slurries. These wastes are mainly constitued of feces and urine.

Schaefer identified 10 molecules as important factors of malodor in piggeries: indole, skatole, phenol, *para*-cresol, acetate, propionate, butyrate, isobutyrate, valerate, and isovalerate *(1)*.

Phenol is a degradation product of L-tyrosine by *Clostridium* sp. *(2,3)* or other bacteria possessing a tyrosine phenol lyase activity *(4)*. This conversion has been observed in anaerobically stored piggery wastes *(4–6)*. Another precursor is 4-hydroxybenzoate (4HBA), which was reported to be decarboxylated to phenol *(7)*.

Two degradation pathways of phenol have been proposed. The first one includes cyclohexanone *(8)*. The latter is a conversion of phenol to benzoate *(9)*. Tschech and Fuchs observed that bicarbonate is necessary to ensure the carboxylation of phenol by a denitrifying Pseudomonads *(10)*. They also observed that radioactivity was transferred from $^{14}CO_2$ to 4HBA in the presence of 4HBA by phenol-grown cells *(10)*. This indicates the reversible cleavage of 4HBA to CO_2 and to enzyme-bound phenol.

Sharak-Genther et al. desmonstrated by the use of fluororophenols that benzoate was formed via a *para*-carboxylation of phenol *(11)*. Interestingly, it was shown that 3-fluoro-4-hydroxybenzoate was converted to 3-fluorobenzoate. This reaction was expected in the case of 4HBA dehydroxylation to benzoate.

Zhang and Wiegel studied the degradation of 2,4-dichlorophenol under methanogenic conditions *(12)*. They proposed that a microorganism included in their consortium was first decarboxylating 4HBA to phenol before converting phenol to benzoate. 4HBA was not detected during phenol conversion to benzoate. These authors suggest that this metabolite could only exist under a enzyme-bound form.

Béchard et al. desmonstrated that the presence of proteose peptone was necessary to observe the conversion *(13)*. This suggests the existence of a cometabolism. Conversely, the presence or absence of H_2 had no influence on the reaction. Bisaillon et al. observed that proteose peptone could be replaced by tryptophan and lysine *(14)*.

In this article, we present results concerning the degradation of L-tyrosine to phenol in piggery wastes. The conversion of phenol to benzoate was investigated with an emphasis on the role of 4HBA.

METHODS

Slurry Sampling and Storage

The slurries were collected in piggeries of Gembloux countryside, Belgium. They were collected in pits under animals. The slurries were stored at 4°C until their use for experiments.

L-Tyrosine Degradation

Three samples of slurry from different piggeries were fed with 5 mM of L-tyrosine. The volume of the samples was 100 mL. Phenol, *para*-cresol, and benzoate were measured during the following days by gas chromatography (GC). The piggery slurries were placed in 100-mL flasks for experiments. The flasks were kept at room temperature during assays. In all experiments, a flask of slurry without chemical added was used as control. The slurries were gently mixed before sampling. The volume of samples was about 5 mL.

The role of 4-hydroxyphenylpyruvate (4HPPyrA) and 4-hydroxyphenylacetate (4HPAA) as intermediate compounds in phenol production pathway was investigated. These metabolites were added to the slurries at the approximative concentration of 5 mM. Phenol concentration was measured during the following days.

Phenol, Benzoate, and 4HBA Degradation

The slurry was complemented with phenol, benzoate, or 4HBA at the approximative concentration of 5 mM. The concentration of these three metabolites was measured during the following days by GC or high-performance liquid chromatography (HPLC). In a second series of experiments, the slurry was complemented simultaneously by 5 mM of phenol and various concentrations of benzoate (0, 5, 10, or 20 mM). The experiments were carried out in the same conditions as those described for L-tyrosine degradation.

Adapted Slurries

A slurry in which degradation activities are already expressed is called an adapted slurry.

Gas Chromatography

Five milliliters of slurry were extracted for 2 h with 2 mL of diethyl-ether in the presence of 400 mg of NaCl and 400 mL of a 50% solution of sulfuric acid. The diethyl-ether contained 500 mg/L of dimethylmalonate as internal standard. The ether extract was dehydrated by addition of anhydrous sodium sulfate. It was then ready for GC analysis.

A Hewlett Packard 5890 series II chromatograph was used with an FFAP column and an FI detector. The column length was 25 m. Its internal diameter was 0.32 mm. The carrier gas was nitrogen. The oven temperature was controlled to obtain the best separation of peaks. It was maintained at 34°C for 30 s, increased to 220°C (15°C/min from 34–100°C; 10°C/min from 100–160°C; 15°C/min from 160–220°C), and finally maintained at 220°C for 10 min.

Five microliters of ether extract were injected with the Hewlett Packard 7673 automatic sampler. Phenol and benzoate were measured by this method. 4HBA was not detected. Accuracy and linearity were confirmed in the concentration ranges of the experiments.

High-Performance Liquid Chromatography

One milliliter of slurry was diluted by 5 mL of distilled water. This diluted sample was centrifuged at 27,000g rpm for 10 min to eliminate the solid particles. Three milliliters of supernatant were diluted with 2 mL of acetonitrile and 1 mL of acetic acid. This solution was then filtered with a 0.45-µ filter. We used a Hewlett Packard series 1050 chromatograph equipped with a 250/8/4 Nucleosil $5C_{18}$ column. Five microliters of solution were injected with a Bio-Rad AS-48 automatic sampler. Metabolites (phenol, benzoate, and 4HBA) were eluted (flow = 4.00 mL/min) with a mixture of water, acetonitrile, and acetic acid (3:2:1). They were measured with a diode array detector near their maximum absorption wavelength (4HBA: 256 nm, phenol and benzoate: 270 nm) with a bandwidth of 4 nm. The nature of the metabolites was confined by their absoption spectrum.

Accuracy and linearity were confirmed in the concentration ranges of the experiments.

RESULTS

Degradation of L-Tyrosine to Phenol, *Para*-cresol, and Benzoate

A transient accumulation of phenol, *para*-cresol, and benzoate was observed in the three samples of piggery wastes (A, B, and C) in which L-tyrosine had been added. This is described in Fig. 1.

The accumulation kinetics of phenol were similar in the three slurries, except that in case C, a steady phase was observed before the degradation of phenol. The average accumulation rates were 0.45, 0.50 and 0.57 mmol/L/d in the cases A, B, and C, respectively. The phenol average degradation rates were 0.28, 0.32 and 0.37 mmol/L/d, respectively. Phenol was produced in each sample in larger amounts than *para*-cresol. This is described in Table 1.

Conversion of L-Tyrosine to Phenol: the Role of 4HPPyrA and 4HPAA

The role of 4HPPyrA and HPAA as intermediate compounds in the conversion of L-tyrosine to phenol was investigated. Phenol was produced in the largest amount when L-tyrosine was used as precursor metabolite.

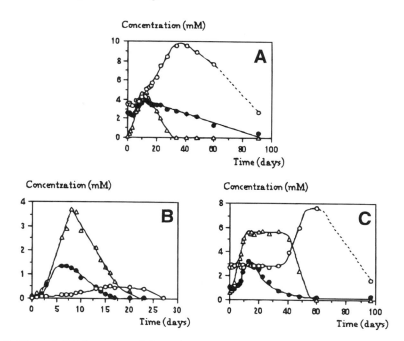

Fig. 1. Phenol (△), benzoate (○), and *para*-cresol (●) were measured after the addition of 5 mM of L-tyrosine. Three samples of piggery wastes were compared in this experiment (**A, B**, and **C**). The accumulation of these metabolites was not observed in the control slurries.

Table 1
Maximal Accumulation of Metabolites in Three Samples of Piggery Wastes

	Phenol	Para-cresol	Benzoate
A	4.6	1.6	6.2
B	3.7	1.26	0.43
C	5.0	2.3	4.8

*Five millimeters of L-Tyrosine were added in three samples of piggery wastes (A, B, and C). This table presents the maximal accumulations (mmol/L) of phenol, *para*-cresol, and benzoate that have been measured. These results represent the substraction of the concentration of the metabolites before their accumulation from their concentration before their degradation.

Figure 2A shows that 4HPPyrA and 4HPAA were nearly inactive as precursors of phenol in the first sample of slurry. Phenol was only significantly produced from L-tyrosine. The maximal concentration observed was 1.88 mmol/L after 17 d.

In another slurry (Fig. 2B), the phenol was produced from L-tyrosine, 4HPPyrA and 4HPAA. A low accumulation of phenol was even observed in the control slurry (the concentration raised from 0.12–0.36 mmol/L at the 20th day). The maximal concentration of phenol was observed when

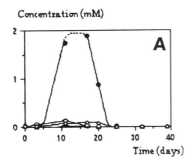 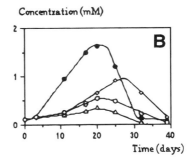

Fig. 2. This figure describes the accumulation of phenol in two samples (**A** and **B**) of piggery wastes. Three precursor metabolites were tested: L-tyrosine (●), 4HPPyrA (○), and 4HPAA (◇). A comparison is made with a control slurry (△).

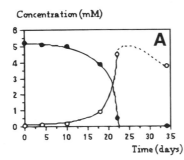 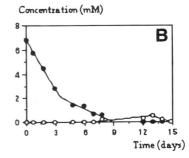

Fig. 3. Phenol was added in two samples (**A** and **B**) of piggery wastes at the approximative concentration of 5 mM. The evolution of phenol (●) and benzoate (○) concentrations was measured. 4HBA concentrations were measured in sample B only: no accumulations have been observed.

L-tyrosine was used as precursor: 1.63 mmol/L after 20 d. The degradation of 4HPPyrA and 4HPAA was observed with a delay compared to L-tyrosine. Maximal concentrations of phenol were 0.53 and 0.90 mmol/L after the addition of 4HPPyrA and 4HPAA, respectively.

Degradation of Phenol

The degradation of phenol was observed in several samples of piggery wastes. Figure 3 presents two different cases. In the case of an unadapted slurry (Fig. 3A), the degradation of phenol was slow in the first day and reached an average rate of 0.58 mM/d between days 18 and 22. In the same time, benzoate was produced at the average rate of 0.90 mM/d. It appears obvious that phenol is completely converted to benzoate. In the case of an adapted slurry (Fig. 3B), the degradation of phenol was achieved through two successive phases. During the first 3 d, the degradation average rate was about 1.36 mM/d. It decreased to 0.26 mM/d during the fol-

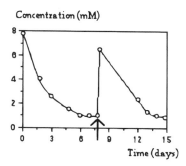

Fig. 4. The degradation of benzoate is observed in a sample of piggery waste. The arrow indicates a second feeding with approx 5 mM of benzoate.

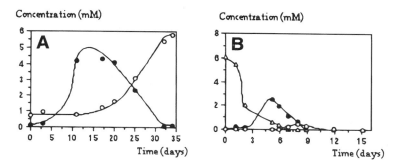

Fig. 5. This figure presents the fate of 4HBA (△) in two samples (**A** and **B**) of piggery wastes. The accumulation of phenol (●) and benzoate (○) was measured.

lowing days. The degradation was complete between the days 9 and 12. In contrast to case A, only weak amounts of benzoate were observed. The maximum was 0.51 mM after 13 d.

Degradation of Benzoate

The degradation of benzoate was also studied. The case presented in Fig. 4 was observed in the adapted slurry that had been used for Fig. 3B. The benzoate degradation began quickly (2.11 mM/d), before slowing and even ceasing between days 6 and 8. A second addition of benzoate—indicated by an arrow on Fig. 4—at day 8 restored the degradation activity.

Degradation of 4HBA

We observed the degradation of 5 mM of 4HBA in the adapted and unadapted slurries previously used for phenol and benzoate degradation experiments. The same slurry was used to obtain the results in Figs. 3B, and 5B. Figure 5A shows that 4HBA was completely converted to phenol and that phenol was consequently completely converted to benzoate. It was impossible to measure directly the disappearance of 4HBA with our

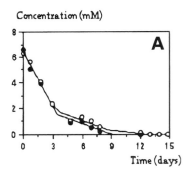 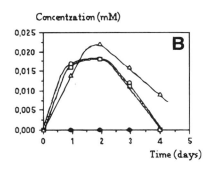

Fig. 6. A sample (**A**) of slurry was fed with both phenol and benzoate at the approximative concentration of 5 mM. Phenol (●) and benzoate (○) degradation were observed. In a second sample (**B**), the addition of the metabolites was repeated, but benzoate was added at various concentration. B shows the accumulation of benzoate in each flask (5 mM ○, 10 mM □, or 20 mM of benzoate △). A comparison is made with a control slurry in which only phenol was added (●).

GC method. The accumulation of phenol reached the average rate of 0.50 mM/d. Between days 20 and 32, benzoate accumulation and phenol disappearance rates were almost the same: 0.30 and 0.34 mM/d.

Case B (Fig. 5B) presents the measurement of 4HBA uptake by HPLC. After a short lag phase, 4HBA concentration decreased at the average rate of 3.46 mM/d. The later diminution was lower: 0.42 mM/d between days 2 and 9. The phenol began to accumulate in the medium after 2 d. The maximal concentration observed was 2.5 mM after 5 d. After this moment, a diminution of 0.60 mM of phenol/d was observed. Phenol had completely disappeared after 10 d. Benzoate accumulated in weak amounts: the maximum observed was 0.46 mM after 8 d.

Simultaneous Degradation of Phenol and Benzoate

When phenol and benzoate were added simultaneously to the slurry, they were degraded at the same rate. This is shown in Fig. 6A. During the first 3 d, 1.46 mM of phenol and 1.33 mM of benzoate disappeared/d. During the later week, the disappearance rate was 0.43 mM/d for phenol and 0.37 mM/d for benzoate. No accumulation of 4HBA was observed in the same time.

In a similar experiment, phenol was added alone (control) or with various concentrations of benzoate (5, 10, or 20 mM). 4HBA could not be detected in the control flask. In the other flasks, 4HBA accumulated in weak concentrations. The maximum was observed in the flask where 20 mM of benzoate were added: 22 µmol/L after 2 d (Fig. 6B).

DISCUSSION

As demonstrated in Fig. 1, L-tyrosine is the precursor metabolite of phenol, benzoate, and *para*-cresol. These degradation products were observed in the three samples studied. This was confirmed in later studies concerning more than 20 samples of piggery wastes from different piggeries (results not published). These reactions can then be considered as ubiquitous.

The degradation pathways of L-tyrosine to phenol were investigated. As indicated in Fig. 2A, a significant accumulation of phenol was observed after the addition of L-tyrosine, although it was not the case after the addition of 4HPPyrA or 4HPAA. This indicates that in this slurry, L-tyrosine is produced through a single reaction, probably by the mean of a tyrosine phenol lyase, as has been described for *Clostridium tetanomorphum* (2).

However, an alternative pathway including 4HPPyrA and 4HPAA possibly exists. It could explain the traces of phenol observed in the case of Fig. 2A and its significant amounts in the case of Fig. 2B after the addition of those two possible intermediate metabolites.

The relationships between phenol and benzoate were investigated. The conversion of phenol to benzoate is described in Fig. 3. We have strong arguments to conclude that 4HBA is not an intermediate compound in the conversion of phenol to benzoate in the piggery waste samples we studied.

Phenol is converted to benzoate without any accumulation of 4HBA, and 4HBA was decarboxylated to phenol, but never directly dehydroxylated to benzoate. If we imagine a two-stages pathway in which 4HBA is the intermediate compound, the fact that 4HBA is not detected must be owing to its quick transformation to benzoate. It was then interesting to observe the effect of an addition of 4HBA in a phenol-adapted slurry. Under such conditions, it was expected that 4HBA would have been directly converted to benzoate. On the contrary, we observed an accumulation of phenol as shown in Fig. 5B. Furthermore, benzoate was only detected after 6 d. It seems impossible then to consider 4HBA as an intermediate metabolite.

In later experiments, slurry was fed simultaneously with 5 mM of phenol and 5 mM of benzoate. It was expected that the presence of benzoate could modify the equilibrium of the reaction and allow the accumulation of 4HBA. As shown in Fig. 6A, phenol and benzoate were curiously degraded at the same rate. This indicates a strict relationship in their uptake and metabolism. It is probable that these metabolites are degraded by the same microorganism.

The effect of an increase in phenol concentration was the accumulation of 4HBA in very low concentration (Fig. 6B). In this experiment, 4HBA was observed with 5 mM of benzoate, although it was not the case previously. We think that it is probably owing to a low reversibility of the reaction of decarboxylation of 4HBA to phenol. Indeed, phenol uptake decreased when 4HBA disappeared from the medium (results not shown). This indicates that 4HBA is transformed to phenol rather then to benzoate.

Zhang and Wiegel *(12)* proposed that 4HBA decarboxylation to phenol and phenol conversion to benzoate was realized by the same microorganism. This is another strong argument for the proposal that 4HBA is not included in the phenol degradation pathway. In such a case, the dehydroxylation of 4HBA is the only way to degrade it.

However, the role of such a conversion is not clear, since 4HBA and phenol are not used as carbon sources. Figures 3A and the 5A effectively show that in unadapted slurries, 4HBA and phenol were completely converted, 5 mM of 4HBA leading to the accumulation of 5 mM of benzoate. Although the phenol may be seen as an electron acceptor, it is not true for 4HBA, as shown by the two reactions below:

$$HO-C_6H_4-COOH \longrightarrow C_6H_5-OH + CO_2 \qquad (1)$$

$$C_6H_5-OH + CO_2 + 2H^+ + 2e^- \longrightarrow C_6H_5-COOH + H_2O \qquad (2)$$

If the benefit from 4HBA conversion was only to obtain phenol as an electron acceptor, we could expect an immediate utilization of phenol. This is not the case. We think that 4HBA decarboxylation to phenol causes its own benefits to the cell.

The fact that these metabolites are not used as carbon sources is also an argument to assume that they are cosubstrates in a cometabolism. This was already previously proposed by Béchard et al. *(13)* and Bisaillon et al. *(14)*.

In conclusion, L-tyrosine is degraded to phenol in one stage, probably by a tyrosine phenol lyase activity. An alternative pathway possibly exists. It includes 4HPPyrA and 4HPAA.

The phenol is converted to benzoate in one stage. 4HBA is not an intermediate compound in this reaction. This metabolite is converted to phenol, and this is probably the final step to phenol and benzoate of an unidentified compound. Since these reactions were observed in all the slurries we have studied, they probably reflect a general phenomenon.

Our further investigations will concern the isolation of the microorganisms implied in the degradation pathways, and especially in the conversion of 4HBA to phenol. We will also investigate the possible conversion of 4HPAA to phenol.

ACKNOWLEDGMENT

This work was supported by IRSIA-IWONL.

REFERENCES

1. Schaefer, J. (1977), *Agriculture and Environ.* **3**, 121–127.
2. Brot, N., Smit, Z., and Weissbach, H. (1965), *Arch. Biochem. Biophys.* **112**, 1–6.
3. Elsden, S. R., Hilton, M. G., and Waller, J. M. (1976), *Arch. Microbiol.* **107**, 283–288.
4. Spoeltra S. F. (1978), *Appl. Environ. Microbiol.* **36, no. 5**, 631–638.
5. Spoeltra S. F. (1977), *J. Sci. Fd. Agric.* **28**, 415–423.
6. Godefroid, J., Antoine, P., Anselme, P., Van Rolleghem, P., De Poorter, M.-P., Van de Woestyne, M., Verstraete, W., and Thonart, P. (1992), *Med. Fac. Landbouww Univ. Gent.* **57/4a**, 1717–1720.
7. Balba, M. T., Clarke, N. A., and Evans W. C. (1979), *Biochem. Soc. Trans.* **7**, 1115–1116.

8. Neufeld, R. D., Mack J. D., and Strakey J. P. (1980), *J. WPCF* **52, no. 9,** 2367–2377.
9. Grbic-Galic, D. and Vogel, T. M. (1987), *Appl. Environ. Microbiol.* **53, no. 2,** 254–260.
10. Tschech, A. and Fuchs, G. (1987), *Arch. Microbiol.* **148,** 213–217.
11. Sharak-Genther, B. R., Townsend, G. T., and Chapman, P. J. (1989), *Biochem. Biophys. Res. Commun.* **162, no. 3,** 945–951.
12. Zhang, X. and Wiegel, J. (1990), *Appl. Environ. Microbiol.* **56, no. 4,** 1119–1127.
13. Béchard, G., Bisaillon, J.-G., Beaudet, R., and Sylvestre, M. (1990), *Can. J. Microbiol.* **36,** 573–578.
14. Bisaillon, J.-G., Lépine, F., and Beaudet, R. (1991), *Can. J. Microbiol.* **37,** 573–576.

The Potential for Intrinsic Bioremediation of BTEX Hydrocarbons in Soil/Ground Water Contaminated with Gas Condensate

ABHIJEET P. BOROLE,[1] KERRY L. SUBLETTE,*,[1]
KEVIN T. RATERMAN,[2] MINOO JAVANMARDIAN,[3]
AND J. BERTON FISHER[4]

[1]Center for Environmental Research and Technology, University of Tulsa, 600 S. College Ave., Tulsa, OK 74104; [2]Amoco Tulsa Technology Center, PO Box 3385, Tulsa, OK 74102; [3]Amoco Corporation, Amoco Research Center, 150 West Warrenville Rd., Naperville, IL 60563; and [4]Gardere and Wynne, 401 S. Boston, Tulsa OK 74103

ABSTRACT

Gas condensate liquids contaminate soil and ground water at two gas production sites in the Denver Basin, CO. A detailed field study was carried out at these sites to determine the applicability of intrinsic bioremediation as a remediation option. Ground water monitoring at the field sites and analysis of soil cores suggested that intrinsic bioremediation is occurring at the sites by multiple pathways, including aerobic oxidation, sulfate reduction, and possibly reduction Fe(III) reduction.

Laboratory investigations were conducted to verify that the water-soluble components of the gas condensate (benzene, toluene, ethylbenzene, and xylene [BTEX]) are intrinsically biodegradable under anoxic conditions in the presence of alternate electron acceptors and soil from the field site. Slurry-phase experiments were conducted in which soil obtained from the field site was mixed with an aqueous phase containing nutrients and electron acceptors (nitrate, Fe[III], sulfate and carbon dioxide) in serum bottles. The aqueous phase also contained soluble components of gas condensate, at two different hydrocarbon concentrations, obtained from the field site. The soil was either pristine (native) soil or soil obtained from a condensate-contaminated region. The aqueous

---

*Author to whom all correspondence and reprint requests should be addressed.

phase was sampled for electron acceptors, hydrocarbons, and possible products of hydrocarbon degradation.

Toluene and xylenes were biodegraded with nitrate or sulfate as the electron acceptor. No degradation of benzene was observed under anoxic conditions.

Index Entries: Intrinsic bioremediation; gas condensate; anoxic; BTEX; hydrocarbon; sulfate reduction; denitrification.

INTRODUCTION

Soil and ground water contamination by petroleum hydrocarbons is an ongoing environmental problem. The BTEX compounds are of special interest because they are relatively water-soluble, and two of these components have been associated with known health risks. Benzene is a confirmed carcinogen, and toluene is a depressant of the central nervous system (1). Recently, researchers have convincingly demonstrated the natural attenuation of hydrocarbon plumes in ground water under anoxic or microaerophilic, as well as aerobic conditions. Nitrate, iron(III) oxides, and sulfate have all been identified as potential electron acceptors for hydrocarbon degradation in the absence of oxygen (2–6).

Gas condensate liquids contaminate the soil and ground water at certain gas production sites operated by Amoco in the Denver Basin. Two of these sites have been closely monitored since July 1993 to determine if intrinsic aerobic or anoxic bioremediation of hydrocarbons occurs at a sufficient rate and to an adequate end point to support a no-intervention decision. The limited migration of the highly soluble BTEX components and the depletion of several potential electron acceptors in the contaminated zone have suggested that intrinsic bioremediation is occurring at the contaminated sites by multiple pathways, including aerobic oxidation, sulfate reduction, and possibly Fe(III) reduction (7,8).

Laboratory investigations have been conducted to accompany field observations in order to verify hydrocarbon degradation by field organisms and identify the primary anoxic biodegradation mechanisms. Two types of experiments were conducted, saturated soil studies with excess hydrocarbon (free phase) and limiting amounts of electron acceptors, and slurry experiments that were hydrocarbon-limiting. The details of the saturated soil experiment are given elsewhere (9). Sulfate reduction was found to be the major mechanism of hydrocarbon degradation under anoxic conditions in the presence of a free hydrocarbon phase. Nitrate and Fe(III) reduction was also observed in the anoxic saturated soil experiments; however, these could not be linked to hydrocarbon degradation. In the presence of an initial limited amount of oxygen, nitrate and Fe(III) reduction was observed and could be linked to the presence of hydrocarbon in the microcosms. It was proposed that utilization of these alternate electron acceptors was stimulated by limited oxygen by generating partially oxygenated hydrocarbons, which were better substrates for anoxic degradation mechanisms.

The slurry-phase, hydrocarbon-limited experiments were conducted to determine which individual water-soluble components of the gas condensate were intrinsically biodegradable under anoxic conditions in the presence of alternate electron acceptors and soil from the field site. Four different electron acceptor conditions were investigated: nitrate, sulfate, Fe(III), and carbon dioxide (methanogenic). The conditions in these experiments simulated the conditions found in the ground water plume down gradient of the residual-free hydrocarbon phase.

MATERIALS AND METHODS

Composition of Slurry-Phase Microcosms

Slurry-phase experiments were carried out in 160-mL serum bottles with 40 g of soil and 80 g of aqueous phase. Microcosms were prepared with two types of soil, one obtained from a condensate-contaminated region and the other obtained from an uncontaminated region (native soil) from the field site. A mineralogical analysis of the native and contaminated soil from these sites has been given previously (9). The native soil was predried for 24 h at 70°C to facilitate handling and sieved through a standard 10-mesh sieve before use. The contaminated soil was collected, transported, and stored anaerobically, and was processed only in an oxygen-free atmosphere in the anaerobic chamber. The contaminated soil contained approx 1000 mg/kg (dry basis) total petroleum hydrocarbons (TPH). These hydrocarbons were shown by gas chromatography to be similar to the gas condensate produced at the site except for reduced concentrations of lighter hydrocarbons. Each microcosm containing contaminated soil, therefore, contained about 40 mg TPH, which could be expected to be significantly depleted in water-soluble components.

The liquid phase consisted of a base medium prepared in deionized water with 1 mL/100 mL trace metals solution + 2.0 mL/100 mL mineral salts solution and 0.35 g/100 mL sodium bicarbonate. The composition of the trace metals solution and mineral stock solution is given by Tanner et al. (10). A low concentration (0.0001%) of resazurin was added to the medium to indicate oxidation-reduction potentials above -0.042 V (11). Resazurin changes color from blue (aerobic) to pink (partially anoxic) at a redox potential of about 0.78 V. This change of color is irreversible (12). The indicator also changes color from pink to colorless (anaerobic) at about -0.042 V. This change is reversible and is generally used to indicate an anaerobic condition.

Nitrate, Fe(III), sulfate, and carbon dioxide were added to microcosms as electron acceptors. The carbon dioxide was present in all microcosms as a part of the buffering system and was, therefore, a potential electron acceptor in all microcosms. The amount of electron acceptor added in each case was in excess of that required for complete mineralization of the hydrocarbon to carbon dioxide. Nitrate and sulfate were added as their sodium salts to the base medium to obtain nutrient medium for the nitrate-amended and sulfate-amended experiments, respectively. Fe(III) was added in the form of an amorphous Fe(III) oxyhydroxide gel. The concentrations of the electron acceptors were as follows: 19.0 mM nitrate in nitrate-amended, 80.0 mM Fe(III) in Fe(III)-amended, and 15.6 mM sulfate in sulfate-amended microcosms. All microcosms contained 20.0% carbon dioxide in the headspace.

The complete microcosms contained soluble components of the gas condensate and were prepared as follows. To simulate the actual contaminant found in the contaminated region, the light hydrocarbons were partially volatilized by heating the gas condensate at 105°C for 1 h to remove about 20 wt% of the hydrocarbon. The heavier retained fraction was referred to as the heavy condensate. The base media amended with electron acceptors (except for the Fe[III]-amended set, where the Fe[III] gel was added directly to microcosms) were contacted with the heavy condensate for a period of 24 h. The organic phase was then separated, giving an aqueous phase saturated with the soluble components of the heavy condensate (approx 32.0 mg/kg total BTEX). The individual components were at the following concentrations in the saturated medium: 6.5 mg/kg benzene, 18.0 mg/kg toluene, 0.4 mg/kg ethylbenzene, 6.0 mg/kg *m,p*-xylene, and 1.0 mg/kg *o*-xylene. The microcosms were prepared with two different total hydrocarbon concentrations, one with 100% hydrocarbon saturation (prepared as given above) and the other with 60% saturation. In the first set of microcosms (100% dissolved hydrocarbon saturation), 80 g of the heavy condensate-saturated solution was used, and in a second set (60% dissolved hydrocarbon saturation), 48 g of the condensate-contacted medium was mixed with 32 g of fresh base medium + electron acceptors resulting in 60% hydrocarbon saturation (approx 19.2 mg/kg BTEX). The microcosms containing 100% dissolved hydrocarbon saturation are referred as "high-BTEX" microcosms, and those with 60% dissolved hydrocarbon saturation are referred as "low-BTEX" microcosms.

Cysteine sulfide (reducing agent) was added to sulfate-amended microcosms to lower the redox potential in the microcosms (1.6 mM sulfide, 0.33 mM cysteine). No cysteine sulfide was added to any other microcosms. All operations were carried out inside an anaerobic chamber. The serum bottles were then stoppered with the Teflon-lined, black, butyl-rubber composite stoppers, removed from the chamber, and immediately crimped with aluminum crimp caps. The headspace in the microcosms was 20% CO_2 + bal N_2.

Sterile controls were prepared in addition to biotic microcosms in order to account for any abiotic losses. Sterile controls were prepared

Table 1
Details of Design of the Slurry-Phase Experiment

soil types	2 (native-soil, contaminated-soil)
microcosm type	2 (biotic, sterile)
carbon source	soluble hydrocarbons, 2 concentrations, 100% saturation of aqueous phase with condensate (32 mg/kg BTEX), and 60% saturation, (19.2 mg/kg BTEX)
terminal electron acceptors	4 (nitrate, Fe(III), sulfate, carbon dioxide)
temperature	1 (30°C)
replicates	2

exactly the same way as the biotic microcosms, except that they were sterilized at 121°C for 60 min. After sterilization, the microcosms were cooled to room temperature and then transferred into an incubator. All microcosms were prepared in duplicate and incubated for 160–200 d at 30°C.

Analysis of samples taken after 15 d of incubation from the nitrate-amended microcosms showed about 90% depletion of toluene in biotic, contaminated-soil microcosms. These microcosms were replenished at this time by replacing about 80% of the liquid phase from the microcosms with fresh medium. Data are reported for these microcosms from this time forward (*see below*). Analysis of all microcosms was done by periodic sampling of the aqueous phase using an anaerobic, sterile syringe and needle. The sampling methodology is given by Borole *(13)*. Details of the design of slurry-phase experiments are given in Table 1.

Analytical

Monoaromatic hydrocarbons (BTEX) were analyzed on an On-Column HP 5890 gas chromatograph with a flame ionization detector. A 15-m long DB-1 megabore (0.32 mm id) capillary column with 1-µ film thickness was used to carry out the separation. Column conditions were as follows: prehold at 35°C for 1 min; oven temperature, 35–170°C; program rate, 7°C/min, "oven track" ON with a temperature difference of 3°C between the oven and injector; detector temperature, 340°C. Calibration standards for the BTEX compounds were made by dissolving pure compounds in a 1:1 (vol:vol) methanol:water. The methods used for analysis of anions and straight-chain organic acids are given elsewhere *(9)*.

RESULTS AND DISCUSSION

Nitrate Amended

Nitrate utilization and degradation of hydrocarbons (primarily toluene and o-xylene) were observed in the biotic microcosms with both contaminated and native soil (Figs. 1 and 2). Although there was little difference in the amount of nitrate reduced or in the removal of toluene or o-xylene in microcosms with contaminated or native soil at the lower BTEX concentration (Fig. 1), significant differences were observed at the higher BTEX concentration (Fig. 2). At the higher BTEX concentration, toluene depletion was incomplete and nitrate utilization lower in the contaminated-soil microcosms, suggesting inhibition of toluene utilization possibly owing to accumulation of an inhibitory product. Alternatively, depletion of a nutrient in the contaminated soil possibly made utilization of the greater concentration of hydrocarbon impossible, or there may have been a combination of nutrient limitation, product inhibition, and inhibition resulting from elevated hydrocarbon concentrations.

o-Xylene was also observed to be utilized in contaminated-soil microcosms at lower BTEX concentrations; however, at the higher BTEX concentration, o-xylene utilization was not complete. Complete o-xylene degradation was observed in both the low- and high-BTEX biotic, native-soil microcosms. Some biodegradation of m,p-xylene was observed in the microcosms with native soil at higher BTEX concentration; however, no m,p-xylene depletion occurred in high-BTEX, contaminated-soil microcosms. No degradation of benzene or ethylbenzene was observed in any nitrate-amended microcosms.

The results given in Figs. 1 and 2 for microcosms with contaminated soil are those in which the aqueous medium was replenished with 80% fresh medium. In the original high-BTEX microcosms, 90% depletion of toluene and 60% depletion of o-xylene was observed in first 15 d; however, in the replenished microcosms, only 50% toluene depletion occurred, and no o-xylene depletion was observed. Depletion of toluene and o-xylene was biologically mediated, since the sterile microcosms did not show a similar decrease. The microcosms prepared with native soil were not replenished, and the results presented in Figs. 1 and 2 are for a first-time exposure to BTEX. Toluene depletion occurred at both the high- and low-BTEX concentrations with no apparent lag at the lower BTEX concentration, but with a lag of about 10 d at the higher BTEX concentration. These observations are again consistent with a nutrient limitation or inhibitory substance associated with the contaminated soil.

The contaminated soil contained black precipitates of iron sulfide produced by sulfate reduction in the field. Sulfate production was observed in nitrate-amended, contaminated-soil microcosms. This production was most likely biologically mediated under denitrifying condi-

BTEX Hydrocarbons

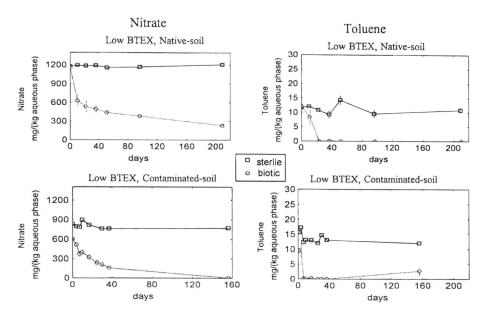

Fig. 1. Nitrate reduction and toluene depletion in low-BTEX, nitrate-amended slurry-phase microcosms. (Vertical bars indicate range of duplicate samples.)

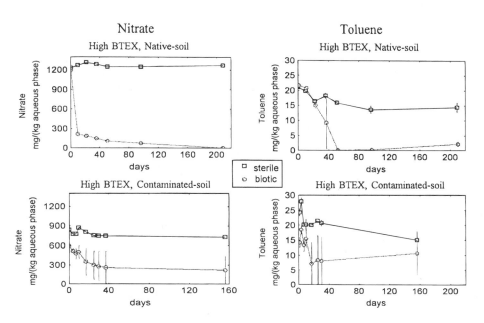

Fig. 2. Nitrate reduction and toluene depletion in high-BTEX, nitrate-amended, slurry-phase microcosms. (Vertical bars indicate range of duplicate samples.)

Table 2
Comparison of Stoichiometric and Observed Utilization of Nitrate as Electron Acceptor for Toluene and o-Xylene Degradation

Microcosm type	NO₃⁻ used/ (toluene + o-xylene) (wt/wt)	
	predicted	observed
Low BTEX, contaminated-soil microcosms	5.04	32.5
High BTEX, contaminated-soil microcosms	5.04	6.17
Low BTEX, native-soil microcosms	5.04	73.6
High BTEX, native-soil microcosms	5.04	48.4

tions, since only biotic microcosms showed sulfate production. The ratio of nitrate consumed to total initial toluene and o-xylene is given in Table 2 for each nitrate-amended microcosms. Nitrate consumed was corrected for nitrate utilized for sulfide oxidation based on sulfate produced.

The stoichiometry for complete degradation of toluene with nitrate as the electron acceptor is given by Eq. (1) (4).

$$C_7H_8 + 7.2\ H^+ + 7.2\ NO_3^- \Rightarrow 7\ CO_2 + 7.6\ H_2O + 3.6\ N_2 \qquad (1)$$

The higher utilization of nitrate observed in these experiments may be owing to (1) incomplete reduction of nitrate (14) or (2) increased electron acceptor demand resulting from indigenous organic carbon present in the soil (1000 mg total organic carbon/kg dry soil). Accumulation of nitrite was observed in both native-soil and contaminated-soil biotic microcosms; however, all the nitrite produced was depleted before the end of the experiment. The amount of nitrite produced in biotic, native-soil microcosms was different from that produced in the corresponding contaminated-soil microcosms. Nitrite concentration reached a peak of about 120 mg/(kg aqueous phase) in low-BTEX and 360 mg/(kg aqueous phase) in high-BTEX biotic, native-soil microcosms. In the contaminated-soil microcosms, the corresponding concentrations were 80 mg/(kg aqueous phase) and 70 mg/(kg aqueous phase), respectively. The peak in nitrite concentration occurred in the first 10 d with both soils; however, no BTEX consumption occurred during this period, which suggests that an alternate carbon source (indigenous organic carbon) may have served as the carbon donor during this time. Moreover, since much larger amounts of nitrite were used and more nitrite was produced in native-soil microcosms, the indigenous organic carbon present in the native soil is inferred to be more readily degradable under denitrifying conditions than that present in contaminated soil (principally hydrocarbon).

Sulfate Amended

Sulfate utilization was observed in both native-soil and contaminated-soil biotic microcosms (Figs. 3 and 4); however, longer lag times were observed with native-soil (100 d) compared to contaminated-soil microcosms (15–37 d). The rate of sulfate depletion was faster and more sulfate was consumed in contaminated-soil microcosms compared to the native-soil microcosms. Sulfate reduction activity was observed in the plume region of the Denver Basin contaminated site (7), therefore, the contaminated soil was expected to be enriched in sulfate-reducing bacteria (SRB). These results suggest that this was the case.

In the contaminated-soil microcosms, toluene depletion accompanied sulfate utilization at the lower and higher BTEX concentrations. However, at the higher BTEX concentration, the two replicates showed appreciable differences in toluene concentrations during the experiment. At both BTEX concentrations, toluene was completely depleted within 60 d (low BTEX) and 100 d (high BTEX). Utilization of xylenes was also observed in microcosms with contaminated soil (Table 3). In native-soil microcosms, utilization of toluene or xylenes was less conclusive, and less sulfate was consumed compared to corresponding microcosms with contaminated soil. Utilization of ethylbenzene was inconclusive at lower BTEX concentrations, but ethylbenzene utilization was documented at the higher BTEX concentration in both native-soil and contaminated-soil microcosms (Table 3). No utilization of benzene was observed in any sulfate-amended microcosms.

Toluene mineralization to CO_2 with sulfate as the electron acceptor (with no cell mass production) is given by Eq. (2) (2)

$$C_7H_8 + 4.5\ SO_4^- + 3\ H_2O \Rightarrow 7\ HCO_3^- + 2.25\ HS^- + 2.25\ H_2S + 0.25\ H^+ \quad (2)$$

The amount of sulfate utilized in these microcosms was far in excess of that required to fully oxidize all of the BTEX in the microcosms. The indigenous organic carbon present in native soil and contaminated soil was capable of producing a sulfate demand, and therefore, part of the sulfate consumed may have been used for oxidation of the indigenous organic carbon. Large amounts of acetate (200 mg/kg aqueous phase) were observed in both low-BTEX and high-BTEX sulfate-amended microcosms in both native-soil and contaminated-soil microcosms during the first 15 d. No detectable sulfate reduction or BTEX depletion occurred during this initial time period, which implies that BTEX compounds did not serve as the electron donor for the production of acetate. No other straight- or branched-chain fatty acids were detected in the aqueous phase of any microcosms.

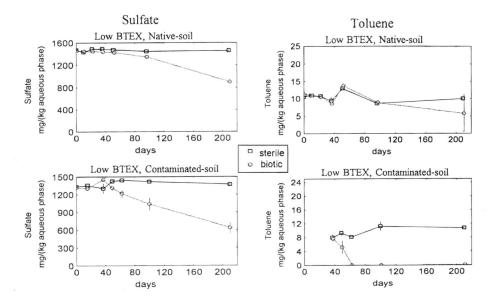

Fig. 3. Sulfate reduction and toluene depletion in low-BTEX, sulfate-amended, slurry-phase microcosms. (Vertical bars indicate range of duplicate samples.)

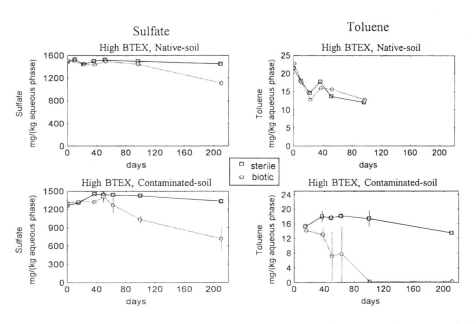

Fig. 4. Sulfate reduction and toluene depletion in high-BTEX, sulfate-amended, slurry-phase microcosms. (Vertical bars indicate range of duplicate samples.)

Table 3
Qualitative Analysis of Biotic and Abiotic Losses
of BTEX in Slurry-Phase Microcosms

Anoxic	Soil type	BTEX amount	Benzene	Toluene	Ethyl-benzene	m,p-Xylene	o-Xylene
			Biodegradation - Abiotic loss - lag period (days)*				
NO_3^-	Native	low	N-N	Y-N-0	?-N	?-Y	Y-Y-0
	Native	high	N-Y	Y-Y-4	?-N	Y-Y-10	Y-Y-0
	Contaminated	low	N-N	Y-Y-7	?-N	N-Y-7	Y-N-4
	Contaminated	high	N-N	Y-Y-10	N-Y	N-Y	?-Y
Fe(III)	Native	low	N-N	Y-N-0	N-N	N-Y	?-Y
	Native	high	N-Y	Y-Y-0	N-N	N-Y	N-Y
	Contaminated	low	N-N	N	N-N	?-Y	Y-?-50
	Contaminated	high	N-?	Y-?-100	Y-?-62	Y-Y-15	Y-?-50
SO_4^{--}	Native	low	N-N	Y-N-100	?-N	Y-Y	Y-Y-10
	Native	high	N-Y	N-N-100	Y-N	Y-Y	Y-Y-0
	Contaminated	low	N-?	Y-?-37	N-N	Y-?-37	Y-?-15
	Contaminated	high	N-?	Y-?-15	Y-Y-62	Y-Y-0	Y-Y-15
Non-amended	Native	low	N-N	N-N	N-Y	N-Y	N-Y
	Native	high	N-Y	N-Y	Y-Y-22	Y-Y	?-Y
	Contaminated	low	N-?	N-N	N-?	N-Y	N-N
	Contaminated	high	N-Y	Y-Y-100	N-Y	N-Y	N-Y

*The 2 letters N and Y stand for no and yes, respectively, and indicate whether biodegradation or abiotic loss took place in the microcosms. A question mark implies unreliable or nonconclusive data. A third numerical character is used in those cases where biodegradation was observed to indicate the lag time (days) observed for biodegradation of the BTEX component.

Fe(III) Amended and Nonamended

Iron analysis could not be done in this experiment because of the nonsacrificial nature of the microcosms, and no analysis was conducted on the liquid phase since a negligible amount of iron was present in the aqueous phase. No benzene depletion occurred in any of the contaminated-soil or native-soil microcosms. Toluene depletion was observed in native-soil microcosms and not in contaminated-soil microcosms amended with Fe(III). Degradation of xylenes was observed in contaminated-soil microcosms, but not in native-soil microcosms amended with Fe(III) (data not shown).

Benzene or toluene biodegradation did not occur in any of the nonamended microcosms. Only some abiotic loss of xylenes was observed in the nonamended contaminated-soil and native-soil microcosms. No attempt was made to analyze for methane in the nonamended microcosms, since sulfate (present as a soil constituent) was also present until the last time event (210 d), and degradation of BTEX components could not be documented.

CONCLUSIONS

Biodegradation of toluene and xylenes has been verified under sulfate-reducing and denitrifying conditions in microcosms with either native soil or soil previously exposed to gas condensate (contaminated soil) from a Denver Basin gas-producing field site as a source of microorganisms. These experiments confirmed sulfate reduction and denitrification as viable anoxic mechanisms for the degradation of compounds in ground water contaminated with gas condensate. Since field observations indicate nitrate is limited at the site relative to sulfate, sulfate can be expected to be the dominant of the two mechanisms in the field. Depletion of BTEX hydrocarbons could not be linked to Fe(III) reduction or methanogenesis in these experiments.

REFERENCES

1. Sittig, M. (1985), *Handbook of Toxic and Hazardous Chemicals and Carcinogens*, 2nd ed., Noyes Publications, Park Ridge, NJ, pp. 868–870.
2. Norris, R. D., Hinchee, R. E., Brown, R., McCarty, P. L., Semprini, L., Wilson, J. T., Kampbell, D. H., Reinhard, M., Bouwer, E. J., Borden, R. C., Vogel, T. M., and Ward, C. H. (1993), "In-Situ Bioremediation of Ground Water and Geological Materials: A Review of Technologies," 68-C8-0058, Robert S. Kerr Environmental Research Laboratory, Ada, OK.
3. Lovley, D. R. and Phillips, E. J. P. (1986), *Appl. Environ. Microbiol.* **51**, 683–689.
4. Hutchins, S. R., Sewell, G. W., Kovacs, D. A., and Smith, G. A. (1991), *Environ. Sci. Technol.* **25**, 68–76.
5. Edwards, E. A., Wills, L. E., Reinhard, M., and Grbic-Galic, D. (1992), *Appl. Environ. Microbiol.* **58**, 794–800.
6. Haag, F., Reinhard, M., and McCarty, P. L. (1991), *Environ. Toxicol. Chem.* **10**, 1379–1389.
7. Barker, G., Fisher, J. B., Raterman, K. T., Corgan, J., Trent, G., Brown, D., Kemp, N., McInerney, M. J., Borole, A. P., Kolhatkar, R. V., and Sublette, K. L. (1995), Proceedings of the Fifth International Conference on Microbial Enhanced Oil Recovery and Related Biotechnology for Solving Environmental Problems, Plano, TX, September.
8. Barker, G. W., Raterman, K. T., Fisher, J. B., Corgan, J., Trent, G., Brown, D. R., Kemp, N., and Sublette, K. L. (1996), *Appl. Biochem. Biotechnol.* **57/58**, 791–801.
9. Borole, A. P., Fisher, J. B., Raterman, K. T., Kemp, N., Sublette, K. L., and McInnerney, M. J. (1996), *Appl. Biochem. Biotechnol.* **57/58**, 817–826.
10. Tanner, R. S., McInerney, M. J., and Nagle, D. P., Jr. (1989), *J. Bacteriol.* **171**, 6534–6538.
11. Balch, W. E. and Wolfe, R. S. (1976), *Appl. Environ. Microbiol.* **32**, 781–791.
12. Norris, J. R. and Robbins, D. W. (1969), in *Methods in Microbiology*, vol. 3B, Academic, London, pp. 117–132.
13. Borole, A. P. (1996), Ph.D. dissertation, The University of Tulsa, Tulsa, OK.
14. Zehnder, A. J. B. (1988), in *Biology of Anaerobic Microorganisms*, Wiley-Interscience, New York, p. 551.

Adsorption of Heavy Metal Ions by Immobilized Phytic Acid

George T. Tsao, Yizhou Zheng, Jean Lu, and Cheng S. Gong*

Laboratory of Renewable Resources Engineering, Purdue University, West Lafayette, IN 47907

ABSTRACT

Phytic acid (myoinositol hexaphosphate) or its calcium salt, phytate, is an important plant constituent. It accounts for up to 85% of total phosphorus in cereals and legumes. Phytic acid has 12 replaceable protons in the phytic molecule, rendering it the ability to complex with multivalent cations and positively charged proteins. Poly 4-vinyl pyridine (PVP) and other strong-based resins have the ability to adsorb phytic acid. PVP has the highest adsorption capacity of 0.51 phytic acid/resins. The PVP resin was used as the support material for the immobilization of phytic acid. The immobilized phytic acid can adsorb heavy metal ions, such as cadmium, copper, lead, nickel, and zinc ions, from aqueous solutions. Adsorption isotherms of the selected ions by immobilized phytic acid were conducted in packed-bed column at room temperature. Results from the adsorption tests showed 6.6 mg of Cd^{2+}, 7 mg of Cu^{2+}, 7.2 mg of Ni^{2+}, 7.4 mg of Pb^{2+}, and 7.7 mg of Zn^{2+} can be adsorbed by each gram of PVP–phytic acid complex. The use of immobilized phytic acid has the potential for removing metal ions from industrial or mining waste water.

Index Entries: Phytic acid; heavy metal ion removal; immobilization; poly 4-vinyl pyridine (PVP).

INTRODUCTION

In nature, phytic acid ($C_6H_{18}O_{24}P_6$) exists as free acid, phytate, or phytin according to the physiological pH and metal salts. For instance, the sodium salt is known as sodium phytate, calcium salt as calcium phytate, and calcium/magnesium salt is known as phytin. In literature, the name

*Author to whom all correspondence and reprint requests should be addressed.

phytic acid has been used interchangeably with the term phytate, which is a salt. Phytic acid has six "phosphate covalent bond groups" that account for up to 85% of the total phosphorus in many cereals and legumes (1).

Phytic acid serves many physiological functions, and influences the functional and nutritional properties of cereals and legumes and their derived foods by forming the complex with essential minerals and proteins. It also provides the starting materials for germinating plants. During seed germination, phytate is hydrolyzed to inorganic orthophosphorus, and a series of lower phosphoric esters of myoinositol and free myoinositol by the seed enzyme, phytase (2). The young seedlings utilize the myoinositol as the substrate for the myoinositol oxidation pathway and for cell-wall polysaccharide formation (2). The presence of phytate in seeds and grains also gives additional benefits for the preservation of foods and protection against some diseases in plants (3).

Phytic acid has a complex structure that has been subjected to intense studies. With potentiometric titration study, Cosgrove (4) demonstrated that phytic acid has 12 replaceable protons in the phytic molecule. Six are strongly dissociated at a pK_a of about 1.84; two are weak acid with pK_a of 6.3; and six are weakly dissociated at a pK_a of 9.7. Using NMR and pH titration methods, Costello et al. (5) reported the similar results with six replaceable protons in the strong acid range with pK_a of 1.1–2.1. One in the weak range of pK_a = 5.7, two with pK_a of 6.8–7.6, and three at range of pK_a = 10.0–12.0. Therefore, phytic acid exists as a strong negatively charged molecule over a wide pH range. With this property, it has tremendous potential for forming a complex with positively charged multivalent cations and positively charged proteins. With heavy metal ion binding capacity, phytic acid can have the potential for removing metal ions from industrial waste water. Furthermore, the adsorbed metals can be recovered for reuse.

As a strong acid, phytic acid forms a variety of salts with metal ions easily and exists as the phytate–metal ion complex. The stability and the solubility of the complex depend on the pH and the concentrations and type of metal ions. The solubility and stability of various phytate–metal complexes have been studied. Vohra et al. (6) indicated the order of stability of phytate–metal complex as $Cu^2 > Zn^{2+} > Ni^{2+} > Co^{2+} > Mn^{2+} > Ca^{2+}$ at pH 7.4, whereas Maddaiah et al. (7) establish the stability order as $Zn^{2+} > Cu^{2+} > Mn^{2+} > Ca^{2+}$.

Although many studies indicated the wide existence of phytate–metal complex in nature, phytic acid has not been considered an adsorbent for the removal of heavy metal ions from contaminated water. In order for phytic acid to be considered the adsorbent, it is necessary to find a solid support material that can be used to immobilize phytic acid with good capacity. The immobilized phytic acid should be stable and should have a good adsorption capacity within a reasonable pH range. Furthermore, the adsorbed metal ions should be able to be separated from the adsorbent readily and recovered in concentrated form that can be easily disposed of or recovered for reuse.

Table 1
Some Properties of PVP–425[a]

Property	PVP-425
Appearance	white spherical beads
Particle size	18-50 mesh
Surface area (m^2/g)	$\cong$ 90
Bulk density (g/cm^3)	0.29
H$^+$ capacity, in water (meq/g)	5.5
pKa	3-4
Swelling (%) in water (free base to HCl form)	52%
Temperature stability (°C)	260
Solubility	insoluble in water, acid, base or organic solvent

[a]Data supplied by Reilley Ind., Indianapolis, IN.

Previously, we reported the use of poly 4-vinyl pyridine (PVP) to remove phytic acid from corn steep water (8). In this article, we study the immobilization of phytic acid onto PVP and use the immobilized phytic acid for heavy metal adsorption.

MATERIALS AND METHODS

Materials

The heavy metal ions used in this study were Cd^{2+}, Cu^{2+}, Pb^{2+}, Ni^{2+}, and Zn^{2+}. The aqueous solutions of 1000 ppm each was prepared from cadmium chloride, cupric sulfate, lead nitrate, nickel chloride, and zinc chloride, respectively. Phytic acid (40% solution in water) was purchased from Aldrich Chemical Co., Milwaukee, WI. A polymeric resin, poly (4-vinyl pyridine) or PVP-Reillex 425, was generously supplied by Reilly Ind., Indianapolis, IN. Some properties of PVP resins are listed in Table 1.

Methods

Preparation of Resins

A column with PVP resins was washed with three bed volumes of 5% HCl, five bed volumes of deionized water, followed by three bed volumes of 4% NaOH and another five bed volumes of deionized water. The flow rate was adjusted to allow at least 30 min of contact time between the resins and washing solution. The final water rinse was performed until the effluent pH was < 8.0.

Batch Adsorption

Composite adsorption isotherms were determined using a 1:10 (w/v) ratio of dry resins and a starting sodium phytate solution using equilibrium method. A sodium phytate solution with concentration ranging from 0.5–16 mg/mL at different pHs (adjusted with sodium hydroxide) was used as the starting solution. The flasks containing the adsorbents and phytate were mixed at 25°C in an incubator-shaker (New Brunswick Scientific, Edison, NJ) and allowed to equilibrate for at least 24 h. The pH of the bulk solution at equilibrium was recorded, and phytic acid concentrations were determined by HPLC.

Fixed-Bed Immobilization of Phytic Acid

The acid-alkaline-washed PVP resins were transferred into a jacked column (25 × 1.0 cm) equipped with adjustable plungers. Air trapped in the resin bed was removed, and the plunger was lowered to the top of the resin bed. The feed solution containing 100 mg of sodium phytate/mL at pH 6.5 was introduced into the column by a peristaltic pump. The flow rate was adjusted to allow at least 30 min of contact time between the resins and feed solution. Samples were collected from the column, and the phytate concentration was analyzed to monitor the breakthrough point. The resins were considered saturated when the phytic acid concentration in the effluent was at least 95% that in the feed. At this point, the column was washed by distilled water until no more phytate could be detected in the effluent. The adsorbed phytate is considered immobilized, since it cannot be removed from the resins by dilute acid or alkaline solution. The adsorption capacity was expressed as g phytic acid adsorbed/g of dry resin.

Adsorption of Metal Ions on Immobilized Phytic Acid

A PVP column (25 × 1.0 cm) with immobilized phytic acid was used to carry out the metal ion adsorption experiments at 25°C. The breakthrough point for column adsorption is defined as the time when the effluent metal ion concentration reached 0.1% of the feed metal ion concentration of 1000 ppm at pH 6.5. The feed rate was maintained at 1.0 mL/min. For comparison, adsorption experiments were also carried out using PVP column without phytic acid.

Analytical Methods

Phytic acid was quantified by high-performance liquid chromatography (Hitachi Instrument, L-6200A) using a Bio-Rad Aminex HPX-87H ion-exclusion column (300 × 7.8 mm) with a refractive index Detector (Hitachi Instrument, L-3350 RI). The column was eluted with dilute sulfuric acid (0.005M) at a column temperature of 80°C and a flow rate of 0.8 mL/min

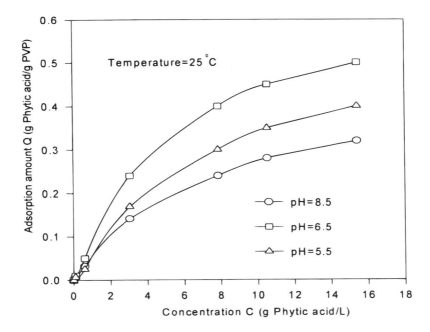

Fig. 1. Adsorption isotherms of phytate at different initial pHs by PVP resins.

over an 8-min period. The concentrations of metal ions were determined using a Perkin Elmer 2280 Atomic Absorption Spectrophotometer.

RESULTS

Similar to other organic acids, pH is one of the important factors determining the adsorption capacity of resins. The adsorption isotherms of phytic acid by PVP-425 at pHs of 5.5, 6.5, and 8.5 are shown in Fig. 1. Resins exhibited highest adsorption capacity for phytic acid at pH 6.5 with 0.51 g phytic acid adsorbed/dry g PVP resin. The capacity dropped to 0.4 g/g at pH 5.5 and 0.31 g/g at pH 8.5.

The feed solution containing 1000 ppm of different metal ions at pH 6.5 was introduced into the columns containing phytic acid by a peristaltic pump at different flow rates. When flow rate was raised to 3.7 mL/min, the shape of the breakthrough curve remained more or less unchanged. Figures 2–6 show the breakthrough data for five different metal ions with or without immobilized phytic acid. With PVP alone, the metal ion adsorption capacities are in the range of 3.1 mg/g PVP for lead ions to 3.7 mg/g PVP for nickel ions. With PVP–phytic acid, the ion adsorption capacity is considerably higher. Table 2 summarizes the capacities of PVP and PVP–phytic acid resins to adsorb different metal ions. The adsorption capacity of as much as 7.7 mg zinc ions/g of immobilized phytic acid was recorded. The order of metal ion's adsorption capacities at pH 6.5 by PVP–phytic acid is as follows: $Ni^{2+} > Zn^2 > Cu^2 > Co^2 > Cd^2 > Pb^{2+}$.

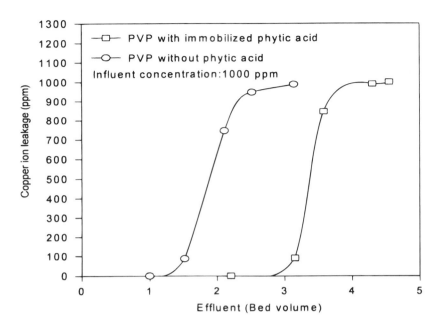

Fig. 2. Adsorption breaktrhough curves of copper ions from aqueous solution using PVP with or without immobilized phytic acid.

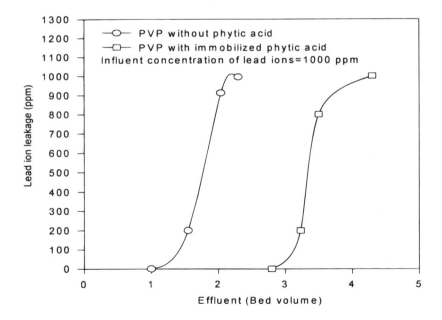

Fig. 3. Adsorption breakthrough curves of lead ions from aqueous solution using PVP with or without immobilized phytic acid.

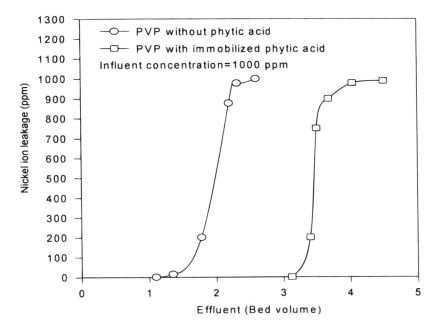

Fig. 4. Adsorption breakthrough curves of nickel ions from aqueous solution using PVP with or without immobilized phytic acid.

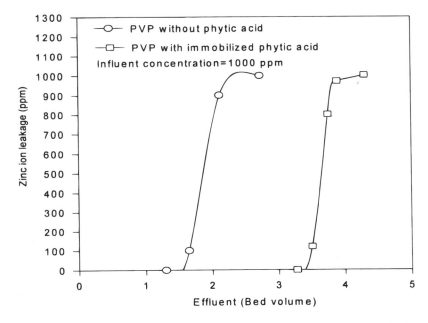

Fig. 5. Adsorption breakthrough curves of zinc ions from aqueous solution using PVP with or without immobilized phytic acid.

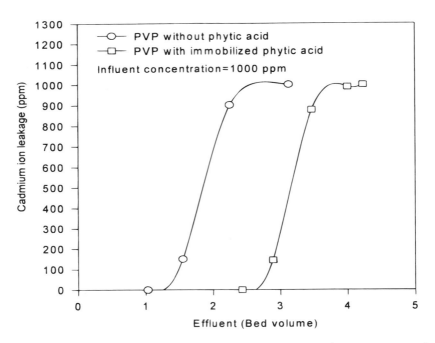

Fig. 6. Adsorption breakthrough curves of cadmium ions from aqueous solution using PVP with or without immobilized phytic acid.

Table 2
Metal Ions Adsorption Capacity of PVP and PVP–Phytic Acid

Metal Ions	Adsorption Capacity		
	PVP	PVP-Phytic Acid	
	mg/g	mg/g	mole/mole
Cd^{2+}	3.3	6.6	0.38
Cu^{2+}	3.6	7.0	0.72
Ni^{2+}	3.7	7.2	0.84
Pb^{2+}	3.1	7.4	0.24
Zn^{2+}	3.6	7.7	0.78

Comparing the breakthrough curves of different metal ions in the same mixture (Fig. 7), the concentration of ions in eluent promptly reached the concentration of feed solution after the breakthrough point. It also shows that the breakthrough curves cross each other. The order of adsorption capacity of different metal ions established is the same as those conducted

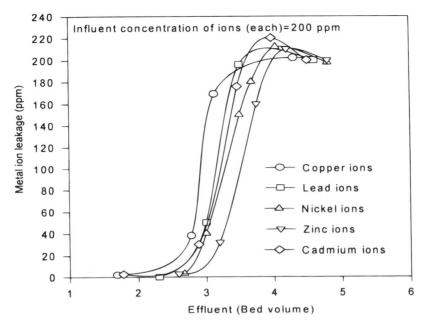

Fig. 7. Comparison of adsorption breakthrough curves of five different metal ions from mixed-ion aqueous solution using PVP with the immobilized phytic acid. Metal ion concentration: 200 PPM; pH = 6.5.

with single metal ions. Although the data are not shown, the PVP–phytate resins can be reused without loss of metal ion's adsorption activity.

DISCUSSION

Growing concern has been given to the potential health hazard presented by heavy metal contamination in drinking water, watershed, and the environment in general. Available processes for heavy metal decontamination/removal include chemical precipitation, evaporation, cementation, flotation, reverse osmosis, and ion adsorption or absorption. Most of these methods suffer from some drawbacks, such as high capital and operational costs and the disposal of residual metal sludge. Ion-exchange resins, which are used in treatment of several pollutants, are usually expensive to use and to regenerate. Therefore, there is a need for research and development of low-cost and readily available materials that can provide an alternative source for removing heavy metals economically. Recently, there has been research conducted, utilizing unconventional and renewable materials for heavy metal removal. The materials studied for heavy metal adsorption and removal include: sawdust *(9)*, modified bark *(10)*, egg shell *(11)*, cyanobacteria *(12)*, and marine algae *(13)*. More recently, there have been efforts to remove heavy metals from contaminated soils using plants. This method is known as "phytoremediation" *(14)*.

Phytic acid is a natural, biodegradable material that can be found in the seeds of cereals and legumes. One of the most readily available source of phytic acid is corn steep liquor, a byproduct from corn wet milling. For example, in a typical concentrated corn steep liquor from corn wet milling process, there is about 4% of phytic acid on a dry basis. Phytic acid in the corn steep liquor is derived from corn germ during steeping process. Considering that over 8 billion bushels of corn are expected to be produced each year in the US and approx 10% of the corn crop will be subjected to a wet milling process *(15)*, the potential amount of phytic acid available is large. It is estimated that 3.6 pounds of corn steep liquor can be generated from each bushel of corn. The total amount of phytic acid that can be recovered is approaching 115 million pounds or 52 million kilograms, a portion can be used for heavy metal removal. Therefore, phytic acid can be a low-cost and readily available material suitable for decontamination of polluted water sources.

PVP is chosen as the adsorbent for the immobilization of phytic acid, because PVP adsorbs phytic acid, a strong acid. Phytic acid is difficult to remove from resins even by alkaline elution *(8)*. This is not surprising, since PVP is known to adsorb acid that is not involved entirely in the ion-exchange mode. The exact mode of phytic acid adsorption by PVP is not yet clear. The adsorption could be in part owing to the interaction between pyridyl group of PVP that obtained a proton from phytic acid and formed the complex.

At stationary condition, the adsorption capacity of phytic acid for metal ions is the function of equilibrium concentration. The information regarding adsorption capacity as a function of product equilibrium concentration in kinetic condition is necessary for process design. This allows the estimation of the solid holdup required to remove as much metal ion from the aqueous solution as possible.

Phytic acid is very stable, particularly under the immobilized form by PVP. Likewise, PVP is very stable and will not deteriorate easily over time under normal application condition (unpublished observation). The adsorbed metal ions can be recovered in concentrated forms, but we have yet to conduct a systematic investigation for the recovery of adsorbed metal ions. A system based on the recovery of organic acid was tried; however a substantial amount of ions remained adhered to the column. Further investigation regarding the adsorption of metal ions in polluted water and the methods for the recovery of metal ions is required.

ACKNOWLEDGMENT

This study was supported in part by The Consortium for Plant Biotechnology Research, Inc. by DOE cooperative agreement no. DE-FC05-920R22072. This support does not constitute an endorsement by DOE or by The Consortium for Plant Biotechnology Research, Inc. of the view expressed in this article.

REFERENCES

1. Reddy, N. R., Pierson, M. D., Sathe, S. K., and Salunkhe, D. K. (1989), in *Phytates in Cereals and Legumes*, CRC, Boca Raton, FL, 152 pp.
2. Loewus, F. A. and Loewus, M. W. (1983), *Annu. Rev. Plant Physiol.* **34,** 137–161.
3. Hall, J. R. and Hodges, T. K. (1966), *Plant Physiol.* **41,** 1459–1464.
4. Cosgrove, D. J. (1966), *Rev. Pure Appl. Chem.* **16,** 209–252.
5. Costello, A. J. R., Glonek, T., and Mayers, T. C. (1976), *Carbohydr. Res.* **46,** 159–171.
6. Vohra, P., Gray, G. A., and Kratzer, F. H. (1965), *Proc. Soc. Exp. Biol. Med.* **120,** 447.
7. Maddaiah, V. T., Kurnick, A. A., and Reid, B. L. (1964), *Proc. Soc. Exp. Biol. Med.* **115,** 391.
8. Moravec, S., Gong, C. S., Lu, J, Yang, C. W., and Tsao, G. T. (1995), Recovery of lactic acid and phytic acid from corn steep liquor using adsorption. IFT, 1995 Annual Meeting, Anaheim, CA.
9. Bryant, P. S., Petersen, J. M., Lee, J. M., and Brounds, T. M. (1992), *Appl. Biochem. Biotechnol.* **34/35,** 777–788.
10. Gaballah, I. and Kilbertus, G. (1994), in *Separation Process Heavy Metals, Ions and Minerals*, Mistra, M. ed., The Minerals, Metals & Materials Society, pp. 15–26.
11. Suyama, K., Fukazawa, Y., and Umetsu, Y. (1994), *Appl. Biochem. Biotechnol.* **45/46,** 871–879.
12. Corder, S. L. and Reeves, M. (1994), *Appl. Biochem. Biotechnol.* **45/46,** 847–859.
13. Holan, Z. R., and Volesky, B. (1994), *Biotechnol. Bioeng.* **43,** 1001–1009.
14. Salt, D. E., Blaylock, M., Kumar, N. P. B. A., Dushenkov, V., Ensley, B. D., Chet, I., and Raskin, I. (1995), *Bio/Technology* **137,** 468–474.
15. Johnson, L. A. (1991), in *Handbook of Cereal Science and Technology*, Lorenz, K. J. and Kulp, K., eds., Marcel Dekker, New York, pp. 55–131.

Observations of Metabolite Formation and Variable Yield in Thiodiglycol Biodegradation Process

Impact on Reactor Design

TSU-SHUN LEE,[1,2] WILLIAM A. WEIGAND,[2] AND WILLIAM E. BENTLEY*,[1,2]

[1]*Center for Agricultural Biotechnology, University of Maryland Biotechnology Institute; and* [2]*Department of Chemical Engineering, University of Maryland, College Park, MD 20742*

ABSTRACT

The complete microbial degradation of thiodiglycol (TDG), the primary hydrolysis product of sulfur mustard, by *Alcaligenes xylosoxydans* ssp. *xylosoxydans* (SH91) was accomplished in laboratory-scale stirred-tank reactors. An Andrews substrate inhibition model was used to describe the cell growth. The yield factor was not constant, but a relationship with initial substrate concentration has been developed. Using a substrate-inhibition and variable-yield kinetic model, we can describe the cell growth and substrate consumption in batch and repeated batch fermentations. Several reactor-operating modes successfully degrade TDG concentration to below 0.5 g/L. According to the experimental results, the two-stage repeated batch operation has the best degradation efficiency, and it also can degrade 500 mM TDG ($\approx$60 g/L) to 5 mM ($\approx$0.7 g/L) in <5 d. A hypothesis for explaining variable-yield and byproduct formation based on the capacity and utilization of metabolic loads is presented.

Index Entries: Biodegradation; thiodiglycol; sulfur mustard; metabolic model.

---

*Author to whom all correspondence and reprint requests should be addressed.

INTRODUCTION

Biodegradation is the method that uses biological means for waste treatment. This technology has become increasingly important, not only for industrial hazardous chemical disposal, but also for chemical warfare agent disposal. This method has several advantages. First, it is the most economic method for toxic waste mineralization. For example, biodegradation processes cost $40–70/t of waste, whereas incineration processes are almost 10 times higher. Second, it is environmental friendly, because the pollutant is naturally degraded to innocuous products, such as CO_2, H_2O, and microorganisms. Finally, biodegradation is a natural process, which makes it easier for the public to accept. Accordingly, our lab has evaluated the biodegradation of thiodiglycol (TDG), the main hydrolysis product of sulfur mustard.

Sulfur mustard, a chemical warfare agent, can be hydrolyzed in water (1), forming hydrochloric acid and TDG. Yang and others have investigated the aqueous hydrolysis reaction kinetics (2). Compared to sulfur mustard, TDG is relatively nontoxic; the LD_{50} (dose at which 50% lethality is effected) of TDG varies from 3000–6610 mg/kg, depending on the species (3). This level of toxicity is similar to isopropanol. TDG has many industrial uses, for example, it is used in elastomers, lubricants, stabilizers, antioxidants, inks, dyes, and so forth. Although TDG is relatively nontoxic and has many industrial uses, it can also be used to manufacture sulfur mustard. TDG is listed in the chemical weapon convention treaty (4) and must be degraded in a sulfur mustard mineralization process.

In this article, *Alcaligenes xylosoxydans* ssp. *xylosoxydans* (SH91), a Gram-negative bacterium that consumes TDG as the sole carbon source, was employed. In previous papers, an Andrews-type inhibition model with constant yield was used to describe SH91 growth on TDG-based media (5,6). In the present work, simulations based on the previous model were compared to experiments with high initial TDG concentration (≥80 mM, 10 g/L). As the initial TDG concentration increased, so did the deviation between model and experiment. By HPLC analysis, the amounts of metabolic byproduct increased with initial TDG concentration. The specific growth rate and cell mass yield were affected by these metabolites, so an improved Andrews inhibition model incorporating a variable cell mass yield was developed and is presented here. Also, several different batch-type operating strategies were used to degrade TDG. An evaluation of these operating modes is included. Finally, a new model is proposed based on the cell's metabolic capacity for degrading TDG and its metabolites. The model qualitatively describes the variable-yield phenomena noted in our experiments.

POSSIBLE DEGRADATION PATHWAY

TDG was injected into rats, and was metabolized into TDG sulfoxide (TDGS), TDG sulfone, and so forth (7). The initial step of TDG degradation by SH91 was determined by Zulty et al. (8), TDG is converted to *S*-(2-

Thiodiglycol Biodegradation Process

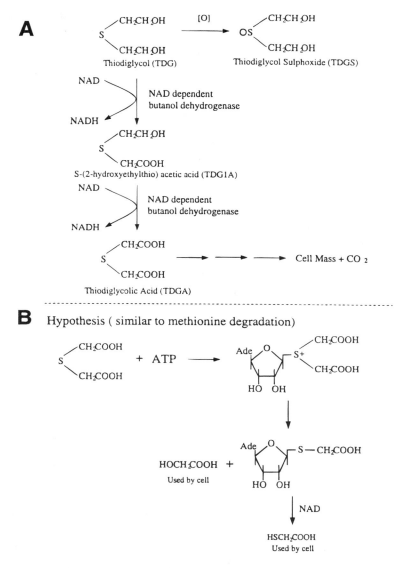

Fig. 1. The metabolic pathway for SH91 to degrade TDG. **(A)** SH91 use NAD-dependent butanol dehydrogenase to convert TDG to TDG1A, and TDG1A to TDGA. The other degradation pathway is TDG oxidized by oxygen to produce TDGS. **(B)** The possible TDGA degradation pathway (hypothesis based on methionine degradation).

hydroxyethylthio) acetic acid (TDG1A). The enzyme used to catalyze this reaction is NAD-dependent butanol dehydrogenase. This enzyme is used again to convert S-(2-hydroxyethylthio) acetic acid to TDG acid (TDGA). Two NADH are produced in these steps. Additionally, we found that TDG is oxidized by molecular oxygen yielding TDGS, which is detected by HPLC. The proposed metabolic pathway is summarized in Fig. 1A. Based on methionine degradation pathways elucidated in *Escherichia coli* bacteria, we suggest an analogous metabolic pathway is used for TDGA in SH91, as shown in Fig. 1B. This is consistent with our metabolic byproduct analyses.

MATERIALS AND METHODS

Microorganism, Media, and Culture Conditions

A. xylosoxydans ssp. *xylosoxydans* (SH91), a Gram-negative bacterium, uses TDG as the sole carbon source for growth. Stock cultures were maintained on TDG medium, which has the following composition (per liter): TDG, 3.66 g (30 mM); ammonium sulfate, 2 g (15.1 mM); potassium phosphate dibasic 2 g (11.5 mM); and modified Wolin salts solution 10 mL (WS42). The composition of WS42 was (per liter): nitrilotriacetic acid, 3 g (15.7 mM; add NaOH until dissolved); $MgSO_4 \cdot 7H_2O$, 6 g (24.3 mM); NaCl, 1 g (17.1 mM); $MnSO_4 \cdot H_2O$, 1 g (5.9 mM); $FeSO_4 \cdot 7H_2O$, 0.5 g (1.8 mM); $CaCl_2 \cdot 2H_2O$, 0.1 g (0.68 mM); $CoCl_2 \cdot 6H_2O$, 0.1 g (0.42 mM); $ZnSO_4 \cdot 7H_2O$, 0.1 g (0.35 mM); H_3BO_3, 0.02 g (0.32 mM) $Na_2MoO_4 \cdot 2H_2O$, 0.01 g (0.04 mM); $CuSO_4$, 0.01 g (0.06 mM). Finally, NaOH was used to adjust the pH of TDG medium to 9.2. The best growth conditions were 30°C and pH-8.0. Cells were transferred to fresh TDG medium every 4 d in order to sustain exponential growth. Inoculum was prepared by adding 4 mL SH91 freezer stock to 250-mL shake flasks with 100 mL TDG medium. This was incubated for 3 d in a reciprocating shaker at 30°C and 150 rpm. Thereafter it was added to the fermenters (New Brunswick Scientific, Hatfield, UK).

Analytical Methods

Cell concentration was determined by optical density (OD_{590}, Milton Roy, FL, Spec21). The absorbance of the background media was used to correct the absorbance readings. Dry cell weights were directly proportional to optical density up to 0.4 OD_{590} (conversion factor = 0.8 g/$OD_{590} \cdot$ L). Samples above 0.4 OD_{590} were diluted into the linear range with deionized water.

TDG, TDG1A, TDGA, TDGS were determined by HPLC (Waters Model 590) at 214 nm (Waters, UV detector) with a Pham-Pak column (9). TDG and TDGA standards were purchased from Sigma Chemical Co. St. Louis, MO). TDGS standard was prepared by addition of 30% H_2O_2 to TDG (7). The HPLC running conditions were as follows: column temperature, 65°C; mobile phase, 3 mM phosphoric acid (flow rate: 1 mL/min.). The linear range for TDG, TDGA, and TDGS was 0.3–10 mM, 0.5–20 mM, and 0.5–15 mM, respectively. All samples taken from the fermenter were filtered (0.2 μm) before injection. *See* Table 1 for typical results.

Reactor Operation

Batch fermentations were run in a 700-mL BioFlo I fermenter with a 500-mL working volume and 2-L BioFlo I fermenter with a 1.5-L working

Table 1
The Accumulated Metabolite Concentrations
in Four Different Batch Final Broths

	30 mM	60 mM	120 mM	170 mM
TDG (mM)	0	0	0	0
TDG1A (mM)	0	0	0.5	17.18
TDGA (mM)	0	3.02	17.4	60.34
TDGS (mM)	2.4	3.89	6.91	6.84
Total (mM)	2.4	6.91	24.86	84.36

volume (New Brunswick Scientific, Inc.). The agitation rate was 400 rpm, and the inlet air was sterilized by a 0.2-μm filter. The inlet air was humidified by passing through a water bath to avoid evaporation. Temperature and pH were controlled at 30°C and 80°, respectively. There are four different fermentation operating policies as shown in Fig. 2. They are batch, repeated batch, two-stage batch, and fed batch with linear feed. For repeated batch, two different TDG feed concentrations were employed (60 and 270 mM). For the 60-mM case, fresh TDG medium was fed into the bioreactor after an initial batch phase (increase from 250–500 mL), and the reactor was run batchwise until all TDG was consumed (≈48 h). Then, 250 mL medium was withdrawn, 250 mL fresh 60 mM TDG medium was added, and a new cycle was begun. This cyclic process was run until 10 L (60 mM) of TDG medium was processed. For the 270-mM case, we followed almost the same procedure. Initially, 1 L of fresh TDG medium was added into the reactor with 0.5 L broth (1.5 L), and a new cycle was begun (cycle time: 60 h). Then, 1 L of fresh TDG medium was withdrawn and replaced with fresh media at every cycle. The two-stage batch was divided into two batch processes as shown in Fig. 2. First, 250 mL fresh 500 mM TDG medium was added into a reactor (from 500 to 750 mL), and the reactor was run batchwise for 2 d. Then, 500 mL fresh 500 mM TDG medium were added, and again the reactor was run batchwise until TDG was consumed to the desired level (3 d). The final batch-type operating mode was fed batch with linear feed. It was divided into three periods. They were feeding, reacting, and withdrawing. In 24 h the reactor volume was increased steadily from 500 to 1500 mL (flow rate: 41.67 mL/h), at which time the reactor was run batchwise for the remaining time (60 h). The withdrawal period was fast (≈10 min), so it was neglected in modeling calculations (Fig. 2). Three complete cycles were run.

MATHEMATICAL MODEL

The model used to describe SH91 growth in TDG medium is an Andrews-type substrate-inhibition model *(10)*:

$$\mu(s) = [\mu_{max}/(1 + (K_s/s) + (s/K_i))] \qquad (1)$$

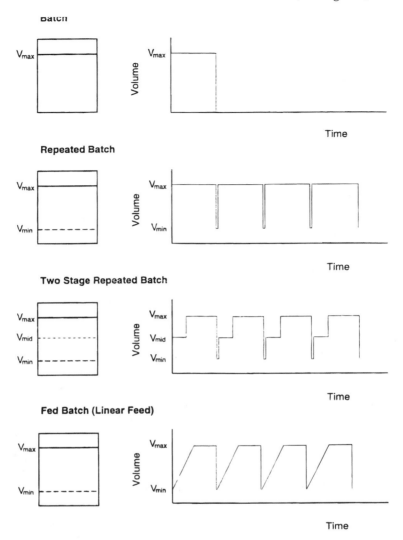

Fig. 2. Four different kinds of batch fermentation. They are batch, repeated batch, two-stage repeated batch, and fed batch.

where s is the growth-limiting substrate (TDG) concentration (g/L), μ_{max} is the maximum specific growth rate (1/h), K_s is the substrate saturation constant (g/L), and K_i is the substrate inhibition constant (g/L). The cell growth is described by an autocatalytic reaction including cell death:

$$(dx/dt) = \mu(s)x - k_d x \quad (2)$$

where k_d is the death rate constant (1/h) as measured previously (5). During cell growth, the substrate is simultaneously consumed:

$$(ds/dt) = (-1/Y_{x/s})\mu(s)x \quad (3)$$

Table 2
Model Parameters

Parameter	Value
Maximum Growth Rate (μ_{max})	0.171 (1/hr)
Substrate Constant (K_s)	2.77 g/L (49.45 mM)
Inhibition Constant (K_i)	9.47 g/L (81.68 mM)
Death Rate Constant (k_d)	7.8×10^{-3} (1/hr)
Yield Factor ($Y_{x/s}$)	Variable
Max. Yield Factor ($Y_{x/s}^M$)	0.345 (g/g)

where ds/dt is the substrate accumulation rate (g/L · h), and $Y_{x/s}$ is the apparent yield factor. In the present work, the yield coefficient is a function of initial TDG concentration:

$$Y_{x/s} = 0.345\, e^{-0.5s} \quad (4)$$

These four equations were used to simulate SH91 growth in TDG. All the parameters are listed in Table 2.

SIMULATION AND EXPERIMENTAL RESULTS

A series of batch fermentations with different initial TDG concentrations was run to determine cell mass yield and specific growth rate. The calculated yield coefficient (obtained from the slope of OD vs S plots) was not constant. Instead, it decreased with increased initial TDG concentration. Interestingly, the yield was constant over the time-course of a fermentation, but varied depending on the initial concentration. For simplicity, an exponential decay function was used to fit the experimental data (Fig. 3). The specific growth rates diverge at initial TDG concentration above 10 g/L, as shown in Fig. 4. In our previous studies, the specific growth rate observed at high TDG was from shake flasks without pH control. The specific growth rate determined here was measured from batch reactors with pH control. Additionally, the bioreactor data were obtained in a single extended repeated batch experiment, where the inoculum for the next experiment was the material left after draw-down of 90% of the reactor volume from the previous experiment. As will be addressed later, this method retains 10% of the metabolic byproducts produced during the batch. This, in addition to the pH control (which in isolated batch cultures yields more metabolites; not shown), provides a basis for why the growth rate is different at high initial TDG (more metabolites produced per TDG). The three calculated kinetic parameters are μ_{max}, K_S, and K_i, and their values are 0.171/h$^-$, 2.77 g/L (22.68 mM/L), and 9.47 g/L (77.51 mM/L), respectively. The parameters from our previous studies were 0.277 h^{-1} (μ_{max}), 6.04 g/L (K_S), and 9.98 g/L (K_i). The new specific growth rate is

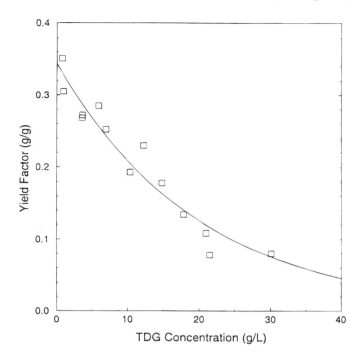

Fig. 3. The yield factor is a function of TDG concentration: $y_{x/s} = 0.345 \times \exp. -0.05s$ ($s = TDG$ conc. g/L).

lower than the previous rate when TDG concentrations are higher than 5 g/L (40 mM). Again this is likely owing to the production of the metabolic byproducts, which affects the TDG consumption rate and, thus, cellular specific growth rate.

The simulation and experimental results for batch reactors are shown in Fig. 5. Four different initial TDG concentration results are shown. Also, the previous and new model simulation results are depicted. As expected, both the new and old models described low TDG ($\leq$60 mM, 7.3 g/L) batch experiments quite well (Fig. 5A,B). The previous model failed to describe high TDG concentration batch experiments, whereas the new model was in good agreement with the experimental data up to 20 g/L (170 mM), as shown in Fig. 5C,D. In Table 1, the metabolite accumulation was more problematic as TDG concentration was increased (refer to Fig. 1A). We further employed the model to simulate 60- and 270-mM repeated batch fermentations (Fig. 6). The new model parameters and variable-yield calculations described the 60-mM repeated batch process as well as or better than the previous model (5). However, the new model still deviated from experimental results for a 270-mM repeated batch process. Note, however, that for the 270-mM repeated batch experiment, we found significantly more TDG1A and TDGA in the broth at the end of each cycle than for the previous batches

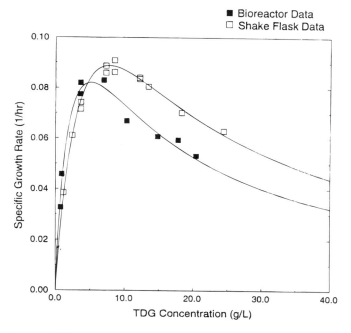

Fig. 4. Andrews inhibition model fit to results of shake flask and bioreactor experimental data: shake-flask model result (K_S = 6.04 g/L [49.45 mM], K_t = 9.98 g/L [81.68 mM], μ_{max} = 0.227 [h^{-1}]), bioreactor model results (K_S = 2.77 g/L [22.67 mM], K_t = 9.47 g/L [77.51 mM], μ_{max} = 0.171 [h^{-1}]).

(30–170 mM; see Tables 1–4). Thus, we hypothesized that several metabolic interactions were at play in this system:

1. High initial TDG resulted in more metabolic byproduct production;
2. High levels of metabolic byproducts in the media were detrimental to cell growth; and
3. Decreased cell growth also meant decreased specific TDG uptake.

Several additional repeated batch operating policies were developed and run to test these interactions. Two-stage batch experimental results are shown in Fig. 7A. Overall, this method degraded TDG from 500 (60 g/L) to 1.1 mM (0.13 g/L) in 120 h. For the first stage, 500 mM TDG was added, increasing the volume from 500 to 750 mL. It took 48 h to degrade TDG from 170 (20.7 g/L) to 9.7 mM (1.2 g/L), when cell mass increased from 2.4 to 3.5 g/L. From HPLC chromatograms, significant levels of metabolite had accumulated (not shown). For the second stage, 500 mM TDG was added to the reactor, increasing the volume from 750 to 1500 mL (maximum value). It took 72 additional hours to degrade TDG from 250 (30.5 g/L) to 1.1 mM (0.13 g/L). The cell mass increased from 1.89 to 4.42 g/L. From HPLC chromatograms, a large amount of both TDG1A and TDGA had accumulated in the reactor (see Table 4).

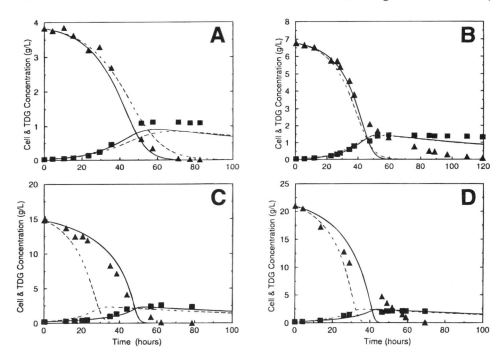

Fig. 5. Four different initial TDG concentration batch experimental and simulation results (dash line: old model; solid line: new model): **(A)** 30-mM batch, **(B)** 60-mM batch, **(C)** 120-mM batch, and **(D)** 170-mM batch.

Table 3
Degradation Index for Three Batch-Type Bioreactors
(First-Calculation Method)

	270mM RB	500mM TSRB	300mM RFB
Influent TDG (mM)	270	512	290
effluent TDG (mM)	0	6.55	4.26
Withdraw volume	1000 ml	1000 ml	1000 ml
Max working volume	1500 ml	1500 ml	1500 ml
Spending time	60 hrs	120hrs	84 hrs
DI (mM/hr)	3.0	2.8	2.3

The results for a linear-feed repeated fed-batch reactor are shown in Fig. 7B. During the 24-h feeding period (41.67 mL/h), TDG concentration increased from 3 (0.36 g/L) to near 130 mM (15.88 g/L), and cell mass decreased slightly owing to dilution. For the remaining batch periods, the TDG decreased from 130 to 4.4 mM, and cell mass gradually increased from 1.9 to 2.5 g/L. As mentioned earlier, metabolites accumulated in the broth (*see* Table 4).

The overall degradation index (DI) was calculated for repeated batch, two-stage batch, and repeated fed-batch reactors. Two different calcula-

Table 4
Degradation Index for Three Batch-Type Bioreactors
(Second-Calculation Method)

	270mM RB	500mM TSRB	300mM RFB
Influent TDG (mM)	270	512	290
effluent TDG (mM)	0	6.55	4.26
Metabolites (mM)	118.27	162.78	58.95
Withdraw volume	1000 ml	1000 ml	1000 ml
Max working volume	1500 ml	1500 ml	1500 ml
Spending time	60 hrs	120hrs	84 hrs
DI (mM/hr)	1.7	1.9	1.8

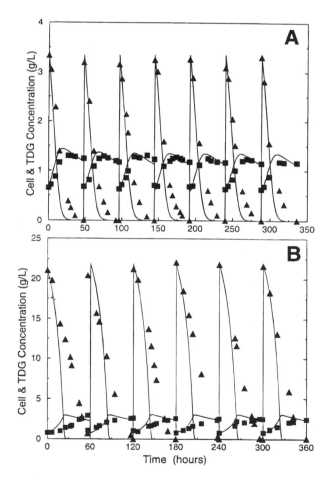

Fig. 6. Two different repeated batch experimental and simulation results: **(A)** 60-mM repeated batch and **(B)** 270-mM repeated batch.

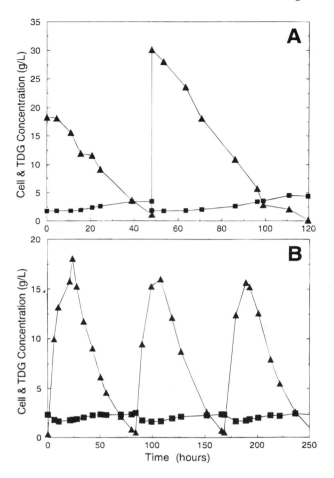

Fig. 7. **(A)** Two-stage batch experimental result and **(B)** repeated fed-batch experimental results.

tion methods are presented here. The first method focuses on TDG degradation, and does not account for the metabolites produced by SH91 during growth. The calculation is as follows:

$$DI = \left[\left(TDG_{inf} - TDG_{eff}\right) \times V_{wd} / V_{max} \times T_f\right] \quad (5)$$

where DI is degradation index, TDG_{inf} and TDG_{eff} are influent and effluent TDG concentration, V_{wd} is the withdrawn volume, V_{max} is the maximum working volume, and T_f is total operating time (*see* 9).

The other method focuses on the amount of TDG completely degraded by SH91. The calculation is as follows:

$$DI = \left[\left(TDG_{inf} - TDG_{eff} - M_{eff}\right) \times V_{wd} / V_{max} \times T_f\right] \quad (6)$$

where M_{eff} is the concentration of accumulated metabolites, including TDG1A, TDGA, and TDGS. The results for these two methods are listed in Table 3 (based on Eq. 5) and Table 4 (based on Eq 6).

DISCUSSION

The simulation results for an Andrews inhibition model with new parameters agree quite well for different batch reactor schemes. The simulation results are significantly improved for high concentrations of TDG (≤120 mM, 14.7 g/L) as shown in Fig. 5C,D. Note at least 600 mM TDG are likely to remain after mustard hydrolysis. Some deviation was found between experimental and simulation results in the 270-mM repeated batch, especially for the TDG consumption rate. The predicted rate was more rapid than the actual consumption rate. For the 270-mM repeated batch, the initial thiodiglycol concentration was near 170–180 mM or was almost the same as the 170-mM batch fermentation. Accordingly, they should have the same kinetic result. The reason for the model failure when describing the 270-mM repeated batch is the influence of metabolite accumulation. From our HPLC chromatograms, we found the TDG1A concentration in 270-mM repeated batch was 2.5× higher than the 170-mM batch reactor. TDG and TDG1A are substrates for the same enzyme, so TDG1A competes with TDG for the NAD-dependent butanol dehydrogenase, as shown in Fig. 1. Thus, the amount of TDG1A directly affects the TDG degradation rate; the more TDG1A is accumulated in the bioreactor, the more the TDG consumption rate is reduced. This phenomenon was also observed in the 500-mM two-phase batch and repeated fed-batch cases.

There are two different goals or paradigms for TDG degradation. The first focuses on TDG alone, and the metabolites produced by SH91 are neglected. For this paradigm, we found the 270 mM repeated batch reactor provided the best degradation efficiency, and the 300 mM repeated fed-batch reactor was the worst. The second paradigm focuses on the amount of TDG completely degraded (i.e., to biomass, CO_2, H_2O, inorganics). In this case, we found the two-stage batch was best, although the 300-mM repeated fed batch was a close second. For the 270-mM repeated batch reactor, much TDC was converted to TDG1A and TDGA, which slowed the overall process down. For the 300-mM fed batch, TDG was continuously introduced, and the TDG concentration increased gradually. Correspondingly, the cells had more time to completely degrade TDG. They were not exposed to as high a TDG concentration in as short a time, so the problem of TDG1A and TDGA accumulation was less serious than with the 270-mM repeated batch. The two-phase batch is midway between a fed-batch and batch scheme, since it divides a batch into two separated batches. It avoids sudden exposure to very high TDG concentration, which has two beneficial effects: first, it reduces the substrate inhibition for cell growth, and second, it produces less metabolite (TDG1A) that competes for the same enzyme as

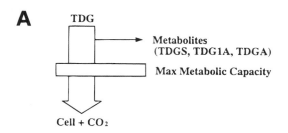

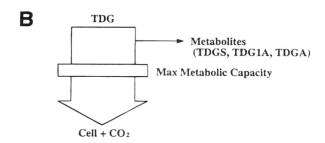

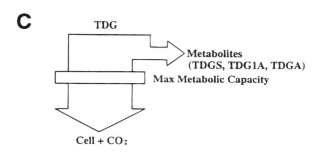

Fig. 8. Illustration of the bottleneck of the oxidative metabolism in SH91. **(A)** Excess oxidative capacity, and completely oxidative TDG to CO_2, TDG1A, TDGA, TDGS (for example, 30-mM batch: 2.4 mM [TDGS only] and 60-mM batch: 6.91 mM [TDGA+TDGS]). **(B)** The oxidative capacity is utilized completely. **(C)** Limit oxidative capacity, and the carbon flux is converted to byproducts by oxido-reductive metabolism (for example, 120-mM batch: 24.86 mM [TDG1A + TDGA + TDGS] and 170-mM batch: 84.36 mM [TDG1A + TDGA + TDGS]).

TDG. For the fed-batch case, with the proper feed rate, we should be able to minimize the TDG concentration, so that the cells consume TDG as it is added. Note, however, that the production of byproducts still occurs, and this method still exposes the cells to potentially high concentrations of byproducts, thus lowering the cell growth rate. The two-stage batch system improves upon this slightly in that the second-stage TDG addition effectively dilutes out the metabolite concentration, but does not expose the cells to too high a TDG level. Note also that this system was run with 500 mM initial TDG compared to 270 mM for the next-best fed-batch case. Thus, we found the 500-mM two-stage batch resulted in high degradation efficiency for both paradigms. According to our experimental results, we suggest a two-stage or multistage fed-batch process may improve the degradation efficiency further, especially for high concentrations of TDG (500 mM, 61 g/L).

Finally, according to our experimental results and HPLC chromatogram analyses, we propose a metabolic capacity model for TDG degradation based on the known metabolic pathways (Fig. 1A). The model is shown in Fig. 8, and the metabolites produced by SH91 are dependent on the TDG concentration (or flux). Little metabolite is produced by SH91 at low initial TDG concentration (as shown in Fig. 8A). This is because the cells have enough capability to convert all TDG to cell mass and CO_2. In Fig. 8B, the cell degradation flux is increased at higher TDG concentrations owing to an increase in capacity. However, metabolites were produced when TDG flux increased further and was higher than the cell maximum metabolic capacity (Fig. 8C). This phenomenon then directly affected the cell mass yield. The yield factor became smaller when TDG flux increased over maximum degradation capacity. A more complex structured metabolic flux model is presently under consideration in our laboratory for describing SH91 growth on TDG.

ACKNOWLEDGMENT

The partial support of the US Army Edgewood, Research Development and Engineering Center is gratefully acknowledged.

REFERENCES

1. Somani, S. M. (1992), *Chemical Warfare Agents*, Academic, London.
2. Yang, Y.-C., Szafraniec, L. L., Beaudry, W. T., and Ward, J. R. (1988), *J. Org. Chem.* **53(14)**, 3293–3297.
3. Sutherland, R. G. (1991), in *Verification of Dual-Use Chemicals Under the Chemical Weapons Convention: The Case of Thiodiglycol*, Lundin, S. J., ed., Oxford University Press, New York, p. 32.
4. Bentley, W. E., Bunnett, J. F., DeFrank, J. J., Fahnestock, M. V., Haley, M. V., Harvey, S. P., Isaacson, J. J., Kilbane, J. E., Kolakowski, M. P., Labare, J.-L., and Ontiveros, J. R. (1994), in Proceedings of the 1993 Scientific Conference on Chemical Defense Research, Aberdeen Proving Ground, Maryland pp. 1133–1138.
5. Lee, T.-S., Pham, M. Q., Weigand, W. A., Harvey, S. P., and Bentley, W. E. (1996), *Biotechnol. Prog.*, **12**, 533–539.
6. Pham, M. Q., Weigand, W. A., Harvey, S. P., and Bentley, W. E. (1995), *Appl. Biochem. Biotechnol.*, **57/58**, 779–789.
7. Black, R. M., Brewster, K., Clarke, R. J., Hambrook, J. L., Harrison, J. M., and Howells, D. J. (1993), *Xenobiotica* **23**, 473–481.
8. Zulty, J. J., DeFrank, J. J., and Harvey, S. P. (1994), in Proceedings of the Scientific Conference on Chemical and Biology Defense Research, Edgewood Research Development and Engineering Center, p. 68.
9. Lee, T.-S. (1994), Using *Alcaligenes xylosoxidans xylosoxidans*, (SH91) to degrade thiodiglycol, the product of sulfur mustard hydrolysis, M.S. thesis. University of Maryland, College Park, MD.
10. Andrews, J. F. (1968), *Biotechnol. Bioeng.* **10**, 707–723.

Pilot-Scale Bioremediation of PAH-Contaminated Soils

S. P. Pradhan,* J. R. Paterek, B. Y. Liu, J. R. Conrad, and V. J. Srivastava

Institute of Gas Technology, 1700 South Mount Prospect Road, Des Plaines, IL

ABSTRACT

The Institute of Gas Technology (IGT) conducted a pilot-scale study at a former manufactured gas plant (MGP) site in New Jersey. The objective of the study was to determine the effectiveness of an innovative chemical/biological treatment process (MGP-REM process) to remediate soils contaminated with polynuclear aromatic hydrocarbons (PAHs). In order to identify the benefits of the MGP-REM process, the system was also operated in the conventional bioremediation mode.

Results showed that the MGP-REM process can effectively treat PAH-contaminated MGP site soils, and it reduced the toxicity of the soil by a factor of 50, as indicated by the Microtox Toxicity Test. The MGP-REM process was 70% more efficient than conventional bioremediation in the removal of the PAHs from the soils. Air emissions data suggest that minimal air pollution control and monitoring are required for the slurry-phase application of both the MGP-REM process and the conventional biological treatment. Process economics indicate that the MGP-REM process in a slurry-phase mode has an estimated treatment cost of $100/cubic yard for remediation of PAH-contaminated soils.

Index Entries: Pilot-scale study; polynuclear aromatic hydrocarbons; slurry-phase mode; chemical/biological treatment (MGP-REM process); Fenton's reaction; manufactured gas plant site.

INTRODUCTION

A pilot study was conducted in Elizabeth, New Jersey to evaluate the slurry-phase bioremediation of former manufactured gas plant (MGP) site soils contaminated with polynuclear aromatic hydrocarbons (PAHs). The

*Author to whom all correspondence and reprint requests should be addressed.

remedial technologies evaluated to treat the soil at this site were Institute of Gas Technology's (IGT's) innovative chemical/biological treatment process (MGP-REM process) and conventional bioremediation. The effect of the treatments on contaminant concentrations, soil microbiology, and toxicity was examined. Air emissions from the system were monitored to ensure compliance with the regulatory limits set by New Jersey Department of Environmental Protection (NJDEP).

Background

In 1950, natural gas replaced manufactured gas as the major gaseous fuel, and hence, the production of manufactured gas came to an end. Operations and residuals management practices of many former manufactured gas plants resulted in the contamination of their site soils with hazardous organic compounds. One of the main contaminants of concern are PAHs. Some of these PAHs are known carcinogens and pose an environmental hazard. As a result, these MGP sites may require cleanup to ensure environmental safety.

IGT has been developing engineering processes to remediate waste media, such as sediment, soil, sludge, ground water, and surface water, at former MGP sites contaminated with hazardous organic compounds (1–3). The ultimate goal is to provide cost-effective waste treatment technologies that furnish an efficient alternative to incineration and landfilling.

After identifying the limitations of conventional bioremediation, IGT has developed approaches to overcome these limitations. As a result of extensive bench-scale studies carried out since 1987, IGT has developed a process for PAH-contaminated soils that is a combination of biological and chemical treatment—the MGP-REM process.

Technology Description: MGP-REM Process

The MGP-REM process combines two remedial steps: (1) biological treatment and (2) chemical treatment. These steps can be applied in different sequences depending on the nature and degree of contamination, and the contaminated matrix.

Biological treatment harnesses the ability of microorganisms to break down or transform organic compounds into less hazardous forms. The two-, three-, and some of the four-ring PAHs are mostly biodegradable and can support the growth of bacteria. However, biodegradation of most of the higher ring (four- to six-ring) PAHs is often slow, incomplete, and/or insufficient (2,4,5). This is when the chemical treatment is beneficial.

Chemical treatment uses Fenton's reagent to transform the organic compounds in the waste matrix into environmentally benign end products. The Fenton's reaction, interaction of hydrogen peroxide (H_2O_2) and ferrous ions (Fe^{2+}), involves formation of highly reactive free hydroxyl radicals that oxidize the recalcitrant contaminants, such as the higher-ring PAHs, into

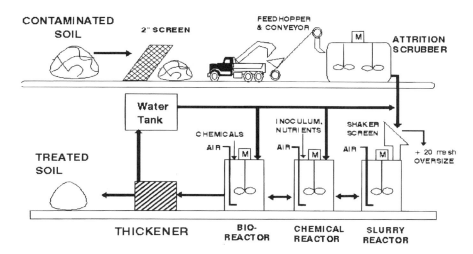

Fig. 1. Schematic of the slurry-phase MGP-REM process.

more readily biodegradable and water-soluble compounds *(1,6)*. Thus, the chemical treatment enhances the rate and extent of degradation of the PAHs in the soil.

METHODS

Process Description

A schematic diagram of the pilot plant is shown in Fig. 1. The treatment train included the following sections: feed preparation, reactor operation, and solids dewatering.

Feed Preparation

Excavated soil was screened to remove objects >2 in. The screened soil was then transported to an attrition scrubber, where soil was mixed with water to make a 50% soil slurry. Slurry from the attrition scrubber was sent to a 20-mesh vibrating screen to remove particles >1 mm, which were sent to the reject stockpile.

Reactor Operation

The reactor system was comprised of a slurry reactor, a bioreactor, and a chemical reactor; each with an identical capacity of 2100 gal. Slurry passing through the 20-mesh vibrating screen was diluted to the desired solids content in the slurry reactor and then pumped to the bioreactor. A startup run was used to promote and develop naturally occurring microbial population in the soil. The supernatant from this run that contained the active microorganisms was used as part of the makeup water to slurry the soil for subsequent tests. During the biological treatment, appropriate conditions for microbial growth were maintained as outlined in Table 1.

Table 1
Operating Conditions During Biological Treatment

Parameter	Operating Range
pH	6.5 - 7.5
Temperature, °F	65 - 85
Dissolved Oxygen (DO), mg/l	2 - 6
Nitrogen, mg/l	≥ 50
Phosphorus, mg/l	≥ 10

For chemical treatment, the slurry was transferred to the chemical reactor. Prior to chemical addition, the slurry was allowed to settle, and the supernatant (about 20% of the total volume) was decanted. This procedure served as a way to oxidize only the contaminants in the soil and to keep the microbial population in the supernatant unaffected. The remaining thick slurry in the chemical reactor was mixed with predetermined concentrations of ferrous sulfate (range of 1–100 mM in slurry) and hydrogen peroxide (range of 0.5–2% by volume of slurry). In order to prevent excessive foam formation, the aeration was temporarily stopped. Following the chemical treatment, the pH of the slurry was adjusted to 7.0, the supernatant with active microbial population was added back for subsequent biological treatment, and aeration was restarted.

Solids Dewatering

Following the termination of each test run, the treated soil slurry was pumped to a thickener for solid/water separation. The overflow from the thickener was transferred to the water management tank for reuse in subsequent tests, and the underflow was discharged to the drying bed.

Operating Protocol

The two test runs were operated in a batch mode. The treatment conditions in the reactor system are shown in Table 2. The MGP-REM process used a treatment sequence of biological followed by chemical and a final biological step. The MGP-REM process was compared with the Conventional Bioremediation Test, which included only biological treatment of the soil.

Table 2
Treatment Conditions in the Reactor System for Pilot-Scale Study

Parameter	Treatment Range
Slurry Volume, gals.	1800 - 2100
Total Solids, %	25 - 35
Slurry Residence Time, days	10 - 20
Air Flow Rate, SCFM	0 - 25
Hydrogen Peroxide, % (vol./vol.)	0.5 - 2.0
Treatment Scheme	1. MGP-REM Process (Biological-Chemical-Biological Treatment) 2. Conventional Bioremediation (Biological Treatment)

Sampling and Analyses

Sampling and analytical methods were consistent with the NJDEP-approved methods. Soil and slurry samples, corresponding to each step of the MGP-REM Process, were collected at suitable time-points.

PAHs in the samples were determined using a modified United States Environmental Protection Agency (USEPA) SW-846 Method 8270B analysis of soxhlet extracts. Soxhlet extraction was performed by EPA SW-846 Method 3540 using a 1:1 mixture of acetone and hexane as the extraction solvent. The modifications of Method 8270B involved a 30-m long XTI-5 (Restek) column (0.25-mm id and 0.5-µm film) installed in a Hewlett Packard 5890 II gas chromatograph/5971 mass selective detector. The injector temperature was 270°C, and the following temperature program was used: 40°C, for 4 min followed by an increase to 300°C at a rate of 10°C/min isothermal at 300°C for 20 min. The PAH analysis was done by Analab, Edison, NJ.

Microbiology tests (total heterotrophic bacterial counts) were conducted on samples using the Method of Most Probable Numbers, at IGT, Des Plaines, IL. The Microtox Test was used to determine the toxicity of the samples. The Microtox Test was used to determine the toxicity of the soil samples. This test system measures the light output of the luminescent bacteria after they have been challenged by a sample of unknown toxicity and compares it to the light output of a control (reagent blank) that contains no sample. The degree of the toxicity of the sample is indicated by the degree of light loss, which is a measure of the metabolic inhibition of the

Table 3
Air-Monitoring Procedures for the Pilot Study

Analytical Parameter	USEPA Method	Air Flow Rate	Cartridge
BTEX	TO-10	25 cc/min	Carbotrap-300
PAHs	TO-1	1000 cc/min	XAD-2 Resin

test organisms. Toxicity levels in the samples were tested at MSL Inc., Pendleton, SC.

Air-Monitoring System

The entire process, which included the feed preparation, biological, and chemical treatment, was conducted in closed systems: attrition scrubber, slurry reactor, bioreactor, and chemical reactor. In order to prevent organic emissions, off-gas from these units was passed through two granular activated carbon (GAC) canisters operated in series.

Air samples were collected before and after the GAC units, and were analyzed by Aqua Air Analytical, Raritan, NJ, for benzene, toluene, ethyl benzene, and xylene (BTEX) and PAHs. The procedures used for collecting and analyzing the air samples are shown in Table 3.

RESULTS AND DISCUSSION

Results of the pilot study are summarized in this section. PAHs were the main contaminants of concern, and hence, the reduction in PAH concentrations in the soil was used to evaluate the performance of the remedial technology.

PAH Results

MGP-REM Process

The complete results of the MGP-REM test are plotted in Fig. 2. PAH concentrations of samples taken at days 0, 20, and 40 are summarized in Table 4. Day 20 sample corresponds to the last day of the reactor operation, and the day 40 sample was taken from the thickener, 20 d after the treatment was completed. Table 4 shows that a 95% removal efficiency was achieved for total PAHs, and the noncarcinogenic and carcinogenic PAHs showed a reduction of 97 and 90%, respectively. Table 4 also shows that PAHs with fewer (2, 3, and 4) aromatic rings were degraded more effectively than those with more (5 and 6) aromatic rings. However, the 87 and 88% removal efficiency of five- and six-ring PAHs, respectively, is notable. Figure 3 compares the carcinogenic PAH concentrations before and after the MGP-REM

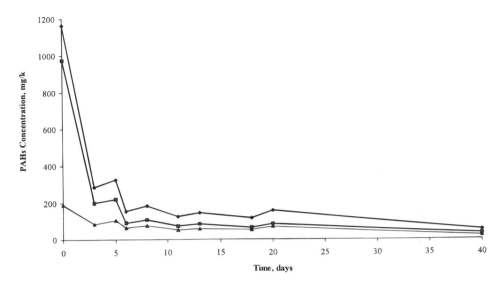

Fig. 2. Time plot of total, noncarcinogenic, and carcinogenic PAHs during the MGP-REM process. ◆ Total, ■ noncarcinogenic, ▲ carcinogenic.

Table 4
PAHs Degradation Data for MGP-REM Process

PAHs, (mg/kg)	Day 1	Day 20	Day 40*	% Degraded
Total	1164	157	53	95
Non-Carcinogenic	974	86	33	97
Carcinogenic	191	71	20	90
Individual Ring PAHs				
2-ring	202	21	7.6	96
3-ring	516	29	13.2	97
4-ring	337	41	18.2	95
5-ring	85	46	11.3	87
6-ring	24	21	2.8	88

*Samples taken from thickener.

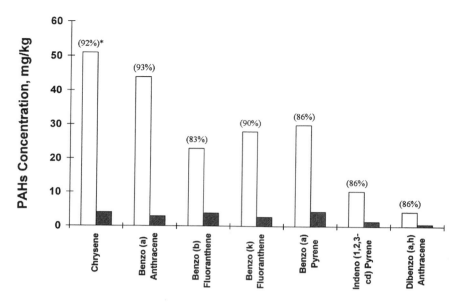

Fig. 3. Comparison of initial and final concentrations of carcinogenic PAHs during the MGP-REM process. ☐ Day 0, ■ day 40 (thickener). *Percent reduction is shown in parentheses.

process. The 86% reduction of benzo(a)pyrene is of special significance, since it is considered one of the most toxic and recalcitrant PAH moieties.

Conventional Bioremediation Test

The complete results of the Conventional Bioremediation Test are plotted in Fig. 4. PAH concentrations of samples taken at days 0, 13, and 35 are summarized in Table 5. Day 13 corresponds to the last sample taken during the reactor operation, and the day 35 sample was taken from the thickener. Table 5 shows that a 60% removal efficiency was achieved for total PAHs, and the noncarcinogenic and carcinogenic PAHs showed a reduction of 68 and 36%, respectively. Surprisingly, the two-ring PAHs showed a relatively lower removal efficiency of 31% compared to the three- and four-ring PAHs, which showed 76 and 69% reduction, respectively. The five-ring PAHs showed a very low removal of 22%, whereas the six-ring PAHs were not removed at all. Higher concentrations of PAHs were observed in the samples for day 35 compared to day 13, probably because of sampling and analytical variations. Figure 5 compares the carcinogenic PAH concentrations, before and after the conventional biological treatment. No reduction in concentration of benzo(a)pyrene was observed.

MGP-REM vs Conventional Bioremediation

The percent reductions in the total, noncarcinogenic, and carcinogenic PAHs for both the MGP-REM Process and conventional biological treatment are presented in Fig. 6. The comparison is made for day 40 and

Pilot-Scale Bioremediation

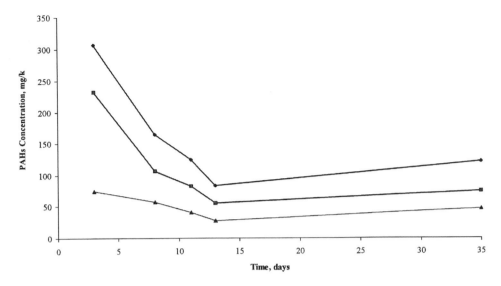

Fig. 4. Time plot of total, noncarcinogenic, and carcinogenic PAHs during the conventional bioremediation process. ◆ Total, ■ noncarcinogenic, ▲ carcinogenic.

Table 5
PAH Degradation Data for Conventional Biological Treatment

PAHs, mg/kg	Day 3	Day 13	Day 35*	% Degraded
Total	306	83	123	60
Noncarcinogenic	232	55	75	68
Carcinogenic	74	28	47	36
Individual Ring PAHs				
2-Rings	26	16	18	31
3-Rings	121	19	29	76
4-Rings	112	25	35	69
5-Rings	40	16	31	22
6-Rings	8	8	10	0

*Samples taken from thickener.

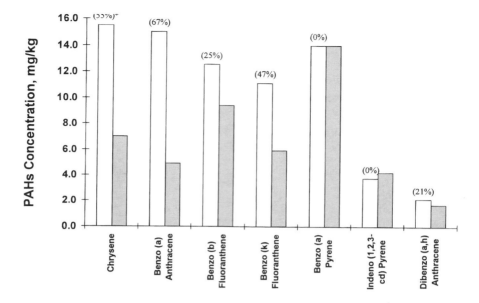

Fig. 5. Comparison of initial and final concentrations of carcinogenic PAHs during the conventional bioremediation process. ☐ Day 3, ■ day 35 (thickener). *Percent reduction is shown in parentheses.

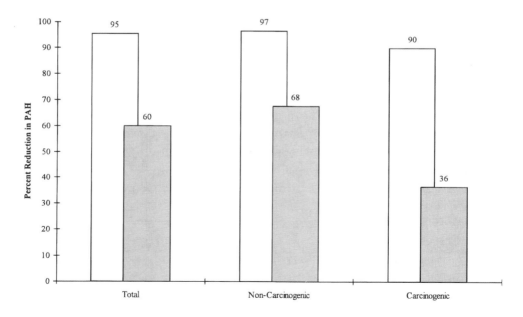

Fig. 6. Comparison of percent reductions in total, noncarcinogenic, and carcinogenic PAHs between the MGP-REM process and conventional bioremediation process. ☐ MGP-REM (day 40), ■ conventional bioremediation (day 35).

35 samples for the MGP-REM process and conventional biological treatment, respectively. The MGP-REM process reduced the total and noncarcinogenic by more than approx 30% over that achieved by the conventional biological treatment. More significant was the 54% higher reduction of the carcinogenic PAHs achieved by the MGP-REM process, compared to the conventional biological treatment, as shown earlier in Tables 4 and 5, respectively. The five- and six-ring PAHs were reduced 87 and 88%, respectively, for the MGP-REM process compared to the only 22 and 0% reduction observed for the conventional biological treatment. Also, more than 95% removal was observed for the two- to four-ring PAHs compared to much lower efficiencies obtained by conventional biological treatment. Thus, the MGP-REM process was clearly superior in the extent of removal of the PAHs from the soil. This is also indicated by the treatment end points of 53 and 123 mg/kg achieved for the total PAHs in the soil by the MGP-REM process and conventional biological treatment, respectively.

The pilot-scale study demonstrated that the MGP-REM process can effectively treat PAH-contaminated soils in slurry-phase bioremediation, and improves the extent (by 70%) of biodegradation of the PAHs (especially the carcinogenic) over the conventional biological treatment.

Microbiology Results

The results of the microbiological testing for the MGP-REM process and conventional biological treatment are shown in Fig. 7. The colony-forming units (CFU)/mL of slurry for the MGP-REM process were in excess of 10^6 through the entire treatment. The data for conventional biological treatment are available only up to day 8. The rest of the data were not generated because of improper sample storage. However, the CFU/mL were above 10^7 up to day 8, indicating a healthy bacterial population. A healthy bacterial population suggests that the chemical treatment did not annihilate the microorganisms during the MGP-REM process.

Toxicity Test Results

Soil samples were analyzed using the Microtox Toxicity Test to determine the effectiveness of the MGP-REM process to reduce the toxicity level in soil. The samples tested were soils before and after treatment. The toxicity results are presented in Table 6. The measure of toxicity used here is EC_{50}, meaning the percentage of soil slurry that will inhibit 50% of a known bacterial activity. Therefore, the lower the EC_{50} reading, the higher the toxicity in the soil. As seen from Table 6, the soil prior to testing was very toxic, and the MGP-REM process decreased the toxicity by more than 50 times.

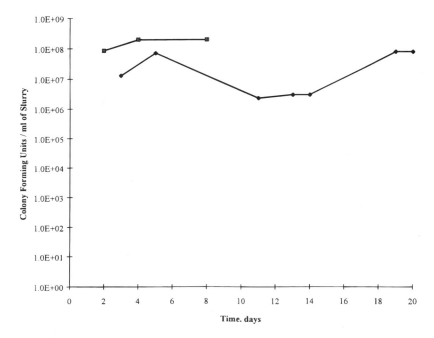

Fig. 7. Total heterotrophic counts in the slurry during the MGP-REM process and conventional bioremediation process. ◆ MGP-REM, ■ conventional bioremediation.

Table 6
Toxocity Test Results for the MGP-REM Process

Soil Sample Description	EC50	Interpretation
Before Treatment	0.212	Very toxic
After Treatment (day 20)	>10	Non toxic

Air-Monitoring Results

The compounds detected in the air emissions included BTEX, and naphthalene, acenaphthene, fluorene, and phenanthrene among the PAHs. These compounds together are termed as total hydrocarbons.

The highest emission rates of benzene and total hydrocarbons (BTEX + PAHs) in the exit airstream observed during the two test runs are shown in Table 7. These numbers are compared with the maximum emission rates permitted by the NJDEP. These results indicate that at any time during the pilot study, the emissions prior to and after the carbon treatment were significantly lower (by five orders) than the NJDEP limits of 0.05 and 0.5 lbs/h for benzene and total hydrocarbons (BTEX + PAHs), respectively.

Table 7
Maximum Emission Rates Obtained During the Pilot Study

Test Description	BTEX and PAHs, lbs/hr	Benzene, lbs/hr
MGP-REM Process	3.4×10^{-7}	Not detected
Conventional	8.1×10^{-6}	4.0×10^{-7}
NJDEP Limits	**0.5**	**0.05**

Table 8
Total Hydrocarbons (BTEX and PAHs) in the Soil During the Pilot Study

Test Description	Cumulative Air Emissions, lbs	Volatilization, %
MGP-REM Process	6.8×10^{-5}	0.003
Conventional Bioremediation	1.3×10^{-3}	0.8

The cumulative air emissions and percent volatilization of total hydrocarbons (BTEX + PAHs) for both the test runs are shown in Table 8. The percent volatilization of the total hydrocarbons was 0.003 and 0.8% for the MGP-REM process and the conventional bioremediation test, respectively. These results suggest that the air emissions from both test runs do not pose a health hazard, and therefore, this process requires minimal air pollution control and monitoring.

Process Economics

The economic evaluation for this pilot-scale soil remediation slurry-phase system was performed as an independent study (7). This evaluation was based on specific site characteristics, which were estimated for a soil volume of 71,500 cubic yards. Shorter treatment times were assumed for the MGP-REM process compared to Conventional Bioremediation. The treatment costs are summarized in Table 9.

SUMMARY OF THE RESULTS

- The pilot study demonstrated that the MGP-REM process can effectively treat MGP site soils and reduce the total PAHs concentration in the soil by 95%.
- The MGP-REM process reduced the toxicity levels present in the soil by approx 50 times, as indicated by the Microtox Toxicity Test.

Table 9
Treatment Costs for Slurry-Phase Bioremediation

Test Description	Cost in $/cubic yard
MGP-REM Process	100-135
Conventional Bioremediation	125

- The MGP-REM process is superior to conventional bioremediation in the extent (by approx 70%) of PAH removal from the soil.
- Microbiological data suggest that sufficient microbial levels were present during both the MGP-REM process and the conventional biological treatment.
- Negligible air emissions were detected during the slurry-phase operation, and off gas emissions from both processes were well in compliance with the NJDEP limits.
- Process economics study indicates that the MGP-REM process has an estimated cost of $100–135/cubic yard for remediation of soil.

Acknowledgments

This pilot study was sponsored by the Elizabethtown Gas Company, Elizabeth, NJ, the Gas Research Institute (GRI), and the Sustaining Membership Program (SMP) of the Institute of Gas Technology. The authors wish to acknowledge IGT's technical committee members, OHM Remediation Services Corp., and Degussa Corporation for their contribution to this project.

REFERENCES

1. Kelley, R. L., Gauger, W. K., and Srivastava, V. J. (1990), Application of Fenton's reagent as a pretreatment step in biological degradation of aromatic compounds. Presented at IGT's Third International Symposium on Gas, Oil, Coal, and Environmental Biotechnology, New Orleans, LA.
2. Liu, B. Y., Pradhan, S. P., Srivastava, V. J., Pope, R., Hayes, T., Linz, D., Proulx, C., Jerger, D., and Woodhull, P. (1994), An evaluation of slurry-phase bioremediation of MGP soils. Proceedings of Gas, Oil, and Environmental Biotechnology VII, Des Plaines, Illinois: Institute of Gas Technology, Des Plaines, IL.
3. Liu, B. Y., Srivastava, V. J., Paterek, J. R., Pradhan, S. P., Pope, J. R., Hayes, T. D., Linz, D. G., and Jerger, D. E. (1993), MGP soil remediation in a slurry-phase system: a pilot-scale test. Presented at Sixth International IGT Symposium on Gas, Oil, and Environmental Biotechnology, Colorado Spring, CO.
4. Park, K. S., Sims, R. C., DuPont, R. R., Doucette, W. J., and Matthews, J. E. (1990), *Environ. Toxicol. Chem.* **9,** 187–195.

5. Lauch, R. P., Herrmann, J. G., Mahaffey, W. R., Jones, A. B., Dosani, M., and Hessling, J. (1992), *Environ. Prog.* **11, no. 4,** 265–271.
6. Ravikumar, J. X. and Gurol, M. D. (1992), Fenton's reagent as a chemical oxidant for soil contaminants. Paper presented at the Chemical Oxidation Technology for the Nineties, Second International Symposium, Vanderbilt University, Nashville, TN.
7. Remediations Technologies Inc. (1995), Bioslurry treatment of MGP impacted soils site specific economic evaluation Elizabethtown Gas Company. Prepared for Gas Research Institute, Chicago, IL.

Relating Ground Water and Sediment Chemistry to Microbial Characterization at a BTEX-Contaminated Site**

S. M. Pfiffner,[1] A. V. Palumbo,*,[1] T. Gibson,[2] D. B. Ringelberg,[3] and J. F. McCarthy[1]

[1]Environmental Sciences Division, Oak Ridge National Laboratory, Oak Ridge, TN 37831; [2]General Motors Research and Development Center, Warren, MI 48090; and [3]Center for Environmental Biotechnology, The University of Tennessee, Knoxville, TN 37932

ABSTRACT

The National Center for Manufacturing Science is investigating bioremediation of petroleum hydrocarbon at a site near Belleville, MI. As part of this study, we examined the microbial communities to help elucidate biodegradative processes currently active at the site. We observed high densities of aerobic hydrocarbon degraders and denitrifiers in the less-contaminated sediments. Low densities of iron and sulfate reducers were measured in the same sediments. In contrast, the highly contaminated sediments showed low densities of aerobic hydrocarbon degraders and denitrifiers, and high densities of iron and sulfate reducers. Methanogens were also found in these highly contaminated sediments. These contaminated sediments also showed a higher biomass, by the phospholipid fatty acids, and greater ratios of phospholipid fatty acids, which indicate stress within the microbial community. Aquifer chemistry analyses indicated that the highly contaminated area was more reduced and had lower sulfate than the less-contaminated area. These conditions suggest that the subsurface environment at the highly contaminated area had progressed into sulfate reduction and methanogenesis. The less-contaminated area, although less reduced, also appeared to be progressing into primarily iron- and sulfate-reducing microbial communities. The proposed treatment to stimulate bioremediation includes addition of oxygen and nitrate to the subsurface. Ground

*Author to whom all correspondence and reprint requests should be addressed.

water chemistry and microbial analyses revealed significant differences that resulted from the injection of dissolved oxygen and nitrate. These differences included an increase in Eh, small decrease in pH, and large decreases in BTEX, dissolved iron, and sulfate concentrations at the injection well. Injected nitrate was rapidly utilized by the subsurface microbial communities, and significant nitrite amounts were observed in the injection well and in nearby down-gradient observation wells. Microbial and molecular analyses indicated an increase in denitrifying bacteria after nitrate injection. The activity and population of denitrifying bacteria were significantly increased at the injection well relative to a down-gradient well for as long as 2 mo after the nitrate injection ended.

Index Entries: Microbial characterization; BTEX or petroleum hydrocarbon; bioremediation; subsurface; ground water; oxygen injection; nitrate injection.

INTRODUCTION

Microbial populations capable of degrading petroleum hydrocarbons (BTEX) *(1–6)* can be found at contaminated sites. Microbial biomass, community structure, and biodegradative activities are limited by properties of the subsurface environment, such as moisture, pH, and the availability of carbon, nutrients, and electron donors/acceptors *(6–8)*, and the microbial community can affect these properties. For example, biodegradation at gasoline contaminated sites has been associated with partial depletion of subsurface oxygen, nitrate, and sulfate *(9–11)*, and the addition of oxygen and nitrate has enhanced the biodegradation of BTEX *(1,2,5,6)*. As part of the National Center for Manufacturing Science (NCMS) petroleum hydrocarbon site bioremediation study, we examined microbial communities to help elucidate biodegradative processes currently active at the bioremediation site. Dissolved oxygen and nitrate were injected for about 3 mo to test their effects on subsurface geochemistry, microbiology, and rates of intrinsic bioremediation from aerobic and denitrification processes. The goal of this demonstration was to determine the presence of extant bacteria capable of BTEX biodegradation and to monitor the bioremediation effort. This article examines some of the microbial populations and degradative activities of sediments prior to remedial efforts, and tests ground waters following oxygen and nitrate additions to the subsurface.

MATERIAL AND METHODS

Site Description, Operations, and Sample Collection

The NCMS (Ann Arbor, MI) Advanced *In Situ* Bioremediation study site at the industrial facility near Belleville, MI, was contaminated prior to 1991 by gasoline from a leaking underground storage tank. The site con-

Ground Water and Sediment Chemistry

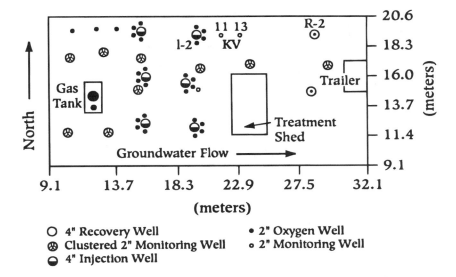

Fig. 1. The site map shows the location of the injection and recovery wells (I-2 and R-2) and the monitoring wells (KV-11 and KV-13). Ground water flows eastwardly from the highly contaminated up-gradient well (I-2) to the less-contaminated down-gradient well (R-2).

tains a shallow perched aquifer with uniform sandy soil isolated vertically by underlying clay till at 3.0–4.3 m below land surface, and has been well characterized for subsurface contaminant distribution, geology, and hydrology. A total of 100 wells and piezometers were installed at the site, and the distribution of contaminants was analyzed in soil cores and ground water recovered from the wells. This study focused on four sampling locations shown in Fig. 1. The injection and recovery wells (4-in. diameter), I-2 and R-2, were fully screened across the aquifer. Well I-2 is approx 9.1 m down-gradient from the source area, and well R-2 is 8.8 m down-gradient of I-2. Monitoring wells (2-in diameter) KV-11 and KV-13 were between the injection and recovery wells, and were 0.8 and 1.7 m, down-gradient from well I-2, respectively. Another fully screened monitoring well, I-2B, is located 0.3 m down-gradient of well I-2.

In spring 1995, subsurface samples for geochemical and microbiological characterization were recovered from core hole I-2 and R-2 using the standard split-spoon sampling technique. The microbial community structure was analyzed in sediments recovered from two depths (1.8–2.1 and 2.4–2.7 m) in both bore holes. For the first bioremediation treatment, oxygen was injected into well I-2 using an innovative passive diffusion technique *(11)*, and wells KV-11, KV-13, and R-2 served as down-gradient monitoring wells. After the end of oxygen injection experiments (June–August 1995, 70 d), nitrate was injected into well I-2. Prior to nitrate injection, nitrate levels in the ground water were below detection limits (0.44 mg/L). Sodium nitrate was added at 5.2 g/d for 35 d (August–October 1995). During the nitrate injection, ground water was monitored for geochemical analyses. In January 1996, 2 mo after nitrate injection ceased, ground water was collected for complete characterization.

Geochemical Analyses

Moisture content, BTEX, total organic carbon, and total petroleum hydrocarbon were measured in sediment and ground water samples at the on-site field laboratory. Ground water temperature and conductivity were measured at the time of sample collection. Also, Eh was measured by platinum electrode (ORP/Hach); nitrate was measured by using a Hach #817 or Hach #353 test methods; and sulfate, ferrous iron, and total iron were measured with other Hach methods.

Microbial Analyses

Denitrifiers, methanogens, iron reducers, and aerobic hydrocarbon degraders were assessed by a turbidimetric most probable number (MPN) three-tube technique (12). Aerobic hydrocarbon degraders were grown in phosphate-buffered mineral salt medium (PBBM) (12) supplemented with 0.1 mL Texaco gasoline. Methanogens were grown on PBBM supplemented with 40 mM acetate and methanol, and with cysteine HCl as a reductant. Denitrifiers in sediments were enumerated with nutrient broth supplemented with KNO_3 (1 g/L), but ground water samples were enumerated on a basal salt medium described by Fries et al. (13). Iron reducers were enriched for with a medium described by Lovley et al. (14). Media pHs-were adjusted to 7.1 for aerobes and 7.3 for anaerobes. These microbial enumerations were performed at the University of Tennessee.

Total heterotrophic bacterial counts in sediment and ground water samples were determined at the field site laboratory by the spread plate enumeration method (11) using plate count agar (Hach, Loveland, CO). Plates were incubated at 25°C and counted at maximum growth (4–14 d).

Biological Activity Reaction Tests (BART, Droycon Bioconcepts, Inc., Regina, Saskatchewan, Canada) were performed at the industrial plant laboratory and were used to determine the specific reactions that are catalyzed by the bacterial enzymes (15). Thus, the activity tests can be used to estimate the capacity of the native microbial population to perform oxidation and reduction reactions that can biodegrade contaminants under the natural subsurface conditions. Total aerobes, sulfate reducers, denitrifiers, iron-related bacteria (aerobic and anaerobic), and fluorescent pseudomonads were enumerated. The test reactor containing nutrient medium was inoculated with 15–20 mL of ground water and was incubated for 48 h at a temperature of 25°C prior to analyses. The tests were used to determine the semiquantitative densities of various physiological populations.

Microbial Phospholipids

Microbial biomass and community structure were estimated using ester-linked phospholipid fatty acids (PLFA) *(16)*. Total PLFA was recovered from 75 g of sediment or 1 L of ground water filtered through 0.2-µm pore size inorganic filters (Anodisc 47, Whatman, Maidstone, England). PLFA was quantitatively extracted from the frozen samples (–50°C) as previously described *(16)*.

The extract was fractionated into specific lipid classes and transesterified to form phospholipid fatty acid methyl esters *(17)*. A gas chromatograph equipped with a mass-selective detector was used to identify and verify individual PLFA *(18)*. Double-bond position in the monounsaturated PLFA was determined as described in Nichols et al. *(19)*. Ground water samples showed biomass levels near the background detection limit of 2.14 pmol/L (or 5.36×10^4 cells/L).

Statistical Analyses

Log-transformed PLFA mole percentages were used in statistical analyses. Ein*sight pattern recognition software (Infometrix, Inc., Seattle, WA) was used for hierarchical cluster (HCA) and principal component analyses (PCA). These analyses were used to determine sample relatedness and factors that may account for variance in the data set *(18)*.

RESULTS AND DISCUSSION

Sediment and Ground Water Analyses Prior to Treatment

The highest total BTEX concentration (2.96 mg/L) was present in I-2 the most western (up-gradient) well in the transect. Concentrations decreased toward the east in the direction of ground water flow. R-2, the most eastern well, contained a very low total BTEX level (0.02 mg/L). Anaerobic conditions were associated with the hydrocarbon contaminant plume. Aquifer chemistry analyses (Table 1) indicated that the highly contaminated area was more reduced (Eh –69 mV) and had lower sulfate (37 mg/L) than the less-contaminated area (Eh –47 mV, sulfate 68 mg/L).

High densities of aerobic hydrocarbon degraders and denitrifiers, but low densities of methanogens, iron reducers, and sulfate reducers were observed in the less-contaminated R-2 sediments (Fig. 2, Table 2). In contrast, the highly contaminated I-2 sediments showed low densities of aerobic hydrocarbon degraders and denitrifiers and high densities of iron and sulfate reducers (Fig. 2, Table 2). Methanogens were also found in these highly contaminated sediments. The chemical and microbial analyses indicate that the subsurface environment at the highly contaminated area had progressed into sulfate reduction and methanogenesis. The less-contaminated area, although less reduced and containing more sulfate,

Table 1
Ground Water Geochemical Characterization

Monitoring well	Prior to treatment				Two months after both oxygen and nitrate injection ended			
	I-2	KV-11	KV-13	R-2	I-2	KV-11	KV-13	R-2
Screen zone or distance from injection well	2-3 m depth	2.7 m depth	2.7 m depth	2-3 m depth	0 m from injection	0.8 m from injection	1.8 m from injection	8.8 m from injection
Eh (mV)	-69.5	-65.2	-55.4	-47.0	-32.0	-101.3	-65.4	+72.7
Fe(II) (mg/L)	21.2	49.9	26.0	2.15	7.33	21.7	18.3	4.6
total Fe (mg/L)	25.8	56.6	32.5	3.25	9.17	31.3	22.1	5.1
Sulfate (mg/L)	37.0	8.0	12.0	68.0	1.0	0.00	0.00	67
Nitrate (mg/L)	<0.88	0	0	<0.44	0.00	0.00	0.00	0.00
Total BTEX (mg/L)	2.96	2.91	2.58	0.022	0.19	1.01	0.909	0.0058

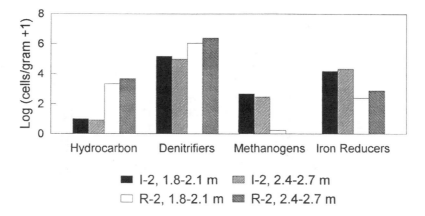

Fig. 2. Physiological types of bacteria present in the four sediments sampled at two depths from the highly contaminated (I-2, 10–16 mg/kg BTEX) and the less-contaminated (R-2, 0.02 mg/kg BTEX) boreholes.

Table 2
Aquifer Microbial Characterization Performed by the Field Site Laboratory
(CFU/mL for Spread Plate Counts and BART)

Analysis	Location I-2	Location R-2
Total heterotrophs	1900	500
Fluorescent pseudomonads	Present	Present
Total aerobic bacteria	~60	~320
Iron-related bacteria	~500	~250
Sulfate-reducing bacteria	~1300	~250
Denitrifying activity	Weak positive	Negative

also appeared to be progressing into primarily iron- and sulfate-reducing microbial communities.

Aquifer material, when examined for physiological types of microorganisms (Table 2), revealed a four to five orders of magnitude difference between the heterotrophic densities observed by the field site laboratory and by the University of Tennessee laboratory. The differences could be attributed to sample heterogeneity, but may also be influenced by sediment shipment and handling, and the types of enumeration media used in the two laboratories. Results between the total aerobic bacteria and the aerobic hydrocarbon degraders were comparable for the two laboratories. I-2 exhibited 7 hydrocarbon degraders and 60 total aerobes/g of sediment, whereas R-2 sediment contained 300 total aerobes and 4600 hydrocarbon degraders/g of sediment (Fig. 2, Table 1). Iron-related and sulfate-reducing bacteria were observed at abundances >250 cells/g of sediment or mL of aquifer material by both laboratories.

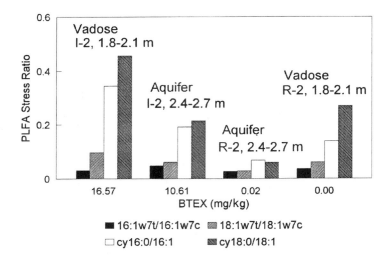

Fig. 3. PLFA stress indicators, based on four types of PLFA ratios, found in the four sediment samples and related to BTEX concentration. Larger stress ratios indicated more stress was experienced in the microbial community. The 1.8–2.1 m depth sediments were from the vadose zone and the 2.4–2.7 m depth sediments were from the aquifer.

Phospholipid fatty acid methyl ester results showed that the highly contaminated I-2 sediments had two to four times the biomass (195–259 pmol/g) exhibited by the less-contaminated R-2 sediments (55–97 pmol/g). A large variety of PLFA were detected in the sediments, but monounsaturated PLFA were the most abundant fatty acid methyl ester representing ~40% of the total PLFA. The abundance of 18:1ω7c and 16:1ω7c indicated the utilization of a pathway in fatty acid synthesis known as anaerobic desaturation, which is typically employed by Gram-negative bacteria. This dominant (~40%) Gram-negative community expressed higher stress ratios (trans/cis, cyclopropyl/monounsaturate) in the two samples with the higher biomass (Fig. 3). Those two samples were from the highly contaminated area. Physiological stress was also indicated in the vadose zone sediments compared to the saturated aquifer sediments (Fig. 3). A substantial level of terminally branched and mid-chain-branched saturates was found, which indicated the presence of sulfate reducers *(20,21)*. However, the ratios of iso/anteiso PLFA are lower than expected for pure cultures of sulfate reducers; these PLFA may be from Gram-positive influences as well. Other evidence for sulfate reducers included the presence of 10me16:0 and i17:1ω7c. R-2 sediment (1.8–2.1 m) showed the largest percentage or relative abundance (20%) of 10me16:0, thus indicating the presence of sulfate reducers.

The HCA of the PLFA profiles indicated that the I-2 sediments were closely related at a similarity index of 0.65, where 0 represents no similarity and 1 represents an identical match. A distant relationship (0.13 similarity index) was exhibited between the R-2 sediment (1.8–2.1 m) and both I-2 sediments. The R-2 sediment (2.4–2.7 m) appeared to be a unique sample with no similarity to the other sediments. These differences between samples demonstrated the spatial heterogeneity seen in the subsurface *(22)*. PCA indicated that 92% of the variance could be explained with a single principal component (PC1). The fatty acid methyl esters under PC1, which were most heavily weighted or assigned the greatest positive correlation coefficients, were 16:1ω7c, 16:0, 18:1ω7c, and 10me16:0 (listed in descending order). These PLFAs indicated the dominance of Gram-negative bacteria and sulfate-reducing bacteria as previously described. Under PC2, sediment R-2 (1.8–2.1 m) was separated from the other sediments by the abundance of 10me16:0, which as already indicated suggests the presence of sulfate reducers. Likewise, PC2 revealed that sediment R-2 (2.4–2.7 m) contained a higher abundance of 16:1ω7c and 18:1ω7c, indicative of the anaerobic desaturation pathways, thus containing a dominant Gram-negative microbial community.

Effect of Oxygen Injection

From June–August 1995, high levels of dissolved oxygen (up to 39 mg/L) were injected from a source of pure oxygen into the ground water at well I-2. Two additional monitoring wells (KV-11 and KV-13) were installed along the ground water flow path between wells I-2 and R-2. Oxygen injection caused other significant changes in the geochemistry for wells in this transect. The oxidation state of the aquifer materials was strongly affected, leading to increases in Eh. Oxygen injection in the test area decreased dissolved ferrous iron in the aquifer ground water while increasing the ferric iron, presumably in suspended and complexed forms *(11)*. Total BTEX, sulfate, and dissolved iron in I-2 decreased, probably because of enhanced aerobic biodegradation processes (Table 1). Carbon dioxide and methane levels were higher in the up-gradient end (I-2) of the transect (data not shown), probably as a result of the aerobic and methanogenic processes coexisting in the subsurface. Ammonium and phosphate both decreased in the transect, probably as a result of enhanced microbial growth during the oxygen injection experiment (data not shown). Total heterotrophic counts and the level of aerobic bacteria shown by BART increased with oxygen addition to the subsurface, but these increases continued only while increased dissolved oxygen was maintained *(11)*.

Effect of Nitrate Injection

The objective of the nitrate addition was to monitor the migration of nitrate and its utilization by denitrifying bacteria, and to look for changes in activity for denitrifying bacteria. Two months after the nitrate injection, ground water samples were collected over a transect from well I-2 to well

Table 3
Ground Water Microbial Characterization Performed
by the University of Tennessee (Cells/mL)

Location	I-2	KV-11	KV-13	R-2
Denitrifiers (NO_3)	28	4	4	28
Ammonium utilizers (NH_4Cl)	1	10	100	1000
Heterotrophs (aerobic)	100,000	100	100	100,000
Heterotrophs (anaerobic)	10	1	100	1000
Sulfate reducers	10	100	100	1
Iron reducers	0	0	0	0
Methanogens	0–1	1	1–10	0–1
PLFA biomass	68	277	84	377

R-2. Ground water chemistry results are shown in Table 1. Interestingly, the KV wells, which had higher BTEX concentrations (Table 1), showed lower abundances of denitrifiers and aerobic heterotrophs than ground water with lower BTEX concentrations (Table 3). When the concentration of BTEX was compared before and after nutrient addition, it was found that >90% of the BTEX was degraded. Similarly, aerobic and anaerobic degradation of diesel fuel has been enhanced by the addition of oxygen and nitrate to microcosms (2).

There were lower Eh values in ground water from the KV wells, indicating anaerobic conditions down-gradient from the injection well, but higher Eh values were observed furthest down-gradient from the contaminated area (Table 1, Fig. 4). Although, ferrous iron decreased because of abiotic interaction with the oxygen addition, the nitrate injection stimulated the biological conversion to ferrous iron as was seen in the KV wells as sulfate was utilized and BTEX was degraded. This data correspond with the work of Beller and Reinhard (23), who showed the enhancement of anaerobic toluene degradation under sulfate-reducing condition by the addition of ferrous iron. Lower sulfate concentrations were observed in ground water samples (KV wells) where higher abundances of sulfate-reducing bacteria were seen (Fig. 4). These wells also contained the highest levels of contamination. Ground water recovered from wells (I-2 and R-2) had lower levels of contamination (Table 1) and exhibited higher Eh values, higher sulfate concentrations, and lower numbers of sulfate reducers (Fig. 3).

The concentrations of nitrate and nitrate were monitored in the injection well (I-2) and in down-gradient wells (I-2B, KV-11). Nitrate concentrations increased rapidly in the injection well (I-2) and rose more slowly (I-2B) or not at all (KV-11) in down-gradient wells (Fig. 5). Significant nitrate could only be tracked about 0.3 m from the injection well after 35 d of injection. In contrast, a conservative tracer (bromide injection at concentrations of at least 260 mg/L) had migrated to all the monitoring points by

Ground Water and Sediment Chemistry

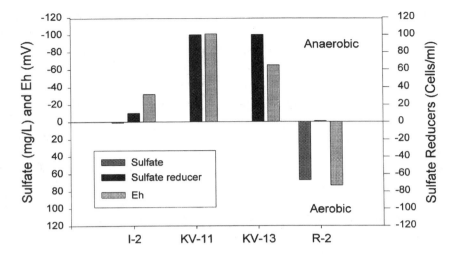

Fig. 4. Sulfate reducers (top portion of right axis), sulfate concentrations (bottom portion of left axis), and Eh values (full scale on left axis) observed in ground water from monitoring wells 2 mo after the oxygen and nitrate treatments ceased. The top portion of the graph represents observed values indicative of anaerobic conditions, whereas the bottom portion of the graph indicates aerobic conditions.

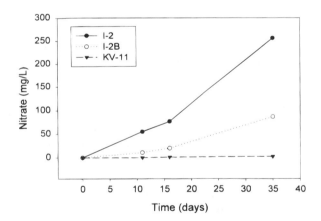

Fig. 5. Changes in ground water nitrate concentrations during the injection of nitrate. Well I-2 was the injection well with down-gradient monitoring wells I-2B and KV-11 at distances from I-2 of 0.3 and 0.8 m, respectively.

11 d after the injection and was detected at 65% of the concentration injected. Using the bromide tracer, the ground water flow velocity was estimated at 0.24 m/d. Nitrite was not detected in any of the ground water at day 11. By day 16, nitrite was detected in I-2 ground water at 1 mg/L, and by day 35, both I-2 and I-2B ground water samples showed 2–3 mg/L nitrite. Nitrate was not detected in KV-11 ground water. The low recovery of nitrite (1–5%) may be attributed to the continued reduction of nitrite by

the subsurface microbial communities to nitric oxide, nitrous oxide, or nitrogen. The utilization of nitrate on injection and the low recovery of nitrite observed at the industrial site was similar to that experienced in laboratory microcosms by Ball and Reinhard (24).

Denitrifying populations increased after nitrate addition, and the subsurface community was still able to utilize nitrate 2 mo after nitrate injection ceased. The denitrifing bacteria, as estimated by BART, were enumerated at <100 cells/mL for day 0 in I-2 ground water, and for days 0, 12, and 35 in KV-11 ground water. For ground water from well I-2 for the sampling days 12 and 35, denitrifiers were estimate at 100–100,000 and 10,000–1,000,000 cells/mL, respectively. Toluene-degrading denitrifiers were isolated from well I-2 ground water and confirmed by PCR amplification of primers specific to *Azoarcus tolulyticus*, a known toluene-degrading denitrifier, and related strains (13). These denitrifiers were not observed in well KV-11 or further down-gradient in KV-13 or R-2. The detection of toluene degradation activity correlated with the presence of BTEX as an electron donor and nitrate as an electron acceptor as observed in well I-2 ground water. Thus, the denitrifying populations were stimulated by nitrate injection, and BTEX degradation was enhanced.

CONCLUSIONS

Microbial characterization of both sediment and ground water revealed an anaerobic microbial community that consisted of sulfate reducers and methanogens in the highly contaminated area, and primarily iron and sulfate reducers in the less-contaminated area. These characterizations were supported by the aquifer and ground water chemistry, which showed more reduced conditions (lower redox potentials) and less sulfate in the highly contaminated area compared with the less-contaminated areas. Microbial analyses of ground water indicated changes in the microbial community composition as a result of oxygen amendment at well I-2. Decreases in strict anaerobic bacteria along with increases in aerobic bacteria were demonstrated in well I-2 ground water. The geochemical data of a higher redox potential and sulfate concentration support these observations. Microbial characterization indicated that several electron acceptors were important in implementation and treatment to achieve effective and efficient bioremediation. Monitoring the microbial community resulted in direct evidence for the changes seen in the geochemical parameters as different electron acceptors were utilized by the subsurface microbial populations.

ACKNOWLEDGMENTS

We thank the staff at the industrial plant for collecting the sediment and ground water samples, as well as for their sample analyses. Appreciation goes to Chuck Payne, Laurel O'Connor, Scott Newton, and

Jennifer Cooper at CEB/UT, who initiated and analyzed the samples for microbial activity and characterization. This research was supported, in part, by an appointment through the Oak Ridge National Laboratory (ORNL) postdoctoral research associate program administered by the Oak Ridge Institute for Science and Education and ORNL. This work was funded by DP/DOE under Cooperative Research and Development Agreement (CRADA) No. DOE 92-0077 between Lockheed Martin Energy Research Corp. and the National Center for Manufacturing Sciences. ORNL is managed by Lockheed Martin Energy Research Corp; for the US Department of Energy under contract DE-AC05-96OR22464.

REFERENCES

1. Aelion, C. M. and Bradley, P. M. (1991), *Appl. Environ. Microbiol.* **57**, 57–63.
2. Bregnard, T. P-A., Hohener, P., Haner, A., and Aeyer, J. (1996), *Environ. Toxicol. Chem.* **15**, 299–307.
3. Siegrist, R. L., Phelps, T. J., Korte, N. E., Pickering, D. A., Mackowski, R., and Cooper, L. W. (1994), *Appl. Biochem. Bioeng.* **45**, 757–773.
4. Phelps, T. J., Siegrist, R. L., Korte, N. E., Pickering, D. A., Strong-Gunderson, J., Palumbo, A. V., Walker, J. F., Morrissey, C. M., and Mackowski, R. (1994), *Appl. Biochem. Biotechnol.* **45**, 835–845.
5. Long, S. C., Aelion, C. M., Dobbins, D. C., and Pfaender, F. K. (1995), *Microbiol. Ecol.* **30**, 297–307.
6. Zhou, E., and Crawford, R. L. (1995), *Biodegradation* **6**, 127–140.
7. Phelps, T. J., Pfiffner, S. M., Sargent, K. A., and White, D. C. (1994), *Microbiol. Ecol.* **28**, 351–364.
8. Palumbo, A. V., Scarborough, S. P., Pfiffner, S. M., and Phelps, T. J. (1995), *Appl. Biochem. Biotechnol.* **55/56**, 635–647.
9. Borden, R. C., Gomez, C. A., and Becker, M. T. (1995), *Ground Water* **33**, 180–189.
10. Wiedemeier, T. H., Swanson, M. A., Wilson, J. T., Kampbell, D. H., Miller, R. N., and Hansen, J. E. (1995), in *Intrinsic Bioremediation,* Hinchee, R. E., Wilson, J. T., Downey, D. C., eds., Battelle, Columbus, OH, pp. 31–51.
11. Gibson, T. L., Abdul, S. A., and Chalmer, P. D., (1996), Annual Conference for Petroleum Hydrocarbons and Organic Chemicals in Groundwater. National Groundwater Association, Houston, TX, Nov. 13–15.
12. Pfiffner, S. M., Phelps, T. J., and Palumbo, A. V. (1995), in *Bioremediation of Chlorinated Solvents,* Hinchee, R. E., Leeson, A., and Semprini, L., eds., Battelle, Columbus, OH, pp. 263–271.
13. Fries, M. R., Zhou, J., Chee-Sanford, J, and Tiedje, J. M. (1994), *Appl. Environ. Microbiol.* **60**, 2802–2810.
14. Lovley, D. R., Chapelle, F. H., and Phillips, E. J. P. (1990), *Geology* **18**, 954–957.
15. Cullimore, D. R. (1993), *Practical Manual of Groundwater Microbiology,* Lewis, Chelsea, MI.
16. White, D. C., Davis, W. M., Nickels, J. S., King, J. D., and Bobbie, R. J. (1979), *Oceologia* **40**, 51–62.
17. Tunlid, A., Ringelberg, D. B., Phelps, T. J., Low, C., and White, D. C. (1989), *J. Microbiol. Methods* **10**, 139–153.
18. Kieft, T. L., Ringelberg, D. B., and White, D. C. (1994), *Appl. Environ. Microbiol.* **60**, 3292–3299.
19. Nichols, P. D., Guckert, J. B., and White, D. C. (1986), *J. Microbiol. Methods* **5**, 49–55.
20. Kohring, L. L., Ringelberg, D. B., Devereux, R., Stahl, D., Mittelman, M., and White, D. C. (1994), *FEMS Microbiol. Lett.* **119**, 303–308.

21. Vainshtein, M., Hippe, H., and Kroppenstedt, R. M. (1992), *Syst. Appl. Microbiol.* **15,** 554–566.
22. Palumbo, A. V., Zhang, C., Phelps, T. J, and Jager, H. (1996). Abstracts of the 96th General Meeting of the American Society for Microbiology, p. 113.
23. Beller, H. R. and Reinhard, M. (1995), *Microbiol. Ecol.* **30,** 105–114.
24. Ball, H. A. and Reinhard, M. (1996), *Environ. Toxicol. Chem.* **15,** 114–122.

Retaining and Recovering Enzyme Activity During Degradation of TCE by Methanotrophs**

A. V. Palumbo,* J. M. Strong-Gunderson, and S. Carroll

Environmental Sciences Division, Oak Ridge National Laboratory, P.O. Box 2008, Oak Ridge, TN 37831-6038

ABSTRACT

To determine if compounds added during trichloroethylene (TCE) degradation could reduce the loss of enzyme activity or increase enzyme recovery, different compounds serving as energy and carbon sources, pH buffers, or free radical scavengers were tested. Formate and formic acid (reducing power and a carbon source), as well as ascorbic acid and citric acid (free radical scavengers) were added during TCE degradation at a concentration of 2 mM. A saturated solution of calcium carbonate was also tested to address pH concerns. In the presence of formate and methane, only calcium carbonate and formic acid had a beneficial effect on enzyme recovery. The calcium carbonate and formic acid both reduced the loss of enzyme activity and resulted in the highest levels of enzyme activity after recovery.

Index Entries: TCE; methanotrophs; biodegradation; enzyme activity; formate.

INTRODUCTION

The cometabolic degradation of trichloroethylene (TCE) by methanotrophs utilizing the methane monooxygenase enzyme has become a major focus of study (e.g., *1–6*), since the discovery of methane induced TCE degradation by Wilson and Wilson (*7*). The prime motivation for studies of the biodegradation of TCE is the high incidence of TCE presence

*Author to whom all correspondence and reprint requests should be addressed.

at superfund sites, Department of Energy sites, and Department of Defense sites across the country.

Maintaining high rates of cometabolic degradation of TCE by methanotrophs is difficult because of reductions in enzyme activity. Loss of enzyme activity may be the result of free radical effects caused by the epoxidation of the TCE by the soluble methane monooxygenase (sMMO) enzyme *(8)*. Maintenance of high levels of sMMO is important, because cultures with high sMMO activity achieve the highest TCE degradation rates (e.g., *9,10*). The addition of formate is apparently one way to maintain sMMO activity. The effects of formate *(11,12)* on TCE degradation have been examined, and the effect of this added reducing power is beneficial to TCE degradation.

The goal of this study was to identify methods for the maintenance or recovery of high levels of enzyme for ultimate use in a multistage bioreactor for methanotrophic degradation of TCE *(13)*. The objective of these experiments was to determine if classes of compounds other than formate, e.g., antioxidant compounds, could either prevent enzyme destruction during TCE degradation or promote enzyme recovery.

MATERIALS AND METHODS

Culture Conditions

These experiments were performed with a mixed culture of *Methylosinus trichosporium* strain OB3b, a Type II obligate methanotroph and a heterotroph. A modification *(4)* of NATE medium *(14)* was used to grow the cultures. Further modifications of the medium in these experiments were substitution of additional nitrate for ammonia and elimination of copper from the trace metal formulation.

Cells for sMMO inhibition and recovery experiments were obtained from the mixed cultures maintained in an airlift bioreactor (Kontes) continuously flushed with 3% methane in air. Optical density ($\lambda = 600$ nm) and sMMO were measured prior to use, and allowed for the standardization and comparisons of experiments. Growth phase appeared to affect sMMO activity (unpublished data). Cultures having an OD of at least 0.8, but not higher than 1.0 expressed the highest activity. It appeared that methane and oxygen were in excess, since a plateau in optical density was reached, but optical density could be increased further with the addition of supplementary inorganic nutrients in the same proportions as in the original media (unpublished data).

Analytical Techniques

Relative sMMO levels were determined by the naphthalene oxidation assay *(9)*. The initial OD (OD_i) is used as a biomass indicator. The change in OD during the sMMO assay (ΔOD) is used as an indicator of the total sMMO activity and the change in OD divided by the initial OD ($\Delta OD/OD_i$) is used as an indicator of biomass specific activity. These units of optical density can be converted to mol of naphthol/h/mg of cells using the following relationships. We worked in a range where there was a linear relationship between OD cell concentration with an OD_i of 0.1, which is equivalent to approx 110 mg cells/L (unpublished data). The relationship between ΔOD and naphthol was also linear with an extinction coefficient of 38,000 mol/cm. Thus, dividing ΔOD by 38,000 and dividing again by the cell concentration (in our experiments, usually about 100 mg/L) and the incubation time (usually 1 h) gives the biomass specific naphthol production rate. Using these relationships, an approximate conversion factor of 18 can be used to convert $\Delta OD/OD_i$ to nmol of napthol/mg cells, and these converted figures are used in this article.

TCE was analyzed using a Sigma 2000 Model (Perkin Elmer, Norwalk, CT) gas chromatograph (GC). The GC was equipped with a capillary column and an electron capture detector (T = 300°C). The oven temperature was set at 150°C. TCE had a retention time of 3.4 min, and was measured in 30-µL samples of the headspace gas. Standards in triplicate consisted of NATE plus TCE added for a final concentration of 0.5, 1.0, and 5.0 ppm. Autoclaved cells plus TCE (1.0 ppm) were used as a control for adsorption to the cellular biomass.

Experimental Design

The enzyme inhibition and recovery experiments were run at 20°C, and consisted of exposing OB3b cells containing high levels of sMMO to TCE in the absence of methane using 40 mL EPA vials with Teflon-lined septa (Supelco, Bellefonte, PA) and a liquid volume of 5 mL. TCE was added as a saturated aqueous solution, and concentrations given are nominal concentrations because all the added TCE did not remain in the liquid phase. Actual concentrations in the liquid phase were lower because of partitioning into the gas phase. Vials were incubated inverted on a shaker. A series of preliminary experiments were performed to determine the appropriate concentrations of formate and TCE and exposure times for use in these recovery experiments. A total of 10 cm^3 of methane (100%) was added to the vials in the recovery experiments. Since the headspace was 35 mL, the methane concentration in the liquid phase approached saturation.

A scoping experiment was performed to document enzyme loss on exposure to 10, 20, and 50 ppm (nominal concentrations) TCE. Vials were incubated inverted on a shaker for 4 h, and residual headspace TCE was measured. Recovery of enzyme activity was initiated in the presence of 10 cm^3 added methane (100%) and 4 mM formate.

In the primary experiment, cells were grown to densities specified above, and sMMO activity measured. Cultures were exposed to 10 ppm TCE for 6 h in the presence of the chemicals being tested for their effects on enzyme levels and recovery (Table 1). Treatments included bacteria (b)

Table 1
Treatments During Primary Experiment

Treatment	Bacteria	Methane 10 mM	TCE, degradation	Addition during TCE
A	Yes	No	No	None
B	Yes	Yes	No	None
C	Yes	Yes	Yes	None
D	Yes	Yes	Yes	2 mM formate
E	Yes	Yes	Yes	2 mM citric acid
F	Yes	Yes	Yes	2 mM ascorbic acid
G	Yes	Yes	Yes	2 mM formic acid
H	Yes	Yes	Yes	$CaCO_3$ (saturated)

[a]Because there was no methane or TCE in treatment A, it is a control examining reduction in sMMO caused only by starvation. Treatment B is a control in which there is no treatment to reduce sMMO. Treatment C represents the baseline TCE effect of 10 mM on sMMO, and the remaining treatments examine the effect of the additions on the sMMO activity and its recovery. Formate and formic acid were added as a source of reducing power and energy. Citric acid and ascorbic acid were added as free radical scavengers, and calcium carbonate was added as a pH buffer.

alone, b + methane, b + TCE, and b + treatment chemicals. The chemical additions were 2 mM formate, 2mM citric acid, 2 mM ascorbic acid, 2 mM formic acid, and $CaCO_3$ to yield a saturated solution. Residual TCE (headspace) was measured, and cultures were then air-sparged to removed from contact with residual TCE. Activity of sMMO was assayed, and pH was measured. Recovery of enzyme activity was initiated in the presence of 4 mM formate and 10 cm^3 methane (100%), and continued for 32 h. At 16 and 32 h, enzyme activity was measured.

RESULTS AND DISCUSSION

TCE at 10, 20, and 50 ppm significantly reduced sMMO activity over that which was seen in unexposed cells (Fig. 1). The oxidation of TCE is apparently detrimental to methanotrophs and results in loss of TCE degradation capacity (15). This toxicity is, in part, likely owing to the inhibition of the sMMO by the TCE epoxide. TCE toxicity is much lower for cells grown on methanol that do not express MMO than it is for cells grown on methane (10), which do express MMO. Although some substrates (i.e., acetylene) appear to be suicide substrates (15) and specifically result in toxicity to methane-oxidizing activity, there is also evidence that TCE toxicity can be nonspecific (15) and thus damage other enzyme systems. In these studies run with 10 ppm TCE, complete degradation was measured (detect limits < 5 ppb).

When added before TCE degradation, the citric acid actually had a negative effect on sMMO activity remaining after the degradation period

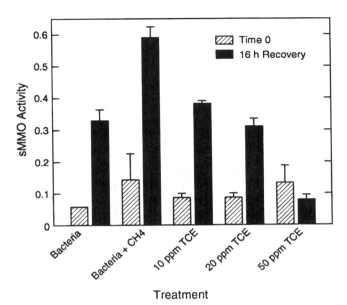

Fig. 1. Results of preliminary experiment showing sMMO activity immediately after a 6 h of exposure to 10, 20, and 50 ppm TCE (Time 0) and 16 h after initiation of recovery with addition of 4 mM formate, compared to levels in cells without TCE exposure, but with added methane (Bacteria + CH_4) and cells starved for methane, but not exposed to TCE (Bacteria).

(Fig. 2A) and prior to the recovery period. After a 16-h recovery period, sMMO activity in the citric acid treatment was still lower than in any other treatment (Fig. 3). During TCE degradation, pH in this treatment fell to 4.9 (Fig. 2B). Measurements in previous experiments had indicated there was no effect by the chemical additions on initial pH. In all other treatments, the pH remained at ~7 (Fig. 2B). Published data have shown that pH can affect TCE degradation *(16)* and pH can decline during TCE degradation in response to release of chloride ions *(17)*.

From our previous experiments and the literature (e.g., *12*), both formate and formic acid promoted recovery of the sMMO to higher levels than with methane alone (Fig. 3). Also, with the formate and formic acid present, sMMO activities did not fall as low during TCE degradation as it did in their absence (Fig. 2). The effect of formate on TCE degradation has been examined in a number of studies and often has shown a beneficial effect *(10,18,19)*. The positive effect has been attributed to provision of reducing equivalents believed to overcome rate limitations *(19)*.

Although it did not affect the loss of enzyme activity during TCE degradation (Fig. 2), the addition of calcium carbonate apparently had a beneficial effect on recovery of sMMO activity (Fig. 3). Ascorbic acid apparently had a slightly positive effect in reducing enzyme loss during

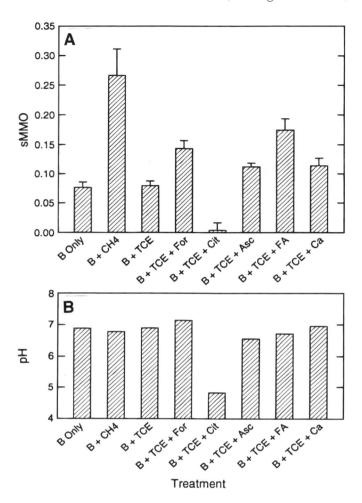

Fig. 2. The effect of TCE degradation on sMMO activity prior to recovery period (**A**) and pH in vials after period of TCE degradation (**B**) in the presence of compounds as noted in Table 1.

degradation (Fig. 2A) and little effect on final enzyme levels after the recovery period (Fig. 3). These results for calcium carbonate and ascorbic acid may be owing to physiological effects on the bacteria.

Addition of calcium carbonate and formate (or formic acid) appears to result in either reduced loss of sMMO activity or enhanced recovery of sMMO activity to higher levels after TCE exposure than without these compounds. However, addition of antioxidant compounds (citric and ascorbic acid) appeared to have no beneficial effect on the recovery of sMMO or in protecting against loss of sMMO. Based on these results, formate addition was included in the bioreactor project (13) to promote recovery of the enzyme activity. The addition of calcium carbonate and ascorbic acid should be considered for future efforts.

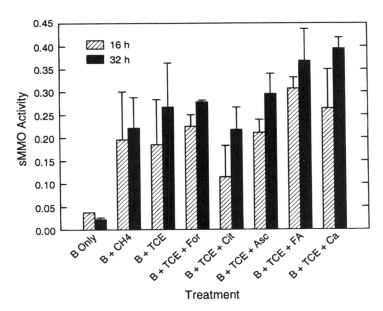

Fig. 3. The effect of 10 ppm TCE on sMMO activity 16 and 32 h after initiation of recovery with the addition of 4 mM formate. Treatments as noted in Table 1.

ACKNOWLEDGMENTS

This research was supported by an agreement between ORNL and Armstrong Laboratories, Environics Directorate, Tyndall Air Force Base, FL. This research was also supported in part by an appointment to the Oak Ridge National Laboratory Postdoctoral Research Program (J. S. G) administered by the Oak Ridge Institute for Science and Education. Oak Ridge National Laboratory is managed by Lockheed Martin Energy Research, Corp. for the US Department of Energy under contract DE-AC 05-96OR22464. The authors would like to thank S. Pfiffner for her review of the manuscript.

REFERENCES

1. Fliermans, C. B., Phelps, T. J., Ringleberg, D., Mikell, A. T., and White, D. C. (1988), *Appl. Environ. Microbiol.* **54,** 1709–1714.
2. Henson, J. M., Yates, M. Y., and Cochran, J. W. (1989), *J. Ind. Microbiol.* **4,** 29–35.
3. Jansen, D. B., Grobben, G., and Witholt, B. (1988), in *Proceedings of the 4th European Congress on Biotechnology*, Vol. 3, Neijssel, O. M., Van der Meer, R. R., and Luyben, K. C. A. M., eds., Elsevier Science Publishers, Amsterdam, pp. 515–518.
4. Little, C. D., Palumbo, A. V., Herbes, S. E., Lindstrom, M. E., Tyndall, R. L., and Gilmer, P. J. (1988), *Appl. Environ. Microbiol.* **54,** 951–956.
5. Tsien, H-C., Brusseau, G. A., Hansong, R. S., and Wackett, L. P. (1989), *Appl. Environ. Microbiol.* **55,** 3155–3161.
6. Zylstra, G. J., Wackett, L. P., and Gibson, D. T. (1989), *Appl. Environ. Microbiol.* **55,** 3162–3166.

7. Wilson, J. T. and Wilson, B. H. (1985), *Appl. Environ. Microbiol.* **49,** 242–243.
8. Ensley, B. D. (1991), *Ann. Rev. Microbiol.* **45,** 283–299.
9. Brusseau, G. A., Tsien, H-C., Hanson, R. S., and Wackett, L. P. (1990), *Biodegradation* **1,** 19–29.
10. Eng, W., Palumbo, A. V., Sriharan, S., and Strandberg, G. W. (1991), *Appl. Biochem. Biotechnol.* **28/29,** 887–906.
11. Alvarez-Cohen, L. and McCarty, P. L. (1991), *Appl. Environ. Microbiol.* **57,** 228–235.
12. Henry, S. M. and Grbic-Galic, D. (1991), *Appl. Environ, Microbiol.* **57,** 236–244.
13. Tschantz, M., Bowman, J., Donaldson, T. L., Bienkowski, P., Strong-Gunderson, J. M., Palumbo, A. V., Herbes, S. E., and Sayler, G. S. (1995), *Environ. Sci. Technol.* **29,** 2073–2082.
14. Whittenbury, R., Phillips, K. C., and Wilkinson, J. F. (1970), *J. Gen. Microbiol.* **61,** 205–218.
15. Oldenhuis, R., Oedzes, J. Y., Van Der Warrde, J. J., and Janssen, D. B. (1991), *Appl. Environ. Microbiol.* **57,** 7–14.
16. Uchiyama, H. (1995), *J. Fermentation Bioengineering* **79,** 608–613.
17. Parvatiyar, M. G. (1995), *Biotechnol. Biol.* **50,** 57–64.
18. Grbic-Galic, D., Henry, S. M., Godsy, E. M., Edwards, E., and Mayer, K. P. (1991), in R. Baker (ed). *Organic Substances and Sediments in Water* Vol. 3, Biological, Baker, R., ed., Lewis Publishers. Chelsea, MI, pp. 239–266.
19. Oldenhuis, R., Vink, J. M., Janssen, D. B., and Witholt, B. (1989), *Appl. Environ. Microbiol.* **55,** 2819–2826.

Spatial and Temporal Variations of Microbial Properties at Different Scales in Shallow Subsurface Sediments

CHUANLUN ZHANG,[1] RICHARD M. LEHMAN,[2]
SUSAN M. PFIFFNER,[1] SHIRLEY P. SCARBOROUGH,[1]
ANTHONY V. PALUMBO,*,[1] TOMMY J. PHELPS,[1]
JOHN J. BEAUCHAMP,[1] AND FREDERICK S. COLWELL[2]

[1]*Oak Ridge National Laboratory, Oak Ridge, TN 37831;
and* [2]*Idaho National Engineering Laboratory, Idaho Falls, ID 83415*

ABSTRACT

Microbial abundance, activity, and community-level physiological profiles (CLPP) were examined at centimeter and meter scales in the subsurface environment at a site near Oyster, VA. At the centimeter scale, variations in aerobic culturable heterotrophs (ACH) and glucose mineralization rates (GMR) were highest in the water table zone, indicating that water availability has a major effect on variations in microbial abundance and activity. At the meter scale, ACH and microaerophiles decreased significantly with depth, whereas anaerobic GMR often increased with depth; this may indicate low redox potentials at depth caused by microbial consumption of oxygen. Data of CLPP indicated that the microbial community (MC) in the soybean field exhibited greater capability to utilize multiple carbon sources than MC in the corn field. This difference may reflect nutrient availability associated with different crops (soybean vs corn). By using a regression model, significant spatial and temporal variations were observed for ACH, microaerophiles, anaerobic GMR, and CLPP. Results of this study indicated that water and nutrient availability as well as land use could have a dominant effect on spatial and temporal variations in microbial properties in shallow subsurface environments.

*Author to whom all correspondence and reprint requests should be addressed.

Index Entries: Aerobic culturable heterotrophs; glucose mineralization rates; community-level physiological profile; seasonal variation; sampling scale effects; water availability.

INTRODUCTION

Release of organic contaminants into aquifers disturbs the *in situ* physical and chemical conditions and the ecology within the subsurface environment. As a result, significant variations in chemical and microbial properties have been observed in contaminated systems. For example, exposure to petroleum contamination altered a microbial community structure by enriching some specific degraders, while suppressing other microbial populations *(1)*. Significant spatial variations in microbial abundance and degradation rates have been observed for contaminated aquifers *(2–5)*. Seasonal changes have also been reported to impact significantly anaerobic microbial activities, such as methanogenesis, sulfate reduction, and Fe(III) reduction in shallow contaminated aquifers *(5–7)*, likely because of changes in temperature, recharge volumes, and chemistry.

Subsurface microbial communities (MCs) in pristine aquifers can often adapt to chemical pollution *8(–9)*. However, because sampling at contaminated sites is usually conducted after the spill or leak has been detected, evaluating the magnitude of MC changes in response to contamination is difficult without knowing the MC status before contamination. To provide background information on microbial changes in uncontaminated shallow aquifers, studies at an Atlantic coastal plain site in Virginia were conducted to investigate spatial and temporal variations in microbial properties and to determine environmental factors controlling microbial variability in uncontaminated subsurface sediments. In this study, we examined microbial variations at the centimeter and meter scales and at varying points in time at the Oyster, VA site. Results showed that temporal changes in water availability and land use properties controlled the observed spatial variability in subsurface sediments within a single field site.

MATERIALS AND METHODS

Site Description and Field Sampling

The field area for this study is located near Oyster, VA (Fig. 1A). Subsurface sediments at the site consist of unconsolidated, fine-to-coarse beach sands and gravels that are clean and well sorted *(10)*. Ground water flow rates at the site are estimated to be approx 20 m/yr with a regional gradient of approx 60 m/km (A. Mills, University of Virginia, personal communication).

Samples were collected during June and August 1994, and during July 1995. Figure 1B shows corehole locations in a soybean field and a corn field. Split-spoon coring tools and a hollow-stem auger system were used during

Microbial Properties

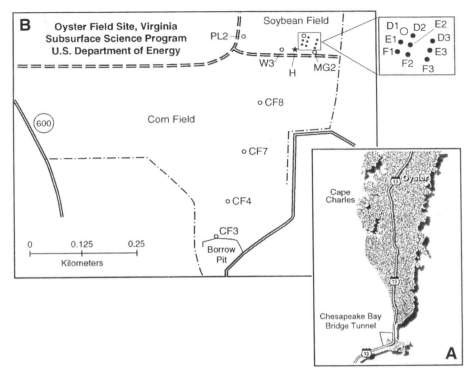

Fig. 1. Geographic location of the study area **(A)** and corehole locations **(B)** at the Oyster site in Virginia. Open circles represent coreholes drilled in June 1994; solid circles represent coreholes drilled in August 1994; the star represents the corehole drilled in July 1995.

the 1994 sampling period, and a sonic drilling system was used during the July 1995 sampling period. Quality assurance and quality-control steps were taken to minimize contamination from drilling operations *(11)*.

To collect undisturbed sediments from the cores, a flame-sterilized handsaw was used to section the core material. The newly exposed core face was pared away with a sterile spatula. The center material was then collected into sterile Whirl-pak bags, and shipped on ice for biological and chemical analyses. At selected depths, sediments were collected 3–5 cm apart for examining small-scale changes in aerobic culturable heterotrophs (ACH), microearophiles, aerobic and anaerobic glucose mineralization rate (GMR), and community-level physiological profiles (CLPP).

Enumeration of Viable Microorganisms

A procedure described by Balkwill *(12)* was used to determine the abundance of ACH. Briefly, sediment samples were blended with sterile 0.1% sodium pyrophosphate (pH 7.0) in the ratio of 1:10 (w/v). Diluted sediment slurries were made with sterile distilled water. An aliquot of 0.1

mL from each dilution (10^{-2}–10^{-5}) was spread onto agar plates in triplicates. After 7 and 14 d of incubation at room temperature (22–25°C), colony-forming units (CFU) were counted by using an automatic bacterial colony counter (Spiral System Instruments, Gaithersburg, MD, Model 500A, countable colony size $\geq$ 0.25 mm).

Microaerophiles were enumerated on dilute-substrate mineral salts in semisolid medium tubes using single-series dilution techniques (13). Tubes were incubated at room temperature, and observations to detect microbial growth in the tube were performed weekly. Microaerophilic growth was judged on the basis of characteristic color bands in the semi-solid medium.

Activity Assays

Experiments examining aerobic and anaerobic GMR were conducted within 30 h of sample collection using anaerobic crimp-top tubes (Bellco Glass, Vineland, NJ). Incubations were at room temperature (22–25°C) for up to 7 d. At t_0 and other appropriate points in time, glucose mineralization was inhibited with 0.5 mL of 2.0M NaOH, and tubes were immediately frozen until analysis. One hour before analysis, tubes were acidified with 0.5 mL of 6M HCl solution, and the CO_2 plus $^{14}CO_2$ that evolved during mineralization was analyzed by gas chromatography and gas proportional counting according to Phelps et al. (11).

Community-Level Physiological Profiles

CLPP integrates the metabolic potential of the heterotrophic community by using the Biolog microplate method (Biolog, Hayward, CA), which tests the utilization of preselected substrates and characterizes environmental isolates. Experiments examining subsurface CLPP were performed according to methods in Garland and Mills (14) and Lehman et al. (15). Briefly, sample slurries, made from blending 10–25 g of sediments with 100 mL of 0.1% sodium pyrophosphate, were flocculated with 0.5 g of a calcium chloride and magnesium carbonate mixture ($CaCl_2 \cdot 2H_2O:MgCO_3$ = 8:5). Then 150 µL of the supernatant were inoculated into each of the 95 wells in Biolog GN microplates (well 96 served as a control). Each well contained lyophilized nutrients and a tetrazolium redox dye as well as one sole carbon source (16). Incubation of these plates was done in the dark at 22°C under a locally humidified atmosphere. Oxidation of the carbon source by the mixed community is indicated by colorimetric reduction of the redox dye. The sum of positive tests for a given sample equals the total number of carbon sources respired.

Statistical Methods

All statistical calculations were performed with SAS (17) on log-transformed values because the majority of microbiological and chemical properties at the Oyster site followed a log-normal distribution (data not shown). Analyses of spatial and temporal effects on ACH and aerobic GMR were performed using a logistic regression analysis with the transformed response regressed on depth. Because more than one corehole was drilled during the June and August 1994 sampling periods, the ACH and GMR data were first analyzed for slope variations between coreholes within a single sampling data. If slopes for the individual coreholes did not differ significantly, these coreholes were grouped to derive a composite slope value for that date for comparison with the slope from a different sampling date. However, if the slopes of a variable in different coreholes differed significantly within a single sampling date, the regression model could not be used for further evaluation of the temporal effect on parameter variation. In this case, the temporal effect on the variable was evaluated on the basis of other information. We excluded coreholes that had samples only below a 1-m depth because of the restricted depth range for the coreholes.

The replicate-to-replicate variability in ACH was estimated from the pooled variance from replicate samples across all coreholes for a given date. This pooled variance estimate was then compared with the error mean square (EMS) from the regression of ACH on depth for the same data set. The EMS provided an estimate of the variability of the ACH observations around the regression line.

RESULTS AND DISCUSSION

Spatial and Temporal Variations at the Centimeter Scale

At the centimeter scale, ACH varied much less, in all three depth zones during the July 1995 sampling period, than during the June and August 1994 sampling periods (Table 1). On the other hand, the Max:Min CFU was highest in the water table zone that transited between the capillary fringe and water-saturated depths for all three sampling periods (Table 1). This indicated a major effect of water table fluctuation on microbial abundance at the centimeter scale. The Max:Min ratio for available aerobic and anaerobic GMR ranged from 1.2 in the zone below the water table to 15.6 in the water table zone, suggesting that greater variability in microbial activity was also because of water level fluctuations.

Variations in microbial abundance and activity at the centimeter scales have been reported for soil environments, aquifer systems, and marine sediments (4,18–20). Microbial abundance and activity reported by Beloin et al. (18) for subsurface sediments appeared to vary by a factor of 3 to >220 at intervals spaced 10 cm apart. However, less variability of microbial abundance at centimeter intervals was reported for a marine sediment study (20). In this study, microbial abundance (ACH) at close intervals appeared to be most variable in the zone transiting between the unsaturated capillary fringe to the water-saturated depth, reflecting the

Table 1
Spatial and Temporal Variations in Aerobic Culturable Heterotrophs
at Close Intervals (3–5 cm Apart)[a]

Depth zone	Max:min CFU		
	June 1994	August 1994	July 1995
Above WT[b]	8 ± 2	10 ± 3	3
At WT	10 ± 2	14 ± 2	4
Below WT	7 ± 8	4 ± 2	1

[a]values are ratios of Max:Min CFU (i.e., $10^4/10^3 = 10$) between sampled intervals. For June and August 1994, each value is a mean +1SD for two to three coreholes in each depth zone where close intervals were selected. The June 1994 data included a cord field and a soybean field; August 1994 and July 1994 data were from the soybean field only.
[b]WT = water table.

effects of water level fluctuation on MC dynamics. On the other hand, the relatively small variation in ACH at close intervals in the other two depth zones, especially the zone below the water table (Table 1), may indicate the uniform texture of the sediments. At the Oyster site, grain sizes changed little with depth; they mainly consisted of fine sands that were clear and well sorted. Physical heterogeneity of subsurface sediments often increases with increasing grain sizes (21). At a different site of the same geological formation, it was shown that significant increases in microbial abundance and activity occurred in a coarse grain zone, and grain size is a major factor controlling microbial abundance and activity (22).

Spatial and Temporal Variations at the Meter Scale

ACH, Aerobic GMR, and Microaerophiles

When ACH data of the soybean field samples were plotted against depth, significant correlations ($P < 0.05$) were found between log-transformed ACH and depth for all three sampling dates (Fig. 2). Using a likelihood ratio test, the slopes of estimated regression lines for log-transformed ACH from three sampling dates were significantly ($P < 0.001$) different (Table 2). The two dates with the greatest difference in slope were June 1994 (S = –1.81) and August 1994 (S = –2.96); thus, ACH decreased with depth more dramatically in August than in June. When we examined the pooled replicate-to-replicate ACH (transformed) variability and compared this with the EMS from the regression of ACH (transformed) on depth, the ratio was >20. This implies that the replicate-to-replicate variability was a trivial component of the overall variability (EMS) noted in the regression analysis.

Abundances of microaerophiles also decreased significantly ($P < 0.05$) with depth from >10^4 cells/g above a 1-m depth to <10 cells/g below a

Microbial Properties

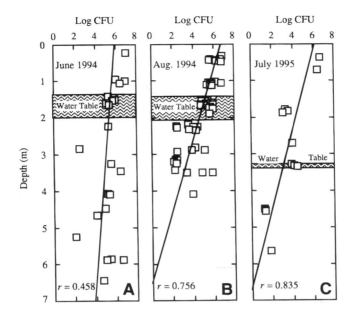

Fig. 2. Depth profiles of aerobic culturable heterotrophs (log CFU) in the soybean field for the June 1994 **(A)**, August 1994 **(B)**, and July 1995 **(C)** sampling dates at the Oyster site in Virginia. r = correlation coefficient. The water table ranges from 1.4–2.0 m in A, from 1.5–2.2 m in B, and is about 3.3 m in C.

Table 2
Maximum Likelihood Estimates of Slopes for Log ACH vs Depth at Different Sampling Dates in the Soybean Field

Date	Number of samples	Number of coreholes	Estimates of slopes, SE
June 1994	13	1[b]	–1.81 (0.69)
August 1994	49	7[c]	–2.96 (0.27)
July 1995	13	1[d]	–2.21 (0.39)

[a]See Fig. 1 for corehole locations. SE = standard error.
[b]Corehole MG2.
[c]Coreholes D3, E1, E2, E3, F1, F2, and F3.
[d]Corehole H.

6-m depth. Large differences in slopes were observed between the soybean field and corn field from the June 1994 sampling date. However, because a significant slope difference was observed between individual coreholes, slopes for microaerophiles could not be pooled for a given site, and thus, the difference in the microaerophile slopes of these two fields could not be evaluated by using the likelihood ratio test.

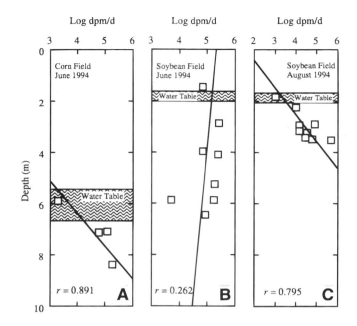

Fig. 3. Depth profiles of anaerobic glucose mineralization rates (log dpm/d) in the corn field (**A**) and soybean field (**B** and **C**) for the June and August 1994 sampling dates. r = correlation coefficient. The water table ranges from 5.3–6.5 m in A, from 1.4–2.0 m in B, and is about 3.3 m in C.

When the likelihood test was applied to aerobic GMR, the slopes did not vary significantly between the two fields or between different sampling dates (data not shown). This finding suggests that aerobic GMR was not sensitive to spatial and temporal variations at this site.

Anaerobic GMR

Trends for anaerobic GMR differed between fields and between two sampling dates (Fig. 3A, B, C). Although the individual slopes for the June 1994 corn field and soybean field data did not significantly differ from 0 ($P > 0.13$ for both), the two slopes did differ significantly ($P = 0.01$) from each other. Anaerobic GMR profiles in the soybean field appeared to be distinct between the June and August 1994 sampling dates: the latter showed a significant ($P < 0.05$) increase in anaerobic GMR with depth (Fig. 3C). The increase in anaerobic GMR with depth in the corn field samples of June 1994 and in the soybean field samples of August 1994 may indicate lower redox conditions with depth. Such conditions are possible, because at lower depths oxygen can be gradually consumed by aerobic microorganisms and result in anoxic conditions, at least in some microenvironments (23–25). Interestingly, bacterial enumeration experiments failed to detect any heterotrophic anaerobes from subsurface sediments collected during the August 1994 sampling period (data not shown). This result suggests that anaerobic glucose mineralization may have been more sensitive in characterizing potential anaerobic activities than other microbiological methods.

CLPP

The CLPP approach has proven to be a rapid, sensitive, and reproducible method for discriminating between microbial communities from a variety of environments, including the subsurface *(14,15,26,27)*. CLPP integrates a measure of the metabolic diversity in heterotrophic communities using a high number of substrate utilization tests to resolve differences in community structure and/or potential function. The CLPP approach is not dependent on isolation. Use of the number of positive tests (sum of binary data) or "community metabolic diversity" has been demonstrated as an effective method of distinguishing communities by Zak et al. *(26)* and Bossio and Scow *(27)*, whereas multivariate analysis of the carbon source utilization profile (continuous data) is a more powerful approach for comparing communities that oxidize a similar number of carbon sources. In either case, the carbon sources may be treated as dimensionless test responses without losing any ability of the CLPP to distinguish communities effectively and reproducibly and assess variability within and among samples.

At the Oyster site, use of the number of positive tests was sufficient to demonstrate community differences. Examination of CLPP data revealed that several features corroborated with microbial abundance and activity analyses. First, CLPP showed decreasing trends with depth for all three sampling dates in the soybean field (Table 3), as did the ACH and microaerophile profiles. Second, large variations in CLPP, as indicated by large standard deviations associated with the mean values of multiple carbon sources utilized by the community, existed for some intervals (1–2, 2–3, and 5–6 m in the soybean field; 6–7 m in the corn field; Table 3). These variations suggest that the MC sometimes changed dramatically within a meter distance at this site. Third, the soybean field appeared to harbor a MC more capable of growth on sole carbon sources than the cornfield within the first 6-m depths. This may be because of the difference in crops between the two fields. Because N_2 fixation occurs more readily with soybeans than with corn, more nutrients may be available in the soil zone in the soybean field than in the corn field. The use of different pesticides or herbicides for soybean and corn crops may also contribute to the difference in CLPP between the two fields. Microbial utilization of multiple carbon sources increased at the water table in the corn field (Table 3). This increase suggests that water availability may be a limiting factor for microbial metabolism at the Oyster site.

Water availability may affect subsurface microbiology in several ways. In arid regions of the US, increased moisture content often stimulates microbial abundance and activity, possibly because of increased bioavailability and transport of sediment-associated nutrients *(28–30)*. Also, deep subsurface sediments containing abundant clays showed

Table 3
Community-Level Physiological Profile for the Soybean Field
and Corn Field Samples Collected During Three Sampling Dates
Between June 1994 and July 1995

Depth interval, m	Community-level physiological profile			
	Corn field, June 1994	Soybean field, June 1994	Soybean field, August 1994	Soybean field, July 1995
0–1[a]	—[b]	87	93 ± 1	83 ± 8
1–2	0	78[c]	57 ± 50[c]	36 ± 50
2–3	0	35 ± 50	7	0
3–4	0	0.5 ± 0.7	3	5 ± 6[c]
4–5	—	0.5 ± 0.7	—	0
5–6	0	0	—	19 ± 26
6–7	36 ± 51[c]	4	—	—
7–8	60	—	—	—

[a]A value in each depth interval indicates the number of positive tests of a sample to the 95 carbon sources used. Average values + 1 SD indicate multiple samples (2–3) in that depth zone.
[b]Not available.
[c]Interval where water table exists.

greater stimulation of microbial growth with nutrient supplements than did sediments dominated by sands (31). In humid areas where shallow subsurface sediments are composed of loosely compacted sands, increased water availability may adversely affect sediment microbial abundance and activity by either decreasing the bacterial attachment capability to sediment particles (32) or transporting nutrients away from the sediments. At the Oyster site, the magnitude of effects of water and phosphate on microbial variability was estimated by using a multiple-constraints model (Palumbo et al., manuscript in preparation).

In summary, significant spatial and temporal variations in microbial abundance (ACH, microaerophiles), activity (anaerobic GMR), and community structures (CLPP) occurred at the Oyster site. Results of this study demonstrated that water abundance, nutrient availability, scale of sampling, and even land use could have dominant effects on microbial variability within a single lithology and field site in shallow subsurface environments.

ACKNOWLEDGMENTS

We thank Aaron Mills for providing field maps and access to the site, and helping with field sampling. Thanks also go to Norman D. Farrow, Charlie A. Lamb, and Tim Griffin for helping with drilling operations. This research was sponsored by the Subsurface Science Program of the US Department of Energy, managed by Frank Wobber. Chuanlun Zhang and Susan M. Pfiffner were supported through Oak Ridge Associated Universities. Oak Ridge National Laboratory is managed by Lockheed Martin Energy Research Corp. for the US Department of Energy under contract DE-AC05-96OR22464.

REFERENCES

1. Long, S. C., Aelion, C. M., Dobbins, D. C., and Pfaender, F. K. (1995), *Microb. Ecol.* **30**, 297–307.
2. McNabb, J. F. and Dunlap, W. J. (1975), *Ground Water* **13**, 33–44.
3. Harvey, R. W., Smith, R. L., and George, L. (1984), *Appl. Environ. Microbiol.* **48**, 1197–1202.
4. Smith, R. L., Harvey, R. W., and LeBlanc, D. R. (1991), *J. Contaminant Hydrology* **7**, 285–300.
5. Adrian, R. N., Robinson, J. A., and Suflita, J. M. (1994), *Appl. Environ. Microbiol.* **60**, 3632–3639.
6. Beeman, R. E. and Suflita, J. M. (1990), *J. Indus. Microbiol.* **5**, 45–58.
7. Vroblesky, D. A. and Chapelle, F. H. (1994), *Water Resour. Res.* **30**, 1561–1570.
8. Wilson, J. T., McNabb, J. F., Cochran, J. W., Wang, T. H., Tomson, M. B., and Bedient, P. D., (1985), *Environmental Toxicol. Chem.* **4**, 721–726.
9. Aelion, C. M., Swindoll, C. M., and Pfaender, F. K. (1987), *Appl. Environ. Microbiol.* **53**, 2212–2217.
10. Mixon, R. B. (1985), *U.S. Geological Survey Professional Paper* 1067–C.
11. Phelps, T. J., Raione, E. G., White, D. C., and Fliermans, C. B. (1989), *Geomicrobiol. J.* **7**, 79–91.
12. Balkwill, D. L. (1989), *Geomicrobiol. J.* **7**, 33–52.
13. Arrage, A. A., Phelphs, T. J., Benoit, R. E., and White, D. C. (1993), *Appl. Environ. Microbiol.* **59**, 3545–3550.
14. Garland, J. L. and Mills, A. L. (1991), *Appl. Environ. Microbiol.* **57**, 2351–2359.
15. Lehman, R. M., Colwell, F. S., Ringelberg, D. B., and White, D. C. (1995), *J. Microbiol. Methods* **22**, 263–281.
16. Bochner, B. R. (1989), *ASM News* **55**, 536–539.
17. SAS Institute Inc. (1989), Cary, NC, p. 846.
18. Beloin, R. M., Sinclair, J. L., and Ghiorse, W. C. (1988), *Microb. Ecol.* **16**, 85–97.
19. Murray, R. E., Feig, Y. S., and Tiedje, J. M. (1995), *Appl. Environ. Microbiol.* **61**, 2791–2793.
20. Litchfield, C. D., Devanas, M. A., Zindulis, J., Carty, C. E., Nakas, J. P., and Martin, E. L. (1979), *Am. Soc. Test. Mater. Special Tech. Pub.* **673**, 128–147.
21. Parsons, B. S. and Swift, D. J. P. (1995) *Contribution No. 9 of the Sediment Dynamics Laboratory*, Department of Oceanography, Old Dominion University, Norfolk, VA.
22. Zhang, C., Palumbo, A. V., Phelps, T. J., Brockman, F., Parsons, B. S., and Swift, D. J. P. (1996), *Annual Meeting of Geological Society of America*, Denver, CO.
23. Fredrickson, J. K., Garland, T. R., Hicks, R. J., Thomas, J. M., Li, S. W., and McFadden, K. M. (1989), *Geomicrobiol. J.* **7**, 53–66.
24. Hirsch, P. (1992), *Progress in Hydrogeochemistry*, Matthess, G., Frimmel, F., Hirsch, P., Schulz, H. D., and Usdowski, H.-E., eds., Springer-Verlag, New York, pp. 308–311.
25. Murphy, E. M., Schramke, J. A., Fredrickson, J. K., Bledsoe, H. W., Francis, A. J., Sklarew, D. S., and Linehan, J. C. (1992), *Wat. Resour. Res.* **28**, 723–740.
26. Zak, J. C., Willig, M. R., Moorhead, D. L., and Wildman, H. G. (1994), *Soil Biol. Biochem.* **26**, 1101–1108.
27. Bossio, D. A. And Scow, K. M. (1995), *Appl. Environ. Microbiol.* **61**, 4043–4050.

28. Brockman, F. J., Kieft, T. L., Fredrickson, J. K., Bjornstad, B. N., Li, S. W., Spangenburg, W., and Long, P. E. (1992), *Microb. Eco.* **23,** 279–301.
29. Kieft, T. L., Amy, P. S., Brockman, F. J., Fredrickson, J. K., Bjornstad, B. N., and Rosacker, L. L. (1993), *Microb. Ecol.* **26,** 59–78.
30. Palumbo, A. V., McCarthy, J. F., Parker, A., Pfiffner, S. M., Colwell, F. S., and Phelps, T. J. (1994), *Appl. Biochem. Biotechnol.* **45/46,** 823–834.
31. Phelps, T. J., Pfiffner, S. M., Sargent, K. A., and White, D. C. (1994), *Microb. Ecol.* **28,** 351–364.
32. Jewett, D. G., Logan, B. E., and Arnold, R. G. (1995), *Abstract with Program*, 1995 annual meeting of Geological Society of America, New Orleans, p. A103.

Development of a Membrane-Based Vapor-Phase Bioreactor

NATHALIE ROUHANA, NARESH HANDAGAMA,
AND PAUL R. BIENKOWSKI*

Department of Chemical Engineering, University of Tennessee, Knoxville, TN 37996-2200

ABSTRACT

A vapor-phase bioreactor has been developed utilizing porous metal membranes in a cylindrical design employing radial flow as opposed to traditional axial flow for the vapor stream. The system was evaluated for the biodegradation of p-xylene (p-xylene) from a water-saturated air stream by *Pseudomonas putida* ATCC 23973 immobilized onto sand. The biocatalyst was placed in the annular space between two cylindrical, porous stainless-steel membranes. Details of the reactor system are presented along with biological data verifying system performance. The feed flow rate and p-xylene concentration were varied between 60 and 130 cm^3/min and 15–150 ppm, respectively. Continuous reactor operation was maintained for 80–200 h with removal efficiencies (based on p-xylene disappearance) between 80 and 95%. The effluent concentration histories were compared to determine the operating range of the bioreactor.

Index Entries: Bioreactor; biocatalytic reactor; biofilter; biodegradation; bioremediation; volatile organic hydrocarbon; p-xylene.

INTRODUCTION

Vapor-phase bioreactors have been used for the degradation of gaseous trichloroethylene [1], for hydrocarbon removal [2], and for p-xylene degradation [3,4]. Currently, over 200 vapor-phase bioreactors are in operation in Holland for the removal of organic contaminants from air in industries, such as ceramics for the treatment of ethanol and isopropanol; resin production facilities to treat phenol; plywood production

---

*Author to whom all correspondence and reprint requests should be addressed.

facilities to treat formaldehyde; and paint production facilities to treat various solvents (5). Since the US is primarily focused on site-specific remediation, vapor-phase bioreactors have not yet gained as widespread use there as they have in Europe. Clearly, expanding bioremediation technology and effectively employing vapor-phase bioreactors require innovation and research to establish the proper conditions under which these types of bioreactors can function.

The principle on which this research is based is that an organic vapor stream of known composition can be degraded through a biodegradative process by an immobilized microbial phase in a radial-flow, packed-bed reactor. This research was motivated by pollution control standards mandated by the Environmental Protection Agency (EPA) with the aim of environmental protection, specifically, air pollution control. The EPA dictates the emission standards of hazardous air pollutants, of which organic compounds comprise a majority. *Para*-xylene (*p*-xylene) was chosen for the test system, because it is on the EPA's list of hazardous chemicals as given by section 112(b) of the Clean Air Act and because from a microbial kinetic standpoint, the system is relatively simple with known kinetics for ATCC 23973 (6). The primary objective of the first phase of this research was to validate the performance of the new reactor design.

Apel *et al.* (4) used a packed-bed bioreactor for *p*-xylene degradation with *Pseudomonas putida* immobilized onto Pall rings. The reactor configuration consisted of a liquid nutrient feed stream flowing in a countercurrent flow to the inlet gas stream. The *p*-xylene feed rate was 16 mL/min with a concentration of 140 mg/min of organic. This system achieved 46% degradation. However, 75 mg/min of organic was detected in the liquid stream leaving the reactor. Vaughn *et al.* (3) employed a similar system, except glass beads and diatomaceous earth were used as packing materials. Nearly 100% removal rates were reported. However, *p*-xylene was detected in the liquid stream from the reactor. The membrane reactor developed in this work does not have a liquid outlet stream.

To provide a contribution to the growing vapor phase bioreactor technology, among the objectives of this research were (1) to build and test a novel packed-bed vapor-phase reactor for the biodegradation of *p*-xylene with an immobilized microbial phase and (2) to evaluate the bioreactor's performance by generating data at different process conditions, i.e., influent concentration and flow rate. The radial-flow reactor was developed with the aim of increasing mass-transfer rates of the organic hydrocarbon to the immobilized cells, or biofilm, which is an important factor when treating pollutants whose water solubilities are low. Radial-flow operation provides a high surface area for contacting of the feed stream and maintains a low pressure drop (2–4 psi), which ensures a uniform distribution of the feed stream along the entire length of the reactor. Additionally, the occurrence of dry spots within the reactor packing is prevented as the flow is uniformly distributed across the bed.

REACTOR DESCRIPTION AND OPERATION

Figure 1 is a schematic of the reactor showing two different crosssectional views. The top view represents a cross-section in the longitudinal direction (xy-plane), and the lower view a cross-section in the transverse direction (xz-plane). The core of the reactor consists of two concentric cylinders made of 316 microporous stainless steel (Pall PMM Metal Membrane Filters, New York), each of which was tungsten-inert gas welded to form the cylinders. The outer diameter of the inner cylinder is 1.63 cm, and the inner diameter of the outer cylinder is 2.54 cm, which gives a reactor thickness of 0.91 cm. This entire assembly is fitted and sealed on top and bottom with o-rings into an outer cylindrical casing made of 304 stainless steel. The reactor plumbing is made from 304 stainless-steel tubing with a diameter of 0.125 in. The reactor was tested for operation up to 60 psi, though the experiments did not produce pressures above 20 psi.

The packing material, which also served as the support medium for the biofilm, consisted of white quartz sand particles (Sigma Chemical Company, St. Louis, MO), which are nonporous and have an average diameter of 0.025 cm. The average crystalline density of the sand was measured to be 2.7 g/mL, and the void fraction of the bed was determined to be 0.30. The reactor volume for this bench-scale model is approx 169 cm^3.

The p-xylene contaminant stream is introduced into the reactor through the inner cylinder core (Fig. 1). The concentration and humidity of the contaminant stream are regulated using an air tank and a three-stage bubbler, as indicated by Fig. 2. Because the two inner cylinders are capped, the vapor mixture flows radially outward through the microporous metal to the inner cylindrical shell containing the sand and biofilm where the p-xylene is enzymatically degraded by the microorganism. The gaseous effluent stream then flows to the outer cylindrical shell by convection and leaves the reactor through the exit port located at the top of the reactor.

The essential elements of the experimental setup are depicted in Figure 2, in which the reactor is integrated into a flow system for proper control of flow rates and effluent analysis. The 700–800 ppm (the units of ppm in this work are on a molar basis) p-xylene/air mixture was obtained premixed in a compressed gas tank (Airco, Port Allen, LA). The air used as the carrying medium was supplied from a second cylinder and was metered through a mass flow controller into a three-stage bubbler, where it became saturated. Mass flow controllers from Brooks Instrument Division (models 5850C, Emerson Electric, Hatfield, PA) were used to maintain volumetric gas flow rates at specified values up to 200 mL/min. The mass flow meters were calibrated using a bubble flow meter and permitted operational ranges of 0–200 and 0–100 cm^3/min, respectively, for the p-xylene/air mixture and the air stream. The effluent from the reactor flowed to a Vista 6000 series gas chromatograph (GC) (Palo Alto, CA) connected to a Hewlett-Packard integrator (Avondale, PA) where the p-xylene composition of the effluent gas was

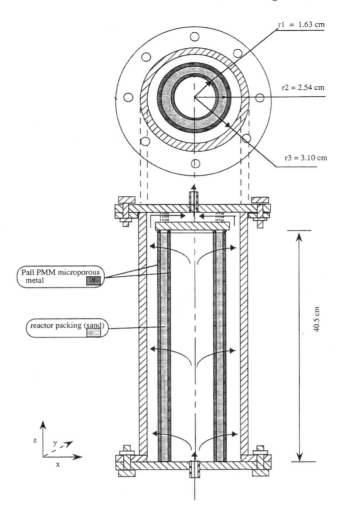

Fig. 1. Schematic of reactor configuration showing concentric cylindrical shell arrangement. Height of filled reactor is 40.5 cm (16 in.).

determined. The GC was equipped with a flame ionization detector for the quantitative analysis of the p-xylene. Two 1-mL sample loops installed in the GC permitted direct sampling of both the introduced gas and the effluent gas. The estimated combined error for control and measurement of the p-xylene flow into and out of the reactor is ±5%.

MATERIALS AND METHODS

Microorganism and Cultivation

The microorganism used in the experiments was *P. putida* ATCC 23973, which was ordered freeze-dried from American Type Culture Collection (Rockville, MD). The medium that was used to culture the cells

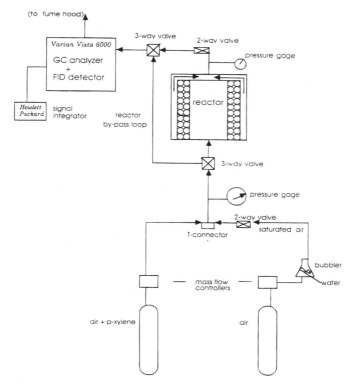

Fig. 2. Schematic line diagram of experimental setup.

was a benzoate medium (medium 1271), the suggested choice of medium for *P. putida* 23973 from ATCC. The composition of the medium on a 1-L solution basis was as follows: NaCl (5.0 g), $(NH_4)_2HPO_4$ (3.0 g), $C_7H_5O_2Na$ (3.0 g), KH_2PO_4 (1.2 g), $MgSO_4·7H_2O$ (0.20 g), all from Mallinckrodt (Paris, KY); and yeast extract (0.50 g), from Difco Labs (Detroit, MI). The cells were cultivated in four 1-L flasks with 250 mL of growth medium and 250 g of white quartz sand, which had been autoclaved. (The sodium benzoate was prepared separately as a filter-sterilized solution and added to the autoclaved growth solution prior to inoculation.) The cell solutions were grown in a shaking flask incubator at 25°C.

The growth state of the cells was monitored by sampling approx 2 mL of the cell-culture solution at various times and measuring the absorbances at a wavelength of 600 nm. Absorbance being directly proportional to the mass of cells in the sample, the growth curve was initially obtained from a graph of the natural logarithm of absorbance as a function of time from a culture grown in a sand-free medium. From this graph, the exponential growth phase of the cells was complete between the fifth and sixth days of growth, after which the onset of the stationary phase was observed. Based on the discussion by Dunbar (7), cells should be harvested during the stationary phase of growth for an active biofilm.

Between the fifth and sixth day of growth, the cell cultures were ready to be transferred to the vapor-phase bioreactor reactor.

Biodegradation Experiments

A controlled-release method of micronutrient delivery in the form of nutrient beads (provided by Grace-Sierra Horticular Products Company, Milpitas, CA) was used for the cells during the biodegradation experiments. The nutrient breakdown of these beads is as follows on a weight percentage basis: ammonium (10.9%), nitrate (9.1%), urea (4.0%), phosphorus (5.5%), sulfur (0.7%), potassium (0.5%), and micronutrients (3.1%), with the balance being a resin coating made from linseed or soybean oil reacted with a cyclic diene. This provided controlled release of nutrients on exposure to moisture. When the cells had reached steady-state growth, nutrient beads in an amount approximately equal to the mass of the nutrients in the original growth medium were added to each of the four flasks containing the cell cultures.

Prior to packing the reactor, the microporous metal membrane cylinders between which the packing was to be situated were autoclaved for 40 min. On cooling of the metal, the cell culture/sand solutions were poured between the two metal membrane cylinders. The solution was drained through the base, the assembly placed within the reactor's metal casing, and the top and bottom of the reactor attached. The biodegradation experiments commenced as the flows of the saturated air and p-xylene/air mixture were passed through the reactor. The flows were adjusted to achieve influent p-xylene concentrations between 15 and 150 ppm at flow rates between 60 and 130 cm^3/min.

Analysis

GC analysis of p-xylene in both the feed and the effluent gas streams was performed at various time intervals during reactor operation. The operating conditions of the GC were: oven temperature set at 130°C, injector temperature set at 140°C, FID temperature set at 150°C, sensitivity set at 10–12 mV, carrier flow rate (He) set at 25 mL/min, with sample loops of 1 mL each. The column was 5% didecyl phthalate plus 5% Bentone 34 on Chromosorb W-HP 80/100 with an inner diameter of 2 mm and a length of 2 m.

To determine the amount of cell mass on the biofilm, several mixed samples of approx 1 g of sand were taken from the packed reactor before and after each experiment. A known volume of a 10% solution of sodium pyrophosphate *(7)* was added to each sand sample to shear the cells off of the sand. The samples were then vortexed, the sand allowed to settle, and the absorbance of the solution at 600 nm measured. Quantifying cell mass based on absorbance measurements was accomplished by prior dry cell weight measurements taken from a culture grown in a solids-free medium. From the correlation of optical density with cell mass obtained, the cell mass removed from the sand was determined. By drying and weighing the sand, estimates of cell mass/g of sand were made and resulted in 9.4 E^{-04}–1.2 E^{-03} g of biomass/g of sand as the initial concentration of biomass within the reactor.

RESULTS AND DISCUSSION

The parameters that were investigated and allowed to vary were the influent *p*-xylene concentration and the flow rates of the influent streams. The relative values of the *p*-xylene/air stream and the saturated air stream were chosen such that a range of different *p*-xylene concentrations would be studied at total flow rates within two ranges: 60–63 and 121–130 cm^3/min. The vapor-phase *p*-xylene concentration that would cause inhibition was estimated from the liquid-phase inhibition constant for the suspended-culture microbial kinetics of the *P. putida* strain used (6) and Henry's law. The influent substrate concentration was chosen to be less than this estimated value. Microbial kinetics developed for suspended growth systems have been shown to apply equally to fixed-film systems (8–10).

The ordinate of the graphs shown in Figs. 3–7 is a normalized concentration representing the ratio of *p*-xylene in the outlet stream to that in the inlet stream. These ratios were determined from the integrated response areas of the GC signals for the *p*-xylene streams entering and exiting the reactor.

To determine the adsorption breakthrough time of *p*-xylene on the biofilm/sand support, experiments were performed to determine the effects of any possible adsorption phenomena in the absence of biodegradation reactions. A representative result is shown in Fig. 3, which shows the elution curve of 284 ppm of *p*-xylene at a flow rate of 60 cm^3/min in an experiment performed with cells that had been autoclaved prior to packing of the reactor. This experiment was preformed at a reactor pressure of 15 psi with only the saturated air flowing through the system prior to starting the *p*-xylene/air mixture flow at time t = 0. Breakthrough occurs within 1 h with an inlet feed of 284 ppm. Thus, any decrease in the outlet concentration of *p*-xylene would subsequently be attributed to biodegradation kinetics.

As Fig. 4 and its inset show, steep variations in the outlet concentration of *p*-xylene occurred when sampling of the outlet stream was done during the first 25 h of operation. During the first 10–25 h of operation, several factors are simultaneously occurring:

1. The pressure inside the reactor is stabilizing as flow to the reactor is begun. Although the pressure drop through the reactor is not >3 psi, the pressure inside the reactor reaches a stable value of 10–14 psi during its first hour of operation.
2. Adsorption and breakthrough of *p*-xylene are occurring during the first hour of operation.
3. Acclimation of the cells to the new carbon substrate is occurring. During packing of the reactor, cells that have been growing from 5–6 d in a growth medium with benzoate as their sole carbon

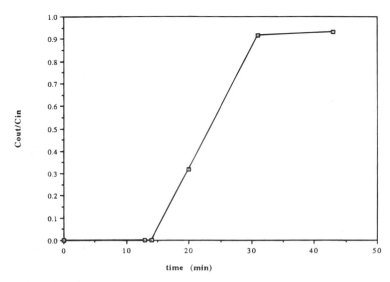

Fig. 3. Elution curve of 284 ppm feed of *p*-xylene at a flow rate of 60 cm^3/min in packed bed experiment performed with immobilized cells that had been autoclaved.

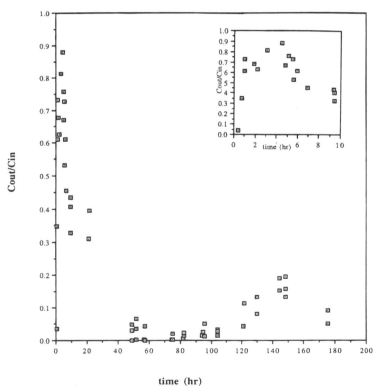

Fig. 4. Normalized concentration of *p*-xylene in the exit stream of the reactor as a function of time at the following conditions: C_o = 36 ppm, flow rate = 121 cm^3/min. The exit concentration of *p*-xylene is normalized with respect to its feed concentration. The inset is blow-up of the initial region of the effluent curve of *p*-xylene.

source were abruptly immersed into an environment initially containing only micronutrient beads; p-xylene as the substrate is introduced to the packed, immobilized cells only after transport to and final assembly of the reactor are completed 1–2 h later. During these initial stages of reactor operation, the cells are then becoming accustomed to a new environment.

Factors 1–3 all occur during the initial stages of reactor operation and have an effect on the effluent profile obtained during the first 25 h, as shown by Figs. 5 and 6.

As time progresses beyond this initial region characterized by high variations, the effluent concentration curves are seen to vary in a more continuous manner (Figs. 4–7). Figure 5 shows the effluent curves obtained with a feed concentration of 15 ppm at flow rates of 62 and 125 cm^3/min. The bioreactor was run for 180 h (7.5 d) at a flow rate of 62 cm^3/min with only 8–16% of the p-xylene feed detected in the reactor outlet. This corresponds to a removal of 84–94% of the p-xylene being fed to the reactor. In a separate experiment with an inlet p-xylene concentration of 15 ppm at a higher flow rate of 125 cm^3/min, 60–88% removal of p-xylene was observed during the 120 h (5 d) of operation. Although there is relatively more scatter in the exit concentrations than in the latter case, the reactor does seem to approach a steady state after 60 h of operation.

The results for the experiments conducted within the 28–36 ppm concentration range at flow rates of 121 and 60 cm^3/min are shown in Figs. 4 and 6, respectively. Both of these experiments produced comparable overall results with the bioreactor reaching 80–90% removal efficiencies after the first 25 h of operation. These removal efficiencies were maintained for 128 h (5.3 d) for the experiment run at 60 cm^3/min, and for 178 h (7.4 d) at a flow rate of 121 cm^3/min.

The experiment run at the lower flow rate of 60 cm^3/min with an inlet feed of 28 ppm of p-xylene (Fig. 6) seemed to reach a steady-state value with the normalized output concentration of p-xylene varying between 0.06 and 0.15. After 60 h of operation, <10% change in the effluent concentration of p-xylene was observed. The experiment run at the higher flow rate of 121 cm^3/min with an inlet p-xylene feed of 36 ppm (Fig. 4) produced removal efficiencies above 95% after the 57th h of operation. This performance continued for the next 48 h, after which the normalized effluent concentration of p-xylene began a slight increase reaching a relative maximum of 0.2. Approximately 30 h after this relative peak, the normalized p-xylene effluent concentration decreased to 0.05–0.1. The slight rise in the effluent profile corresponded to an experimental observation of an increase in pressure to 18 psi beyond the 105th h of reactor operation. This increase in pressure may have been caused by an increase in biomass within the packed bed, because removal efficiencies were in the range of 90–95% directly preceding this pressure increase.

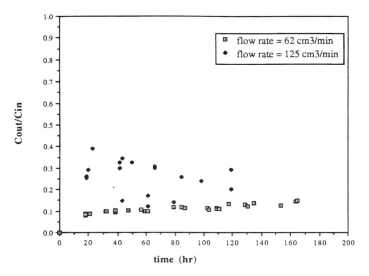

Fig. 5. Normalized concentration of *p*-xylene in the exit stream of the reactor as a function of time at an influent *p*-xylene concentration of 15 ppm.

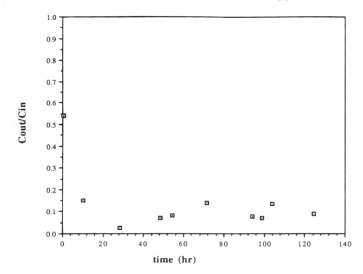

Fig. 6. Normalized concentration of *p*-xylene in the exit stream of the reactor as a function of time at an influent *p*-xylene concentration of 28 ppm and a flow rate of 60 cm^3/min. The exit concentration of *p*-xylene is normalized with respect to its feed concentration.

A graphical compilation of other results is shown in Fig. 7, in which the results of the experiments performed with *p*-xylene feed concentrations of 52, 100, and 150 ppm at comparable flow rates are shown. With increasing levels in *p*-xylene feed concentrations, the removal efficiencies decrease. The experiment conducted with the 52 ppm feed is seen to reach a steady state based on the exit concentrations of *p*-xylene. The other two cases run

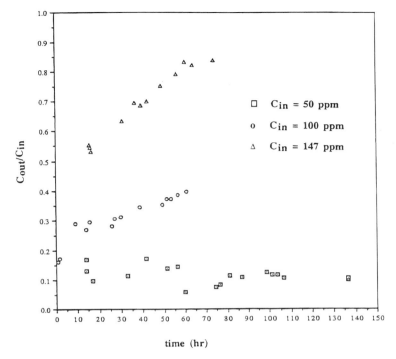

Fig. 7. Normalized concentration of p-xylene in the exit stream of the reactor as a function of time at an influent p-xylene concentrations of 50 ppm (squares), 100 ppm (circles), and 147 ppm (triangles) at flow rates of 63, 62, and 62 cm³/min, respectively. The exit concentration of p-xylene is normalized with respect to its feed concentration.

at feed concentrations of 100 and 150 ppm show a steady increase in the amount of p-xylene in the effluent stream over time. Per Fig. 7, overall p-xylene removal efficiencies in this range successively decreased as the influent concentration increased. This was probably owing to organism toxicity resulting from increased p-xylene concentrations in the reactor bed.

Figure 8 is a plot of the average removal rate of p-xylene in mg/h vs the mass flow rate to the reactor. As the mass flow rate increases, the removal rate increases up to an observed maximum at 0.02215 mg/min corresponding to an inlet concentration of 50 ppm and a flow rate of 125 cm³/min. Higher mass flow rates of p-xylene result in a decrease in productivity of the reactor, which is attributed to organism inhibition at the higher liquid-phase orgainc concentrations. The reactor productivity ranged from 0.0021 mg/(h cm³) up to 0.013 mg/(h cm³)

CONCLUSIONS

The radial-flow vapor-phase bioreactor functioned very well for the biodegradation of p-xylene and could achieve high removal rates (based on p-xylene disappearance). The region of operation for the defined reactor

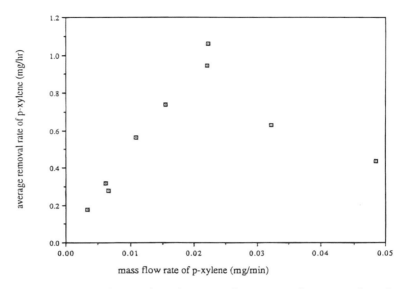

Fig. 8. Average removal rate of *p*-xylene vs influent mass flow rate of *p*-xylene for all nine experimental runs.

system/immobilized *P. putida* microorganism was found to be within the range of 15–150 ppm for the influent *p*-xylene concentration and flow rates between 60 and 130 cm^3/min. Continuous biodegradation was demonstrated within this defined operating region, which resulted in removal efficiencies of 80–90% of the influent *p*-xylene for operating times up to 200 h.

Mass-transfer limitations did not appear to be significant for this reactor system within the 15–36 ppm range, once removal efficiencies were comparable and in both cases remained above 60% when the total feed flow rate was doubled. As the influent feed concentration increased from 50–100–147 ppm, the removal efficiencies successively decreased, while still remaining as high as 60 for the 100-ppm case. The system did appear to become saturated with *p*-xylene after 60 h of operation at the 147 ppm and 62 cm^3/min conditions.

The gas-phase bioreactor developed in this study does support and verify the novel design for the vapor-phase biodegradation of *p*-xylene with the particular microbial system utilized. Modeling and numerical validation of the reactor system studied are a currently ongoing endeavor.

ACKNOWLEDGMENTS

This work was supported by the Waste Management Research and Educational Institute (WMR&EI) and the Department of Chemical Engineering, University of Tennessee. The authors would like to acknowledge the technical support of John Bowman and Gary Sayler of the Center for Environmental Biotechnology (CEB), University of Tennessee.

REFERENCES

1. Uchiyama, H., Oguri, K., Yagi, O., and Kokufuta, E. (1992), *Biotechnol. Lett.* **14(7),** 619–622.
2. Davison, B. H. and Thompson, J. E. (1994), *Appl. Biochem. Biotechnol.* **45/46,** 917–923.
3. Vaughn, B., Jones, W., and Wolfram, J. (1994), in *Proceedings of the 48th Industrial Waste Conference 1993*, Purdue Research Foundation, Purdue University, West Lafayette, IN, pp. 393–405.
4. Apel, W. A., Dugan, P. R., Wiebe, M. R., Johnson, E. G., Wolfram, J. H., and Rogers, R. D. (1993), in *Emerging Technologies in Hazardous Waste Management III*, Tedder, D. W. and Pohland, F. G., eds., American Chemical Society, Washington, DC, pp. 411–428.
5. Atlas, R. M. (1995), *Chem. Eng. News*, **April 3,** 32–42.
6. Lee, J. Y., Choi, Y. B., and Kim, H. S. (1993), *Biotechnol. Prog.* **9(1),** 46–53.
7. Dunbar, P. D. (1993), Ph.D. thesis, University of Tennessee, Knoxville, TN.
8. San, H. A., Tanik., A., and Orhon, D. (1993), *J. Chem. Tech. Biotechnol.* **58,** 39–48.
9. Gujer, W. and Wanner, O. (1985), *Water Sci. Technol.* **17,** 27–44.
10. Harremoes, P. and Arvin, E. (1990), *Water Sci. Technol.* **2,** 171–192.

Intrinsic Bioremediation of Gas Condensate Hydrocarbons

Results of Over Two Years of Ground Water and Soil Core Analysis and Monitoring

KERRY L. SUBLETTE,*[,1] RAVINDRA V. KOLHATKAR,[1] ABHIJEET BOROLE,[1] KEVIN T. RATERMAN,[2] GARY L. TRENT,[2] MINOO JAVANMARDIAN,[3] AND J. BERTON FISHER[4]

[1]*Center for Environmental Research and Technology, University of Tulsa, 600 S. College Avenue, Tulsa, OK 74104;* [2]*Amoco Tulsa Technology Center, PO Box 3385, Tulsa, OK 74102;* [3]*Amoco Corporation, Amoco Research Center, 150 West Warrenville Road, Naperville, IL 60563; and* [4]*Gardere and Wynne, 2000 Mid Continent Tower, 01 S. Boston, Tulsa, OK 74103*

ABSTRACT

Condensate liquids have been found to contaminate soil and ground water at two gas production sites in the Denver Basin operated by Amoco Production Co. These sites have been closely monitored since July 1993 to determine whether intrinsic aerobic or anaerobic bioremediation of hydrocarbons occurs at a sufficient rate and to an adequate end point to support a no-intervention decision. Ground water monitoring, soil gas analysis, and analysis of soil cores suggest that bioremediation is occurring at these sites by multiple pathways, including aerobic oxidation, sulfate reduction, and methanogenesis. Results of over two years of monitoring of ground water and soil chemistry at these sites are presented to support this conclusion.

Index Entries: Intrinsic bioremediation; gas condensate; hydrocarbons; ground water; sulfate reduction; hydraulic gradient.

---

*Author to whom all correspondence and reprint requests should be addressed.

INTRODUCTION

Amoco operates more than 800 natural gas wells within the Denver Basin, Colorado, which each average about 10^5 std ft^3/d (2830 std m^3/d) of gas and less than about 3 barrels/d of associated water and condensate liquids. Condensate has been found to contaminate soil and ground water at certain sites, and Amoco has sought a low-cost alternative to active remediation of these sites wherein acceptable environmental conditions would be restored. Natural or intrinsic bioremediation is one such option. This option recognizes that indigenous microorganisms in the subsurface are capable of hydrocarbon degradation when critical environmental factors are not limiting (e.g., nutrients, temperature, moisture, pH, salinity, and electron acceptor). Recently, researchers have convincingly demonstrated the natural attenuation of hydrocarbon plumes in ground water under both aerobic and anaerobic conditions (1,2). Oxygen, nitrate, Fe(III) oxides, sulfate, and carbon dioxide have all been identified as terminal electron acceptors for the biochemical oxidation of hydrocarbons (3–7).

Amoco has initiated a study to determine whether intrinsic aerobic or anaerobic bioremediation of hydrocarbons occurs at the Denver Basin sites at a sufficient rate and to an adequate end point to support a no-intervention or intrinsic remediation option. Tasks specific to this objective are:

1. Long-term ground water and soils monitoring (initiated July 1993) to document field hydrocarbon losses and bioactivity over time (quarterly sampling events for 5–6 yr for ground water; annual sampling events for soils);
2. Laboratory verification of hydrocarbon degradation by field microorganisms and identification of primary biodegradation mechanisms (initiated September 1993); and
3. Microbiological characterization of the sites to understand the spatial utilization of electron acceptors in and around the hydrocarbon plume (initiated November 1995).

We report in this article preliminary results (July 1993 to November 1995) from the ground water and soils monitoring program. The implications of these data to natural attenuation of hydrocarbons are discussed. An additional three years of sampling are planned. Further laboratory investigations and the microbial characterization of the sites will be reported at a later date.

SITE CHARACTERIZATION AND MONITORING

In July 1993, two sites situated near the Platte River in agricultural areas near Ft. Lupton, Co, were chosen for in-depth site assessments. Preliminary evaluations had shown that both soil and shallow ground water beyond the storage tank containment area were contaminated with gas condensate. The

Gas Condensate Hydrocarbons

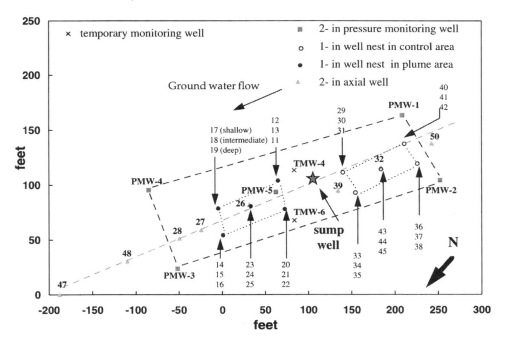

Fig. 1. Site map of KPU-2 showing types of wells and their locations. (Numbers shown are well identification numbers.)

aquifer material (gravely sands, sands, and silty sands) is highly permeable, and the water table elevations fluctuate greatly with seasonal irrigation. Further, potential surface water receptors are near both sites. Therefore, the potential for contaminant transport was deemed high, and both sites were placed in a high-priority category for further investigation.

Ground Water Monitoring

Based on an initial soil gas survey (8), permanent ground water monitoring wells were installed to determine the extent of hydrocarbon loss and the degree of bioactivity over time. Figure 1 shows the locations of 1-in. (2.54-cm) OD vertically nested monitoring wells that were installed in five-spot patterns within both the plume and control areas of the KPU-2 site. This monitoring arrangement was adopted to define areal and vertical variations of hydrocarbon and electron acceptor concentrations in ground water. Each vertical well nest consisted of three wells screened over 18-in. (45.7-cm) intervals and placed 0, 5, and 10 ft (0, 1.52, and 3.05 m) below the water table at the time of installation. At the time of installation, the water table was high. However, the water table has never been lower than just below the bottom of the topmost screened interval. Additional 2-in (5.1-cm) OD monitoring wells were placed along the longitudinal axis of predominant ground water flow to monitor plume migration and electron acceptor transport. These axial wells were arranged along a path extending from upgradient of the control area, through the original source area and downgradient of the plume. The 2-in. (5.1-cm) wells were screened over a 10-ft (3.05-m) interval to allow for seasonal ground water fluctuations.

To address hydraulic modeling requirements adequately and thereby ultimately to assess the role of abiotic mechanisms (e.g., dispersion, advection) in hydrocarbon loss, both the downgradient area with hydrocarbons and the upgradient area without hydrocarbons were contained within a larger hydraulic five-spot monitoring pattern. Hydraulic monitoring wells were completed as the 2-in. (5.1-cm) monitoring wells. Pressure transducers were permanently installed at a fixed depth in each well. Average, maximum, and minimum water table fluctuations were recorded daily. These measurements have a resolution of 0.5 in (1.3 cm). In addition to these pressure monitoring wells, the axial 2-in. wells were also used to measure water level during ground water sampling.

Soil Core Sampling

Soil cores were obtained from each site in November 1993 to document the initial soil hydrocarbon and electron acceptor distributions. Four cores were taken from within each control and plume area. Coring locations were situated approximately halfway between the center and corner well clusters in each pattern quadrant. Continuous cores were obtained with a 2-in. (5.1-cm) split-spoon sampler from the surface to a total depth of 15 ft (4.6 m). Core samples were composited at 1.5-ft (45.7-cm) intervals and stored in a reduced oxygen environment at 4°C until requisite analyses could be performed. A second set of cores was collected in November 1994 to document changes in the hydrocarbon concentrations. These cores were taken within 1 m of the 1993 coring locations. To minimize disturbance, the cores taken in November 1994 were obtained by vibracoring. In the vibracoring operation, continuous cores were obtained in a 3-in (7.6-cm) acrylic sleeve from surface to a total planned depth of 10 ft (3.0 m). The coring done in November 1995 was again done with a split-spoon sampler to a total depth of 15 ft (4.5 m). Material from these cores was composited and analyzed in the same manner each year.

Analytical

Baseline ground water samples were collected during the first week of November 1993. Fresh water samples were obtained by producing approximately three well volumes from each monitoring well prior to sampling. Individual samples were collected and analyzed within 24 h for inorganic constituents, such as nitrate, sulfate, total alkalinity (as mg/L $CaCO_3$), and Fe(II) (8). Samples for BTEX and TPH were collected in clean VOA vials, immediately extracted with Freon and shipped to Amoco's

Groundwater Management Section Laboratory, Tulsa, OK for analysis by Amoco-modified EPA method 8015.

Soil solids were analyzed for moisture content, acid extractable Fe(II) and Fe(III), porosity, bulk density and saturated paste pH, nitrate, and sulfate. Details of the preparation of saturated soil paste and the individual analyses are described in ref. *(9)*. Soil solids were also analyzed for BTEX and TPH after extracting the soil with methylene chloride by EPA method 8020 *(8)*.

RESULTS

Ground Water

Based on the baseline ground water data collected in the first week of November 1993, it was inferred that the contamination was largely confined to shallow depths and that sulfate reduction was an important mechanism in hydrocarbon attenuation at this site *(8,9)*. The importance of sulfate as an electron acceptor was also corroborated by microcosm studies undertaken in the laboratory *(10,11)*. Additional sampling since that time continues to support sulfate reduction as an important pathway for the attenuation of hydrocarbons at the site.

Initial ground water data indicated that BTEX and TPH were confined primarily to the shallow well depth (3.5 to 5 ft below grade) *(12)*. At this depth, various electron acceptors including sulfate, nitrate and dissolved oxygen (DO) exhibited diminished concentrations in the plume the control zone. Similarly, concentrations of Fe(II), a product of Fe(III) reduction, and total alkalinity (mainly bicarbonate) were higher. Lower hydrocarbon concentrations were detected at the intermediate depth (6.5–8 ft below grade) in the plume. Trends in electron acceptor utilization at the intermediate depth within the plume were similar to those at the shallow depth, but less pronounced. Deep well data indicated nondetectable levels of hydrocarbons in both plume and control. Likewise, electron acceptor data showed no appreciable differences. Since most of the activity is occurring at shallow depths, further discussion will be restricted to only the shallow-screened wells.

BTEX data are plotted in Fig. 2 for shallow-screened, 1-in. wells using complete data collected from November 1993 through November 1995. BTEX was completely absent from wells upgradient of sump (i.e., control zone), whereas considerable concentrations of BTEX (20–40 mg/L) were detected in the downgradient wells (i.e., plume zone). Extremely high BTEX concentrations seen in wells 14 and 17 in February 1994 were owing to sample collected from a vapor point, since the water table was below the screen. These were point samples as opposed to volume averaged samples over the 18-in. screened interval for the other samples. Lower values of BTEX were observed in the sump compared to plume, indicating that

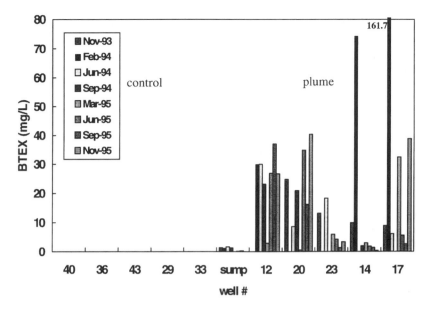

Fig. 2. BTEX in 1-in. shallow wells.

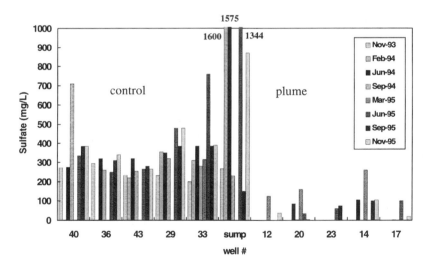

Fig. 3. Sulfate in 1-in. shallow wells.

this region did not have any further source of contamination, such as a non aqueous-phase liquid (NAPL).

A similar plot for sulfate concentrations (Fig. 3) indicates considerably diminished sulfate in the plume zone associated with the presence of hydrocarbons, whereas wells in the control zone had sulfate on the order of 230 mg/L. In addition, all the ground water samples from the wells in plume zone had a strong hydrogen sulfide smell.

Similarly, increased Fe(II) and total alkalinity as well as reduced DO were observed in the plume zone compared to in the control zone (data not shown). Increased levels of Fe(II) in the plume (4–6 mg/L compared to near zero concentrations in control) indicate Fe(III) reduction occurring in this zone, although the utilization of Fe(III) as electron acceptor cannot be quantified because of its low solubility. Fe(III) reduction may be biotic or abiotic. Hydrogen sulfide produced by biological sulfate reduction can reduce Fe(III) to Fe(II) and precipitate it as iron sulfide *(6)*.

DO in the plume ranged from 0.25–1.4 mg/L in the absence of sulfate. Bicarbonate, the major component of total alkalinity, is a product of biodegradation of hydrocarbons. Considerably higher values of alkalinity in the plume (average 600 mg/L $CaCO_3$ compared to 300 mg/L in control) strongly suggest that biodegradation of hydrocarbons is occurring in the plume. The diminished sulfate concentrations in the plume suggest that sulfate reduction is a major contributor to biodegradation of the hydrocarbons.

As seen in Fig. 2, no clear decrease in BTEX concentrations in ground water has been observed to date (November 1995) owing presumably to replenishment of dissolved hydrocarbons from a sink of sorbed hydrocarbons and water table fluctuations. Individual water level measurements from the 2-in. axial wells showed water table fluctuations of 2.5–3 ft (0.76–0.91 m) with the water table being higher during the summer (June to September) and lower during winter (November to March). Based on the water level data obtained from these wells, it was evident that the local gradient exhibited fourfold seasonal variations (Fig. 4). Higher gradient was seen during winter and lower gradient during summer. This could be the result of flooding in a downgradient stream (that discharges into the Platte River) during summer when meltdown of snow increases recharge in this area. During the winter months, when the stream level is low and less recharge occurs owing to freezing temperatures, higher gradient is established. This has important bearing on relative magnitudes of hydrocarbon mobility and attenuation rate. During winter, when biological activity would be slower, mobility of dissolved hydrocarbons would be higher because of higher ground water velocity, causing the dissolved plume to migrate at a higher rate. The reverse phenomena would take place during summer, resulting in increased attenuation.

Finally, axial monitoring well data indicated that highly soluble BTEX components had migrated to well 28, i.e., a distance of only 165 ft (50.3 m) from the source over an estimated 20-year time period (Fig. 5). Sulfate and Fe(II) concentrations were at background levels at this distance (data not shown).

In summary, the following observations are made on the basis of ground water data acquired to date. The aerobic biodegradation potential of hydrocarbon appears limited owing to uniformly low DO concentrations (1.4 mg/L or less) throughout the investigated area; however, the role of oxygen at the edge of the plume is potentially important. Nitrate is

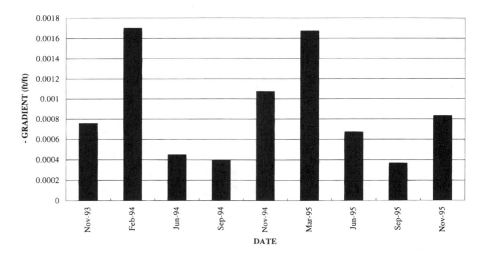

Fig. 4. Hydraulic gradients in 2-in. axial wells.

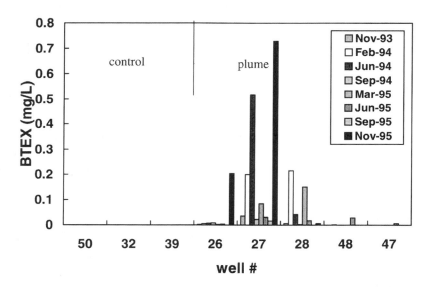

Fig. 5. BTEX in 2-in. axial wells.

present as a result of agricultural applications of fertilizer, but because of its low concentration, has limited potential for hydrocarbon degradation (9). In contrast, the utilization of sulfate appears significant. Background concentrations are on the order of 230 mg/L, whereas sulfate is practically absent in the shallow area containing hydrocarbons. Given the large initial concentrations of sulfate and its favorable stoichiometric utilization for hydrocarbon degradation, it appears that sulfate reduction is a major means of hydrocarbon remediation at these sites. Moreover, concentrations of sulfate observed in June 1995 sampling show that a substantial amount of sulfate is transported by infiltration of rainwater and that the rate of sulfate consumption in the hydrocarbon-contaminated ground water exceeded the rate of supply by infiltration. This sampling was conducted immediately after a period of persistent heavy rainfall. In the control area at KPU-2, background sulfate levels in the shallow wells had nearly doubled (average of 5 wells = 418 mg/L sulfate), whereas sulfate concentrations for intermediate and deep wells were comparable to those observed previously. In contrast, sulfate concentrations in shallow wells from plume area showed an absence of sulfate.

Soil Cores

Soil core analyses for total iron, Fe(II), BTEX, and TPH were performed on composited samples from each 1.5-ft (45.7-cm) interval. Within the plume area (Figs. 6 and 7), BTEX and TPH were consistently present in soil cores, but were confined to an approx 3-ft (0.91-m) interval at the water table/air interface. Soil samples acquired from upgradient cores within the plume showed less BTEX and TPH than their downgradient counterparts. Because upgradient cores were situated nearest the original hydrocarbon source, this evidence supports the contention that the original source of the existing hydrocarbons in soil and ground water has been effectively eliminated at these sites. Total petroleum hydrocarbon concentrations decreased in the first year, and then were more or less constant; however, BTEX concentrations observed in soil cores collected in November 1994 and 1995 are greatly reduced compared to total BTEX concentrations observed in November 1993. Since BTEX concentrations observed in ground water monitoring wells have remained essentially unchanged, this is interpreted to indicate that the total hydrocarbon inventory at this site is being decidedly reduced by natural attenuation. As discussed earlier, BTEX concentrations in the ground water did not show a temporal trend because of the presence of a residual hydrocarbon phase (Fig. 2). Also, no specific movement of the dissolved hydrocarbon plume was observed (Fig. 5). The plume appears to be stable based on the available data; however, once the BTEX inventory diminishes, ground water BTEX concentrations may start decreasing and the plume also may start shrinking.

Soil sulfate concentrations and Fe(II)/Fe(III) ratios obtained by saturated paste extract method (3) showed distinct differences between plume and control zone samples (data not shown). Sulfate concentrations were lower in the plume region (4–100 mg/kg compared to 100 mg/kg and above for control). Although sulfate was absent in many of the 1-in. shallow-screened wells in the plume, soil still contained sulfate in mineral form. Dissolution of sulfate minerals might have controlled the supply of sulfate to the ground water. No specific change was observed in sulfate concentrations in the soil cores from November 1993 to November 1994 (data not shown).

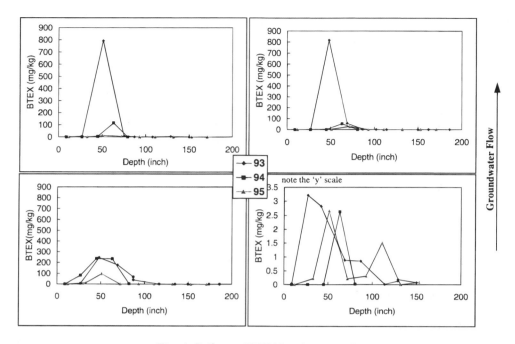

Fig. 6. Soil core BTEX in plume region.

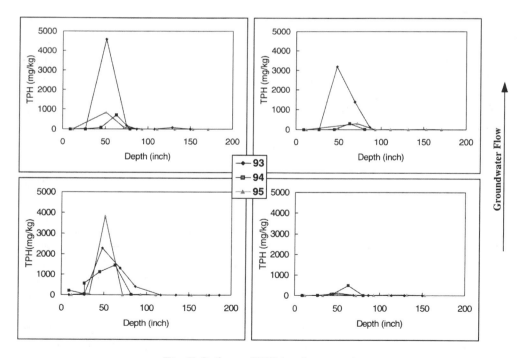

Fig. 7. Soil core TPH in plume region.

When hydrocarbons were present, the ratio of Fe(II) to Fe(III) in soils was increased. Assuming both iron species were initially distributed uniformly across the site, it appears that Fe(III) was subsequently reduced to Fe(II) within the zone of significant hydrocarbon presence. The reduction of iron in the presence of hydrocarbons is indicative of anaerobic biodegradation and further supports the hypothesis that intrinsic bioremediation of hydrocarbons is occurring at these sites by multiple pathways.

Finally, visual inspection of soil cores showed a significant accumulation of a black precipitate (acid volatile sulfide) associated solely with the presence of hydrocarbons. The accumulation of iron sulfide (FeS) in the presence of hydrocarbon is consistent with the anaerobic biodegradation of hydrocarbons by sulfate reduction *(6)*.

CONCLUSIONS

At the sites investigated, intrinsic aerobic and anaerobic bioremediation play a strong role in attenuating hydrocarbons. Even though the sites are old (>20 yr), hydrocarbons are laterally and vertically confined to a small portion of the total aquifer. Further, there is strong geochemical evidence of both aerobic and anaerobic bioactivity. Oxygen is much depleted and carbon dioxide is much elevated in the soil gas immediately overlying the ground water associated with the hydrocarbon plume. Ground water that is immediately associated with the hydrocarbon plume shows depleted levels of both nitrate and sulfate compared to ground water from the upgradient uncontaminated zone. Moreover, based on anecdotal observations of the response of the uppermost level of ground water to the infiltration of sulfate-laden rainwater, the rate of sulfate consumption in ground water immediately associated with hydrocarbons is very high compared to that of ground water not associated with hydrocarbons. Soils associated with hydrocarbons show abundant iron sulfide production. Although sulfate is depleted in ground water associated with hydrocarbons, sulfate remains available as an electron acceptor by means of diffusion from the surrounding aquifer, by dissolution of sulfate from the solid phase, and by importation of sulfate by infiltration of rainwater. Potentially, sulfate reduction could be accelerated in the hydrocarbon-impacted areas through the addition of sulfate.

REFERENCES

1. Norris, R. D., Hinchee, R. W., Brown, R., McCarty, P. L., Semprini, L., Wilson, J. T., Kampbell, D. H., Reinhard, M., Bouwer, E. J., Borden, R. C., Vogel, T. M., Thomas, J. M., and Ward, C. H. (1993), In-Situ *Bioremediation of Groundwater and Geological Material: A Review of Technologies*, 68-C8-0058, Robert S. Kerr Environmental Research Laboratory, Ada, OK.
2. Bennett, P. C., Siegel, D. E., Baedecker, M. J., and Hult, M. F. (1993), *Appl. Geochem.* **8,** 529–549.

3. Barker, J. F., Patrick, G. C., and Major, D. (1987), *Groundwater Monitoring Rev.*, 64–71.
4. Hutchins, S. R., Sewell, G. W., Kovacs, D. A., and Smith, G. A. (1991), *Environ. Sci. Technol.* **25(1)**, 68–76.
5. Lovley, D. R. and Lonergan, D. J. (1990), *Appl. Environ. Microbiol.* **56(6)**, 1858–1864.
6. Edwards, E. A., Wills, L. E., Reinhard, M., and Grbic-Galic, D. (1992), *Appl. Environ. Microbiol.* **58(3)**, 794–800.
7. Grbic-Galic, D. and Vigel, T. M. (1987), *Appl. Environ. Microbiol.* **53(2)**, 254–260.
8. Raterman, K. T., Barker, G. W., Corgan, J. M., Fisher, J. B., and Trent, G. L. (1994), in *Proceedings of 1994 Rocky Mountain Symposium on Environmental Issues in Oil and Gas Operations*, pp. 229–238.
9. Barker, G. W., Raterman, K. T., Fisher, J. B., Corgan, J. M. Trent, G. L., Brown, D. R., Kemp, N. P., and Sublette, K. L. (1996), *Appl. Biochem. Biotechnol.*, **57/58**, 791–802.
10. Borole, A. P. (1996), Intrinsic bioremediation of gas condensate hydrocarbons, Ph.D. thesis, The University of Tulsa, Tulsa, OK.
11. Borole, A. P., Fisher, J. B., Raterman, K. T., Kemp, N., Sublette, K. L., and McInerney, M. J. (1995), *Appl. Biochem. Biotechnol.*, **57/58**, 817–826.
12. Fisher, J. B., Sublette, K. L., Kolhatkar, R. V., Brown, D. R., Barker, G. W., Raterman, K. T., Trent, G. L., and Corgan, J. M. (1995), in *Proceedings of 1995 Rocky Mountain Symposium on Environmental Issues in Oil and Gas Operations*, pp. 235–249.

Copyright © 1997 by Humana Press Inc.
All rights of any nature whatsoever reserved.
0273-2289/97/63-65—0835$10.25

Effects of a Nutrient-Surfactant Compound on Solubilization Rates of TCE

M. T. GILLESPIE AND J. M. STRONG-GUNDERSON*

P.O. Box 2008, Building 1505, MS 6038, Oak Ridge National Laboratory, Oak Ridge, TN 37831-6038

ABSTRACT

BioTreat™, a commercially available nutrient-surfactant compound, was investigated for its ability to solubilize TCE. Potential mechanisms for enhancing biodegradation rates by the use of nutrient-surfactant mixtures are: increased solubilization of TCE into the aqueous phase, and increased nutrients for the bacteria and greater numbers of colony forming units (CFUs). In aqueous systems, no measured solubilization of 0.1 and 1.0 ppm TCE from the headspace into the liquid phase was observed with BioTreat added at concentrations <0.5%. However, at BioTreat concentrations in excess of the CMC (≥0.5%), increased solubilization of TCE was measured. A second question was the nutrient effect of BioTreat on the growth of the TCE-degrading bacterium, *Burkholderia cepacia* G4 $PR1_{301}$. The added nutrients provided by BioTreat was evident and lead to increased cell numbers. The effect of BioTreat on the expression of ortho-monooxygenase, the enzyme necessary for TCE degradation by *B. cepacia* was also investigated. Enzyme expression as detected by a colorimetric assay was inhibited for BioTreat concentrations >0.05%.

Index Entries: TCE; degradation; surfactants; enzyme; biodegradable.

*Author to whom all correspondence and reprint requests should be addressed. *"The submitted manuscript has been authorized by a contractor of the U.S. Government under contract No. DE-AC0S-960R22464. Accordingly, the U.S. Government retains a nonexclusive, royalty-free license to publish or reproduce the published form of this contributuion, or allow others to do so, for U.S. Government purposes."

INTRODUCTION

Soils contaminated with volatile organic compounds (VOC) can pose a problem for remediation when the contaminant is sorbed to the soil, is present as a nonaqueous-phase liquid, or volatilized within the pore spaces. For bioremediation to occur, the contaminant must be bioavailable, i.e., solubilized, into the aqueous phase. Surfactants can facilitate this process of solubilization by reducing the interfacial tension between the soil–liquid interface. Surfactants can also facilitate solubilization from the pore spaces into the liquid phase by reducing the interfacial tension at the gas–liquid interface *(1)*.

Aqueous-phase surfactant concentrations can easily be quantified by measuring their ability to reduce the interfacial tension of the liquid. Typically, surfactants used for contaminant solubilization are used at or above the critical micelle concentration (CMC). The CMC is the concentration at which surfactant monomers self-aggregate into micelles with the hydrophobic tails oriented to the inside and the hydrophilic heads oriented outside. At concentrations above the CMC, monomers and micelles are present, but below the CMC, only surfactant monomers are in solution *(1)*.

Fertilizers have been found to increase remediation rates by stimulating bacterial growth. The oleophilic fertilizer, Inipol EAP22, was used in the remediation of the Exxon Valdez oil spill in Prince William Sound, AK to increase natural biodegradation rates of stranded oil by the indigenous microorganisms *(2)*. In contrast to this fertilizer, we examined the effects of an aqueous-based nutrient plus surfactant mixture and its application to hazardous waste remediation. This bioenhancing compound, Bio Treat™, along with two nonionic synthetic surfactants, Poly-Tergent 42® and Tween-80® were used in these experiments in order to investigate enhanced TCE solubilization for biodegradation by *Burkholderia cepacia* G4 PR1$_{301}$, a constitutive TCE degrading bacterium *(3)*. Potential mechanisms for enhanced biodegradation rates are: (1) increased solubilization of TCE into the aqueous phase, and (2) increased nutrients or alternate carbon source for the bacterium, thus increased cell numbers. In this study, we investigated the effects of this nutrient-surfactant product on TCE solubilization, *B. cepacia* growth, and expression of the enzyme necessary for TCE degradation.

EXPERIMENTAL PROCEDURES AND MATERIALS

Bacterial Strain, Culture Conditions, and Chemicals

The bacteria used throughout these experiments was *B. cepacia* G4 PR1$_{301}$, a nongenetically engineered constitutive TCE degrader (M. Shields, University of West Florida, Pensacola, FL; 3–4). The bacteria was grown in continuous culture in basal salts media (BSM) *(5)* with 20 mM

glucose as the sole carbon source. Liquid cultures were routinely started from nonselective agar plates of R2A growth media (Difco Laboratories, Detroit, MI) or a selective growth media of BSM + 1.7% noble agar, and either 20 mM glucose or 20 mM sodium lactate. The plates were scraped after 7 d, and the bacteria resuspended in 10 mL of BSM + 20 mM glucose in a 15-mL sterile centrifuge tube (Corning, Corning, NY). After a 2-d incubation on a rotary shaker (250 rpm) at ambient temperature, the optical density (OD) increased to 0.2–0.5 at 600 nm (Gilford Response UV-Visible spectrophotometer, Oberlin, OH). These cultures were transferred into 90 mL of fresh media in a 250-mL Erlenmeyer flask and returned to the shaker until OD ≥ 2.0.

Expression of the enzyme responsible for TCE degradation by *B. cepacia*, *ortho*-monooxygenase *(3)*, was measured using the triflouromethyl phenol or *m*-hydroxy benzotrifluoride (TFMP) oxidation assay. The rate of production of TFHA (7,7,7-trifluoro-2-hydroxy-6-oxo-2,4-heptadienoic acid), a yellow product, from TFMP correlates to the potential rate of TCE degradation by the enzyme *(4)*.

The synthetic surfactants used in these experiments were Poly-Tergent 42 (Olin, Stamford, CT), Tween 80 (Sigma Chemical, St. Louis, MO), and a commercially available nutrient–surfactant mixture, BioTreat (Rem-Tec, Clemmons, NC). All surfactant stock solutions were made in Milli-Q water and filter-sterilized through a 0.2-µm filter (Nalgene, Rochester, NY). The CMC values for these surfactants are: 0.01% BioTreat, 0.05% Poly-Tergent 42, and 0.03% Tween 80. The surfactants were tested at concentrations below, equal to, or above their aqueous CMC.

Growth of *B. cepacia* on Surfactants

B. cepacia G4 PR1$_{301}$ was assayed with 0.05% BioTreat and 0.05% Poly-Tergent 42 as the sole carbon sources. Poly-Tergent 42 was chosen as the nonionic surfactant, because it was reported to be biodegradable (Olin Corporation, personal communication). Bacteria were initially cultured on agar plates for 7 d, scraped, and resuspended in sterile minimal salts media *(6)* to OD = 1.5 at 600 nm. The bacterial suspension (300 µL) was inoculated into 100 mL of minimal salts media in a 250-mL Erlenmeyer flask. BioTreat and Poly-Tergent 42 that were filter-sterilized were added to a final concentration of 0.05%. The sterility of the surfactants was verified by streaking on nutrient agar plates (Becton Dickson Company, Cockeysville, MD) and observing for growth. The flasks were shaken at room temperature at 250 rpm on a rotary shaker and daily optical density measurements were recorded at 600 nm. The cultures were amended with an additional 0.05% surfactant until the OD = 2.

B. cepacia was also grown with BioTreat provided as a secondary carbon source. The bacteria were cultured as described previously with 20 mM glucose as the primary carbon source, and 0.01%, 0.05%, and 0.1%

BioTreat as an additional carbon/nutrient source. Optical density measurements at 600 nm were recorded at days 3, 4, and 5. The specific activity of the enzyme was assayed at day 5 using the TFMP oxidation assay.

Abiotic TCE Solubilization

TCE solubilization experiments utilized sterile 15-mL glass vials (EPA vials) containing 5 mL of sterile phosphate-buffered solution (PBS, 1.2 g Na_2HPO_4, 2.2 g NaH_2PO_4, 8.7 g NaCl/L water). Filter-sterilized BioTreat was added at concentrations of 0.001, 0.01, 0.05, 0.5, 1 and 5%. Tween 80 was tested at concentrations of 0.01, 0.03, 0.1, 0.5, 1 and 5% and was chosen based on its ability to solubilize low concentrations of carbon tetrachloride (7). The vials were sealed with Teflon-coated septa prior to the addition of 0.1 or 1.0 ppm TCE, which was added through the septa via syringe. The vials were inverted and equilibrated overnight at ambient temperature on a rotary shaker at 250 rpm. Headspace measurements (30 µL) were performed on a gas chromatograph (GC).

Analytical Methods

A GC (Hewlett Packard 5890 Series II Plus, San Fernando, CA) equipped with an electron capture detector and a megabore capillary column DB624 (30 m 0.53 mm inner diameter) (Alltech, Deerfield, IL) was used to analyze headspace gas samples. Argon-methane was the carrier gas. The injector temperature was 100°C. The oven temperature started at 70°C and increased to 110°C at a rate of 10°C/min. The detector temperature was 300°C, and column carrier gas flow was 7.0 mL/min. A headspace gas volume of 30 µL was injected for each sample.

RESULTS AND DISCUSSION

Previous work has shown that surfactants can enhance solubilization of chlorinated solvents in soil systems (8–10). The effect of surfactant-enhanced partitioning, i.e., solubilization, of TCE from the headspace into the liquid phase is important when determining the bioavailability of chlorinated solvents and other VOCs. This work examined abiotic TCE solubilization in aqueous systems. The amount of TCE partitioned into the headspace was followed throughout the surfactant treatments. All solubilization data presented here are expressed in GC area units of TCE in the gas phase.

Initial experiments examined low surfactant concentrations and their affect on TCE solubilization. Surfactants facilitate solubilization of VOCs from the headspace into the aqueous phase by reducing the interfacial tension at the gas–liquid interface. Relatively low surfactant concentrations (<0.1%) had no effect on the enhanced TCE solubilization for either 0.1 or 1.0 ppm in an aqueous system (Figs. 1 and 2). Data presented

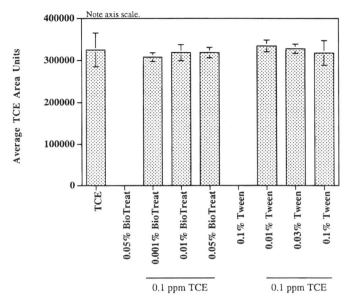

Fig. 1. Gas-phase measurements for the abiotic solubilization of 0.1 ppm TCE. Standard deviations are within bars, $n = 3$. BioTreat concentrations represent below (0.001%), at (0.01%), and above (0.05%) aqueous CMC. No enhanced partitioning of TCE from headspace into liquid phase visible. Tween concentrations represent below (0.01%), at (0.03%), and above (0.1%) aqueous CMC. No enhanced partitioning of TCE is detected.

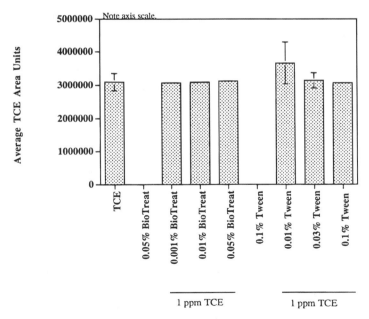

Fig. 2. Gas-phase measurements for the abiotic solubilization of 1.0 ppm TCE. Standard deviations are within bars, $n = 3$. No enhanced solubilization of TCE at 0.001, 0.01, or 0.05% BioTreat. No enhanced solubilization of TCE at 0.01, 0.03, and 0.1% Tween 80.

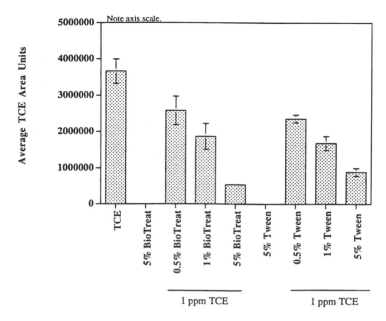

Fig. 3. Gas-phase measurements for the abiotic solubilization of 1.0 ppm TCE. Standard deviations are within bars, $n = 3$. Samples contain surfactant in excess of the CMC (0.5, 1, 5% BioTreat; 0.5, 1, 5% Tween 80). Increased partitioning of TCE from headspace into aqueous phase as surfactant concentration increased.

here showed that this was true for both Tween 80 and the nutrient–surfactant compound, BioTreat. This is expected based on previous work by other investigators, which showed that below the CMC, surfactants exist only as monomers, which do not enhance TCE solubilization (15). However, as the surfactant concentrations increased in excess of the CMC (>0.5%), there was a decrease in TCE headspace concentration compared to controls. These data support an increased solubilization of TCE into the aqueous phase (Fig. 3). This enhanced solubilization of TCE from the headspace into the aqueous phase makes the contaminant ultimately more bioavailable to the microorganisms for degradation.

An important consideration in choosing a surfactant for remediation is surfactant persistence and toxicity (11). The surfactant chosen for primary consideration in this study was BioTreat. Furthermore, the addition of nutrient solutions to contaminated soil has also been shown to increase remediation rates (2,12,13). For example, Inipol EAP22, an oleophilic fertilizer used in the cleanup of the Exxon Valdez oil spill, also has surfactant properties (14,15). BioTreat, although an aqueous-based fertilizer, is similar to Inipol EAP22 in that it demonstrates surfactant properties in addition to its nutrient-providing capabilities. Thus, the ability of *B. cepacia* to metabolize BioTreat as its sole carbon source (Fig. 4) was measured. An initial concentration of 0.05% BioTreat supported microbial growth to an OD = 0.32. All OD measurements were performed with a blank con-

Nutrient Surfactants

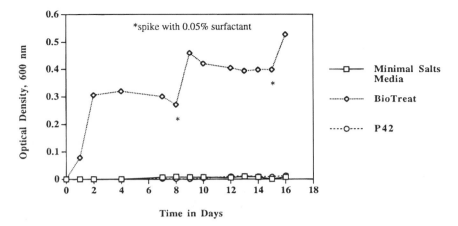

Fig. 4. Growth of *B. cepacia* G4 PR1$_{301}$ with surfactants as sole carbon source. Optical density measurements over time. Samples were spiked with additional 0.05% surfactant at days 8 and 15 in order to provide more carbon to the system. Bacteria can metabolize BioTreat, but not Poly-Tergent 42.

sisting of surfactant and water to account for the surfactant's possible effect on absorbance. Subsequent additions of BioTreat resulted in sharp increases in cell growth, followed by stationery growth periods. Stationery phases represent a depletion of the carbon source, inorganic nutrients, or a buildup of toxic cellular byproducts *(16)*. In TCE degradation experiments in our lab (data not shown), CFUs were increased by 34% over controls when BioTreat was added. Previous experiments indicated that Poly-Tergent 42 was not actually biodegradable by *B. cepacia* and therefore was used as a control. No growth of *B. cepacia* was observed after 16 d of exposure to 0.15% Poly-Tergent 42.

Although BioTreat at concentrations higher than the aqueous CMC can increase surfactant partitioning and act as a nutrient source for *B. cepacia*, the bacteria does not produce the enzyme necessary for TCE degradation when grown in the presence of high concentrations of BioTreat. *B. cepacia* was cultured with 20 m*M* glucose and varying concentrations of BioTreat as a secondary carbon source. Enzyme expression was inhibited at 0.05 and 0.1% BioTreat, but was not affected by BioTreat concentrations of 0.01% (Fig. 5). However, these low surfactant concentrations are not effective at increasing the TCE solubilization. BioTreat does appear to increase cell numbers, which would increase rates of TCE degradation by changing the TCE equilibrium concentrations between the aqueous and headspace phases. It is apparent that *B. cepacia* growth is not inhibited by the surfactant properties of BioTreat. However, the additional source of carbon does not correspond to an increase in OD over that of controls with glucose only. It is currently unclear if there may be other factors within the system that would slow cell growth, such as limited trace elements in the BSM media. Additional data from the TFMP

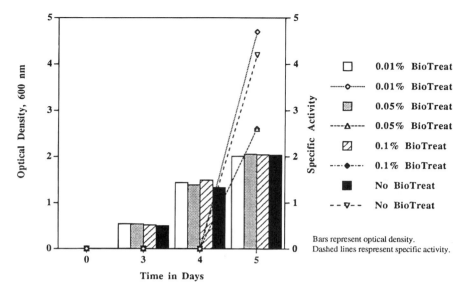

Fig. 5. Growth of *B. cepacia* G4 PR1$_{301}$ on glucose in the presence of BioTreat. Optical density and specific activity measurements over time. Glucose 20 mM present as carbon source. BioTreat (0.01, 0.05, 0.1%) did not inhibit growth of bacteria. Concentrations of 0.1% BioTreat inhibited enzyme production.

assays confirm that the bacteria does not produce the enzyme when grown on BioTreat only. An alternative explanation is the interference of BioTreat with the colorimetric TFMP enzyme assay. Work is currently being done to verify that no enzyme is produced vs the inability to detect enzyme production by the TFMP oxidation assay.

The BioTreat product used in this study acted as a nutrient to the bacteria, thus leading to higher cell numbers, but did not increase TCE bioavailability or solubilization at low concentrations. Although utilization of the compound at concentrations greater than the CMC increased TCE headspace partitioning, i.e., solubilization, and thus increased bioavailabilty, these same concentrations inhibited the expression of *ortho*-monooxygenase, which is essential for TCE degradation by *B. cepacia* G4 PR1$_{301}$. These results suggest no additional value in adding BioTreat to aqueous systems when increased solubilization of TCE for biodegradation by *B. cepacia* G4 PR1$_{301}$ is the only desired effect. Current work is examining the solubilization and nutrient effects of BioTreat in soil-slurry systems in preparation for use in a potential field demonstration.

ACKNOWLEDGMENTS

We would like to thank Malcolm Shields at the University of West Florida for the *B. cepacia* G4 PR1$_{301}$, Tom Smith at Rem-Tec, Inc., for the BioTreat, Eva Krieger at Olin Chemical for the Poly-Tergent surfactant, and Sue Carroll at the Oak Ridge National Laboratory (ORNL) for her expert technical assistance.

This research was sponsored by the *In-Situ* Remediation Technology Development Program of the Office of Technology Development, US Department of Energy (US DOE). ORNL is managed by Lockheed Martin Energy Research Corporation, under contract DE-AC05-96OR22464 with the US Department of Energy (ESD Publication #4654).

REFERENCES

1. West, C. C. and Harwell, J. H. (1992), *Environ. Sci. Tech.* **26**, 2324–2330.
2. Lindstrom, J. E., Prince, R. C., Clark, J. C., Grossman, M. J., Yeager, T. R., Braddock, J. F., and Brown, E. J. (1991), *Appl. Environ. Microbiol.* **57**, 2514–2522.
3. Shields, M. S., Montgomery, S. O., Cuskey, S. M., Chapman, P. J., and Pritchard, P. H. (1991), *Appl. Environ. Microbiol.* **57**, 1935–1941.
4. Shields, M. S. and Reagin, M. J. (1992), *Appl. Environ. Microbiol.* **58**, 3977–3983.
5. Hareland, W., Crawford, R. L., Chapman, P. J., and Dagley, S. (1975), *J. Bacteriol.* **121**, 272–285.
6. Little, C. D., Palumbo, A. V., Herbes, S. E., Lindstrom, M. E., Tyndall, R. L., and Gilmer, P. J. (1998), *Appl. Environ. Microbiol.* **54**, 951–956.
7. Bowden, S. R. (1996), Master's thesis. University of Tennessee, Knoxville, TN.
8. Jackson, R. E. (1994), *Remediation*, **Winter 1993/94**, 77–91.
9. Pennel, K. D., Jin, M., Abriola, L. M., and Pope, G. A. (1994), *J. Cont. Hydrol.* **16**, 35–53.
10. Shiau, B., Sabatini, D. A., and Harwell, J. H. (1994), *Groundwater* **32**, 561–569.
11. Ankley, G. T. and Burkhard, L. P. (1992), *Environ. Toxicol. Chem.* **11**, 1235–1248.
12. Sveum, P., Fakness, L. G., and Ramstad, S. (1994), in *Hydrocarbon Bioremediation*, Hinchee, R. E., Alleman, B. C. Hoeppel, R. E., and Miller, R. N., eds., Lewis, Boca Raton, FL, pp. 163–174.
13. Sveum, P. and Ramstad, S. (1995), in *Applied Bioremediation of Petroleum Hydrocarbons*, Hinchee, R. E., Kittel, J. A., and Reisinger, J. H., eds., Battelle, Columbus, OH, pp. 201–218.
14. Basseres, A., and Ladousse, A., (1993) in *ACS Preprints, Bioremediation and Bioprocessing*, Vol. 38, II, pp. 246–253.
15. West, C. C. (1992), in *Transport and Remediation of Subsurface Contaminants: Colloidal, Interfacial, and Surfactant Phenomena*, Sabatini, D. A., and Knox, R. C., eds., American Chemical Society, pp 159–168.
16. Churchill, S. A., Griffin, R. A., Jones, L. P., and Churchill, P. F. (1995), *J. Environ. Quality* **24**, 19–28.
17. Atlas, R. M. (1988), in *Microbiology Fundamentals and Principles*, 2nd ed., MacMillan, New York, p. 107.

Porphyrin-Catalyzed Oxidation of Trichlorophenol

SALEEM HASAN AND KERRY L. SUBLETTE*

Center for Environmental Research & Technology, University of Tulsa, 600 S. College Avenue, Tulsa, OK 74104

ABSTRACT

Porphyrin–metal complexes are potentially useful to catalyze redox reactions, which convert toxic and biologically recalcitrant compounds to compounds that are less toxic and more amenable to biotreatment. Porphyrins, in the absence of proteins as in ligninases, peroxidases, and oxidases, are potentially more robust than enzymes and microbial cultures in the treatment of inhibitory substances.

2,4,6-Trichlorophenol was used as a model compound for chlorinated phenols and as a substrate for various porphyrin-metal complexes acting as oxidation catalysts. t-Butyl hydroperoxide was the oxidizing agent. TCP was shown to be at least partially dechlorinated and the aromatic ring broken in reaction products. All porphyrins exhibited saturation kinetics with regard to the initial TCP concentration in reaction mixtures. Electron-withdrawing substituents on the porphyrins were observed to increase stability of the catalysts to inactivating ring-centered oxidation.

Index Entries: Trichlorophenol; porphyrin; biomimetic; heme; dechlorination.

INTRODUCTION

Extensive use of chlorinated phenols, such as pentachlorophenol, as fungicides and wood preservatives has resulted in significant contamination of ground water, surface water, and soils with these compounds. Chlorinated phenols are also found in the effluents of Kraft paper mills formed as a result of lignin degradation *(1)*. The toxicity of chlorophenols in aquatic systems and to humans is well established *(2–4)*.

Many microorganisms have been shown to mineralize pentachlorophenol. However, the same organisms do not necessarily degrade other struc-

---

*Author to whom all correspondence and reprint requests should be addressed.

turally similar chlorinated phenols. Mixtures of chlorinated phenols can escape treatment because of crossinhibition (5). Anaerobic/aerobic mixed cultures in series have been shown to mineralize several chlorinated phenols (6). The maintenance of two dissimilar cultures in series clearly adds a level of complexity to the process. The white-rot fungus *Phanerochaete chrysosporium* has been shown to degrade a variety of chlorinated phenols via a system of exogenous, nonspecific peroxidases or ligninases (7,8). However, these ligninases are produced only when the organism is starved for a key nutrient like nitrogen (9). Therefore, the utilization of this organism for treatment of chlorinated phenols and other substituted aromatics requires a feed-starve approach to maintain viability and ligninase production.

Several investigations have been demonstrated that porphyrins alone, in the absence of protein as in ligninases, can catalyze oxidation and reduction reactions of lignin model compounds. Porphryin–metal complexes have been shown to catalyze the reductive dehalogenation of halogenated hydrocarbons and the reduction of nitroaromatics to corresponding amines (10–13). Shimada et al. (14) have also reported the oxidative cleavage of C—C bonds in lignin model compounds catalyzed by Fe^{+3} protoporphyrin IX in the presence of an active oxygen donor, such as *t*-butylhydroperoxide (TBHP). Hickman et al. (15) have reported that oxidation of metalloporphyrins may be metal-centered or ring-centered. Ring-centered oxidation results in inactivation of the porphyrin as a catalyst. Metal-centered oxidation allows the oxidized metal ion to abstract an electron from a susceptible functional group and be reduced to the original oxidation state. Nanthakumar and Goff (16) have shown that porphyrins with electronegative substituents in the *o*-aryl positions are more stable toward ring-centered oxidation and, therefore, conceivably more useful as redox catalyst.

Porphyrin–metal complexes are potentially useful to catalyze redox reactions, which convert toxic and biologically recalcitrant compounds to compounds that are less toxic and more amenable to conventional biological treatment. Porphyrins, in the absence of protein as in ligninases and other peroxidases and oxidases, are potentially more robust than enzymes and microbial cultures in the treatment of inhibitory substrates. In this work, 2, 4, 6-trichlorophenol (TCP) was used as a model compound for chlorinated phenols and as a substrate for various porphyrin–metal complexes acting as oxidation catalysts. The objective was the catalytic oxidation of TCP to less-chlorinated and less-toxic products.

MATERIALS AND METHODS

Chemicals

All porphyrins used in this work, including Fe^{3+}-protoporphyrin IX (Fe-pro-Ph), Fe^{3+}-meso-tetra(*o*-diclorophenyl) porphine sulfonate (Fe-diCl-Ph), and Fe^{3+}-meso-tetra(*o*-difluorophenyl) porphine sulfonate (Fe-diF-

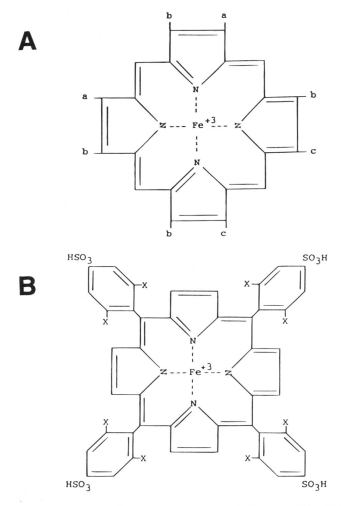

Fig. 1. Structures of **(A)** Fe^{3+}-protoporphyrin IX (a = —CH=CH; b = —CH$_3$; c = —CH$_2$CH$_2$COOH) and **(B)** Fe^{3+}-meso-tetra(*o*-dichlorophenyl)porphine sulfonate (x = Cl) and Fe^{3+}-meso-tetra(*o*-difluorophenyl)porphine sulfonate (x = F).

Ph), were obtained from Porphyrin Products (Logan, UT). TCP was obtained from Aldrich Chemical Co. (Milwaukee, WI). A 100 mg/L standard of TCP in methanol was obtained from Chem Service (West Chester, PA) and TBHP was obtained from Sigma Chemical Co. (St. Louis, MO). All chemicals used were reagent-grade.

Porphyrin-Catalyzed Oxidation of TCP

Three porphyrins were used in this study, Fe-pro-Ph, Fe-diCl-Ph, and Fe-diF-Ph (*see* Fig. 1). Fe-pro-Ph was chosen as an example of a porphyrin susceptible to ring-centered oxidation. Fe-diCl-Ph and Fe-diF-Ph both have electronegative substituents in the *o*-aryl positions and should be more stable toward ring-centered oxidation. The sulfonate groups enhance water solubility.

Stock solutions of porphyrins were prepared in $0.1M$ phosphate buffer at pH 7.0 and stored in flasks covered with aluminum foil to prevent exposure to light. Porphyrin concentrations were verified using a calibration curve based on UV-VIS absorbance at a characteristic wavelength of each porphyrin. A stock TCP solution (5.0 mM) was also prepared in the $0.1M$ phosphate buffer at pH 7.0.

Reactions were typically conducted in 100-mL sealed septum bottles containing 3.0 mM TCP, 0.0010–0.0020 mM porphyrin, and 45 mM TBHP in a reaction volume of 35.0 mL. The temperature was 55°C and the pH was 7.0, which was shown to be optimal. The reaction was initiated by adding the oxidizing agent. Samples (1.0 mL) were removed periodically with a syringe and placed in 0°C/water bath to quench the reaction. Samples were analyzed for TCP, chloride, and other reaction products as detailed below. Control experiments were identical to the reaction experiments, except for the absence of porphyrin in the reaction mixture.

In order to determine the effect of pH on the reaction, a number of batch experiments were conducted at 55°C using Fe-diCl-Ph at various pH values. These experiments were conducted as described above, except that the pH was varied from 4.5–10.1. A $0.1M$ phosphate buffer was used at each pH. Sodium hydroxide was used to adjust the pH to the desired value. In each case, the pH remained constant during the course of the reaction. Other initial reaction conditions were as follows. Fe-diCl-Ph (0.0020 mM), TCP (1.5 mM) and TBHP (45.0 mM).

A comparison of the kinetics of TCP oxidation catalyzed by Fe-pro-Ph, Fe-diCl-Ph, and Fe-diF-Ph was conducted as follows. The effect of initial TCP concentration on the reaction catalyzed by each of the three porphyrins was conducted at a temperature of 55°C, pH 7.0, porphyrin concentration of 0.0010 mM, and a TBHP concentration of 45.0 mM. The initial TCP concentration was varied from 0.20–3.00 mM. Comparison of the amount of TCP oxidized/mol of porphyrin for different initial TCP concentrations was also used to demonstrate the catalytic nature of the reaction.

The effect of porphyrin concentration on the reaction rate was studied by conducting a series of TCP oxidation reactions using TBHP as oxidizing agent and Fe-diCl-Ph as the porphyrin catalyst. The temperature was 55°C, the TBHP concentration was 45.0 mM, and the initial TCP concentration was 1.6 mM in all the experiments, whereas the concentration of Fe-diCl-Ph was varied from 0.0–0.0014 mM.

A series of batch experiments was also conducted to study the effect of temperature on dechlorination of TCP. Reaction mixtures contained 2.1 mM TCP, 45.0 mM TBHP, and 0.0080–0.033 mM porphyrin. The reactions were conducted at 55, 70, and 100°C and pH 7.0, and were otherwise conducted as described above.

Analytical

A Hewlett Packard model 1090L high-pressure liquid chromatograph (HPLC), with UV-VIS diode array detector was used to measure TCP concentration in samples of the reaction mixture. The column used was (100 mm long and 46 mm id) Hypersil ODS (5 µm) and was supplied by Hewlett Packard (Palo Alto, CA). The solvent was methanol/water (45:55) at a flow rate of 1.5 mL/mm. The operating temperature was 40°C and the injection volume was 0.3 µL. Under these conditions, the retention times of TCP and TBHP were 4.0 and 1.0 min, respectively. A TCP standard solution (100 mg/L) from Chem Service (West Chester, PA) was used to quantify TCP concentration.

Chloride ion concentration in reaction mixtures was measured using an Orion Model 94-17B chloride electrode (Cole-Parmer Instruments, Niles, IL) using the method of standard addition. A 1000 mg/L standard chloride solution was also obtained from Cole-Parmer.

All reaction mixtures contained a fine precipitate that was filtered out and analyzed separately. A known volume of the reaction mixture was filtered through a membrane filter (pore size 0.45 µm) produced by Belman Instrument Co. (Ann Arbor, MI). The precipitate was dried and dissolved in 4.0 mL of HPLC-grade methanol.

Product identification was initially attempted using HPLC. The HPLC operating conditions were the same as those used for TCP quantification. Compounds that were possible products of the reaction were injected into the HPLC to identify the unknown product by matching the retention time and UV-VIS spectrum. The filtrate was also analyzed on Dionex 200li series liquid ion chromatograph to determine the concentration of organic acids that may have been formed.

Infrared analysis was carried out on the precipitate (in methanol) to get some indication of the functional groups on the reaction product. Infrared analysis was conducted on a Nicolet 510P FT-IR spectrometer manufactured by Nicolet (Madison, WI). A background subtraction was done to eliminate the effect of the methanol solvent. In addition, gas chromatograph/mass spectrometric (GC/MS) analysis of the reaction medium, as well as the precipitate dissolved in methanol, was done using a VG 70-25OHF mass spectrometer (VG Analytical, Manchester, England) in conjunction with a model 5790 Hewlett Packard GC. The GC/MS operating conditions were: column, DB-1 capillary thin film (60 m); He as carrier gas; column pressure 25 psi and split ratio 20:1; column and oven temperature 35°C for 2 min, then 7 C/min to 325°C; injector temperature, 260°C.

RESULTS AND DISCUSSION

Porphyrin-Catalyzed Oxidation of TCP

The optimum pH for Fe-diCl-Ph-catalyzed oxidation of TCP was determined to be pH 7.0. The reaction rate was reduced to 40% at pH 5.0,

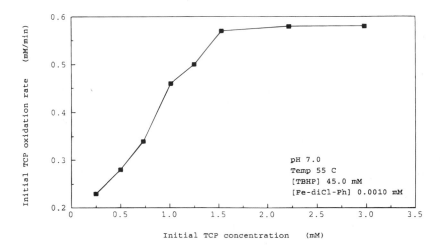

Fig. 2. Effect of initial TCP concentration on Fe-diCl-Ph-catalyzed TCP oxidation with TBHP.

but was reduced only to 80% at pH 9.0 compared to the rate at pH 7.0. All subsequent reactions with all three porphyrins were conducted at this pH.

Under assay conditions described above, complete conversion of TCP was observed with Fe-diCl-Ph and Fe-diF-Ph as catalyst. However, TCP conversion was incomplete (35%) in reactions catalyzed by Fe-pro-Ph. With Fe-pro-Ph, TCP conversion could be increased by increasing the initial porphyrin concentration, indicating that TCP conversion was reduced by ring-centered oxidation and inactivation of the porphyrin. Fe-diCl-Ph and Fe-diF-Ph apparently exhibited greater conversion because of greater stability to ring-centered oxidation.

Batch experiments, with Fe-pro-Ph, Fe-diCl-Ph, and Fe-diF-Ph, were conducted to study the effect of initial TCP concentration on TCP oxidation. In each case, the initial rate increased linearly with initial TCP concentration and then leveled off. Further increase in initial TCP concentration did not result in higher initial oxidation rates. With Fe-diCl-Ph (0.0010 mM) as catalyst, no further increase in reaction rate was seen at TCP concentrations above 1.5 mM (Fig. 2). This reaction exhibited saturation kinetics, a characteristic typical of enzyme-catalyzed reactions (such as ligninase-catalyzed reactions). For comparison, the initial rates of reaction for each system at an initial TCP concentration of 1.0 mM and porphyrin concentration of 0.0010 mM were: Fe-pro-Ph, 0.10 mM/min; Fe-diCl-Ph, 0.50 mM/min; and Fe-diF-Ph, 0.72 mM/min.

In addition to demonstrating that porphyrin-catalyzed oxidation reactions exhibit saturation kinetics, the initial rate vs initial TCP concentration data were used to show that stoichiometric amounts of TCP were not consumed for reaction mixtures containing the same amount of porphyrin. Table 1 shows in that reactions catalyzed by Fe-diCl-Ph the mmol of TCP

Table 1
Effect of Initial TCP Concentration on TCP Conversion[a]

Initial TCP mM	Final TCP mM	TCP Converted mM	mmoles TCP/mmole Fe-diCL-Ph
0.25	0.00	0.25	250
0.50	0.09	0.41	410
0.73	0.15	0.58	580
1.01	0.31	0.70	700
1.25	0.47	0.78	780
1.53	0.62	0.91	910
2.22	1.27	0.95	950
2.98	1.88	1.10	1100

[a]The Fe-diCl-Ph-catalyzed reaction was conducted at 55°C, pH 7.0. The Fe-diCl-Ph concentration was 0.0010 mM and TBHP concentration was 45.0 mM, the TCP concentration was caried from 0.20–3.00 mM.

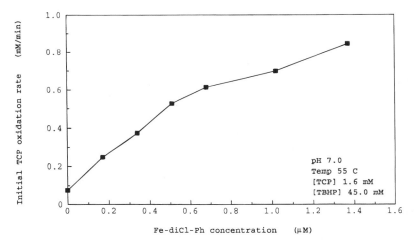

Fig. 3. Effect of Fe-diCl-Ph concentration on TCP oxidation rates. TBHP was the oxidizing agent.

oxidized/mmol of porphyrin is different in each run and increases with initial TCP concentration, thus supporting a catalytic role for porphyrins in the TCP oxidation reaction.

The effect of porphyrin concentration on TCP oxidation rate was also investigated. Porphyrin concentration was varied from 0.0–0.0014 mM. As expected, increasing the catalyst concentration in each case resulted in a roughly linear increase in the initial oxidation rates (Fig. 3).

The extent of dechlorination was determined for all three catalysts (Fe-pro-Ph, Fe-diF-Ph, and Fe-diCl-Ph) at temperatures of 55–100°C

(Table 2). Almost complete dechlorination (93.7%) of the oxidized fraction of initial TCP present occurred when Fe-pro-Ph was used as catalyst at a reaction temperature of 55°C. The extent of dechlorination by Fe-pro-Ph dropped to 75.3% at 100°C possibly because the higher temperature increased the rate of ring-centered oxidation. The ability to dechlorinate TCP was found to be lower in the case of both Fe-diCl-Ph and Fe-diF-Ph, however, the extent of dechlorination did not change significantly with increased reaction temperature, suggesting again that both these porphyrins were much more stable than Fe-pro-Ph with regard to ring-centered oxidation.

The filtrate from reaction mixtures was analyzed for organic acids by ion chromatography for organic acids. Very small amounts of formate, acetate, and propionate (all <0.01 mmol/mmol TCP) were detected. Infrared analysis of the precipitate indicated the presence of one or more carbonyl groups as indicated by a major absorbance at 1670 cm^{-1}. GC-MS analysis of the filtrate and precipitate led to the following observations:

1. The major products formed during TCP oxidation were the same irrespective of the choice of Fe-diF-Ph or Fe-diCl-Ph as the catalyst.
2. Except for one product, which had two chlorine atoms, isotopic analysis showed that all other chlorinated products contained only one chlorine atom, indicating that two of the three chlorine atoms had been removed from the TCP. This further confirms the data obtained from the chloride balance, which indicated that 60–70% dechlorination occurred in the Fe-diF-Ph or Fe-diCl-Ph catalyzed oxidation of TCP.
3. Based on the interpretation of the fragmentation patterns, the absence of aromatic compounds indicates that in addition to dechlorination, the benzene ring was also broken.

CONCLUSIONS

Three porphyrins, Fe^{3+}-protoporphyrin IX, Fe^{3+}-meso-tetra(o-dichlorophenyl) porphine sulfonate and Fe^{3+}-meso-tetra(o-difluorophenyl) porphine sulfonate have been shown to catalyze the oxidation 2, 4, 6-trichlorophenol. TCP conversion was incomplete in reactions catalyzed by Fe^{3+}-protoporphyrin IX owing presumably to ring-centered oxidation and inactivation. The two substituted porphyrins were observed to give 100% conversion of TCP under assay conditions used in these studies. Reaction rates were significantly greater for the substituted porphyrins. All three porphyrins exhibited saturation kinetics with TCP, as substrate and reaction rates increased linearly with porphyrin concentration.

Fe^{3+} protoporphyrin-catalyzed reactions were characterized by complete dechlorination of TCP oxidized, although TCP conversion was incomplete. All reaction products in reactions catalyzed by the substi-

Table 2
Dechlorination Activity of Porphyrin Catalysts[a]

Porphyrin	Temperature C	Porphyrin Conc mM	TCP Initial mM	TCP Final mM	TCP Converted mM	Measured Cl mM	TCP Dechlorination[b] %
Fe-pro-Ph	55	0.0080	2.10	1.36	0.74	2.09	93.7
Fe-diCl-Ph	55	0.0080	2.11	0.00	2.11	4.00	63.3
Fe-diF-Ph	55	0.0080	2.13	0.00	2.13	4.00	62.5
Fe-pro-Ph	55	0.0330	1.92	0.45	1.48	3.61	81.3
Fe-diCl-Ph	55	0.0330	1.93	0.00	1.93	3.61	62.2
Fe-pro-Ph	100	0.0330	1.89	0.44	1.46	3.30	75.3
Fe-diCl-Ph	100	0.0330	1.90	0.00	1.90	4.11	72.1
Fe-diF-Ph	100	0.0330	1.90	0.00	1.90	4.11	72.1

[a]Batch reactions were conducted at two temperatures, 55 and 100°C, at pH 7.0, and two porphyrin concentrations, 0.0080 and 0.0330 mM. Reactions were carried out in 0.10M phosphate buffer at pH 7.0 and a TBHP concentration of 45.0 mM.
[b]Based on complete dechlorination of TCP converted, not an initial TCP present.

tuted porphyrins were at least partially dechlorinated. Most were monochlorinated compounds with no indication that the benzene ring had survived intact.

Porphyrin-catalyzed oxidation has been shown to be a potentially viable process concept for partial oxidation and detoxification of chlorinated phenols.

REFERENCES

1. Hackman, E. E. (1978), *Toxic Organic Chemical: Destruction and Waste Treatment*, NOYES Publications, Park Ridge, NJ.
2. Konemann, H. and Musch, A. (1981), *Toxicology* **19(3)**, 223–228.
3. Holcombe, G. W., Flandt, J. T., and Phipps, G. L. (1980), *Water Res.* **14(8)**, 1073–1077.
4. Devillers, J. and Chambon, P. (1986), *Bull. Environ. Contam. Toxicol.* **37(4)**, 599–605.
5. Chaudhry, G. R. and Chapalamadugu, S. (1991), *Microbiol. Rev.* **55**, 59–79.
6. Armenante, P. M., Kafkewitz, D., Lewandowski, G., and Kung, C.-M. (1992), *Environ. Prog.* **11(2)**, 113–122.
7. Bumpus, J. A. and Aust, S. D. (1987), *Appl. Environ. Microbiol.* **53**, 2001–2008.
8. Mileski, G. J., Bumpus, J. A., Jurek, M. H., and Aust, S. D. (1988), *Appl. Environ. Microbiol.* **54**, 2885–2889.
9. Kirk, T. K. (1987), Biochemistry and genetics of cellular degradation, paper presented at the Federation of Microbiological Societies Symposium, Paris, Sept 7–9.
10. Holmstead, R. L. (1976), *J. Agric. Food. Chem.* **24**, 620–624.
11. Marks, D. (1976), Removal of chlorinated hydrocarbon pesticides from industrial wastewater, paper delivered at the ACS Meeting, New York, NY, April 7.
12. Cho, J-G., Potter, W. T., and Sublette, K. L. (1994), *Appl. Biochem. Biotechnol.* **45/46**, 861–870.
13. Habeck, B. D. and Sublette, K. L. (1995), *Appl. Biochem. Biotechnol.* **51/52**, 747–759.
14. Shimada, M., Habe, T., Higuchi, T., Okamoto, T., and Panijpan, B. (1987), *Holzforschung* **41**, 277.
15. Hickman, D. L., Nanthakumar, A., and Goff, H. M. (1988), *J. Am. Chem. Soc.* **110**, 6384–6390.
16. Nanthakumar, A. and Goff, H. M. (1990), *J. Am. Chem. Soc.* **112**, 4047–4049.

Copyright © 1997 by Humana Press Inc.
All rights of any nature whatsoever reserved.
0273-2289/97/63–65—0855$10.50

Bacterial Reduction of Chromium

ERIC A. SCHMIEMAN,[1] JAMES N. PETERSEN,*,[2]
DAVID R. YONGE,[1] DONALD L. JOHNSTONE,[1]
YARED BEREDED-SAMUEL,[2] WILLIAM A. APEL,[3]
AND CHARLES E. TURICK[3]

[1]Civil & Environmental Engineering Department, Washington State University, Pullman, WA; [2]Chemical Engineering Department, Washington State University, Pullman, WA; and [3]Idaho National Engineering and Environmental Laboratory, Center for Industrial Biotechnology, P. O. Box 1625, Idaho Falls, ID

ABSTRACT

A mixed culture was enriched from surface soil obtained from an eastern United States site highly contaminated with chromate. Growth of the culture was inhibited by a chromium concentration of 12 mg/L. Another mixed culture was enriched from subsurface soil obtained from the Hanford reservation, at the fringe of a chromate plume. The enrichment medium was minimal salts solution augmented with acetate as the carbon source, nitrate as the terminal electron acceptor, and various levels of chromate. This mixed culture exhibited chromate tolerance, but not chromate reduction capability, when growing anaerobically on this medium. However, this culture did exhibit chromate reduction capability when growing anaerobically on TSB. Growth of this culture was not inhibited by a chromium concentration of 12 mg/L. Mixed cultures exhibited decreasing diversity with increasing levels of chromate in the enrichment medium. An *in situ* bioremediation strategy is suggested for chromate contaminated soil and groundwater.

Index entries: Bioremediation; chromium; chromium-reduction; chromium-tolerance; chromate; mixed culture; diversity.

*Author to whom all correspondence and reprint requests should be addressed.

INTRODUCTION

More than 300,000 metric tons of chromite ore are processed each year in the US for use in stainless-steel production, electroplating, leather tanning, wood preservation, drilling muds, magnetic recording tape, photographic films, pigments, and many other purposes *(1)*. These uses of chromium have resulted in chromium-contaminated soils and ground water at some production sites *(2)*. At such sites, chromium exists primarily in two valence states: Cr^{3+} and Cr^{6+}. Of these two states, trivalent chromium is the dominant species in the natural environment *(3)*. This species is relatively immobile in ground water owing to complexation with organics or precipitation as an hydroxide *(4)*. This form of chromium is also an essential trace element in the human diet: the glucose tolerance factor has been identified as a complex of Cr^{3+} with three amino acids *(5)*.

In contrast to trivalent chromium, hexavalent chromium (Cr^{6+}) is found in the natural environment almost exclusively as a result of human activities *(6)*. Further, unlike the trivalent form, this species is soluble and very mobile in ground water *(7)*. Hexavalent chromium is typically found in ground water systems as chromate, which is an oxyanion that is actively transported into cells by the sulfate transport system *(8)*. Because chromate is a carcinogen *(9)* capable of directly causing lesions in DNA as well as indirectly generating oxygen radicals, the maximum allowable concentration of chromium in drinking water is 0.1 mg/L (100 ppb) *(10)*. Effective remediation methods must be developed for sites contaminated with chromium. Methods that have previously been tested to remediate these sites include excavation, pump and treat, *in situ* vitrification, and chemical treatment with a reductant *(11)*. We are exploring biological reduction of chromate to the less hazardous trivalent state as an alternative remediation strategy.

In this article, the diversity of bacteria enriched in chromate-containing medium under denitrifying conditions and the relative rates of chromate reduction of two mixed cultures are reported. Further, the application of a bacteriological treatment strategy for *in situ* remediation is discussed.

METHODS

Two enrichment cultures, designated culture A and culture B, were employed in this study. Mixed Culture A was enriched from surface soil obtained by Idaho National Engineering and Environmental Laboratory (INEEL) from an eastern US site that had been contaminated by chromite ore tailings used as landfill. The soil Cr^{6+} concentration was approx 250 mg Cr/kg soil. Soil dilutions (10^{-1} g/mL) were made using isotonic phosphate buffer *(12)* and were inoculated into sealed serum vials containing tryptic soy broth (Difco Laboratories, Detroit, MI) and nitrogen gas in the headspace. The mixed culture was harvested and added to a packed-bed reactor filled with sterile polymeric biocatalyst support beads (Bio-Sep™, DuPont, Newark, DE). After operation of the reactor under anaerobic conditions with various Cr^{6+} concentrations, the beads were removed and stored at 4°C prior to their use.

Mixed culture B was enriched from subsurface soil obtained by Pacific Northwest National Laboratory (PNNL) from the edge of a dichromate contamination plume (13) on the US Department of Energy's Hanford facility in southeastern Washington. The soil used was a 30.5-cm long, 10 cm diameter (1-ft long, 4 in. diameter), core from 14.5 m (47.5 ft) below the surface in well 199-H5-15. The water table is 12.2 m (40 ft) below the surface at this location. After the core was removed, ground water analyses from the well had a mean value of 0.055 mg/L (55 ppb) total chromium (14). The core was obtained aseptically on November 8, 1995, and stored anaerobically (in an inert argon atmosphere) in a plastic cylinder at 4°C until March 29, 1996. The cylinder was opened in a laminar flow hood and 10 g of soil transferred into each of 12 165-mL serum bottles that had been autoclaved (121°C for 15 min). During the soil-transfer process, the serum bottles were continuously purged with filtered (0.2-μm) prepurified nitrogen. Ninety milliliters of a filter-sterilized minimal salts medium augmented with 200 mg/L acetate and 100 mg/L nitrate were added to each of the serum bottles. Potassium chromate was added to three bottles to obtain a concentration of 20 mg/L Cr^{6+}, to three bottles to obtain a concentration of 5 mg/L Cr^{6+}, and to three bottles to obtain a concentration of 0.5 mg/L Cr^{6+}. Three autoclaved bottles containing only 100 mL of the filter-sterilized minimal salts medium augmented with 200 mg/L acetate and 100 mg/L nitrate were used as a control. The 15 serum bottles were capped and pressurized to 5 psig with filtered prepurified nitrogen. After 3 d, 100 μL were aseptically withdrawn from each serum bottle and serially diluted in 0.1% peptone sterile blanks to perform heterotrophic plate counts on tryptic soy agar (Difco Laboratories, Detroit, MI) under aerobic conditions at 30°C (15). Diversity, based on colony morphology and cultural characteristics, was noted for each countable plate. Two distinctly different colonies were found in each bottle containing 20 mg/L Cr^{6+}, and they were labeled culture B strain B and culture B strain C.

Ten milliliters of tryptic soy broth (Difco Laboratories, Detroit, MI) were autoclaved (121°C for 15 min) in each of 18 test tubes. After cooling to room temperature in a laminar flow hood, filtered (0.2 μm) prepurified nitrogen gas was bubbled through the broth for 15 min to remove oxygen. Potassium chromate was added to nine of the tubes to a concentration of approx 15 mg/L. Six tubes (three with potassium chromate and three without) were inoculated with the mixed culture A by placing five polymeric biocatalyst support beads (Bio-Sep™, DuPont, Newark, DE) removed from the INEEL packed-bed reactor into each of the tubes and vortexing for 30 s. Six tubes (three with potassium chromate and three without) were inoculated with pure culture B, strain B, by transferring a single, isolated colony growing aerobically on tryptic soy agar into each of the tubes and vortexing for 30 s. Six tubes (three with potassium chromate and three without) were inoculated with pure culture B, strain C, in a similar manner. The tubes were capped, pressurized to 5 psig with filtered prepurified nitrogen, and incubated at 27°C until used as sources for inoculation of the batch reactors.

Batch reactors consisted of 125-mL serum vials. One hundred milliliters of tryptic soy broth (Difco Laboratories) were autoclaved (121°C for 15 min) in each of 18 125-mL serum vials (batch reactors). Immediately after removal from the autoclave, each reactor was purged with filtered (0.2 µm) prepurified nitrogen gas in the headspace for 5 min in a laminar flow hood. Potassium chromate was added to nine of the reactors to a concentration of approx 15 mg/L. The reactors were capped and pressurized to 5 psig with filtered prepurified nitrogen. Three reactors with potassium chromate and three reactors without potassium chromate were not inoculated and used as abiotic controls. Five milliliters were transferred from each mixed culture A tube described above (with or without potassium chromate as appropriate) into each culture A reactor (with or without potassium chromate, as appropriate). Mixed culture B reactors were inoculated by transferring 2 1/2 mL from a culture B, strain B, tube and 2 1/2 mL from a culture B, strain C, tube described above (with or without potassium chromate, as appropriate) into each culture B reactor (with or without potassium chromate, as appropriate). Each transfer was performed in a laminar flow hood by disposable sterile hypodermic syringe and needle after sterilizing the tube septum or reactor septum with 70% ethanol.

The 18 reactors in the preceding paragraph constitute a completely random experimental design with a 2×3 factorial arrangement of treatments. The treatments were randomly applied to the reactors, which were then identified by the numbering scheme summarized in Table 1.

The 18 reactors were placed on a gyratory shaker operated at 200 RPM in a 27°C incubator. Each reactor was vortexed for 30 sec and sampled at approx 6-h intervals. Each 5-mL sample was obtained by disposable sterile hypodermic syringe and needle after sterilizing the reactor septum with 70% ethanol. The samples were transferred to sterile 10-mL disposable test tubes. Three and one-half milliliters were then transferred to a plastic cuvet, and absorbance was immediately measured spectrophotometrically at 600 nm (Milton Roy Genesys 5 spectrophotometer) using type I reagent-grade water as a blank.

Chromate (Cr^{6+}) concentration in each sample was measured spectrophotometrically. Two hundred microliters of each sample were transferred by digital pipet (Eppendorf) to a sterile 10-mL disposal test tube containing 9.80 mL of type I reagent-grade water. The contents of a Hach Chrom Ver3 pillow (0.09 g diphenylcarbazide) was added to the tube, the tube shaken for 30 s, and absorbance measured between 5 and 20 min later

Table 1
Experimental Design[a]

Treatment B Initial Concentration Cr^{6+} (mg/L)	Treatment A, Inoculum								
	Sterile			Culture A			Culture B		
0	1	7	13	3	9	15	5	11	17
12	2	8	14	4	10	16	6	12	18

[a]The completely random experimental design employed a 2 × 3 factorial arrangement of treatments, Treatment A, inoculum, was applied at three levels: sterile; mixed culture A from DuPont Bio-Sep™ beads; and mixed culture B enriched from Hanford subsurface soil. Treatment B, initial hexavalent chromium concentration, was applied at two levels: 0 and ~12 mg/L. Treatments were randomly assigned to 125 mL reactors containing 100 mL of TSB, and the reactors were then identified with numbers from 1–18 as indicated in the matrix above.

at 542 nm (Milton Roy Genesys 5 spectrophotometer) using similarly treated samples from the aseptic reactors without chromate as a blank. Chromate concentration (mg/L Cr^{6+}) was calculated for each sample based on a five-point calibration curve developed over the range 0.1–20 mg/L Cr^{6+} undiluted sample in a TSB matrix.

In the final sample taken from each reactor, total Cr in the supernatant of the centrifuged (10,000g for 5 min) samples was measured by direct aspiration into an inductively coupled argon plasma simultaneous emission spectrometer (Thermo Jarrell Ash ICAP-61).

Results

Soil from the highly contaminated eastern site was enriched in TSB under anaerobic conditions with one level of chromate concentration, 20 mg/L. As reported previously by Turick and Apel (16), three species were recovered from the enriched mixed culture (culture A). This mixed culture exhibited chromate reduction during the enrichment process.

Soil from the Hanford subsurface was enriched on minimal salts medium augmented with acetate and nitrate under anaerobic conditions with four different chromate concentration levels. None of the enrichment cultures exhibited significant chromate reduction during the enrichment process. Heterotrophic plate counts indicated that both the stationary-phase populations (Fig. 1) and the population diversity (Fig. 2) decrease with increasing Cr^{6+} concentration over the range 0–20 mg/L. The two species constituting the mixed cultured enriched at 20 mg/L Cr^{6+} are referred to as mixed culture B.

When grown on TSB under anaerobic conditions in the presence of Cr^{6+} mixed culture A and mixed culture B exhibit significantly different growth rates (Fig. 3) and chromate reduction rates (Fig. 4). Mixed culture A

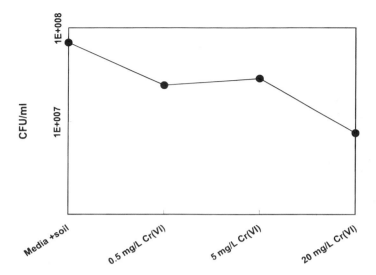

Fig. 1. Heterotrophic plate count from Hanford core 199-H5-15. 10^{-1} dilution of soil into minimal salts medium augmented with 200 mg/L acetate and 100 mg/L nitrate, followed by 3 d of anaerobic growth at 30°C. Spread plates on full-strength TSA were counted after 2 d of aerobic growth at 30°C. All dilutions and spread plates were done in triplicate.

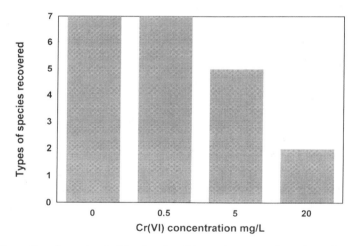

Fig. 2. Diversity of species in Hanford soil inoculum as a function of medium chromate concentration. The same diversity was noted on all replicates of countable spread plates. Diversity was based on colony morphology and cultural characteristics.

reduced all chromate in <12 h, whereas mixed Culture B required more than 24 h to reduce all of the Cr^{6+} to Cr^{3+}. After the Cr^{6+} concentration was reduced to levels below the detection limits, the total Cr in the supernatant of centrifuged samples was found to be equal to the initial Cr^{6+} concentration, within the limits of the methods employed. Thus, although Cr^{3+} was

Bacterial Reduction of Chromium

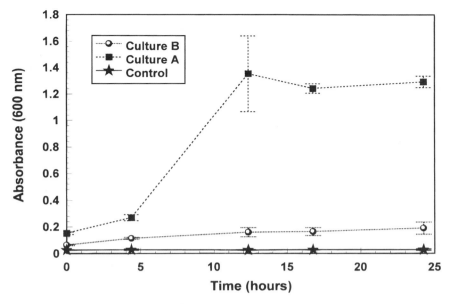

Fig. 3. Microbial growth rates of two mixed cultures growing anaerobically on TSB. ★ Abiotic control, ■ culture A, ○ culture B. Error bars are 1 SD.

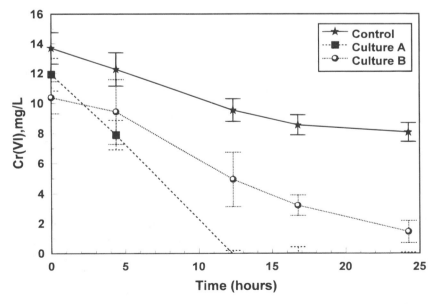

Fig. 4. Chromate reduction rates of two mixed cultures grown anaerobically on TSB. ★ Abiotic control, ■ culture A, ○ culture B. Error bars are 1 SD.

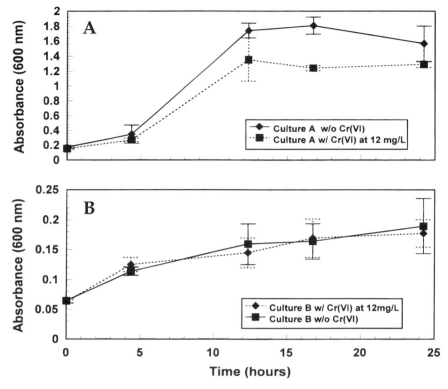

Fig. 5. Chromate inhibition. (A) mixed culture A, (B) mixed culture B. ♦ No chromate in medium, ■ Chromate present in medium. Error bars are 1 SD.

not directly measured, a mass balance on Cr^{6+} suggests that it had been reduced to Cr^{3+} and that the Cr^{3+} remained soluble at neutral pH.

In Fig. 5, the microbial concentration of each of the cultures, as indicated by absorbance at 600 nm, is shown as a function of time. Further, the bacteria were either grown with or without Cr^{6+} in solution during growth. Mixed culture A exhibited inhibition by 12 mg/L Cr^{6+} (Fig. 5A), whereas mixed culture B did not exhibit inhibition at this concentration of Cr^{6+} (Fig. 5B).

DISCUSSION

This research demonstrates that it is possible to enrich for chromium-reducing bacteria from soil at chromium-contaminated sites. This suggests an *in situ* bioremediation strategy: determine conditions limiting growth of the autochthonous chromium-reducing bacteria, and then provide conditions that support their growth. When the hazardous Cr^{6+} has been reduced to the relatively benign Cr^{3+} withdraw support and allow the bacterial populations to return to their prior state.

As indicated above (Fig. 3), absorbance at 600 nm was used as an indicator of the microbial concentration in the reactor. Other methods that would have allowed direct comparison of the biomass concentration in the various reactors were considered. For example, a portion of the reactor broth could have been filtered and the total suspended solids could have been determined. However, production of exopolysaccharides by both mixed cultures prohibited measurement of the biomass concentration using such filtration-based techniques. Alternatively, total protein concentration was considered a measure of biomass concentration. However, the use of TSB as the carbon source precluded the use of protein assay for measurement of biomass. Therefore, although absorbance at 600 nm is indicative of the biomass concentration, this parameter cannot be directly related to dry cell weight.

Because of this inability to measure the total biomass concentration directly in the reaction vessels, it is possible that the chromate reduction rate exhibited by the two cultures, when considered on a per unit dry cell weight basis, is different than the relative chromate reduction rates indicated by absorbance. That is, were it possible to express the chromate reduction rate in terms of mg of Cr reduced/mg dry cell weight/hour, culture B might be found to reduce Cr at a rate greater than culture A.

Acetate was used as the carbon source for the enrichment from Hanford soil because of prior success in the use of acetate as a nutrient for other *in situ* bioremediation activities at this site. Mixed culture B exhibited chromate resistance, but not chromate reduction, when grown on acetate. Although chromate reduction is exhibited when growing on TSB, this complex carbon source is not suitable for *in situ* application for economic and technical reasons.

In addition to biomass measurement limitations described above, the complex carbon source also precludes the use of ion chromatographic methods for chromium speciation. Future research on the biological reduction of chromium will be facilitated by finding a simple carbon source on which the cells will exhibit chromate reduction. The success of the *in situ* strategy for application of this bioremediation technique is also dependent on the determination of such a simple, inexpensive, and pumpable carbon source.

ACKNOWLEDGMENTS

This work was supported through the INEEL University Research Consortium, Contract No. CC-S-622890 administered by Lockheed Martin Idaho Technologies Company under DOE Idaho Operations Office Contract DE-AC07-94ID13223. Eric Schmieman is on educational leave of absence from Pacific Northwest National Laboratory, which is operated by Battelle Northwest Laboratories for the DOE. We are thankful to Yuri Gorby of Pacific Northwest National Laboratory for providing the cores from Hanford, which were used in this study.

REFERENCES

1. Papp, J. F. and Bureau of Mines (1996), *Mineral Industry Surveys: Chromium, Monthly*, US Department of the Interior, Washington, D.C.
2. Turick, C. E., Apel, W. A., and Carmiol, N. S. (1996), *Appl. Microbiol. Biotechnol.* **44,** 683–688.
3. Johnson, C. A. and Sigg, L. (1996), in *Metal Compounds in Environment and Life*, E. Meriam and W. Haerdi, Northwood, Charlottesville, NC, **vol. 4,** pp. 73–80.
4. Fendorf, S. E. and Sparks, D. L. (1994), *Environ. Sci. Technol.* **28,** 290–297.
5. Mertz, W. (1975), *Nutr. Rev.* **33,** 129–135.
6. Bartlett, R. J. (1991), *Environ. Health Perspect.* **92,** 17–24.
7. Davis, J. A., Kent, D. B., Rea, B. A., Maest, A. S., Garabedial, S. P., (1993), in *Metals in Groundwater*, Allen, H. E., Perdue, E. M., and Brown, D. S., eds., Lewis, Boca Raton, FL, pp. 223–273.
8. Sugiyama, M. (1992), *Free Radical Biol. Med.* **12,** 397–407.
9. Bianchi, V. L. (1987), *Toxicol. Environ. Chem.* **15,** 1–24.
10. U.S. Environmental Protection Agency (1995), *Maximum Contaminant Levels for Inorganic-Contaminants*, 40 CFR 141.62, US National Archives and Records Administration, Washington, DC.
11. Vermeul, V. R., Gorby, Y. S., Teel, S. S., et al. (1995), *Geologic, Geochemical, Microbiologic, and Hydrologic Characterization at the* In Situ *Redox Manipulation Test Site*, PNL-10633, Pacific Northwest Laboratory, Richland, WA.
12. APHA, AWWA, WEF (1995), in *Standard Methods for the Examination of Water and Wastewater*, 19th ed., Eaton, A. D., Clesceri, L. S., Greenberg, A. E., eds., American Public Health Association, Washington, DC, Part 9000, page 39.
13. Dresel, P. E., Thorne, P. D., Luttrell, S. P., et al. (1995), *Hanford Site Ground-Water Monitoring for 1994*, PNL-10698, Pacific Northwest Laboratory, Richland, WA.
14. Teel, S. S. Fax to Schmieman, E. A. (1996), Preliminary baseline sample data.
15. APHA, AWWA, WEF (1995), in *Standard Methods for the Examination of Water and Wastewater*, 19th ed., Eaton, A. D., Clesceri, L. S., and Greenberg, A. E., ed., American Public Health Association, Washington, DC, Part 9000, pp. 35–37.
16. Turick, C. E., Camp, C. E., and Apel, W. A. (1996), *Appl. Biochem. Biotechnol.* **63–65,** 871–877.

Degradation of Polycyclic Aromatic Hydrocarbons (PAHs) by Indigenous Mixed and Pure Cultures Isolated from Coastal Sediments

MAHASIN G. TADROS*,[1] AND JOSEPH B. HUGHES[2]

[1]Department of Biology, Alabama A&M University, P.O. Box 14004, Huntsville, AL 35815; and [2]Department of Environmental Science and Engineering, Rice University, P.O. Box 1892, Houston, TX 77251

ABSTRACT

The goal of this paper was to quantify and characterize microorganisms (bacteria) in sediment samples contaminated with polycyclic aromatic hydrocarbons (PAHs: fluorene and naphthalene). The isolated organisms were evaluated for their ability to degrade PAHs compounds. The results indicated that the total number of recovered heterotrophic colony forming units was higher than zone forming units produced by the PAHs compounds. There was a relationship between the biomass of the bacteria recovered from the sediment and the degradation of the compounds. This indicated the utilization of the compounds by the bacteria as a carbon source. Two bacterial species were isolated from the contaminated sediments and identified as *Pseudomonas* sp. and *Ochrobactrum* sp.

Index Entries: Polycyclic aromatic hydrocarbons (PAHs); indigenous cultures; degradation; mixed species; single species.

INTRODUCTION

Polycyclic aromatic hydrocarbons (PAHs) are priority pollutants, and some are known to be chemical carcinogens and mutagens *(1,2)*. PAHs are widespread in the environment. Because of their hydrophobic nature, PAHs sorb to organic-rich soils and sediments, and can accumulate in fish and other aquatic organisms. Microbial degradation of PAHs is considered

---

*Author to whom all correspondence and reprint requests should be addressed.

to be the major decomposition process for these contaminants in nature and is of great practical interest for implementation of bioremediation. Microorganisms capable of degrading PAH compounds are common in soil and sediments previously contaminated with PAHs. A great deal of effort has been directed to the use of indigenous microorganisms to accomplish bioremediation of sites contaminated by such compounds. The use of specially selected microorganisms to enhance bioremediation efforts has proven effective in a number of applications (3,4). Aquatic sediments act as important sinks for PAH derivatives that are discharged either directly or indirectly into the aquatic environment. In this article, degradation of PAH compounds was tested with indigenous mixed and pure cultures isolated from coastal sediments.

PROCEDURE AND METHODS

Chemicals were obtained from Sigma Chemical Company (St. Louis, MO).

Sediment Sampling and Extraction

Sediment samples were obtained from the basin of Houston Ship Channel. Sediment samples were taken with sterilized shovels after removing 20 cm of the surface layer. The materials were collected in sterilized 1-L jars. The jars were transferred to the laboratory within a few hours, stored aerobically at 4°C, and were shipped from Rice University. The microorganisms were isolated from the sample by mixing 1 g of sediment with 10 mL of sterile sodium pyrophosphate solution ($Na_2P_2O_7$, 2.8 g/L) and 3-g glass beads (3 mm diameter) in a 50-mL plastic centrifugation tube. The tube was closed and shaken for 2 h in a horizontal position on a rotary shaker (350 rpm). The solid particles were allowed to settle for 30 min, and aliquots of the supernatant phase were used as inoculum. Dilutions of the inoculum were prepared with pyrophosphate solution. The sample extract (0.1 mL) was plated on the media, which were either blank (reference plates) or coated with one PAH compound/plate as a sole carbon source. The plates were incubated for up to 14 d at 25°C, and monitored regularly for growth or zone formation.

Isolation and Purification of Single Species

The isolation of single species was started by a repeated streaking of individual zone-forming colonies on solidified nutrient broth medium without PAH. Then the nutrient-grown colonies were transferred to the original PAH containing mineral medium. Colonies that were still able to grow and to form clearing zones on PAH were picked again and checked for purity by microscopical means. The purification steps were repeated several times. Species were identified by using the Biolog Microbial Identification System.

Culture Conditions and Degradation Experiment

Batch cultures were carried out in shake flasks. The composition of the mineral salt medium (4) was as follows: (mg/L) $(NH_4)_2SO_4$ (1000), K_2HPO_4 (800), KH_2PO_4 (200), $MgSO_4$ (1000), $CaCl_2\text{-}H_2O$ (100), sea salt (400), $FeCl_3\text{-}6H_2O$ (5) and $(NH_4)_6Mo_7O_{24}\text{-}4H_2O$. A 200-mL sample of culture medium in a 500-mL conical flask was placed in a shaker at 30°C, 125 rpm, in the dark for growth and degradation of PAH components. The flasks were treated with PAH compounds in the range of 50 mg/L in acetone. Twelve conical flasks were inoculated, and three were taken at certain periods for analysis of dry weight (dry biomass) and residual PAH component. Parallel sterile controls were prepared similarly, except that no culture was added. Killed cell controls were obtained by autoclaving the flask containing the added culture. After an appropriate period of incubation time, three replicate cultures and controls were sampled following the introduction of 10 mL of acetonitrile (HPLC-grade, Fisher Scientific, Norcross, GA) into each of the Erlenmeyer flasks to stop growth and to solubilize the PAH. The flasks were placed back on the shaker and shaken overnight.

Gas Chromatography

The samples were centrifuged, and supernatants were analyzed by using HPLC, according to specifications of Tsomides et al. (2). Aqueous samples were solvent-extracted by shaking 10–15 mL of aqueous solution with 1 mL of dichloromethane for 30 min. The solvent phase was then removed and analyzed on the Gas Chromatograph (GC). A Hewlett Packard Model 5890 Series II Gas Chromatograph equipped with a photoionization detector and HP-5 column was used to perform the organic analysis of fluorene, phenanthrene, and naphthalene. The run parameters for the GC were as follows: initial temperature of 75°C, held isothermally for 1 min, increased at 20°C/min to 230°C, and held for 2 min. Injection port temperature is 250°C, and the detector temperature is 250°C. The carrier gas was helium at a flow rate of 40 cm/s. A series of standards was prepared by dissolving PAH crystals in dichloromethane. The standards were then analyzed on the GC, and a linear calibration line was plotted. This plot was used to measure the concentration of the experimental samples. Standards were run prior to analysis of each set of experimental samples.

Dry Weight

For estimation of dry biomass, the total batch (200 mL culture) was centrifuged at 2500 rpm for 15 min. The supernatant was decanted into another flask for analysis of the PAH residual. The residue was quantitatively transferred to a thin aluminum foil of known weight, dried in an oven overnight at 105°C, and then transferred to a desiccator at room temperature. It was weighed again, and from the difference between the final and initial weights, the amount of dry biomass was estimated. All analyses were in triplicate and averaged.

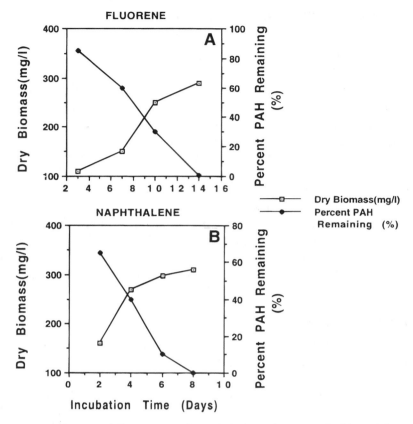

Fig. 1. Degradation of fluorene and naphthalene (0.05 mg/mL) and biomass of microflora in 200 mL of mineral medium. Standard deviation did not exceed 2% of the mean for three replicates.

RESULTS AND DISCUSSION

The ability of mixed organisms isolated from the sediments to degrade the PAH compounds is shown in Fig. 1. The biomass of the microorganism in the media containing PAH compound as sole carbon was also determined. The degradation percentage of the compounds and the biomass of microflora expressed, as a dry weight, are both shown as a function of the time. The degradation of the PAH components as well as the biomass of the sediment bacterial consortium was investigated during the growth of cultures as media containing 0.05 mg/mL naphthalene, phenanthrene, and fluorene as sole carbon sources (Fig. 1A, B). The biomass of the microflora grew with increased degradation percentage of the PAH component, which served a as source of carbon. The degradation of fluorene took a longer time than naphthalene. In sterile control tests containing no inoculum, the concentration of PAH's chemicals remained unchanged during incubation.

Table 1
Degradation of PAH compounds Fluorene and
Napthalene by mixed and Isolated Microorganisms
Enriched from the Sediments[a]

Culture	Fluorene	Napthalene
Mixed	90.4	98.7
Ochrobactrum sp.	85.3	77.2
Pseudomonas sp.	74.7	87.4

[a]Individual PAH degradation % used as sole growth substrate, by indigenous mixed and isolated cultures after 8 days of incubation.

The individual strains and the mixed cultures were compared for their ability to remove the PAH compounds (Table 1) after 8 d of incubation period. Individual strains were adjusted to the same optical density. The mixed cultures were prepared by mixing equal volumes of the individual strains, so that the mixture has the same optical density of the individual components. The results indicated that the degradation by the mixed culture was faster than the isolated strains. The degradation percentage of PAH components was almost complete when mixed indigenous cultures were tested, but when the isolated pure cultures were tested, the degradation percentage was slow and less than that mediated with mixed cultures. The isolated strains were able to degrade all the tested components, but each species degraded certain components at a higher rate when compared to the other components. For example, *Ochrobactrum* degraded fluorene at a much higher percentage than the other components. As is typically observed. (6,7), low-mol wt compounds were degraded at faster rates than high-mol wt structures in time-course studies. The bioavailability and biodegradability of these compounds depend mainly on the complexity of the chemical structures and the corresponding physicochemical properties (8).

Two bacterial strains were isolated from the original mixed culture, identified as *Pseudomonas* sp, and *Ochrobactrum* sp. *Pseudomonas* and *Flavobacterium* species have been found to degrade different PAHs in numerous studies. Mueller and coworkers (3) demonstrated that *Pseudomonas paucimobilis* EPA 505 degraded a range of aromatic compounds, including naphthalene, fluorene, anthracene, phenanthrene, and fluoranthene. *P. paucimobilis* and *Pseudomonas versicularis* have been shown to metabolize phenanthrene and fluorene, respectively. A *Flavobacterium* species was found in a mixed culture degrading anthracene oil (8).

In biodegradation studies with either fluoranthene alone or a defined PAH mixture, better degradation was observed using the mixed culture than using individual strains. Walter and coworkers (8) demonstrated that degradation of anthracene oil, which contains a complex mixture of PAHs resulting from fractional distillation of coal tar, involved a 15-member mixed culture, including species of *Pseudomonas, Achromobacter, Alcaligenes*, and *Flavobacterium*.

An interesting and significant conclusion from these studies is that bioaugmentation of remediation systems should focus on the addition of mixed cultures and not pure ones. As evidenced in these studies, individual strains could degrade several PAHs, but preferred one. Thus, bioaugmentation with pure culture may not be particularly effective in field systems, where multiple contaminants are present. The collective metabolism by mixed cultures of microorganisms may result in an enhanced PAH utilization, since intermediary biotransformation products from one microorganism may serve as substrates for catabolism and growth by others *(9–11)*. Future research should focus on identifying the factors that enhance the biodegradation of PAH's compounds by the indigenous species of the sediments. Such an understanding of these limitations will lead to ecologically based and cost-effective strategies for stimulating indigenous microbial species to degrade pollutants, unless special circumstances demand inoculation with specific degrading organisms *(12)*. Several reports have been successful with *in situ* bioremediation using inoculation with specific biodegrading organisms. Biodegradation of added pollutants is often more rapid in samples from sites that have previously been exposed to pollutants than from uncontaminated sites *(2, 12–14)*.

ACKNOWLEDGMENT

This work was supported by a grant from the Hazardous Substance Research Center South/Southwest.

REFERENCES

1. Wilson, C. S. and Jones, C. K. (1993), *Environ. Pollut.* **81**, 229–249.
2. Tsomides, H. J., Hughes, J. B., Thomas, J. M., and Ward, C. H. (1995), *Environ. Toxicol. Chem.* **14**, 953–959.
3. Mueller, J. G., Chapman, P. J., Blattmann, B. O., and Pritchard, P. H. (1990), *Appl. Environ. Microbiol.* **56**, 1079–1086.
4. Mueller, J. G., Chapman, P. J., and Pritchard, P. H. (1989), *Environ. Sci. Technol.* **23**, 1197–1201.
5. Weissenfels, W. D., Beyer, M., and Klein, J. (1990), *Appl. Microbiol. Biotechnol.* **32**, 479–484.
6. Bossert, I. D. and Bartha, R. (1986), *Bull. Environ. Contam. Toxicol.* **37**, 490–495.
7. Wodzinski, R. S. and Johnson, M. J. (1968), *Appl. Microbiol.* **16**, 1886–1891.
8. Walter, U., Beyer, M., Klein, J., and Rehm, H. J. (1991), *Appl. Microbiol. Biotechnol.* **34**, 671–676.
9. Kiyohara, H., Nagao, K., and Yana, K. (1982), *Appl. Environ. Microbiol.* **43**, 454–457.
10. Bhatnagar, L. and Fathepure, B. (1991), in *Mixed cultures in Biotechnology*, Zeikus, J. G. and Johnson, E. A., (eds.), McGraw-Hill, New York, pp. 293–340.
11. Keck, J., Sims, R. C., Coover, M., Park, K., and Symons, B. (1989), *Water Res.* **23**, 1467–1476.
12. Spain, J. P., Pritchard, P. H., and Bourquin, A. W. (1980), *Appl. Environ. Microbiol.* **40**, 726–734.
13. Thomas, J. M., Lee, M. D., Scott, M. J., and Ward, C. H. (1989), *Indust. Microbiol.* **4**, 109–120.
14. Cerniglia, C. E. and Heitkamp M. A. (1989), in *Metabolism of Polycyclic Aromatic Hydrocarbons in the Aquatic Environment*, Varanasi, U. ed., CRC, Boca Raton, FL., pp. 41–68.

Reduction of Cr(6⁺) to Cr(3⁺) in a Packed-Bed Bioreactor

CHARLES E. TURICK,*,[1] CARL E. CAMP,[2] AND WILLIAM A. APEL[1]

[1]*Idaho National Engineering Laboratory, Idaho Falls, ID, 83415-2203; and [2]DuPont Company, Wilmington, DE*

ABSTRACT

Hexavalent chromium, Cr(6⁺), is a common and toxic pollutant in soils and waters. Reduction of the mobile Cr(6⁺) to the less mobile and less toxic trivalent chromium, Cr(3⁺), can be achieved with conventional chemical reduction technologies. Alternatively, Cr(6⁺) can be biochemically reduced to Cr(3⁺) by anaerobic microbial consortia which appear to use Cr(6⁺) as a terminal electron acceptor. A bioprocess for Cr(6⁺) reduction has been demonstrated using a packed-bed bioreactor containing ceramic packing, and then compared to a similar bioreactor containing DuPont Bio-Sep beads. An increase in volumetric productivity (from 4 mg Cr(6⁺)/L/h to 260 mg Cr(6⁺)/L/h, probably due to an increase in biomass density, was obtained using Bio-Sep beads. The beads contain internal macropores which were shown by scanning electron microscopy to house dense concentrations of bacteria. Comparisons to conventional Cr(6⁺) treatment technologies indicate that a bioprocess has several economic and operational advantages.

Index Entries: Hexavalent chromium; bacterial reduction; bioprocess; chromate; bioreduction.

INTRODUCTION

Hexavalent chromium (Cr[6⁺]) has been used extensively in the industrial and government sectors throughout this century *(1,2)*. Consequently, Cr(6⁺) is present in soils and ground waters, and presents a considerable health risk as a toxic, mutagenic, and carcinogenic pollutant

*Author to whom all correspondence and reprint requests should be addressed.

(3). Without remedial activities, $Cr(6^+)$ has been projected to persist in the environment for 1000 years (4).

Health risks significantly diminish on the reduction of $Cr(6^+)$ to trivalent chromium ($Cr[3^+]$) owing to decreased solubility and bioavailability. Previous work indicates that anaerobic bacteria capable of $Cr(6^+)$ reduction may be ubiquitous in soils (5). These findings suggest that a bioprocess operating in conjunction with soil washing and pump and treat technologies could be developed to reduce $Cr(6^+)$ to the more benign $Cr(3^+)$. Turick and Apel (6) demonstrated a packed-bed, anaerobic bioprocess, incorporating a mixed culture of $Cr(6^+)$ reducers from soil. This bioprocess has been shown to operate with a continuous stream of $Cr(6^+)$ ranging from 140–750 mg/L at a dilution rate of 0.5 d^{-1}. $Cr(6^+)$ reduction occurred at a rate of 4 mg/L/h. Although these $Cr(6^+)$ reduction rates were low, the results demonstrated the possibility of developing a bioprocess using a mixed culture of soil isolates to reduce $Cr(6^+)$ continuously.

In the present study, an attempt was made to increase bacterial density in the bioreactor and subsequent volumetric productivity of the bioprocess. Solid supports (porcelain saddles) from the previous study were replaced with Bio-Sep beads. The increased porosity of the beads allows for potentially higher densities of bacteria owing to entrapment and immobilization.

An estimation of economics relative to operational costs and size of a scaled-up $Cr(6^+)$ reducing bioreactor is calculated based on data presented here.

MATERIALS AND METHODS

Bioreactor Description

The packed-bed bioreactor consisted of a 1.9-L glass cylinder fitted with a water jacket and sealed with Teflon stoppers at either end. Liquid volume was 0.750 L with the solid supports making up the remaining volume. The operational temperature in the reactor was maintained at 30°C. $Cr(6^+)$ concentration was maintained at 200 mg/L using a syringe pump. $Cr(6^+)$ contacted the Tryptic Soy broth (TSB) via an in-line mixer prior to both the TSB and $Cr(6^+)$ entering the bioreactor. Oxygen-free TSB containing $Cr(6^+)$ was circulated through the reactor with a peristaltic pump positioned upstream of the reactor (Fig.1). The bioreactor was operated anaerobically at a dilution rate of 2 h^{-1}.

Solid Supports

The remaining volume of the bioreactor contained porous, adsorptive, spherical supports (2–3 mm diameter) (Fig. 2) (total dry wt of beads: 1.250 kg) consisting of a polyarimid fibrous matrix and powdered activated carbon (Bio-Sep beads). Inoculation of supports was accomplished in batch by

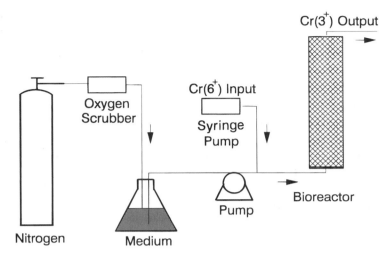

Fig. 1. Schematic of Cr(6+)-reducing bioprocess. Oxygen-free liquid nutrients (TSB) were pumped into a packed-bed bioreactor after contacting a Cr(6+) stream in an in-line mixer. The process was operated at 30°C with a dilution rate of 2 h^{-1}.

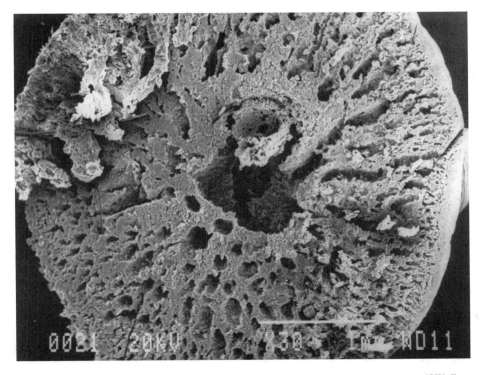

Fig. 2. Cross-section of Bio-Sep bead with scanning electron microscopy (SEM).

contacting the supports with effluent (5 L) from a $Cr(6^+)$-reducing bioreactor (7) containing a mixed culture of $Cr(6^+)$ reducing bacteria and 200–500 mg/L $Cr(6^+)$. Inoculation occurred over 10 d at 20°C with a total of 2175 mg $Cr(6^+)$.

Adsorption

Solid supports were added to 9-mL glass tubes and contacted continuously at 20 and 30°C with distilled water containing 200 mg/L $Cr(6^+)$, at a dilution rate of 0.5 h^{-1}. $Cr(6^+)$ concentrations were analyzed to determine the degree of adsorption. Rate constants for $Cr(6^+)$ adsorption (k) (h^{-1}) and total $Cr(6^+)$ adsorbed (a_1) (mg/g) were calculated using least-squares fit of the data (Statgraphics version 5 STSC, Inc.) to a first-order kinetic model described by the following equation:

$$a = a_1(1 - e^{-kt}) \quad (1)$$

Cr and pH Analysis

Influent samples were taken periodically and analyzed for pH and $Cr(6^+)$. $Cr(6^+)$, total Cr, and pH of the effluent were monitored in the reactor effluent. $Cr(6^+)$ concentrations in the samples were measured by clarifying via centrifugation, diluting the clarified solution 1:100 or 1:1000, adding 0.09 g of ChromaVer 3 Chromium Reagent Powder (Hach Chemical, Loveland, CO), and measuring the absorbance of the mixed solution at 542 nm on a Shimadzu 160U UV VIS spectrophotometer. Total chromium was analyzed using inductively coupled plasma emission spectroscopy (Model 3410, ARL).

Scanning Electron Micrographs

Supports were removed from the bioreactor after operation, thin-sectioned, and gold-sputter-coated. They were photographed during inspection under a scanning electron microscope.

Carbon Source Study

An equal number of inoculated solid supports from the bioreactor were transferred to 10-mL serum vials and incubated anaerobically with equal volumes of TSB or mineral salts medium containing: 10 g/L sucrose; 10 g/L K_2HPO_4; 3.5 g/L $Na(NH_4)HPO_4 \cdot 4H_2O$; and 0.2 g/L $Mg\,SO_4 \cdot 7H_2O$ at 30°C. $Cr(6^+)$ concentrations were measured over time as described above. Results were used to determine the economic utility of using sucrose as an inexpensive sole carbon source in future scale-up studies.

RESULTS AND DISCUSSION

The $Cr(6^+)$ adsorptive capacity of the supports was estimated (Eq. [1]) to be 0.18 and 0.77 mg/g at 20 and 30°C, respectively. Similarly, $Cr(6^+)$ adsorption rates were determined as 0.18 and 0.25 h^{-1}. Based on these

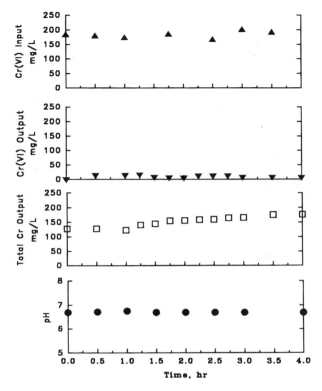

Fig. 3. Results of Cr(6+)-reducing bioreactor study.

values, 90% of the adsorption capacity of the supports was attained within four reactor volumes. These data indicated that Cr(6+) adsorption onto the beads during experimental operation was not a significant factor, since adsorption occurred during bioreactor inoculation, as described above.

Approximately 95% of the Cr(6+) entering the bioreactor was reduced to Cr(3+) (Fig. 3), resulting in a reduction efficiency of 260 mg/L/h at a dilution rate of 2 h^{-1}. Throughout the experiment, the total Cr concentrations in the effluent increased with time and were similar to Cr(6+) influent concentrations (Fig. 3), indicating that sorption in the reactor was minimal.

Throughout this study, pH values of the bioreactor effluent remained circumneutral (Fig. 3), varying little from the input values of 7.0. Therefore, additional pH treatment of the effluent is not required.

Inoculated Bio-Sep beads demonstrated an increase of Cr(6+) reduction rates by nearly 65 times relative to previous work (6). Improved volumetric productivity was probably achieved by increasing bacterial density in the bioreactor through immobilization into beads. High bacterial density is evident in the solid supports used in this study and is exhibited in Fig. 4. This improvement in the rate of Cr(6+) reduction, relative to previous studies (6), can be presumed to be owing to increased bacterial density

Fig. 4. Cross-section of inoculated Bio-Sep bead with SEM, demonstrating high bacterial density.

obtained during the present work. Analysis of bacterial density in the beads was not performed in this preliminary study owing to the difficulty in obtaining accurate values. The beads were found to interfere with protein analysis, and their robust physical properties did not allow for blending, eliminating the potential of plate counts of immobilized bacteria. The rich nutrient medium reduced the reliability of bacterial activity measurements by analysis of carbon utilization. Future studies are planned to deal with this problem by using a simplified carbon and energy source more amenable to physiological measurements of bacterial activity.

Batch studies amended with sucrose (10 g/L) demonstrated rates of $Cr(6^+)$ reduction similar to batch experiments using TSB (data not shown). These findings support the use of sucrose-rich feedstocks, such as molasses, as suitable, inexpensive sources of carbon and energy for future studies. The above data allow for the calculation of size and operational costs of a scaled-up bioprocess. The resulting estimation, based on 200 mg/L of $Cr(6^+)$ at a flow rate of 37.85 L/min into a bioprocess (10 gal/min), indicates a size range of 2436–2832 L (86–100 cubic feet) with an operating cost range from $0.20–0.50/3785 L (1000 gal), based on economic evaluations of sucrose-rich feedstocks, such as molasses (7). Based on operational costs and size, a $Cr(6^+)$-reducing bioreactor using sucrose as a carbon and energy source would compare favorably to conventional chemical processes for $Cr(6^+)$ reduction. Advantages of a bioprocess for $Cr(6^+)$ reduction would include low capital and maintenance costs, as well as potential modular design.

ACKNOWLEDGMENT

This work was supported in part under contract no. DE-AC07-941D13223 for the US Department of Energy to the Idaho National Engineering Laboratory.

REFERENCES

1. Riley, R. G. and Zachara, J. M. (1991), *Nature of Chemical Contaminants on DOE Lands and Identification of Representative Contaminant Mixtures for Basic Subsurface Science Research*. OHER Subsurface Science Program, PNL, Richland, WA.
2. Witmer, C. (1991), *Environ. Health Perspect.* **92,** 139–140.
3. Olson, P. A. and Foster, R. F. (1956), *Effect of Chronic Exposure of Sodium Dichromate on Young Chinook Salmon and Rainbow Trout*. Annual Report for 1955/56, HW 41500:35, Hanford Biological Research, Richland, WA.
4. Xing, L. and Okrent, D. (1993), *J. Hazardous Materials* **38,** 363–384.
5. Turick, C. E., Apel, W. A., and Carmiol, N. S. (1996), *Appl. Microbiol. Biotechnol.* **44,** 683–688.
6. Turick, C. E. and Apel, W. A. (1997), *J. Ind. Microbiol.*, in press.
7. Leeper, S. A. and Andrews, G. F. (1991), *Production of Organic Chemicals via Bioconversion: A Review of the Potential*, DOE Report No. EGG-BG-9033.

Managed Bioremediation of Soil Contaminated with Crude Oil
Soil Chemistry and Microbial Ecology Three Years Later

KATHLEEN DUNCAN,[1] ESTELLE LEVETIN,[1]
HARRINGTON WELLS,[1] ELEANOR JENNINGS,[1]
SUSAN HETTENBACH,[1] SCOTT BAILEY,[1] KEVIN LAWLOR,[2]
KERRY SUBLETTE,*,[2] AND J. BERTON FISHER**,[2]

[1]*Department of Biological Sciences,* [2]*Department of Chemical Engineering, University of Tulsa, Tulsa, OK; and Amoco Technology Center, Tulsa, OK*

ABSTRACT

Analysis of samples taken from three experimental soil lysimeters demonstrated marked long-term effects of managed bioremediation on soil chemistry and on bacterial and fungal communities 3 yr after the application of crude oil or crude oil and fertilizer. The lysimeters were originally used to evaluate the short-term effectiveness of managed (application of fertilizer and water, one lysimeter) vs unmanaged bioremediation (one lysimeter) of Michigan Silurian crude oil compared to one uncontaminated control lysimeter. Three years following the original experiment, five 2-ft-long soil cores were extracted from each lysimeter, each divided into three sections, and the like sections mixed together to form composited soil samples. All subsequent chemical and microbiological analyses were performed on these nine composited samples.

Substantial variation was found among the lysimeters for certain soil chemical characteristics (% moisture, pH, total Kjeldahl nitrogen [TKN], ammonia nitrogen [NH_4-N], phosphate phosphorous [PO_4-P], and sulfate [SO_4^{-2}]). The managed lysimeter had 10% the level of total petroleum hydrocarbons (TPH-IR) found in the unmanaged lysimeter. Assessment of the microbial community was performed for heterotropic bacteria, fungi, and aromatic hydrocarbon-degrading bacteria (toluene, naphtha-

---

*Author to whom all correspondence and reprint requests should be addressed.
**Current affiliation Gardere & Wynne, Tulsa, OK.

lene, and phenanthrene) by dilution onto solid media. There was little difference in the number of heterotrophic bacteria, in contrast to counts of fungi, which were markedly higher in the contaminated lysimeters. Hydrocarbon-degrading bacteria were elevated in both oil-contaminated lysimeters. In terms of particular hydrocarbons as substrates, phenanthrene degraders were greater in number than naphthalene degraders, which outnumbered toluene degraders. Levels of sulfate-reducing bacteria seem to have been stimulated by hydrocarbon degradation.

Index Entries: Crude Oil; bioremediation; TPH; sulfate-reducing bacteria; nematodes; soil; fungi; hydrocarbon.

INTRODUCTION

It is well established that most of the hydrocarbons of crude oils are amenable to biodegradation by microorganisms indigenous to soil provided sufficient aeration, moisture, and mineral nutrients are available. The ultimate objective of *in situ* bioremediation of petroleum-contaminated soil is to return the economic and/or aesthetic value of the site. Clearly the long-term effects of the original contamination and the bioremediation process on the soil ecosystem are important determinants to the eventual postremediation use of the soil.

In 1992 Amoco Production Company initiated a study of managed and unmanaged bioremediation of soil contaminated with Michigan Silurian crude oil. Three lysimeters were filled with topsoil. One was left uncontaminated as a control, whereas the other two were contaminated with crude oil. Of the contaminated lysimeters, one received fertilizer and water (managed), and the other did not (unmanaged). The rates and extent of bioremediation in the contaminated lysimeters were then studied for 6 mos. After the conclusion of this work, the lysimeters were unattended and exposed to the elements.

Three years after the original contamination of the lysimeters, the work presented in this article was initiated with the objective of evaluating the long-term effects of crude oil contamination and bioremediation on soil chemistry and microbial populations. This study consisted of a comparison of the three lysimeters on the basis of soil chemistry and bacterial and fungal populations at different depths.

MATERIALS AND METHODS

Soil Lysimeters

The three soil lysimeters were originally set in place in 1992 at the Amoco Production Research Environmental Test Facility in Rogers County, OK. Each lysimeter measures 9.1 ft (2.8 m) by 9.1 ft (2.8 m) by 3 ft (0.91 m) (depth) and consists of a shallow, reinforced concrete container

Table 1
Analysis of Original Lysimeter Soil

Soil Moisture (%)	9.6
pH	7.0
Saturated Paste Moisture (%)	26.4
CEC (meq/100g)	4.8
NO_3 - N (ppm)	0.3
PO_4 - P (ppm)	30.2
EPTA K (ppm)	61.7
Soluble Cations (mg/L)	
Na	0.2
Ca	1.9
Mg	0.9
Exchangeable Sodium (%)	<1.0
TPH-IR	17.5
TPH-GC	4.1
Exchangeable Cations (meq/100g)	
Na	<0.1
Ca	2.1
Mg	1.0

Table 2
Type Analysis of Michigan Silurian Crude Oil

Carbon Number	Mole %
C4 - C10	48.98
C11 - C20	36.66
C21 - C30	10.26
C31 - C40	3.05
C41 - C53	1.05

containing 9.2 yd^3 (7.1 m^3) of soil. Each lysimeter is drained by a pipe into a collection pond. The soil used to fill the lysimeters was collected from northwest Tulsa County, and was representative of the loamy Okay series found in that area (1). This soil was not contaminated with salt or hydrocarbon. An analysis of this soil as given by Fisher and King (2) is given in Table 1.

Twenty-one gallons of Michigan Silurian Reef crude oil (Table 2) were applied to two of the lysimeters by hand-spraying evenly over the surface and tilling to a depth of 1 ft (0.305 m). This gave an initial oil loading of 1.7% by weight. Fertilizer was added to one of the contaminated lysimeters (managed bioremediation) to provide nitrogen (0.73 kg, in the form of urea), phosphate (0.18 kg, as P_2O_5), and potassium (0.18 kg as K_2O). The managed lysimeter was watered as needed to maintain soil moisture at 80% of container capacity. Total petroleum hydrocarbons (TPH) were monitored for 190 d. By the end of this period, TPH in the unmanaged lysimeter showed a 26% decrease (as determined by infrared absorbance) and an 88% decrease in the managed lysimeter.

Soil Gas Measurements

Three years after the original contamination with crude oil, the current project was initiated with a soil gas analysis in the control (C), oiled (O), and oiled and fertilized (OF) lysimeters. Soil gases (O_2, CO_2, volatile organic compounds, volatile organic compounds minus methane) were measured 12 in. (30.5 cm) below the surface at three (C, OF) or four (O) well-separated locations in each lysimeter. Soil vapors were sampled with an AMS Soil Gas Vapor Probe (SGVP; Forestry Suppliers, Jackson, MI). SGVP dedicated sampling tips, perforated with vapor inlet holes, were driven 12 in. (30.5 cm) into the subsurface. The SGVP drive tubes were removed, leaving the vapor tip probe imbedded at the desired sampling depth. A Teflon vapor tube, connected to the tip and extending to the surface, was used to sample soil gases near the tip. The VOCs in soil gases were measured using a Gastech Trace-Techtor hydrocarbon analyzer with range settings of 100, 1000, and 10,000 ppm. The analyzer was calibrated against hexane calibration gas (4350 ppm). Soil gas concentrations of CO_2 and O_2 were measured using a Gastech model 32520 × CO_2/O_2 analyzer. The CO_2 calibration was performed against atmospheric CO_2 concentration (0.05%) and a 2.5% standard. The O_2 was calibrated using an atmospheric standard (20.9%). Both analyzers had an internal vacuum pump for sampling soil gases.

Soil Sampling

Soil core probes were made of 3-cm (id) stainless-steel tubing. The inside surface of all probes was washed with methylene chloride to remove hydrocarbon residues. The probes were then sterilized by autoclaving at 300°C for 90 min. Soil cores were taken in a five-spot pattern from each lysimeter from the surface to 24 in. (61 cm) below the surface. The probes were brought back to the laboratory to be processed immediately after collection. Using a tubing cutter, each probe was cut into three sections: surface to 3 in. (7.6 cm) below surface (L1); 3 in. (7.6 cm) below surface to 12 in. (30.5 cm) below surface (L2); 12 in. (30.5 cm) below surface to 24 in. (61 cm) below surface (L3). Sterilized spoons were used to remove soil from each section of a probe. Soil samples were placed into pails that had previously been washed with methylene chloride and then sterilized by autoclaving. Composite soil samples were made by mixing with a sterile spoon the five like sections (in terms of sample depth) from each lysimeter in the sterilized pails, giving a total of nine samples, three from each lysimeter. The composited soil samples were then subdivided for chemical and microbial analysis.

Soil Chemical Analysis

The pH, % moisture, carbonate, bicarbonate, chloride, sulfate, sodium, calcium, magnesium, potassium, sodium adsorption ratio (SAR), cation-exchange capacity (CEC), total Kjeldhal nitrogen (TKN), nitrate nitrogen (NO_3-N), ammonia nitrogen (NH_4-N), total phosphorous, available phosphate phosphorous (PO_4-P), and total carbon (%) were determined for each of the composited soil samples. Composited soil samples were also analyzed for total petroleum hydrocarbons by both gas chromatography (TPH-GC) and by infrared absorbance of a solvent extract (TPH-IR), and for benzene, toluene, ethylbenzene, and xylenes (BTEX). Samples were shipped by overnight delivery in completely filled glass jars with Teflon-lined lids to Soil Analytical Services, College Station, TX for analysis.

Fungi

Nine milliliters of sterilized water were added to 1 g of soil, mixed thoroughly by vortexing, then diluted, and spread (three replicates) onto Malt Extract Medium (Difco, Detroit, MI) containing 300 µg/mL of the antibacterial antibiotic streptomycin (Sigma Chemical, St. Louis, MO). The plates were incubated at room temperature for 2 wk before colonies were counted.

Heterotrophic and Hydrocarbon-Degrading Bacteria

Ten milliliters of sterilized water were added to 1 g of soil, mixed thoroughly by vortexing, then diluted, and spread (three replicates) onto Plate Count Agar (PCA, Difco Co.) containing 40 µg/mL of the antifungal antibiotic cycloheximide (Sigma Chemical Co.), and onto a mineral salts agar with trace metals containing either toluene (TOL), naphthalene (NAP), or phenanthrene (PHE), to select for toluene-, naphthalene-, or phenanthrene-degrading bacteria, respectively. PCA media were incubated at room temperature for 48 h before colonies were counted. Each of the three hydrocarbons was separately administered in the vapor phase as crystals (NAP, PHE) or liquid on filter paper (TOL) placed on the inside of the Petri plate lid. Each type of hydrocarbon medium was sealed separately in a Rubbermaid container, and incubated for 36 at room temperature in an active chemical fume hood in order to prevent mixing of the vapors.

Sulfate-Reducing Bacteria

Three grams of composited soil were added to a bottle containing 9 mL of SRB medium (Bioindustrial Technologies, Austin, TX). The soil and medium in the tube were then thoroughly mixed by vortexing for 1 min. A 1-mL sample was then withdrawn with a sterile needle and syringe, and used to inoculate another 9-mL bottle of SRB medium. The procedure was

Table 3
Soil Gas Measurements[a]

Lysimeter	O_2 (%)	CO_2 (%)	VOC (ppm)	VOC-Me (ppm)
C	18-19	0.5-0.6	12-16.5	10-11.5
O	14.5-15.8	2.1-2.9	36-40.5, 119	30-37
OF	11.7-14	2.2-4.2	130-145	37-40

[a]Soil gases were measured 12 in. below the surface at three (C, OF) or four (O) well-separated locations in each lysimeter. The numbers span the range of values obtained.
ppm = parts per million; VOC = volatile organic compounds; VOC-Me = volatile organic compounds minus methane.

repeated for a total of 10 dilutions of the original soil sample, and performed in triplicate for each of the nine composited samples. The tubes were incubated for 25 d and then scored for growth and formation of a black precipitate (iron sulfide), indicating the presence of sulfate-reducing bacteria (SRB). The most-probable number (MPN) of SRB in the original samples was estimated from the characteristic number for a three-tube MPN (3).

RESULTS AND DISCUSSION

Soil Gas

Results of soil gas analysis (at 1 ft or 30.5 cm) are summarized in Table 3. Total VOCs were highest in the fertilized and oiled (OF) lysimeter compared to the oiled (O) and control (C) lysimeters. However, as shown in Table 3, most of the VOC in the soil gas of the OF lysimeter was methane, suggesting that the fertilizer originally applied to this lysimeter has had a stimulating effect on methanogenesis in anaerobic zones or microenvironments. High levels of methane were also detected at one location in the O lysimeter. The substrates for methanogenesis could be products of aerobic degradation of the hydrocarbons. Elevated nonmethane VOCs in the O and OF lysimeters were accompanied by reduced oxygen (O_2) and elevated carbon dioxide (CO_2) concentrations, indicative of aerobic biodegradation of petroleum hydrocarbons. The greater CO_2 concentrations and lower O_2 concentrations in the OF lysimeter compared to the O lysimeter again suggest a stimulation of bioactivity 3 yr after application of the fertilizer.

Hydrocarbon Analysis

As expected, no BTEX was detected in any composited soil samples. Results of analysis of composited soil samples for TPH are given in Table 4. For comparison, the TPH-IR levels in 12-in. (30.5-cm) composited samples 190 d after oil application were (as given by Fisher and King [2]): 17.5

Table 4
Total Petroleum Hydrocarbon Analysis of Composited Soil Samples[a]

Lysimeter	Level	TPH-IR (mg/kg)	TPH-GC (mg/kg)
C	L1	< 10	< 25
	L2	< 10	< 25
	L3	< 10	< 25
O	L1	17237	6000
	L2	1452	249
	L3	30.7	60
OF	L1	1633	84
	L2	863	75
	L3	< 10	< 25

[a]C = control lysimeter; O = oiled lysimeter; OF = oil and fertilized lysimeter; L1 = 0–3 in. depth composite; L2 3–12 in. depth composite; L3 = 12–24 in depth composite.

mg/kg (C), 8440–16,277 mg/kg (O), and 1511–2169 mg/kg (OF). The data given in Table 4 suggest that although the soil in each oiled lysimeter was tilled to 1 ft after application, most of the oil remained near the surface. If the TPH-IR data in Table 4 for L1 and L2 are averaged over a 1-ft (30.5-cm) depth, the weighted averages are 5400 mg/kg (O) and 1060 mg/kg (OF). Therefore, in both the O and OF lysimeters, additional TPH-IR reduction has been realized since the 190-d analysis. It is interesting to note though that if the TPH-GC is taken to be the lighter fraction of the total hydrocarbons in the soil, then the hydrocarbons in the oil and fertilized lysimeter (OF) have been significantly enriched for the heavier components compared to the oiled lysimeter (O). The soil gas analysis suggests that these heavier components or residual partially oxidized products of the original oil in place are still undergoing active aerobic biodegradation.

Soil Chemistry

Results of the chemical analysis of composited soil samples from the three lysimeters are given in Table 5. The following observations are made based on the comparative soil chemistry of the lysimeters:

1. The capacity of the soil for holding moisture correlated somewhat with TPH. Higher TPH-IR resulted in greater moisture retention in the soil.
2. There was a reduction in pH, which correlated to greater hydrocarbon biodegradation. This is attributed to the production of organic acids as intermediates of hydrocarbon degradation, perhaps because of local oxygen limitations.

Table 5
Lysimeter Soil Chemistry

Lysimeter	Level	pH	SAR	% Moisture	CEC (meq/100g)	TKN (mg/L)	NO_3-N (mg/L)	CO_3^{-2}	HCO_3^-	SO_4^{-2}	Soluble cations and anions (meq/L) Cl^-	Na^+	Ca^{+2}	Mg^{+2}	K^+
C	L1	6.3	0.1	5.0	4.1	155	<1.0	<0.1	2.0	0.3	0.7	0.1	1.2	0.5	0.1
	L2	6.6	0.6	11.5	4.2	218	<1.0	<0.1	1.3	0.5	1.4	0.5	0.8	0.5	0.1
	L3	6.7	0.5	15.8	4.0	112	<1.0	<0.1	1.2	0.5	0.8	0.4	0.8	0.5	0.1
O	L1	6.0	0.1	7.8	4.7	1160	<1.0	<0.1	0.6	0.7	0.2	0.1	1.1	0.4	<0.1
	L2	6.1	0.1	15.4	5.7	290	<1.0	<0.1	1.0	0.4	0.2	0.1	0.9	0.4	<0.1
	L3	6.8	0.1	18.1	5.8	154	<1.0	<0.1	2.4	0.3	0.2	0.1	1.6	0.5	0.1
OF	L1	5.5	0.1	4.9	5.7	554	<1.0	<0.1	0.9	0.4	0.4	0.1	1.0	0.4	0.2
	L2	5.8	0.1	12.5	5.4	359	<1.0	<0.1	0.9	0.5	0.2	0.1	0.9	0.3	0.2
	L3	6.3	0.2	15.2	5.3	308	<1.0	<0.1	1.4	0.4	0.3	0.2	1.0	0.4	0.1

Lysimeter	Level	NH_4-N (mg/L)	Total P (mg/L)	Avail-PO_4-P (mg/L)	LECO Total C (%)
C	L1	5.9	252	39.1	0.2
	L2	3.1	198	39.0	0.2
	L3	8.1	302	42.4	0.1
O	L1	10.6	404	26.8	1.2
	L2	9.7	301	36.2	0.3
	L3	8.3	283	35.7	0.2
OF	L1	14.7	284	69.0	0.6
	L2	14.7	246	51.8	0.4
	L3	12.9	276	33.8	0.3

3. The oiled and fertilized (OF) lysimeter still contained elevated levels of NH_4-N, available PO_4-P, and TKN relative to the C and O lysimeters 3 yr after application.
4. The OF lysimeter still contained significant levels of total carbon compared to the O lysimeter, out of proportion to the relative TPH levels. This suggests that high concentrations of partially degraded hydrocarbons that are not measured as TPH remain in the OF lysimeter that continue to fuel biological activity as indicated by the soil gas analysis.

Fungi

Fungi were must abundant in the uppermost level of all lysimeters, with higher concentrations in the O and OF lysimeters ($5-7 \times 10^4$ CFU/g soil) than in C (2×10^4 CFU/g soil), suggesting that one effect of crude oil contamination and subsequent bioremediation was to promote the growth of soil fungi. The positive effect seen in O and OF may be partially owing to the crude oil acting as a carbon and energy source for fungi that can degrade oil, since fungi were abundant on media containing aromatic hydrocarbons. Also, the low pH in the O and OF lysimeters (Table 5) may have favored fungi at the expense of bacteria, since fungi are more tolerant of low pH.

Bacteria

Mean Heterotrophic Bacteria

Aerobic heterotrophic bacteria were more abundant in the two uppermost levels of the lysimeters, but differed little among the lysimeters. The mean concentration of viable bacteria (CFU)/g soil (dry wt) for each level of the three lysimeters was approx 10^6. Crude oil contamination seemed to have relatively little effect after 3 yr on densities of bacteria assayed on this medium, unlike the soil fungi.

Hydrocarbon-Degrading Bacteria

Toluene-degrading bacteria were present in all lysimeters and all levels at approximately the same low concentration ($4 \times 10^3 - 10^4$ CFU/g soil). The concentration of naphthalene degraders was somewhat elevated in OL2 and OL3 (10^3 CFU/g soil) above the corresponding levels in C ($7 \times 10^1 - 8 \times 10^2$ CFU/g soil), whereas the concentration in OF ($4 \times 10^4 - 2 \times 10^6$ CFU/g soil) was elevated for all levels. Phenanthrene degraders were much more abundant in OF ($4 \times 10^5 - 2 \times 10^7$ CFU/g soil) and O ($10^5 - 2 \times 10^6$ CFU/g soil) than in C ($3 \times 10^3 - 7 \times 10^4$ CFU/g soil). The relative densities of these hydrocarbon degraders (PHE > NAP > TOL) probably reflect the current level of these classes of compounds as the hydrocarbon is enriched for the heavier components. Fungi were also abundant on these plates, but have not yet been examined.

Table 6
Concentration (Most Probable Number/g Soil)
of Viable Sulfate-Reducing Bacteria in Soil from Each Lysimeter

Level	Lysimeter C	Lysimeter O	Lysimeter OF
L1	2.5×10^2	9×10^5	1.5×10^6
L2	2.5×10^2	5×10^6	5×10^6
L3	10^5	10^9	10^9

Sulfate-Reducing Bacteria

The mean concentration of sulfate-reducing bacteria (most probable number/g) in composited samples from each level of each lysimeter is shown in Table 6. A higher concentration of SRB was found in L2 and L3 of each lysimeter than in L1, as expected for these anaerobic bacteria. Higher concentrations of SRB were found in each level in the lysimeters contaminated with crude oil than in the control. It appears that aerobic degradation of hydrocarbons in L1 has stimulated the growth of SRB, especially in L3.

CONCLUSIONS

Three soil lysimeters, originally used to study the short-term degradation of crude oil in soil, have been re-examined 3 yr after the application of oil. The application of crude oil and subsequent bioremediation (managed and unmanaged) have had marked effects on soil chemistry and microbial populations. Soil analysis showed that the fertilized lysimeter still contained elevated levels of NH_4-N, available PO_4-P, and TKN relative to the control lysimeter (C) and the lysimeter (O), which received oil only. The application of fertilizer to a crude oil-contaminated lysimeter (managed bioremediation—OF) compared to unmanaged bioremediation (O) has resulted after three years in:

1. Greater degradation of hydrocarbons (TPH);
2. Greater rates of oxygen consumption in aerobic zones; and
3. Increased methanogenesis.

However, although the OF lysimeter TPH was much reduced compared to the O lysimeter, the OF lysimeter still contained significantly higher levels of total carbon than the O or C lysimeters, suggesting high concentrations of partially degraded hydrocarbons. The soil gas analyses strongly suggest that these compounds continue to fuel high levels of bioactivity. Both the O and OF lysimeters exhibited reduced pH and increased levels of fungi, SRB, and hydrocarbon degraders compared to C.

In summary, three years after crude oil contamination and subsequent application of fertilizer (managed bioremediation), microbial populations and microbial activity still remain stimulated, and bioremediation of the soil continues. However, more time and/or active intervention will be required for this soil to return to normal as defined by the uncontaminated lysimeter.

REFERENCES

1. Cole, E., Bartolina, D., and Swafford, T. (1977), Soil Survey of Tulsa County, OK, US Department of Agriculture Soil Conservation Service, p. 140.
2. Fisher, J. B. and King, G. (1994), Monitoring of Hydrocarbon Bioremediation Progress with Microtox, Amoco Production Co. Report 940870004-APR.
3. Rodina, A. G. (1972), *Methods in Aquatic Microbiology*, University Park Press, Baltimore, MD.

Author Index

A
Adney, W. S., 315, 585
Akanbi, F., 423
Antoine, P., 707
Apel, W. A., 855, 871

B
Bailey, S., 879
Bajpai, R. K., 495, 511
Baker, J. O., 315, 585
Barnard, S., 637
Barnes, J. M., 667
Baron, M., 257, 305, 327
Beauchamp, J. J., 797
Beck, M. J., 667
Bentley, W. E., 743
Bereded-Samuel, Y., 855
Bienkowski, P. R., 483, 809
Bon, E. P. S., 87, 203, 409
Bonfim, T. M. B., 305
Borole, A. P., 719, 823
Brainard, A., 243
Bredwell, M. D., 637
Burr, B., 3

C
Camp, C. E., 871
Cao, N., 129, 387, 541
Carroll, S., 789
Carvajal, E., 203
Chang, V. S., 3
Chen, C., 387
Chen, R., 435
Chen, Z. D., 243
Cheong, T. K., 213
Cheryan, M., 423
Chocial, M. B., 305
Chua, H., 627, 669
Colwell, F. S., 797
Conrad, J. R., 759

D
Dale, B. E., 625
Davison, B. H., iii, 483, 565
De Moraes, F. F., 527
Dominguez, J. M., 117
Donnelly, J. A., 375
Donnelly, M. I., 153
Dorsch, R. R., 349
Dronawat, S. N., 363, 365
Du, J., 387, 541
Duncan, K., 879

E
Ehrman, C. I., 585
Eley, M. H., 35

F
Felipe, M. G. A., 557
Ferreira, A. G., 327
Finkelstein, M., iii
Fisher, J. B., 719, 823, 879
Florêncio, J. A., 305
Foidl, G., 457
Foidl, N., 449, 457
Fontana, J. D., 257, 305, 327
Fontana, M. A., 327
Freire, D. M., 409
Fürlinger, M., 173

G
Gaddy, J. L., 597
Ghirardi, M. L., 141
Gibson, T., 775
Gillespie, M. T., 835
Gollhofer, D., 173
Gong, C. S., 117, 129, 387, 541, 731
Gregg, D. J., 609
Griessler, R., 159
Gübitz, G. M., 449, 457
Guimarães, M. F., 257, 305, 327

H
Haltrich, D., 159, 173, 189
Handagama, N., 809
Hanley, T. R., 363, 375
Harris, B. G., 153
Hasan, S., 845
Hatzis, C., 349
Haug, I., 173
Hettenbach, S., 879
Himmel, M. E., 315, 585
Hinman, N. D., 599
Ho, L. Y., 627
Ho, N. W. Y., 243
Holtzapple, M. T., 1, 3
Hughes, J. B., 865

I
Ingledew, W. M., 59

J
Javanmardian, M., 719, 823
Jeffries, T. W., 97, 109
Jennings, E., 879
Joerke, C. G., 327
Johnstone, D. L., 855

K
Karplus, P. A., 315
Kaufman, E. N., 625, 677
Kelley, S. S., 469
Klasson, K. T., 339
Kolhatkar, R., 823
Kulbe, K. D., 159, 173, 189
Kulkarni, G., 153

L
Lafferty, R. M., 457
Lawford, H. G., 221, 269, 287
Lawlor, K., 879
Lee, T.-S., 743
Lee, Y. Y., 21, 435
Lehman, R. M., 797
Levetin, E., 879

Little, M. H., 677
Liu, B. Y., 759
Loha, V., 395
Lokshina, L. Ya., 45
Lu, J., 731
Lumpkin, R. E., 243

M
Magasanik, B., 203
Maraschin, M., 305, 327
Markov, S. A., 577
McCarthy, J. F., 775
McMillan, J. D., 269, 469
Monceaux, D., 597
Myers, M. D., 469

N
Nakamura, H., 659
Narita, M., 659
Nascimento, H. J., 87
Nghiem, N. P., 565
Nidetzky, B., 159, 173, 189

O
Ohuchi, S., 659
Oriel, P., 213

P
Padukone, N., 469
Palumbo, A. V., 775, 789, 797
Paterek, J. R., 759
Pereira-Meirelles, F. V., 73
Petersen, J. N., 855
Pfiffner, S. M., 775, 797
Phelps, T. J., 797
Philippidis, G. P., 243
Pradhan, S. P., 759
Prokop, A., 395

Q
Qiang, H., 649

R
Raterman, K. T., 719, 823

Index

Réczey, K., 351
Ribeiro, J. D., 557
Richardson, G. R., 565
Richmond, A., 649
Rieth, T. C., 375
Riley, C. J., 243
Ringelberg, D. B., 775
Rocha-Leão, M. H. M., 73
Rouhana, N., 809
Rousseau, J. D., 221, 269, 287
Rowen, D., 203
Rytov, S. V., 45

S

Sachslehner, A., 189
Saddler, J. N., 609
Sakon, J., 315
Sant' Anna, Jr., G. L., 73, 409
Sarisky-Reed, V., 71
Savithiry, N., 213
Scarborough, S. P., 797
Schmidt, S. L., 469
Schmieman, E. A., 855
Sebghati, J. M., 35
Seibert, M., 141, 577
Selvaraj, P. T., 677
Shah, M. M., 423
Silva, J. G., Jr., 87
Silva, S. S., 557
Soares, M. B., 327
Soares, V. F., 87
Sode, K., 659
Sosulski, F. W., 59
Sosulski, K., 59
Souza, A. M., 327
Spurrier, M. A., 483
Sreenath, H. K., 109
Srivastava, V. J., 759
Stanbrough, M., 203
Staubmann, R., 449, 457
Steiner, W., 449, 457

Stols, L., 153
Strong-Gunderson, J. M., 789, 835
Sublette, K. L., 695, 719, 823, 845, 879
Sugiura, H., 659
Sun, M. Y., 483
Suttle, B. E., 565
Svihla, C. K., 363, 375
Szengyel, Zs., 351

T

Tadros, M. G., 865
Taillieu, X., 707
Tang, J., 59
Tanner, R. D., 395
Teles, E. M. F., 409
Telgenhoff, M. D., 637
Thomas, S. R., 315
Thonart, P., 707
Togasaki, R. K., 141
Toon, S. T., 243
Torget, R., 1
Torriani, I., 327
Trent, G. L., 823
Tsao, G. T., 117, 129, 387, 541, 731
Turick, C. E., 855, 871

U

Ulhoa, C., 305

V

Valencia Arbizu, V. M., 457
Vavilin, V. A., 45
Vinzant, T. B., 585
Vitolo, M., 557

W

Wang, C.-J., 495, 511
Wang, S., 59
Weaver, P. F., 577
Webb, O. F., 483
Weigand, W. A., 743
Weinhäusel, A., 159
Wells, H., 879

Winkler, E., 449
Woodward, J., 71, 257
Worden, R. M., 637
Wu, Z., 21
Wyman, C. E., iii

X

Xia, Y., 129

Y

Yancey, M. A., 599
Yang, V. W., 97
Yonge, D. R., 855
Yu, P. H. F., 627, 669
Yuan, N., 541

Z

Zacchi, G., 351
Zanin, G. M., 527
Zhang, C., 797
Zheng, Y., 731

Subject Index

A
Abrasion, 59
Acetic acid, 423
Acetobacter, 327
Activated sludge, 627
Aerobic biofilm, 669
Airlift, 541
Ammonia steeping, 129
Anabaena variabilis, 577
Anaerobic, 707
Anaerobic digestion, 45
Anaerobiospirillum succiniciproducens, 565
Aromatic hydrocarbons, 759
Asparaginase, 203
Aspergillus awamori, 87
Astaxanthin, 305

B
Bacillus megaterium, 257
Bacterial cellulose, 327
Benzoate, 707
Biodiesel, 449
Biofilm, 327, 387
Biofilter, 809
Biogas, 457
Biomass, 35, 609
Bioremediation, 719, 775, 823, 879
BTEX, 719, 775
Bubble fractionation, 395
2,3-butanediol, 129
Butyribacterium methylotrophicum, 637

C
Candida lipolytica, 73
Candida shehatae, 97
Cellulase, 315, 585
Cellulase adsorption, 21
Cellulase production, 351
Cellulose, 45
Cereal grains, 59
Cheese whey, 495
Chemostat, 511
Chlamydomonas, 141
Chromium, 855, 871
Clostridium thermoaceticum, 423
Corn steep liquor, 287, 305
Corynebacterium callunae, 159
Cost analysis, 695
Crude oil, 879
Cybernetic model, 511

D
D-xylose, 117
Debaryomyces hansenii, 117
Dechlorination, 845
Degradation, 835
Delignification, 21
Desulfurization, 677
Dilute-Acid, 21

E
Energy, 59
Enzyme activity, 789
Escherichia coli, 153, 221
Ethanol, 287, 483, 609
　production, 109
　recovery, 469

F
Fermentation scale-up, 243
Filamentous suspensions, 363, 375
Fluidized-bed reactor, 483, 527
Fumaric acid, 387, 541
Furfural inhibition, 351

G
Gas condensate, 719, 823

Gas-liquid mass transfer, 363
α-1,4-D-Glucan phosphorylase, 159
Glucoamylase, 87
Gluconolactonase, 173
Glucose dehydrogenase, 257
Glucose-fructose oxidoreductase, 173
Glucuronic acid, 221
Gypsum, 677

H
Hardwood, 585
Heavy metal, 731
Helical impellers, 375
Heterotrophs, 797
Hollow fiber membrane, 435
Hydrogen, 141, 577
Hydrolytic bacteria, 45
4-Hydroxybenzoate, 707

I
Immobilization, 257, 731
Immobilized enzyme, 527
In situ separation, 435
Investment analysis, 599

J
Jatropha curcas, 449, 457

K
Kinetics, 495
Klebsiella oxytoca, 129
Kluyveromyces marxianus, 495

L
Lactic acid, 435
Lignocellulose, 3
Lime, 3
Limonene, 213
Lipase production, 73, 409

M
Malic enzyme, 153
Mannanase, 189
Manufactured gas plant, 759
Mass transfer coefficient, 541
Membrane separations, 469
Membrane reactor, 585
Methane production, 457
Methanotrophs, 789
Microbial characterization, 775
Microbubbles, 637
Mineralization, 797
Mixed species, 865
Municipal solid waste, 35

N
Nitrogen regulation, 203

O
Oil, 449
Organophosphorous, 659
Oxygen, 141

P
Pachysolen tannophilus, 97
Packed-bed bioreactor, 871
Pectinase, 395
Penicillium restrictum, 409
Pervaporation, 469
Phaffia, 305
Phenol, 707
6-Phosphofructokinase, 97
Photoautotrophs, 649
Photobioreactors, 649, 577
Phytic acid, 731
Pichia stipitis, 109
Pilot-scale, 759
Poly-hydroxy-alkanoates, 627
Polycyclic aromatic hydrocarbons, 865
Polyvinyl pyridine, 731
Porphyrin, 845
Pretreatment, 3, 21

Index

Protein separation, 395
Public/private partnerships, 599
Pyrolysis, 35

R
Racemic phosphotriesterase, 659
Reduction, 855, 871
Regulation, 97
Respiration, 109
Rhizopus, 541
Rhizopus oryzae, 387
Rotary biofilm contactor, 387

S
Saccharomyces, 97, 203, 243
Sampling, 797
Sclerotium rolfsii, 189
Seed production, 269
Simultaneous saccharification
    and fermentation, 243, 483
Smoothing data, 339
Starch, 527
Succinic acid, 153, 565
Sugar cane bagasse, 557
Sulfate-reducing bacteria, 677, 879
Sulfate reduction, 719
Sulfide, 695

Surfactants, 835
Switchgrass, 3
Synthesis gas, 637, 677

T
TCE, 789, 835
Techno-economic modeling, 609
α-Terpineol, 213
Thermostable, 213
Thiobacillus denitrificans, 695
Thiodiglycol biodegradation, 743
Trichlorophenol, 845
Trichoderma reesei, 351, 585
Two-phase, 213

V
Volatile organic hydrocarbon, 809

W
Waste water, 669

X
X-ray structures, 315
Xylanase, 189, 315
Xylitol, 117, 557

Z
Zymomonas mobilis, 173, 269, 287